Ozone in Water Treatment

Application and Engineering

Cooperative Research Report

Edited by

Bruno Langlais
David A. Reckhow
Deborah R. Brink

Second Printing 1991

Library of Congress Cataloging-in-Publication Data

Ozone in water treatment: application and engineering / edited by
Bruno Langlais, David A. Reckhow, Deborah R. Brink.
p. cm.
"American Water Works Association Research Foundation: Compagnie
Générale des Eaux."
Includes bibliographical references and index.
1. Water—Purification—Ozonation. 2. Drinking water—
Purification. I. Langlais, Bruno, II. Reckhow, David A.
III. Brink, Deborah R. IV. AWWA Research Foundation. V. Compagnie
Générale des Eaux (Paris, France)
TD461.0962 1991
628.1′662—dc20 90-13476
ISBN 0-87371-474-1

Bottom cover photograph courtesy of Compagnie Générale des Eaux

AMERICAN WATER WORKS ASSOCIATION RESEARCH FOUNDATION
6666 West Quincy Avenue, Denver, Colorado 80235

LEWIS PUBLISHERS, INC.
121 South Main Street, Chelsea, Michigan 48118

PRINTED IN THE UNITED STATES OF AMERICA

Contents

Disclaimer

Foreword

The AWWA Research Foundation is a non-profit corporation that is dedicated to the implementation of a research effort to help local utilities respond to regulatory requirements and traditional high priority concerns of the industry. The research agenda is developed through a process of grass roots consultation with members, utility subscribers, and working professionals. Under the umbrella of a Five-Year Plan, the Research Advisory Council prioritizes the suggested projects based upon current and future needs, applicability, and past work; the recommendations are forwarded to the Board of Trustees for final selection.

This publication is a result of one of those sponsored studies and it is hoped that its findings will be applied in communities throughout the world. The following report serves as a means of not only communicating the results of the water industry's centralized research program but also as a tool to enlist the further support of the nonmember utilities and individuals.

Projects are managed closely from their inception to the final report by the Foundation's staff and a large cadre of volunteers who willingly contribute their time and expertise. The Foundation serves a planning and management function and awards contracts to other institutions, such as water utilities, universities, and engineering firms. The funding for this research effort comes primarily from the Subscription Program. The program, through which water utilities subscribe to the research program and make an annual payment proportionate to the volume of water they deliver, offers a cost-effective and fair method for funding research in the public interest.

A broad spectrum of water supply issues are addressed by the Foundation's research agenda: resources, treatment and operations, distribution and storage, water quality and analysis, toxicology, economics, and management. The ultimate purpose of the coordinated effort is to assist local water suppliers in providing the highest possible quality of water economically and reliably. The true benefits are realized when the results are implemented at the utility level. The Foundation's Trustees are pleased to offer this publication as a contribution toward that end.

This book is the culmination of a joint cooperative effort between the AWWA Research Foundation and Compagnie Générale des Eaux, a private French water company. Compagnie Générale des Eaux has developed into a decentralized but consolidated group of 950 different companies representing a work force some 150,000 strong. World leader in water distribution, CGEaux supplies drinking water to more than a third of the population in France (it is responsible for the distribution of water to over five million people in Paris and the greater Paris area and supplies 7,000 out of 36,000 listed French municipalities).

If water distribution and treatment remains its main activity, CGEaux also places priority on expanding its companies' service capabilities, including hydraulic works; heat production; waste collection, disposal, and recycling; housing construction; and public works, as well as audiovisual and recreational activities.

Because of its know-how in the field of services, CGEaux has extended its activities abroad. Over the last decade, manufacturing subsidiaries and other operational facilities have been set up in Europe, the United States, and Canada, as well as in developing countries. These local companies have become full-fledged members of the industrial community.

In line with its strategy, CGEaux is resolved to maintain an active presence in all aspects of advanced scientific issues relating to its traditional and newly diversified companies. Although most of these companies, as well as their subsidiaries and branches, conduct their own research and development work, a high-technology research center has also been set up on the Seine River a few miles west of Paris, at Maisons-Laffitte, through an original organization called Anjou Recherche. Anjou Recherche acts as a catalyst where all functions of research and development are concerned. Areas of research conducted there range from water, energy, and waste recycling technologies to physics, chemistry, biotechnology, and microelectronics. Moreover, the laboratory is pursuing extensive programs in fundamental research. Another obvious responsibility of the research center is to ensure that relevant data are quickly collected and dashed off to:

- the local communities with whom CGEaux's companies are involved;
- universities, national research institutes, and agencies cooperating with CGEaux in joint research and development programs;
- the industry at large; and
- universities, research centers, and organizations abroad.

This book is the perfect demonstration of the research center's philosophy in this field.

These two organizations, with the assistance of 35 expert authors, have collaborated to produce a state-of-the-science manual on the application of ozone for drinking water treatment. The use of ozone is increasing worldwide, and particularly in North America. New disinfection and disinfection by-product regulations are driving the recent interest in ozone. This book is intended for all individuals interested in ozone technology. Topics covered include ozone chemistry, generation, toxicology, practical applications, design, operation, and economics.

Richard P. McHugh
Chairman, Board of Trustees
AWWA Research Foundation

James F. Manwaring, P.E
Executive Director
AWWA Research Foundation

Acknowledgments

This book would not have been possible without the dedicated efforts of the 35 authors who have volunteered many long hours of their time to successfully produce it. The AWWA Research Foundation and Compagnie Générale des Eaux thank all authors for their significant contributions to this effort.

To produce this book, a select group of ozone experts from both Europe and North America was assembled to assimilate the vast body of knowledge on the subject into a comprehensive state-of-the-science manual on the use of ozone in drinking water applications. Initially, primary authors and reviewers were assigned for each chapter subsection. Both North American and European experts were involved with each subsection to maximize the cross-fertilization of their respective ozone experiences. Because of the significant contributions to the work by both authors and reviewers we will not distinguish between the two. All participants played a critical role in the development and quality of this document. The following list defines the contributions of each participant by chapter and subsection.

North American Participants

David A. Reckhow, Associate Professor, Department of Civil Engineering, University of Massachusetts, Amherst, MA (Co-editor and North American Committee Chairperson)

Introduction
Fundamental Aspects
  Chemistry—Introduction
Practical Application of Ozone: Principles and Case Studies
  Introduction; Iron and Manganese Removal; Color Abatement; Control of Tastes and Odors; Elimination of Synthetic Organic Chemicals; Particulate Removal; Algae Removal; Disinfection By-product Control; Biological Stabilization
Engineering Aspects
  Treatability Studies

Deborah R. Brink, Senior Project Manager, AWWA Research Foundation, Denver, CO (Co-editor and AWWARF Project Manager)

Introduction
Ozone System Terminology, Measurements, and Conversions

William D. Bellamy, Assistant Director of Water Supply and Treatment, CH2M Hill Consulting Engineers, Denver, CO

Fundamental Aspects
Practical Application of Ozone: Principles and Case Studies
  Control of Tastes and Odors; Biological Stabilization
Engineering
  Treatability Studies; Design of Contact Chambers and Diffusion Systems; Retrofit of Ozone Systems
Economics of Ozone Systems: New Installations and Retrofits
  "Order-of-Magnitude" Estimates of Ozone System Costs; Considerations for Estimating the Cost of an Ozonation Facility; Influence of the Designed Ozonation

System on the Capital Cost; Influence of the Designed Ozonation System on O&M Costs; Case Studies

F. Bernard Daniel, Chief, Biochemical and Molecular Toxicology Branch, Health Effects Research Laboratory, U.S. Environmental Protection Agency, Cincinnati, OH

Fundamental Aspects
Toxicology
Practical Application of Ozone: Principles and Case Studies
Minimization of Adverse Health Effects

Gilbert Gordon, Professor, Department of Chemistry, Miami University, Oxford, OH

Fundamental Aspects
Chemistry—Introduction; Theoretical Aspects of Ozone Analysis—Introduction; Ozone Generation; Ozone Gas Transfer
Practical Application of Ozone: Principles and Case Studies
Iron and Manganese Removal

William R. Knocke, Professor, Department of Civil Engineering, Virginia Polytechnic Institute and State University, Blacksburg, VA

Practical Application of Ozone: Principles and Case Studies
Iron and Manganese Removal

Benjamin Lykins, Jr., Chief, Systems and Field Evaluation Branch, U.S. Environmental Protection Agency, Drinking Water Research Division, Cincinnati, OH

Practical Application of Ozone: Principles and Case Studies
Disinfection By-product Control
Economics of Ozone Systems: New Installations and Retrofits
"Order-of-Magnitude" Estimates of Ozone System Costs; Considerations for Estimating the Cost of an Ozonation Facility; Influence of the Designed Ozonation System on the Capital Cost; Influence of the Designed Ozonation System on O&M Costs; Case Studies

Richard S. Miltner, Environmental Engineer, U.S. Environmental Protection Agency, Drinking Water Research Division, Cincinnati, OH

Practical Application of Ozone: Principles and Case Studies
Disinfection By-product Control; Biological Stabilization

Gilbert Pacey, Professor, Department of Chemistry, Miami University, Oxford, OH

Fundamental Aspects
Chemistry—Introduction; Theoretical Aspects of Ozone Analysis—Introduction

Michèle Prévost, Research and Development Group Manager, Gendron LeFebvre Consultants, Laval, Canada

Practical Application of Ozone: Principles and Case Studies
Biological Stabilization

Kerwin L. Rakness, Principal, Process Applications, Inc., Fort Collins, CO

Engineering Aspects
Treatability Studies; Feed Gas Preparation; Ozone Generation; Design of Contact Chambers and Diffusion Systems; Instrumentation and Control Systems; Ozone Destruction Systems; Corrosion Considerations and Ozone-Resistant Materials; Retrofit of Ozone Systems; Performance Evaluation; Other Considerations; Checklists for Ozone System Design
Operating an Ozonation Facility
Operating Methods; Ozone System Reliability; Required Plant Maintenance; Employee Training; Safety
Economics of Ozone Systems: New Installations and Retrofits
"Order-of-Magnitude" Estimates of Ozone System Costs; Considerations for Estimating the Cost of an Ozonation Facility; Influence of the Designed Ozonation System on the Capital Cost; Influence of the Designed Ozonation System on O&M Costs; Case Studies
Ozone System Terminology, Measurements, and Conversions

C. Michael Robson, Vice President–Facilities Engineering, Heritage Environmental Services, Inc., Indianapolis, IN

Engineering Aspects
Treatability Studies; Feed Gas Preparation; Ozone Generation; Design of Contact Chambers and Diffusion Systems; Instrumentation and Control Systems; Ozone Destruction Systems; Corrosion Considerations and Ozone-Resistant Materials; Retrofit of Ozone Systems; Performance Evaluation; Other Considerations; Checklists for Ozone System Design
Operating an Ozonation Facility
Operating Methods; Ozone System Reliability; Required Plant Maintenance; Employee Training; Safety
Economics of Ozone Systems: New Installations and Retrofits
"Order-of-Magnitude" Estimates of Ozone System Costs; Considerations for Estimating the Cost of an Ozonation Facility; Influence of the Designed Ozonation System on the Capital Cost; Influence of the Designed Ozonation System on O&M Costs; Case Studies

Philip C. Singer, Professor, Department of Environmental Science and Engineering, University of North Carolina, Chapel Hill, NC

Practical Application of Ozone: Principles and Case Studies
Particulate Removal

Otis J. Sproul, Dean, College of Engineering and Physical Sciences, University of New Hampshire, Durham, NH

Practical Application of Ozone: Principles and Case Studies
Control of Tastes and Odors; Algae Removal; Disinfection

Special thanks are also due to **E. Marco Aieta,** (Vice President, James M. Montgomery Consulting Engineers, Inc., Pasadena, CA), who provided the first draft of the Treatability Studies section of the Engineering Aspects chapter.

European Participants

Bruno Langlais, Research Engineer, Compagnie Générale des Eaux, Research Center, Maisons-Laffitte, France (Co-editor, and European Committee Chairperson)

Introduction
Fundamental Aspects
Theoretical Aspects of Ozone Analysis—Introduction; Ozone Generation; Ozone Gas Transfer
Practical Application of Ozone: Principles and Case Studies
Introduction; Iron and Manganese Removal; Color Abatement; Control of Tastes and Odors, Algae Removal; Disinfection, Biological Stabilization
Engineering Aspects
Treatability Studies; Feed Gas Preparation; Ozone Generation; Design of Contact Chambers and Diffusion Systems; Instrumentation and Control Systems; Ozone Destruction Systems; Performance Evaluation; Other Considerations; Checklists for Ozone System Design
Operating an Ozonation Facility
Operating Methods; Ozone System Reliability; Required Plant Maintenance; Employee Training; Safety
Economics of Ozone Systems; New Installations and Retrofits
"Order-of-Magnitude" Estimates of Ozone System Costs; Considerations for Estimating the Cost of an Ozonation Facility; Influence of the Designed Ozonation System on the Capital Cost; Influence of the Designed Ozonation System on the O&M Costs; Case Studies
Ozone System Terminology, Measurements, and Conversions

Guy Bablon, Engineering Development Group Manager, Compagnie Générale des Eaux, Paris, France

Fundamental Aspects
Chemistry—Introduction
Practical Application of Ozone: Principles and Case Studies
Biological Stabilization

Gilles Billen, Professor, Microbiology of the Aquatic Environment, Free University of Brussels, Brussels, Belgium

Practical Application of Ozone: Principles and Case Studies
Biological Stabilization

Marie-Marguerite Bourbigot, Head of the Research Center, Compagnie Générale des Eaux, Maisons-Laffitte, France

Fundamental Aspects
Toxicology
Practical Applications of Ozone: Principles and Case Studies
Minimization of Adverse Health Effects

François Damez, Research Coordinator, Compagnie Générale des Eaux, Paris, France

Engineering Aspects
Corrosion Considerations and Ozone-Resistant Materials; Retrofit of Ozone Systems
Operating an Ozonation Facility
Operating Methods; Ozone System Reliability; Required Plant Maintenance; Employee Training; Safety

Marçel Doré, Professor, Laboratory of Water Chemistry and Pollution, University of Poitiers, Poitiers, France

Fundamental Aspects
Chemistry—Introduction

Françoise Erb, Professor, Pharmaceutical University of Lille, Lille, France

Fundamental Aspects
Toxicology
Practical Application of Ozone: Principles and Case Studies
Minimization of Adverse Health Effects

Cyril Gomella, Board Member of Setude, Paris, France

Practical Application of Ozone: Principles and Case Studies
Disinfection

Philippe Hartemann, Professor, Public Health Laboratory, Medical University, Vandoeuvre, France

Practical Application of Ozone: Principles and Case Studies
Disinfection

Jean-Claude Joret, Research Engineer, Compagnie Générale des Eaux, Research Center, Maisons-Laffitte, France

Practical Application of Ozone: Principles and Case Studies
Disinfection

Alain Laplanche, Assistant Professor, Laboratory of Pollution Chemistry and Environmental Engineering, National High School of Chemistry, Rennes-Beaulieu, France

Fundamental Aspects
Chemistry—Introduction; Ozone Generation; Ozone Gas Transfer
Practical Application of Ozone: Principles and Case Studies
Elimination of Synthetic Organic Chemicals

Bernard Legube, Professor, Laboratory of Water Chemistry and Pollution, University of Poitiers, Poitiers, France

Fundamental Aspects
Chemistry—Introduction
Practical Application of Ozone: Principles and Case Studies
Particulate Removal

Guy Martin, Professor, Laboratory of Pollution Chemistry and Environmental Engineering, National High School of Chemistry, Rennes-Beaulieu, France

Fundamental Aspects
Chemistry—Introduction
Practical Application of Ozone: Principles and Case Studies
Elimination of Synthetic Organic Chemicals

Nathalie Martin, Research Engineer, Compagnie Générale des Eaux, Research Center, Maisons-Laffitte, France

Practical Application of Ozone: Principles and Case Studies
Color Abatement

Willy J. Masschelein, Director of Laboratories and Technical Services, Brussels Intercommunal Waterboard, Brussels, Belgium

Fundamental Aspects
Theoretical Aspects of Ozone Analysis—Introduction; Ozone Generation; Ozone Gas Transfer

Antoine Montiel, Head of the Quality Department, Societé Anonyme de Gestion des Eaux de Paris, Paris, France

Practical Application of Ozone: Principles and Case Studies
Biological Stabilization
Engineering Aspects
Instrumentation and Control Systems

Daniel Perrine, Assistant Professor, Laboratory of Zoology and Parasitology, Pharmaceutical University of Caen, Caen, France

Practical Application of Ozone: Principles and Case Studies
Disinfection

Pierre Schulhof, Senior Vice President, Compagnie Générale des Eaux, Paris, France

Economics of Ozone Systems: New Installations and Retrofits
"Order-of-Magnitude" Estimates of Ozone System Costs; Considerations for Estimating the Cost of an Ozonation Facility; Influence of the Designed Ozonation System on the Capital Cost; Influence of the Designed Ozonation System on the O&M Costs; Case Studies

Pierre Servais, Professor, Microbiology of the Aquatic Environment, Free University of Brussels, Brussels, Belgium

Practical Application of Ozone: Principles and Case Studies
Biological Stabilization

Claire Ventresque, Engineer, Compagnie Générale des Eaux, Paris, France

Fundamental Aspects
Chemistry—Introduction
Practical Application of Ozone: Principles and Case Studies
Biological Stabilization

Special thanks are also due to **Marie-Francoise Morin** (Research Engineer, until October, 1989, Compagnie Générale des Eaux, Research Center, Maisons-Laffitte, France) who supported the European chairperson in the administration portion of this work and who provided the first and second drafts for the following:

Practical Application of Ozone: Principles and Case Studies
Iron and Manganese Removal; Color Abatement; Control of Tastes and Odors; Elimination of Synthetic Organic Chemicals; Algae Removal

We thank Mary Ann Sherrie, Karen Potts, and Cheryl Mannion of the AWWA Research Foundation for their hard work and support throughout this project. We also thank Margaret Sandle, the translator for the French committee, and Sylvie Chaillou, who supported the European co-chairperson in the administrative part of this work.

Mary Kay Kozyra, Senior Technical Editor of the American Water Works Association, deserves our thanks for initially editing over half the book, and for her help in the early stages of the book's development.

Finally, we would like to thank Brian Lewis of Lewis Publishers, Inc., who undertook this project under severe time constraints, and Robin Berry, who edited the remaining chapters of the book. Thanks are also due to Beth Tobin and Terry Cobb of Braun-Brumfield, Inc., who shepherded the book through the typesetting and printing process in record time.

Without the commitment of all these individuals, this book would not have been completed.

I

Introduction

Deborah R. Brink
Bruno Langlais
David A. Reckhow

I.A PURPOSE OF BOOK

This book on ozone for the treatment of drinking water is intended for practicing engineers, water treatment plant managers and personnel, and others interested in ozonation. The purpose of this book is to provide guidance on the various applications of ozone and appropriate system design and operation. Ozone chemistry and physics are also covered to increase the reader's understanding of ozone fundamentals, which should aid in design and operation. While many of the principles presented here can be applied to the treatment of other aqueous process streams (for example, municipal or industrial wastewater, industrial process waters, swimming pools), other references (for example, Rice and Browning 1981; Masschelein 1982; Blogoslawski and Rice 1975; Rice 1981) should be consulted for specifics on these other applications.

The book has been designed so that each chapter can stand alone, although related sections in other chapters are referenced when appropriate. Readers with a strong interest in chemistry may focus on chapters II (Fundamental Aspects) and III (Practical Applications), while readers whose interest lies in engineering design may find chapters III, IV (Engineering), V (Operations), and VI (Economics) of greatest benefit.

Chapter II is devoted to the review of aqueous ozone chemistry, toxicology, analytical methods, and the physics of ozone production and gas transfer. Questions such as how ozone decomposes in water and what types of pollutants are most likely to be attacked are addressed in sec. II.A. Section II.B discusses the toxicology of ozone. Section II.C covers analytical techniques that can be used to measure ozone concentrations. Then, in sec. II.D and II.E, fundamental principles of ozone generation and gas transfer are presented. The chapter assumes a basic knowledge of chemistry and physics.

Chapter III provides the all-important bridge between fundamental chemistry and physics (chapter II) and the specifics of design (chapter IV). Here 10 different ozone applications are discussed in detail: iron and manganese removal (sec. III.B), color removal (sec. III.C), taste and odor control (sec. III.D), destruction of synthetic organic chemicals (sec. III.E), particulate removal (sec. III.F), algae control (sec. III.G), disinfection (sec. III.H), control of disinfection by-products (sec. III.I), minimization of adverse health effects (sec. III.J), and improvement in biological stability in the distribution system (sec. III.K). In each case, the applicable chemistry of aqueous ozone is briefly reviewed to explain field and laboratory results. Where possible, actual case studies are presented. And finally, design considerations for each application are discussed to provide a link to the engineering chapters.

Chapter IV presents the basics of ozone system design. The first section (sec. IV.A) addresses questions of how treatability studies should be conducted and how the data can be used. Later sections focus on specific system components or engineering aspects, including feed gas preparation (sec. IV.B), ozone generation (sec. IV.C), contacting and diffusion (sec. IV.D), instrumentation (sec. IV.E), ozone destruction (sec. IV.F), corrosion considerations (sec. IV.G), system retrofits (sec. IV.H), performance evaluations (sec. IV.I), and other miscellaneous concerns (sec. IV.J). A checklist for ozone system design (sec. IV.K) is included to identify issues that must be addressed during different phases of a project. This material is intended to provide a framework upon which intelligent engineering design can be based.

Aspects of system operation are discussed in chapter V. Principles of operation are discussed in sec. V.A. Of particular concern are reliability (sec. V.B), maintenance (sec. V.C), training (sec. V.D), and safety (sec. V.E).

Chapter VI presents information on economics. Cost curves for order-of-magnitude cost estimates of both capital and operation and maintenance costs are presented along with worksheets that can be used to quickly obtain budget estimates (sec. VI.A). The remainder of the chapter expands upon the factors needed to create exact cost estimates (sec. VI.B) and focuses on the influence a designed ozonation system has on capital costs (sec. VI.C) and on operation and maintenance costs (sec. VI.D). Some case studies are presented at the end of this chapter (sec. VI.E).

Much of the material presented here relates to applications of ozone without the use of activating agents such as hydrogen peroxide or ultraviolet light. However, some consideration is also given to processes dominated by secondary radical oxidants, where such activating agents have been added (that is, advanced oxidation processes). This material is integrated into the various chapters and subsections of the book. For a more specialized treatment of such advanced oxidation processes, the reader should consult Doré (1989) and Peyton et al. (1990).

This book is a comprehensive compilation of the state-of-the-science of ozone technology, as applied to drinking water production. Equipment manufacturers or trade names are cited only for illustrative purposes. Such references do not in any way constitute an endorsement.

I.B CURRENT ROLE OF OZONE IN DRINKING WATER TREATMENT

I.B.1 History of Ozone Use

The ability of ozone to disinfect polluted water was recognized in 1886 by de Meritens (Vosmaer 1916). A few years later, the German firm Siemens & Halske, manufacturers of electrical equipment, contacted local Prussian officials who were willing to test ozone's application for the disinfection of drinking water. Accordingly, a pilot plant was constructed at Martinikenfelde, Germany. Froelich reported in 1891 that tests at this site showed ozone to be effective against bacteria (Vosmaer 1916).

In 1889, the French chemist Marius Paul Otto began studying ozone at La Sorbonne University in Paris. In 1897, he obtained a doctorate degree for his thesis on ozone, the first of its kind. He also created the first specialized company for the manufacture and installation of ozonation equipment in 1897. This company, the Compagnie Provençale de l'Ozone, was renamed in 1919 the Compagnie Générale de l'Ozone, and in 1929 renamed the Compagnie des Eaux et de l'Ozone, thereby associating ozone application with drinking water treatment.

The first full-scale application of ozone in drinking water treatment was in 1893 at Oudshoorn (Netherlands). Later, plants were constructed in Paris, France (1898),

Wiesbaden, Germany (1901), Paderborn, Germany (1902), Niagara Falls, N.Y. (1903), Saint-Petersbourg (Leningrad), USSR (1905), Nice, France (1906), Chartres, France (1908), Paris/St. Maur, France (1909), and Madrid, Spain (1910). The number of ozone installations grew rapidly prior to 1914. Vosmaer (1916) counted at least 49 European plants using Siemens, de Frise, Marmier, Abraham, and Otto ozone generators by 1915. However, the first four types of generators would soon disappear.

This growth in ozone usage soon slowed in the early part of the century when research on poisonous gases conducted during World War I led to the development of inexpensive chlorine. In much of the world, this stimulated the use of chlorine as a disinfectant and severely constrained the spread of ozone. Nevertheless, the construction of new ozone plants continued at a slow pace, especially in France. There was also some construction of ozone plants elsewhere, such as in the Belgian Congo where 11 plants were built before 1939 (Pascal 1986). By 1936, there were close to 100 ozone plants in France and 30–40 in other parts of the world (Evans 1972). However, it was not until after World War II that the rate of construction of ozone facilities fully returned to its earlier level.

Applications of ozone. All of these first-generation plants, the majority of which were located in France, used ozone as a disinfectant. In addition, it was realized quite early that improvements in taste and odor accompanied the disinfection of water by ozone (Vosmaer 1916). Thus, most ozone applications (since ozone's first use around the turn of the century until the 1950s) were for both disinfection and taste and odor control. However, application of ozone for purposes other than disinfection and taste and odor control were emerging even during the pre–World War I era. Vosmaer (1916) reported observations by treatment engineers of color removal and iron and manganese oxidation following ozonation. In Britain, a few ozone plants had been installed prior to 1960 to treat high-quality groundwater in areas where it was considered inappropriate to "contaminate" such supplies by using marginal chlorination (Lowndes 1985).

The 1960s saw widespread introduction of several new applications for ozone. Many of these new applications required that ozone be added during the early stages of treatment, hence the term preozonation. Until this time, ozonation was, in most cases, the last stage of treatment. In France and Germany during the early 1960s, ozone was used specifically to oxidize iron and manganese. While the ability of ozone to oxidize these metals was long recognized (Vosmaer 1916), widespread full-scale application came much later. Some of the first applications for this purpose were at Dusseldorf, Germany (1957 and 1980), Sitterdor, Switzerland (1963), Montjean, France (1967), and Rouen-la-Chapelle, France (1974). At about this same time, several Scottish and Irish plants that employed ozone for color removal were designed and constructed (O'Donovan 1965; Greaves and Lowndes 1987). Water from the Scottish and Irish highlands is characteristically high in color and of low turbidity. Plants in Baerum, Norway (1961), Loch Turret, Scotland (1967), Sligo, Ireland (1970), Loch Lomond, Scotland (1971), and Manchester, England (1970) represent early examples of this application.

During the mid-1960s, the coagulating effects of ozone were first exploited (Guillerd 1968). The observations in Scotland and France that spontaneous flocculation occurred in some ozone contact chambers led to development of the "micellisation–demicellisation" process, an early application of ozone to enhance particulate removal (see chapter III, sec. III.F). Plants that incorporated this process were constructed in Roanne, France (1964), Lausanne, Switzerland (1967), and Constance, Germany (1966). Processes employing the coagulating effect of ozone have since evolved into modern preozonation applications for turbidity control. Also at this time, ozone was used in West Germany and Switzerland to oxidize certain specific micropollutants. In particular, phenolic compounds and several pesticides were early targets of this treatment.

The ability of ozone to control algal growth and the improved treatment in the presence of an algal bloom was first exploited in France in the late 1970s. This is exemplified by the plant at Fromne.

The most recent applications of ozone are for disinfection by-product (DBP) control and biological stabilization, or minimization of the microbiological growth potential of the water. In the case of DBP control, most applications have been in the United States. The discovery of trihalomethanes (THMs) as chlorination by-products in 1973 (Rook 1974), the promulgation of an enforceable standard in the United States in 1979 (*Federal Register* 1979, 1980), and the promulgation of the European Economic Community (EEC) directive in 1980 provided the impetus for this particular application. Biological stabilization was recognized in Europe as a benefit in treatment trains incorporating ozone prior to granular activated carbon (GAC) filtration. The practice of suppressing prechlorination and using ozone prior to GAC, often referred to as biological activated carbon (BAC) filtration, afforded significant extension of carbon bed life. The early full-scale applications of this technology were in Germany. Although biological stabilization was originally thought of as an additional treatment step for organics removal, it is now recognized as a necessary treatment to assure the absence of excessive biological growth in the distribution system when surface waters with high dissolved organic carbon (DOC) levels are used as drinking water sources.

Two of the more recent applications of ozone are unique in that they were proposed based on fundamental chemical considerations. Based on current knowledge of ozone's reactions with synthetic organic chemicals (SOCs) it was proposed that ozone could be used to effectively control certain types of organic compounds. Also, based on the similarities between ozone electrophilic attack (not the only pathway for ozone reaction) and chlorine electrophilic attack, the use of ozone for the control of chlorination by-products was proposed. These uses contrast with other applications that originated from empirical observations.

I.B.2 Current Status in North America

United States. Several water treatment plants in New York, Pennsylvania, and Indiana experimented with ozone during the pre–World War II era, including a plant in Niagara Falls, N.Y., in 1903 (Vosmaer 1916; Consoer 1941). However, the oldest currently operating U.S. drinking water ozonation system, located in Whiting, Ind., was not brought on-line until 1940. Since that time, very little new activity occurred in the United States in terms of ozone usage until the 1970s. In the 1980s, concern over chlorination by-products has generated widespread and accelerating United States interest in ozone.

As of 1990, there are close to 40 U.S. water treatment plants that are equipped with ozonation facilities. Among the plants is the third largest plant in the world, located in Los Angeles (Los Angeles Department of Water and Power Aqueduct Filtration Plant). About 20 additional plants are under construction or in the design phase. Figure I–1 shows the current geographical distribution of North American water treatment plants using ozone. About 40 percent of U.S. ozone plants use lake water as a source, and the remaining plants are evenly split (approximately 20 percent each) between river, reservoir, and groundwater sources (Fujikawa et al. 1989).

Figure I–2 shows the size distribution (in terms of population served) of U.S. community water systems. The height of the bars in this figure represents the ratio of the total population served by the particular size system to the total U.S. population served by all community water systems. Figure I–3 shows the percent of plants within each size category that use ozone. Note the pronounced increase in the use of ozone with increasing plant size. This was not always the case; early U.S. applications were typically at small plants trying to solve site-specific problems (Robson et al. 1990). Only recently has the U.S. water treatment industry fully accepted ozonation, opening the way to larger-scale applications.

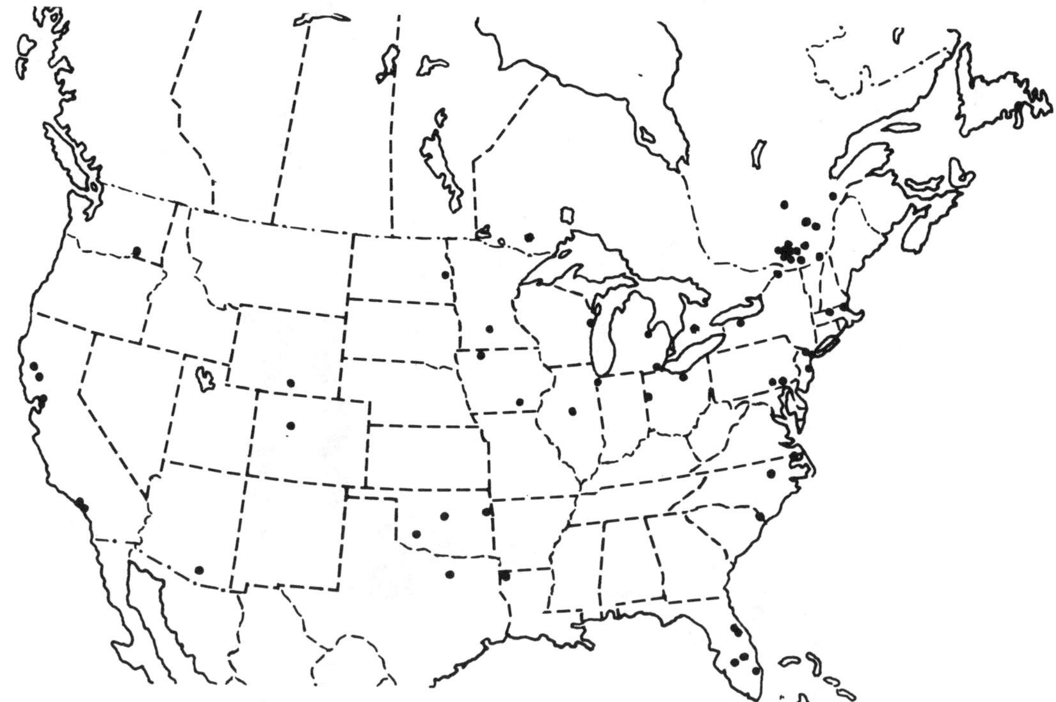

Figure I–1 Location of North American Water Treatment Plants Using Ozone

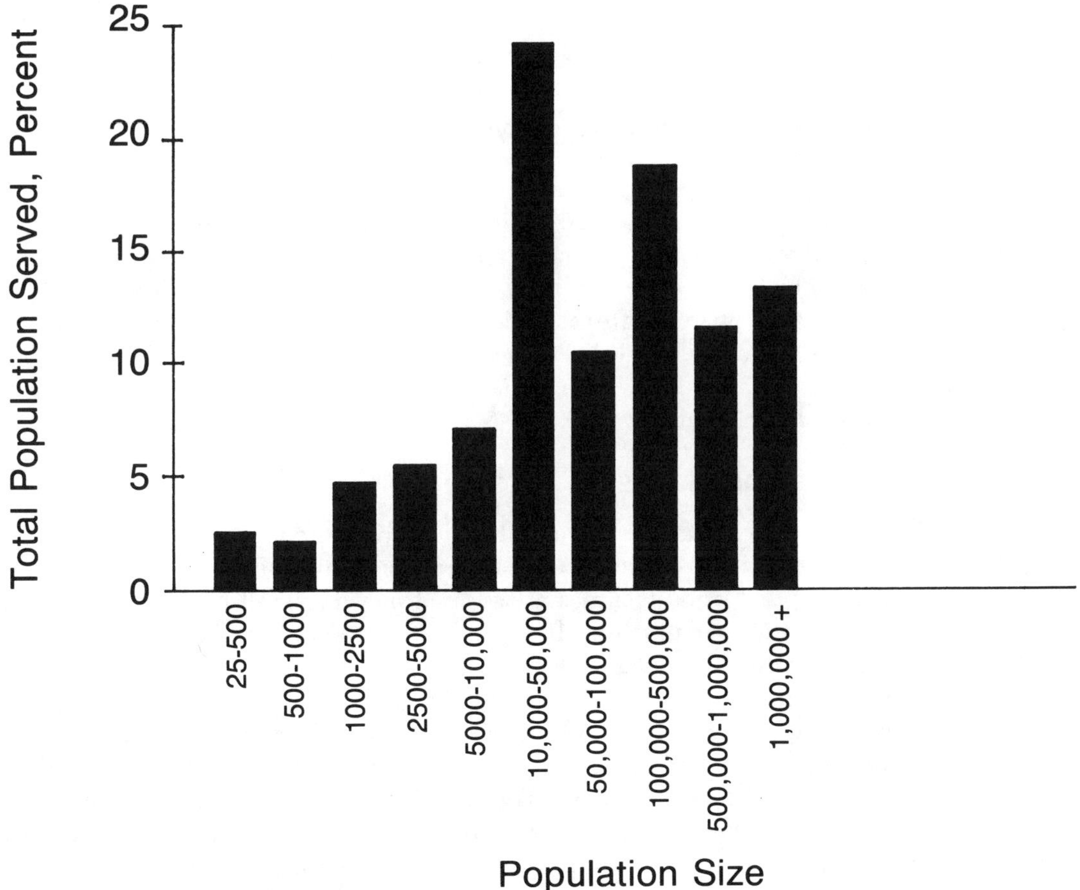

Figure I–2 Size Distribution of U.S. Community Water Systems

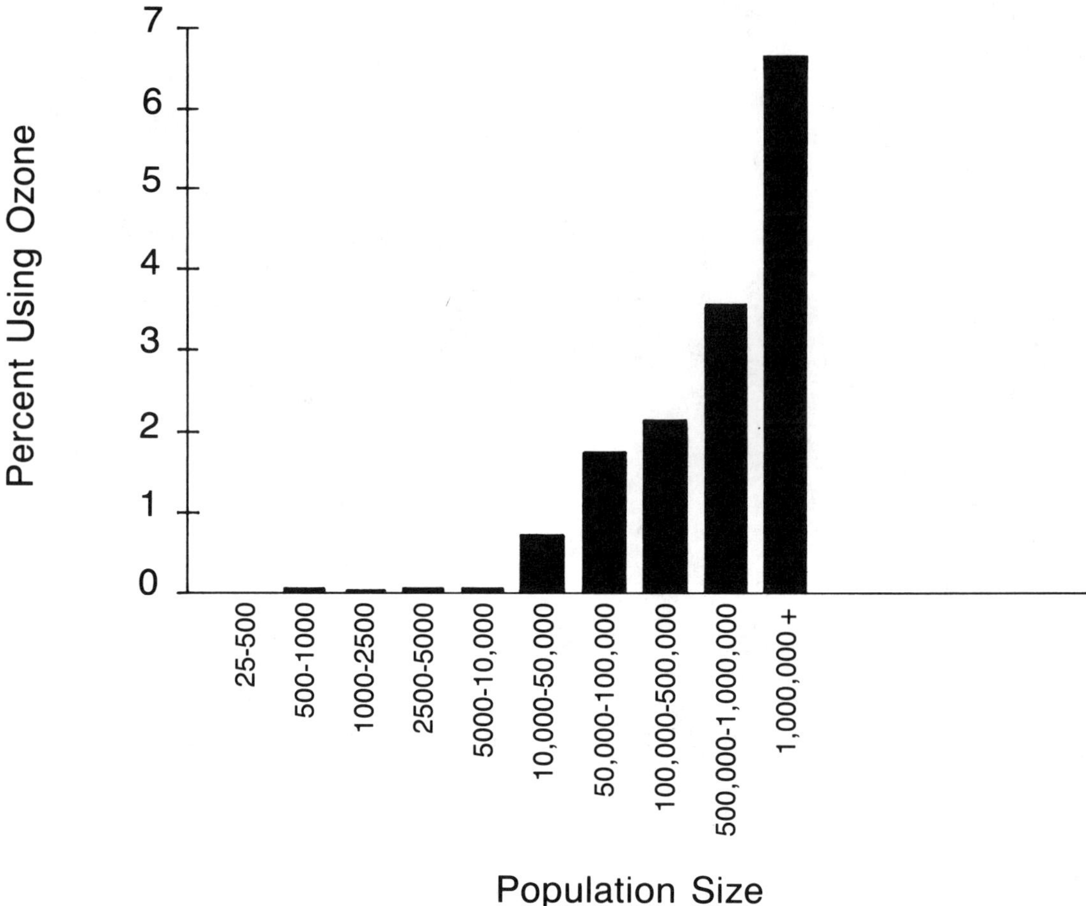

Figure I–3 Percent of U.S. Community Water Systems Using Ozone

Canada. The first Canadian ozone plant was constructed in Sainte Therese, Quebec, in 1956. In 1980, 26 Canadian plants were using ozone for disinfection and taste and odor control. There are currently close to 35 plants in operation, and several more are in the design phase. The vast majority of the operating plants are located in the Province of Quebec (Figure I–1). There are, in fact, so many in the area around Montreal that all sites cannot be shown on a map of the scale of Figure I–1. The world's fourth largest plant is in Montreal.

I.B.3 Current Status in Europe

Western Europe. The first full-scale ozone applications started in western Europe about 100 years ago (see sec. I.A.1) and were primarily located in France, Germany and Switzerland. Ozone facilities are located in very small or medium-sized plants as well as very large plants (from 10 m^3/h to over 30,000 m^3/h [0.06 mgd to over 190 mgd]). Each year more than 40 new ozonation facilities are installed in new water treatment plants or as retrofits in existing treatment trains. In addition, an increasing number of old ozone facilities are being replaced by updated ozone equipment.

France. As of 1990, over 700 French water plants are equipped with operating ozone systems; of these, 50 ozone facilities are over 30 years old and 250 facilities are over 20 years old. The largest ozonation plants are located in the greater Paris area where 12 water plants (total ozone output capacity greater than 500 kg O_3/h [26,500 lb/day]) and using air as the feed gas) are able to treat a daily flow of 3,000,000 m^3 (about 792 mgd) to serve more than 10 million people. Among these facilities is the Choisy-le-Roi ozone facility which is presently the second largest ozone plant in the world with an output of 160 kg O_3/h (8466 lb/day), and the Neuilly-sur-Marne facility, which holds the world record for ozone output per generator—30 kg O_3/h (1587 lb/day)—using air as the feed gas. Total output for this plant is 140 kg O_3/h (7407 lb/day).

Switzerland. Since the 1950s, ozone has been used in over 80 Swiss water plants to disinfect slightly contaminated groundwater and spring water and to oxidize the organics in heavily contaminated surface water. Since 1980, there has been a 25 percent increase in the number of ozone installations in Switzerland (Geering 1988). The largest installations are located in Zürich, Geneva, and Lausanne. About 40 percent of the plants using ozone treat surface water. Of these facilities, about 80 percent are now equipped with an activated carbon filtration step (Geering 1987).

Germany. Full-scale ozonation plants were built in Germany beginning in the 1950s. Today about 70 installations are in operation. Some plants ozonate surface water before it is reinjected into the aquifer, including plants at Mülheim (Jekel 1982; Sontheimer et al. 1978) and Essen.

Other western European countries. Only a few ozonation facilities exist in Austria, Belgium, The Netherlands, United Kingdom, and Ireland. Following is a summary of the ozone activities in these countries.

- In Austria, after several successful ozone installations for iron and manganese removal, ozonation is being considered for oxidation of trace organic contaminants in surface water prior to filtration and prior to activated carbon filtration (Frischherz 1982).
- In Belgium, two plants use ozone to treat the drinking water supplies of Brussels (the largest plant) and Antwerp (the oldest plant). The Tailfer plant in Brussels, with an output of 24 kg O_3/h (1267 lb/day), was constructed in 1970. The plant in Antwerp, which was constructed in 1967, has an output of 6 kg O_3/h (317 lb/day).
- In The Netherlands, the largest ozonation plant was built to supply water for Amsterdam in 1973 (25 kg O_3/h [1323 lb/day]). The other plants in this country were also installed or retrofitted in the 1970s and the 1980s (Dyer-Smith et al. 1985).
- In the United Kingdom, ozone is used in potable water treatment primarily for color removal. In Britain, 5 of the 10 ozone plants commissioned in the past 20 years were designed specifically for color removal, and two were designed for the control of tastes and odors together with color reduction (Greaves and Lowndes 1987). Recently, two plants were designed with two-stage ozonation: Bristol, with intermediate ozonation for taste and odor removal and postozonation for disinfection, and Campion Hills, with preozonation and intermediate ozonation for reduction of organics and disinfection.

Northern Europe. In the Scandinavian countries ozone is not usually used. However, five ozone facilities are in operation: one in Baerum, Norway, built in 1961; two in Helsinki, Finland, built in 1978; one in Espoo, Finland, built in 1980; one in Iggesund, Sweden, built in 1977; and one in Hudiksvall, Sweden, built in 1987 (Westin 1987).

Eastern Europe. Since the mid-1960s, about 80 cities in Eastern Europe have included ozonation in their water treatment process. Ozone is used primarily in large cities that supply potable water to large populations. Such cities include:

- in the USSR—Moscow, with the largest ozonation plant in the world (200 kg O_3/h [10,582 lb/day]), Kiev, Gorki, and Minsk.
- in Poland—Katowice, Lodz, Mokry Dwor, and Wroclaw.
- in Czechoslovakia—Prague.
- in Hungary—Budapest (Varszegi et al. 1987) and Debrecen.

- in Yugoslavia—Belgrade, Sisak, Arilje, Strumica, and Stip.
- in Bulgaria, for the supply of smaller towns such as Burgas, Targovichte, and Preslav.

Southern Europe. In Italy, 20 water utilities with ozone disinfection were built before 1940, essentially for groundwater treatment. Since the late 1970s, ozone has been used in surface water treatment as a preoxidant, for example, in Torino (Merlo and Verde 1983), or in two stages, for example, in Bologna (Magagnoli 1983), Pesaro (Melchiorre 1983), and Florence. In Palermo, postozonation is used for disinfection (Delcominette 1989).

In Spain, ozone has been used only very recently. Since the 1980s, over 10 plants have been equipped with ozonation facilities for pre- and/or intermediate ozonation. Among the largest cities supplied are Bilbao (Le Paulouë 1987), Madrid (two plants), Añarbe, and Valencia.

I.B.4 Current Status in the Middle East and Far East

In the Middle East, very few ozonation plants for drinking water treatment exist. The town of Aleppo, Syria, is supplied by four plants using ozone, the oldest dating back to 1951.

In the Far East, the first large ozonation facilities were installed in Japan and Singapore during the 1970s. They were used primarily for disinfection at the end of a conventional treatment process. During the following decade, Japanese water utilities using ozone, such as Chiba Prefecture (which operates the largest ozone plant in Japan—38 kg O_3/h [2010 lb/day]), installed GAC filters downstream of the ozonation step. For the 1990s, Japanese scientists and local governments plan to modify the existing treatment at large and medium-sized operating plants by adding ozonation and GAC filtration.

New plants built in Singapore and Korea use pre- or intermediate ozonation. In Malaysia, the first plant will be installed in 1991. In China, three plants are reportedly equipped with an ozonation step.

I.B.5 Current Status in the World at Large

It should be noted that in Africa, Zaire has been using ozone since the 1930s (more than 10 plants were in existence), due to the availability of cheap hydraulic power. In the 1980s, several additional small installations have become operational; and in South Africa, three medium-size installations exist, mainly in the Transvaal region.

In Latin America and the Pacific region, the use of ozone is just beginning.

I.C THE FUTURE

Interest in the use of ozone, particularly in the United States, has been steadily increasing over the last several years. New regulations on disinfection and disinfection by-products, which are emerging as a result of the 1986 Amendments to the Safe Drinking Water Act (PL 99–339), are expected to further increase the water industry's interest in this technology. In Europe, the EEC regulations on drinking water quality (Journal Officiel des Communautés Europeénes—80/778/CEE) also lead to an increasing interest in ozone use.

Much research is currently under way to further improve the efficiency of ozonation for various applications. For example, uses of ozone in combination with other oxidants such as hydrogen peroxide or ultraviolet light and in combination with solid-phase catalysts are currently being evaluated in numerous studies. In addition, research is under way to improve ozone generation efficiency, as well as mixing and dissolution in contactors. Advances in ozone residual measurement, especially at low-level aqueous-phase concentrations, will continue.

References

Blogoslawski, W.J. & Rice, R.G. 1975. *Aquatic Applications of Ozone*. Pan American Committee/International Ozone Association, Norwalk, Conn.

Consoer, A.W. 1941. Use of Ozone for Water Purification. *Civil Engrg.*, 11:12:701.

Delcominette, A. 1989. Ozone in Italy. *Ozonews*, 17:6:9.

Doré, M. 1989. *Chimie des Oxidants et Traitement des Eaux*. Technique et Documentation, Lavoisier, Paris.

Dyer-Smith, P. et al. 1985. Uprating Existing Facilities With Medium Frequency Ozone Generators. Proc. Intl. Conf.: The Role of Ozone in Water and Wastewater Treatment. Edited by R. Perry and A.E. McIntyre. Selper Ltd., London (pp. 62–72).

Evans, F.L. 1972. *Ozone in Water and Wastewater Treatment*. Ann Arbor Science Publishers, Inc., Ann Arbor, Mich.

Federal Register. 1979. 44, No. 231, 68624-68707 (Nov. 29).

_____. 1980. 45, No. 49, 15542–15547 (March 11).

Frischherz, H. 1982. Application of Ozone in Drinking Water Treatment in Austria. *Gas. Wasser. Warme*. 36:3:81–86.

Fujikawa, E.G., et al. 1989. Ozonation in America: An Evolution of Success. *Wtr. Engrg. Mgmt.,* Oct.: 20–24.

Geering, F. 1987. Experiences With Ozone Treatment of Water in Switzerland. Proc. 8th Ozone World Congress. International Ozone Association, Zürich, Switzerland (pp. B59–B75).

_____. 1988. Experiences With Ozone Treatment of Water in Switzerland. *Ozone News*, 16: 5:18–24.

Greaves, G.F. & Lowndes, M.R. 1987. Ozone Use at British Water Treatment Plants. Proc. 2nd Intl. Conf.: The Role of Ozone in Water and Wastewater Treatment. Edited by D.W. Smith and G.R. Finch. TekTran Intl. Ltd., Kitchener, Ont. (pp. 245–262).

Guillerd, J.R. 1968. L'Evolution dans le Traitement des Eaux par l'Ozone au Cours des Quinze Dernières Années. *Tech. Sci. Munic.*, 63:10:279–312.

Jekel, M.R. 1982. The DOHNE Treatment Plant in Mülheim (Ruhr): The Mülheim Process. In *Ozonation Manual for Water and Wastewater Treatment*. Edited by W.J. Masschelein. John Wiley and Sons, Inc., New York (pp. 306–308).

Journal Officiel des Communautés Europeénnes 1980. Directive du Conseil du 15 Juillet 1980 Relative à la Qualité des Eaux Destineés à la Consommation Humaine (80/778/CEE) L 229/11-L229/29 (30 Août 1980).

Le Pauloué, J. 1987. New Ozonation Facilities in Poland and Spain. *Ozonews*, 15:3:10.

Lowndes, M.R. 1985. Ozone and the (British) Consumer—Organoleptic Considerations. Proc. Intl. Conf.: The Role of Ozone in Water and Wastewater Treatment. Edited by R. Perry and A.E. McIntyre. Selper Ltd., London (pp. 189–194).

Magagnoli, M. 1983. L'Approvvigionamento Idropotabile dei Comuni Bolognesi. Memorie e Note. Edited by A.M.G.A.–Bologna. Officine Graphiche Pitagora–Technoprint. Bologna, Italy (5, 5–11).

Masschelein, W.J. 1982. *Ozonization Manual for Water and Wastewater Treatment*. John Wiley and Sons, Inc., Chichester, England.

Melchiorre, I. 1983. Impianto di Potabilizzazione del Fiume Metauro (Pesaro). Memorie e Note. Edited by A.M.G.A.–Bologna. Officine Graphiche Pitagora–Technoprint. Bologna, Italy (5, 76–82).

Merlo, G. & Verde, L. 1983. Primi Parziali Risultati Relativi all' Impiego dell'Ozono nell' Impianto di Potabilizzazione PO_3 dell'Azienda Acquedotto Municipale di Torino. Memori e Note. Edited by A.M.G.A.–Bologna. Officine Graphiche Pitagora–Technoprint. Bologna, Italy (5, 63–69).

O'Donovan, D.C. 1965. Treatment with Ozone. *Jour. AWWA*, 57:9:1167–1194.

Pascal, O. 1986. Ozone Disinfection in Drinking Water. Presented at Workshop on Water and Wastewater. USEPA Water Engrg. Res. Lab., Cincinnati, Ohio (pp. 63–84).

Peyton, G.R. et al. 1990. Selection Between Various Advanced Oxidation Processes on the Basis of Water Composition. AWWA Ann. Conf., Cincinnati, Ohio.

Rice, R.G. 1981. Ozone Treatment of Water for Cooling Applications. Sem. of the International Ozone Association Pan American Committee, Norwalk, Conn.

Rice, R.G. & Browning, M.E. 1981. *Ozone Treatment of Industrial Wastewater*. Noyes Data Corp., Park Ridge, N.J.

Robson, P.M. et al. 1990. Status of U.S. Drinking Water Treatment Ozonation Systems. Unpublished report.

Rook, J.J. 1974. Formation of Haloforms During Chlorination of Natural Water. *Wtr. Trmt. Exam.*, 23:234–243 (Part 2).

SONTHEIMER, H. ET AL. 1978. The Mülheim Process. *Jour. AWWA*, 70:7:393–396.

VARSZEGI, C. ET AL. 1987. Experiences With the New Ozone Generating Plant of the Budapest Waterworks. Proc. 8th Ozone World Congress. International Ozone Association, Zürich, Switzerland (pp. B1-B18).

VOSMAER, A. 1916. *Ozone: Its Manufacture, Properties, and Uses*. Van Nostrand Publishers, New York.

WESTIN, S.O. 1987. Ozone for Potable Water Treatment at Hudiksvall, Sweden. Proc. 8th Ozone World Congress. International Ozone Association, Zürich, Switzerland, (pp. D57-D61).

II

Fundamental Aspects

Guy Bablon
William D. Bellamy
Marie-Marguerite Bourbigot
F. Bernard Daniel
Marcel Doré
Françoise Erb
Gilbert Gordon
Bruno Langlais
Alain Laplanche
Bernard Legube
Guy Martin
Willy J. Masschelein
Gilbert Pacey
David A. Reckhow
Claire Ventresque

II.A CHEMISTRY—INTRODUCTION

In an aqueous solution, ozone may act on various compounds (M) in the following two ways (Hoigné and Bader 1977a, 1977b, 1978a):

- by direct reaction with the molecular ozone, and
- by indirect reaction with the radical species that are formed when ozone decomposes in water.

These two basic reactions of ozone in water are illustrated in Figure II–1.

II.A.1 Molecular Ozone Reactivity

The extreme forms of resonance structures in ozone molecules can be represented as follows:

$$O{=}\overset{\delta+}{O}{-}O^{\delta-} \longleftrightarrow O^{\delta+}{-}O{-}O^{\delta-} \qquad \text{(II–1)}$$

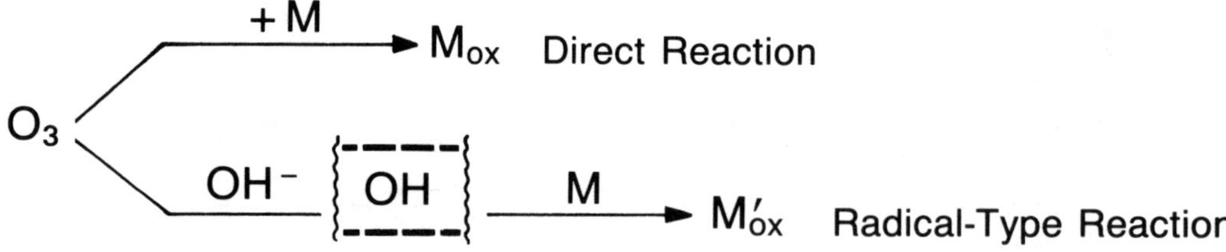

Figure II–1 Reactivity of Ozone in Aqueous Solution

This structure illustrates that the ozone molecule will act as a dipole, as an electrophilic agent, and as a nucleophilic agent. These three types of action, which are described in the following paragraphs, have usually been evidenced in organic solvents (Bailey 1978).

Cyclo addition (Criegee mechanism). As a result of its dipolar structure, the ozone molecule may lead to 1–3 dipolar cyclo addition on unsaturated bonds, with the formation of primary ozonide (I) corresponding to the following reaction:

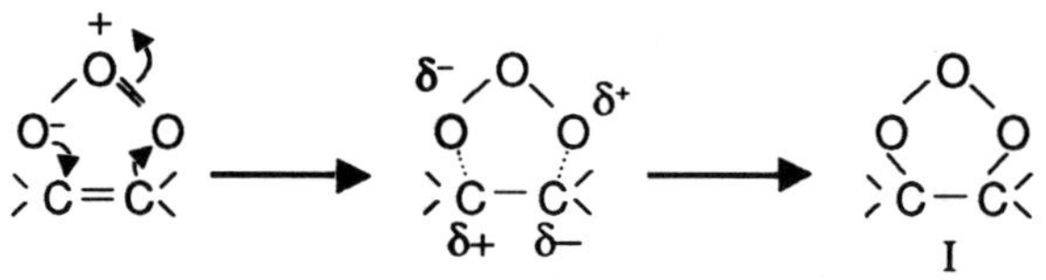

I (II–2)

In a protonic solvent such as water, this primary ozonide decomposes into a carbonyl compound (aldehyde or ketone) and a zwitterion (II) that quickly leads to a hydroxy–hydroperoxide (III) stage that, in turn, decomposes into a carbonyl compound and hydrogen peroxide (see the following reactions).

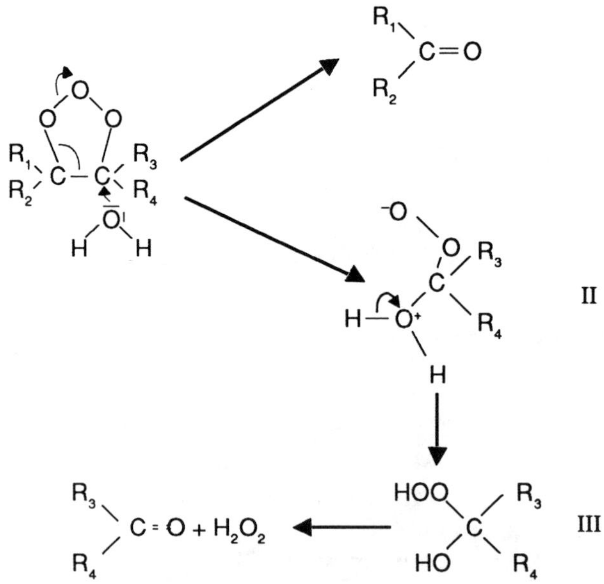

II (II–3)

III

Electrophilic reaction. The electrophilic reaction is restricted to molecular sites with a strong electronic density and, in particular, certain aromatic compounds. Aromatics substituted with electron donor groups (OH, NH_2, and similar compounds) show high electronic densities on carbons located in the ortho and para positions, and so are highly reactive with ozone at these positions. On the contrary, the aromatics substituted with electron-withdrawing groups (−COOH, $-NO_2$) are weakly ozone reactive. In this case, the initial attack of the ozone molecule takes place mainly on the least deactivated meta position.

The result of this reactivity is that the aromatic compounds bearing the electron donor groups D (for example, phenol and aniline) react quickly with the ozone. This reaction is schematically represented as follows:

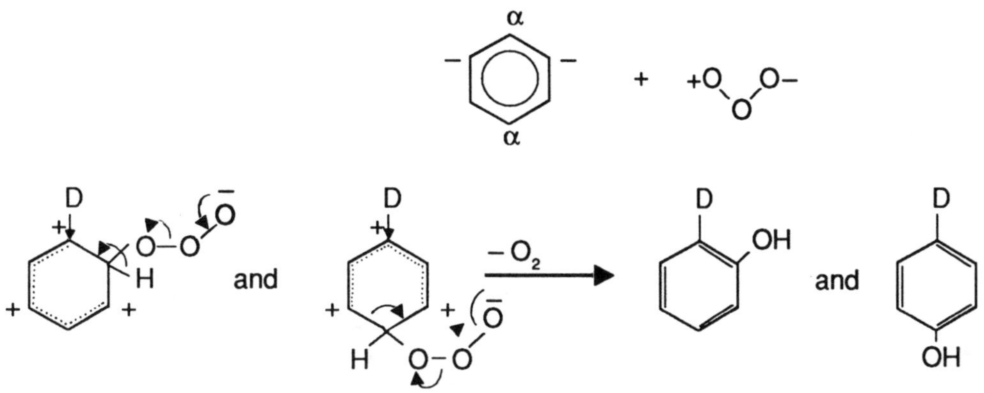

(II–4)

This initial attack of the ozone molecule leads first to the formation of ortho- and para-hydroxylated by-products. These hydroxylated compounds are highly susceptible to further ozonation. The compounds lead to the formation of quinoid and, due to the opening of the aromatic cycle, to the formation of aliphatic products with carbonyl and carboxyl functions.

Nucleophilic reaction. The nucleophilic reaction is found locally on molecular sites showing an electronic deficit and, more frequently, on carbons carrying electron-withdrawing groups.

In summary, the molecular ozone reactions are extremely selective and limited to unsaturated aromatic and aliphatic compounds as well as to specific functional groups.

II.A.2 Decomposition of Ozone

The stability of dissolved ozone (or its half-life as measured in the comparative studies) is readily affected by pH, ultraviolet (UV) light, ozone concentration, and the concentration of radical scavengers (Tomiyasu et al. 1985).

The decomposition rate, measured in the presence of excess radical scavengers, which prevent secondary reactions, is expressed by a pseudo first-order kinetic equation of the following configuration:

$$-\left(\frac{d\,[O_3]}{dt}\right)_{pH} = k'\,[O_3]^1$$

and

$$-\left(ln\frac{[O_3]}{[O_3]_0}\right)_{pH} = k't$$

Where:

k' = pseudo first-order rate constant for a given pH value.

The pseudo first-order constant is a linear function of pH, as shown in Figure II–2. This evolution reflects the fact that the ozone decomposition rate is first order with

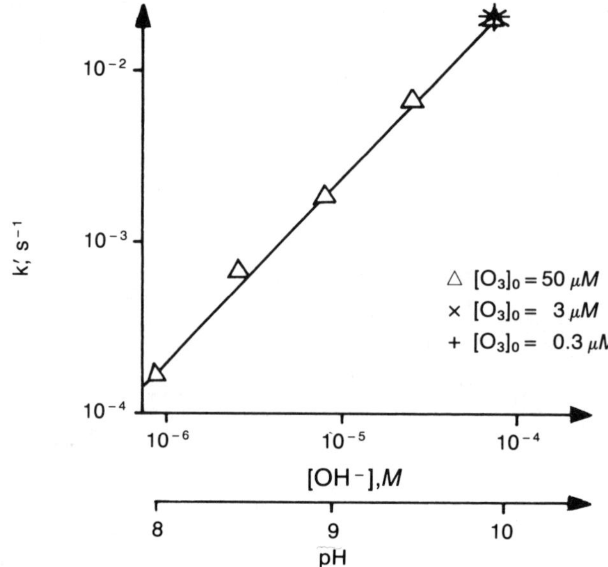

Source: Reprinted with permission from *Envir. Sci. Technol.*, 16:10:678, Staehelin, J. and Hoigné, J. © 1982 American Chemical Society.

Variation of pseudo first-order rate constant k' versus $[OH^-]$.

Figure II–2 Stability of Ozone

respect to both ozone and hydroxide ions, resulting in an overall equation of the following form:

$$-\frac{d[O_3]}{dt} = k[O_3][OH^-]$$

Where:

$$k = \frac{k'}{[OH^-]}$$

Hoigné, Staehelin, and Bader mechanism. Ozone decomposition occurs in a chain process that can be represented by the following fundamental reactions (Weiss 1935; Staehelin et al. 1984), including initiation step 1, propagation steps 2 to 6, and break in chain reaction steps 7 and 8.

(1) $O_3 + OH^- \xrightarrow{k_1} HO_2 + O_2^-$ $k_1 = 7.0 \times 10^1\ M^{-1}\ s^{-1}$
HO_2 : hydroperoxide radical

(1′) $HO_2 \overset{k_2}{\rightleftarrows} O_2^- + H^+$ k_2 (ionization constant) $= 10^{-4.8}$
O_2^-: superoxide radical ion

(2) $O_3 + O_2^- \xrightarrow{k_2} O_3^- + O_2$ $k_2 = 1.6 \times 10^9\ M^{-1}\ s^{-1}$
O_3^- : ozonide radical ion

(3) $O_3^- + H^+ \underset{k_{-3}}{\overset{k_3}{\rightleftarrows}} HO_3$ $k_3 = 5.2 \times 10^{10}\ M^{-1}\ s^{-1}$
$k_{-3} = 2.3 \times 10^2\ s^{-1}$

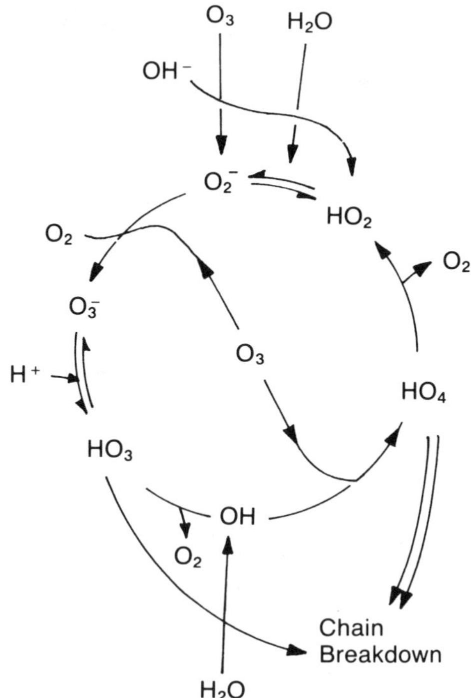

Source: Reprinted with permission from *Jour. Phys. Chem.*, 88:6000, Staehelin, J. et al. © 1988 American Chemical Society.

Figure II–3 Reaction Diagram and Rate Constants for Ozone Decomposition Process

(4) $$HO_3 \xrightarrow{k_4} OH + O_2 \qquad k_4 = 1.1 \times 10^5\ s^{-1}$$

(5) $$OH + O_3 \xrightarrow{k_5} HO_4 \qquad k_5 = 2.0 \times 10^9\ M^{-1}\ s^{-1}$$

(6) $$HO_4 \xrightarrow{k_6} HO_2 + O_2 \qquad k_6 = 2.8 \times 10^4\ s^{-1}$$

(7) $$HO_4 + HO_4 \rightarrow H_2O_2 + 2O_3$$

(8) $$HO_4 + HO_3 \rightarrow H_2O_2 + O_3 + O_2$$

The overall pattern of the ozone decomposition mechanism is shown in Figure II–3.

The first fundamental element in the reaction diagram and in the rate constant values is that the free-radical initiating step constitutes the rate-determining step in the reaction. The second is that the regeneration of the superoxide radical ion O_2^-, or its protonic form HO_2, from the hydroxyl radical OH implies that 1 mol of ozone is consumed. As a result, all the species capable of consuming hydroxyl radicals without regenerating the superoxide radical ion will produce a stabilizing effect on the ozone molecule in the water.

Gordon, Tomiyasu, and Fukutomi mechanism. In the scheme of Hoigné, Staehelin, and Bader, the initiation step is characterized by either an oxygen radical transfer from the ozone to the hydroxide or an oxygen atom transfer from the ozone to the hydroxide with an accompanying one-electron transfer from the hydroxide to the ozone.

$$O_3 + OH^- \rightarrow O_2^- + HO_2$$

Table II–1 Typical Initiators, Promotors, and Inhibitors for Decomposition of Ozone by Radical-Type Chain Reaction

Initiator	Promotor	Inhibitor
OH^-	$R_2-CH(OH)$ (R₂–C with H and OH)	CH_3-COO^- (CH₃–C with =O and O⁻)
H_2O_2/HO_2^-	Aryl–(R)	Alkyl–(R)
Fe^{2+}		HCO_3^-/CO_3^{2-}
Formate	Formate	
Humics	(Humics)	(Humics)
(TOC)	O_3	(TOC)

Source: Reprinted with permission from *Envir. Sci. Technol.*, 19:121, Staehelin, J. and Hoigné, J. © 1985 American Chemical Society.

Kinetic studies and mechanistic descriptions of the decomposition of aqueous ozone have been investigated extensively in recent works (Tomiyasu et al. 1985; Grasso 1987; Gordon 1987). The proposed mechanism involves a two-electron transfer process or an oxygen atom transfer from ozone to hydroxide ion. Following are the steps for the mechanism of ozone decomposition.

$$O_3 + OH^- \xrightarrow{k_9} HO_2^- + O_2 \quad (9) \qquad k_1 = (40 \pm 2)\ M^{-1}\ s^{-1} \text{ (initiation step)}$$

$$HO_2^- + O_3 \xrightarrow{k_{10}} O_3^- + HO_2 \quad (10) \qquad k_{10} = 2.2 \times 10^6\ M^{-1}\ s^{-1}$$

$$HO_2 + OH^- \rightleftarrows O_2^- + H_2O \quad (11) \qquad k_a = 10^{-4.8}$$

$$O_2^- + O_3 \xrightarrow{k_2} O_3^- + O_2 \quad (2) \qquad k_2 = 1.6 \times 10^9\ M^{-1}\ s^{-1}$$

$$O_3^- + H_2O \xrightarrow{k_{12}} OH + O_2 + OH^- \quad (12) \qquad k_{12} = 20\text{–}30\ M^{-1}\ s^{-1}$$

$$O_3^- + OH \xrightarrow{k_{13}} O_2^- + HO_2 \quad (13) \qquad k_{13} = 6 \times 10^9\ M^{-1}\ s^{-1}$$

$$O_3^- + OH \xrightarrow{k_{14}} O_3 + OH^- \quad (14) \qquad k_{14} = 2.5 \times 10^9\ M^{-1}\ s^{-1}$$

$$OH + O_3 \xrightarrow{k_{15}} HO_2 + O_2 \quad (15) \qquad k_{15} = 3 \times 10^9\ M^{-1}\ s^{-1}$$

$$OH + CO_3^{2-} \xrightarrow{k_{16}} OH^- + CO_3^- \quad (16) \qquad k_{16} = 4.2 \times 10^8\ M^{-1}\ s^{-1}$$

$$CO_3^- + O_3 \rightarrow \text{products } (CO_2 + O_2^- + O_2) \quad (17)$$

In this mechanism, the two intermediates, HO_3 and HO_4, are not proposed. In Hoigné's mechanism, if these species agree with the mechanistic pathway, some additional experiments may be necessary to confirm their existence.

Initiators, promotors, and inhibitors of free-radical reactions. The fundamental role played by the hydroxide ions in initiating the ozone decomposition process in water has been described. In fact, a wide variety of compounds are able to initiate, promote, or inhibit the chain-reaction processes (Hoigné and Bader 1977a; Staehelin and

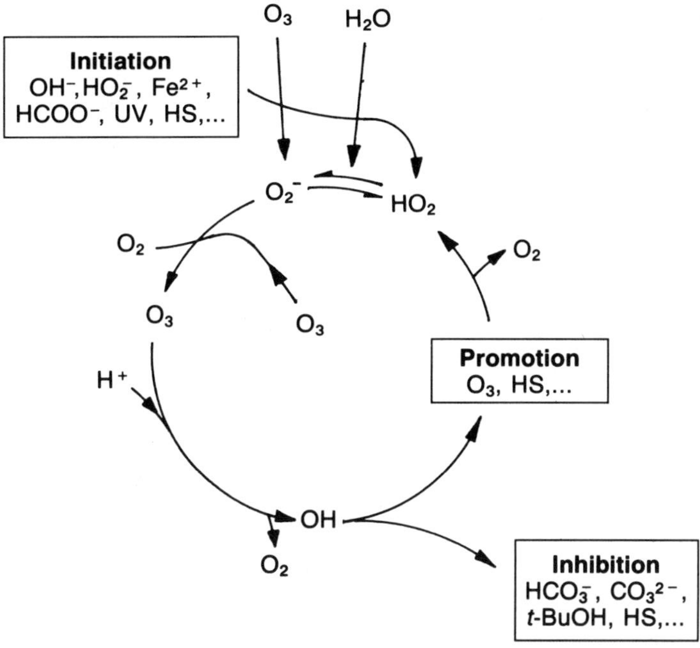

Figure II–4 Mechanism of Ozone Decomposition—Initiation, Promotion, and Inhibition of Radical-Type Chain Reaction

Hoigné 1983). For Hoigné and co-workers, the initiators, promotors, and inhibitors are defined in Table II–1 and Figure II–4 (Hoigné and Bader 1985).

Initiators. The initiators of the free-radical reaction, that is, the compounds capable of inducing the formation of a superoxide ion O_2^- from an ozone molecule (see Figure II–4), are inorganic compounds (for example, hydroxyl ions [OH^-], hydroperoxide ions [HO_2^-], and some cations) and organic compounds (for example, glyoxylic acid, formic acid, and humic substances). Ultraviolet radiation at 253.7 nm is also capable of initiating the free-radical process. This activation of ozone by UV light, plus the combination of H_2O_2/HO_2^-, constitutes the basis of the advanced oxidation processes that will be described later in this section.

Promotors. The promotors of the free-radical reaction are all organic and inorganic molecules capable of regenerating the O_2^- superoxide anion from the hydroxyl radical (see Figure II–4). The rate at which O_2^- reacts with ozone is very great compared with the latter's reactions with other aqueous solutes present in the water. Consequently, the superoxide anion can promote the decomposition of ozone. The regeneration of O_2^- by a promotor is therefore as follows:

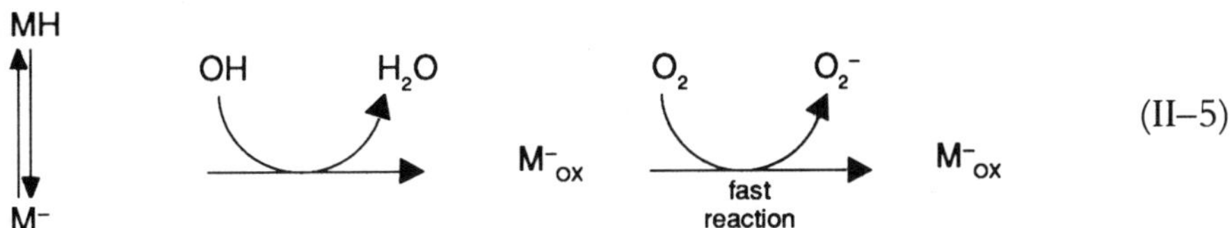

(II–5)

where MH represents inorganic or organic compounds.

Common promotors that are also organics include aryl groups, formic acid, glyoxylic acid, primary alcohols, and humic acids. As an example, the regeneration of O_2^- by formic acid will correspond to the following diagram (Staehelin and Hoigné 1985):

$$HCOOH \rightleftarrows HCOO^- \xrightarrow{OH \rightarrow H_2O} COO^- \xrightarrow[\text{fast reaction}]{O_2 \rightarrow O_2^-} CO_2 \qquad \text{(II–6)}$$

Among the inorganic compounds that are promotors, phosphate species are worth special mention (see Figure II–4). The O_2^- promotion process in the presence of inorganic promotors goes through the following stages (Hoigné and Bader 1985):

$$H_2PO_4^- \rightleftarrows HPO_4^{2-} \xrightarrow{OH \rightarrow OH^-} H_2PO_4 \rightleftarrows HPO_4^- \xrightarrow{RCH_2OH \rightarrow H_2PO_4^-} RCHOH \xrightarrow{O_2 \rightarrow O_2^-} X \qquad \text{(II–7)}$$

This reaction implies that a proton has been extracted from an organic molecule such as alcohol.

Inhibitors. The inhibitors of the free-radical reaction are compounds capable of consuming OH radicals without regenerating the superoxide anion O_2^- (see Figure II–4). Some of the more common inhibitors include bicarbonate and carbonate ions, alkyl groups, tertiary alcohols, and humic substances (Hoigné and Bader 1985).

In the case of bicarbonate and carbonate ions, the inhibiting reactions are as follows (Hoigné and Bader 1985):

$$HCO_3^- \rightleftarrows CO_3^{2-} \xrightarrow{OH \rightarrow OH^-} HCO_3 \rightleftarrows CO_3^- \xrightarrow{O_2 \not\rightarrow O_2^-} X \qquad \text{(II–8)}$$

With *t*-butanol, the following reaction is obtained:

$$(CH_3)_3COH \xrightarrow{OH \rightarrow H_2O} (CH_3)_2C(CH_2)OH \xrightarrow{O_2} (CH_3)_2C(CH_2OO)OH \xrightarrow{\not\rightarrow O_2^-} X \qquad \text{(II–9)}$$

Discussion. In the ozonation of natural water, the first-order rate law is not always correct (Gurol and Singer 1982; Yurteri and Gurol 1988; Tomiyasu et al. 1985). For example, in some cases at pH 8–11, a combined first- and second-order rate law most correctly describes the results (Tomiyasu et al. 1985). Thus,

$$-\frac{d\,[O_3]}{dt} = k\,[O_3] + k'\,[O_3]^2$$

However, the second-order term is not observed if a radical scavenger (Na_2CO_3) is present in the solution. The rate law becomes nearly first-order with respect to ozone as the concentration of added Na_2CO_3 is increased. This fact agrees with the results of Hoigné and co-workers.

The rate at which hydroxyl radicals react with the various organic solutes contained in raw water can be expressed as follows:

$$-\frac{d\,[\mathrm{M}]}{dt} = k_{\mathrm{OH}}\,[\mathrm{M}]\,[\mathrm{OH}]$$

The k_{OH} rate constants are $10^8 \leq k_{OH} \leq 10^{10}\ M^{-1}\ s^{-1}$ (Farhataziz and Ross 1977). Moreover, the reaction of the hydroxyl radicals with the organic solutes is not selective and is probably the only reaction capable of degrading the saturated aliphatic molecules.

Typical rate constants for the hydroxyl radical reactions with the primary radical traps, carbonates, bicarbonates, and *t*-butanol include the following:

CO_3^{2-}	$k_{OH} = 4.2 \times 10^8\ M^{-1}\ s^{-1}$
HCO_3^-	$k_{OH} = 1.5 \times 10^7\ M^{-1}\ s^{-1}$ (per Anbar and Neta 1967)
$(CH_3)_3{-}C{-}OH$	$k_{OH} = 5 \times 10^8\ M^{-1}\ s^{-1}$

These rate constants are, therefore, slightly less than those of OH with organic solutes. However, with regard to bicarbonates in particular, concentrations of scavengers may be much higher than concentrations of the micropollutants to be oxidized. Hence, total inhibition of the free radical chain may occur (Doré et al. 1987), in accordance with the following competitive reaction scheme:

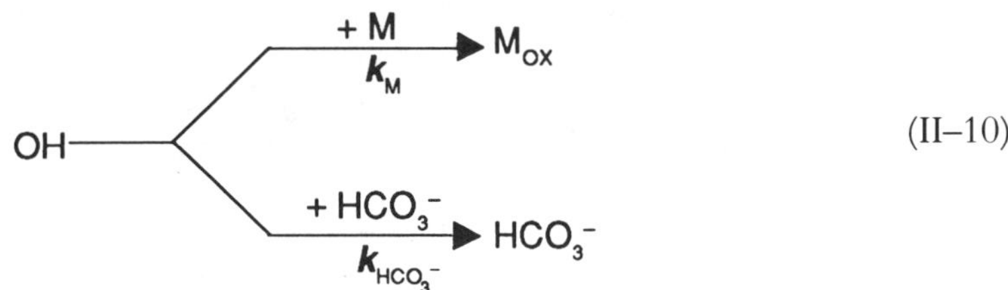

(II–10)

Where:

$k_M\,[M] < k_{OH}\,[HCO_3^-]$

The presence of bicarbonates in the water may, therefore, actually inhibit the free-radical reaction chain, hence slowing down decomposition of ozone in the water. This gives two different results: (1) more ozone will become available for direct and more selective reaction, and (2) less OH radical–induced oxidation will occur. Thus, from a practical point of view, removal of carbonate by softening or remineralization of the water prior to ozonation will influence the nature of the ozonation reactions.

The humic substances that are complex macromolecules seem to be capable, by their various functions, of playing the role of initiator, promotor, or even radical trap (Staehelin and Hoigné 1985). More details are given concerning the role of natural organic compounds in the radical chain reaction of ozone in water in sec. II.A.5.

Advanced oxidation processes. Hydrogen peroxide and UV radiation are among the factors likely to induce decomposition of ozone in water, generating highly reactive hydroxyl radicals. These two entities are used to activate ozone in a neutral pH water and, when combined with ozone, provide advanced oxidizing treatment techniques.

Hydrogen peroxide. Hydrogen peroxide (IV) is a weak acid. When combined with water, it partially dissociates into hydroperoxide ion (V). Thus,

$$\underset{\mathrm{IV}}{H_2O_2} + H_2O \rightleftarrows \underset{\mathrm{V}}{HO_2^-} + H_3O^+ \qquad k_a = 10^{-11.6}$$

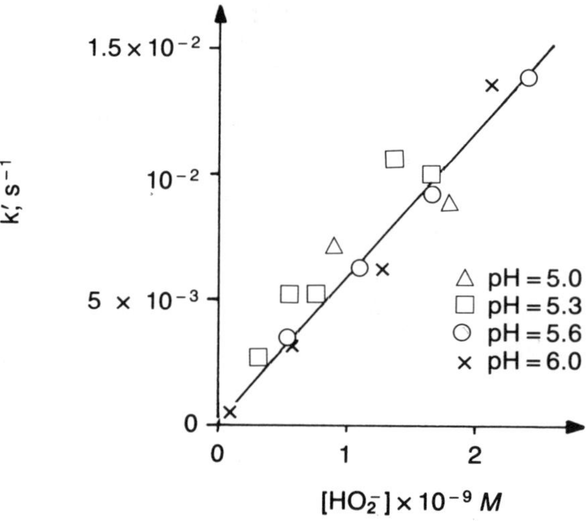

Source: Reprinted with permission from *Envir. Sci. Technol.*, 16:10:678, Staehelin, J. and Hoigné, J. © 1982 American Chemical Society.

Figure II–5 Measurement of Pseudo First-Order Ozone Decomposition Versus Concentration of HO_2^- (methyl mercury = 1.5 mM, $[PO_4]$ t = 50 mM)

The hydrogen peroxide molecule reacts very slowly with ozone (Taube and Bray 1940), whereas the hydroperoxide anion is highly reactive. As a result, the ozone decomposition rate by hydrogen peroxide increases with increasing pH.

For each increase in pH value, the concentration in HO_2^- can be expressed as a function of k_a and of the dissociation coefficient α of the hydrogen peroxide. Therefore,

$$k_a = \frac{[HO_2^-][H_3O^+]}{[H_2O_2]}$$

$$\alpha = \frac{[HO_2^-]}{[H_2O_2] + [HO_2^-]} = \frac{[HO_2^-]}{[H_2O_2]_t}$$

Where:

$pH << pk_a$
$[H_2O_2]_t = [H_2O_2] +]HO_2^-] \approx [H_2O_2]$
$k_a = \alpha\,[H_3O^+]$
$\log \alpha = pk_a - pH.$

Given a known concentration of HO_2^-, it is seen that the O_3 decomposition rate is first order with respect to ozone and depends on the HO_2^- concentration. This is shown by the development of the kinetic constant, k', corresponding to the following kinetic equation (Figure II–5):

$$-\frac{d[O_3]}{dt} = k'[O_3]$$

Where:

k' = pseudo first-order rate constant of ozone decomposition with H_2O_2.

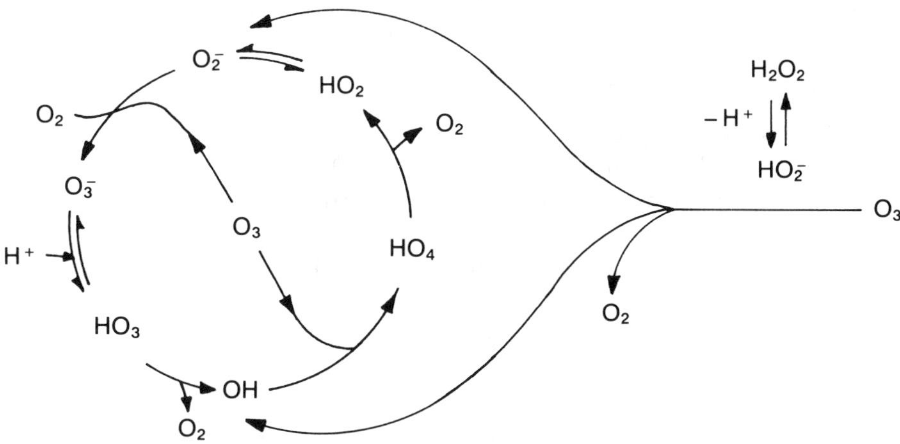

Figure II–6 Ozone Decomposition Process by Hydroperoxide Ions

This development reflects an overall kinetic equation of the following form:

$$-\frac{d\,[O_3]}{dt} = k''[O_3]\,[HO_2^-]$$

Where:

k'' = second-order rate constant of ozone decomposition with H_2O_2.

This rate constant ($k'' = 5.5 \pm 1.0 \times 10^6\ M^{-1}\ s^{-1}$) is therefore vastly superior to that of ozone decomposition initiated by hydroxyl ions. Therefore, the result is that very low concentrations of HO_2^- are kinetically effective in initiating O_3 decomposition. In this context, it should be noted that HO_2^- is not normally measured directly, and it is the kinetic law that explains the role of HO_2^-.

According to the mechanism of ozone decomposition, the initiation and propagation reactions could be as follows (equation numbers are from sec. II.A.2):

(0)	$H_2O_2 + H_2O \rightleftarrows HO_2^- + H_3O^+$	k_a	$= 10^{-11.6}$
(10)	$O_3 + HO_2^- \rightarrow OH + O_2^- + O_2$	k_{10}	$= 2.2 \times 10^6\ M^{-1}\ s^{-1}$
(1′)	$O_2^- + H^+ \rightleftarrows HO_2$	$1/k_a$	$= 10^{4.8}$
(2)	$O_3 + O_2^- \rightarrow O_3^- + O_2$	k_2	$= 1.6 \times 10^9\ M^{-1}\ s^{-1}$
(3)	$O_3^- + H^+ \rightleftarrows HO_3$	k_3 k_{-3}	$= 5.2 \times 10^{10}\ M^{-1}\ s^{-1}$ $= 2.3 \times 10^2\ s^{-1}$
(4)	$HO_3 \rightarrow OH + O_2$	k_4	$= 1.1 \times 10^5\ s^{-1}$

This is schematically displayed in Figure II–6.

The hydroperoxide ions (HO_2^-) consumed by ozone are very quickly replaced by shifting the equilibrium (reaction 1). The rate at which the H_2O_2 is consumed by O_3 then takes the following form:

$$-\frac{d\,[H_2O_2]}{dt} = k\,[H_2O_2]_t\,[O_3]$$

UV radiation. Both in a gaseous phase and in an aqueous solution, ozone absorbs UV radiation. Maximum absorption is at 253.7 nm.

In a gaseous phase enriched with water vapor, the mechanism of photolysis involves the release of a molecule of oxygen and an atom of oxygen as ^1D (McGrath and Norrish 1960). The latter may react with water to produce hydroxyl radicals.

$$O_3 + h\nu \rightarrow O_2 + O(^1D)$$
$$O(^1D) + H_2O \rightarrow 2OH$$

In an aqueous phase, the hydroxyl radicals may combine to form hydrogen peroxide (Taube 1956). The direct production of H_2O_2 by the following reaction is also mentioned (Peyton and Glaze 1983):

$$O_3 + H_2O \xrightarrow{h\nu} O_2 + H_2O_2$$

The hydrogen peroxide formed in this way may be either photolyzed or decomposed by ozone, as shown in the following reaction:

$$H_2O_2 \underset{+H^+}{\overset{-H^+}{\rightleftarrows}} HO_2^- \qquad \text{(1)}\; \xrightarrow{+h\nu} 2OH \qquad \text{(2)}\; \xrightarrow{+O_3} HO_2 + O_3^- \rightarrow OH \qquad \text{(II–11)}$$

Considering H_2O_2 photolysis is very slow (Guittonneau et al. 1988a, 1988b) compared with the rate at which O_3 is decomposed by HO_2^-; it seems that at a neutral pH path 2 is the main pathway.

During ozone reactions with organic MH in aqueous solutions, hydrogen peroxide molecules may be regenerated. The overall pattern is shown in Figure II–7 (Peyton and Glaze 1988).

A continuous ozonized air or oxygen diffusion reactor constitutes a heterogeneous system, and initiation of the free radical process could result from three different processes: the photolysis of the ozone in solution, the photolysis of the ozone in the gaseous phase (inside the gas bubble), and the photolysis of the ozone in the liquid film at the gas–liquid interface.

Measuring the photon flux. Applying UV radiation requires determining the flux of photons emitted by the lamps. The photon flux must be measured for three reasons: the characterization of the lamps, the determination of the quantum yield, and the optimization of reactor geometry.

Determination of the photon flux can be performed using actinometry or radiometry (Braun et al. 1986). Radiometry consists of installing sensors that convert radiant energy into an electric signal, the intensity of which depends on the energy of the incident light. The classical radiometric systems in use are thermal detectors and photodetectors.

Actinometry is based on the determination of a number of molecules (actinometers) that undergo photochemical transformation when submitted to a given period of irradiation. In fact, if the quantum yield ($\phi_{AC,\lambda}$) of the actinometer is known for the wavelength under study, the number of photons absorbed at a wavelength λ over a

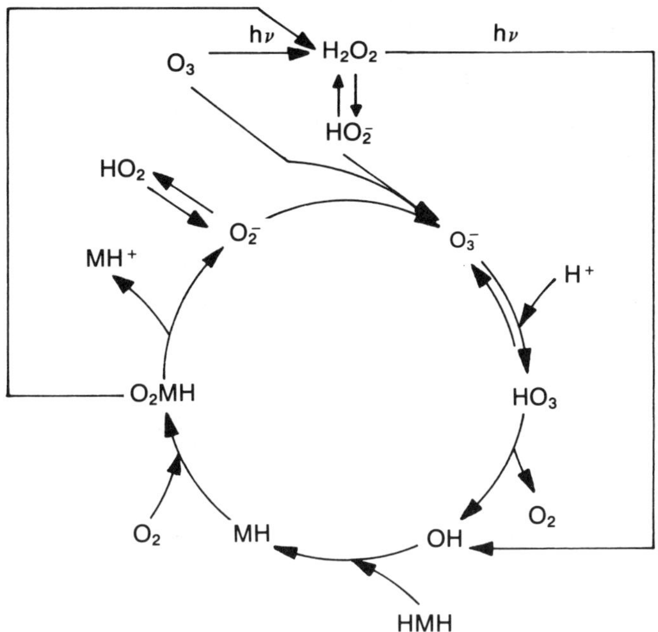

Source: Reprinted with permission from *Envir. Sci. Technol.*, 22:766, Peyton, C.R. and Glaze, W.H. © 1988 American Chemical Society.

Figure II–7 Ozone Decomposition Process by Photolysis at 253.7 nm

given time $N_{AC,\lambda}$ can be obtained from the number of actinometer molecules that have reacted $\Delta\, n_{AC}$ by applying the following ratio:

$$N_{AC,\,\lambda} = \Delta n_{AC} / \phi_{AC,\,\lambda}$$

The incident photon flux is then obtained from the photon flux absorbed according to the Beer–Lambert law.

Two of the most widely used actinometers for measuring the photon flux of UV lamps are uranyl oxalate $UO_2C_2O_4$ and potassium ferrioxalate $K_3Fe\ (C_2O_4)_3$, $3H_2O$. In the first case, the UV radiation induces the decomposition of the oxalate ions through the following reaction:

$$H_2C_2O_4 \xrightarrow{h\nu} H_2O + CO_2 + CO$$

The oxalate is titrated by potassium permanganate before and after irradiation. In the second case, the photochemical reaction is reflected by the oxidation of the oxalate ions and the reduction of ferric iron into ferrous iron.

$$2Fe^{3+} + C_2O_4{}^{2-} \rightarrow 2Fe^{2+} + 2CO_2$$

The Fe^{2+} ions are measured by the *o*-phenanthroline method. The application and effectiveness of these combined oxidation processes are discussed in chapter III (sec. III.C)

II.A.3 Reaction With Inorganics

The reaction of ozone with inorganic compounds found in water generally follows a first-order kinetic law with respect to both ozone and the oxidizable compound. The ozone disappearance rate can typically be expressed by equations of the following form:

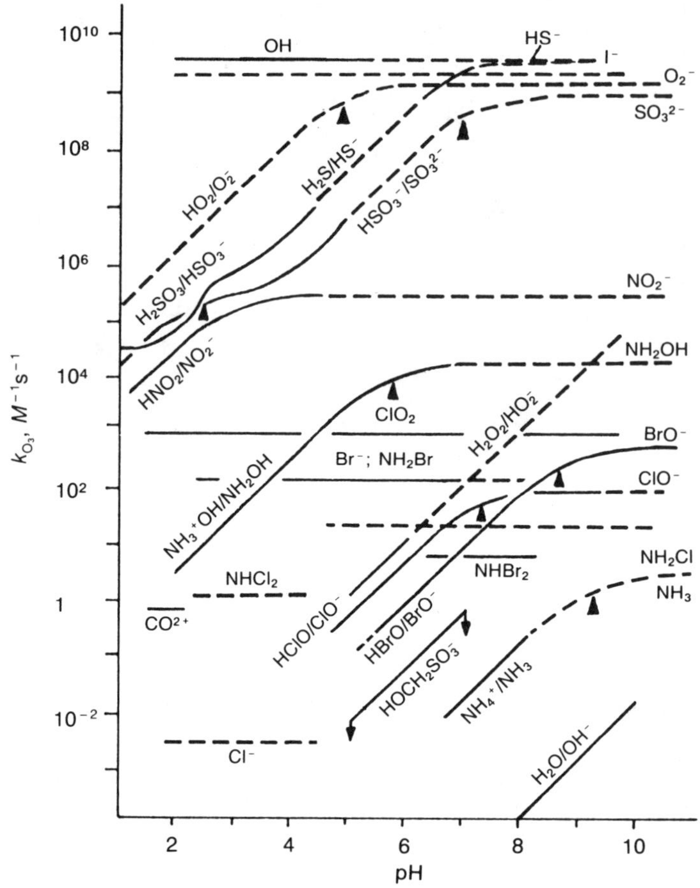

Source: Reprinted with permission from *Wtr. Res.*, 19:8:1000, Hoigné, J. et al. © 1985 Pergamon Press PLC.

Figure II–8 Ozonation of Inorganic Compounds: Apparent Rate Constants Versus pH (pk_{HB} Value)

$$-\frac{d\,[O_3]}{dt} = k_{O_3}\,[O_3]\,[M]$$

Where:

M = solute.

The rate constant k_{O_3} varies according to the pH, and Figure II–8 points to the fact that the reactivity of molecular and anionic forms is greater than protonated forms.

Action on metals: iron and manganese. Iron and manganese are two of the most abundant elements found in the earth's crust. The oxidized forms Fe^{3+} and Mn^{4+} are most prevalent, but it is the reduced forms (rhodochrosite [$MnCO_3$] and siderite [$FeCO_3$]) that are present in water (Pouvreau 1984). The existence of iron and manganese in drinking water generates certain unpleasant side effects, such as color and precipitates in water and stains on clothing and plumbing fixtures.

The conversion of these carbonates ($MnCO_3$ and $FeCO_3$) into bicarbonate forms can take place when CO_2 is present, via the following reactions:

$$FeCO_3 + CO_2 + H_2O \rightarrow Fe(HCO_3)_2$$

$$MnCO_3 + CO_2 + H_2O \rightarrow Mn(HCO_3)_2$$

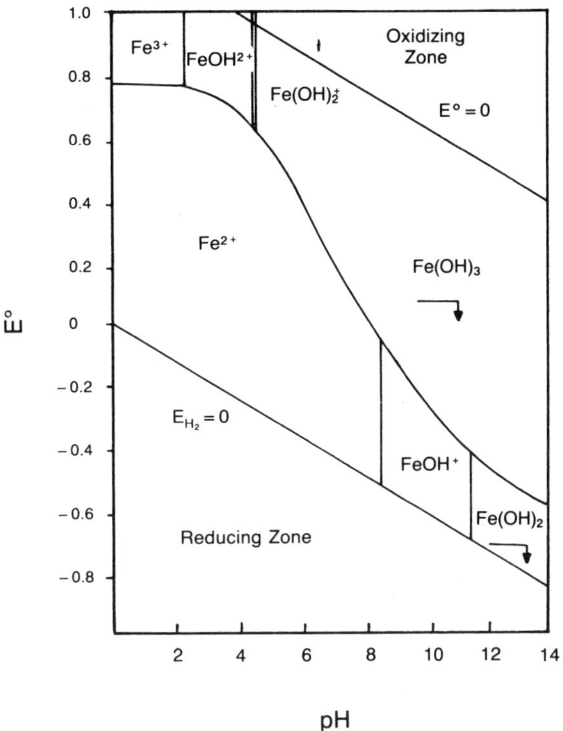

Source: Reprinted with permission from *Chimie des Oxydants et Traitement des Eaux,* Doré, M. © 1989 Technique et Documentation–Lavoisier.

Figure II–9 Diagram of Iron Equilibrium in Water

Large concentrations of CO_2 are generally found in groundwaters and at the bottom of surface water impoundments where anaerobic conditions are prevalent.

Removal of iron and manganese consists of transforming the soluble matter into insoluble compounds that can be filtered out of the water. The transformation of the reduced soluble forms (Fe^{2+} and Mn^{2+}) into insoluble oxides (Fe^{3+} and Mn^{4+}) offers a solution to the problems connected with these compounds.

Iron removal. Figure II–9 displays the distribution of the different forms of iron in equilibrium with water (Doré 1989). It is shown that in a neutral environment the passage from a reduced to an oxidized form is thermodynamically favorable.

The ferrous iron is, therefore, easily oxidized by ozone by the following reactions:

$$2\,Fe^{2+} \xrightarrow{O_3 \;\rightarrow\; O_2} 2\,Fe^{3+} \xrightarrow{H_2O} Fe(OH)_3 \downarrow \qquad \text{(II–12)}$$

This reaction implies a consumption of 0.43 mg of O_3/mg of Fe^{2+}. With regard to the process involved, since Fe^{2+} is an O_3 decomposition initiator, it is likely that the reaction proceeds by the transfer of electrons.

The mechanism for this reaction, proposed by Hart et al. (1983), is as follows:

$$\text{(1)} \qquad Fe^{2+} + O_3 \rightarrow Fe^{3+} + O_3^-$$

$$\text{(2)} \qquad O_3^- \rightleftarrows O^- + O_2$$

$$\text{(3)} \qquad O^- + H_2O \rightarrow OH + OH^-$$

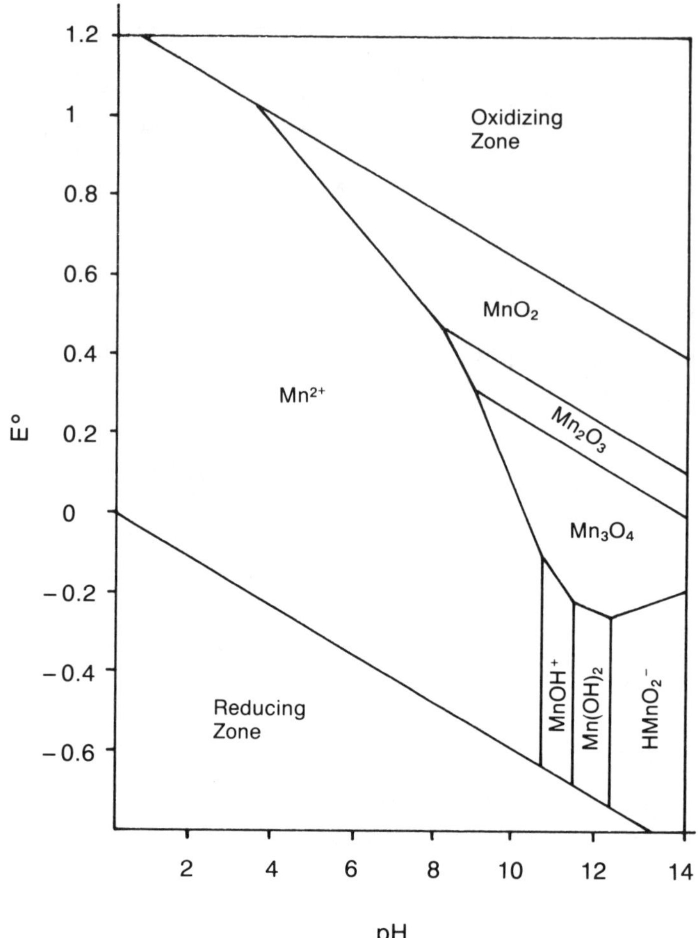

Source: Reprinted with permission from *Chimie des Oxydants et Traitement des Eaux,* Doré, M. © 1989 Technique et Documentation–Lavoisier.

Figure II–10 Diagram of Manganese Equilibrium in Water

$$(4) \qquad OH + Fe^{2+} \rightarrow OH- + Fe^{3+}$$

$$(5) \qquad OH + O_3 \rightarrow HO_2 + O_2$$

However, reaction 2 is not always dominant, and a direct oxidation of Fe^{2+} with O_3^- is possible (Yang and Neely 1986). Thus,

$$O_3^- + Fe^{2+} + H_2O \rightarrow O_2 + Fe^{3+} + 2OH^-$$

Manganese removal. The distribution of the various forms of manganese is shown in Figure II–10 (Doré 1989). Compared with the different forms of iron, this diagram shows that the conversion of Mn^{2+} into MnO_2 requires more oxidizing power than the conversion of Fe^{2+} into $Fe(OH)_3$ because it requires more energy.

The oxidation reaction of manganese by ozone takes the following form:

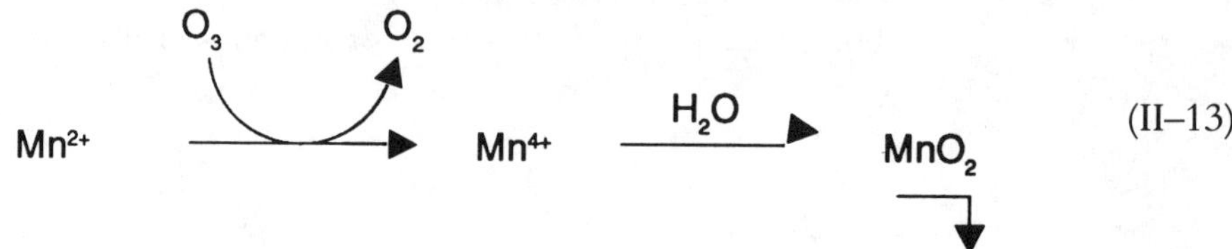

(II–13)

Manganese is often more difficult to remove than iron (Stoebner and Rollag 1981; Pouvreau 1984). When organics and humic acids are present in the water, the reaction yield is severely diminished. The consumption of ozone by the organic compounds is probably responsible for this phenomenon (Giannissis et al. 1985), and the ozonation conditions may act on the yield of the manganese oxidation (Paillard et al. 1989a). More details about the removal of iron and manganese are given in chapter III (sec. III.B)

Action on reduced nonmetal species. The interaction between ozone and hydrogen peroxide has been discussed in an earlier section. The interaction between ozone and reduced nonmetal species will be discussed in the following sections.

Halogens. The standard oxidation–reduction potential (E^0) values applicable to the combination O_3/O_2 and the various halogen/halide couples are as follows:

	E^0 (V)
O_3/O_2	2.07
$HClO/Cl^-$	1.49
$HBrO/Br^-$	1.33
HIO/I^-	0.99

Therefore, it is thermodynamically possible for ozone to oxidize chloride, bromide, and iodide. In fact, when bromide or iodide ions are present in waters being ozonized, free bromine (Br_2) and/or hypobromous acid (HOBr), hypobromite ion (BrO^-), bromate ion (BrO_3^-), free iodine (I_2), and/or hypoiodous acid (HOI), hypoiodite ion (IO^-), and iodate ion (IO_3^-) are some of the halogen-containing oxidants that can be produced (Crecelius 1978; Haag and Hoigné 1983). The rate of reaction is very high in the case of the iodides. It is slower for bromides and practically zero for chlorides.

Bromide ion is oxidized into hypobromite according to the following reaction:

$$O_3 + Br^- \xrightarrow{k_1} O_2 + BrO^-$$

The hypobromite formed can be partially oxidized into bromate ion, while a part of the bromide is regenerated through the following reactions:

$$BrO^- + O_3 \xrightarrow{k_2} (O_2 + BrOO^-) \rightarrow Br^- + 2O_2$$

$$BrO^- + 2O_3 \xrightarrow{k_3} BrO_3^- + 2O_2$$

The overall reaction is as follows:

$$2BrO^- + 3O_3 \rightarrow Br^- + BrO_3^- + 4O_2$$

The reaction pattern (Haruta and Takeyama 1981; Haag and Hoigné 1983) together with the rate constants can be represented by Figure II–11 (Haag and Hoigné 1983). It is interesting to note that for lower pH values, less bromate is formed. Figure II–11 shows that the combination of the following reactions causes the catalytic destruction of ozone:

$$Br^- \xrightarrow{O_3 \;\curvearrowright\; O_2} BrO^- \quad \text{and} \quad BrO^- \xrightarrow{O_3 \;\curvearrowright\; 2\,O_2} Br^- \qquad \text{(II–14)}$$

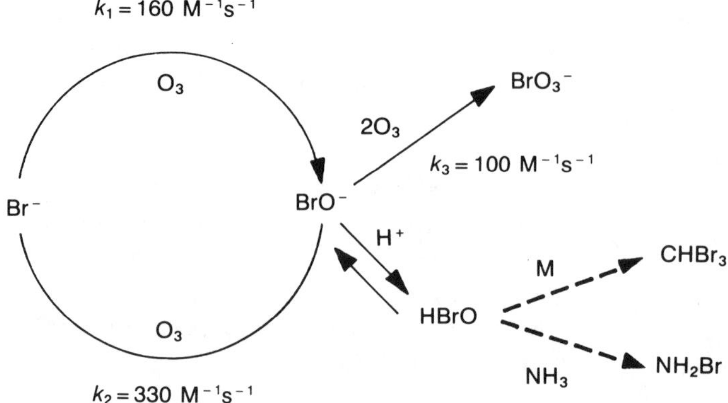

Source: Reprinted with permission from *Envir. Sci. Technol.* 17:266, Haag, W.R. and Hoigné, J. © 1983 American Chemical Society.

Figure II–11 Diagram of Bromide Oxidation Cycle

Oxidation of ammonia. Ammonia can be oxidized by the following reaction (Haag et al. 1984):

$$4O_3 + NH_3 \rightarrow NO_3^- + 4O_2 + H_3O^+$$

with

$$-\frac{d\,[NH_3]}{dt} = k\,[NH_3]\,[O_3]$$

The ammonia oxidation rate is slow, especially at pH levels below the pK_a (about 9.3). The curves diagrammed in Figure II–12 show the influence of pH (part A) and the ammonia-to-nitrate conversion rate in terms of quantity (part B).

The oxidation of ammonia most likely results from the superposition of the direct reaction produced by molecular ozone and the indirect reaction of the OH radicals formed by the decomposition of ozone (Figure II–13) (Hoigné and Bader 1978b). The share of indirect reaction is inversely proportional to the pH level (see Figure II–13).

Influence of bromide. The existence of bromide in water may catalyze the degradation of ammonia. The hypobromous acid formed by the oxidation of the bromide quickly leads to the formation of monobromamine, which in turn is oxidized by ozone resulting in the release of bromide and nitrate. This is shown in the following reaction diagrams (Haag et al. 1984; Hoigné and Bader 1985):

$$Br^- + O_3 \rightarrow BrO^- + O_2 \qquad k_1 = 160\ M^{-1}\ s^{-1}$$

$$BrO^- + H_3O^+ \rightleftarrows HBrO + H_2O \qquad k_a = 2.06 \times 10^{-9}\ M^{-1}\ s^{-1}$$

$$HBrO + NH_3 \rightarrow NH_2Br + H_2O \qquad k_2 = 7.5 \times 10^7\ M^{-1}\ s^{-1}$$

$$NH_2Br + 3O_3 + 2H_2O \rightarrow NO_3^- + Br^- + 2H_3O^+ + 3O_2 \qquad k_3 = 40\ M^{-1}\ s^{-1}$$

$$NH_3 + 4O_3 \xrightarrow{Br^-/HBrO} NO_3^- + 4O_2 + H_3O^+ \text{ (overall)}$$

As bromide is regenerated by the system, it, in effect, plays the part of a catalyst. Moreover, the monobromamine oxidation rate is somewhat faster than that of ammonia and, above all, is independent of the pH. This allows indirect oxidation of the ammonia

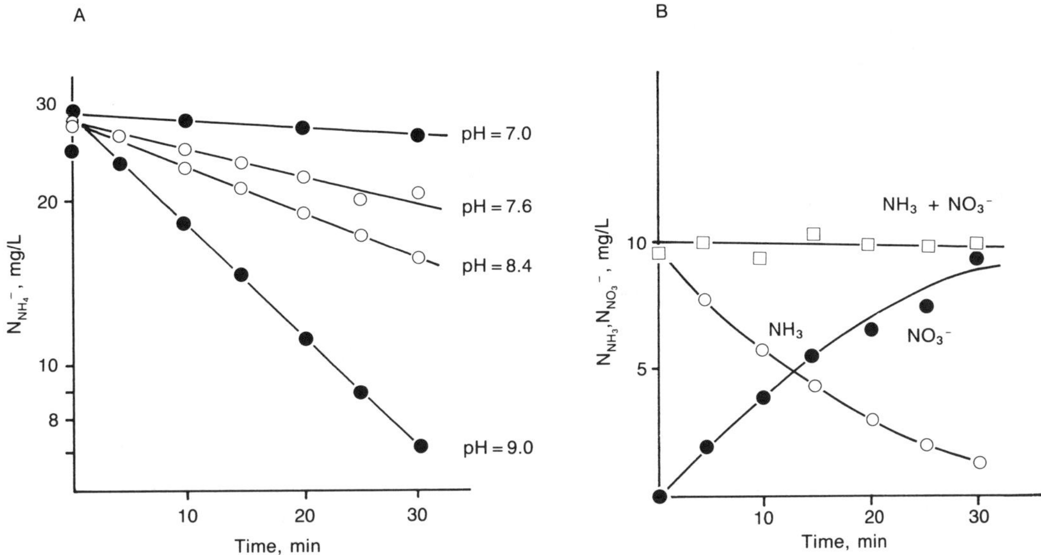

Source: Reprinted with permission from *Wtr. Res.* 9:129–130, Singer, P.C. and Zilli, W.B. © 1975 Pergamon Press PLC.

(A) Influence of pH on ammonia oxidation rate; (B) Conversion of NH_3 into NO_3^- at pH 9.

Figure II–12 Action of Ozone on Ammonia

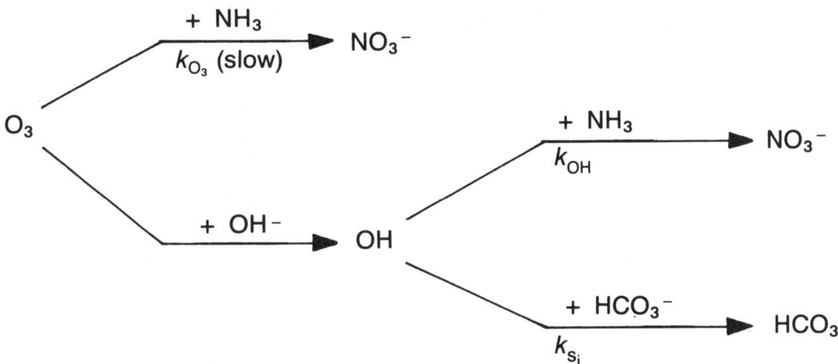

Figure II–13 Flow Diagram of Ammonia Ozonation Process

in a neutral pH environment to occur at a faster rate than would occur during direct ozonation.

Oxidation of nitrite. Nitrite is quickly oxidized by ozone into nitrate at a rate constant of 3.3–3.7 × 10^5 M^{-1} s^{-1} (Garland et al. 1980; Haag et al. 1984).

$$NO_2^- + O_3 \rightarrow NO_3^- + O_2$$

The stoichiometry of the reaction corresponds to a consumption of 1.04 mg ozone/mg NO_2^-.

Oxidation of inorganic forms of sulfur. Sulfide is oxidized into sulfate via the following reactions:

$$S^{2-} + 3O_3 \rightarrow SO_3^{2-} + 3O_2$$

$$SO_3^{2-} + O_3 \rightarrow SO_4^{2-} + O_2$$

giving a total of

$$S^{2-} + 4O_3 \rightarrow SO_4^{2-} + 4O_2$$

This reaction implies a consumption of 6 mg ozone/mg S^{2-}.

As with other compounds, the reaction rates decrease with the degree of anion protonization. This is shown by the following rate constants:

S^{2-}	$k_{O_3} = (3 \pm 1) \times 10^9\ M^{-1}\ s^{-1}$	per Hoigné and Bader (1985)
HS^-	$k_{O_3} = (1.1 \pm 0.4) \times 10^6\ M^{-1}\ s^{-1}$	
H_2S	$k_{O_3} = (3 \pm 2) \times 10^4\ M^{-1}\ s^{-1}$	
SO_3^{2-}	$k_{O_3} = 10^9\ M^{-1}\ s^{-1}$	per Hoigné and Bader (1978a, 1983a, 1983b)
HSO_3^-	$k_{O_3} = 3.2 \times 10^5\ M^{-1}\ s^{-1}$	per Erickson et al. (1977)
H_2SO_3	$k_{O_3} = (2 \pm 2) \times 10^4\ M^{-1}\ s^{-1}$	per Penkett (1972)

Action on other oxidants—reactivity with different forms of chlorine.

Free chlorine: hypochlorous acid and hypochlorite. Under normal pH conditions found in water intended for potable water treatment, free chlorine is in the form of hypochlorous acid or hypochlorite. These two species are formed from the hydrolysis of molecular chlorine in the water.

$$Cl_2 + 2H_2O \rightleftarrows HClO + Cl^- + H_3O^+$$

$$HClO + H_2O \rightleftarrows ClO^- + H_3O^+$$

Like hypobromites, hypochlorites are oxidized by ozone resulting in chlorates and chlorides as shown below.

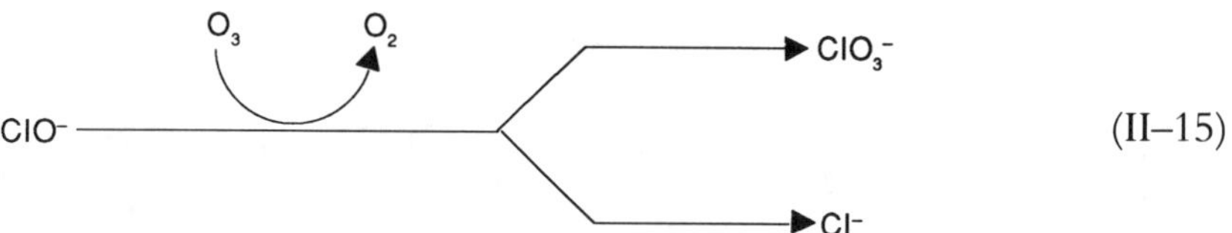

(II–15)

At a given pH, the rate of free chlorine oxidation by ozone is greater than the rate of free bromine oxidation. For example, the half-life of chlorine at pH 7 in the presence of 1 mg/L O_3 is expected to be about 15 min, whereas the bromine half-life should be about 1 h.

The basic difference between hypochlorite and hypobromite is the fact that, during hypobromite oxidation, the bromide that is formed initially is eventually oxidized to hypobromite, whereas the chloride formed is not oxidized to chlorate ion.

Combined chlorine. Inorganic chloramines and bromamines are degraded by ozone, although the degradation rates are not very high. This is illustrated by the rate constants listed in Table II–2.

Chlorine dioxide, chlorite and bromite, and chlorate and bromate. Chlorine dioxide, chlorite, and bromite are very reactive with ozone. This is shown by the rate constants listed in Table II–3. In comparison, chlorates and bromates are not reactive with ozone, as can be judged from the following rate constants:

Table II–2 Rate Constants for Ozonation of Ammonia and Its Main By-Products

Compound	pk_B	Rate Constants $k_{HB}{}^{+\star}$ ($M^{-1}s^{-1}$)	$k_B{}^{\dagger}$ ($M^{-1}s^{-1}$)
$NH_4{}^+/NH_3$	9.2	0.00	20 ± 2
NH_2Cl			26 ± 4
$NHCl_2$			1.3 ± 0.5
NH_2Br			160 ± 40
$NHBr_2$			40 ± 20
NH_3OH^+/NH_2OH	6.0	<2	$2.1 \pm 0.2 \times 10^4$

Source: Reprinted with permission from *Wtr. Res.*, 19:8, J. Hoigné et al. © 1985 Pergamon Press PLC.
$^\star k_{HB}{}^+$ is the rate constant of the protonic form.
$^\dagger k_B$ is the rate constant of the molecular form.

Table II–3 Rate Constants for Ozonation of Oxygenated Chlorine and Bromide Compounds

Compound	pk_B	Rate Constants $k_{HB}{}^{\star}$ ($M^{-1}s^{-1}$)	$k_B{}^{\dagger}$ ($M^{-1}s^{-1}$)
ClO_2		$1.1 \pm 0.1 \times 10^3$	
$HClO_2/ClO_2{}^-$	1.9		10^4
$HBrO_2/BrO_2{}^-$			10^5

Source: Reprinted with permission from *Wtr. Res.*, 19:8, J. Hoigné et al. © 1985 Pergamon Press PLC.
$^\star k_{HB}$ is the rate constant of the molecular form.
$^\dagger k_B$ is the rate constant of the anionic form.

$$ClO_3{}^- \quad k << 10^{-4}$$

$$ClO_4{}^- \quad k << 2 \times 10^{-5}$$

$$BrO_3{}^- \quad k < 10^{-3}$$

II.A.4 Oxidation of Synthetic Organic Chemicals

In the drinking water treatment process, ozone may react with dissolved organic micropollutants. Testing for the presence of these contaminants requires some knowledge of both the micropollutants' concentrations and their structure and reactivities.

Nature and origin of micropollutants. Aqueous organic constituents that are present in water prior to the treatment process are derived from either natural or human sources. In the latter case, the constituents may have undergone certain abiotic or biotic changes prior to entering a water treatment plant.

It should be noted that influent water may carry large particles (>0.45 μm) and some dissolved organic matter. This dissolved organic matter can include natural organic compounds (for example, amino acids, fulvic acids, and carbohydrates), and untransformed micropollutants (for example, aromatics and aliphatic hydrocarbons, chlorinated solvents, phenols, substituted or nonsubstituted polyphenols, pesticides, plasticizers, and surfactants), as well as some micropollutants formed during passage through natural media, including products from hydrolysis, photolysis, or biological transformation.

In general, only those micropollutants found dissolved in water will be discussed here. Although many products are sparingly soluble in water they are highly soluble in oils, and while their concentrations are low in water, their concentrations may be increased in the biomass and in sediments.

Organic carbon is found dissolved in surface water at concentrations commonly ranging from 2 to 10 mg/L; only 5 percent is in the form of micropollutants of human origin (Thurman 1985; Le Cloirec Renaud 1984). Consequently, the latter are not sensitive markers of total organic carbon (TOC) occurrence. The analytical techniques used to quantify TOC should therefore be more specific. Gas and liquid chromatography and gas chromatography–mass spectrometry (GC–MS) are commonly accepted methods.

An analysis of the nitrogen micropollutants present in surface water in Brittany, France (Elmghari-Tabib 1981; Le Cloirec Renaud 1984), has resulted in the identification of approximately 20 compounds in raw water, primarily pesticides. The compounds include substituted ureas (monuron, linuron, chloroxuron [1–10 μg/L]), *s*-triazines (simazine, atrazine [4–45 μg/L]), nitrocompounds (*o*-nitrophenol, nitrotoluene, dinoterbe, nitrobenzene [3–40 μg/L]), chloramphenicol (2–20 μg/L), anilines (usually chlorinated [1–5 μg/L]), and nitrosamines (2–10 μg/L).

In another study, Brauch and Kuehm (1988) showed that the concentration of synthetic organic human-related micropollutants in Rhine River water is less than 1 μg/L, except where EDTA (ethylenediaminetetracetic acid) and NTS (nitrilotriacetic acid) are present. Most of the identified substances were halogenated compounds, aromatics (chloro and nitro compounds), phthalates, pesticides, and chelating agents. Systematic analysis of micropollutants at all steps of drinking water treatment shows the seasonal character of various species along with the occurrence or non-occurrence of these compounds according to the quality of raw water and the location of the water treatment plant.

The chief source of micropollutants is probably related to human activity, either by direct contamination or by formation (hydrolysis, metabolism, or similar process) into the aqueous medium.

The above examples of human-derived micropollutants are based on the determination of nitrogenous organic compounds. For the most part, these molecules have aromatic or heterocyclic structures, which simplifies the study of their reactivities.

Other detected micropollutants may include some aromatic derivatives such as phthalates (at levels of a few hundred ng/L), acids, and hydrocarbons (at levels of a few μg/L). Although some of these substances are related to human activities (phthalates), others are primarily the result of the degradation of natural organic matter (lignin, humic substances). However, some micropollutants originate exclusively from industry. Among these are some chlorinated solvents such as trichloroethylene and tetrachloroethylene (at levels of a few hundred μg/L, especially in groundwater). The treatment of drinking water, for example, chlorination and ozonation, can also generate such micropollutants as trihalomethanes (THMs) and aldehydes (Rice 1980). Lastly, mention should be made of the occurrence of dangerous micropollutants that are generated in polluted natural media. These include toxic substances produced by hypertrophic freshwater *cyanophyta*.

To conclude, drinking water is polluted by natural substances, by untransformed and transformed products related to human activity, and by compounds generated in situ in waters containing high concentrations of organic matter. Table II–4 summarizes types and amounts of dissolved organic compounds found in natural waters that are used to produce drinking water. (For additional information see Thurman [1985].)

Behavior toward these micropollutants. Analysis of organic pollutants is limited both by their diversity and the analytical tools necessary for quantification, especially at very low concentrations. Raw water can be analyzed to study priority pollutants or discharged wastewater can be used to survey specific pollutants. For example, in 1982, the European Economic Community Commission (EECC) recorded 429 micropollutants. Cotruvo and Vogt (1987) indicate that the U.S. Environmental Protection Agency

Table II–4 Types and Amounts of Dissolved Organic Compounds

Organic Substance	Typical Molecule	Chemical Oxygen Demand (%)	Typical Range (mg C/L) River Water	Groundwater
Humic substances	$CH_3-(CH_2)_{16}-COOH$ Stearic acid	75	0.5–4	0.01–0.1
Carboxylic acids	CH_3-COOH Acetic acid	8		
Nonvolatile fatty acids		4	200 (50–500)	20 (5–50)
Volatile fatty acids		2	100	50
Hydroxy acids	$CH_3-CH(OH)-COOH$ Lactic acid	1	50	—
Dicarboxylic acids	$HOOC-(CH_2)_2-COOH$ Succinic acid	0.5	10 (3–100)	—
Aromatic acids	 CO_2H, OH *p*-Hydroxy benzoic acid	0.5	10 (3–100)	—
Phenols	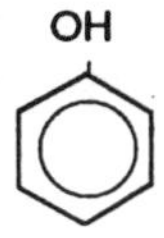 OH Phenol	0.5	(0.2–15)	(0.2–1)
Amino acids	NH_2, CH_2-COOH Glycine (gly)	2–3	300 (50–1000)	50 (20–350)
Free amino acids Humic amino acids Combined amino acids	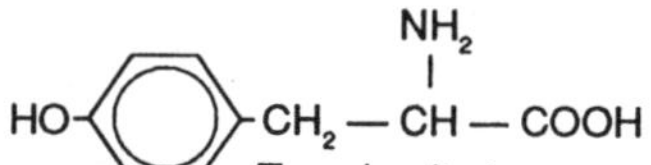 HO–C_6H_4–$CH_2-CH(NH_2)-COOH$ Tyrosine (tyr)			
Carbohydrates	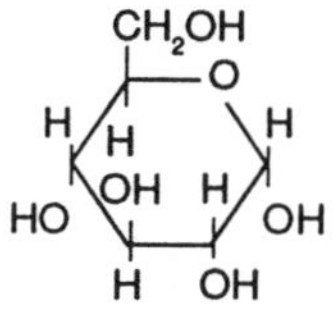 CH_2OH, O, H, H, H, OH, H, HO, OH, H, OH Monosaccharide (glucose)	10	500 (100–300)	100 (65–105)
Monosaccharides		1		
Polysaccharides	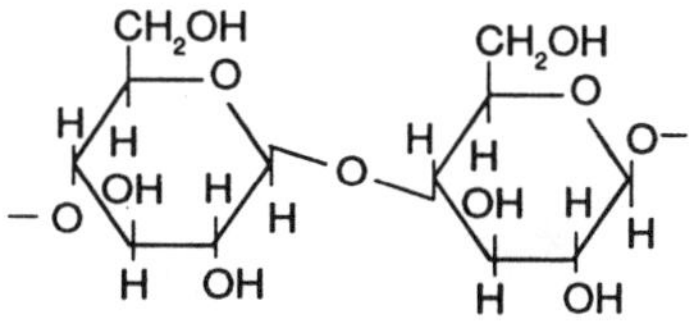 CH_2OH, O, H, H, OH, H, H, –O, H, OH, O, CH_2OH, O, H, H, O–, OH, H, H, H, OH Polysaccharide (cellulose)	5		
Humic saccharides	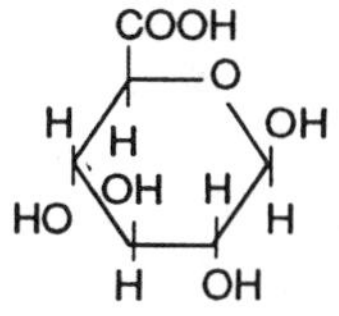 COOH, O, H, H, OH, OH, H, HO, H, H, OH Sugar acid (glucuronic)	2		
Sugar acids		1		

Table II–4 Types and Amounts of Dissolved Organic Compounds (Continued)

Organic Substance	Typical Molecule	Typical Amount (range, μg/L)
Amino sugars	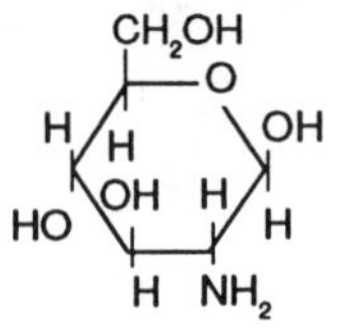 Amino sugar (glucosamine)	1
Hydrocarbons		
Methane	CH_4 Methane	10–10,000 in lake and groundwater
Halogenated methane	$CHCl_3$ $CHBrCl_2$ $CHBr_2Cl$ $CHBr_3$ Halogenated methanes	10–100 produced by chlorination
Chlorinated hydrocarbons	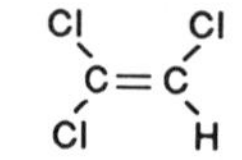 Trichloroethene 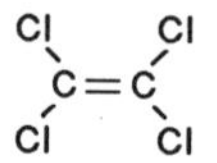 Tetrachloroethene	0.1–5
Aliphatic hydrocarbons	$CH_3—(CH_2)_n—CH_3$ *n* – Alkanes	0.1–5
Aromatic hydrocarbons	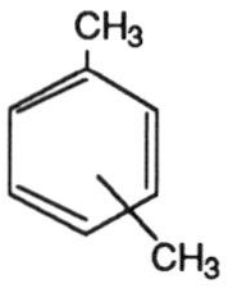  Xylene, 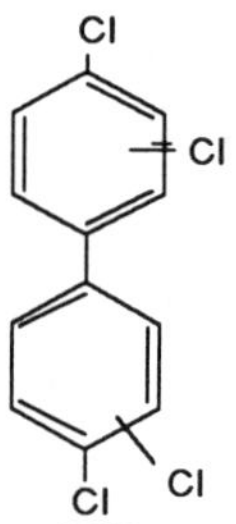 Benzene, PCB	0.1–5
Polyaromatic hydrocarbons	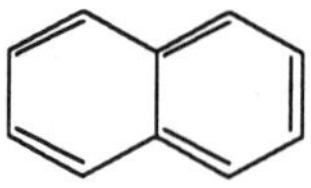 Naphthalene, 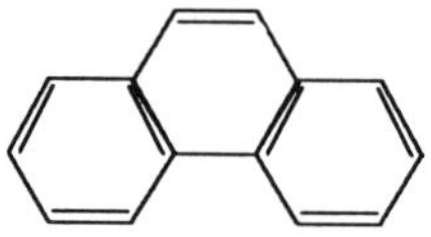 Phenanthrene	0.1–5
Aldehydes	$CH_3—(CH_2)_n—CHO$ where n = 0 - 9 *n* – Alkyl aldehydes; 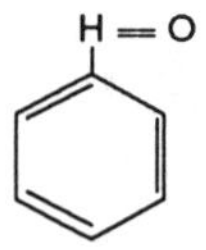 Benzaldehyde	0–1 after ozonation
Organic bases and nucleotides	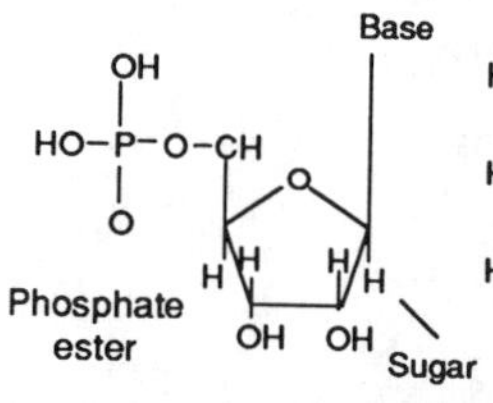 Structure of a mononucleotide (Base, Sugar, Phosphate ester)	0–1
ATP	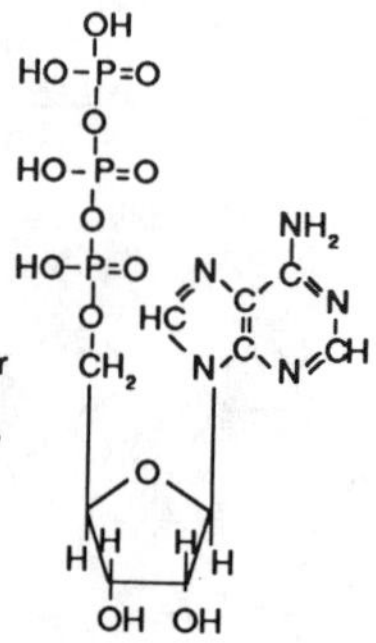 ATP	0–0.5

Table II–4 Types and Amounts of Dissolved Organic Compounds (Continued)

Organic Substance	Typical Molecule	Typical Amount (range) μg/L
Sulfur compounds	CH_3-S-CH_3 Dimethylsulfide $CH_3-CH_2-CH_2-CH_2SH$ *n* - Butylmercaptan	0–10^{-2}
Alcohols	CH_3CH_2OH $CH_2OH-CHOH-CH_2OH$	0–5
Amines	$(CH_3)_3N$ Trimethylamine $HOCH_2CH_2NH_2$ Ethanolamine	0.1–6
Sterols	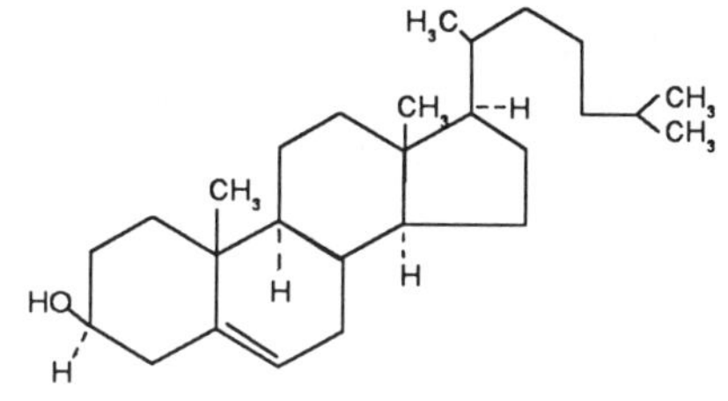 Cholesterol	0.1–6
Pesticides		
s-Triazines	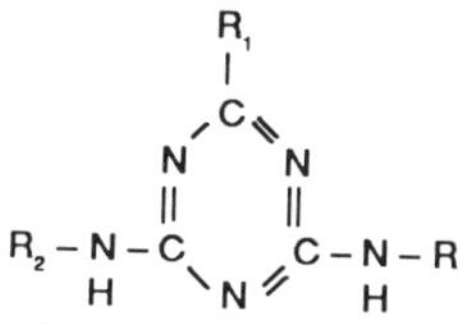 *s* - Triazines	0–50
Substituted ureas	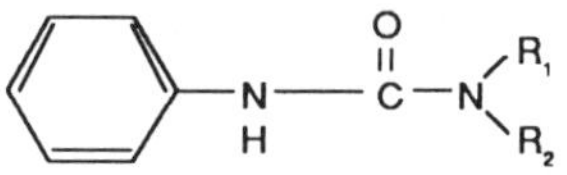 Substituted ureas	0–10
Organochlorous	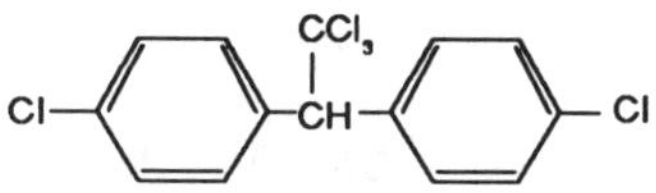 DDT	0–10^{-1}
Organophosphorous	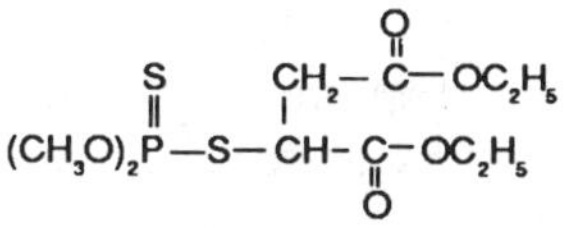 Malathion	0–10^{-1}
Carbamates	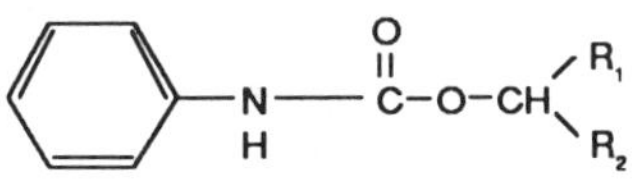 Phenylcarbamate	0–10^{-1}
Phenoxy acetic acid derivatives	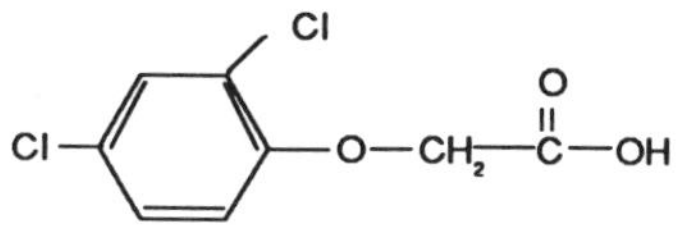 2.4 - D	0–10^{-1}
Phthalates	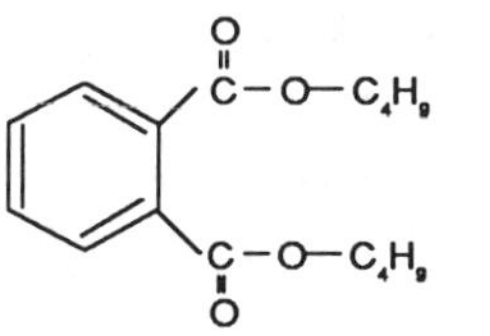 Dibutylpthalate	10^{-1}

(EPA) in its National Primary Drinking Water Regulations proposes to rank 83 specific contaminants according to threshold occurrence levels in water. The regulations also identify the best available technology (BAT) for the elimination of these products. Many of these 83 substances are volatile organic compounds (VOCs), such as benzene, benzene derivatives, chlorinated ethane, and ethylene. Other substances include heavier aromatic compounds, phenols, and chlorinated pesticides. For these substances, very low guide levels (maximum contaminant level goals [MCLGs]) have been set, although the enforceable standards (maximum contaminant levels [MCLs]) are generally higher.

Molecular structure and reactivity with ozone. The two ozonolysis pathways, which were discussed in sec. II.A, are

- direct attack by electrophilic or dipolar cyclo addition, and
- indirect attack by free radicals produced by reaction with water and water constituents.

Ozone reactions with the substrates M are first-order reactions with respect to both M and ozone:

$$- d[M]/dt = k[O_3][M]$$

In complex media, the reactivity of the micropollutants with ozone must be very high for degradation to occur. This potential for reaction is usually more pronounced with ozone than with other oxidants.

Initial molecular reaction sites are either multiple bonds (C=C, C=C−O−R, −C=C−X) or atoms carrying a negative charge (N, P, O, S, and nucleophilic carbons). A strong initial reactivity is therefore predicted for ortho-activated aromatics by substituents such as OH, CH_3, or OCH_3. A weaker initial reactivity is predicted for molecules with NO_2, CO_2H, or CHO groups. It should be noted that this simple reasoning applies to O_3 as well as other electrophilic oxidants. Radical reactions (for example, OH) are unselective. Depending on the structure of the first oxidation byproduct, additional reactions may occur. Some examples are reviewed in the following paragraphs.

Initial reactivity is related to electron density at the site. By studying this electron density, one is able to predict the reactions. Thus, using the CNDO method, Laplanche et al. (1984) and Decoret et al. (1984) studied reactive sites on organo-phosphorus compounds and aromatic compounds, respectively.

A simpler approach might be to define molecular electron densities using a simple physical technique. As an example, semi-empirical relationships have been obtained from nuclear magnetic resonance (NMR) experiments (Martin 1966), which link the electron density to C_{13} and proton chemical shifts to magnetic susceptibility and to Taft and Hammett constants. The constant σ is related to the substituent's nature for a reference structure.

Therefore, as shown in Figure II–14, excellent linear free energy relationships ($\log_{10}k/k_0 = \sigma\rho$) can be formulated. This may reflect the initial reactivity of ozone and the aromatic derivatives. k_o is the rate constant of the reference structure (benzene), k is the rate constant of the tested product, and ρ is a constant characteristic of the tested substituent (Shorter 1972; Gould 1987; Hoigné and Bader 1983a, 1983b).

Stock–Brown (March and Exner 1982) have proposed another approach based on the free enthalpy of the molecule.

These approaches are valuable to predict the orientation of the initial reaction, but they do not provide any information on the chemical metabolites that will be formed.

Kinetics of ozonation of dissolved organic micropollutants. The oxidation rate of micropollutant substrates (M_i) in a solution may be illustrated by the following equations:

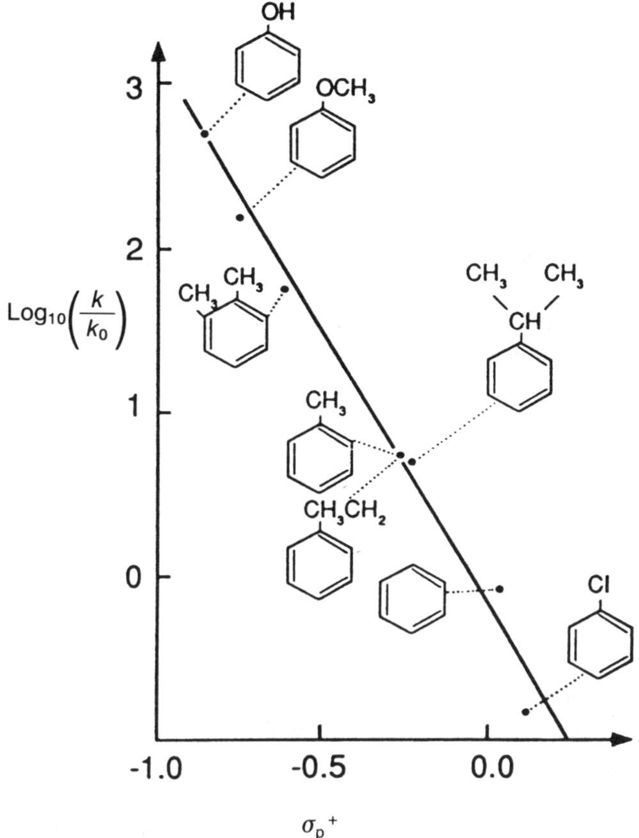

Source: Reprinted with permission from Proc. IOA Symp. (Toronto), p. 16, Hoigné, J. and Bader, H. © 1977 IOA.

Values for ssp$^+$ are based on the solvency of dimethylphenolcarboxychloride.

Figure II–14 Stock–Brown Plot of k_o Values of Substituted Benzenes

$$O_3 + M_i \begin{array}{l} \nearrow \; k_{1_i} \rightarrow \text{products} \\ \searrow \; k_{2_i} \rightarrow OH \pm \text{products (initiation step)} \end{array}$$

$$M_i + OH \begin{array}{l} \nearrow \; k_{3_i} \rightarrow \text{products (termination step)} \\ \searrow \; k_{4_i} \rightarrow OH_2 + \text{products (propagation step)} \end{array}$$

Let

$$\frac{d\,[M_i]}{dt} = (k_{1_i} + k_{2_i})\,[O_3]\,[M_i] + (k_{3_i} + k_{4_i})\,[OH]\,[M_i]$$

where $[O_3]$ and $[M_i]$ are the concentrations of dissolved ozone and micropollutant i, respectively. The OH radicals are generated by ozone attack on M_i OH^- organic and inorganic initiators. According to Yurteri and Gurol (1988),

$$[OH] = \Psi\,[O_3]$$

Where:

$$\Psi = \frac{\sum_i k_{2_i} [M_i] + 2\, k_1 [OH^-]}{\sum_i k_{si} [S_i]}$$

where k_1 is the rate constant of the initiation step in the decomposition of ozone by OH^-, and S_i and k_{si} represent the concentrations and rate constants of the scavengers, respectively.

The total oxidation rate of the particular substrate is

$$\frac{d\,[M_i]}{dt} = \text{Rate}_i = (k_i\star + k_i\star\star\, \Psi)\, [O_3]\, [M_i] = k_i\, [O_3]\, [M_i]$$

Where:

$$k_i\star = k_{1_i} + k_{2_i}$$

$$k_i\star\star = k_{3_i} + k_{4_i}.$$

It should be noted that k_i varies as a function of $[M_i]$.

The ozonation rate for a substrate M_i is therefore dependent on the aqueous matrix through ψ, and more specifically on the alkalinity of the water.

In most cases, the value of the ψ factor is about $1 \times 10^{-10} - 1 \times 10^{-8}$; the value of $k_i\star\star$ is about $10^8 - 10^{10}\ M^{-1}s^{-1}$. So the value of the product of $k_i\star\star\ \psi$ is between 10^{-2} and $10^2\ M^{-1}s^{-1}$. Note that $k_i\star$ varies substantially and is dependent on the molecular structure.

If operating a completely mixed-flow ozonation reactor, it can be estimated, on the basis of the total rate ($d[M_i]/dt$), that the effluent micropollutant concentration M_i will be

$$[M_{io}]/[M_i] = 1 + k_i\, [O_3]\, \theta$$

where θ is the mean O_3 contact time in the reactor. If $\theta = 10$ min and $[O_3] = 0.5$ mg/L (or 10^{-5} mol/L), a 90 percent decrease of M_i requires an overall rate constant of $k_i = 1.5 \times 10^3\ M^{-1}s^{-1}$. Thus, the reaction between the substrates and ozone must have rate constants in the range of $10^3\ M^{-1}s^{-1}$ or higher in order to obtain a significant degradation.

This model implies that:

- When $k_i\star\star\ \psi > k_i\star$, that is, when molecules are poorly reactive under direct-attack conditions, the radical mechanism is preponderant. In this case, k_i stays low, and under normal drinking water treatment conditions, the M_i pollutant will cross the water processing system.
- When the reaction by the direct attack of ozone on the chemical species M_i is fast (for example, phenols), M_i will be degraded but the radical mechanism will not be an important factor.

Except for soft water, at high pH, the radical-mediated pathway will be of little importance without the use of an advanced oxidation process.

Micropollutants and their reactivity to ozone.

Aliphatic hydrocarbons and halogenated derivatives. Halogenated compounds can be industrial products such as solvents; they can also be generated during the process of water chlorination. The electron densities attached to C or H atoms of alkanes will cause this organic family to be unreactive. Alkenes are characterized by a π cluster that

Table II–5 Ozonation of Hydrocarbons

Compound	Ozonation k, $M^{-1}s^{-1}$ In CCl_4 or $CHCl_3$ (25°C)	In Water (25°C)
Cyclopentane	2.6×10^1	
$CH_2=CH_2$	2.5×10^4	
$CH_2=CHCl$	1.1×10^3	
$CH_2=CCl_2$	2.2×10^1	1.1×10^2
$CHCl=CCl_2$	3.6×10^{-1}	1.7×10^1
$CHCl=CHCl$	5.8×10^2	5.7×10^3
$CCl_2=CCl_2$	1×10^0	$<10^{-1}$
Hexene 1	7.6×10^4	
Cyclopentene	2×10^5	
Styrene	1×10^5	

Source: Hoigné & Bader (1976, 1983a).

Table II–6 Summary of Half-Lives ($t_{1/2}$) for Volatile Organic Contaminants in a Semibatch System

Compound		VOC Influent Concentration (μg/L)	$t_{1/2}$ (min) Ozone and Air*	VOC Influent Concentration (μg/L)	$t_{1/2}$ (min) Air Only†
Alkanes	Range	175–240	130–>250	180–250	139–>150
	Average	203	>105	178	>147
	Number of compounds	4		4	
Aromatics	Range	60–203	1–11	50–225	68–150
	Average	108	4	142	>111
	Number of compounds	12		12	
Alkenes	Range	125–140	3–6	118–130	46–60
	Average	133	5	124	53
	Number of compounds	2		2	

*Reaction & stripping.
†Stripping only.

aids in promoting electrophilic ozonolysis by dipolar cyclo addition. Such reactivity is depressed when the electron density undergoes modification by way of the electrophilic group. Table II–5 presents data relating to the ozonation of hydrocarbons (Hoigné and Bader 1976, 1983a). For example, attack on the ethylene-type double bond is less readily achieved if the H atoms carried by the double bond are replaced by Cl atoms. A splitting of the double bond gives rise to aldehydes and acids. It is obvious that under conventional ozone application conditions, both ethylene and vinyl chloride will be adequately removed.

Kuo (1984) compared the ozonation of aqueous solution of alkenes (cyclohexene, cyclopentene, pentene-1) to the reactivity of the corresponding alkanes. It was found that the former are much more reactive ($k = 8.5 \times 10^5\ M^{-1}s^{-1}$). Fronk (1987) compared the half-lives of alkanes, alkenes, and aromatic hydrocarbons, and confirmed the relative inertness of the alkanes (Table II–6).

Unreactive species such as $CHCl_3$, $BrCHCl_2$, tetrachloroethylene, and trichloroethylene can be destroyed by OH radicals through the application of combination

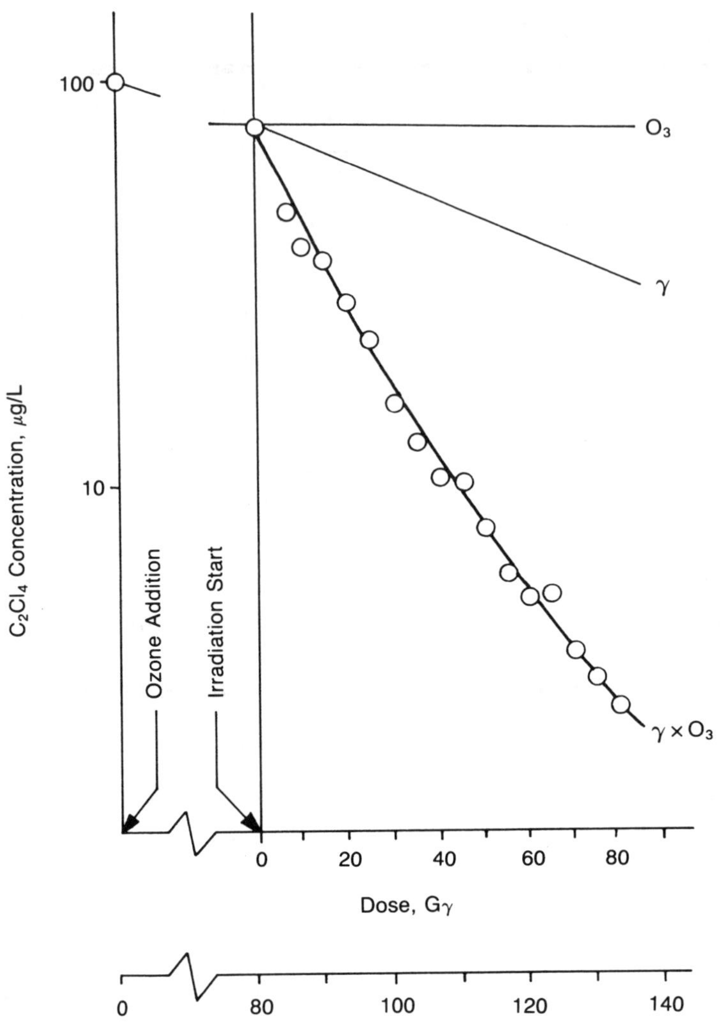

Source: Reprinted with permission from *Wtr. Res.*, 22:5:645, Gehringer, P. et al. © 1988 Pergamon Press PLC.

Initial ozone concentration is equal to 5 ppm.

Figure II–15 Decomposition of Perchloroethylene in Drinking Water

systems such as O_3/UV, for which the degradation rate would be increased onefold to fiftyfold (Glaze et al. 1986; Paillard et al. 1987), or O_3/gamma radiation (Gehringer et al. 1988) (Figure II–15), or by using solid catalysts (Ginocchio 1979).

Alcohols, ethers, aldehydes, carboxylic acids. Alcohols, ethers, aldehydes, and carboxylic acids are characterized by the occurrence of a nucleophilic group −O−. Dalouche (1981) has investigated the ozonation of alcohols. The initial attack on O gives rise to a peroxy compound, which is thought to be the starting point for the formation of aldehydes or ketones, acids, esters, and peresters (Figure II–16). Figure II–17 depicts the formation of rather stable compounds such as carbonyl compounds and esters.

Alcohols are strongly nucleophilic but they are solvated in water. The p*k* value for alcohols is approximately 20. Alcohols occur in the non-ionized, solvated form in water at pH 7, where they are poorly accessible to electrophilic reagents.

Hoigné and Bader (1983b) have indicated that the occurrence of an acidic CO_2H group deactivates a molecule. Acetic and oxalic acids have extremely low reaction rates. The formate ion is fairly reactive ($k = 1.4 \times 10^2\ M^{-1}s^{-1}$), while non-ionized acid is practically inert. For the two aromatic compounds, benzoic acid and salicylic acid,

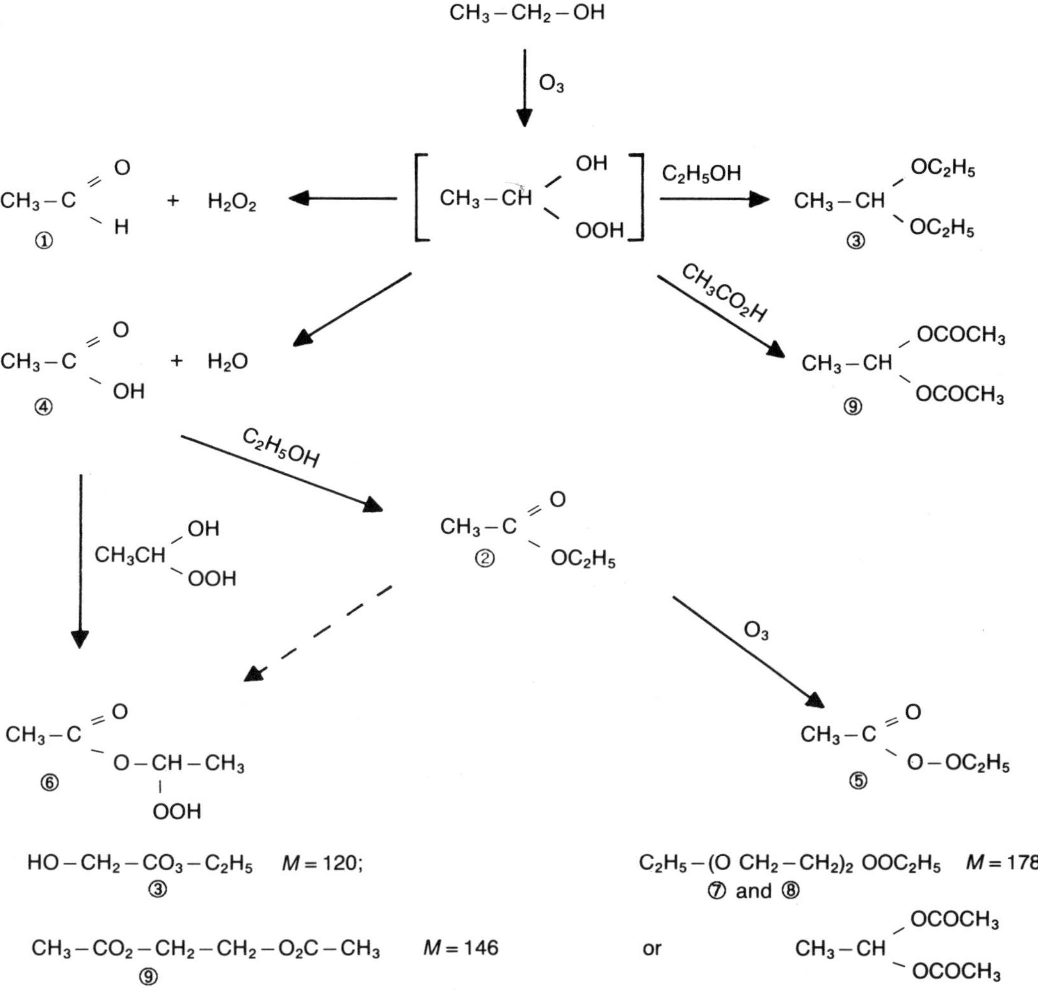

Source: Reprinted with permission from Wasser Berlin '81, p. 588, Elmghari-Tabib, M. et al. © 1981 IOA.

Figure II–16 Schematic of Ozonation of Ethanol

reactivity to ozone is depressed by the CO_2H group. Likewise, in the aliphatic series, maleic acid and fumaric acid ($CO_2H - CH = CH - CO_2H$) are less reactive than their alkene analogues.

One may predict a low reactivity for polyols, polyethylene glycols (PEG), sugars, polysaccharides, and ethers. Conventional catalytic systems (O_3/UV) are effective in promoting a better degradation of the PEGs (Suzuki et al. 1983).

Aromatic compounds. Simple or complex aromatic hydrocarbons (symbolized by AH or PAH) constitute a major family of chemicals encountered in raw waters. They are released either from the natural environment (for example, humic and fulvic acid monomers) or from organic synthesis. Ozonation of humic substances will be discussed in sec. II.A.6.

Ozonolysis of aromatic compounds may involve either the aromatic ring itself (resulting in decyclization and loss of the aromatic structure) or one of the side-chain substituents.

As a general rule, simple aromatic products fall into the following two categories (Doré and Legube 1983):

- those with electron-donating groups (OH, NH_2) (these react readily with ozone); and

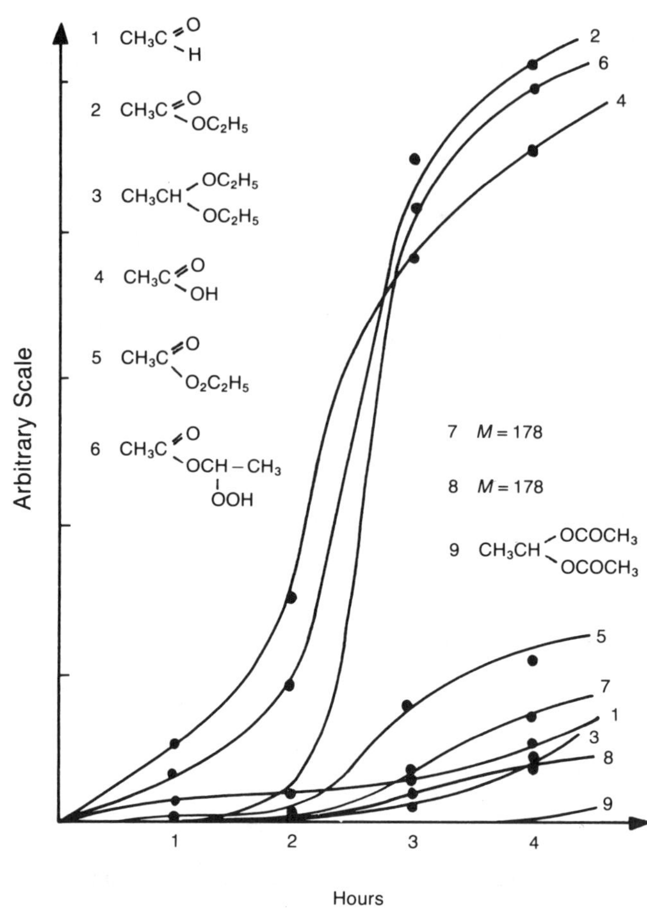

Source: Reprinted with permission from Wasser Berlin '81, p. 590, Elmghari-Tabib, M. et al. © 1981 IOA.

Figure II–17 Formation of Metabolites as a Result of the Ozonation of Ethanol

- those with electron-withdrawing groups (NO_2, Cl), for which the reaction is much slower.

Direct ozonolysis is an electrophilic reaction of the oxidant, whereby ozonolysis occurs at the ortho or para position of a donor substituent, or at the meta position in the case of an acceptor substituent. Attack with a reactive side chain may occur in competition with attack on the aromatic nucleus.

Figure II–18 diagrams the general reaction of ozonation of aromatics (Legube 1983). Simple ozonation provides for the degradation of mononuclear and polynuclear aromatics (Marley et al. 1987; Dreher and Klamberg 1988), and of benzopyrene compounds (Chedal 1976; Morozzi et al. 1986; Nazarov et al. 1979). As an example, the ozonation of naphthalene by Legube et al. (1985, 1986) shows 1–3 dipolar cyclo addition of ozone on the 1,2 bond of naphthalene (Figure II–19). With substituted naphthalenes (1-chloro and 2-methyl), the initial attack of the oxidant on the 1,2 bond of naphthalene is electrophilic in nature. Oxidation products derived from polynuclear compounds are summarized in Table II–7.

Degradation of the less reactive aromatic compounds requires the application of oxidation systems with free-radical-producing ozonation. Dieter et al. (1988) reported better degradation performances with polychlorinated dibenzo *p*-dioxins and polychlorinated dibenzo furanes at a pH of 10 rather than 5, because the high pH encouraged OH formation. Anselme et al. (1988) used the $O_3{-}H_2O_2$ system to eliminate nitrobenzenes and chlorobenzenes from groundwater.

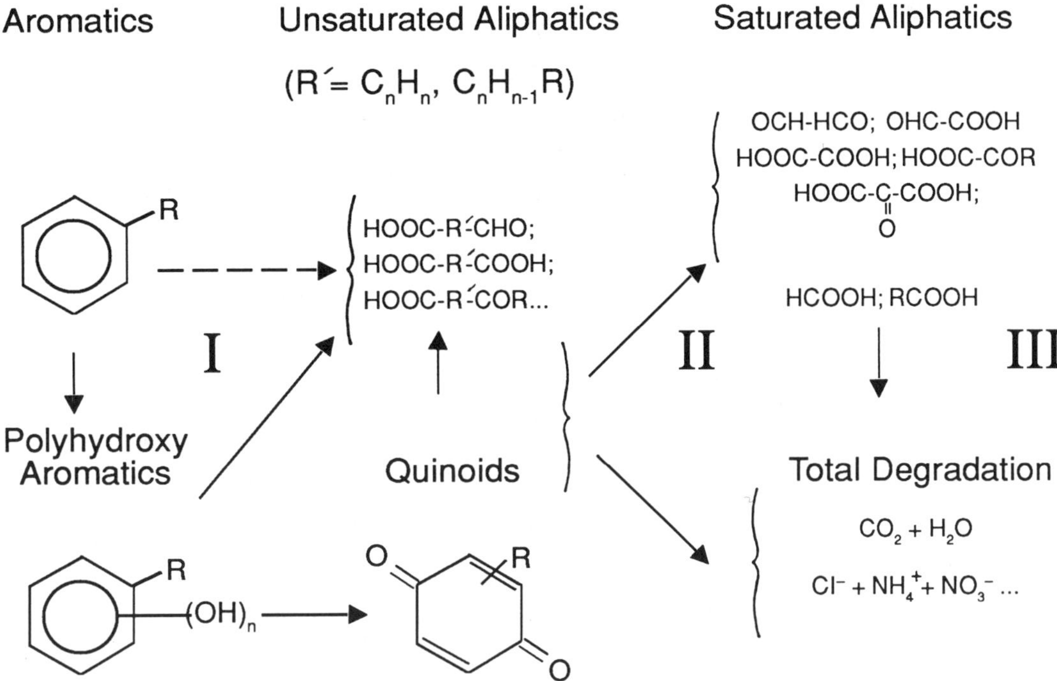

Source: Reprinted with permission from *Envir. Technol. Lett.* 5:210, Decoret, C. et al. © 1984 Science and Technology Letters.

Figure II–18 Schematic of Ozonation of Polar and Simple Aromatics

The degradation scheme for polynuclear products is similar to that observed for simple compounds. Figure II–20 depicts the general degradation pathway of aromatic compounds. The overall reaction follows second-order kinetics (Kuo 1985; Dieter et al. 1988).

$$-\frac{d\,(AH)}{dt} = k\,[O_3]\,[AH]$$

Figure II–21 shows the amount of ozone needed for 90 percent destruction of the starting aromatic compound. This value reflects the rate of oxidant attack as determined by the nature of the substituents.

Phenols and derivatives. Phenolic products (phenol, substituted phenols, quinones, and polyphenols) are characterized by the presence of the OH donor group on the aromatic nucleus. These compounds are strongly reactive with ozone, and their oxidation as it relates to drinking water treatment and wastewater management has been extensively studied.

The following two mechanisms of attack on phenols by ozone have been described:

- At acidic or neutral pH, researchers (Decoret et al. 1984; Gurol and Vatistas 1987; Doré and Legube 1983; Eisenhauer 1968, 1971) have evoked an electrophilic attack of the oxidant on reactive carbons (ortho or para positions) (Figure II–22).
- At neutral or basic pH, some researchers favor, as an adjunct to the above-mentioned mechanism, either electrophilic attack on phenate ions (Singer and Gurol 1983) or an OH radical mechanism that would be initiated both by hydroxyl and phenate ions (Gurol and Vatistas 1987).

The oxidation sequence yields the following intermediates:

- products resulting from aromatic nuclear hydroxylation, such as diphenols and quinones;

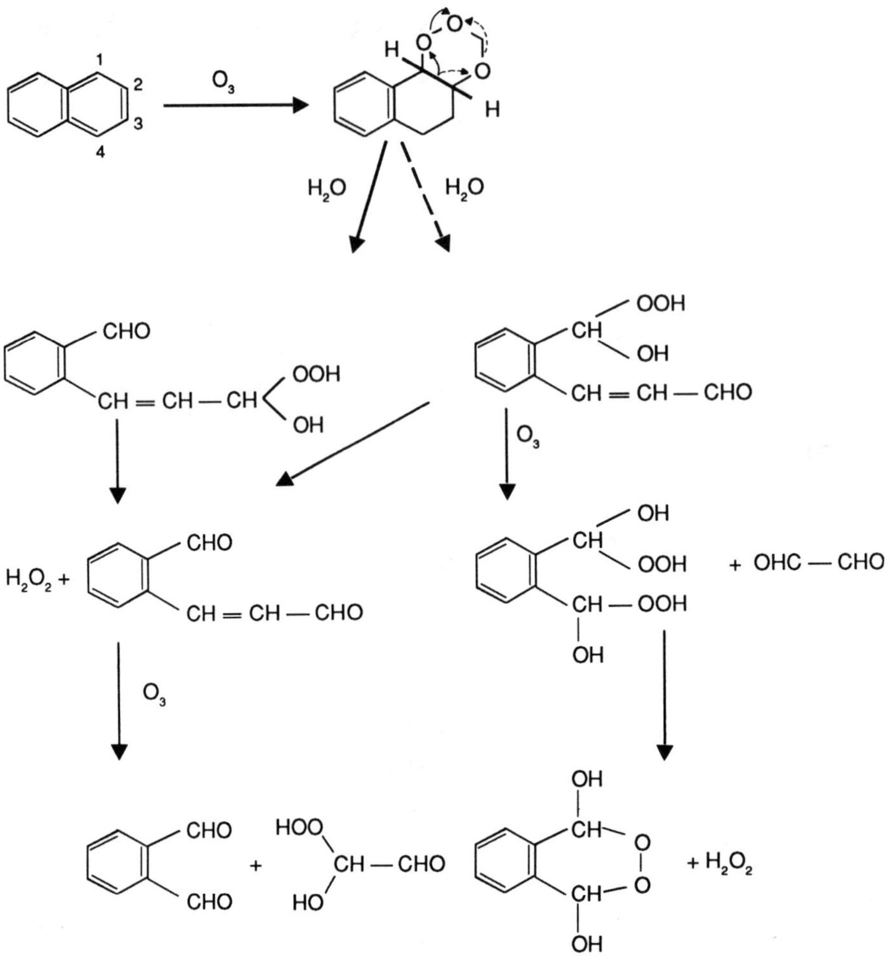

Source: Legube et al. (1985, 1986).

Figure II–19 Schematic of Ozonation of Naphthalenes

- products issued from ring breakdown, such as muconic acids and derivatives; and
- final oxidation products, such as glyoxylic, oxalic and formic acids, and glyoxal.

The formation of condensation products (polymerization) has been described on certain occasions (Duguet et al. 1985).

Whatever the mechanism involved, ionic or radical, the oxidation products are the same. Nevertheless, performance is improved with the advanced oxidation processes. Gurol and Vatistas (1987) propose the following order:

$$O_3 + UV > O_3 > UV \text{ alone.}$$

The electrophilic attack takes place according to first-order kinetics with respect to ozone and phenol as follows (Hoigné and Bader 1983a, 1983b; Gurol and Nekouinaini 1984):

$$-\frac{d\,[\text{phenol}]}{dt} = k\,[O_3]\,[\text{phenol}]$$

Depending on the structure, the reaction rate of phenolic compounds increases when the electron density on the aromatic ring is augmented and decreases with

Table II–7 Oxidation Products From Polynuclear Compounds

Starting Aromatic Taking Part in Ozonation	Initial Intermediates
Naphthalene	Cyclic peroxide
	cis- and trans- isomers of formyl cinnamaldehyde
	Oxalic acid
	Oxomalonic acid
	Formic acid
	Orthophthaldialdehyde acid
	Phthalaldehydic acid
	Hydrogen peroxide
	1,4-Naphthoquinone
	Phthalic acid
Fluorene	Phthalic acid
	9-Fluorenone-1-carboxylic acid
	Fluorene-1,9-decarboxylic acid
	Homophthalic acid
Benzothiophene	*o*-Hydroxybenzene sulfonic acid
	o-Sulfobenzoic acid

increasing substituent size (Gould 1987). There is also a consensus concerning increased reaction rates at higher pH values due to the ionic dissociation of the phenols to phenate ion. However, reported rate constants vary according to the source.

Hoigné and Bader (1983a, 1983b) give the following data for phenol in the pH range of 2–6:

$$k \text{ nondissociated phenol} = 1.3 \pm 0.2 \times 10^3 \ M^{-1}s^{-1}$$
$$k \text{ phenate ion} = 1.4 \pm 0.4 \times 10^9 \ M^{-1}s^{-1}$$

These authors have determined the rate constants for other phenolic compounds to be $10^3 - 10^4 \ M^{-1}s^{-1}$ for nondissociated species, and in the order of 10^9 for dissociated ones. Their conclusion is that the overall reaction rate increases tenfold with each pH unit. On the other hand, Anderson (1977) reported that the rate of ozone reaction with phenate ion was only 27 times that of undissociated phenol. By varying the pH, Joshi and Shambaugh (1982) arrived at the overall constants listed in Table II–8.

The opening of the phenol ring requires 4–6 mol of ozone/mol of phenol (Gould and Weber 1976; Doré et al. 1978; Gilbert 1974).

Lastly, the toxic effects of solutions obtained from the degradation of chlorophenols by ozone (Benoît-Guyot et al. 1984) on *Antenia salina* showed increased toxicity levels at the onset of ozonation, minimum levels during treatment, and a second toxic peak prior to total destruction of the molecules.

In summary, phenols are easily and rapidly oxidized by ozone. Initial intermediates of oxidation show greater toxicity than the starting products. However, owing to rapid degradation, adequate ozone dosages must be applied for a sufficiently long contact time, in order to get through the periods of quinone occurrence. These conditions exist in conventional ozonation treatment facilities.

Amines and derivatives. Amines constitute a highly reactive family of compounds with respect to ozone; this is due to the fact that the electrophilic attack is promoted by the presence of a negatively charged nitrogen atom. However, when the nitrogen atom is deactivated by an electron-withdrawing group (nitrosoamines), the attack by the oxidant is slower.

Upon examination of the different compounds isolated during the ozonation of amines, the following four reaction types arise (Table II–9):

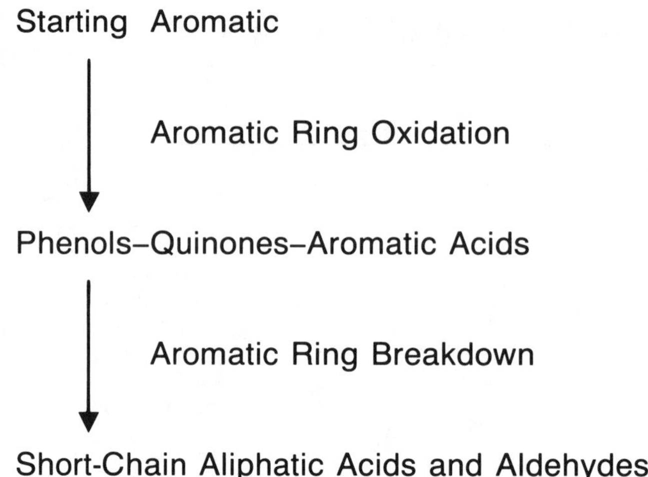

Figure II–20 Degradation of Polynuclear Compounds

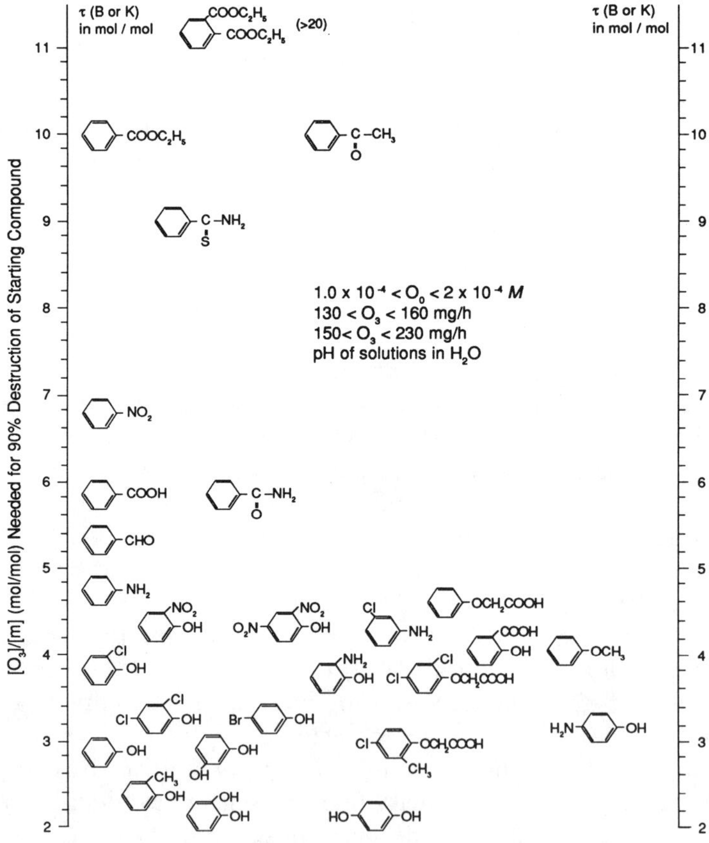

Source: Reprinted with permission from *Jour. Fr. d'Hydrol.*, 1:40:18, Doré, M., and Legube, B. © 1983 *Journal Français d'Hydrologie.*

Figure II–21 Ozone Dosage Needed to Obtain 90 Percent Aromatic Clearance

- direct oxidation of the nitrogen atom resulting in formation of hydroxylamine, oxime, and aminoxide-type derivatives;
- oxidation of the carbon positioned in the α position with respect to N, with formation of amide and formylamine functional groups;

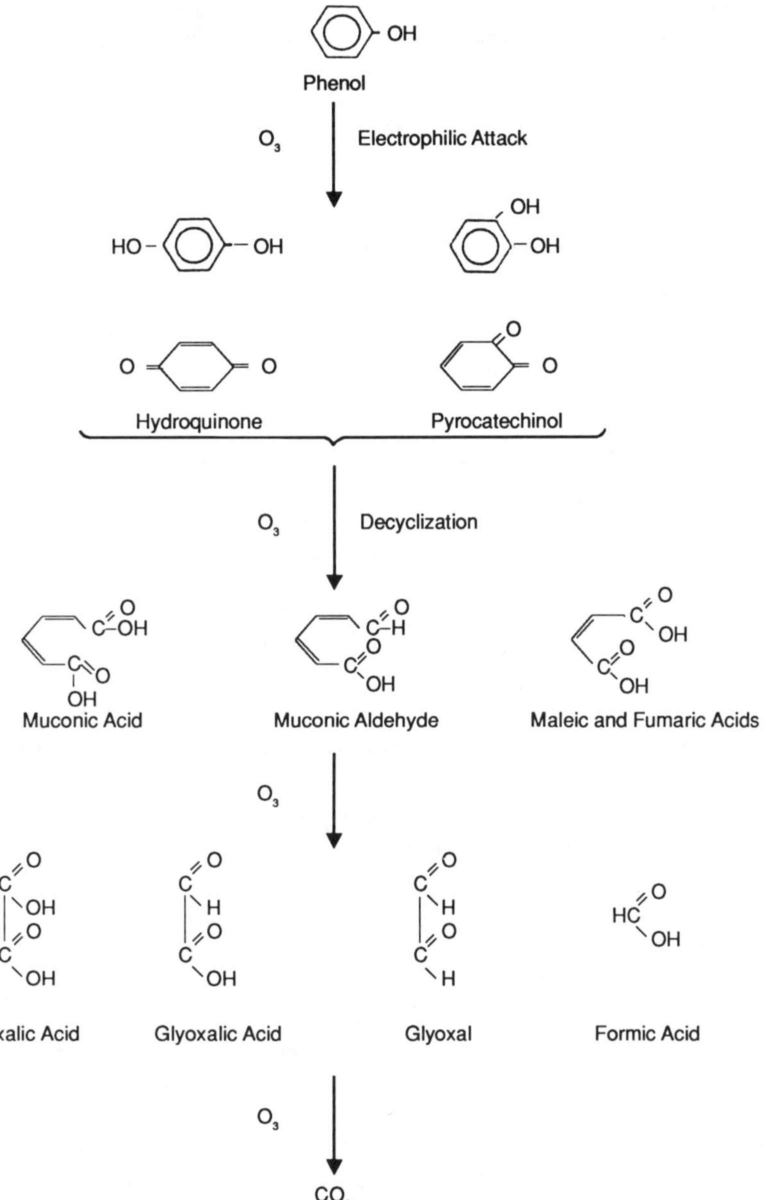

Figure II–22 Schematic of Ozonation of Phenol

- splitting of the C–N bond (the nitrogen may then be found in the inorganic form NH_4^+ or NO_3^-, and the carbon gives rise to oxygenated compounds [aldehydes, acids]); and
- secondary condensations involving the starting product as well as already oxidized intermediates.

Based on existing studies, the reaction of ozone on the amines follows first-order kinetics with respect to both the oxidant and the amine.

$$-\frac{d\,[\text{amine}]}{dt} = k\,[O_3]\,[\text{amine}]$$

Table II–8 Oxidation of Phenol and Effect of pH on Reaction Kinetics*

pH	S (mol phenol/mol ozone)	k_2 ($M^{-1}s^{-1}$)
6.0	0.240	0.09×10^3
7.0	0.240	0.140×10^3
8.0	0.226	0.180×10^3
9.0	0.420	0.246×10^3

*Conditions for the experiments: temperature = 20°C; initial phenol concentration = 1.0632 μM/mL; ozone concentration in feed gas = 3 percent by weight; total gas flow rate to the reactor = 1.4331 NTP/min.

Hoigné and Bader (1979) identified many reaction rate constants and showed the influence of pH on them (Figure II–23). A decrease in rate paralleling that of pH may be explained by protonation of the amine. Based on the compounds identified, Elmghari-Tabib et al. (1982) and Laplanche (1982) have proposed an ozonation scheme (Figure II–24). After the electrophilic ozone reaction on the nitrogen atom, the intermediate adduct formed may evolve along the following two reactional schemes:

- In the first instance, oxygen loss leads to monohydroxylamine formation. A second attack gives rise to dihydroxylamine with formation of oximes and degradation products. Simulation tests carried out using the MNDO method seem to favor this pathway (Laplanche et al. 1985).
- The second scheme involves the alpha carbon atom. Subsequent rearrangement of the unstable hydroxyaminoalcohol obtained leads to the formation of C–O compounds such as amides.

Urea (Orlov et al. 1983), ethylene diamine (Marchenko and Obolonchir 1979), hydrazine (Sierka and Cowen 1981), and dimethyl nitrosoamine (Kobylinski and Peterman 1979) are oxidized very slowly by ozone. Nevertheless, some of the aforementioned researchers note that these compounds are oxidized at a much faster rate when production of free radicals is promoted (basic pH; O_3/UV system).

Pesticides. Numerous studies have been conducted to assess the action of ozone on pesticides. Recently, a review on aqueous ozonation of pesticides has been presented by Reynolds (1989). Chlorinated organic pesticides are slowly reactive and incompletely destroyed by direct oxidation (Kroyer et al. 1986). For example, aldrine is transformed into dieldrine and heptachlor is transformed into heptachlorepoxide. On the other hand, Prengle and Mauk (1978) have demonstrated the efficiency of the O_3/UV system. Figure II–25 depicts the chain reaction for dichlorodiphenyltrichloroethane (DDT) studied by those researchers.

Organophosphorus pesticides are much more susceptible to direct ozonation. Numerous products have been studied (Laplanche et al. 1984). The principal reactions encountered are

- oxidation of the P=S bond with formation of P=O; and
- splitting of the molecule at one of the bonds involving the phosphorus atom, yielding simple esters of phosphoric acids. Thus, H_3PO_4 is the ultimate product of ozonation concerning the phosphorated fraction of these insecticides.

For parathion, the initial oxidation kinetics, according to Laplanche et al. (1976), are

$$-\frac{d\,[\text{parathion}]}{dt} = k\,[\text{parathion}][O_3]$$

Where:

$$k = 70\ M^{-1}\ s^{-1}$$

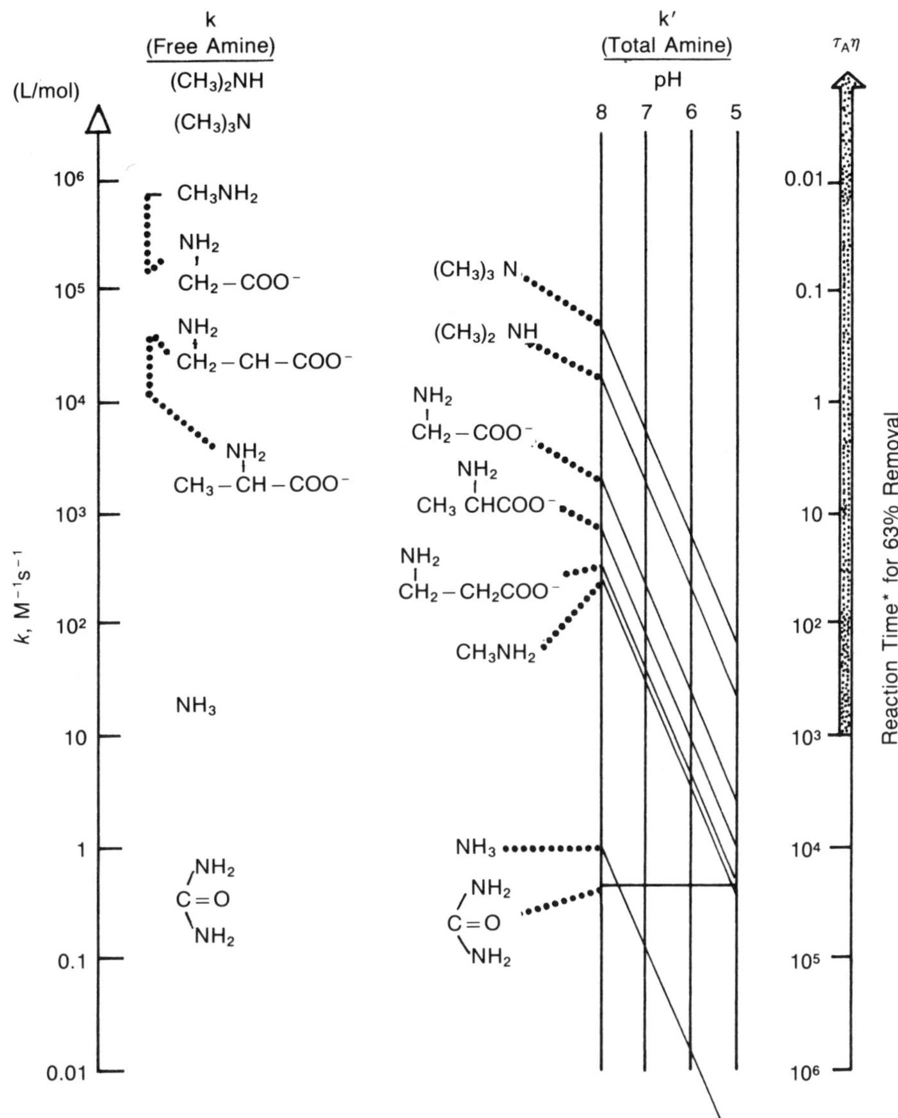

Source: Reprinted with permission from *Ozone Sci. Engrg.* 1:73, Hoigné, J. and Bader, H. © 1979 IOA.

*Assuming a constant aqueous ozone concentration of 5 mg/L.

Figure II–23 Rate Constants for Amine–Ozone Reactions

The same kinetic equation applies to paraoxon, with a rate constant slightly lower ($k = 50\ M^{-1}s^{-1}$), causing an accumulation of paraoxon prior to complete destruction (Figure II–26). In this type of slow degradation, complete removal of toxic compounds requires that a longer contact time be maintained than generally used for disinfection.

Herbicides of the phenoxyacetic class are readily degraded (Doré et al. 1978, 1980; Quentin et al. 1978). These compounds have the same degradation scheme as phenols, with the first intermediates being hydroxylation products of the aromatic cycle.

Compared to other heterocyclic nitrogenous compounds, atrazine is not very reactive (6-atom heterocycles). Still, an attack can be noted on the various nitrogen atoms. Products isolated from atrazine at ozonation are itemized in Table II–10 (Legube et al. 1987b).

Surfactants. Surface-active agents are the chief active products found in detergents. They include four families: anionic, cationic, non-ionic, and ampholyte surfactants. Their molecular weight is in the range of 350 g/mol. They occur at low

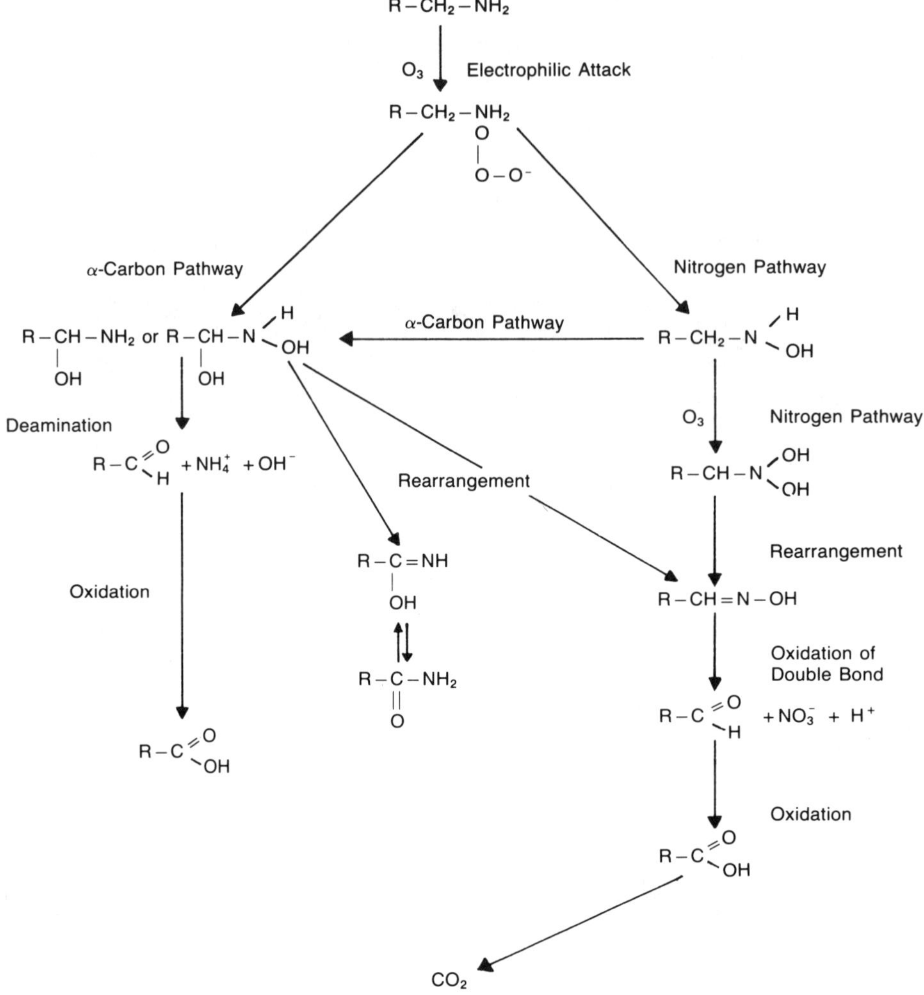

Figure II–24 Schematic of Ozonation of Primary Amines

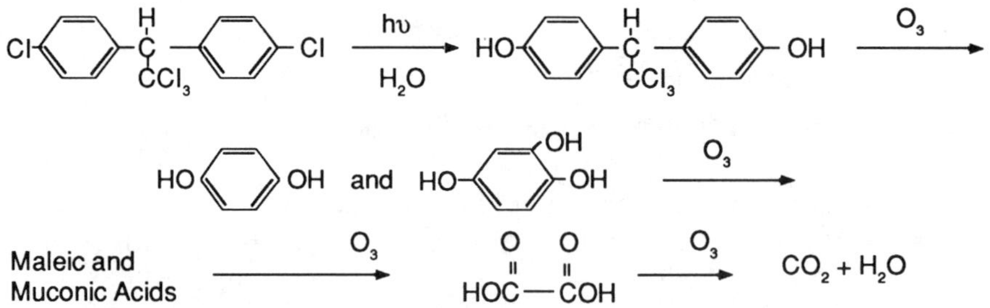

$$\text{Cl-C}_6\text{H}_4\text{-CH(CCl}_3\text{)-C}_6\text{H}_4\text{-Cl} \xrightarrow[H_2O]{h\nu} \text{HO-C}_6\text{H}_4\text{-CH(CCl}_3\text{)-C}_6\text{H}_4\text{-OH} \xrightarrow{O_3}$$

$$\text{HO-C}_6\text{H}_4\text{-OH and HO-C}_6\text{H}_3\text{(OH)-OH} \xrightarrow{O_3}$$

$$\text{Maleic and Muconic Acids} \xrightarrow{O_3} \text{HOC(=O)—C(=O)OH} \xrightarrow{O_3} CO_2 + H_2O$$

Source: Reprinted with permission from *Ozone/Chlorine Dioxide Oxidation Products of Organic Materials,* p. 309. Prengle, H.W. and Mauk, C.E. (R.G. Rice and J.W. Cotruvo, eds.). © 1978 IOA.

Figure II–25 Efficiency of O_3/UV System: Chain Reaction Starting With DDT

concentrations in surface waters. Surfactants are adsorbed at the solutions' interface and are not readily oxidized by ozone. They are eliminated from water by adsorption onto floc, activated carbon, biomass, and similar solid surfaces.

Alkylbenzene sulfonate is an important representative of the anionic series. Ozone attack occurs on the aromatic ring and the rate is dependent on the nature of R.

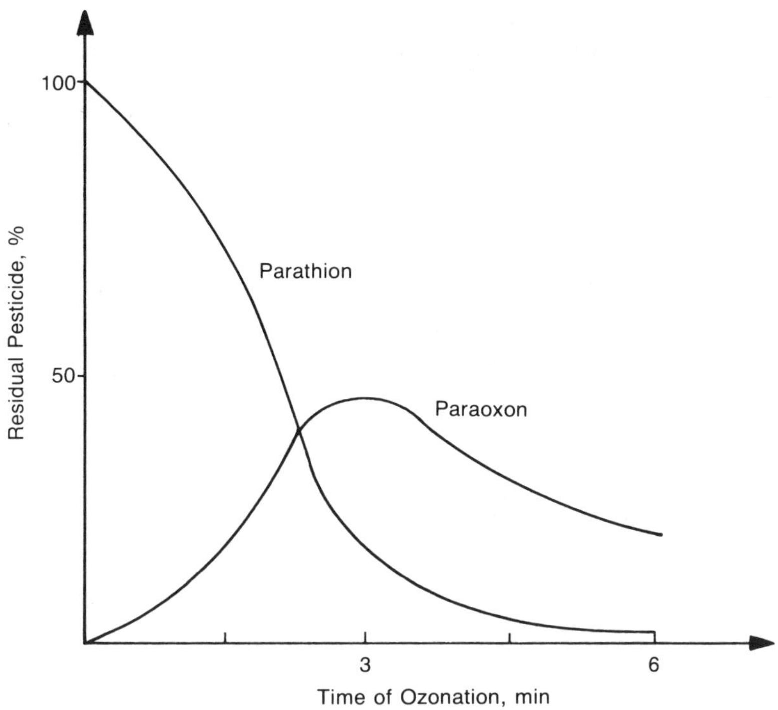

Source: Reprinted with permission from *Ozonation Manual for Water and Wastewater Treatment,* p. 75, Laplanche, A. and Martin, G. (W.J. Masschelein, ed.). © 1982 John Wiley & Sons Inc.

Figure II–26 Pilot Plant Results of Ozonation of Parathion: 3 mg/L O_3, 80 μg/L Parathion

It is important to note that when aromatic ring cleavage occurs, biodegradability is enhanced. Anionic forms are more rapidly oxidized than neutral forms (for example, with R = C_9H_{19}, $C_{10}H_{21}$). Joy et al. (1980) have shown that ozonolysis of *p*-toluene sulfonic acid occurs five times less rapidly at pH 12 than pH 3. Total loss of the parent compound and accompanying release of sulfate requires 16 mol O_3/mol of substrate. Still, at this point only 28 percent of TOC is eliminated, and the chemical oxygen demand (COD) is reduced (in an aqueous solution of the product alone) by 74 percent.

No data are available for the reaction of ozone with cationic surfactants, but it is predicted that they should be relatively inert.

In the case of non-ionic surfactants, ozonation products include polyethylene glycols and polyethers. They are slowly reactive and require that high ozone doses be applied for long contact periods to achieve significant removal. The presence of these by-products is attributed to the formation of short polyethylene glycol chains as intermediates. With regard to phenol ethoxylates, aromatic ring cleavage is slow (Narkis et al. 1980).

Park and Lee (1986) studied the degradation of dodecylbenzene sulfonate (DBS) in the presence of humic acids, Fe, and Mn at 20°C and pH 7. With an ozone dose of 20 mg/L and a 6-min contact time, 80–100 percent of the DBS (340 mg/L starting concentration) was removed. They found that Fe, humic acids, Mn, and DBS were oxidized in that order.

Dyes. The color of treated drinking water is primarily related to the presence of fulvic materials, although certain industrial micropollutants occurring in trace quantities may also color the water. These organic substances are partially ionized, are always present as conjugated aromatic structures, and are therefore strongly reactive.

Studies on the reactivity of dyes to ozone have been carried out by numerous researchers (Gould and Groff 1987; Matsui et al. 1981). Most of the works referenced in

Table II–9 Ozonation Products for Amines

Starting Product	Nitrogen Oxidation	Carbon Oxidation	Split Products	Condensation Products
Ethylamine $C_2H_5-NH_2$	CH_3CH_2-NHOH $CH_3CH=NOH$	CH_3CONH_2	CH_3CHO CH_3CO_2H	C_2H_5NHCHO $C_2H_5NHCOCH_3$ $CH_3CONHCOCH_3$
Dimethylamine CH_3 NH CH_3		CH_3 NH CHO	HCHO CH_3NHOH	CH_3 N – CHO CH_3 CH_3 CHO N – N CH_3 CHO
Trimethylamine CH_3 N – CH_3 CH_3	CH_3 CH_3 — N – O CH_3	CH_3 N – CHO CH_3	HCHO CHO NH CH_3	
Cyclohexylamine NH₂	NHOH N – OH		O O	NH – CHO
Aniline NH_2		OH NH_2	OH H C O C C O H C OH O C O C C O HO OH O O C—C HO OH O H—C OH O H_2N—C O—NH_4^+	NH H – C_6H_5 C H C H O N – C_6H_5 HC N NH NH—C O H

Source: Reprinted with permission from *Wtr. Res.*, 16:226, Elmghari-Tabib, M. et al. © 1982 Pergamon Press PLC.

this book report on studies that were carried out on concentrated solutions. Nevertheless, the first-order kinetics is worthy of discussion.

The initial attack on the double bonds C = C and N = N and on the aromatic rings

Table II–10 Products Isolated During Ozonation of Atrazine

Structure	Chemical Name
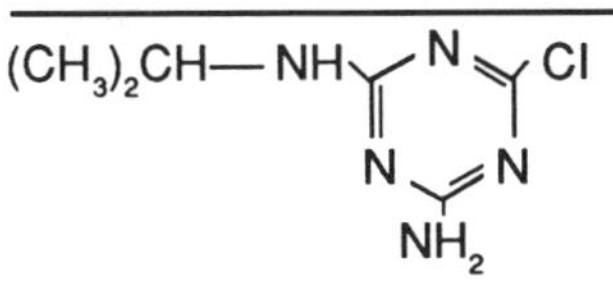 $(CH_3)_2CH$—NH, N, Cl, N, N, NH_2	2-Chloro-4-amino-6-isopropylamine-1,3,5-triazine
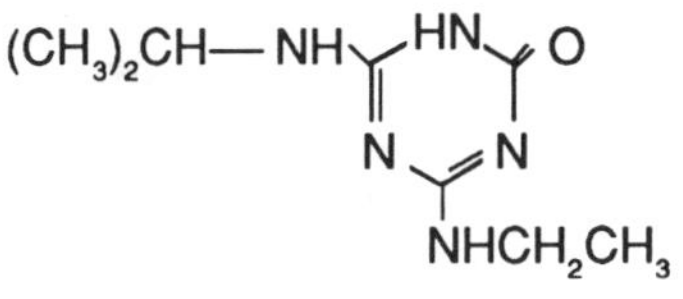 $(CH_3)_2CH$—NH, HN, O, N, N, $NHCH_2CH_3$	2-Oxo-4-ethylamino-6-isopropylamine-1,2-dihydro-1,3,5-triazine or
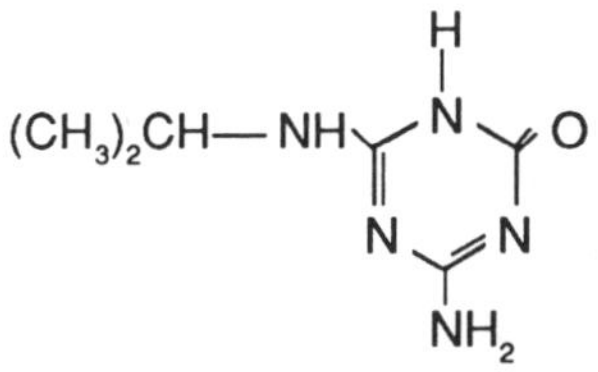 H, $(CH_3)_2CH$—NH, N, O, N, N, NH_2	2-Oxo-4-amino-6-isopropylamine-1,2-dihydro-1,3,5-triazine
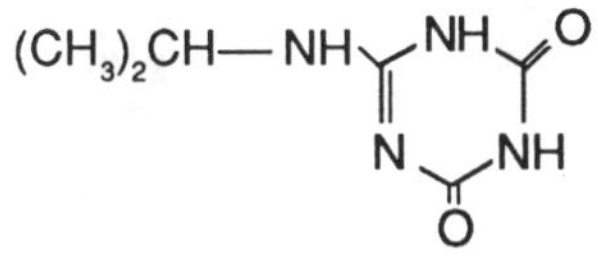 $(CH_3)_2CH$—NH, NH, O, N, NH, O	2,4-Dioxo-6-isopropylamine-1,2,3,4-tetrahydro-1,3,5-triazine
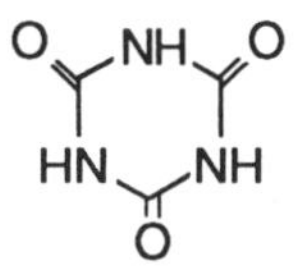 O, NH, O, HN, NH, O	2,4,6-Trioxo-hexahydro-1,3,5-triazine

Source: Reprinted with permission from *Ozone Sci. Engrg.*, 9:240, Legube, B. et al. © 1987 IOA.

leads to the production of cleavage and/or oxidation products. The complexity of the molecules entails the formation of a large number of intermediates, usually still colored, whereby longer ozonation times are likely to be used in order to obtain discrete color reductions. For example, in the azoic series, Nishina and Motoyuki (1978) have shown that 1- to 2-h contact time is needed to obtain a 50–80 percent absorbance reduction at 500–700 nm.

Conclusions and summary. Examples of ozonation mechanisms and by-products obtained have been discussed. Lists of oxidation compounds are found in several publications, including Masschelein et al. (1980). Legube (1983) should be referenced for aromatic hydrocarbons and Laplanche (1982) for pesticides.

The following conclusions can be made regarding the oxidation of synthetic organic compounds:

- Compounds related to human activity occur at very low concentrations and do not show up or do so very weakly when tested by general analytical methods (for example, TOC). Their identification is often based on advanced analytical technology. During ozonation and in conventional water treatment plants, only substrates having ozonation kinetic constants greater than 10^3 $M^{-1}s^{-1}$ will be degraded significantly.

- Compounds susceptible to reaction with ozone are primarily those containing nucleophilic sites such as O, N, S, P, and activated C. Aromatics with mesomeric substituent groups are particularly reactive substrates.
- In the degradation process, many molecular cleavage products are formed. These are more oxidized, more polar, and more biodegradable. This can occur without significant changes in TOC levels.
- It is possible that ozonation could cause certain products to occur in trace amounts due to the degradation of other complex substances.
- In the case of poorly reactive substances (those with structures presenting no strong nucleophilic sites and possibly with very symmetrical structures), it would be necessary to combine either UV, H_2O_2, or some catalysts with O_3 to eliminate these organic compounds.
- Considering the complexity of the organic matrix of a water and the diversity of ozonation by-products, treated water quality assessments should incorporate toxicity testing with chemical analyses.

II.A.5 Reaction of Ozone on Cell Constituents In Vitro

Cell constituents are macromolecules that often have a high molecular weight. A special feature of macromolecules is that their specific structural conformations are dictated by low-energy bonds. The basic chemical groups that make up these macromolecules represent the majority of the functionality of organic chemistry.

A study of ozone effects on cell constituents is very complicated, even in the case of in vitro studies. The major difficulties result from the analytical techniques that are required to quantify the components to be ozonated and the ozonation products. Moreover, the large number of reactive sites in cell constituents can induce different types of direct ozone reaction mechanisms as well as radical reactions, as discussed earlier in this chapter.

The main ozonation reactions of the chemical compounds that compose cell constituents will be discussed in the following paragraphs. This will be done by identifying the most sensitive structures, that is, those structures whose destruction or modification can induce cell deactivation.

Structure of cell constituents. An important class of chemical compounds in cell constituents is carbohydrates. Monosaccharides are cyclic polyalcohols (five to eight carbon atoms) and include hexose- and pentose-type structures, which are shown in Table II–11. The combination of two hexose molecules (by way of a glycoside bond) results in disaccharides, and a combination of several molecules results in polysaccharides, which account for the majority of carbohydrates. Other cell constituents may include sugars, principally glycolipids (or liposaccharides) and glycoproteins, in their structure.

Proteins (polypeptides) are one of the most complex and essential classes of compounds found in living matter. Simple proteins are amino acid polymers, linked by peptide chains, whereas conjugated proteins are composed of amino acids and lipids (lipoproteins) or glucids (glycoproteins). Each basic polypeptide structure has a specific conformation in space (for example, coiled to form a helix), imposed upon it by low-energy bonds.

Simple lipids are glycerol and fatty acid (saturated and unsaturated) esters. Combined lipids include phospholipids (esterification of glycerol phosphate and fatty acids) and glycolipids (a polysaccharide–fatty acid combination). Unlike those mentioned above, some lipids, such as steroids, pigments, and vitamins, present a more complicated structure.

Another very important class of chemical components in cell constituents is the nucleobases derived from purine (adenine and guanine) and pyrimidine (uracil, thymine,

Table II–11 Classes of Organic Compounds Included in Cell Constituents

Class	Typical Structure	Name
Saccharides		
Hexoses	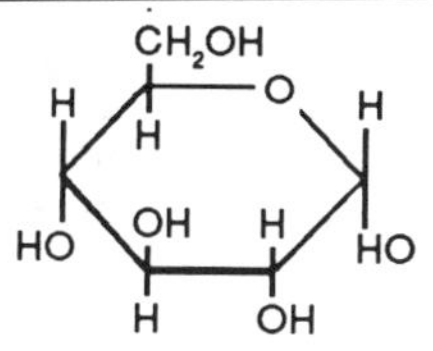	Glucose
Pentoses	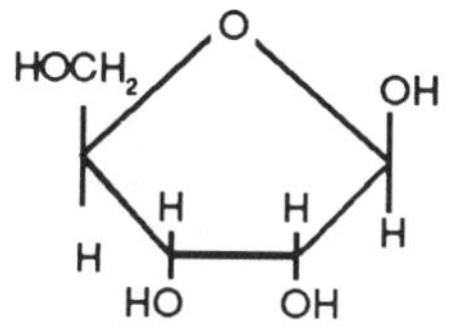	Ribose
Amino acids	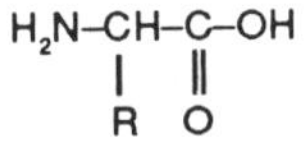 $H_2N\text{-}CH(R)\text{-}C(=O)\text{-}OH$	R = alkyl, aromatic, sulfurated alkyl
Lipids	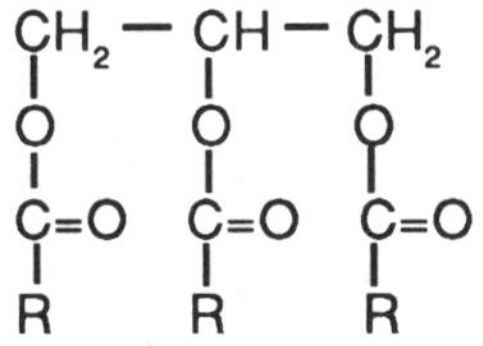 $CH_2(OC(=O)R)\text{-}CH(OC(=O)R)\text{-}CH_2(OC(=O)R)$	R = saturated or unsaturated alkyl groups
Nucleobases		
Purine type	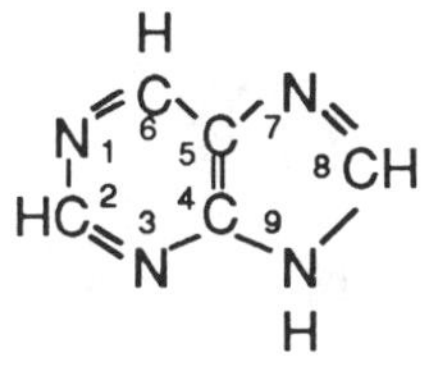	Basic structure of adenine and guanine
Pyrimidine type	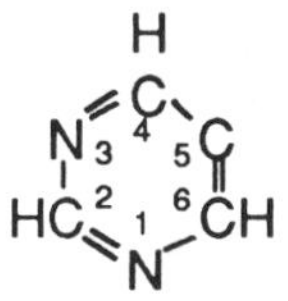	Basic structure of uracil, thymine, and cytosine

and cytosine). One of these bases, a ribose (or deoxyribose) and phosphoric acid linked together by carbon–nitrogen and carbon–oxygen bonds, constitutes a class of nucleotide precursors of the nucleic acids, ribonucleic acid (RNA) and deoxyribonucleic acid (DNA). Lastly, adenine is a part of the structure of other molecules of interest, such as adenosine triphosphate or other substituted nucleotides, including three essential coenzymes (NAD^+, FAD, and CoA).

Reaction of ozone with main cell-forming components. With regards to the reaction of ozone with the main cell-forming components and other reactions already described, the rate constants described here have almost all been obtained in a batch reactor by mixing a solution of the component under study with a solution of ozone. Some studies on the consumption of ozone (stoichiometry) or ozonation products were performed in semibatch or bubble-column-type units.

Carbohydrates. Owing to their chemical structure, which contains no strongly nucleophilic sites, it is likely that carbohydrates will react only slightly with ozone, as is

Table II–12 Rate Constants (k_{O_3}) ($M^{-1}s^{-1}$)* for Ozonation of Carbohydrates at Acid pH and 20°C

Compound	Hoigné and Bader (1983a)†	Perez et al. (1987)‡	Doré et al. (1989)§
D-glucose	0.45 ± 0.05 (1)	1.92 ± 0.05	0.17
D-xylose		1.43 ± 0.06	0.16
D-ribose		2.24 ± 0.05	
Saccharose	0.12 ± 0.02 (2)		0.13
Cellobiose		3.2 ± 0.1	0.21
Lactose		4.0 ± 0.1	0.24
D-gluconic acid			0.24
D-glucuronic acid		1.33 ± 0.07	0.09

*k_{O_3} = constant obtained from measuring ozone depletion versus time in batch reactor (k_{O_3} = k, when stoichiometry = 1).
†pH = 2; [compound] = 5–100 nM; (1) indicates without *t*-butanol; (2) indicates with *t*-butanol = 20–200 mM.
‡pH = 4; [compound] = 0.02–0.5 mM; without radical scavenger.
§pH = 2; [compound] = 2–5 mM; *t*-butanol = 10 mM.

the case with aliphatic alcohols (Hoigné and Bader 1983a). The attack by the ozone molecule on primary and secondary aliphatic alcohols may, however, lead to formation of hydrogen peroxide, aliphatic aldehydes, acids, or ketones, through an unstable hydroxyhydroperoxide (discussed in sec. II.A.4). Consequently, the reaction of ozone with carbohydrates can be a precursor to hydroxyl radicals that, in turn, react strongly with the hydrocarbons (Anbar and Neta 1967). Since carbohydrates promote the radical chain by which ozone decomposes in water (sec. II.A.1), it is important to distinguish between the studies performed with and without the presence of a radical scavenger in the reactive environment, even at low pH. This is particularly true in the case of carbohydrates because there is a great difference between the kinetic constants of the reaction produced by molecular ozone (Table II–12) and those observed with hydroxyl radicals (for glucose, $k_{OH} = 10^9\ M^{-1}s^{-1}$, according to Anbar and Neta [1967]).

Table II–12 gives the rate constants for the action of molecular ozone on certain monosaccharides (and related molecules) and disaccharides. These values are very low when compared, for instance, with those usually obtained on unsaturated sites (sec. II.A.4). Table II–12 illustrates that the presence of butanol significantly reduces the reaction speed, which points to the possibility of a reaction with hydroxyl radicals, even with an acidic pH. Increasing the pH causes a significant increase in the kinetic constant, although the constant remains below 10 $M^{-1}s^{-1}$, at a pH of 6 and in the absence of a radical scavenger (Hoigné and Bader 1983a; Perez et al. 1987). At a neutral pH, Doré et al. (1989) demonstrate that the increase in bicarbonate ions in the environment induces a decrease in the reactivity of glucose.

The reaction of ozone on carbohydrates leads to ozone consumption levels of over 2 mol/mol of monosaccharide for total removal (Bonnet 1988), without substantial losses in TOC (Duguet et al. 1987; Bonnet 1988).

Regarding the products of ozonation, research performed by Bonnet (1988) on the ozonation of glucose, xylose, and cellobiose at an acid pH (without a radical scavenger) has evidenced a large increase in total acidity (1–1.5 meq/mmol of oxidized carbohydrate), 20 percent of which is due to the formation of formic acid. This same author has identified, by GC–MS analysis, gluconic acid and arabinose from the ozonation of glucose; xylonic acid from xylose and glucose; and arabinose and gluconic acid from a disaccharide (cellobiose). In addition, Yamada and Somiya (1980) found a significant amount of formaldehyde and pyruvic acid after ozonating glucose. As a final note, by applying ozone to a secondary municipal effluent, Legube et al. (1985) detected an increase in monosaccharides, at the expense of the initially present polysaccharides.

The conclusion can be drawn from these results, when taken as a whole, that the action of ozone on polysaccharides leads to the rupture of the glycosidic bonds, followed by the oxidation of the alcoholic functions of the free monosaccharides, eventually leading to the formation of aliphatic acids and aldehydes. This slow reaction includes not only the action of molecular ozone but also the action of hydroxyl radicals, especially in the absence of radical scavengers.

Amino acids. Amino acids [R–CH(NH_2)COOH] can react with ozone at two levels of their structure—the amine function and the R group. It is known that, with regard to the primary amine function, its reactivity with ozone is largely dependent on pH. This is shown by the kinetic constants obtained by Hoigné and Bader (1983a, 1983b) with respect to ammonia and some amines (Table II–13).

In the case of saturated aliphatic primary amines, the action of ozone (or hydroxyl radicals) apparently leads to the formation of hydroxylamine, then oxime and/or amide, and finally aldehydes, acids, and nitrate ions, according to Laplanche (1982) (see Figure II–24).

The R group of amino acids can be an alkyl group (saturated aliphatic), a sulfur alkyl group, or even an aromatic or unsaturated heterocyclic group (imidazole or indole). In the first case of the alkyl substituted amino acid R group, ozone is only slightly reactive at an acid pH, as is the case with alkylamines. However, an increase in pH can greatly increase the speed of the reaction, as with glycine (Table II–13; Duguet et al. 1982), alanine (Table II–13; Le Cloirec Renaud et al. 1984), and leucine (Narkis et al. 1977; Yamada and Somiya 1980). As with alkylamines, the ozone should react by the electrophilic process on nitrogen and finally lead to partial decarboxylation plus a buildup of acids, aldehydes, nitrate ions, and ammonia (Table II–14).

In the second case of the amino acid R group (such as a sulfur alkyl group), the reactivity of ozone is very high, even at an acidic pH (Table II–13), and increases proportionally with the increasing pH (Mudd et al. 1969). The center of reaction for these sulfur amino acids is apparently the sulfur atom oxidizing into a sulfoxide or sulfonic group. Table II–14 presents some ozonation by-products of methionine, cysteine, and cystine.

In the third case of the amino acid R group (such as an aromatic or unsaturated heterocyclic group), the attack of ozone can occur preferentially either on the R group or on the amine function. It would seem that the reactivity with ozone depends on pH (Mudd et al. 1969; Le Cloirec Renaud et al. 1984), and that the reaction can result in several products (aromatic aliphatic acids and aldehydes, and condensation products), some of which are shown in Table II–14.

To conclude, ozone can react significantly with amino acids, especially at neutral and basic pH, either on the nitrogen atom or on the R group (alkyl sulfur or unsaturated) or on both at once. So the reactivity of polypeptides and proteins will, in all probability, depend on the nature of their constituent amino acids. It is worth noting that Brette et al. (1987) obtained total removal of glycyl-L-tyrosine by ozonating this peptide at a dose rate of 0.7 mol ozone/mol carbon.

Fatty acids. Saturated fatty acids, the main constituent of lipids, react only slightly with ozone, as can be seen from the few kinetic constants obtained by Hoigné and Bader (1983b) (Table II–13). However, their reactivity increases when there are several ethylenic bonds in the carbonated chain. When this occurs, the main ozonation products are aldehydes, acids, and hydrogen peroxide, demonstrating a probable ozone attack by 1–3 dipolar cyclo addition on the double bond. A common example is that of oleic acid ozonation, resulting in the formation of nonanal and hydrogen peroxide (Spanggord and McClurg 1978; Guyon 1984).

Nucleobases. Purine and pyrimidine are the basic structures of the five nucleobases that make up nucleosides, nucleotides, and nucleic acids (Table II–15). Despite the low reactivity of ozone on purine and pyrimidine, the corresponding nucleic bases

Table II–13 Rate Constants ($M^{-1}s^{-1}$) at 20°C for Ozonation of Amino Acids, Aliphatic Amines, and Aliphatic Acids*

Compound	Value of k_{O_3}† at pH = 2	Value of k_{O_3}† at pH = 8
Amino acids (HOOC–CH(NH$_2$)–R)		
Glycine (R = –H)	50×10^{-3}	1600
Alanine (R = $-CH_3$)	3×10^{-3}	640
Methionine (R = $-CH_2-CH_2-S-CH_3$)	$>500 \times 10^{3}$‡	
Cysteine (R = $-CH_2-SH$)	$(30 \pm 10) \times 10^{3}$‡	
Cystine (R = $-CH_2-S-S-CH_2-CH(NH_2)-COOH$)	550 ± 10‡	
Aliphatic amines		
Dimethylamine	$<0.13 \pm 0.2$	20×10^{3}
Butylamine	<0.02‡	340
Aliphatic acids (HOOC–R)		
Acetic acid (R = $-CH_3$)		$\sim 3 \times 10^{-5}$
Butyric acid (R = $-CH_2-CH_2-CH_3$)		$\sim 4 \times 10^{-2}$
Oxalic acid (R = –COOH)		$\sim 4 \times 10^{-2}$
Malonic acid (R = $-CH_2-COOH$)	<4	7 ± 2
Maleic acid (R = $-CH_2-COOH$)	1×10^{3}‡	$>5 \times 10^{3}$

*Values are according to Hoigné, J. & Bader, H. 1983. Rate Constants of Reactions of Ozone With Organic and Inorganic Compounds in Water. II. Dissociating Organic Compounds. *Wtr. Res.*, 17:2:185.
†k_{O_3} = constant obtained by measuring the depletion of ozone versus time in a batch reactor ($k_{O_3} = k$ with stoichiometry = 1).
‡In the presence of radical scavenger (*t*-butanol or HCO_3^-).

react quickly with ozone, as has been demonstrated in kinetic studies by Ishizaki et al. (1984) and Legube et al. (1987b), followed by Doré et al. (1989) (Table II–16).

The kinetic study conducted in a batch reaction vessel by Doré et al. (1989) has resulted in the following reactivity sequence: T > U > G > A > C. During an earlier study performed in a semibatch reactor and without radical traps, Ishizaki et al. (1984) obtained another reactivity sequence: G,T > U > C > A. The difference between these two sequences is perhaps due first to the presence or absence of radical traps. Secondly,

Table II–14 Amino Acid Ozonation Products

Amino Acids ($HOOC-CH(NH_2)-R$)	Ozonation Products	Reference
Glycine $R = -H-$	Formic acid, nitrate	Duguet et al. (1982)
Alanine $R = -CH_3$	Formaldehyde, acetic acid, ammonium, nitrate	Le Cloirec Renaud et al. (1984)
Leucine $R = -CH_2-CH(CH_3)-CH_3$	Formaldehyde, butanol, pyruvic acid, ammonium, nitrate	Yamada and Somiya (1980); Narkis et al. (1977)
Methionine $R = -CH_2-CH_2-S-CH_3$	Sulfoxide group on R	Mudd et al. (1969)
Cysteine $R = -CH_2-SH$	Cystine, cysteic acid	Mudd et al. (1969)
Cystine $R = -CH_2-S-S-CH_2-CH(NH_2)-COOH$	Cysteic acid	Mudd et al. (1969)
Phenylalanine (1985) $R = -CH_2-C_6H_5$	Phenylacetaldehyde, phenylacetic acid, phenyl acetamide, phenylhydroxylamine, hydroxy-4 phenylacetic acid, hydroxy-2 benzoic acid, ammonium, nitrate, condensation products	Laplanche et al. (1985)
Tryptophane $HO-C_6H_4-CH_2-CH(NH_2)-COOH$	N-formyl-kynurenine ammonium	Mudd et al. (1969)
Tyrosine $R = -CH_2-$(indolyl, N–H); $R = -CH_2-C_6H_4-OH$	Di-hydroxyphenylalanine, hydroxy-tyrosine condensation products	Mudd et al. (1969); Verweij et al. (1982)
Histidine $R = -CH_2-$(imidazolyl, N, N–H)	Ammonium	Mudd et al. (1969)

it may be due to the fact that measurements of the remaining component versus ozone-bubble-formation time in a semicontinuous reactor led to an evaluation of reactivity that includes the gaseous ozone transfer rate in an aqueous environment.

With regard to nucleotides, Ishizaki et al. (1984) determined in a batch reactor (without radical traps) the kinetic constants of four deoxyribonucleoside-5′-monophosphates (Table II–16). It should be noted that the sequence of kinetic constants obtained by these authors (dGMP > dTMP > dCMP > dAMP) is different from those obtained in the case of nucleobases by Doré et al. (1989). Here again the lack of components trapping the radicals is important, since Ishizaki et al. (1984) pointed out that the

Table II–15 Chemical Formulas of Nucleobases and Selected Containing Structures

Nucleobases

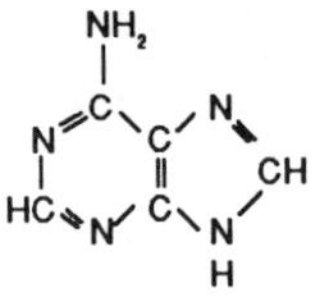 Adenine (A)	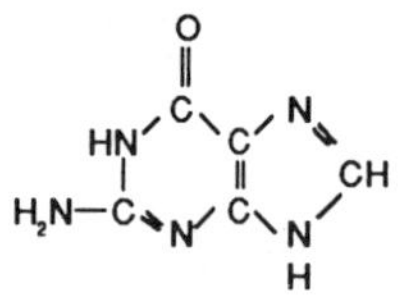 Guanine (G)	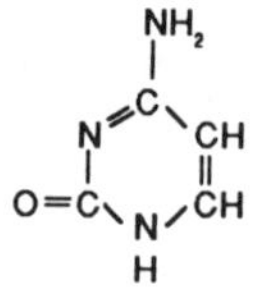 Cystosine (C)	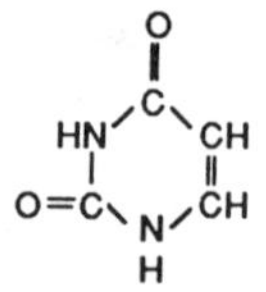 Uracil (U)	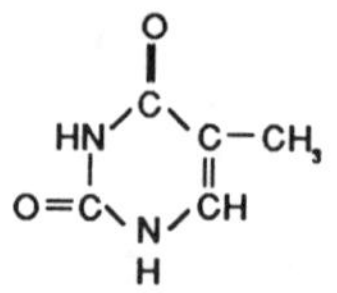 Thymine (T)

Nucleosides		Nucleotides	
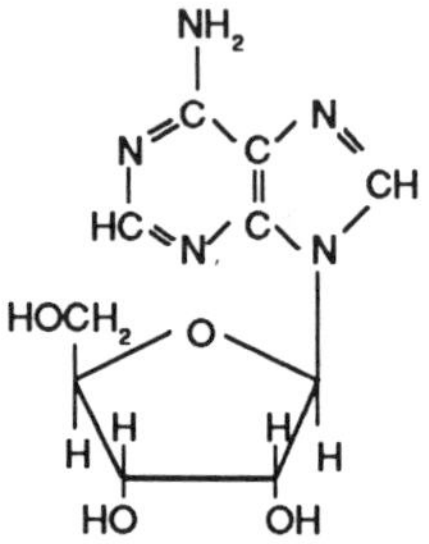 Adenosine	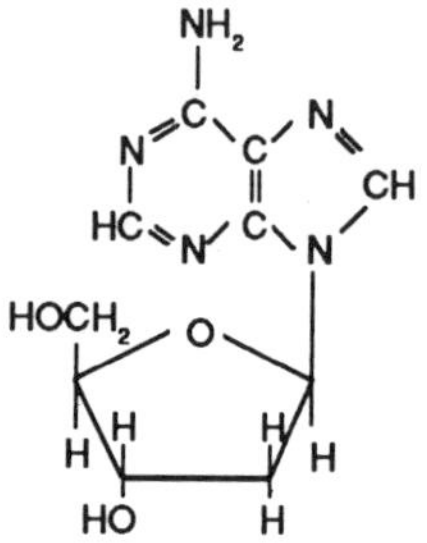 Deoxyadenosine	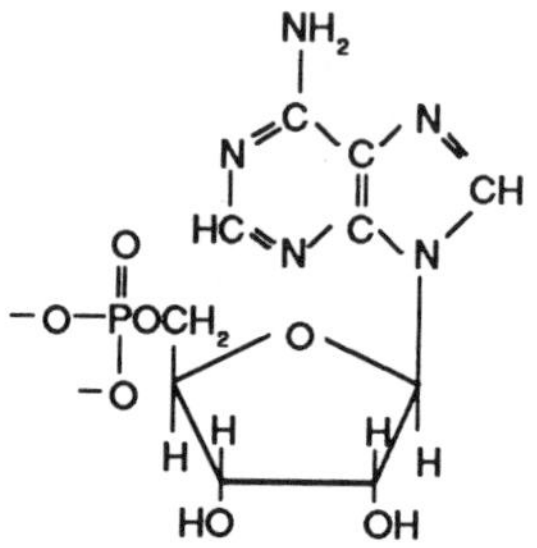 Adenosine-5′-phosphate (AMP)	AMP, GMP CMP, UMP dAMP, dGMP dCMP, dTMP

Nucleic Acids

Phosphoric acid 2-Deoxyribose Thymine Cytosine Adenine Guanine	DNA	Phosphoric acid Ribose Uracil Cytosine Adenine Guanine	RNA

Table II–16 Rate Constants of Puric and Pyrimidic Bases and Selected Nucleotides (Deoxyribonucleoside-5′-monophosphates) at a Neutral pH

Compound	Stoichiometry (mol O_3/mol compound, in semibatch reactor)	k ($M^{-1}s^{-1}$, in batch reactor)	Reference
Purine	4	7.9 ± 0.7 (at 20°C)	Legube et al. (1987b)★
		3.9 ± 0.7 (at 10°C)	
		1.4 ± 0.2 (at 2°C)	
Adenine (A)	2	$(2.2 \pm 0.5) \times 10^3$ (at 2°C)	
Guanine (G)	1	$(3.5 \pm 1.1) \times 10^3$ (at 2°C)	
Pyrimidine		2.2 ± 0.4 (at 20°C)†	Doré et al. (1989)★
Cytosine (C)	1	$(7.5 \pm 0.85) \times 10^2$ (at 2°C)	Doré et al. (1989)★
		9.3×10^2 (at 15°C)	Ishizaki et al. (1984)‡
Uracil (U)	1	$(4.5 \pm 0.8) \times 10^3$ (at 2°C)	Doré et al. (1989)★
Thymine (T)	1	$(6.8 \pm 0.8) \times 10^3$ (at 2°C)	Doré et al. (1989)★
		2.3×10^4 (at 15°C)	Ishizaki et al. (1984)‡
dGMP	1	5×10^4 (at 15°C)	Ishizaki et al. (1984)‡
dAMP	2	2×10^2 (at 15°C)	
dCMP	1	1.4×10^3 (at 15°C)	
dTMP	1	1.6×10^4 (at 15°C)	

★pH = 7.6; [compound] = 0.01–0.15 mM; (HCO_3^-) = 10 mM.
†k_{O_3} and not k (stoichiometry unknown).
‡pH = 6.9; [compound] = 0.1–0.3 mM; without radical scavengers.

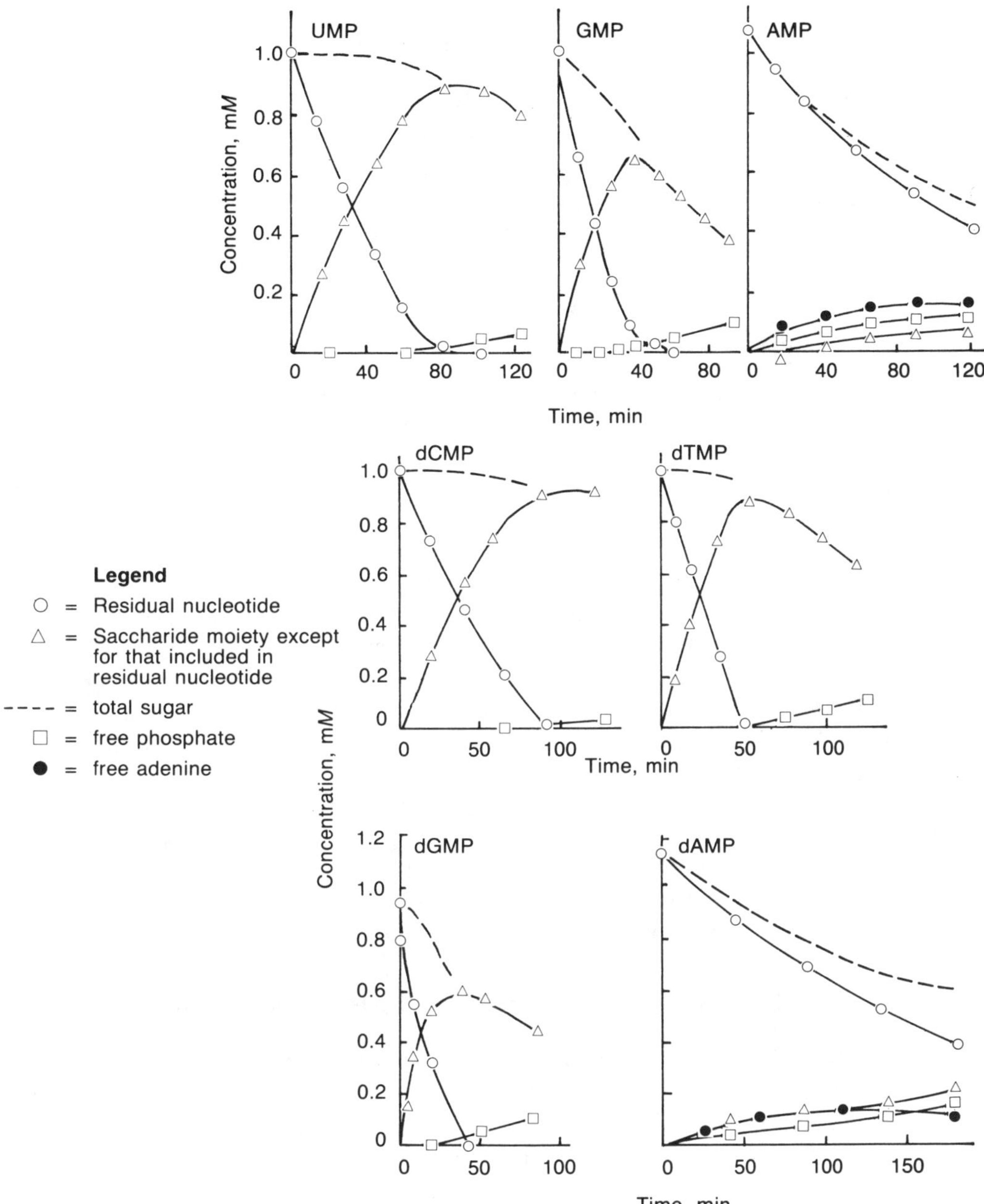

Source: Adapted with permission from *Chem. Pharm. Bull.*, 29:868, Ishizaki, K. et al., and 32:3603, Ishizaki, K. et al. © 1981 and 1984 Pharmaceutical Society of Japan.

Figure II–27 Ozone Degradation of Ribonucleoside-5′-monophosphates and Deoxyribonucleoside-5′-monophosphates

presence of *t*-butanol significantly modified the value of some kinetic constants. Finally, Shinriki et al. (1981) showed that the ozonation of a mixture of four ribonucleotides led mainly to the degradation of the GMP.

The ozonation products of the nucleic bases have not been determined at the time of this writing (1990). However, Ishizaki et al. (1981; 1984) showed an increase in appearance of free ribose and phosphate ions through ozonation at a neutral pH of the three ribonucleoside-5′-monophosphates (UMP, GMP, and AMP), as well as free deoxyribose and phosphate ions from the four deoxyribonucleoside-5′-monophosphates. At high ozone doses, deoxyribose and ribose are then oxidized. In the case of AMP and dAMP, the appearance of adenine has been measured. The results are given in Figure II–27.

In conclusion, ozone reacts quickly with nucleobases, especially thymine, guanine, and uracil, as well as with the corresponding nucleotides, by releasing the carbohydrate and phosphate ions.

Consequence for microorganism inactivation. Knowledge regarding ozone reactions with various chemical components constituting biological cells makes it possible to locate a few sensitive sites during the deactivation of bacteria (and viruses) by ozonation.

It can be considered that a bacterium is schematically composed (from the outside to the inside) of a cell wall surrounded by exo-polysaccharides, then a cytoplasmic membrane, and finally a cytoplasm containing the genetic-information-carrying chromosome (Block 1982). The cell liquid offers a near-neutral pH and a high concentration of bicarbonate ions. It is, therefore, probable that the radical action of ozone is inhibited inside the cell. As a result, the chemical reaction of the molecular ozone on the cell wall is weak, bearing in mind the composition of the cell wall, which is rich in polysaccharides, phospholipids, and amine sugars. The cytoplasmic membrane can provide a site for ozone reaction, due to the numerous proteins among its constituents. If residual ozone crosses this membrane, the cytoplasm and the chromosome become preferred sites, since the nucleic acids (mainly guanine and thymine) are quickly degraded by the ozone. This has been shown in some in vitro studies on nucleic acids (Shinriki et al. 1983, 1984) and other studies carried out on *Escherichia coli* (for example, Ishizaki et al. 1987). These considerations may be useful for a better understanding of ozone disinfection (see chapter III, sec. III.H).

Natural waters. Simple and polymerized carbohydrates, free and combined amino acids, and even nucleobases (at lower level) are natural organic compounds found in surface waters at highly variable concentrations. Finally, it is important to note that some ozonation by-products of amino acids, especially aldehydes and acids, are commonly found in ozonated natural waters (see chapter III, sec. III.I).

II.A.6 Reactions With Aquatic Humic Substances

Natural waters contain varying concentrations of numerous organic and inorganic natural compounds. These compounds are present in dissolved and suspended forms. The dissolved organic load, or dissolved organic carbon (DOC), is the organic fraction that can be analyzed after the water has been filtered (0.45-μm filter); the remainder of the TOC represents only 10 percent of the total (Thurman 1985). The concentration of TOC is usually low in groundwater, but it is variable in rivers (a few milligrams per liter) and can be very high in certain eutrophic lakes (as much as 30 mg/L and greater).

This natural organic matter results from elutriation of the soil and also from biological, chemical, and photochemical reactions in the water that occur due to the presence of algae and of animal and vegetable by-products. The main organic constituents in natural water are a collection of polymerized organic acids called humic substances. Other types of carboxylic acids, amino acids, and carbohydrates can also be added to this group.

It is essential that the reaction of these natural compounds with ozone be studied. Such study is important to understanding ozone consumption in water treatment. Above all, such study is necessary to understand the formation of ozonation products, some of which can create tastes and odors in water or possibly cause adverse health effects.

There are two possible methods of study. One method is to identify the oxidation products present after the different ozonation stages in treatment plants (or pilot units) and to correlate the results with observations on the subsequent quality of the treated water. The second method is to study the behavior of aqueous solutions of isolated natural compounds toward ozone. In addition, these same solutions would be studied for the effects of ozonation on toxicity and biodegradability, and the effects on stages of treatment. In fact, these two methods of study are complementary. The first method is discussed in chapter III.I. The second laboratory study method will be discussed in this chapter after an overview of the different dissolved natural organic compounds found in water, particularly humic substances.

Natural organic compounds in water. Most types of functional groups known in organic chemistry have been found in unpolluted natural water. A simple classification of natural aquatic organic components is not easy to make because such a classification would have to reflect the numerous separation, concentration, and analysis techniques used by analytical chemists and geochemists. However, a distinction can be made between classes of compounds based on methods of isolation. For example, humic substances can be distinguished from compounds such as simple carboxylic acids, phenols, carbohydrates, amino acids, hydrocarbons, and others that can be isolated and identified as to their exact structure.

Aquatic humic substances. Humic substances have been used to represent natural dissolved organic components in water because of their high concentration (30 to 50 percent DOC; Thurman [1985]) and their truly natural origin. Several theories in the literature explain the presence of humic substances in natural water. Whether they are formed in the ground (then washed out) or directly in the water itself, humic substances are the result of the microbiological, chemical, and photochemical reactions that occur during the degradation and polymerization of vegetable matter.

There are many techniques for extracting humic substances from water. In fact, there are so many methods that comparisons between the results obtained cannot be easily made. For this reason, the International Humic Substance Society (IHSS) (McCarthy et al. 1986) has made available reference humic substances, isolated from the Suwannee River, near Fargo, Georgia. The society has also proposed a method of isolation, which is essentially the same as the one used by Thurman and Malcolm (1981). This method basically includes microfiltration of the water and adsorption of organics on XAD-8 resin at a pH of 2, followed by sodium hydroxide elution and separation by precipitation at an acid pH (pH = 1, COD = 500 mg/L). The precipitated fraction is humic acid and the part remaining in the pH 1 solution is fulvic acid.

Fulvic acids, which are more soluble than humic acids (mainly in colloidal form), always represent the larger fraction. A series of extractions performed according to Thurman and Malcolm's method (1981) on 10 different French surface water samples taken from lakes and rivers showed that the fulvic acid/humic acid mass ratios were between 1.5 and 9 (Legube et al. 1989a). These results explain why most of the investigations on solutions of humic substances have been carried out with the fulvic acid fraction.

The typical elemental composition of aquatic humic substances is generally close to that of reference fulvic and humic acids given by IHSS (Table II–17). Note that the commercial humic substances may be of a slightly different elemental composition, and their use for evaluating the behavior of humic substances in water treatment is not always appropriate (Malcolm and McCarthy 1986).

Aquatic humic substances possess many chemical properties that arise from the presence of several functional groups in their complex structure about which little is known. A description of these functional groups and quotations from the relevant bibliographic references is beyond the scope of this book; however, Table II–18 lists some of the structural characteristics of humic substances. The main physicochemical properties include:

Table II–17 Elemental Composition of Reference Aquatic Humic Substances*

Reference Aquatic Humic Substance	%C	%H	%O	%N	%S	%P	Total	%ASH
Fulvic acid (average of four values)	53.75	4.29	40.48	0.68	0.50	0.01	99.71	0.82
Humic acid (average of four values)	54.22	4.14	39.00	1.21	0.82	0.01	99.40	3.19

*Reference aquatic humic substances are provided by the International Humic Substances Society, Denver, Colo.

- a strong reactivity with halogens, leading to a high consumption of chlorine in potable water treatment and production of several volatile and nonvolatile halogenated organic compounds, some with a strongly mutagenic nature;
- a significant adsorbability on solids such as activated carbon and alumina, but also on mineral colloids, which is likely to modify the performance of the coagulation/flocculation process;
- possible complexation with trace metals, which is likely to make them more soluble and hinder their removal, or a possible complexation with coagulant metals, inducing the precipitation of the humic matter during the clarification process; and
- the possibility that the humic substances might combine with organic micropollutants, including some pesticides, which can cause them to be more soluble in natural water.

Other natural organic compounds. In addition to the humic substances defined as the fraction retained by XAD resin at an acid pH, Lecnher and Huffman (1976) have characterized the nonretained fraction, including hydrophilic acids (30 percent, according to Thurman [1985]). Little is known about this class of humic substances, which is composed mainly of acid sugars and simple and polymerized organic acids.

In addition to these two main classes of natural compounds, there are others that can be specifically analyzed, some of which are considered to be humic substances. Table II–19 summarizes the data collected by Thurman (1985) on natural organic compounds that can account for as much as 20 to 25 percent of the DOC in natural water. Carbohydrates, carboxylic acids, and amino acids are the most important classes.

Reaction of ozone with humic substances. As opposed to chlorination, only a few studies have been carried out on the ozonation of humic substances in an aqueous solution. The following paragraphs will summarize the results obtained from this research. The research work presented here includes only experiments performed on solutions of humic substances extracted from the natural environment. The methods of Thurman and Malcolm (1981) and proposed by the IHSS (McCarthy et al. 1986) were primarily used.

Ozone consumption. Little work has been done regarding kinetic studies of the ozone consumed by humic substances. In this book, the results found in the literature are expressed in terms of mass or molar ozone consumption versus humic mass or mole of carbon, rather than as a kinetic constant. There are, however, a few studies on the determination of kinetic constants applicable to ozone consumption. The constants measured are of apparent first order, bearing in mind the difficulty (if not the impossibility) of analyzing precisely the concentration of ozone consumption sites on humic substances during ozonation.

Anderson et al. (1986) showed that semibatch ozonation of a phosphate-buffered fulvic acid (extracted from Black Lake in North Carolina; pH = 7.2) induced a large consumption of ozone that increased with the ozone dosage applied (Figure II–28). Using lower concentrations of the same fulvic acid, Reckhow (1984) observed the same phenomenon. Moreover, the consumption of ozone was shown to be dependent on the

Table II–18 Main Structural Characteristics of Aquatic Humic Substances

Structural Characteristic	Comments	Reference
Apparent molecular weight	Depends on analytical method (UF, gel, X-ray): HA* = 2–5 kD FA† = 0.5–2 kD	Thurman (1985)
UV absorption	No characteristic bands HA absorbs more than FA For 11 FAs (at 254 nm): 0.025–0.040 AU‡/mg C	Oliver and Thurman (1983); Legube et al. (1989a)
Aromatic carbon content	By NMR ^{13}C: 20–40 percent of total carbon	Thurman (1985)
Carboxylic functions	By titrimetry or NMR ^{13}C: 4–7 meq/g HA or FA By titrimetry for 11 FAs: average of 12 meq/g C	Thurman (1985); Thurman and Malcolm (1983); Legube et al. (1989a)
OH—phenolic functions	By titrimetry or NMR ^{13}C: 1–4 meq/g HA or FA	Thurman (1985); Thurman and Malcolm (1983)
Carbohydrate content	2–4 percent of total carbon	Thurman (1985)
Amino acid content	0.5–1 percent of total carbon	

*HA = humic acid.
†FA = fulvic acid.
‡AU = absorbance unit.

Table II–19 Classes of Specifically Analyzable Natural Compounds in Water

Compound	Types
Carbohydrates	Monosaccharides, oligosaccharides, polysaccharides, sugar acids, amino sugars
Carboxylic acids	Fatty acids, diacids, hydroxy acids, aromatic acids, phenols, tannins
Amino acids	Free amino acids, polypeptides, proteins
Hydrocarbons	Saturated and unsaturated aliphatics, simple and polyaromatics
Trace compounds	Aldehydes, sterols, purines and pyrimidines, organosulfides, alcohols, ketones, ethers, chlorophyll and other pigments

presence of bicarbonate ions in the solution, as shown in Figure II–29. Bicarbonate stabilizes the ozone in the water, even in the presence of fulvic acid (Reckhow et al. 1986). The fact that bicarbonate ions have this effect on the consumption of ozone has been shown by other authors (Croué 1987).

With the same fulvic acid (extracted from Black Lake), Anderson et al. (1986) established for two identical initial molar ratios (O_3/C) that the relative consumption of ozone is independent of the initial fulvic acid content (Figure II–28). Legube et al. (1989b) observed the same phenomenon for a solution of fulvic acid (isolated from pond water in a French forest) in the presence of *t*-butanol (OH scavenger) at a slightly acid pH (Figure II–30). These same authors showed that the consumption of ozone versus the initial concentration in humic matter depended on the nature of the substances studied and, in particular, on the UV absorbance of the fulvic acids (Table II–20).

Attempts at a kinetic approach. Bearing in mind the complex chemical structure of humic substances, their reactivity, and their importance in natural waters, it is not

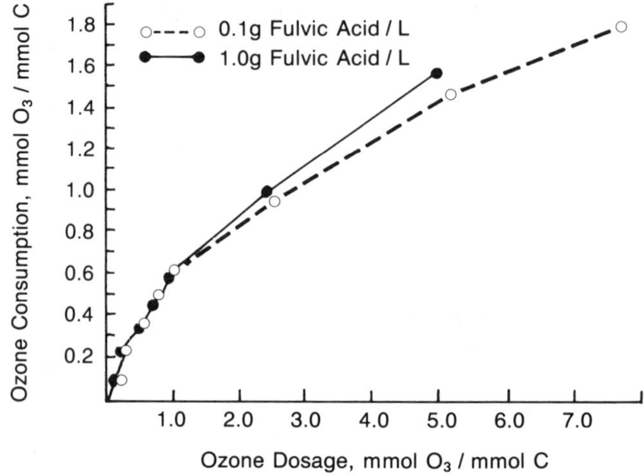

Source: Reprinted with permission from *Envir. Sci. Technol.*, 20:740, Johnson, D. et al. © 1986 American Chemical Society.

Figure II–28 Ozone Consumption for an Aquatic Fulvic Acid

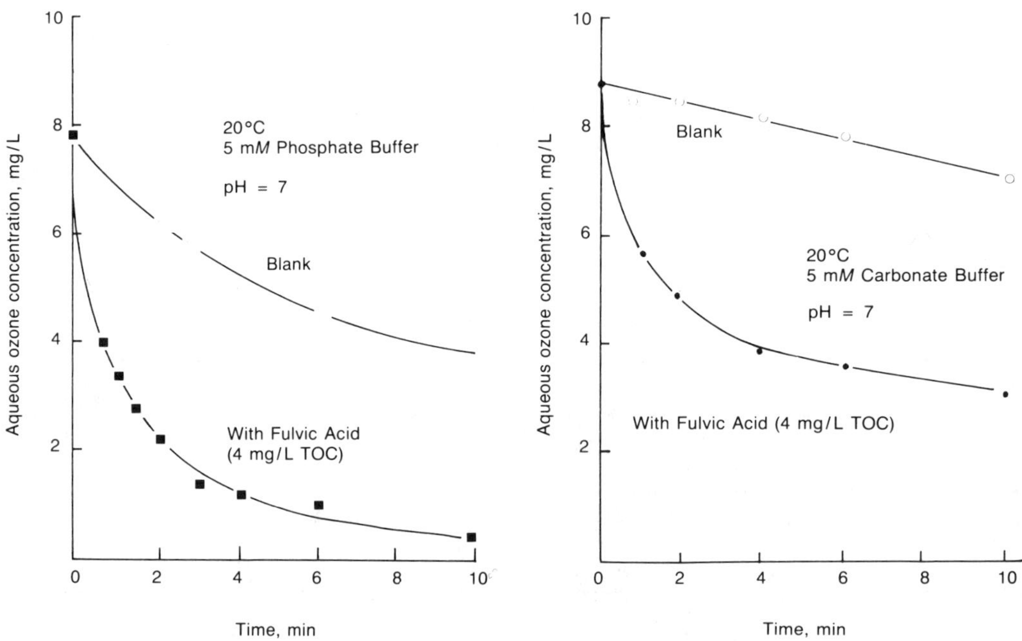

Source: Reprinted with permission from *Wtr. Res.*, 20:8:989, Reckhow, D.A. et al. © 1986 Pergamon Press PLC.

Figure II–29 Effect of Bicarbonate Ions on Ozone Stability With and Without Aquatic Fulvic Acid

unreasonable to assume that solutions of humic substances react with ozone in the same way as natural waters (Figure II–31). That is to say they react like a solution containing different concentrations of molecular ozone consumer sites $(HS)_i$ and $(HS)_d$, leading to the formation of radical or non-radical species (reactions 2 and 1), and different concentrations of hydroxyl radical consumer sites $(HS)_p$ and $(HS)_s$, possibly leading to a radical chain propagation (reactions 3 and 4). In addition to these principal reactions, there is also the reaction of ozone on hydroxide ions, leading to the formation of the hydroxyl radical (reactions 5a, 5b), and the reaction of ozone with OH, resulting in the automatic decomposition of the ozone in the water (reaction 6). These reactions have

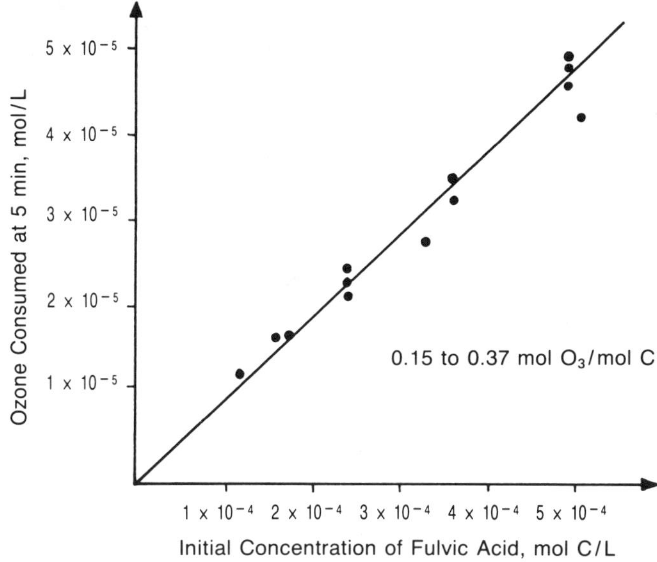

Source: Reprinted with permission from *Ozone Sci. Eng.*, 11:1:73, Legube, B. et al. © 1989 IOA.

Figure II–30 Ozone Consumption Versus Initial Fulvic Acid Concentration, Reaction Time 5 min

Table II–20 Consumption of Molecular Ozone (at 5 min) and UV Absorbance of Aquatic Fulvic Acids*

Source of Fulvic Acid	Ozone Consumption (mol O_3/mol C)	Initial UV Absorbance (AU/mol C)
Pinail (forest ponds)	0.094	432
Moulin Papon (reservoir)	0.087	372
Villejean (reservoir)	0.064	308
Les Landes (lake)	0.055	264

*Fulvic acid extracted using Thurman and Malcolm method (1981). Ozonation in batch reactor for 5 min $[O_3]_0/[C]_0 = 0.15–0.37$ mol/mol; pH = 4.7; I = 10^{-3} (phosphates); *t*-butanol = 10^{-2} M; 2°C.

been described earlier in this chapter. The formation of the superoxide anion O_2^- and the hydroperoxyl radical HO_2 by reactions 3, 5a, and 6 involves further ozone consumption in order to produce more hydroxyl radicals (reaction 5b).

The overall kinetic equation of this complex reactive pattern was developed by Staehelin and Hoigné (1985) for an initial dose of ozone much below the initial humic content.

$$-\left(\frac{d\,[O_3]}{dt}\,\frac{1}{[O_3]}\right) \text{tot} = k'_t = \text{total apparent rate constant of ozone depletion}$$

Where:

$$k'_t = k_d\,[HS]_d + k'_c$$

$$k'_c = k_1\,[OH^-] + (2k_1\,[OH^-] + k_i\,[HS]_i)\left(1 + \frac{k_p\,[HS]_p}{k_s\,[HS]_s}\right)$$

Reaction 1	$(HS)_d + O_3 \xrightarrow{k_d}$ products + O_2 + H_2O_2	(k_d : variable)
Reaction 2	$(HS)_i + O_3 \xrightarrow{k_i}$ products. . . $\longrightarrow$ OH	(k_i : variable)
Reaction 3	$(HS)_p$ + OH $\xrightarrow{k_p}$ products + ($O_2^- \rightleftharpoons HO_2$)	($k_p = 10^8$–$10^9/M^{-1}s^{-1}$)
Reaction 4	$(HS)_s$ + OH $\xrightarrow{k_s}$ products	($k_s = 10^7$–$10^9/M^{-1}s^{-1}$)
		(e.g. : $k_{t_{BuOH}} = 5 \times 10^8/M^{-1}s^{-1}$)
Reaction 5 (5a)	$OH^- + O_3 \xrightarrow{k_1} (O_2^- \rightleftharpoons HO_2)$	($k_1 = 70/M^{-1}s^{-1}$)
		($k_1 = 40/M^{-1}s^{-1}$)
(5b)	$O_2^- + O_3 \xrightarrow{k_2}$ OH	($k_2 = 1.6 \times 10^9/M^{-1}s^{-1}$)
Reaction 6	OH + $O_3 \xrightarrow{k_3} (O_2^- \rightleftharpoons HO_2)$	($k_3 = 3 \times 10^9/M^{-1}s^{-1}$)

Source: Reprinted with permission from *Envir. Sci. Technol.*, 19:1211, Hoigné, J. and Staehelin, J. © 1985 American Chemical Society.

Figure II–31 Possible Reactions of Ozone Consumption in a Natural Aquatic Environment

At acidic pH, or even neutral pH, reaction 5a from Figure II–31 can be ignored. Hence $2k_1[OH^-] << k_i[HS]_i$. In this case, the adequate kinetic model is

$$k'_t = k_d\,[HS]_d + (k_i\,[HS]_i)\left(1 + \frac{k_p\,[HS]_p}{k_s\,[HS]_s}\right)$$

If humic substances are initiators and promotors of ozone decomposition in water, that is, $[HS]_i > 0$ and $[HS]_p >> [HS]_s$, the addition to the water of components that trap the OH radical will increase the total amount of scavenger sites $[HS]_s$. As a result, the consumption rate of ozone (measurable by k'_t) will decrease, since the second term of the kinetic model will approach the value $k_i[HS]_i$ as $[HS]_s$ becomes large. If these humic substances are not initiators, that is, $[HS]_i = 0$, or if they are at the same time both initiators and inhibitors, that is, $[HS]_i > 0$ and $[HS]_s >> [HS]_p$, the addition of a radical trap in the reactional mixture should not significantly modify the total apparent rate constant k'_t.

The experiments conducted by Legube et al. (1989c) using a fulvic acid taken from an impounded water confirmed the above kinetic model at pH 2 to 8. In the presence of *t*-butanol (Figure II–32), the change of the total apparent rate constant k'_t versus pH was found to be similar to the change of the rate constant of certain aromatic acids. The aquatic fulvic acid behaves like an organic diacid with a reactivity toward ozone that increases when the dissociation of carboxyl groups by OH-phenolic groups increases; it must be noted that Staehelin and Hoigné (1985) found similar values ($0.02\ s^{-1}$ and $0.035\ s^{-1}$ at pH 3.5 and 4.5, respectively) for k'_t determined with 2 mg/L of commercial humic acid in the presence of *t*-butanol.

In the absence of *t*-butanol, the total apparent rate constant k'_t was always greater than in the presence of a radical scavenger, except at pH 2 (Figure II–32). This is expected with the kinetic model presented above, under conditions where the humic substance is both initiator and promotor of ozone decomposition in water.

As a result of similar experiments performed on preozonated humic acid at pH 8, Staehelin and Hoigné (1985) pointed out that the value of k'_t shows a linear decrease as

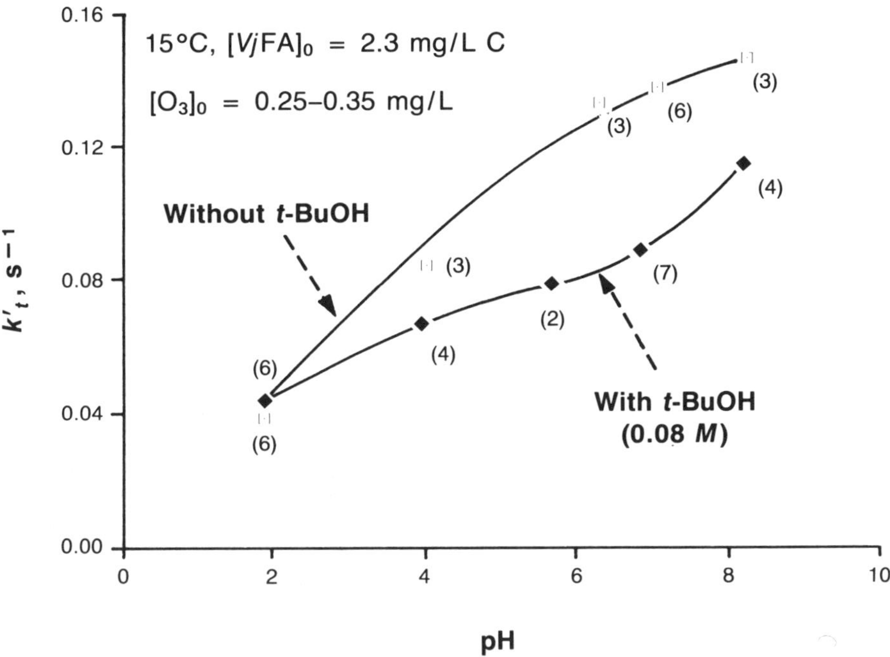

Source: Legube et al. (1989c).

Figure II–32 Effects of pH and Presence of Radical Scavenger on Total Apparent Rate Constant

the concentration of t-butanol added to the water is increased (Figure II–33). Therefore, elimination by preozonation of sites $(HS)_d$ results in a large participation of the humic substances, during a second ozonation stage, in the cycle producing the radical-type decomposition of ozone. This points to the probable presence of initiator sites $(HS)_i$ and promotor sites $(HS)_p$ in preozonated humic substances. These sites may also be by-products of the direct reaction.

When the ozonation of humic substances is performed with initial $[O_3]/[HS]$ concentration ratios greater than in the tests described in Figures II–32 and II–33, a large portion of the ozone may be consumed by radical processes. This latter quantity must be proportionately larger as the initial $[O_3]/[HS]$ ratio increases, as observed in the above-mentioned works of Reckhow (1984) and Anderson et al. (1986). In this case, reaction 6 (Figure II–31) is certainly not negligible. Consequently, the addition of a radical scavenger, such as *t*-butanol (or bicarbonate ions), enables the consumption of ozone to be significantly reduced by the limitation of reactions 3 and 6, as observed in Figure II–29.

Conclusions. The following conclusions regarding ozone consumption can be made:

- At neutral or acid pH, with or without radical scavengers, the consumption of ozone versus the initial concentration in humic substances is constant when the ratio of initial concentrations is also constant.
- At a constant initial concentration of fulvic acid, the relative ozone consumption grows very significantly when the initial ozone dosage is increased (at a neutral pH and without radical scavengers).
- The presence of bicarbonate ions (as radical scavengers) stabilizes the ozone in fulvic acid solutions at a neutral pH, causing a lower rate of consumption compared with results obtained in the absence of bicarbonates.
- The kinetic data obtained show that the reaction probably occurs in two stages—the first by direct action and the second by the radical process.

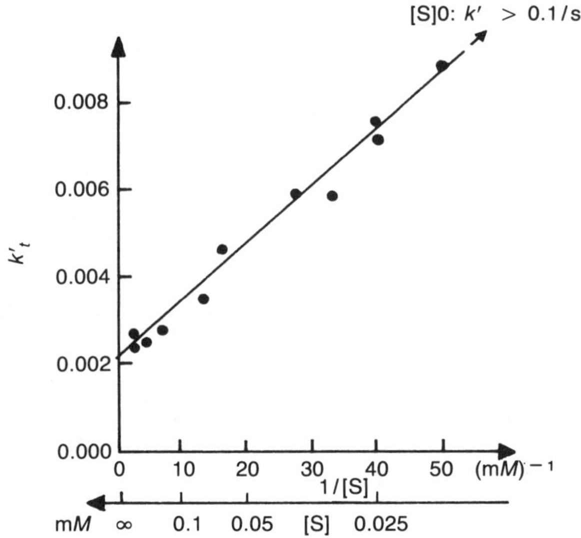

Source: Reprinted with permission from *Envir. Sci. Technol.*, 19:1208, Staehelin, J. and Hoigné, J. © 1985 American Chemical Society.

Figure II–33 Pseudo First-Order Rate Constant for Depletion of Ozone in the Presence of Preozonated Humic Acid Versus Dilution of *t*-butyl Alcohol (= 1/[S])

- Direct consumption is greater when the UV absorption of the humic substances is high.
- The radical process can be initiated by certain sites originally present in the humic substances or formed during direct reaction and by hydroxide ions. The propagation of radicals can be initiated by sites on humic substances (and by ozone itself).

Structural evolution. Based on what is known at this time, the action of ozone on aquatic humic substances in a neutral pH solution leads primarily to:

- a slight abatement of TOC;
- a strong degradation of color and of UV absorbance;
- a slight decrease in the high apparent molecular weight fractions, and a slight increase in the smaller fractions;
- a significant increase of the carboxylic functions; and
- the formation of some identifiable ozonation products (see following paragraph).

Through analysis of TOC before and after ozonation of humic substances at neutral pH, one can observe small changes as a function of ozone dose. For an ozonation dosage of 1 mg O_3/mg C, there is a fall of approximately 10 percent (Reckhow 1984; Croué 1987), which increases as the ozone dose increases (Anderson et al. 1986). From time to time, an apparent rise in TOC content is also observed (Killops 1986; Paillard et al. 1989b; Lefebure et al. 1990); the ozone may slightly increase the rate at which macromolecular organics are oxidized into carbon dioxide during analysis.

The degradation of color by ozone has been found to be almost total when the dosages applied are between 1 and 3 mg O_3/mg C (Watts 1985; Killops 1986). Ultraviolet absorption is significantly degraded by ozonation at a neutral pH (Anderson et al. 1986; Reckhow et al. 1986; Legube et al. 1989b), especially when bicarbonate ions are present in the solution, as shown in Figure II–34 and Table II–21.

Despite a slight decrease in TOC, most of the literature is in agreement that the ozonation of humic substances at a neutral pH leads to a decrease in the organic carbon contained in the high apparent molecular weights, resulting in an increase in the organic carbon of low molecular weights. Intermediate fractions develop either way, depending on ozonation conditions. This effect of ozonation has been evidenced either

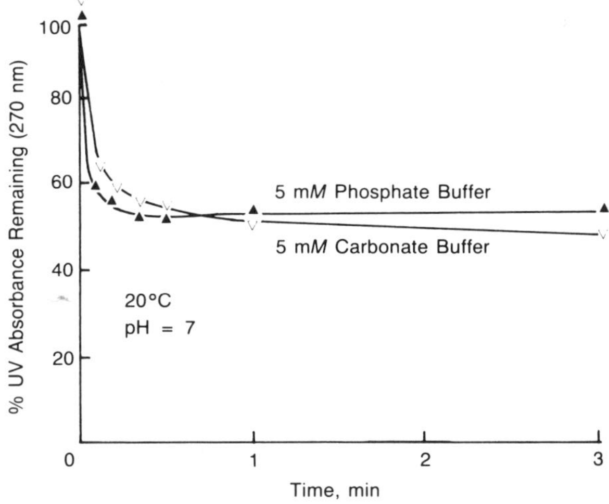

Source: Reprinted with permission from *Wtr. Res.*, 20:8:989, Reckhow, D.A. et al. © 1986 Pergamon Press PLC.

Figure II–34 Effect of Bicarbonate Ions on UV Absorbance Degradation by Ozone for an Aquatic Fulvic Acid

Table II–21 UV Absorbance (270 nm) Reduction by Ozonation of Aquatic Fulvic Acids*

	Without Carbonates		With Carbonates (4 mM)	
Source of Fulvic Acid	Ozone Dosage (mg O_3/mg C)	UV (270 nm) Reduction (%)	Ozone Dosage (mg O_3/mg C)	UV (270 nm) Reduction (%)
Pinail (forest pond)	0.36	25	0.42	32
	0.74	42	0.71	47
	0.88	52	0.91	57
Les Landes (lake)	0.78	54	0.78	61

*Fulvic acid extracted using Thurman and Malcolm method (1981). Batch reactor ozonation: FA = 5 mg/L; pH = 7.5; I = 10^{-2} (phosphate or phosphate-carbonate).

by ultrafiltration (Flogstad and Odegaard 1985; Legube et al. 1989b) or by gel filtration (Veenstra et al. 1983; Anderson et al. 1986).

With regards to the evolution of the main functional groups among humic substances, relevant literature mentions an increase in the rate of acidity and in the carboxylic functions analyzed by titrimetry (Figure II–35). It is worth mentioning that Killops (1986) has reported an increase in the aliphatic/aromatic proton ratio (by H-NMR) during the ozonation of an aquatic humic substance.

By-products of ozonation. It is known that humic substances can trap organic components of a low molecular weight, either in a free state or physically linked by low-energy bonds, within their matrices.

The identification of oxidation products in a diluted aqueous environment always requires numerous operations, including a removal step from the aqueous support (using an absorbant resin and/or an organic solvent), a concentration step (sometimes in vacuum conditions), a chemical derivation step, and, lastly, analysis by gas (or liquid) chromatography. These different stages weaken the existing bonds between the macromolecular matrix and the low molecular weight components and may cause the release of the latter. Ozone can also release or form components with a low molecular weight. It is therefore

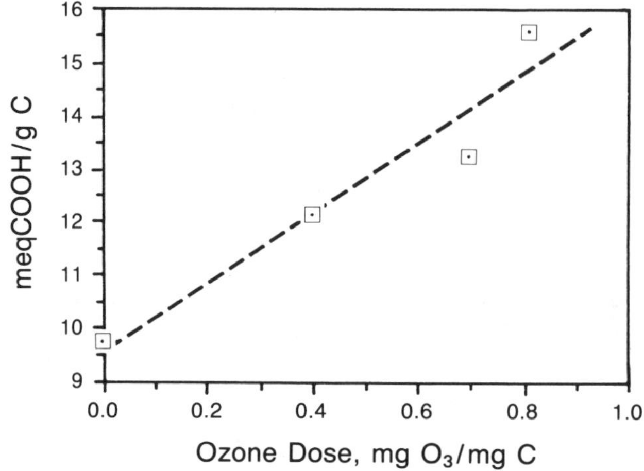

Source: Legube et al. (1989b).

Figure II–35 Increase in the Carboxylic Functions of an Aquatic Fulvic Acid by Ozonation

not surprising that the literature on ozonation by-products is rare and difficult to interpret. It is always difficult to distinguish actual ozonation by-products from those that are simply released by ozone and by the analytical treatment of the sample, as mentioned by Lawrence et al. (1980), Watts (1985), Killops (1986), and Croué (1987).

Lawrence et al. (1980), in their study of soil fulvic solution, identified many compounds by GC–MS following sequential extraction with petroleum ether, ethyl acetate, or hexane. These compounds primarily included mono- and dicarboxylic aliphatic acids, some cyclic ketonic compounds (isobenzofuranedione), and phthalates (combined with humic structures and released during the ozonation process).

Using XAD-2 adsorption and sequential diethyl ether and methanol elution followed by GC–MS and high-performance liquid chromatography (HPLC) analyses, Watts (1985) showed that nonvolatile ozonation products derived from humic acids extracted from river and reservoir waters include:

- humic acid—aromatic compounds (acids, carbonyl derivatives, and similar compounds; and
- fulvic acid—aliphatic compounds (alcohols, acids, diacids, linear alkanes, substituted aromatics, and similar compounds).

It would seem that a number of compounds already exist prior to ozonation. These compounds are not detected and become apparent only as they are released after ozonation. Despite high ozone doses (3 mg/mg HA), very few new products are actually formed.

Killops (1986) confirmed the fact that many compounds that are identified are present prior to ozonation and that 50 percent of the compounds remain unaffected by ozonation. XAD-2-isolated volatile ozonation products detected by GC–MS include:

- humic acid—aldehydes (C_9-C_{10}), vanillin, alkyltetrahydrofuran, tetrahydrofuran dicarboxylic acid, and similar compounds; and
- fulvic acid—alkanals; alkanoic acid esters; hepta-, hexa-, and monodecane; tetracosane to nonacosane; and similar compounds.

The author concludes that very few of the compounds identified are truly ozonation products and that, in fact, they are released in the oxidation process. Consequently, ozonation by-products would chiefly be carboxylic acids and aldehydes.

Table II–22 shows a few classes of components found in ozonated solutions containing humic substances and in unozonized controls. The primary components found after ozonation are saturated aliphatic acids and diacids, aliphatic aldehydes, alkanes, and

Table II–22 Classes of Compounds Identified by GC-MS Analysis in Ozonated and Unozonated Solutions of Humic Substances

Compound	References for Compounds Detected Only in Ozonated Sample*	References for Compounds Detected in Unozonated Samples*
Saturated aliphatic acids/diacids (or esters)	1, 2, 4, 5, 6, 7	1, 2, 4, 5
Unsaturated aliphatic acids (or esters)		5
Aliphatic hydroxy acids (or esters)	4, 5	
Aliphatic ceto-acids (or esters)	5, 6, 7	
Aliphatic aldehydes	4, 5, 6, 7	5
Aliphatic cetones	4	
Aliphatic alcohols	4	5
Aliphatic ethers	4	
Aromatic acids (or esters)	2, 4, 7	4, 5
Alkylbenzenes	2	4, 5, 6
Chlorobenzenes	1, 2	
Phenols	2, 3	4, 6
Phthalates	1, 2	
Naphthalenes or indenes		4, 5, 6
Linear alkanes	4, 5, 7	
Cyclic alkanes	3, 7	
Furans and derivatives	4, 6	4, 6
Nitrogen compounds		4, 5, 6

*References are as follows:
(a) 1, 2, and 3: Lawrence et al. (1980). Fulvic acid extracted from a soil; ozonation 1 mg O_3/mg C in pure water; liquid-liquid extraction by petroleum ether (1), ethyl acetate (2), or hexane (3); vacuum evaporation, diazomethane derivatization.
(b) 4 and 5: Watts (1985). Humic acid (4) and fulvic acid (5) extracted from river and reservoir waters; ozonation at 2–4 mg O_3/mg saturated aliphatic acid in pure water; extraction by XAD-2 at acid pH; trimethylsilylation.
(c) 6 and 7: Killops (1986). Humic acid (6) and fulvic acid (7) extracted from river water; ozonation at 3 mg O_3/mg SH at pH 6-7; extraction by XAD-2 at acid pH.

some aromatic acids. These classes of compounds, also identified during oxidation with $KMnO_4$ or by hydrolysis with NaOH (Liao et al. 1982), are probably not all true products of ozonation but rather compounds released by combinations of primary molecular structures and components through the oxidizing action of molecular ozone or hydroxyl radicals. Among the compounds that disappear under the effects of ozonation, particular mention must be made of aromatics, such as alkylbenzenes and naphthalenes. It is also pointed out that the compounds presented in Table II–22 account for only a small amount of the initial carbon contained in humic substances.

Aliphatic aldehydes represent an important class of by-products generated by the ozonation of organic compounds (see chapter III, sec. III.I). These compounds may be formed by the ozonation of humic substances. Table II–23 shows the concentrations in formaldehyde and acetaldehyde obtained by the ozonation of an aquatic fulvic acid (Legube et al. 1989b). Note that aliphatic acids and aldehydes represent the main ozonation by-products found in ozonated natural waters (chapter III, sec. III.I).

It would appear that the ozonation of humic substances generates less undesirable oxidation by-products than chlorination. Nevertheless, in the presence of bromide ions, ozonation leads to the formation of hypobromites (sec. II.A.3) that can, with humic matter, ultimately form organohalogen by-products. Several authors have observed the formation of bromoform through the ozonation of a humic substance in the presence of bromide ion at a neutral pH (Merlet et al. 1980; Haag and Hoigné 1983; Legube et al. 1989b). By GC–MS analysis, Croué (1987) identified different organobromide products after ozonation of a fulvic acid solution enriched with bromides. These products, primarily bromoform, bromoacetic and dibromoacetic acid, and other saturated and unsaturated brominated acids, are comparable to certain of those obtained by the chlorination of humic matter.

Table II–23 Short-Chain Aliphatic Aldehydes in an Ozonated Fulvic Acid Solution*

Ozone Dosage (mg O_3/mg C)	Formaldehyde (HCHO) (μg/mg C)	Acetaldehyde (CH_3CHO) (μg/mg C)
0	0.15	<0.15
0.3	2.3	<0.15
0.6	4.4	0.6

*Fulvic acid extracted from forest pond using Thurman and Malcolm method (1981). Ozonation in batch reactor: FA = 7 mg/L C; pH = 7.5; I = 10^{-2} (phosphates).

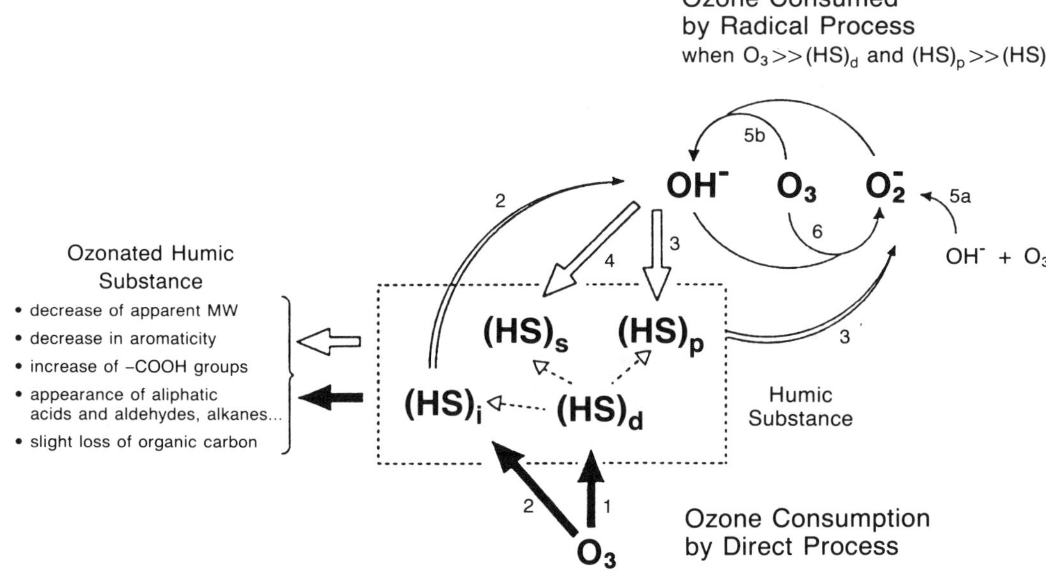

Figure II–36 Ozone Action on Humic Substances Containing Several Reactive Sites ($[HS]_d$, $[HS]_i$, $[HS]_p$, and $[HS]_s$)

Reactional pattern. Ozone can react (Figure II–36) on humic substances in an aqueous environment by the following two processes:

- direct process corresponding to the action of ozone:
  —on nucleophilic centers $(HS)_d$ (for example, aromatic nuclei, other unsaturated sites); and
  —on sites initiating the formation of radicals $(HS)_i$, originally present or formed during ozonation of nucleophilic sites (for example, organic acids, hydrogen peroxide).
  The initiation phase follows the phase in which the nucleophilic centers are ozonated.
- by a radical-type process corresponding to the action of hydroxyl radicals and depending on several factors, including:
  —the initiating capacity of the radical system—pH, $[HS]_i$;
  —the quantity of ozone applied—$[O_3]/[HS]_d$; and
  —propagator and initiator site content—$[HS]_p$, $[HS]_s$.

The propagator sites are not necessarily nucleophilic centers (OH is not a selective oxidant). They can be of various types, such as oxygenated compounds, aromatics, and ozone itself. Inhibitor sites may be included in the structure of humic substances, such as the alkyl groups, or be incorporated in the solution, as are carbonate and bicarbonate ions or *t*-butanol.

These reactions as a whole lead to a slight loss of organic carbon and a decrease in the average apparent molecular weight of the humic matter. Simultaneously, ozonation induces a distinct increase in acidity through the carboxylic functions; a formation (or the release) of trace amounts of some aliphatic acids, aldehydes, alcanes, and other products; and finally the removal of color and UV absorption. For this latter point, the addition of bicarbonate ions to humic solutions can limit ozone consumption by a radical-type process and afterwards the removal of UV absorption by the consumption of molecular ozone on nucleophilic sites.

Influence on other potable water treatment processes. One consequence of ozone reactivity with humic substances is the removal of color from the host water (see chapter III, sec. III.C). Other effects result from the above overall pattern shown in Figure II–36. For instance, the increase in the number of carboxylic functions and the decrease in the average molecular mass of the humic substances through ozonation will affect the adsorption of these compounds on solids. It is known that this development of the two parameters concerned generally leads to a decrease in their propensity to adsorb on activated carbon (De Laat 1988). Additionally, the precipitation of these humic substances by coagulation with hydrolyzable metals (Al III and Fe III) depends on the molecular mass of the humic material and may also depend on the carboxylic group content. Consequently, preozonation may have an effect not only on the removal of humic substances by coagulation (Paillard et al. 1989b) but also on the flocculation of colloids in the presence of dissolved humic matter (Sheng and Singer 1988).

For water containing high concentrations of humic substances, the removal of manganese by ozonation will be difficult, especially in the presence of weak concentrations of bicarbonate ions, which will cause a large consumption of ozone by humic substances, directly or by radical effect (Legube et al. 1988) (see chapter III, sec. III.B).

These "secondary" effects of ozonating humic substances are discussed in chapter III. Two other important effects—the degradation (and formation) of chlorine reactivity precursor sites and the increase in biodegradability—are described below.

Effect of ozone on the reactivity of humic substances with chlorine. Several studies have been conducted on the effect ozonation has on the reactivity of humic substances with chlorine, particularly the formation of chloroform. The results are often contradictory. It is difficult to compare them in view of the great differences in operating conditions, such as nature of the humic substances, ozone dosage, pH and ozonation time, chlorination rate, chlorination time and pH, and the presence or absence of bicarbonates in the solutions. However, a decrease in the chloroform formation potential generally results from ozonation of humic substance found in natural waters (see chapter III, sec. III.I).

Using an aquatic fulvic acid, Reckhow et al. (1986) showed that ozonation at a neutral pH (phosphate buffer solution) decreased the formation potential of chloroform, total organic halides (TOX), trichloroacetic acid, and dichloroacetonitrile. Nevertheless, the formation potential of dichloroacetic acid and of trichloroacetone may increase, together with the chlorine consumption potential. Through an addition of bicarbonate ion in the solution to be ozonated, the same experiments will lead systematically to a better result, whatever the parameter analyzed. Figure II–37 illustrates the case of chloroform and TOX compounds.

Comparable results have been obtained under similar conditions by Legube et al. (1989b, 1989c) on numerous fulvic acids taken from surface waters (Table II–24). Note that in the absence of bicarbonates, ozonation generally leads to an increase in the potential for chlorine demand and to an increase in the chloroform formation potential for one of the studied fulvic acids.

Since chlorine and molecular ozone are electrophilic reactives, it is not surprising that preozonation at a neutral pH can degrade the precursor sites of organohalide compounds

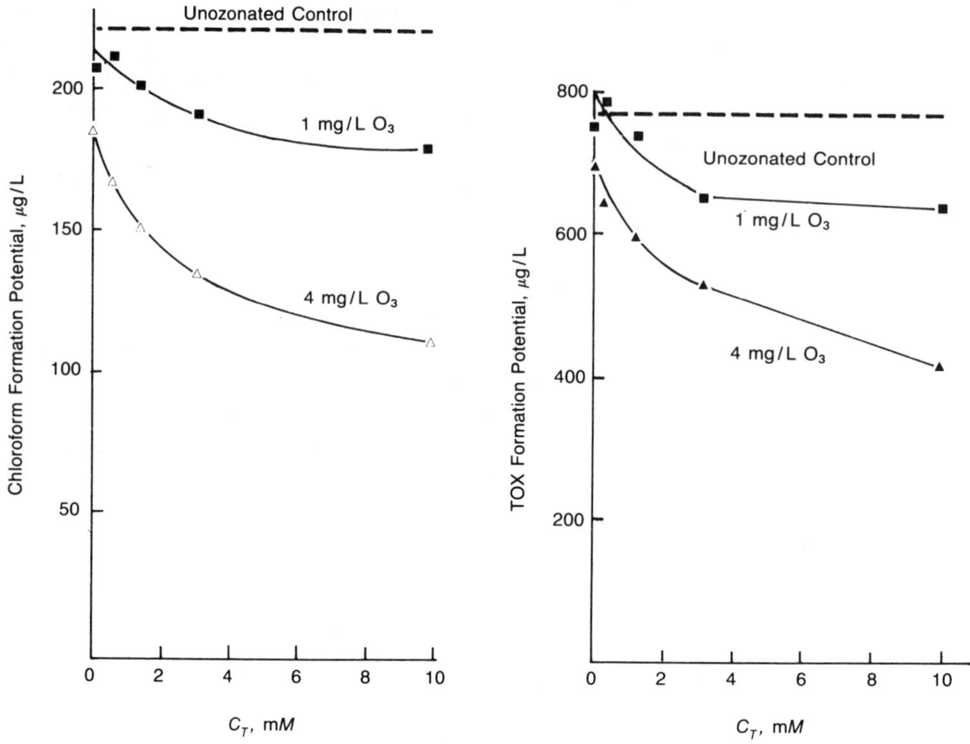

Source: Reprinted with permission from *Wtr. Res.*, 20:8:989–990, Reckhow, D.A. et al. © 1986 Pergamon Press PLC.

Figure II–37 Removal of Organohalide Precursors as a Function of Bicarbonate Concentration in an Ozonated Aquatic Fulvic Acid

on fulvic acids. This phenomenon has already been demonstrated on simpler molecules, such as resorcinol (Merlet et al. 1980). The action of molecular ozone can indeed lead to the degradation of aromatic sites (probably precursors of organochlorinated compounds), but the action of the hydroxyl radicals does not cease at that point. Hydroxyl radicals can induce, in addition to an increase in the decomposition of molecular ozone, the possible formation, by hydroxylation, of sites that are more strongly precursor sites than the initial structures. As discussed in relation to ozone consumption (Figure II–29), the presence of bicarbonate ions in the system can partly trap the radical flux. Bicarbonate ions, therefore, reduce the formation of new precursors and improve the degradation of the existing ones by increasing the action of molecular ozone in the environment, which is normally reflected in an enhanced degradation of UV absorption (Figure II–34).

As a final note, in the presence of bromides, ozonation alone is liable to form organobrominated compounds. This reaction can, however, be attenuated by the presence of bicarbonate (Croué 1987).

Effects of ozone on the biodegradability of humic substances. Various methods can be used to evaluate the biodegradability of a chemical substance under conditions that simulate the biological phenomena occurring in the natural environment. The methods usually consist of determining the growth of microorganisms in contact with the chemical substance by counting or by dosing with specific constituents (such as adenosine triphosphate [ATP], enzymes), by the amount of oxygen consumed (biochemical oxygen demand [BOD]), or by the amount of carbon dioxide formed. In addition, when evaluating simple substances, the concentration of the compound to be tested can be tracked at the same time. It is not always easy to compare results to those found in the literature because the operating conditions applied are not always the same; bacterial

Table II–24 Effect of Ozonation on the Reactivity of Fulvic Acids With Chlorine*

Reactivity Potential With Chlorine	Experimental Conditions	Aquatic Fulvic Acids: FA, Gartempe (river)	FA, Landes (lake)	FA, Pinail (forest pond)	FA, Moulin Papon (reservoir)	FA, Vienne (river)
Potential formation of $CHCl_3$ (μg/mg C)†		58	28	66	37	56
Development with ozonation at 1 mg O_3/mg C‡	P	−2%	+55%	−15%	−25%	0%
	PC	−22%	−5%	−30%	−46%	−45%
Potential formation of CCl_3COOH (μg/mg C)†			22	50	38	
Development with ozonation at 1 mg O_3/mg C‡	P		−20%	−20%	−35%	
	PC		−50%	−40%	−55%	
Potential formation of $CHCl_2COOH$ (μg/mg C)†			13	25	20	
Development with ozonation at 1 mg O_3/mg C‡	P		0%	−20%	−6%	
	PC		0%	−40%	−31%	
Potential formation of TOX (μg Cl/mg C)†		165	118	194	105	194
Development with ozonation at 1 mg O_3/mg C‡	P	−25%				−17%
	PC	−41%				−37%
Potential formation of Cl_2 (μg/mg C)†		1.45	0.90	1.7	1.00	1.65
Development with ozonation at 1 mg O_3/mg C‡	P	−1%				+2%
	PC	−19%				−22%

*General experimental conditions: FA = 5 mg/L; pH = 7.5; I = 10^{-2}; phosphates (P) or carbonates + phosphates (PC), total carbonates = 4 mM; 20°C. Ozonation conditions: batch reactor; 1 mg O_3/mg C. Chlorination conditions: Cl_2 = 10 mg/L; 72 h.
†Mean values given.
‡(+) indicates enhancement, (−) indicates reduction.

inoculum, incubation time, concentration of the chemical substance, and other conditions can vary.

In general, the results obtained prove that the ozonation of organic compounds tends to increase their biodegradability (Gilbert and Zinecker 1980; Brunet 1981) by modifying the structures generated by the oxidizing treatment. For a few compounds, the reverse effect may be observed at very low ozonation dose rates (Gilbert 1983).

Similar methods are used to measure the biodegradability of potable water undergoing treatment, determined by biodegradable organic carbon (BDOC) analysis. In this field, the effect of ozone leads to a significant increase in biodegradability and an increase in the efficiency of downstream activated carbon filters (Van der Kooij et al. 1987).

It is believed that the few studies performed on humic substances have been based on measuring development of the BOD (5 days, BOD_5) of humic substances versus the amount of ozone consumed during ozonation. Table II–25 shows the results obtained. These results indicate that humic substances are only slightly or not at all biodegradable and that ozonation induces a slight increase in their biodegradability for doses of ozone corresponding to a consumption of 1 or 2 mg O_3/mg C.

By using a different technique (measuring the amount of CO_2 formed), De Laat (1988) showed that ozonation on fulvic acid and humic acid extracted from natural water

Table II–25 Biodegradability Development (BOD_5 Measurement) of Humic Substances by Ozonation

Humic Matter	Ozone Consumed (mg O_3/mg C)	BOD_5/TOC (mg O_2/mg C)	BOD_5/COD	Reference
Humic substance from soil (by NaOH extraction)	0		0	Larose et al. (1982)*
	0.16		0.02	
	0.24		0.05	
	0.68		0.11	
	1.44		0.12	
	2.08		0.12	
Aquatic humic substance (ion exchange extraction)	0	0.15	0.07	Gilbert (1983)†
	1	0.4	0.15	
	1.4	0.5	0.25	
Commercial humic substance	0	0.1		Somiya et al. (1986)‡
	1	0.2		
	2	0.3		
	4	0.3		
	6	0.3		

*250 mg/L C; pH = 7.5–8 (ammonium phosphate).
†67 mg/L C; pH = 3.7.
‡15 mg/L C in pure water.

induced a significant increase in biodegradability. The results given in Table II–26 indicate that 40 to 50 percent of the organic carbon is biodegraded after 28 days of incubation (mixed bacteria) for an ozonation rate of 2 or 3 mg O_3/mg C. This same author has shown that preozonated fulvic acid (0.5 mg O_3/mg C) is better removed on an activated carbon biological filter than the same fulvic acid without preozonation (30 percent increase in DOC removal after treatment with 13,000 v/v of granular activated carbon [GAC]).

Conclusions. Humic substances, a typical class of natural compounds dissolved in water, react quickly with ozone when applied directly and by radical process. The rate constant of direct molecular ozone consumption depends on pH. Ozone consumption by the radical process depends on the concentration of radical traps, present or injected (for example, HCO_3^-/CO_3^{2-}), in the reactive environment, the pH, and the amount of directly consumed ozone.

The action of ozone on humic substances leads to significant changes in their structure. The changes include decrease of apparent molecular weight (MW), increase of –COOH groups, degradation of color, and degradation of UV absorption. These changes occur without, however, causing a sizable occurrence of ozone products (mainly aldehydes, aliphatic acids, and alkanes) and carbon dioxide.

The effects of ozonation on humic substances and on their behavior during other steps of treatment are numerous (see chapter III). Based on current research, the most significant effects of ozonation are a decrease in the potential reactivity with chlorine, particularly in the presence of HCO_3^-/CO_3^{2-}, and increased biodegradability.

Consequently, better knowledge of the effects of ozonation on humic substances requires further research, especially regarding the principal ways in which ozone acts on humic substances and the identification of ozonation by-products.

Reaction of ozone with other natural compounds. For all classes of natural compounds listed in Table II–19, one or several examples of the effects of ozonation are given earlier in this chapter. The ozonation of amino acids, carbohydrates, carboxylic acids, hydrocarbons, and other trace elements has already been described.

Table II–26 Effect of Ozone on Humic Substance Biodegradability by Measuring the CO_2 Formed*

Ozonation Rate	Fulvic Acid				Humic Acid	
($mg\ O_3/mg\ C$)	0	1	2	3	0	2
Incubation Time (min)	Mineralization Efficiency (%): CO_2 formed/initial DOC					
4	<1	13	–	38.5	<1	–
7	<1	15	35	46.2	<1	30–45
28	<1	20.5	44	51	<1	40–45

*Fulvic and humic acids extracted from forest pond water using Thurman and Malcolm method (1981). Ozonation in bubble column at pH = 7.7 (phosphate buffer). Biodegradability test from 20 mg/L TOC in presence of secondary effluent bacteria (urban sewage).

One natural compound, lignin, has been the subject of several studies. With sugars in general, lignin accounts for the chemical nature of lignocellular waste, hence contributing to the formation of aquatic humic substances. An in-depth study of this compound was performed by Bonnet (1988), and the principal results were published (Bonnet 1989). Commercial lignin behaves like humic substances toward ozone. The reaction leads to removing color from the solutions, a drop in UV absorbance, a decrease in DOC (15 to 20 percent for 2 mg O_3/mg C), a significant increase in the carboxylic function (six times greater, for 1 mg O_3/mg C), and the formation of many ozonation by-products, including volatile acids, aliphatic diacids, substituted aromatic acids, and furans.

II.B TOXICOLOGY

II.B.1 Principles and Limitations of Toxicological Testing

The significance of toxicological studies on drinking water or on drinking water contaminants is not well understood by the public and the water treatment industry. There are several reasons for this lack of understanding. It is related in part to a general misunderstanding about the principles and goals of toxicology. Some clarification regarding these principles and goals, from a toxicologist's point of view, may help resolve this issue.

The goal of any toxicological study on drinking water contaminants is generally to protect human health. However, research is generally conducted on either laboratory test animals (for example, rats or mice) or on cultured cells (in vitro) of either bacterial or mammalian origin. Thus, the test results are usually not directly applicable to human health risk, and their significance with regards to potential human health effects must be interpreted or interpolated by professional health scientists.

It should be noted that all toxicological tests are not "equal." For example, a positive result from a bacterial mutagenesis assay, such as the Ames test, on a water sample or concentrate from a water sample will generally not be considered as significant as finding that this sample will produce a toxicological effect in a whole animal bioassay (for example, an effect confirmed in rodents drinking the sample).

Compounding the confusion surrounding the field of toxicology is the fact that many basic toxicological procedures, such as the administration of extremely high test dosages to laboratory animals, are viewed with skepticism by many outside of the field. However, the administration of high dosages of a toxicant to laboratory test animals is necessary in order to compensate for the inability to test large numbers of animals. For example, in a typical cancer study, only 200 animals (half mice and half rats) may be tested. If the purpose of the test is to study the health effects that might result from widespread human exposure to some environmental contaminant in a population of 200 million people, then each rodent in the cancer bioassay "represents" one million persons. In a population of 200 million, a 0.01 percent incidence rate for induction of cancer by the contaminant in question would be 20,000 excess cancers. Few, if any, would consider this an acceptable risk. Considering again the animal bioassy, it might be assumed that the rodents used in the toxicological test are identical to humans in their sensitivity to the contaminant being tested. Thus, the treatment of the rodents with the contaminant at environmental dose levels would also be expected to result in a 0.01 percent incidence of cancer. However, for a population of only 200 (the number of test animals), this would amount to 0.02 cancers over the two-year life of the animals. This quantity is not amenable to either analysis or statistical validation. Thus, to compensate for the small number of test animals, the existence of a linear (or other quantifiable) relationship between the dose of the contaminant and the incidence of cancer is assumed. This much higher dose is then applied to the animals in order to induce cancer incidence rates that are in the range of the statistical methods (10 to 90 percent). It should be noted that it currently costs $300,000 to $1,000,000 (1,700,000 to 5,600,000 French francs) to conduct a complete cancer bioassay on 200 rodents. Obviously, there are real limits to the number of chemicals that can be tested in chronic animal bioassays.

The above example does not deal with other militating factors that commonly further complicate this issue. Other examples include the sensitivities of rodents and humans, which are generally not identical, the fact that rodents live for less than three years while humans commonly live to 70 or 75, and, even more significant, the relationship between dose and incidence, which actually may *not* be linear. This example serves to illustrate why there is confusion over toxicological procedures and test results.

Toxicological test levels. The remainder of this section will describe, in general terms, some of the principals of toxicological testing. In addition, it will define some of the more commonly used tests and discuss limitations in the use of toxicological test data.

It is useful to recall that toxicological tests can be ordered in hierarchical levels, with each level having a particular degree of significance to human health. A description of those levels is given in the following paragraphs.

At the highest level are epidemiological studies. In this type of study, health data on individuals exposed to a particular environmental situation are collected and compared, via statistical analysis, with an appropriate control population. These studies are time- and labor-consuming and require highly specialized personnel (epidemiologists) to design and conduct the studies as well as to interpret the results. Despite the limitations and difficulties in conducting epidemiological studies, they yield very relevant information that is usually directly applicable to human health risk. For example, Craun (1988) found that individuals who drink chlorine-disinfected water (for more than 30 years) have a slightly higher risk of developing bladder and colon cancer than comparable individuals who drink chloramine-disinfected water. This finding is of concern and requires further study. This particular epidemiological study has detected a possible effect on humans who are consuming the contaminant of interest in an actual ("real world") situation.

At a level below epidemiological studies are toxicological studies in laboratory animals. These studies are less significant than epidemiological studies, since the test animals (other mammalian species, such as rodents or more rarely dogs or primates), although often similar to humans, are not identical to humans in all aspects of their physiology or metabolism. Likewise, toxicological animal tests are generally carried out on relatively small numbers of animals (rarely more than 50 and usually much fewer) and using much higher dosages of the contaminant than would ever be encountered by humans in their environment. Therefore, the results of these studies must be interpreted (extrapolated) to account for species differences (for example, rodent versus human) as well as the dosage disparities. Imperfect as it is, this kind of animal experimentation provides the backbone of current risk assessment practice. This is true not only for assessment of environmental contaminants but also for product evaluation in the drug, food, and chemical industries.

The third tier of modern toxicological testing involves tests on cultured cells or tissues (in vitro techniques), or on isolated enzymatic or biochemical systems (also referred to as "cell-free" studies). This lowest tier, in turn, has sublevels. For example, a test on cultured rodent cells may have more direct relevance to human health than a test on bacterial cells. Likewise, a test on human cells would have more significance than one on rodent cells, and a study conducted on the specific cell or organ type (for example, hepatocytes) thought to be at risk in the population may be even more appropriate.

In vitro test data. In vitro test data must generally be interpreted by experienced experts. Even then, the following two factors limit or temper test results relative to predicting potential health effects from this data:

- While cultured cells often retain many of the biological properties of the tissue from which they were derived, they are obviously not perfect representations of that tissue.
- In vitro techniques do not allow any estimation of the whole animal pharmacokinetics (that is, the metabolism, absorption, distribution, and elimination) of the chemical being tested. For example, testing of a chemical on a colon cell culture may not register a relevant result, because the toxicity of the chemical to the whole animal may be due to the metabolism by gastrointestinal tract bacteria, which are not present in the cultures. Alternatively, a cultured lung cell may show a toxic response to a contaminant; however, that contaminant may never reach the lung in the whole animal.

Short-term in vitro (or in vivo) tests (STTs) have been widely used in the field of genetic toxicology. Accordingly, the reliability of STTs based on genetic toxicological endpoints as a tool for determining whether environmental complex mixtures are either hazardous or safe, with respect to their potential to cause human health effects (namely cancer or heritable mutation), has been an area of considerable scientific (and political) debate. Misguided arguments on both sides of this issue are not uncommon. For example, the belief that all positive STT results obtained from analysis of environmental mixtures constitute an imminent human hazard is as irrational as the often expressed view that STTs have no significance to human health. The fact is that the significance of a positive result in an STT must be interpreted by knowledgeable scientists on a case-by-case basis. Thus, STTs can play an important role in assessing possible human health effects. While a complete discussion of this issue is beyond the scope of this book, a few general points, discussed in the following paragraphs, should be noted.

The use of STTs is predicated on the logical argument that at the molecular level analogous biochemistry is occurring. For example, in genetic toxicology it is assumed that DNA is the same in all organisms and the mechanisms for DNA damage and repair are at least similar (McCann et al. 1988). Thus, since many (but not all) carcinogens are mutagens and cancer can, on occasion, result from mutation of somatic cells, it is reasonable to expect that a correlation between carcinogenicity and mutagenicity exists. Further, it is reasonable to expect that STTs might be a useful tool for determining which chemicals or mixtures should undergo long-term cancer bioassays.

By the same token, there are several theoretical reasons why STTs may *not* always accurately predict the potential carcinogenicity of an environmental sample. It is clear that chemicals can exhibit genotoxicity by at least several distinct processes, ranging from mutations (induced changes in the DNA sequence) to clastogenic events (chromosomal breaks and aberrations). Since most chemicals are prone to induce predominantly one type of damage, and most STTs (with the possible exception of the mouse lymphoma assay) detect predominantly only one form of damage, the failure to detect a positive response in an STT may be due to the failure of the assay to detect the type of damage being induced by the particular chemical or mixture of environmental agents.

An STT may also fail to detect the carcinogenic potential in an environmental sample for other mechanistic reasons. For example, some carcinogens are not genotoxic per se but rather appear to be capable of causing genotoxic events indirectly via the induction of specific cellular or molecular processes. For example, peroxisome proliferators appear to be a special class of hepatocarcinogens that may induce genotoxicity in rodent liver by virtue of their propensity to induce high levels of active oxygen species (for example, hydroxyl radical), which in turn produce DNA damage (Rao and Reddy 1987). Most peroxisome proliferators would not, for example, produce mutations in *Salmonella* and thus would not be detected in the Ames assay.

Similarly, tumor promotors represent another class of carcinogens that are generally not detected by STTs based on genotoxic endpoints (for example, DNA strand breakage). Tumor promotors increase the tumor incidence (thus, are carcinogens) by virtue of their ability to confer a growth advantage upon the tumor or upon the preneoplastic lesion that precedes the tumor. It has become obvious that tumor promotors are important factors in environmental carcinogenesis. Tumor promotors, like peroxisome proliferators, would not be identified as potential carcinogens in bioassays that use mutations or DNA damage as the measured endpoint.

There are also practical limitations to the ability of STTs (for genotoxicity) to detect carcinogens. For example, cancer is a "whole animal" disease, and the induction of cancer by chemicals is often a unique event, occurring only in a specific organ of a specific species. The reasons for this are complex and due to a combination of pharmacokinetic and pharmacodynamic factors, such as absorption, metabolism (activation versus detoxification), and cellular repair and defense mechanisms (for example, DNA repair), as well as the susceptibility of particular organs to tumor promotion. Clearly, no one STT is likely to mimic all of these parameters simultaneously. Thus, it does not seem probable that STTs will ever replace two-year whole animal cancer bioassays. Likewise, it does not seem highly feasible that risk assessments will be made solely on the basis of these STTs. Nevertheless, genotoxicity-based STTs still may provide useful information on the possible existence of mutagens and other genotoxic compounds in environmental samples. Therefore, these tests can be useful for comparing the relative merits of various control or treatment technologies, such as drinking water disinfection.

In summary, it should be apparent that considerable caution must be exercised when interpreting toxicological results obtained from any water sample. Accordingly, these data should be interpreted by health scientists using the best methods currently available.

II.B.2 Short-Term Mutagenicity and Genotoxicity Tests

Short-term tests for mutagens and carcinogens have evolved substantially since they were first introduced. This evolution is due to the fact that mutagens and many carcinogens share a common cellular target—DNA. For the *Salmonella* assay, this relationship was originally validated at 90 percent by McCann et al. (1975) and has been recently reevaluated by Malaveille and Bartsch (1989). Mutagenicity and genotoxicity tests may be used in vitro for preliminary screening expected for epigenotoxic agents. When results are positive, the implications for human genetic or carcinogenic hazard must be confirmed by another test, such as in vitro cell transformation studies (for example, primary hamster embryo cells), or by in vivo animal experimentation whenever possible. From among the 100 or so tests available, only 6 or 7 are routinely and widely practiced in many laboratories and are generally considered to be validated (Marzin 1984). Several of these are described in detail below.

Mutagenesis assays.

Ames test (Salmonella back-mutation assay). The Ames test is based on the use of mutants of *Salmonella typhimurium* that are deficient in the histidine operon. A reverse mutation, resulting from the effect of some mutagenic agent, enables the bacteria to retrieve their histidine synthesis capacity. The five *Salmonella* strains generally used provide a means for detecting different mutation mechanisms. Strains TA1535 and TA100 are sensitive to base-pair substitutions within DNA, whereas TA1537, TA1538, and TA98 detect frameshift mutations due to a shift at the DNA base code reading frame level (Ames et al. 1985). Some new strains that are even more sensitive (TA97 and TA102) have recently been proposed.

Research dealing with procarcinogens, such as polynuclear hydrocarbons, that are metabolized by animals and do not exert any direct mutagenic effect on cell culture systems that lack metabolic capacity requires the use of a metabolic activation system, namely an S_9 mix (rat liver microsomes + nicotinamide adenine diphosphonucleotide [NADPH] + glucose 6-phosphate). This mix can mimic hepatic metabolism and thereby detect the potential mutagenicity of procarcinogens.

The maximum water volume that can be incorporated into the Ames test medium is 2 mL. This limits the concentration of tested organic matter to approximately 10 μg/L from typical water containing 0–5 mg/L TOC. Thus, these small quantities of organic matter would have to be very potent mutagens in order to be detected by the Ames test.

In order to avoid the concentration step, which is often needed in natural water mutagenicity studies and which may induce losses or extract contamination, a modified fluctuation assay has been proposed (Green et al. 1976). This assay uses strains TA98 and TA100, which are characterized by high spontaneous mutation rates (Tye and Waite 1981). The fluctuation assay may, in some cases, be hundredfold more sensitive than the conventional Ames test, and it has been applied by Harrington et al. (1983) and Fielding and Horth (1986) to unconcentrated drinking water analysis.

Test on yeast. *Sacharomyces cerevisiae* is a diploid strain of yeast, which is based on a mutation at the adenine-synthesis enzyme locus. As such, it allows for the detection of mutagens. This is done by differentiating between red mutants and nonpigmented white colonies capable of synthesizing their own adenine. A technique that provides for the detection of abnormal mitotic recombinations or for the loss of a homologous chromosome (aneuploidy) has been developed in parallel with the Ames test (Cotruvo et al. 1977) and applied to show ozone mutagenicity (Dubeau and Chung 1982).

Point mutation tests in cultured mammalian cells. In order to reveal direct mutations affecting initially normal cells, a test is carried out to detect a mutation at some enzyme locus that might be reflected by suppression of enzyme synthesis and or activity, as exemplified by the mutation occurring at the hypoxanthine–guanine phosphoribosyltransferase (HGPRT) locus. This test has also been adapted to Chinese hamster V79 cells to detect tumor promotors in the presence of tetradecanoyl phorbol acetate (TPA) (Pottenger 1983; Hartemann et al. 1987; Bourbigot et al. 1986).

Chromosomal aberrations.

Metaphase chromosome analysis. Analysis of chromosome structure and number of chromosomes can be carried out after cell division has been blocked at the metaphase stage of mitosis by addition of spindel poisons, such as colcemide or colchicine. This test consists of a cytogenetic analysis of the fixed and stained chromosomes by evaluation of the numbers of breaks, translocations, and anomalies. Highly trained personnel must conduct the test and, for better results, the test should be performed directly in vivo, which will minimize the chance for false positives (Thompson 1986).

Micronucleus test. The micronucleus test is often preferred because it can be more easily performed than the chromosome analysis. The tissue used in the test is bone marrow or peripherical lymphocytes of exposed animals (rats, mice). The test involves the detection of persisting micronuclei that result from chromosomal breaks and non-disjunctive events that occur within the polychromatic erythrocytes following exposure to clastogenic agents. These micronuclei are usually identifiable after mitosis and are relatively easy to count. The micronucleus test has been modified for application to aquatic species. Newts were exposed for 12 days to the tested water prior to carrying out the micronucleus test on blood smears (Siboulet 1984; Grinfeld et al. 1986; Jaylet et al. 1986, 1987; Gauthier et al. 1989a, 1989b; AFNOR 1987).

A preconcentration of the micropollutants is not required, because the larvae are bred and grown in the tested water, allowing the micropollutants to be metabolized by the animals. After a 10-day treatment, the micronucleated red blood cells are evaluated in blood smears. Results are expressed as a comparison between the number of micronucleated cells obtained for a group of treated larvae and the number of micronuclei-positive tests obtained for a control group bred in "high-grade reconstituted water." As an example, benzo(a)pyrene elicits a positive response (0.025 ppm within 8 days).

DNA damage and repair assays.

Chromotest. The SOS Chromotest uses the PQ 37 *E. coli* strain containing a gene coding for β-galactosidase that has been incorporated into one of the SOS genes (Sfia). (The SOS terminology refers to a so-called "last chance" or emergency DNA repair system that appears operational in numerous bacterial species.) The damage produced by DNA adducts (or reaction between DNA and the toxicants) elicits the SOS response with resulting induction of both Sfia and β-galactosidase synthesis. The activity of β-galactosidase is measured in relation to that of noninducible enzyme alkaline phosphatase (Quillardet et al. 1985). This test has been validated and has been used to assess the genotoxicity of chlorinated water (Urien et al. 1986) and ozonated water (Bourbigot et al. 1986).

Sister chromatid exchange. Sister chromatid exchanges (SCEs) occur during DNA synthesis and are related to the level of repair of DNA recombination (a form of DNA repair) phenomena. Sister chromatid exchanges are tested in continuous rodent cell lines or on human lymphocytes treated with the tested chemical and cultivated in the presence of bromodeoxyuridine for two DNA replication cycles. Appropriate treatment allows for differentiation of the two chromatids from the chromosome and visualization of the exchange phenomena. This test may be performed in vivo and has been applied to the study of THMs (Morimoto and Koizumi 1983) and haloacetonitriles.

Unscheduled or unprogrammed DNA synthesis. Unprogrammed DNA synthesis (UDS) generally uses primary rat hepatocyte cultures grown in the presence of tritiated thymidine. DNA incorporation, as evidenced by autoradiography, indicates the level of DNA repair that results from the chemically induced DNA damage (Waters 1984).

II.B.3 Short-Term Bioassay for Carcinogenic/Mutagenic By-Products

Ozone. Ozone has been found to produce chromosome and chromatid aberrations in the lung cells of rodents following inhalation exposure (Zelac et al. 1971; Tice et al. 1978; Gooch et al. 1976). Using cultured Chinese hamster V79 cells, Rasmussen (1986) demonstrated that 1–10 ppm ozone in the gas phase inhibited DNA replication in a manner identical to that exhibited by chemical mutagens (for example, ethylmethane sulfonate) and radiation.

Specific organic chemicals. Several studies have documented the dichotomous ability of aqueous ozone to either destroy or, in contrast, activate (to a mutagenic state) specific classes of organic chemicals. Burleson et al. (1979) reported that ozonation of polyaromatic amines, such as acriflavine, proflavine, and β-napthylamine, and polycyclic aromatic hydrocarbons (PAH), such as benzo(a)pyrene, 7,12-dimethylbenz(a)anthracene, and 3-methylcholanthrene, in water (5 mg/L ozone, pH 6.6, 3–15 min) eliminated or reduced the mutagenic activity of these compounds in several strains of

Salmonella, including TA98 and TA100. Comparable treatment with molecular oxygen had no effect on the mutagenic potential of these chemicals. These same investigators demonstrated that ozone treatment (relative to an equimolar concentration of molecular oxygen) destroyed or reduced the carcinogenicity of PAH to the skin of Lobund CFW mice. The authors considered these chemicals to be either known or potential water contaminants or to be compounds that were structurally related to known water contaminants. This point is probably somewhat specious. Nevertheless, these experiments point out that ozonation may destroy the genotoxic and oncogenic potential of some polycyclic aromatic compounds in aqueous media. In addition, these workers reported that ozone destroys the activity of several other structurally diverse carcinogens (aflatoxin B_1, bis[2-chloroethyl]amine) and some pesticides (for example, captan and Dexon).

In subsequent studies, Burleson and Chambers (1982) reported on the effect of ozonation on 36 additional carcinogens and mutagens. In this study, it was demonstrated that ozonation reduced or eliminated the mutagenic activity of polyaromatic amines (2-aminoanthracene, 2,4-diaminoanisole, and similar compounds) but did not reduce the activity of some simple alkylators (for example, N-nitrosomethylurea, methylmethane- and propane-sulfonate, and β-propiolactone) or of some nitroaromatics, such as nitrofurantoin and 4-nitroquinoline-1-oxide.

The reaction of ozone with hydrazines in aqueous media was found to be complex. The colon carcinogen 1,2-dimethylhydrazine (DMH) was converted by ozone in a pH-dependent reaction from an indirect-acting to a direct-acting mutagen (Chambers and Burleson 1982; Sierka and Cowen 1978). The ozonation of DMH in acidic aqueous media (pH 2.5 to 7) results in the production of an intermediate that reverts *Salmonella* strain TA100 in the absence of a mammalian liver homogenate. In contrast, under alkaline conditions (pH 8.5 to 11.5), ozone does not activate the DMH.

Sayato (1987) and Sayato et al. (1989) demonstrated that the reaction of naphthoresorcinol with ozone in aqueous solution at pH 7.4 produced a series of aldehydes, aldehydic acids, ketoacids, and dicarboxylic acids. These ozone–naphthoresorcinol byproducts were shown to possess direct-acting mutagenic activity in several strains of *Salmonella*, including TA98, TA100, TA102, and TA104. The addition of mammalian microsomal enzymes decreased the activity in all strains, except TA102. Two mutagenic by-products, glyoxal and glyoxylic acid, were identified and appeared to be responsible for a portion of the ozone-induced mutagenicity.

In another study, Cotruvo and co-workers (1977) concluded that the oxidation of 28 selected organic chemicals did not result in the production of mutagens to either *Salmonella* or *Sacharomyces*.

Aqueous humic substances. Many studies pertaining to ozonation of humic materials have been conducted under a variety of operational conditions. Additionally, many of these studies were carried out on extremely diversified samples originating from various sources (for example, aquatic versus terrestrial humic acid), which led to some attempts to develop some synthetic models that could be used to simulate the reactions of ozone with naturally derived humic substances.

Cotruvo et al. (1977), operating under more drastic ozonation conditions than those encountered in a treatment line, have assessed the mutagenicity of soil-extracted humic acids following a 20-min ozonation treatment using both the Ames test and the test on yeast (*Sacharomyces*). According to these authors, there was no evidence of any mutagenicity.

Kowbel et al. (1982) have studied ozone reactions with fulvic acids extracted from two different sources, soil and water, by varying the fulvic acid/ozone mass ratio. The mutagenicity of the extracts was determined on five *Salmonella* strains, both with and without metabolic activation. All of the ozonated samples proved negative, except for the products from the most strongly ozonated sample, which after metabolic activation had a slightly decreased mutagenic potency.

In light of these results and the importance of the ozone concentrations used, it is not possible to reach a systematic conclusion concerning the toxicity hazards generated by ozonation of humic materials, nor is it possible to make a comparison between the two types of fulvic acids used, since the ozonation conditions that were applied varied from one sample to the other.

On the other hand, Kowbel et al. (1984) have shown that chlorination of a soil-extracted fulvic acid solution induces a mutagenicity toward the *Salmonella* TA100 strain (fluctuation assay). This activity shows a dependence on the applied ozone dosages and on the pH (it was higher in an alkaline environment); furthermore, this activity was significant when compared to that measured for the nonchlorinated samples. In a subsequent study (Kowbel et al. 1986), these authors demonstrated that residues of aquatic fulvic acid that have been ozonated, chlorinated, or ozonated and chlorinated had activity in TA100 using the fluctuation assay. All samples that were first ozonated and then chlorinated proved to be mutagenic except when ozonation was carried out at pH 8. In addition, increased ozonation rates did not affect mutagenicity, except at higher pH where the mutagenicity was decreased. It appears that the superoxide radical formed at pH 8 contributes to organic matter destruction by oxidizing high molecular weight fulvic acids into smaller molecules susceptible to volatilization or into products that are devoid of any toxic effect.

Urien (1986) studied the cytotoxicity and mutagenicity of ozonation on a unique model, which was based on a polymerized gentisic acid. The assay was run in parallel with that of a natural fulvic acid extracted from lake water. Cytotoxicity was measured on cultured human (Hela) cells, in terms of the inhibition of the RNA synthesis (Fauris et al. 1985). The results of these experiments showed the following trends:

- No cytotoxicity was elicited by the untreated gentisic acid.
- In the ozonation of the gentisic acid polymer, increased cytotoxicity occurred for the first 15 min of ozonation, and subsequently declined over the following 60 min.
- The behavior of natural fulvic acid was similar to the gentisic acid polymer.

There was no evidence of any genotoxic activity detectable in the SOS Chromotest following the ozonation of gentisic acid, the gentisic acid polymer, or natural fulvic acid. However, positive results were recorded with chlorination products obtained both from gentisic acid polymer and the fulvic acids (Pommery et al. 1989). Ozonation decreased the capacity for both synthetic and natural fulvic acids to inhibit alkaline phosphatase. The ozonation products that were identified by HPLC showed no cytotoxic activity. The depolymerization of the fulvic material may be evoked, with a resulting modification of the radical-type interactions and remodeling, and the progressive loss of the free radicals detected by electronic paramagnetic resonance (EPR).

Sayato et al. (1987) evaluated the mutagenic potential of ozonation products derived from naphthoresorcinol toward several strains of *Salmonella*, including strains TA97, TA98, TA100, TA102, and TA104. The evaluations were conducted both with and without metabolic activation owing to the structural analogy of this diphenol material with the gentisic acid material used in the elaboration of the synthetic humic model described above.

These naphthoresorcinol-derived ozonation products are mutagenic towards *Salmonella* strains TA97, TA98, TA100, and TA104. Moreover, after metabolic activation, the mutagenic effect on strains TA98 and TA100 is suppressed, but it is unchanged in strain TA102.

The HPLC analysis of ozonation products demonstrates the formation of glyoxal following derivatization with 2,4-dinitrophenylhydrazine, as well as occurrence of phthalic, muconic, maleic, mesoxalic, glyoxalic, and oxalic acids as *p*-bromophenacyl derivatives. A study on mutagenicity of the identified products showed that glyoxal and glyoxlic acid have a direct mutagenic effect, the former on *Salmonella* strains TA100,

TA102, and TA104, the latter on strains TA97, TA100, and TA104. Glyoxylic acid is mutagenic on strain TA102 in the presence of the liver homogenate. This is indicative of the fact that these two chemical reaction products are contributors to the mutagenicity of ozonated naphthoresorcinol.

Backlund et al. (1985) submitted water containing high concentrations of humic acid (TOC = 21 mg/L) prior to and following flocculation (TOC = 6.4 mg/L) to ozonolysis using dosages of 10 and 2.9 mg/L, respectively. They showed that such processing did not generate any mutagenic activity toward *Salmonella* strains TA100, TA97, and TA98, while chlorination was highly mutagenic to strain TA100.

Later, Backlund et al. (1988) compared the effect of different disinfectants on water with a high humic matter content with regard to the production of active oxygen species, $CHCl_3$, and chlorohydroxyfuranones. The researchers confirmed that the highest mutagenic activity and highest chlorohydroxyfuranone concentration were found in chlorinated water. Preozonation followed by chlorination induced a slight decrease in both mutagenicity and in chlorohydroxyfuranone levels.

Fielding and Horth (1987) evaluated the formation and elimination of mutagenic chemicals during drinking water treatment processes. They demonstrated that the mutagenic activity (XAD-2 extract, pH = 2) was primarily in *Salmonella* strains TA98 and TA100, both with or without metabolic activation as assessed in the fluctuation assay. The most potent mutagenic agent active on strain TA100 at 1 ng/L concentration was 3-chloro-4-dichloromethyl-5-hydroxy- 2(5H)furanone (MX), which is considered responsible for 22 to 46 percent of the activity of these drinking water extracts at pH 2.

II.B.4 Subchronic Bioassays on Nongenotoxic Ozone Disinfection By-Products

Ozone. Ozone has long been recognized as a hazardous air pollutant, and its toxicological properties toward mammalian species following inhalation exposure have been extensively studied. The acute toxicological effects of ozone are due almost entirely to its extreme reactivity and to its redox potential of + 2.07 V. Thus, ozone is capable of oxidizing numerous biochemical components, including unsaturated fatty acids, amino acids, pyrimidine nucleotides, flavins, and sulfhydryl-containing proteins (Carmichael et al. 1982). The most significant reaction of ozone in vivo is the production of highly reactive, short-lived free radicals, which can react with unsaturated bonds in membrane lipids and with sulfhydryl groups located at critical functional sites in essential membrane-bound enzymes (for example, glutathione reductase).

Airborne ozone. As would be expected, the primary target of airborne ozone is the lung. The initial, acute toxicological effect of ozone inhalation is pulmonary edema accompanied by capillary hemorrhage and inflammation of the respiratory tract. With prolonged exposure, ozone has been shown to cross the alveoli, with toxicological consequences to blood cells and serum proteins. Erythrocyte sphering accompanied by increased membrane fragility (Jaffee 1967) and increases in membrane acetylcholinesterase activity have been observed (Buckley et al. 1975).

Ozone has also been shown to cause adverse effects on the eye and nervous system (Carmichael et al. 1982) and to induce fetal resorptions in rats at doses higher than 1.26 ppm (Kavlock et al. 1980). A brief summary of the toxicological effects of ozone following inhalation exposure has appeared (Carmichael et al. 1982).

Aqueous ozone. In contrast to airborne ozone, the toxicological effects in mammalian systems induced by exposure to aqueous solutions of ozone and its associated reaction by-products have not been extensively studied. However, the toxicities of ozone and/or ozone reaction by-products in aqueous media have been well studied in several fish species.

Paller and Heidinger (1979) observed that the LC_{50} (concentration that is lethal to half the test animals) for a 24-h exposure of bluegills (*Lepomis macrochrius*) to ozone was 0.06 mg/L. Treatment of this same species with 0.01 mg/L ozone over a four-week exposure resulted in a 60 percent mortality. Analogous results were obtained by Roselund (1975), who found that rainbow trout (*Salmo gairdneri*) were killed after several hours exposure to 0.01–0.06 mg/L ozone. Later, Wedemeyer et al. (1979) determined the 96-h LC_{50} for rainbow trout to be 0.0093 mg/L. This study, which included a histopathological examination of the dead fish, concluded that death was due to massive destruction of the gill lamellar epithelium and the accompanying hydromineral imbalance.

The rainbow trout appears to be particularly sensitive to ozone toxicity, evidenced by the need to lower the ozone concentration to 0.002 mg/L in order to avoid significant pathology upon long-term (more that three months) exposure. Paller and Heidinger (1980) subsequently demonstrated that bluegills surviving the initial acute exposure to ozone often died two to seven days later and that the delayed death was due to the destruction and loss of the epithelium covering the gill filaments. This loss in the lamellar structure was accompanied by a rapid drop in serum osmolality from a normal value of 280 mOsm to 225 mOsm, at which point mortality occurred. The authors point out that destruction of the gill tissue also renders the animals highly susceptible to microbial infections.

The white perch (*Morone americana*) appears to be more resistant to ozone, with the 24-h LC_{50} being 0.38 mg/L (Richardson et al. 1983). In this study, however, the authors recognized the fact that some of the toxicity associated with aqueous exposure to ozone may be due to the simultaneous exposure to the ozone reaction by-products. Thus, these authors report their treatment concentrations in terms of "ozone-produced oxidant." The ozone-induced toxicity in the white perch was also mediated via the destruction of the gill epithelia.

Several authors have considered the question of the relative toxicity of ozone and chlorine toward fish species. While a detailed discussion of this subject is beyond the scope of this chapter, several generalizations should be noted.

First, it is often difficult to separate the toxicity associated with the oxidant from that of the accompanying oxidant reaction by-products. Consequently, some of the differences in the reported toxicity of ozone may be related to the variance in the levels of ozone reaction by-products formed.

Second, the actual LC_{50}s (measured in milligrams per liter) for ozone and chlorine in some fish species appear to be similar, but in other species ozone is more toxic. For example, Hall et al. (1981a) found that the 96-h LC_{50} for striped bass (*Morone saxatilis*) larvae was 0.06 mg/L total residual chlorine, while the 96-h LC_{50} for these juvenile forms was 0.08 mg/L total ozone-produced oxidants. In a later study, this same group of researchers concluded that the toxic potential of ozone and chlorine toward yellow perch is quantitatively similar, and that the mechanism of action (the destruction of the gill epithelium) is similar for both oxidants (Hall et al. 1981b). In contrast to these observations, Paller and Heidinger (1979) conclude that ozone is toxic to bluegills (0.06 mg/L) at significantly lower concentrations than is chlorine (1 mg/L).

Third, it has been suggested that due to their chemical nature, oxidants such as chlorine and chloramine might be more easily absorbed across the gill membrane than ozone. Therefore, these oxidants exert systemic toxicological effects (for example, methemoglobinemia induced by chloramine), whereas the effects of aqueous ozone might be confined to the body surface (for example, gill epithelia) (Paller and Heidinger 1980).

In conclusion, studies of fish species are of limited value for assessing potential human health effects, since they appear to be associated more with the toxicological effects of ozone in water and less with the toxic effects of ozone reaction by-products. Nevertheless, it would be logical to assume that fish species could be of substantial value in assessing the toxicological potential of disinfectant (including ozone) by-products since these species can be induced to live in the treated water. Thus, it might be anticipated that fish and other aquatic species could be very useful for comparative studies of the disinfectant reaction by-products of ozone and other disinfectants. For example, the Japanese Medaka (*Oryzias latipes*), a small aquarium fish, has proven very sensitive to hepatocarcinogenesis by nitrosamines (Nakazawa et al. 1985) and other carcinogens (Hatanaka et al. 1982). Presently, it appears that chronic bioassays in aquarium fish species may have great use in assessing the potency of waterborne carcinogens.

Ozone by-products. In contrast to fish species, the toxicity of aqueous ozone by-products in mammalian species is largely unstudied. Recently, Daniel et al. (1989) conducted a 90-day study on the toxicity of aqueous humic acids disinfected with ozone and with ozone followed by chlorine in order to evaluate, in a relative sense, the toxic effects of the various disinfection by-products. This study is considered to be an evaluation of the toxicological potential of ozone reaction by-products and not ozone per se. However, since it is most likely that it would be these disinfection reaction by-products that reach the consumer, these studies may have relevance to the questions concerning the safety of ozone-disinfected drinking water. Since the study of Daniel et al. (1989) is currently the only complete subchronic study assessment of the toxicity of ozone reaction by-products in mammalian species, the study will be described in some detail.

A commercial sample of humic acid was used to prepare solutions of 0.25 and 1.0 g/L. A total of 120 Sprague-Dawley rats, randomly assigned to six dose groups, received as drinking water either untreated humic acid solution (1.0 g/L TOC), ozonated solution (0.25 or 1.0 g/L TOC), ozonated/chlorinated solution (0.25 or 1.0 g/L TOC), or distilled water. The humic acid solutions were ozonated with a carbon:ozone ratio of 1:1 (w/w; pH 7 to 8) and chlorinated at a carbon:chlorine ratio of 1:0.7 (w/w). The humic acid solutions receiving both disinfectants (ozone and chlorine) were ozonated before chlorination.

A complete clinical and histopathological examination of the animals, during and after the exposure period, was performed. Several minor, though statistically significant, hematological and biochemical changes were observed. In female rats, these changes included increased segmented neutrophils in the ozonated/chlorinated humic acids (1 g/L) group, increased serum calcium levels in both the ozonated humic acids (1 g/L) and ozonated/chlorinated humic acids (1 g/L) groups, and increased serum phosphate levels in the ozonated/chlorinated humic acids (1 g/L) group. Male rats had decreased serum glucose levels in almost all humic acid groups and slightly decreased hemoglobin in those receiving untreated humic acids (1 g/L) and ozonated humics (1 g/L) compared to controls. The authors concluded that these were minor changes and, though statistically significant, they exhibited no pattern to indicate a relationship to treatment and may not have biological significance. In particular these effects seemed associated more with the humic acid per se than with the disinfectant types.

Organ weight analysis revealed a significant variation in liver weights in males; however, the magnitude differences were not statistically significant between any single disinfectant(s) treatment group and the control values. A significant difference was found in the relative liver weight (organ-to-body weight ratios) in the male rats treated with the ozonated/chlorinated humics (1 g/L) compared to controls, but the same parameter in the male animals of analogous ozonated-only group were not significantly altered. The authors concluded that the latter finding may warrant further consideration, but that on the whole there was no convincing evidence for any effect related solely to ozonated humic acid treatment. Some thyroid lesions (colloid depletion and follicular cysts) were observed in seven males of the 1-g/L humic acids group but not those treated with 1 g/L ozonated humic acids. Two males in the 1-g/L ozonated/chlorinated humic acids group had thyroid lesions.

In summary, minor toxicological effects were observed. The authors concluded that

1.0 g/L TOC in humic acid either with ozonation or with ozonation/chlorination might be considered the no observable adverse effect level (NOAEL) for a 90-day exposure in rats. A NOAEL such as this must be interpreted cautiously, because the study did not include a higher dose. If, indeed, the various minor findings from the study can be discounted as unrelated to exposure, the actual NOAEL could be higher.

However, a more conservative view of this study could be taken. For example, further research at higher levels of humics might reveal one or more of the present findings at 1 g/L to be a dose-related effect. These possibilities are important when comparison is made to the 0.5-g/L NOAEL previously established for chlorinated humic acids, administered to rats via drinking water in a study similar to this one. This NOAEL was demarcated by effects observed at the 1.0-g/L level (Condie et al. 1985).

Summary. The chemistry of ozone in aqueous solutions and the health effects are complex. It is clear that ozone reacts with products in the water supply (for example, humic acids) to form numerous disinfection by-products. However, the general pattern that emerges from most studies is that the reaction by-products of ozonation appear to be less toxic than those produced by chlorination (for example, chlorohydroxyfuranones, THMs). Many reactions with ozone are dose- and pH-dependent, and this explains the differences between results obtained under various operational conditions.

Two mutagenic by-products, glyoxal acid and glyoxylic acid, were identified after ozonation of naphthoresorcinol, which has some structural analogy with the humic model. On the other hand, it is shown that several carcinogens and pesticides can be destroyed by ozone. Ozonation of polyaromatic amines and polycyclic aromatic hydrocarbons eliminated or reduced the mutagenic activity of these compounds.

II.C THEORETICAL ASPECTS OF OZONE ANALYSIS—INTRODUCTION

From the standpoint of analytical chemistry methodology in water treatment (*Standard Methods* 1985, 1989; Gordon et al. 1987a, 1987b), when ozone is used as a chemical oxidant, measurement of a specific concentration of dissolved ozone rarely is used to monitor the process, except when the materials react rapidly with ozone. Satisfaction of the immediate ozone demand of a particular water is assumed to occur when a measurable residual ozone concentration is obtained.

In slower oxidations, other analytical parameters must be used that depend on the specific purpose of ozonation. When removing manganese by ozone oxidation, for example, the pale purple color of permanganate in an aqueous phase measurement can be taken as the ozone control point. Also, ozone dose can be controlled by monitoring the purple color. However, this approach does not take into account variations in parameters such as transfer coefficients. For color removal, the elimination of color can be the endpoint.

In contrast, by maintaining a constant gas-to-liquid flow ratio and by varying the power input to the ozone generator, the increase in the amount of ozone transferred into solution will be almost in direct proportion to the higher ozone concentration in the inlet gas, up to the theoretical limit defined by Henry's law. A higher exhaust gas ozone level also will occur, but only after more ozone has been transferred to the water.

The technique of monitoring ozone concentrations in contactor exhaust gases is quite promising as a way to ensure the production of adequate quantities of ozone, thus preventing the generation of wasteful excesses (Venosa et al. 1985). This, in turn, provides considerable savings in electrical energy costs for ozone generation. If a dissolved ozone residual is not measurable, and without an exhaust gas control procedure such as this, the ozone dosage is adjusted manually for operating water or wastewater treatment plants.

For removal of cyanide ion, sulfide ion, and/or nitrite ion, the amount of ozone that will be required for total destruction can be calculated from the known concentrations of these materials and their oxidation chemistries. The requisite amount of ozone can then be applied, supplemented by increases to compensate for less than 100 percent ozone transfer efficiency in the contactor system, plus some excess ozone for undefined ozone-demanding components, which may be present.

It is clear that selection of the specific analytical method depends on the control procedure used for ozonation of water supplies, which depends on the specific purpose(s) for which ozone is being applied. This selection process is complicated further if ozone is applied at multiple points for different treatment purposes. Specific analytical control methodologies may be appropriate at different ozonation points in the treatment plant. Concurrently, various materials may be present at the multiple ozonation points that can interfere with specific analytical ozone determinations.

II.C.1 Ozone Measurements in the Gas Phase

Measurements of ozone concentration other than dissolved residual ozone are made in the gas phase. Applied ozone dosage is determined by analyzing the amount of ozone in the gases being fed to the ozone contactor coupled with a knowledge of the (ozone-containing) gas and liquid (to be treated) flow rates. However, the transfer efficiency in a contactor depends on many parameters, such as the type of diffusion system, the immediate ozone demand of the water, temperature, and pH. Consequently, the applied dosage cannot be controlled by simply monitoring the exhaust gas.

Some water treatment plants move a step beyond the applied ozone dosage and rely on the absorbed ozone dosage, or utilized ozone dose. This is determined by subtracting the concentration of ozone in the ozone contactor exhaust gases from the concentration of ozone in the contactor feed gases. As before, knowledge of the gas and liquid flow rates is important.

In recent years, monitoring devices capable of measuring ozone concentration in both of these gas phases have been installed in order to control the addition of ozone, as well as to minimize ozonation costs. The devices are appropriate for applications that do not depend on the level of dissolved ozone residual. Procedures for monitoring levels of dissolved organics that are amenable to ozone oxidation also have been developed for controlling oxidation of organics in water treatment plants.

Points in an ozone water treatment process where gas-phase ozone is monitored include:

- downstream of the ozone generator to measure the output of the ozone generator;
- at the exhaust gas outlet of the ozone contacting chamber (subtracting this value from the ozone generator output gives the amount of ozone consumed or utilized by reaction in the aqueous medium);
- at the outlet of the gas-phase ozone-destruction apparatus, to assure that the ozone destruction is complete;
- in occupied areas within the plant facility; and
- in the environment surrounding the plant, to assure that possible leaks of ozone into the atmosphere are detected.

The ranges of ozone concentrations to be measured in the gas phase are quite large, varying from less than 0.1 ppm by volume (0.2 mg/m^3 NTP) in the ambient atmosphere and at the outlet of the ozone destructor, up to 10 g/m^3 NTP (0.76 percent by weight) in the ozone contactor gases, and up to 100 g/m^3 NTP (6.8 percent by weight) in the gases exiting the ozone generator.

The principle of contactor gas-phase measurements is that the presence of a given level of ozone in the gas phase above the ozone contactor indicates that the ozone demand of the aqueous phase has been satisfied. If this level of ozone in the gas phase changes (that is, a drop means increased ozone demand and an increase means that excess ozone is being generated), the analytical instrument monitoring the gas-phase ozone level can be programmed to increase or decrease the output of ozone from the ozone generator. In this way, the output of ozone can be held constant to just those levels required for the particular application. Consequently, energy savings can be considerable due to the conservation of ozone use.

II.C.2 Interferences Affecting Dissolved Ozone Measurements

Because ozone and many of its decomposition products are such powerful oxidizing agents, a specific analytical technique must be selected carefully. The analytical reagents used for ozone determination are frequently oxidized by ozone, and these same reagents can also be oxidized by many of the oxidation/decomposition products of ozone itself. In addition, many impurities normally found in water are oxidized by ozone to produce oxidizers capable of reacting with the ozone reagents. The effects that these interferents have on ozone analytical measurements are discussed in detail in this section.

Ozone decomposition products acting as oxidants (see sec. II.A.2). The decomposition of ozone involves very reactive and catalytic intermediates that influence the stability of ozone solutions and readily alter the precision and accuracy of various analytical methods for dissolved ozone. These species also influence the role and magnitude of various potential interferents on the ozone decomposition process. From this point of view, it is important to note that many of the old methods for determining residual ozone in solution, such as the iodometric method, are not species-specific methods. For example, the iodometric method measures any and all of the oxidizing species present in a decomposing ozone solution, such as ozone itself, O_3^-, HO_2^-, O_2^-, OH, HO_2, and H_2O_2. Under normal operating conditions, the total concentration of these highly reactive intermediates is less than 1 percent of the applied ozone.

It is also clear that the rate of ozone decomposition in aqueous solution is a complex function of pH and solutes present in the solution, such as CO_3^{2-}, HCO_3^-, PO_4^{3-}, and similarly related species. These species act as scavengers for various free radical intermediates and either enhance or decrease the rate of decomposition (Tomiyasu et al. 1985).

Inorganic oxidants produced upon ozonation (see sec. II.A.3). When bromide or iodide ions are present in waters being ozonized, free bromine (Br_2) and/or hypobromous acid (HOBr), hypobromite ion [BrO^-], free iodine (I_2) and/or hypoiodous acid (HOI), hypoiodite ion [IO^-] and iodate ion [IO_3^-] are some of the halogen-containing oxidants that can be produced (Haag and Hoigné 1984; Haag et al. 1984).

One additional question that must be addressed is the purpose for the ozone measurement. It is clear that before a method is chosen to differentiate between ozone and chlorine-containing compounds, the chemistry involved in the treatment must be understood. Assuming that a specific method for the speciation of O_3 from any and/or all of the various chlorine-containing species in solution is available, the measurement that is actually needed must determine: (a) the initial concentration of ozone prior to reaction with the chlorine species, (b) the ozone concentration at a fixed point in time (and at a specific point in the system), or (c) the concentration of ozone after all ozone–chlorine species reactions are over. These problems just discussed make the prospects of determining ozone in the presence of many of the chlorine species difficult at best.

Organic oxidants produced upon ozonation (see sec. II.A.1). When unsaturated organic compounds are ozonized, products such as ozonides, peroxides, diperoxides, triperoxides, hydroalkylperoxides, and peroxy acids can be produced (Bailey 1978). The limited research that has been conducted on solutions containing water as a participating solvent indicates that these types of intermediate oxidation products decompose readily in water to form simpler products such as aldehydes, ketones, and carboxylic acids. Again, many of these peroxygen-containing compounds (that is, ozonide, diperoxide, and peroxy acids) can easily oxidize many reagents commonly used for the analytical determination of ozone.

Depending on whether the analytical method is colorimetric or electrochemical for the determination of dissolved ozone, these other oxidants produced during ozonation are likely to interfere with the analytical procedure or to cause high ozone readings to be obtained. Because these secondary oxidants are present, many current analytical procedures for ozone will measure the total residual oxidants that may be present rather than simply molecular ozone.

II.C.3 Constraints of Dissolved Ozone Measurements

It should be noted that most methods for dissolved ozone measurement are modifications of chlorine residual methods that determine the total of all oxidants in the solution. Therefore, ozone decomposition products, such as hydrogen peroxide and various free radicals, are also measured. In Europe, the Committee of Standardization of the International Ozone Association (IOA–Europe) has proposed various usable, existing methods of measurement, which at present are imprecise. However, while waiting for the best solution, this proposal allows everyone to use the same methodology, which should lead to comparable results.

Sampling. Sampling must be accomplished without agitation. The sample vessel should not have any headspace (which would allow the ozone to diffuse from the sample). The ozone solution should always be added to the reagent in order to minimize ozone loss. Sample handling must be minimized, since the continuous and fast decomposition of ozone at the pH found at most water treatment plants makes it critical that the time lag between sampling and measurement be short. The best way to accomplish this is to automate the sample measurement into an on-line device. In this way, the measurement is made on a fresh sample and the result reported in near real-time sequence. The potential for accuracy and cost-effectiveness due to process control is significantly increased.

Standard sample preparation. Under ideal conditions, ozone standard samples can be made at pH 2 with ozone-demand-free water and stored for a few hours. If the treatment facility has highly qualified staff to perform standard preparations using spectrophotometry and to make dilutions with shrinking syringes, it would be possible to standardize the analytical system. However, it is easier to use reaction chemistry, which is well-behaved, can be standardized, and requires a minimum of calibration.

In some cases, the relative quantity of ozone is more important than the actual ozone concentration. However, the expensive nature of the generation process makes the accurate determination of ozone necessary.

Reasons for imprecision. Much of the imprecision in measuring ozone is caused by improper sample handling. Since ozone is volatile, any operation that allows the ozone to diffuse from the water will produce random error. The use of stripping techniques, while isolating the ozone from potential interference, creates imprecision that is the result of variations in conditions and enhancement of the decomposition rates.

II.C.4 Reporting Gas-Phase Ozone Concentration

Whatever method is used to determine the concentration of ozone in a carrier gas, all results obtained by various operators must be comparable. As a general rule, results are expressed in percentage weight, percentage volume, grams of ozone per cubic meter of gas, and grams of ozone per normal cubic meter (NTP) of gas, which corresponds to 0°C (32°F) at 1 atm. For control purposes, the concentration of ozone in the gas phase and flow rates must be measured with comparable accuracies. The results must indicate the thermodynamic conditions (temperature and pressure) and the specific units such as percentage by weight or g O_3/m^3 NTP.

II.C.5 Ozone Measurement in the Aqueous Phase (Ozone Residual)

Indigo trisulfonate batch method (colorimetric method). The indigo method of ozone measurement was developed by Hoigné and Bader at the Swiss Federal Institute for Water Resource and Water Pollution Control (EAWAG) (Bader and Hoigné 1979, 1981, 1982a, 1982b). The method is very sensitive, precise, fast, and more selective for residual ozone than other methods. Manganese ions, chlorine, hydrogen peroxide, and ozone decomposition products, and the products of ozonolysis of organic solutes (which interfere with iodometric and similar methods) exhibit less interference with the indigo procedure. However, the masking of chlorine in the presence of ozone can make the indigo method problematic.

When using this method, the change in absorbance is followed most accurately by using spectrophotometry, although a manual procedure is available (Bader and Hoigné 1979, 1981, 1982a). In all indigo procedures, the ozone dose must be adjusted to decolorize between 20 and 90 percent of the indigo reagent. For one fixed concentration of the reagent and one predetermined relative ratio of volume of reagent to volume of aqueous ozone, the dynamic range is restricted to a factor of about 4.5. Precision is usually between 1 and 5 percent, depending on technique, method, and equipment used. The stoichiometry is 1:1 at pH 2, and the decrease in absorbance is linear with increasing concentration over a wide range.

An indigo stock solution is prepared using analytical-grade phosphoric acid and potassium indigo trisulfonate. This stock solution is stable when stored in the dark for periods up to four months. It should be discarded when the absorbance of a hundredfold dilution falls below 0.16 absorbance units/cm.

The spectrophotometric procedure for samples containing between 0.01 and 0.1 mg/L ozone is to add 10.0 mL of indigo reagent I (20 mL indigo stock solution, 10 g sodium dihydrogen phosphate, and 7 mL concentrated phosphoric acid diluted to 1 L) to each of two 100-mL volumetric flasks (A and B). The different ranges are achieved by using different dilutions of the stock solutions or smaller samples. Flask A (the blank) is filled to the mark with distilled water. Flask B is filled to the mark with the ozone-containing sample. The ozone solution is added in such a manner that completely decolorized zones are quickly eliminated by efficient stirring, but such that no ozone degassing occurs by the formation of gas bubbles.

The absorbance of both solutions is measured in 10-cm cuvettes at 600 ± 5 nm as soon as possible, but within at least 4 h. The ozone concentration is calculated from the absorbance difference (Abs) between sample A (the blank) and sample B, by using the equation:

$$\text{ozone concentration (in mg } O_3/L) = (\Delta \text{Abs} \times 100)/(f \times b \times v)$$

Where:

- b = the path length of the cuvette, cm
- v = the volume of the sample added, mL (normally 90 mL)
- f = the slope of the calibration curve at 600 nm (0.42 ± 0.1 cm^{-1} per mg/L, with ϵ (molar absorptivity) = 20,000 $M^{-1} \cdot cm^{-1}$.

The factor of 100 converts the milliliters of sample to 100 mL for all calculations. When high accuracy is required, it is necessary to have the indigo purity verified. The method is normally calibrated against iodometric titrations. However, the UV absorbance of ozone in pure water (ϵ_{258} = 2950 $M^{-1}cm^{-1}$) is a better choice.

To determine ozone in the presence of manganese using the indigo method, a blank solution is prepared using the sample water in which the ozone is selectively destroyed by the addition of glycine before it is added to the indigo reagent. The absorbances of the two solutions (one containing manganese, the other containing manganese plus ozone) are measured, and the difference is taken as a measure of the concentration of ozone. However, the precision of this difference procedure is poorer if the manganese content is much greater than that of ozone.

Automated indigo method. The gas diffusion-flow injection analysis (GD-FIA) technique can eliminate most, if not all, indigo method interferences (in this case, chlorine, bromine, and manganese). The GD-FIA indigo method exhibits an improvement over the manual method in terms of higher linear range, greater precision among samples, higher sampling frequency, and lower reagent consumption (Straka et al. 1984, 1985).

In GD-FIA, the volatile ozone diffuses through a gas-permeable hydrophobic membrane into an acceptor stream where the Indigo reaction takes place and the resulting product is detected. Microporous 0.45-μm pore size Teflon® membranes frequently are used because they are chemically inert, an important characteristic when considering the high oxidation potential of ozone. Gas diffusion cells are commercially available with a 75-mm long × 2-mm wide diffusion area.

The GD-FIA procedure eliminates the interference of oxidized forms of manganese and reduces the interference of chlorine to less than 25 percent of its value in the flow injection analysis method without gas diffusion. This is equivalent to an error in the measurement of 1 mg/L of dissolved ozone of 0.8 percent for each milligram per liter of Cl_2 present (Straka et al. 1985).

Iodometric methods. In the 1985 *Standard Methods* description for the determination of residual ozone in water, iodide ion is oxidized to iodine by ozone in an unbuffered potassium iodide solution. The pH is then adjusted to 2 with sulfuric acid, and the liberated iodine is titrated with sodium thiosulfate to a starch endpoint. The ozone/iodine stoichiometry for this reaction has been studied extensively (Gordon and Grunwell 1983; Gordon et al. 1987a, 1987b) and found to range from 0.65 to 1.5. Factors affecting the stoichiometry include pH, buffer composition, buffer concentration, iodide ion concentration, sampling techniques, and reaction time. The pH during the initial ozone/iodide ion reaction and the pH during the iodine determination have been shown to alter the ozone/iodine stoichiometry.

There have been several modifications of the method in an attempt to obtain the necessary one-to-one stoichiometry. However, when the desired stoichiometry is observed, it appears to be the result of a balance among the various reactions. Ideally, the reaction of ozone with potassium iodide is as follows:

$$O_3 + 2I^- + H_2O = I_2 + O_2 + 2OH^- \quad \text{(II–16)}$$

The released iodine is titrated with sodium thiosulfate:

$$I_2 + 2S_2O_3^{2-} = 2I^- + S_4O_6^{2-} \quad \text{(II–17)}$$

The IOA in Europe (IOA 1989a) has proposed the oxidation of nitrite ion to nitrate ion by ozone as a comparison method for the iodometric procedures. However, the success of the indigo method makes the use of iodometric methods for the measurement of residual ozone inappropriate.

Other colorimetric methods. Several other colorimetric methods have been proposed for the determination of ozone and are described below. Although some color reactions, such as the N,N-diethyl-*p*-phenylenediamine (DPD) or the *o*-tolidine, may be used, they should be considered as qualitative tests for oxidants, rather than quantitative methods for the determination of ozone.

The Leuco crystal violet method. Leuco crystal violet (LCV) is oxidized to crystal violet by oxidants such as ozone (Layton and Kinman 1970). The absorbance maximum is at 592 nm in acidic solution, exhibits a molar absorptivity of 1×10^6 $M^{-1}cm^{-1}$, and is stable for at least 44 days. Since the LCV method is subject to many of the same interferences as iodometry, the procedure cannot be recommended.

Acid chrome violet K method. Acid chrome violet K [1,5-bis-(4-methylphenyl-amino-2-sodium sulfonate)-9,10-anthraquinone] (ACVK) is readily bleached by ozone. The method has been in use since 1968 at the Tailfer water utility, which supplies the city of Brussels, Belgium. Decreases in absorbance at 550 nm can be related directly to the ozone concentration. This procedure is remarkably free of interferences. Free and combined chlorine and chlorate ion concentrations below 10 mg/L do not interfere (Masschelein et al. 1989). The detection limit of the ACVK procedure approaches 25 μg of O_3/L. The apparent advantages of this method are its ease of execution, stability of the dye reagent solution, a linear calibration curve over the range of 0.05 to just over 1.0 mg/L of ozone, and the apparent lack of interferences from low levels of manganese, chlorine, organic peroxides, and other organic oxidation products produced upon ozonation.

Carmine indigo procedure. The carmine indigo procedure has been in use for at least 15 years in water treatment plants in the province of Quebec, Canada. Ozone-containing water is titrated with a solution of carmine indigo until a faint blue color persists. This color will not persist until all of the ozone has been destroyed by oxidizing the blue carmine indigo color.

This procedure has been checked against the indigo method and good agreement is reported. The primary advantage of this procedure is its simplicity, rapidity, and ready adaptability to water treatment plant conditions.

UV absorption. Ultraviolet absorption measurements also can be used for residual aqueous ozone at the wavelengths of 258–260 nm or 253.7 nm, which can be obtained by the use of low-pressure mercury lamps and spectral filters. There is uncertainty with respect to the molar absorptivity for aqueous ozone. In the literature, values from 2900 to 3600 $M^{-1}cm^{-1}$ are reported (Gordon and Grunwell 1983; Hart et al. 1983).

If the molar absorptivity for residual ozone in aqueous solution is known unambiguously, UV absorption, in principle, is an absolute method for the determination of residual ozone, which is not dependent on calibration or standardization against other analytical methods. In fact, water that does not contain UV interferences is very unusual. Thus, for the measurement of residual ozone in drinking water, UV absorption is generally unacceptable.

Electrochemical devices. In drinking water treatment plants, amperometers are used for continuous measurement of the concentration of oxidizing agents, employing a simplified method of amperometric analysis (Masschelein 1982b). They are commonly used in drinking water treatment plants that use ozone for disinfection by installing the devices at the outlets of the ozone contact chambers.

Commercial amperometers measure the concentration of dissolved ozone by two fundamentally different approaches: by bare electrodes, which measure dissolved ozone directly, or by membrane electrodes (Stanley and Johnson 1979), in which ozone diffuses through a thin film of microporous plastic to the cathode.

The electrochemical method of measuring ozone utilizes a measuring cell that acts like a galvanic element with electrodes made of selected metals. The oxidizing agent–containing water flows through the cell and serves as the electrolyte. As long as the water does not contain any oxidizing compounds, the electrode is almost completely polarized and only a very small current flows, which can be compensated electrically.

When oxidants are present in the water, they are reduced on the cathode; the polarization is disturbed, and the electrode supplies an electrical current proportional to the concentration of the oxidant. Depending on the electrodes chosen and the current applied between the electrodes, the resulting measurement can be more or less selective with respect to the measurement of ozone itself.

The galvanic couples commonly used include Au/Cu, Au/Ag, Pt/Cu, and Ni/Cu. These systems are not ozone-selective and they all respond to chlorine (Masschelein et al. 1979). Some couples such as NiO/O_2–Ag/AgCl enable selective ozone measurement, although the results may be sensitive to variations in pH and temperature.

The primary disadvantage of other nonselective amperometric systems as a monitor is that they measure the total of all oxidizing agents present and can be used most effectively when there is a variable concentration of a single substance in solution. Under these conditions, the effect of any other substances present in constant concentration can be cancelled out by adjusting the zero setting of the apparatus. In practice, this is acceptable (for control) if the rate of ozone generation is constant, since the interferents probably would also be generated or produced as by-products at a constant rate.

Measurement after stripping of dissolved ozone into the gas phase. In indirect procedures, dissolved ozone is stripped from the aqueous media using an inert gas such as nitrogen, argon, or air that does not react with ozone. The amount of ozone present is determined directly in the gas phase using techniques such as UV absorption, chemiluminescence, calorimetry, and isothermal pressure differentials, or the gaseous ozone is reabsorbed into aqueous solution and determined by iodometry or other ozone colorimetric methods. This purging, or stripping, technique was developed to minimize the complications caused by the presence of other oxidants in aqueous solution, which interfere with direct iodometric or colorimetric procedures.

The success of this procedure initially depends on the ability to strip ozone as the sole oxidizing agent from the treated water without any decomposition (*Standard Methods* 1985; Verein 1978). The fact that many of the aqueous decomposition products of ozone are ionic or free radicals in nature and that most of the inorganic and organic oxidants produced during the ozonation of water are either ionic in nature or are very water-soluble suggests that, in most cases, only ozone and some nitrogen oxides will be stripped from treated water by the carrier gas. Nitrogen oxides can be removed from the gas stream by adsorption onto a column containing solid potassium permanganate.

On the other hand, ozone does readily decompose even when the gas phase is very dry (see sec. II.E.1). Therefore, application of the gas stripping technique requires close attention to conducting all operations rapidly in the absence of light and at very reproducible time intervals. Even then, and especially at low concentrations of dissolved ozone, the efficiency of gas purging can be expected to be lower than at the higher concentrations.

In addition, passage of a carrier gas through an aqueous solution can be expected to strip out small droplets of water (aerosol size) along with the ozone. These aerosol-sized water droplets will contain the soluble interfering materials. If the stripped gas is passed through a desiccant to remove the water vapor, it is possible that some of the ozone will either be adsorbed by the desiccant (such as silica gel or molecular sieves), which is effective for separating ozone from oxygen and/or water, or be partially catalytically decomposed by surface impurities. Finally, glass frits, which are normally used in standard wash bottles and might be used to disperse the carrier gas, can cause ozone decomposition (Gordon and Grunwell 1984).

Titration of a sample of ozone-containing water that is added directly to the potassium iodide solution (without first passing the ozone through the gaseous phase), as compared to titrating after the ozone stripping procedure, will give some indication whether interfering oxidants are present. If such interferences are absent or negligible, ozone transfer by means of inert gas stripping should not be considered, and the ozone should be determined directly in solution. Since stripping conditions such as temperature, pH, and salinity can vary, the reliability of this method is suspect.

II.C.6 Gas Phase Measurements

Iodometric methods. Iodometric procedures have been used for all of the ozone concentration ranges encountered in water treatment plants. This includes measurement of ozone directly from the generator and measurement of ozone as stripped from aqueous solution. For the iodometric method, the ozone-containing gas is passed into an aqueous solution containing excess potassium iodide, in which the ozone oxidizes iodide ion (Maier and Kurzmann 1977). The difficulties regarding the ozone dosage based on iodometry have been partially described earlier in this chapter. Additional concerns are described below.

Other oxidants in the ozonated gas. All other oxidizing materials act as interferences with the iodometry. The influence of nitrogen oxides, which may be formed when ozone is generated in air, may be determined by absorption in sodium hydroxide solution with subsequent determination of nitrite ion. Alternatively, the effects of nitrogen oxides may be eliminated by passing the ozone-containing gas through absorbents such as potassium permanganate that are specific for nitrogen oxide gases.

KI versus NBKI methods. In all absorption methods, the ozone-containing gas is passed through two washing bottles filled with 200–400 mL of 1 percent or 5 percent potassium iodide solution. The reading of the wet test meter is recorded. The total gas volume is determined accurately and should be more than 1 L and not more than 10 L. After finishing the measurement, the gas flow is switched to the destruction mode and the wet test meter reading is recorded again. The barometric pressure and the temperature of the measured gas are read and recorded.

The three specific variations of this method that are commonly used are a modification of the neutral buffered potassium iodide (NBKI) method, the unbuffered potassium iodide (KI) method, and a recent modification of the NBKI method accepted in Europe by the IOA ($NBKI_e$).

The reaction of ozone with iodide ion, upon which the iodometric method is based, produces two moles of hydroxide ion for each mole of iodine. The presence of hydroxide ion in the reagent solution creates an inherent problem with the iodometric determination of ozone, since the reaction of ozone with hydroxide ion constitutes the initiation step of the ozone decomposition mechanism in solution. With the KI method, the iodide reagent rapidly becomes basic, and the ozone decomposition mechanism is expected to be a factor in the determination. Likewise, in the $NBKI_e$ modification, which is very weakly buffered, the local concentrations of hydroxide ion might be expected to enhance the decomposition of ozone. The NBKI method is strongly buffered at a neutral pH. This helps to minimize ozone decomposition, which occurs rapidly in basic medium. However, iodate ion formation has been shown to be significant at a neutral pH. For this reason, before the titration of iodine by thiosulfate, it is necessary to acidify the buffered solution to transform any iodate that may have been formed back to iodine.

All of these difficulties are evidenced by large differences between the NBKI method, which yields values approximately 15 percent higher than those of the weakly buffered $NBKI_e$ modification, and the unbuffered KI method, when carried out under the same conditions. Clearly, the buffering of the iodide ion–containing reagent has an effect on the determination. The $NBKI_e$ method, which is very weakly buffered, and the unbuffered KI method produce statistically equivalent results (Wood et al. 1989).

Ideally, the iodometric determination of ozone can be reproducible when carried out under very strictly controlled conditions. When defined procedural techniques are used, the KI or $NBKI_e$ method may be useful as an independent check of a UV monitor such as within ± 3 to 5 percent. However, the evidence that even microscopic details of the sample bubble passing through the reagent solution can effect the determination makes the iodometric determination of ozone not an ideal candidate as a standard method for ozone determination (Wood et al. 1989).

Direct amperometric analysis (gas phase). Amperometric analyzers for gas-phase measurement are commercially available (Paré 1982). Ozone is sensed in the air sample by the oxidation–reduction reactions in the sensing solution, which contains potassium iodide. These reactions take place on the cathode portion of the electrode support. In this region, any ozone in the air sample reacts with the iodide ion in the sensing solution. When no voltage is applied between the electrodes, a thin layer of hydrogen gas is produced at the cathode by a polarization current. When a voltage is applied to the electrodes, the hydrogen layer builds to its maximum and the polarization current ceases to flow. When free iodine is produced by reaction with ozone, it reacts immediately with the hydrogen. The removal of hydrogen from the cathode causes a repolarization current to flow in the external circuit, reestablishing equilibrium. The current is directly proportional to the mass of ozone entering the sensor per unit time.

The same concerns raised earlier regarding iodometric methods as applied to the determination of ozone are applicable to this method. The electrochemical detection does not eliminate these concerns.

Direct UV absorption measurements (253–260 nm). Gaseous ozone absorbs light in the short–UV wavelength region with a maximum absorption at 253.7 nm. At 253.7 nm, the generally accepted value for the gas-phase absorption coefficient for ozone is 3000 ± 30 $M^{-1}cm^{-1}$ at 273 K and 1 atm (Duguet et al. 1983; Langlais 1985; Kogelschatz 1987; Mauersberger et al. 1986; Molina and Molina 1986; Zurer 1987).

Instruments for measuring ozone by the absorption of UV radiation are supplied by several manufacturers for gas concentrations below 1 g/m^3 NTP (0.076 percent by weight). In general, these instruments measure the amount of light when no ozone is present and the amount of light when ozone is present. The meter output is the difference of the two readings, or the actual amount of ozone present.

The meters can be calibrated such that the registered number equals the concentration of ozone. In all cases, the temperature and pressure in the cell must be known. Then, the registered number is proportional to the concentration of ozone. In general, the meter is set (adjusted) such that the registered number is equal to the concentration of ozone by taking into account the temperature and pressure in the cell.

If the molar absorptivity of 3000 ± 30 $M^{-1}cm^{-1}$ for gaseous ozone is unambiguously accepted as recommended by the IOA European Standardization Committee (IOA 1989b), UV absorption becomes an absolute method (± 1 percent) for determining gaseous ozone, which is not dependent on calibration or standardization against other analytical methods. Therefore, it can be used for calibration of other analytical methods for ozone. It is specific to the determination of molecular ozone, and is applicable to measurements in gaseous phases. The molar absorptivity used for each instrument must be available and verified in order to ensure accuracy. Ultraviolet absorption has been specified by the U.S. EPA as the method to be used for calibrating atmospheric ozone monitors and analyzers based on other analytical procedures such as chemiluminescence.

Calorimetry. The theoretical basis of calorimetry (Caprio et al. 1982; Maier 1982) for ozone measurement is based on the enthalpy of the catalyzed decomposition of ozone (H = 142.3 kJ/mol). This value for the enthalpy of decomposition has been confirmed by numerous determinations (Leitzke 1977). In this technique, the calorimetric determination of ozone is calibration-independent. The observed temperature difference between the inlet gas and the gas reacting in the catalyzer is measured by means of thermocouples located in a well-isolated cell. This temperature difference is proportional to the concentration of ozone in the gas.

A constant flow of gas (air or oxygen) containing ozone is directed to a heat-insulated measurement cell containing a proprietary catalyst. This cell is fitted with thermocouples that measure the gas temperatures at the inlet and the outlet. Any nitrogen oxides in the carrier gas, liable to poison the catalyst, are retained on the adsorbent, which does not destroy ozone. The adsorbent has a service life of four to five weeks. The technique is specific to the determination of molecular ozone, but is applicable to measurement only in the gas phase. However, the higher the concentration of ozone in the gas phase, the more accurate the method appears to be, since a greater temperature difference is observed. Potential interferents have not been reported.

Isothermal pressure changes. The isothermal pressure change procedure (Caprio and Lignola 1980) is independent of chemical comparative tests and does not require calibration. The procedure is based on the generation of an increased number of gas molecules during the destruction of ozone at constant temperature. The reaction is:

$$2O_3 \rightarrow 3O_2 \quad \text{(II–18)}$$

When this reaction is carried out isothermally in a closed vessel, the increase in pressure of the contained gas is proportional to the ozone concentration.

In principle, this pressure differential procedure achieves a totally physical ozone measurement without requiring calibration using a chemical method. Various automated instrumental checks such as the stored molar absorptivity, the age of the UV light source, the zero point reading, measurement of the flow of the test gas and the flushing gas, and the reading of the diagnostic display are possible. The method has very good potential.

II.C.7 Conclusions

In conclusion, when dissolved ozone is indirectly measured in the gaseous phase after stripping, the methods of measurement are still subject to the same sources of error caused by the stripping itself. The iodometric method remains by far the most widely used, and with some modifications is still proposed today as the accepted method by the International Ozone Association in Europe. The UV method using 3000 ± 30 $M^{-1}cm^{-1}$ at NTP is currently also recommended in Europe for gas-phase ozone measurements. The method of choice in the United States for the measurement of residual ozone, as recommended in the 1989 edition of *Standard Methods*, is the indigo method.

II.D OZONE GENERATION

The existence of ozone can be traced back to ancient times. In book XII, verse 417 of the *Odyssey*, Zeus strikes a ship with a thunderbolt "quite full of sulphurous odor." In 1785, Van Marum observed that air submitted to electrical sparks generated a typical odor (Rideal 1920). Ozone was discovered by Schönbein using electrolysis in 1840 (Schönbein 1840), and in 1848, Hunt (1848) formulated the triatomic oxygen structure of ozone. It was established in 1860 by Andrews and Tait (1856) that ozone thermally decomposes to oxygen.

In 1857, von Siemens developed the first industrial ozone generator, which was based on corona discharges (von Siemens 1857). Two concentrical glass tubes were used; the outer tube was covered externally by a layer of tin, and the inner tube was covered internally by a layer of tin. Air was circulated through the annular space. This technology was later improved by the addition of circulating cooling fluids along the discharge air or oxygen gap, resulting in lower generation temperatures and less thermal destruction of the ozone.

Since 1868, ozone has been considered to be an endothermic compound (Hollman

1868), and the enthalpy of formation has been better understood since 1908 (Jahn 1908). It should be noted that all attempts to generate ozone by thermal activation have failed.

The remainder of this section will highlight the theoretical aspects underlying the design of ozone generation systems.

II.D.1 Basic Principles

Ozone is a metastable molecule produced from elemental oxygen. As such, its production presents conflicting aspects from a thermodynamic viewpoint. It has been called "an allotropic form" of oxygen, which is an inadequate terminology.

The overall reaction for ozone formation is described by an endothermic reaction:

$$3O_2 \rightleftharpoons 2O_3 \ (\Delta H^\circ \text{ at 1 atm}, \ +284.5 \text{ kJ/mol})$$

Also, the entropy of formation is large and unfavorable:

$$\Delta S^\circ \ (1 \text{ atm}) = -69.9 \ (\text{J/mol})/\text{degree}$$

Clearly, ozone cannot be generated by thermal activation of oxygen, since the standard free energy of formation ΔG° (1 atm) = +161.3 kJ/mol. To summarize, ozone can only be decomposed easily by heating, and adequate temperature control of the process gas is an important factor in ozone generation efficiency.

Based on the fundamental thermodynamic principles described above, it is apparent that ozone cannot be liquified by compression. Ozone dissolved in liquid oxygen up to 30 percent by weight is relatively safe, while spontaneous explosions occur at more than 72 percent by weight ozone in liquid oxygen. Ozone has a tendency to separate and concentrate during evaporation due to the higher volatility of oxygen. When this occurs, the composition becomes unavoidably explosive (Waller and McTurk 1965). Conservation of ozone in liquified freons has been attempted, but application of the process to water treatment is a problem (L'Air Liquide, French Patent 1,246,273). Also, ozone decomposes even when dissolved in a liquified matrix. Consequently, in water treatment, ozone must be generated onsite.

The generation of ozone involves the intermediate formation of atomic oxygen radicals, which can react with molecular oxygen. Starting from molecular oxygen, this formation involves energies of 493.3 kJ/mol for the production of $O(^3P)$ radicals and 682.8 kJ/mol for $O(^1D)$ radicals (Figure II–38).

The main reactions by which ozone is considered to be formed are:

$$O(^3P) + O_2X^3\Sigma g^- + M = O_3$$

and

$$O(^1D) + O_2X^3\Sigma g^- = O_3$$

All processes that can dissociate molecular oxygen into oxygen radicals are potential ozone generation reactions. Energy sources that make this action possible are electrons or photon quantum energy. Electrons can be used from high-voltage sources in the silent corona discharge, from chemonuclear sources, and from electrolytic processes. Suitable photon quantum energy includes UV light of wavelengths lower than 200 nm and γ-rays.

Electronic activation of oxygen leading to the formation of monoatomic oxygen ions, O^-, are, at present, considered not to contribute significantly to ozone formation (Kogelschatz et al. 1987).

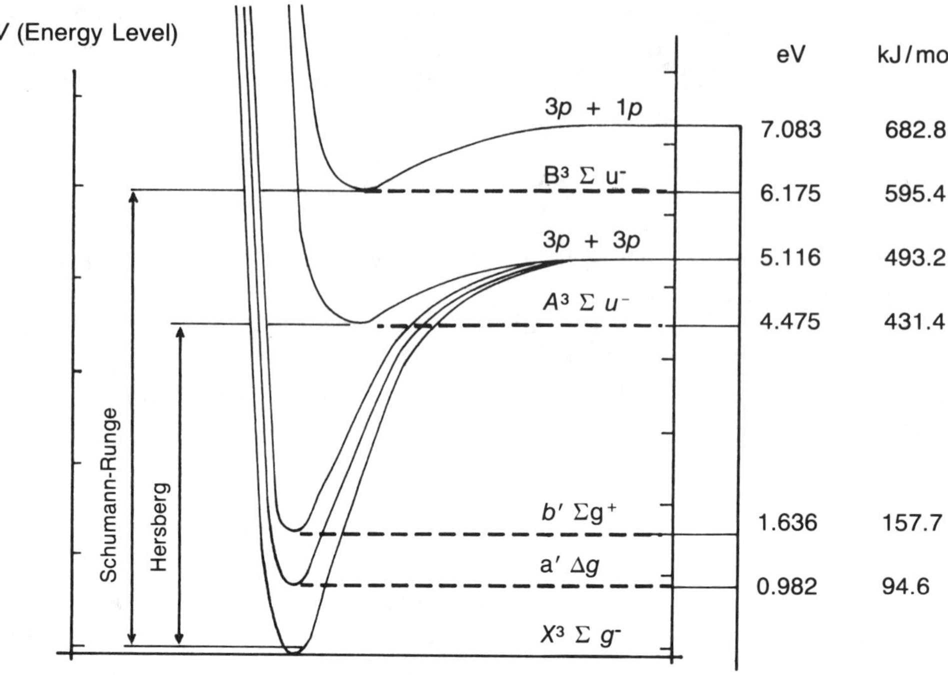

Source: Masschelein (1977).

Figure II–38 Potential Energy Curve of Oxygen

$$O_2\ (X^3\Sigma g^-) + e \begin{cases} \nearrow O^- + 1.510\ \text{eV} \\ \searrow O(^3P) - 5.116\ \text{eV} \end{cases}$$
$$(-\ 3.606\ \text{eV or}\ -\ 387.8\ \text{kJ/mol})$$

Moreover, the monoatomic oxygen ion can promote the destruction of ozone according to the reactions:

$$O_3\ (^1A) + O^- \rightarrow O_2\ (X^3\Sigma g^-) + O_2^-$$
$$(+\ 2.98\ \text{eV or}\ +\ 287.4\ \text{kJ/mol})$$

The intermediate oxygen radicals can recombine to reform oxygen with the liberation of considerable heat as shown by the following reaction:

$$O(^3P) + O(^3P) + M = O_2(X^3\Sigma g^-) + M$$
$$(+\ 5.1\ \text{eV or}\ +\ 491.6\ \text{kJ/mol})$$

Consequently, if the concentration of oxygen radicals becomes too high, the relative yield of ozone generation will drop. At a relative mole fraction $[O]/[X_2]$ of 10^{-3} and lower (X_2 being, for example, O_2), a conversion efficiency $[O_3]/[O]$ of 80 percent and more can be claimed. This efficiency drops very quickly at increasing radical mole fractions, thus establishing an upper limit to the electronic activation. Moreover, electrons can decompose ozone; this decomposition also limits the acceptable discharge strength.

$$O_3\ (^1A) + e \rightarrow O_2(X^3\Sigma g^-) + O(^3P) + e$$
$$(-\ 1.084\ \text{eV or}\ -\ 104.5\ \text{kJ/mol})$$

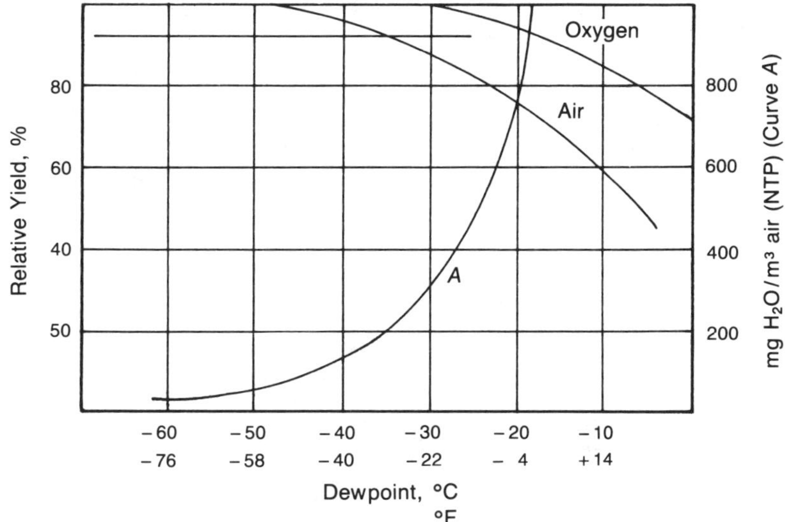

Source: Masschelein (1977).

Figure II–39 Effect of Water Content of Process Gas on Ozone Production

$$O_3\ (^1A) + e \rightarrow O_2(X^3\Sigma g^-) + O^-$$
$$(+\ 0.426\ \text{eV or } +\ 41.1\ \text{kJ/mol})$$

Similar principles apply to photochemical generation processes with mercury lamps, since the more important UV emission at 254 nm is absorbed by ozone and promotes decomposition along with generation, thus establishing an equilibrium with generation.

As a conclusion, ozone generation is an equilibrium process in which the conditions for generation also involve reaction schemes for destruction. An ozone generator is always a compromise in that the designer has considered the relative importance of several factors in view of the objective to be reached, the relative local costs of the different components, and the technical skill available for system operation.

II.D.2 Ozone Generation by Corona Discharge

Corona discharge in a dry process gas containing oxygen is presently the most widely used method of ozone generation for water treatment. A classical production line is composed of the following units: gas source (compressors or liquified gas), dust filters, gas dryers, ozone generators, contacting units, and offgas destruction.

It is of utmost importance that a dry process gas is applied to the corona discharge. Figure II–39 illustrates the effect of moisture on the relative ozone production yield. Limiting nitric acid formation is also important in order to protect the generators and to increase the efficiency of the generation process. With air as the feed gas and under good operating conditions, about 1 mol of N_2O_5 may be produced per 100 mol of ozone.

The process is described by the following reactions:

$$O_2 + N_2 \xrightarrow{e} 2NO$$
$$2NO + O_3 \rightarrow N_2O_5$$
$$N_2O_5 \rightarrow 2NO_2 + 1/2\ O_2 \text{ or}$$
$$N_2O_5 \xrightarrow{e} NO + NO_2 + O_2 \text{ and}$$
$$N_2O_5 + H_2O \rightarrow 2HNO_3$$

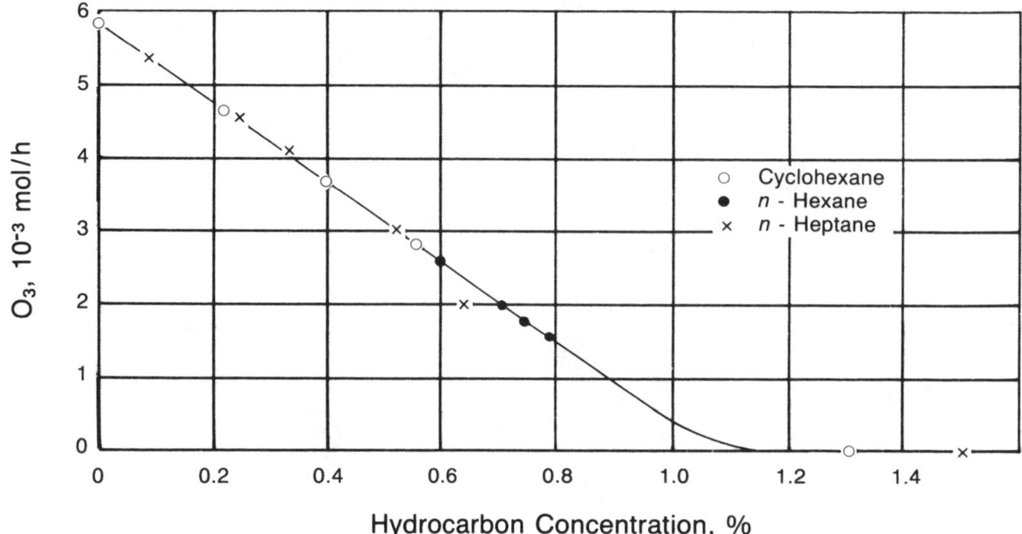

Source: Reprinted with permission from *Ozone Chemistry and Technology* (Advances in Chemistry Series 21), p. 313, Inoue, E. and Sugino, K. © 1959 American Chemical Society.

Figure II–40 Influence of Hydrocarbons on the Generation Yield of Ozone

In normal operation of properly designed systems, a maximum of 3 to 5 g nitric acid is obtained per kilogram ozone produced with air.

If increased amounts of water vapor are present, larger quantities of nitrogen oxides are formed when spark discharges occur. Also, hydroxyl radicals are formed that combine with oxygen radicals and also ozone. As indicated by the following equations, both reactions reduce the ozone generation efficiency.

$$OH + O(^1D) \rightarrow HO_2$$
$$OH + O_3 \rightarrow HO_2 + O_2\ (X^3\Sigma g^-)$$

Direct reactions with NO_2 are also possible.

$$NO_2 + OH + M \rightarrow HNO_3 + M$$
$$NO_2 + O(^1D) \rightarrow O_2 + NO$$

Consequently, the dryness of the process gas is of relevant importance to obtain a reliable yield of ozone. Moreover, with air, nitrogen oxides can form nitric acid, which can cause corrosion.

The presence of organic impurities in the feed gas should be avoided, including impurities arising from engine exhaust, leakages in cooling groups, or leakages in electrode cooling systems. Sample data are presented in Figure II–40. The drop in production yield is almost linear with the increasing hydrocarbons content. At concentrations higher than 1 percent volume of hydrocarbons, practically no ozone is formed.

Chlorofluorocarbons and hydrogen have a negative effect on the ozone production yield. However, some other gases such as carbon monoxide or nitrogen can, when present in low concentrations, slightly increase the generation yield (by 3 or 10 percent) (Figure II–41).

The formation of ozone through electrical discharge in a process gas is based on the nonhomogeneous corona discharge in air or oxygen. There are numerous distributed microdischarges by which the ozone is effectively generated. It appears that each individual microdischarge lasts only several nanoseconds, lasting about 2.5 to 3 times longer

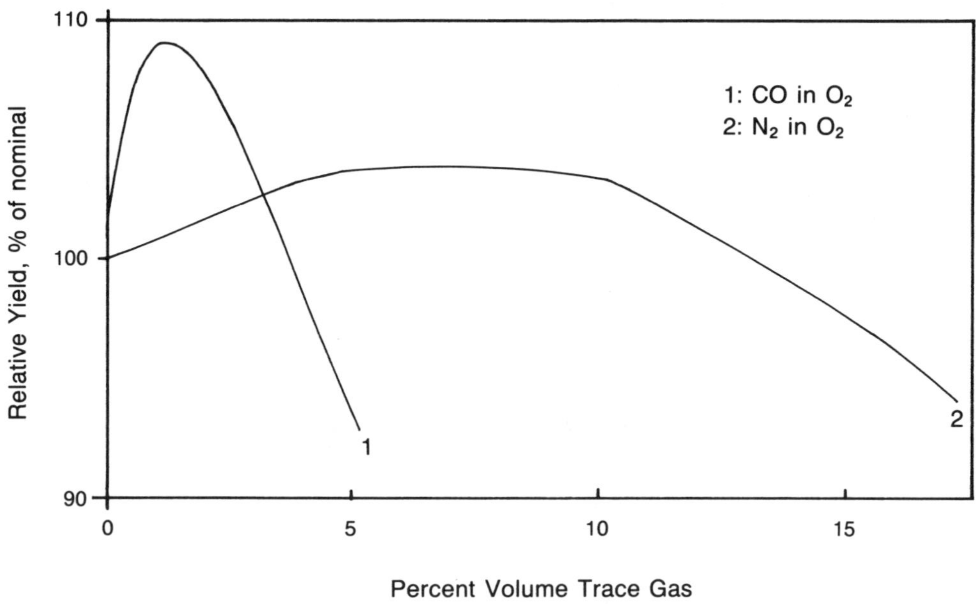

Source: Reprinted with permission from *Ozone Chemistry and Technology* (Advances in Chemistry Series 21), p. 304, Cromwell, W.E. and Manley, T.C. © 1959 American Chemical Society.

Figure II–41 Influence of Trace Gases on Ozone Generation Yield

in air than in oxygen (Kogelschatz et al. 1987). The current density ranges between 100 and 1000 A/cm^2.

Industrial corona cells have a capacitance due to both the discharge gap and the dielectric material (Figure II–42). The formula for the total cell capacitance is given as:

$$C\ (f) = (2\pi\ \epsilon\epsilon_0 L)/\ln(r_e/r_i) \text{ and } C\ (f) = \epsilon\epsilon_0\ A/d$$

for cylindrical geometry and parallel plates, respectively.

Where:

r_e, r_i = outer and inner concentric electrode radius
d = distance between electrodes
L = electrode length
A = electrode area
ϵ_0 = absolute dielectric constant (8.854×10^{-12}f/m)
ϵ = relative dielectric constant (for example, equals 6 for glass, 1 for air/oxygen)
C = capacitance, f.

The driving potential is a function of the current frequency, ν, and so:

$$V = V_0 \sin\ (2\ \pi\ \nu \times t)$$

in which V_0 is the peak potential of the alternative current applied.

At a value V_s, the corona breaks down due to sparking. Values for the sparking potential (in volts) have been reported in the literature (Cobine 1958) as:

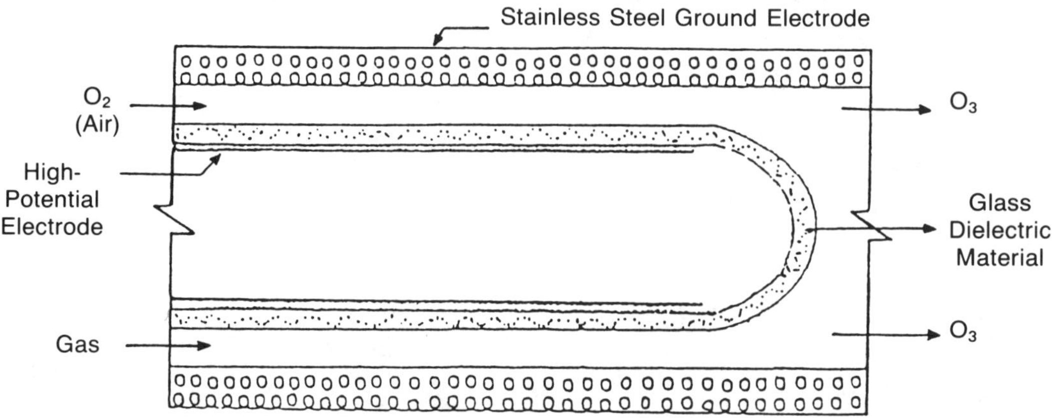

Figure II–42 Schematic of Corona Discharge Gap

$$V_s = 29.64 \text{ P dg} + 1350, \text{ for air}$$
$$V_s = 26.55 \text{ P dg} + 1480, \text{ for oxygen}$$

Where:

P = the absolute gas pressure, kPa
dg = the thickness of the gas gap, mm.

This indicates that for equal values of dg the sparking potential for oxygen is somewhat lower than for air. This can be corrected by using a slightly smaller gap in the design of oxygen-based generators. All other conditions being constant, the sparking potential, when operating at high voltage, is lower when the discharge gap is smaller.

The different parameters are correlated by the following equation, verified experimentally (Reynolds 1955) for the average power input:

$$P = 4 \text{ Cd } V_s \nu [V_0 - ((\text{Cd} + \text{Cg})/\text{Cd}) V_s]$$

Where:

Cd = the capacitance of the dielectric
Cg = the capacitance of the gas layer.

The power taken off by the corona discharge is increased by thinner dielectrics with high dielectric constants. This is done by increasing the peak voltage and by increasing the current frequency.

The expressions for ozone yield can be variable. Theoretically, 2960 J (0.82 kW) are necessary to produce 1 g of ozone. The yield is sometimes expressed by G values, that is, molecules of ozone produced by consumption of 100 eV energy. Reference values for ozone generation at 100 percent yield are as follows: 142.3 kJ/mol O_3; 2965 J/g O_3; 38.5 Wh/mol O_3/h; 0.803 Wh/g O_3/h; and 1.47 eV/molecule O_3, or G = 68. Practical yields obtained with air are illustrated in Figure II–43 (Masschelein 1977).

By using oxygen or enriching the process air in oxygen, the generating capacity of a given ozone generator can be increased by a factor ranging from 1.7 to 2.5 versus the production capacity with air, depending on the design parameters (for example, gas discharge gap and current frequency). Some examples are given in Figure II–44.

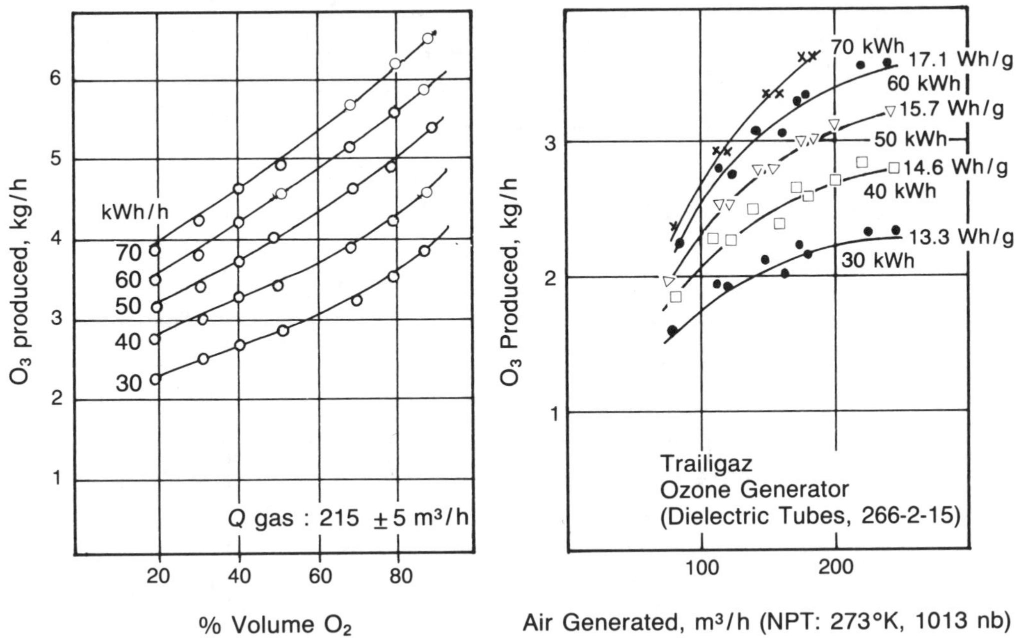

Source: Masschelein (1977).

Figure II–43 Yield Parameters in Ozone Generation With Air (Low Frequency: 50 Hz)

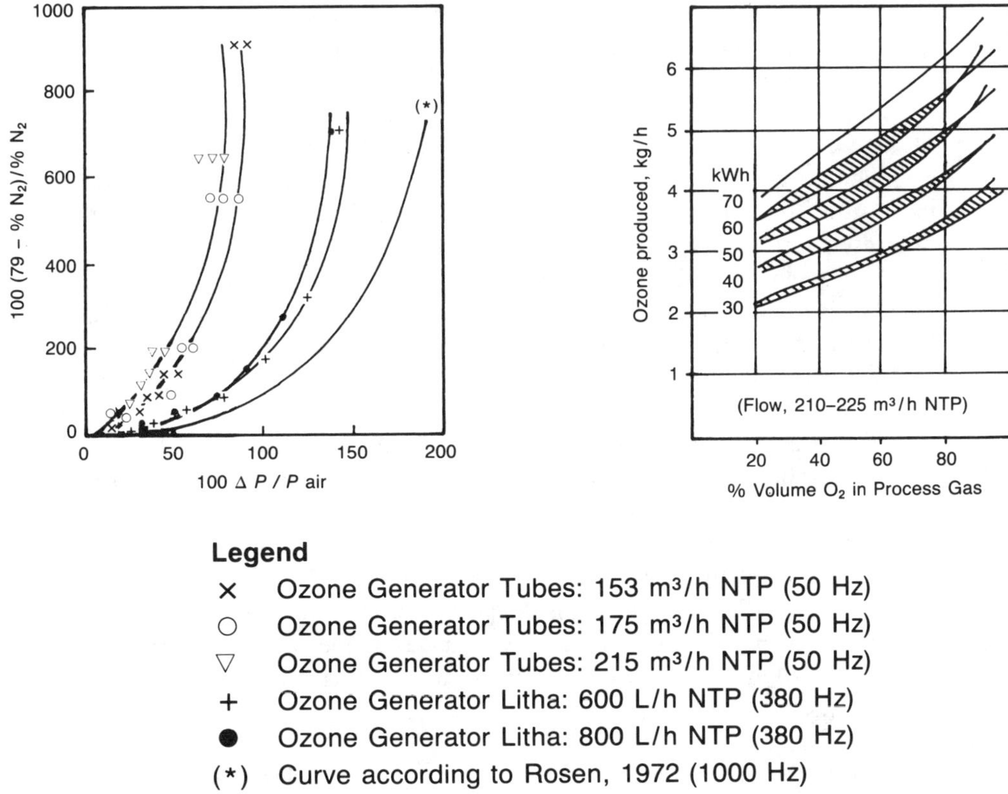

Source: Masschelein (1977).

Figure II–44 Increase in Production Capacity As a Function of Oxygen Content of the Process Gas

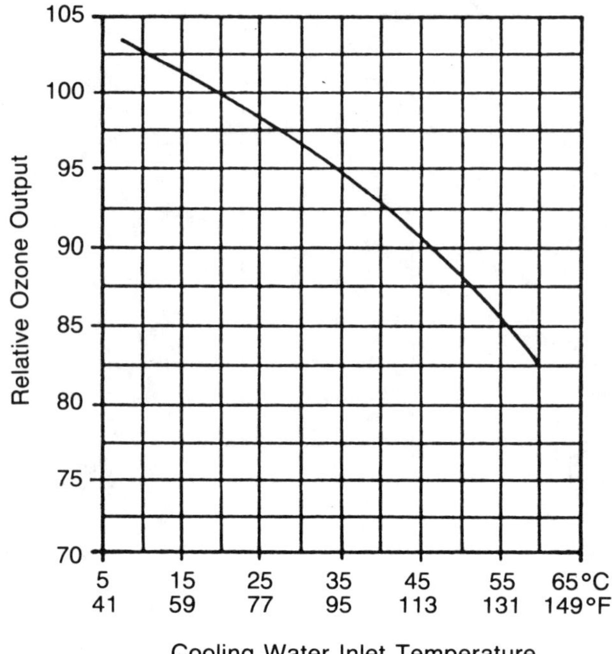

Source: Reprinted with permission from *Envir. Prog.*, 1:2:113–118, Carlins, J.J. © 1982 AIChE.

Figure II–45 Effect of Cooling on Ozone Generation Efficiency

A distinction must be made between the uprating of ozone concentration and ozone quantity produced with oxygen versus air at variable flow of the process gas. This means that the production capacity can be increased by a factor of 1.7 by running a classical air-based design with oxygen, while the production can be decreased by a factor of 2.5 when operating an oxygen-based design with air.

The nominal design capacity at which operation can be performed on a permanent basis must be considered to be at least 20 to 30 percent. Note that the costs at this decreased rate are increased (see Figure II–43).

The yield obtained when using an oxygen-enriched process gas is increased with a smaller gas space and an increased electrical current frequency. Since all variations result in energy loss in the form of heat, cooling of the process gas is very important. Figure II–45 illustrates a typical effect of cooling water temperature on the yield of ozone generation.

The most efficient form of cooling is the "both-side" cooling system, which is a system that has cooling on both the high-voltage side and on the ground side. However, in case of accidental breakage of the dielectric, the cooling liquid (for example, water) enters the discharge gap and causes short-circuiting of the entire system. Therefore, cooling only on the ground side is the safer design.

The space distance in the gas discharge gap is important for cooling. Classical ozone generators using air are designed with a gas space of 2.5–3.2 mm. Ozone generators using oxygen and/or medium–high frequency current are based on gas spaces of 1–1.5 mm thickness, and more heat must be evacuated from a given volume of hardware in the latter case.

Summary. Several parameters must be considered when designing ozone generating systems based on the corona discharge. Parameters related to the process gas include:

- dryness and purity,
- pressure,

- temperature, and
- residence time in the discharge gap space.

Electrical parameters include:

- voltage,
- current intensity,
- current frequency, and
- frequency modulation.

Flexibility in operation includes flow and concentration ranges.

In summary, a corona discharge ozone generator is a chemical reactor that is self equilibrating as a function of external operation parameters.

II.D.3 Alternative Methods

Photochemical ozone generation. The formation of ozone from oxygen exposed to UV light at 140–190 nm was first reported in 1900 (Lenard 1900) and fully assessed in 1903 (Goldstein 1903). It was soon recognized that the active wavelengths for technical generation are below 200 nm, for example, the 185-nm resonance line of mercury. The method has been reviewed more recently in an overview by Du Ron (1982) and in state-of-the-art papers (Dohan and Masschelein 1987).

Considering the energy required for the endothermic formation of ozone from oxygen (that equals 142 kJ/mol), theoretically light of wavelengths shorter than 842 nm could form ozone from oxygen (Du Ron 1982). However, at least ^3P radicals must be formed as intermediates; this means that when starting from the lowest ground state of oxygen $X^3\Sigma g^-$, 5.116 eV/molecule or 493.8 kJ/mol is necessary. This dissociation energy corresponds to a wavelength of 242.4 nm (Dohan and Masschelein 1987).

With the existence of singlet oxygen species in the fundamental state, $A^1\Delta g$ at 94.3 kJ/mol and $B^1\Sigma g^+$ at 157 kJ/mol, respectively, higher energy states than the ground state can be formed by absorption of longer wavelengths. This could decrease the necessary photoquantum energy for the subsequent dissociation of molecular oxygen (see Figure II–38, sec. II.D.1).

The energy at the convergence limit corresponds to the dissociation of molecular oxygen into radicals ^3P and ^1D (+ 684 kJ/mol). The latter can form ozone directly as follows:

$$O\ (^1D) + O_2\ (X^3\Sigma g^-) = O_3\ (^3A)$$

However, the 185-nm wavelength corresponds to 647 kJ/mol, whereby vibrational activation of the O_2 in the ground state remains necessary.

Other reaction pathways are quenching of $O^\star{}_2\ B^3\Sigma u^-$ by ground-state oxygen.

$$\begin{array}{c} O^\star{}_2\Sigma u^- + O_2\Sigma g^- = O_3\ (A^1) + (^3P)\ O \\ 6.175\ \text{eV} + 0 - 1.474\ \text{eV} + 2.558\ \text{eV} \end{array}$$

Predissociation occurs through repulsive $^3\pi u$ states, which possibly dissociate into ^3P radicals. However, quenching of $^3\pi u$ oxygen with $O_2\Sigma u^-$ can be another pathway for ozone generation activated by the 185-nm wavelength.

The quantum yield of ozone formation from oxygen is 2. The absorption index of oxygen at 185 nm is about 0.1/cm atm. Ozone absorbs light at 254 nm in the Hartley-band region, and the absorption index is approximately 135/cm atm. In view of present technologies with mercury-based UV-emission lamps, the 254-nm wavelength is transmitted along with the 185-nm wavelength, and photolysis of ozone is simultaneous with its generation. Moreover, the relative emission intensity is 5 to 10 times higher at 254 nm compared to the 185-nm wavelength.

The photolysis process is a first-order reaction. The relative quantum yield ranges from 4 to 16 in dry oxygen and oxygen saturated with water vapor, respectively (Laforge et al. 1982).

Attempts to reach a suitable photostationary state of ozone formation with mercury lamps have failed (Dohan and Masschelein 1987). The main reason for this failure is that thermal decomposition is concomitant with ozone formation. With annular reactors and mercury lamps, the best results have been obtained at irradiation times lasting from 500 to 800 s. At longer irradiation times, apparent first-order decay of ozone occurs. Increasing the gas pressure increases the yield of ozone formation. To increase the yield, the reactor walls are best cooled to maintain temperatures below 10–15°C (50–59°F).

The practical yields obtained in early experiments were 0.3–0.4 g/kWh (electrical energy), with an occasional yield at 1.8 g/kWh (Lenard 1900; Goldstein 1903). In the present state of reactor development, 16–27 g/kWh can be obtained when using oxygen as the process gas.

Except for small-scale uses or synergic effects, the UV-ozone process (the UV-photochemical generation of ozone) has not reached maturity. Important phases requiring additional development include the development of new lamp technologies with less aging and higher emission intensity at wavelengths lower than 200 nm.

Electrolytic ozone generation. Electrolytic generation of ozone has historical importance because synthetic ozone was first discovered by Schönbein in 1840 by the electrolysis of sulfuric acid (Schönbein 1849). The simplicity of the equipment can make this process attractive for small-scale users or users in remote areas.

Many potential advantages are associated with electrolytic generation, including:

- the use of low-voltage DC current,
- no feed gas preparation,
- reduced equipment size,
- possible generation of ozone at high concentrations, and
- generation in the water, eliminating the ozone-to-water contacting processes.

Problems and drawbacks of the method include:

- corrosion and erosion of the electrodes (Fabjan 1977; Foller and Goodwin 1984),
- thermal overloading due to anodic over-voltage and high current densities,
- need for special electrolytes or water with low conductivity, and
- with the in-site generation process, incrustations and deposits are formed on the electrodes, and production of free chlorine is inherent to the process when chloride ions are present in the water or the electrolyte used.

Significant progress in this technology can be expected (Fabjan 1977) through the development of electrolytic cells with reduced ohmic resistance; the selection of electrolytes with high electrical conductivity to increase the over-voltage of oxygen generation; the development of new electrodes, including membrane technologies (Stucki et al. 1985); and the use of modified current, for example relaxed and superimposed current pulses.

The fundamental principles and reactions involved in the electrochemical generation of ozone are formulated by the following redox equations in "normal" acid solutions (potentials listed in volts, with the standard hydrogen electrode [SHE] as a reference.

$$(1)\quad 2H^+ + 2e = H_2 \qquad E^0 = 0.0$$
$$(2)\quad 2H_2O = O_2 + 4e + 4H^+ \qquad E^0 = +1.23$$
$$(3)\quad Cl^- + H_2O = HOCl + 2e + H^+ \qquad E^0 = +1.49$$
$$(4)\quad 3H_2O = (O_3)(g) + 6e + 6H^+ \qquad E^0 = +1.51$$
$$(5)\quad 2H_2O = H_2O_2 + 2e + 2H^+ \qquad E^0 = +1.77$$

(6) $O_2 + H_2O = (O_3)(aq) + 2e + 2H^+$ $E^0 = +2.07$
(7) $H_2O = OH + e + H^+$ $E^0 = +2.85$

In practice, the following results will be noted:

- Over-voltages are necessary to limit the formation of oxygen, according to reaction 2.
- In the presence of chloride ion, simultaneous formation of chlorine cannot be avoided according to reaction 3. On electrolysis of tap water with an electrical potential in the range of 2–2.5 V and a PbO_2 electrode (1250 cm^2/L), near equal amounts of chlorine and ozone result (Wabner and Grambow 1983). The minimum current density is, in this case, 5 mA/cm^2. The hydroxyl radical is considered to be the intermediate product. It can react in the water to, for example, destroy phenols present in the water when the water is submitted to the electrolytic process. The disinfecting action of the direct electrolytic processes is considered to be due to the secondary formation of chlorine (Borneff 1981).
- The cathodic process can be either the reduction of oxygen (reaction 2) or the evolution of hydrogen (reaction 1). In the latter case, pure water must be supplemented to the cell to compensate for the water loss. The oxygen cathode process also has less thermal electrode stress (Foller and Goodwin 1984).
- A minimum electrical conductivity is necessary if pure water is chosen as the electrolyte. This is obtained, for example, at 35–40 μmho/cm by dissolving CO_2 (approximately 1–1.1 g/L).

The probable pathway of the reaction may involve adsorbed oxygen layers or oxides that split off ozone in consecutive electrochemical steps (Fabjan 1977). Therefore,

$$MO\,(nO)_{(ads)} + H_2O \rightarrow MO\,((n-2)O)_{(ads)} + O_3 + 2H^+ + 2e$$

Costs. The following energy costs are for electrolytic generation of ozone:

Direct hydrolysis:	50–100 kWh/kg ozone (Wabner and Grambow 1983)
H_2SO_4/Pt:	50–100 kWh/kg ozone (Fabjan 1977)
$HClO_4$/Pt:	40–80 kWh/kg ozone (Fabjan 1977)
HBF_4/C:	50–60 kWh/kg ozone (Foller and Goodwin 1984)

The higher energy consumption results when operating at high current density. At present, the energy consumption for electrolytic generation is higher by a factor of 2–5 compared to that for ozone generation by corona discharge.

Radiochemical ozone generation. High-energy irradiation of oxygen by radioactive rays can promote the formation of ozone. The ozone generation yield appears to be independent of the ionization source: γ and β or neutrons emitted by single isotopes. The waste-fission isotope ^{137}Cs is a potential irradiation source that is applicable to ozone generation in water treatment, but ^{60}Co and ^{90}Sr can also be used. In static systems with oxygen at atmospheric pressure and at −78°C (−108°F), 0.5 percent by volume ozone (10 g/m^3 NTP) can be obtained with an energy expense ranging from 10 to 15 kWh/kg. The efficiency of energy conversion is about 30 percent (Kircher et al. 1960; Johnson and Warman 1964).

In static liquid oxygen, formation of highly concentrated ozone can be obtained, for example, 6 percent by weight or 70–75 g/L (liquid), at an energy expense of about 25–30 kWh/kg under laboratory conditions. In flow-through air systems at atmospheric pressure, ozone concentrations formed remain low with ^{60}Co-γ irradiation, for example, 3–4 mg/m^3 NTP (Shah and Maxie 1966).

The best information on the feasibility of cheminuclear ozone generation for water treatment results from the Brookhaven project (Steinberg and Beller 1970).

In a water-cooled closed-loop system in which air is circulated and irradiated, a steady-state ozone concentration is obtained that is almost a direct function of the irradiation intensity. The system is operated under pressure up to 60 bar (870 psi). Cooling (for example, at −20°C [−4°F]) improves the yield in ozone. A few percent of nitrogen injected in the circuit does enhance the yield. Concentrations as high as 500 ppm by weight or about 6 g/L (liquid) are obtainable at an overall thermal yield of 10–12 percent. Major problems with the process are the difficulty of removing fission products and the difficulty of conditioning a suitable gas stream containing the ozone, making the technology complex.

The main nonvolatile reaction products are Cs, Ba, and Sr isotopes, which must be precipitated and removed by filtration. Gaseous reaction products are Xe, Kr, and $C^{14}O_2$. Differential dissolution in freon 11 and 12 has been proposed for absorption stripping (Shah and Maxie 1966). The solubility of ozone in freon is between that of xenon and krypton and requires a series of three contact towers: one for absorption in freon, one for krypton stripping, and one for ozone stripping. Freon is recovered by distillation. An oxygen-separation plant to produce the process gas from air completes the scheme.

Even with the favorable thermodynamic yield of the process and the interesting use of waste fission isotopes, the cheminuclear ozone generation process has not yet become a significant application in water or wastewater treatment. This fact is due to its complicated process requirements.

II.E OZONE GAS TRANSFER

II.E.1 Solubility of Ozone in Water

The dissolved concentration of ozone in water at saturation, C_s, can be expressed either in terms of the solubility S or as an absorption coefficient β. S is defined by the volume of gas dissolved per unit volume of liquid at the temperature and pressure under consideration and in the presence of the equilibrating gas at 1 atm pressure. An alternative expression of S is the so-called solubility ratio, which expresses the relative concentration as milligrams per liter ozone in water to milligrams per liter ozone in the gas. Often called the Bunsen absorption coefficient, β is the volume of gas expressed at NTP, which is dissolved at equilibrium by a unit volume of liquid at a given temperature when the partial pressure of the gas is the unit atmosphere. This is equal to the pressure of the gas itself minus the vapor tension of the liquid (Table II–27).

It is generally agreed in the literature that when ozone is dissolved in water Henry's law is obeyed. This means that the C_s values are proportional to the partial pressure of ozone P_γ at a given temperature.

The expression for the saturation concentration of a dissolved gas under thermodynamic ideal conditions is as follows:

$$C_s\ (\mathrm{kg/m^3}) = \beta\, M \times P_\gamma$$

Table II–27 Solubilities of Gases Associated With Ozonation (at NTP)

Gas Solubility	Ozone	Oxygen	Nitrogen	Carbon Dioxide	Chlorine	Chlorine Dioxide
β, v/v	0.64	0.049	0.0235	1.71	4.54	±60
$\beta\ (O_3)/\beta$ gas	1	13.3	27.7	0.38	0.14	±0.01
C_s, mg/L for $P_\gamma = 1$	1,400	70	30	3,360	14,400	180,000

Source: Masschelein (1982a)

Table II–28 Selected Solubility Data of Ozone as a Function of Water Temperature†

Temperature		Caprio et al. (1982) ■	Khadraoui (1988) (bidistilled) ●	Khadraoui (1988) (HCO_3^- : 200 mg/L) ★	Rawson (1953) ▲	Matrozov et al. (1975) ○
°C	°F					
0.5	32.9	0.54	—	—	—	—
5.6	42.1	—	—	0.39	—	0.41
6.9	44.4	—	0.46	—	—	—
9.6	49.3	—	0.41	—	0.39	—
12	53.6	0.34	—	—	—	—
12.4	54.3	—	0.31	—	—	—
14.5	58.1	—	—	—	0.29	—
14.7	58.5	—	—	0.26	—	0.28
15.9	60.6	—	0.25	—	—	—
20.3	68.5	—	—	—	0.21	—
20.7	69.3	—	0.20	—	—	—
21	69.8	0.24	—	—	—	—
23.6	74.5	—	0.18	—	—	—
25.5	77.9	—	—	0.14	0.17	0.16
30.5	86.9	—	—	0.10	0.14	—
31	87.8	0.17	—	—	—	0.14
33.5	92.3	—	0.10	—	—	—
34.8	94.6	—	—	0.07	—	—
35	95.0	—	—	—	0.12	—
39	102.2	—	—	—	0.07	—
41	105.8	0.12	—	—	—	0.17
43	109.4	—	0.07	—	—	—

†Solubility ratios given as S: mg/L in water to mg/L in gas. Symbols correspond to those diagrammed in Figure II-46.

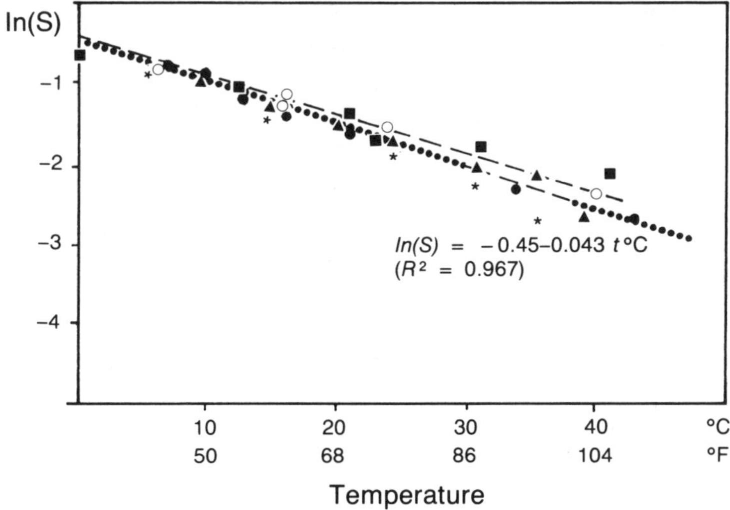

Figure II–46 Dissolution Energy of Ozone in Water

Where:

M = mass volume of the gas, kg gas/m^3 gas NTP (2.14 for ozone)
P_γ = partial pressure of the gas in the given gas phase

Table II–29 Henry-Dalton Constants for Ozone Residuals in Water (Concentration Range of 0.1–0.65 mg/L)

	Temperature, °C (°F)							
	0 (32)	5 (41)	10 (50)	15 (58)	20 (68)	25 (77)	30 (86)	35 (95)
S (water/gas)	0.64	0.50	0.39	0.31	0.24	0.19	0.15	0.12
K_H*	34,990	45,600	59,450	76,180	100,000	128,600	164,800	210,400
H_a	1,945	2,490	3,190	4,200	5,190	6,555	8,302	10,375

*$K_H = (n_i/n_T) \times (n_T \times R \times T)/(V_T \times C_L)$; $H_a = P_\gamma/x$

Where:

K_H = Henry's constant, atm · m^3/mol
H_a = apparent Henry's constant, atm/O_3 molar fraction in the liquid
n_i, n_T = fractional and total number of moles in the gas phase, respectively
R = universal gas constant, 82 atm · m^3/mol, K
T = equilibrium temperature, K
V_T = total volume of the gas phase, m^3
C_L = equilibrium liquid-phase concentration, mol/m^3
P_γ = partial pressure of ozone in the gas phase, atm
x = molar fraction of ozone in the liquid phase.

Influence of temperature. In addition to physical dissolution, several side effects may affect the practical equilibrium. For example, pH affects the decomposition rates of dissolved ozone, and dissolved products can react with ozone during the measurements (see secs. II.A, II.C). The most important parameter is probably the effect of water temperature; however, some controversy does exist on this point (Morris 1988) (Table II–28).

Most data taken from older literature are of questionable accuracy, since less reliable analytical methods, even those not reported, have been used. Sampling is often an uncontrolled source of error of measurement (see sec. II.C).

The data presented in Figure II–46 provide a means to estimate thermodynamic characteristics of ozone dissolution (S is measured in mg/L water to mg/L gas). The following equation can be derived from this relationship:

$$d\,(\ln S)/dT = \Delta E/RT^2$$

where ΔE is the molar energy of dissolution, and

$$\Delta E/RT^2 = -0.048$$

and, at 20°C (68°F), $\Delta E_{293} = (-0.048) \times (8.31) \times (293)^2 = -34.2$ kJ/mol or an enthalpy change of exothermic dissolution as follows:

$$\Delta H_{293} = \Delta E + RT = -34.2 + \frac{293 \times 8.31}{1000} = -31.8 \text{ kJ/mol}$$

Equivalent data are obtained with liquid film contactors and a semibatch contact column as described in the literature (Masschelein and Goossens 1984). Henry-Dalton constants can be computed from the solubility ratios reported in Figure II–46; values are given in Table II–29.

Analysis is best made *immediately* after sampling. If photometric methods are used for the gas phase, appropriate corrections for gas temperature must be made. Corrections may be made according to IOA standards (European IOA Standardisation Committee 1989).

From the data in Table II–29, the following equation can be computed for the apparent Henry's constant:

$$\ln H_a = 22.3 - 4030/T$$

This equation expresses the higher solubility of ozone in water at lower temperatures.

Influence of pH and ionic strength. In addition to temperature, several parameters can influence the Henry's constants. The most relevant are pH and ionic strength.

pH. Data from the literature (Roth and Sullivan 1981) lead to the following equation:

$$H_a = 3.84 \times 10^{-7} \times [OH^-]^{0.035} \exp(-2428/T)$$

which covers the temperature range of 3.5–60°C (38.3–140°F) and pH range of 0.65–10.2.

More recently, the following two correlations have been published (Ouederni et al. 1987) for pH 2 and pH 7 and for a range in temperature of 20–50°C (68–122°F):

$$\text{pH 2:} \quad \ln H_a = 20.7 - 3547/T$$
$$\text{pH 7:} \quad \ln H_a = 18.1 - 2876/T$$

The scatter in the experimental data is considerable. More investigation is still necessary since the stability of ozone in water at higher pH values is lowered, thus causing interference with the solubility of molecular ozone.

Ionic strength. An empirical relation to account for ionic strength on ozone partition between gas and water has been proposed as

$$\ln H = 12.19 - 2297/T + 2.659 \times I - 688\ (I/T)$$

Where:

H = is expressed in atm/L/mol
I = the ionic strength, mol/L
T = temperature, K

In medium-strength mineralization of drinking water (less than 1000 total dissolved solids [TDS] in mg/L) the effect of ionic strength on ozone solubility appears to remain marginal.

Effects of thermal decomposition in the gas phase. In experiments with semibatch contact columns operated to reach the equilibrium of ozone dissolution, ozone consumption or decomposition can be observed in the gas phase even when the dissolved ozone concentration remains stable (Matrozov et al. 1975). The drop in ozone concentration in the gas phase at the exit of a contacting column increases with increasing temperature.

The thermal decomposition of ozone contained in dry oxygen at atmospheric pressure and without any catalyst or activator initiating the formation of radical oxygen is a second-order reaction of the form

$$O_3 + O_2 \rightarrow O_2 + O_2 + O$$

The kinetic constant associated with the reaction is described as:

$$k_2 = 2.02 \times 10^{-2} \exp(24{,}000/RT)$$

and is expressed in $M^{-1}s^{-1}$.

Table II–30 Data on Decomposition of Ozone in Water

Order	Concentration Range, mg/L	Temperature, °C (°F)	pH	Ions	E_{act}, kJ/mol	Reference
$\|O_3\|^1$	0.8–1.8	3.5–60 (38.3–140)	0.5	NaOH		Gordon 1987
	Batch reactor 0.01–2.4	3.5–60 (38.3–140)	0.5–10.2			Roth and Sullivan 1983
	Batch reactor		9.6–11.9			Staehelin and Hoigné 1982
	flow reactor 0.5–5	25 (77)	10–13	NaOH		Rogozhkin 1970
	satur. vs. 0.5 g O_3/L	0.27 (32.5)	12–13		41	Czapski et al. 1968
						Alder and Hill 1950
	oxygen 0.5–0.6 (+02)	20–40 (68–104)	5–6	—	62	*
$\|O_3\|^2$	35–75	0 (32)	2–4			Rothmund and Burgstaller 1913
	5–40	20 (68)	2–10	Acids		Gurol and Singer 1982
$\|O_3\|^{1.5}$	14	25 (77)	0–7	$HClO_4$		Kilpatrick et al. 1956
	10	25 (77)	2.1–10.2			Li 1977
$\|O_3\|^0$						Sheffer and Esterson 1982
$\|O_3\|^1\|OH\|^1$	0.7–6	20 (68)	11–13	CO_3^{2-}/HCO_3^-		Forni et al. 1982
	1.2 (column)	18–27 (64.4–80.6)	8.5–13.5	CO_3^{2-}/HCO_3^-	96	Rizzuti et al. 1976
$\|O_3\|^1\|OH\|^{0.75}$	0.06–1.5	1–20 (33.8–68)	7.6–10	CO_3^{2-}	112	Stumm 1954
$\|O_3\|^1+\|O_3\|^{1.5} \times \|OH\|^{.5}$	Satur. vs. 86 g/L O_2	10–40 (50–104)	2.5–9	(PO_4)	41.3	Sotelo et al. 1987
$\|O_3\|^1\|OH\|^1 + \|O_3\|^2\ \|OH\|^1$	Stop-flow		12			Tomiyasu et al. 1985

*Preliminary data from Brussels' Waterboard.

For low ozone concentrations in pure oxygen a pseudo first-order constant results: $k_1 \sim 9 \times 10^{-14}s^{-1}$. The reaction is strongly temperature dependent.

When the formation of radical atomic oxygen is obtained by an activation source, the decomposition of ozone according to the reaction $O_3 + O \rightarrow 2O_2$ is very fast ($k_2 = 3.37 \times 10^{10} \exp(5700/RT)\ M^{-1}s^{-1}$).

Effect of ozone decomposition in water on gas transfer. The data from literature on ozone decomposition in water are summarized in Table II–30. Under conditions representative of water treatment, that is, for low residual concentrations, the auto-decomposition of dissolved ozone is an apparent first-order reaction with respect to ozone. The hydroxyl ion concentration influences the decomposition rate in an approximately first-order relationship too.

The activation energy in such representative conditions ranges from 100 to 40 kJ/mol and is higher at lower pH values. A classical half-life time (pH 7–8) is from 20 to 30 min for relatively pure waters. The mechanism involved in ozone decomposition in water is discussed in more detail in secs. II.A and II.C.

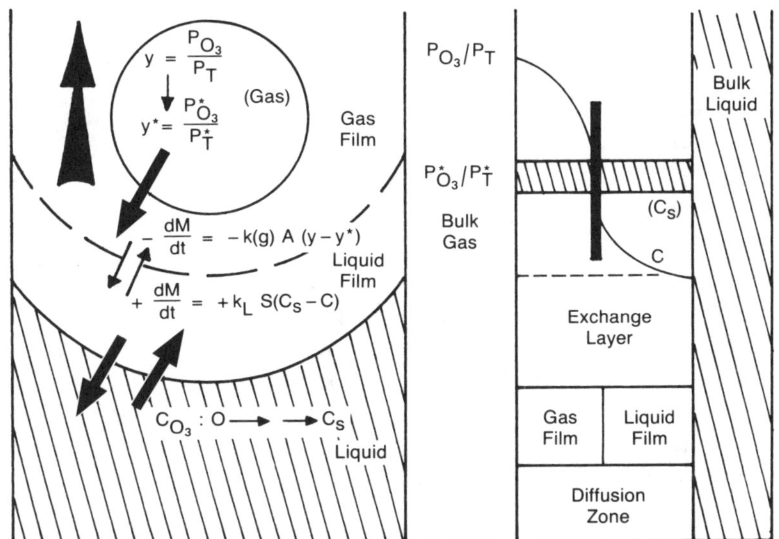

Source: Masschelein (1982a)

Figure II–47 Schematic of Double-Film Transfer

II.E.2 Contacting of Ozone With Water

Diffusion of ozone in water obeys Fick's law for molecular diffusion. At 20°C (68°F), the diffusion constant is $D_{O_3} = 1.74 \times 10^{-9}$ m^2/s, and can be corrected for different physical conditions according to the Nernst–Einstein relationship, which states $[(D \times \mu)/T]$ = constant, where μ is the dynamic viscosity of water and T is the absolute temperature. Consequently, ozone transport is much faster in the gas phase than in the liquid phase; the latter determines the overall transport rate.

Transfer of ozone to water without chemical reaction. The transfer of ozone to water without reaction is currently accepted as occurring according to the double-film model (Figure II–47). The driving force is $(C_s - C)$.

$$(dC/dt) = +k_L S\ (C_s - C);\ \ln((C_s - C)/C_s) = k_L St$$

In this equation, S is the specific exchange surface in the liquid film and depends on practical conditions such as agitation, pressure, and total gas and liquid volumes.

On the gas side, the concentration of ozone is expressed in terms of partial pressure:

$$y = (P(_{O_3})/P_{total})$$

Part of y is transferred to the liquid phase to build up C. If y★ is the part transferred, then:

$$y^\star = (P^\star(_{O_3})/P_{total})$$

and $(y - y^\star)$ is the driving force in the gas phase for ozone transfer into the liquid. The practical transfer rate is given by:

$$(-\ dM/dt) = k(g)\ S(g)\ (y - y^\star)$$

The equilibrium condition, that is, what the gas releases and the liquid receives, is given by:

$$(+\ dM/dt) = +\ k_L S\ (C_s - C) = N_t$$

(the amount transferred per unit time into a unit volume).

In a gas–liquid system it is often difficult to determine the specific exchange area S; as a result, estimations for k_L are variable. By considering the total volume of a reactor V, the result is $d(N_t) = k_L S\ (C_s - C) \times dV$ for a small volume portion of the reactor. For a total reactor, several approximations are necessary, depending on the type of reactor (Roustan and Mallevialle 1982).

Numerous empirical, sometimes very complex, equations have been formulated to evaluate the gas transfer constant as a function of different operational parameters. Only a simplified model can be given here, since few approaches concern the transfer of ozone. The most reliable value of k_L for ozone is in the order $2 - 3 \times 10^{-3}$ m/s, which is about 2.5 times lower than for oxygen (Mallevialle et al. 1975).

In a static water column through which ozonated air is bubbled, the following simple equation applies (Masschelein 1982a):

$$k_L = 1.13 \sqrt{D_{O_3} \times U_{SG}/d_B}$$

Where:

U_{SG} = the superficial gas velocity, m/s
d_B = the diameter of the bubble, m

Bubble column reactors. The superficial gas velocity is equal to the gas flow per cross section of the reactor as follows:

$$U_{SG} = Q_G/A$$

The approximation of Hughmark can be applied to bubbles that are free moving and not coalescent.

$$\epsilon_G = U_{SG}/(0.3 + 2U_{SG})$$

where ϵ_G is the volumic fraction of gas in the reactor.

Values of U_{SG} can be estimated by the relation:

$$U_{SG} = (T \times H)/(C_g \times D_t)$$

Where:

T = treatment doses, g/m^3
H = column height, m
C_g = ozone concentration of gas, g/m^3 NTP
D_t = residence time of water in reactor, s.

Various relationships of the type shown below have been developed for application under several operating conditions and for practical reactor designs.

$$k_L S = \alpha U_{SG}^{\gamma}$$
$$k_L S\ (s^{-1}) = 2.32 \times 10^{-4}\ U_{SG}^{0.82}\ (5°C)\ \text{(Roustan et al. 1987)}$$
$$k_L S\ (s^{-1}) = 3.26 \times 10^{-4}\ U_{SG}^{0.95}\ (12°C)\ \text{(Roustan et al. 1989)}$$
$$k_L S\ (s^{-1}) = 7.91 \times 10^{-4}\ U_{SG}^{0.54}\ (20°C)\ \text{(Laplanche et al. 1989)}$$

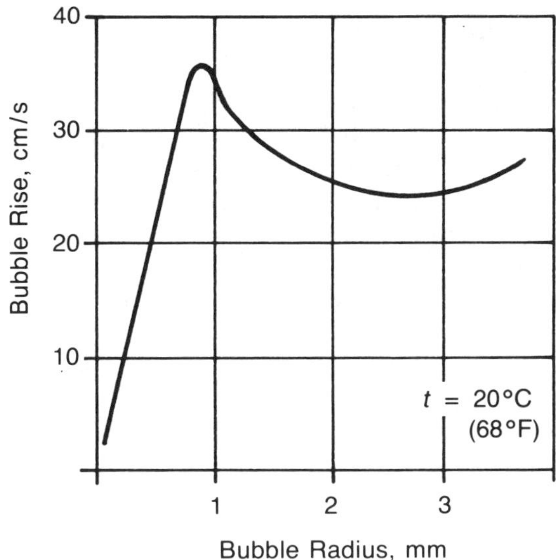

Source: Masschelein (1982a).

Figure II–48 Bubble Rise Velocity as a Function of Dimensions

The free bubble rise velocity also depends on the bubble size (Figure II–48).

Mechanically stirred reactors. Only few equations relate to the transfer of ozone in stirred reactors.

$$k_LS = 2.57 \times 10^{-2} \times (Q_G \times N)^{0.67} \text{ (Ouederni et al. 1987)}$$

Q_G is expressed in 10^{-3} m^3/s, and N is the number of rotations per second. Values of k_LS (s^{-1}) appear to be about four times higher in stirred vessels than in simple bubble-rising systems. More work is still required in this field.

Another relation to consider is the power input per volume, as displayed in the following equation:

$$k_LS = F \times (W/V)^{1/2} \times (Q_G/V)^{1/2}$$

F is an experimental factor, depending on the ozone concentration, and is in the order of 0.02 under the most usual conditions (Mallevialle et al. 1975).

Packed towers. The equation of Onda et al. (1968) and Puranik et al. (1974) is often considered to be most reliable for predicting gas transfer in packed towers.

$$k_L/(a_pD_{O_3}) = 5.1 \times 10^{-3} \times (a_pd_p)^{0.4} \times (U_{SL}\rho_L/a_p\mu_L)^{4/3} \times (U_{SL}a_p/g)^{-1/3} \times (\mu_L/\rho_L D_{O_3})^{1/2}$$

Where:

a_p = the theoretical specific surface of the packing material
d_p = the nominal diameter of the packing material
ρ = the specific weight of the liquid, kg/m^3
μ_L = the dynamic viscosity of the liquid, $kg\ m^{-1}\ s^{-1}$
U_{SL} = the superficial liquid velocity, m/s.

The wet exchange surface of packed columns is increased by at least a factor of 3 compared to bubble columns.

Alternative theories. In Higbie's (1935) penetration theory, it is supposed that the transfer is obtained by renewal of the liquid in the vicinity of the gas phase, that is, a transitional regime of mass transfer. In Danckwerts' (1970) theory of renewal of the interface, a variable stay of the renewed exchange liquid is considered.

These theories present considerable mathematical difficulties when considering the practical conditions of ozonation. Therefore, they are generally not applied to the field, although they could be used as the basis when fast reactions in the liquid phase are in competition with the ozone transfer.

Absorption with chemical reaction. If the ozone transferred to the liquid is consumed by a chemical reaction, the specific transfer coefficient k_L is no longer influenced by only the diffusivity, since a significant part of the ozone dissolved in the liquid phase is exhausted continuously. Therefore,

$$k_L\ (R) > k_L \text{ and } (k_L(R)/k_L) = B$$

where k_L (R) is the transfer coefficient in the presence of chemical reactions. An approximation for B is given by Danckwerts (1970).

$$B = [1 + (D_{O_3}\, k_1)/k_L]^{1/2}$$

B is an acceleration coefficient often called enhancement factor (E), for ozone transfer, while k_1 is the first-order rate constant of the oxidation.

In the practical conditions of postozonation of drinking water, the enhancement factor is always about unity.

If the reaction is very fast, for example, oxidation of a solution of iodide ion at $k_1 \sim 10^4\ s^{-1}$, the oxidation takes place only at the bubble surface and no ozone is transferred into the bulk of the liquid phase ($B \sim 2.3$).

For a k_1 value of $10^2\ s^{-1}$, which is in the range of easily oxidized organic compounds of concentrations of 0.1–0.2 M, B is still about 1.2. For $k_1 = 1\ s^{-1}$ and slower, the direct effect of reacting dissolved compounds on the gas transfer can be neglected and the reaction is that of predissolved ozone. Also occurring in this range and at the pH values for drinking water, the oxidation by OH radicals enters into competition with direct oxidation by ozone (Hoigné 1988).

Competitive inhibition effects in ozone-transfer-controlled reactions. In ozone "gas-transfer controlled reaction rates," the reaction kinetics observed, for example, those in a bubble column, are often of apparent zero order. This has been studied in more detail for nitrophenols oxidized with low ozone concentrations, simulating the conditions of preozonation (Masschelein and Goossens 1984). Compounds that do not react with ozone in similar conditions (for example, those with low pH values) can interfere by competitive inhibition mechanisms. Typical inhibitions studied are acetic acid, ethanol, and ammonium chloride, all of which slow down the oxidation of picric acid in a semibatch reactor (Masschelein et al. 1987).

Specific energy of ozone contacting. The power for ozone dissolution strongly depends on local conditions and plant hydraulics. Numerous systems have been used. Some general observations of these systems are summarized below (see sec. IV.D and sec. V.B.2).

- Dispersion with porous disks or rods (eventually equipped with packing material)—the system is static, but can be subject to clogging. Plug-flow patterns should be approached in the detention zones. Average specific energy equals 2–3 Wh/g O_3.
- Static mixing—has the advantage of static systems, but ozone losses occur. Average specific energy equals 4–5 Wh/g O_3.

- Direct injection in a pipe—is not expensive, but subject to bumping and eventually to corrosion. Dissolution of accompanying gases such as O_2 or N_2 can be at oversaturation, and subsequent degassing can be a problem. Average specific energy equals 0–5 Wh/g O_3.
- Slow downflow injection shaft—has no moving parts, can be subject to channeling, and needs a given water-to-gas flow ratio. Average specific energy equals 4 –5 Wh/g O_3.
- Eductor with total water flow—shows high turbulence and good mixing, but has risk of clogging. Average specific energy equals 4–5 Wh/g O_3, but can range from 10 to 45 Wh/g O_3 in high counter-pressure systems.
- Recirculating propeller or turbine—shows good mixing and high bubble contacting rate, which approaches the theoretical contact time, but is based on moving construction parts. Average specific energy equals 5–7 Wh/g O_3.
- Packed or plate columns (substream saturation)—have a good exchange rate and high ozonation yields, but are subject to clogging and are water-pressure dependent. Average specific energy equals 15–40 Wh/g O_3.

Summary. Ozone-to-water contacting is important because of the relatively low partial pressure of ozone in the process gas. The dissolution of ozone in water is slightly exothermic, and a large variety of gas transfer systems are applicable.

References

AFNOR Normalisation Française. 1987. Detection in an Aquatic Medium of the Genotoxicity of a Substance to Larvae of Batracians. *Fascicule AFNOR*, T 90–325.

L'Air Liquide, *French Patent*, FR. 1,246,273.

Alder, M.G. & Hill, G.R. 1950. The Kinetics and Mechanism of Hydroxide Ion Catalyzed Ozone Decomposition in Aqueous Solution. *Jour. Amer. Chem. Soc.*, 72:1884.

Ames, B.N. et al. 1985. Methods for Directing Carcinogens and Mutagens With the *Salmonella*/ Mammalian-Microsome Mutagenicity Test. *Mut. Res.*, 31:347–363.

Anbar, M. & Neta, P. 1967. A Compilation of Specific Bimolecular Rate Constants for the Reactions of Hydrated Electrons, Hydrogen Atoms, and Hydroxyl Radicals With Inorganic and Organic Compounds in Aqueous Solutions. *Intl. Jour. Appl. Radiation & Isotopes*, 18:493.

Anderson, F.G.L.Q. 1977. Ozonation of High Levels of Phenol in Water. *AIChE Symp. Secr.*, 73:166–265.

Anderson, L.J. et al. 1986. Extent of Ozone's Reaction with Isolated Aquatic Fulvic Acid. *Envir. Sci. Technol.*, 20:7:739.

Andrews, T. & Tait, P.G. 1856. On the Volumetric Relations of Ozone, and the Action of the Electrical Discharge on Oxygen and Other Gases. *Philosophical Transactions*, 113.

Anselme, C. et al. 1988. Development of an Integrated Analytical Approach for Evaluating Toxic Compounds Removal From Groundwater Treated by Ozone Hydrogen Peroxide. Proc. AWWA Wtr. Qual. Technol. Conf., St. Louis, Mo.

Backlund, P. et al. 1985. Mutagenic Activity in Humic Water and Alum Flocculated Humic Water Treated With Alternative Disinfectants. *Sci. Total Envir.*, 47:257.

_____. 1988. Formation of Ames Mutagenicity and of the Strong Bacterial Mutagen 3-chloro-4-(dichloromethyl)-5-hydroxy-2(5H)-furanone and Other Halogenated Compounds During Disinfection of Humic Water. *Chemosphere*, 17: 1329–36.

Bader, H. & Hoigné, J. 1979. Analysis of Ozone in Water and Wastewater by an Indigo Method. 4th Ozone World Congress, IOA, Houston, Texas.

_____. 1981. Determination of Ozone in Water by the Indigo Method. *Wtr. Res.*, 15:449.

_____. 1982a. Determination of Ozone in Water by the Indigo Method: A Submitted Standard Method. *Ozone Sci. Engrg.*, 4:169.

_____. 1982b. Colorimetric Method for the Measurement of Aqueous Ozone Based on the Decolorization of Indigo Derivatives. In *Ozonization Manual for Water and Wastewater Treatment.* Edited by W.J. Masschelein. John Wiley & Sons, Inc., New York.

Bailey, P.S. 1978. *Ozonation in Organic Chemistry*. Academic Press, Inc., New York.

Benoît-Guyot, J. et al . 1984. Chlorophenols: Degradation and Toxicity. *Jour. Français d'Hydrologie*, 15:3:249.

Block, J.C. 1982. Mécanismes d'Inactivation des Microorganismes par les Oxydants. *TSM l'Eau*, 77:11:521.

Bonnet, M.L. 1988. Contribution à l'Etude de la Valorisation des Déchets Lignocellulosiques. Thèse de l'Université de Poiters, France.

_____. 1989. Etude des Sous-Produits d'Ozonation de la Lignine et des Glucides en Milieu Aqueux. *Envir. Technol. Letters*, 10:577–590.

Borneff, M. 1981. Anodische Oxydation zur Trinkwasserdesinfection. I. Mitteilung: Bakteriologische Untersuchungen bei Praxiseinsatz. *GWF.*, 128:4:141.

Bourbigot, M.M. et al. 1986. Role of Ozone and Granular Activated Carbon in the Removal of Mutagenic Compounds. *Envir. Health Perspec.*, 69:159–163.

Brauch, H.J. & Kuehm, W. 1988. Organic Micropollutants in the River Rhine and in Drinking Water Treatment. *Wasser/Abwasser*, 129: 3:186.

Braun, A.M. et al. 1986. *Technologie Photochimique.* Presses Polytechniques Romandes, Lausanne, Switzerland.

Brette, B. et al. 1987. Influence de la Polymérisation des Matières Organiques sur l'Efficacité de la Coagulation. Proc. 67th Congrès AGHTM, Nice, France.

Briner, E. & Perrottet, E. 1939. Détermination des Solubilités de l'Ozone dans l'Eau et dans une Solution Aqueuse de Chlorure de Sodium; Calcul des Solubilité de l'Ozone Atmosphérique dans les Eaux. *Helv. Chim. Acta*, 22: 397.

Briner, E. & Yalda, A. 1941. La Production de l'Ozone par Electrolyse et la Surtension Anodique. *Helv. Chim. Acta*, 24:1328.

Brown, D.W. & Wall, L.A. 1964. Irradiation of Liquid and Solid Oxygen. *Jour. Phys. Chem.*, 65:915.

Brunet, R. 1981. Evolution Chimique de la Micropollution Organique en Cours d'Ozonation d'Eaux de Surface Filtrées. Thèse de 3ème cycle, Université de Poitiers, France.

_____. 1985. Rapport d'Etudes: Oxydants Combinés. Convention de Recherche Compagnie Générale des Eaux–Laboratoire de Chimie de l'Eau et des Nuisances, Université de Poiters, France.

Buckley, R.D. et al. 1975. Ozone and Human Blood. *Arch. Envir. Health*, 30:40–43.

Burleson, G.R. & Chambers, T.M. 1982. Effect of Ozonation on the Mutagenicity of Carcinogens in Aqueous Solution. *Envir. Mutagen.*, 4:469–476.

Burleson, G.R. et al. 1979. Ozonation of Mutagenic and Carcinogenic Polyaromatic Amines and Polyaromatic Hydrocarbons. *Wtr. Cancer Res.*, 39:6 (Pt.1):2149.

Caprio, V. & Lignola, P.G. 1980. Gas Phase Ozone Evaluation by Thermal Decomposition Technique. *Anal. Chem.*, 52:1123.

Caprio, V. et al. 1982. A New Attempt for the Evaluation of the Absorption Constant of Ozone in Water. *Chem. Engrg. Sci.*, 37:1:122.

Carlins, J.J. 1982. Engineering Aspects of Ozone Generation. *Envir. Progress*, 1:113.

Carmichael, N.G. et al. 1982. The Health Implications of Water Treatment With Ozone. *Life Sciences*, 30:117–129.

Chambers, T.M. & Burleson, G.R. 1982. pH-Sensitive Mutagenic Activity of Ozone-Treated 1,2-Dimethylhydrazine in the *Salmonella/* Microsome Assay. *Mut. Res.*, 94:23–29.

Chedal, J. 1976. Effect of Ozone on Micropollutants. Proc. Forum Ozone Disinfection, Intl. Ozone Inst., 178.

Cobine, J.D. 1958. *Gaseous Conductors.* Dover Publications, New York.

Condie, L. et al. 1985. Subchronic Toxicology of Humic Acid Following Chlorination in the Rat. *Jour. Toxicol. & Envir. Health*, 15:305–314.

Cotruvo, J.A. & Vogt, C.D. 1987. Treatment Processes for Control of SOCs. Status of Drinking Water Regulations. Communication à une Journée à la Société Lyonnaise des Eaux, France.

Cotruvo, J.A. et al. 1977. Investigation of Mutagenic Effects of Products of Ozonation Reactions in Water. *Ann. N.Y. Acad. Sci.*, 298:124.

Craun, G.F. 1988. Surface Water Supplies and Health. *Jour. AWWA*, 80:40–52.

Crecelius, E.A. 1978. The Production of Bromine and Bromate in Sea Water by Ozonation. *Ozonews*, 5, part 2,1.

Cromwell, W.E. & Manley, T.C. 1959. Effect of Gaseous Diluents on Energy Yield of Ozone Generation from Oxygen. *Ozone Chem. Technol.* (Advances in Chemistry Series), 21:304.

Croué, J.P. 1987. Contribution à l'Etude de l'Oxydation par le Chlore et l'Ozone d'Acides Fulviques Naturels Extraits d'Eaux de Surface. Thèse de l'Université de Poitiers, France.

Dalouche, A. 1981. Degradation de Composés Organiques Malodorants par Traitement Biologique et Ozonation. Thèse Rennes, no. 696, France.

Danckwerts, P.V. 1970. *Gas-Liquid Reactions.* MacGraw-Hill Publishers, New York.

Daniel, F.B., et al. 1989. Subchronic Toxicity of Ozonated/Chlorinated Humics in Sprague-Dawley Rats. *Envir. Toxicol. Chem.* (submitted).

Decoret, C. et al. 1984. Experimental and Theoretical Studies of the Mechanism of the Initial Attack of Ozone on Some Aromatics in Aqueous Medium. *Envir. Technol. Letters*, 5:5: 207.

De Laat, J. 1988. Contribution à l'Etude du Mode d'Elimination de Molécules Organiques Modèles sur Charbon Actif en Grains: Interactions entre les Processus d'Adsorption et de Biodégradation. Doctorat d'Etat, Université de Poitiers, France.

Dieter, V.D.M. et al. 1988. Dioxins: Treatment with Ozone. Proc. Indl. Waste Conf., 499.

Dohan, J.M. & Masschelein, W.J. 1987. The Photochemical Generation of Ozone: Present State of Art of Technology. *Ozone Sci. Engrg.*, 9:315.

Doré, M. 1989. Chimie des Oxydants et Traitement des Eaux. Technique et Documentation–Lavoisier, Paris.

Doré, M. & Legube, B. 1983. Mechanism of the Ozonation of Monocyclic Aromatic Compounds. *Jour. Français d'Hydrologie*, 14:1:11.

Doré, M. et al. 1978. Ozonation of Phenols and Phenoxyacetic Acids. *Wtr. Res.*, 12:6:413.

_____. 1980a. Mecanisme de l'Ozonation des Herbicides Derives de l'Acide Phénoxyacetique 2,40 at M.C.P.A. *Wtr. Res.*, 14:767.

_____. 1980b. Mechanism of the Reaction of Ozone with Soluble Aromatic Pollutants. *Ozone Sci. Engrg.*, 2:1:39.

_____. 1987. The Role of Alkalinity in the Effectiveness of Processes of Oxidation by Ozone. Proc. 2nd Intl. Conf., The Role of Ozone in Water and Wastewater Treatment. (D.W. Smith and G.R. Finch, editors). Teletran International Ltd., Kitchener, Ontario.

_____. 1989. Ozonation of Molecules Constituting Cellular Matter. Proc. 9th Ozone World Congress, IOA, New York.

Dreher, W. & Klamberg, H. 1988. Degradation of Polycyclic Hydrocarbons. III. Reaction Products of the Ozonolysis of Additional Polycyclic Aromatic Hydrocarbons in Aqueous Systems. *Fresenius Z. Anal. Chem.*, 331:3/4:290.

Dubeau, H. & Chung, Y.S. 1982. Ozone Response in Wild Type and Radiation-Sensitive Mutants of *Saccharomyces cerevisiae*. *Gen. Genetics*, 126:393–398.

Duguet, J.P. et al. 1982. Etude Expérimentale de l'Action de l'Ozone sur les Acides Aminés. *Cebedeau*, 35:71.

_____. 1983. Monitoring an Ozone Process Through UV Measurements. 6th Ozone World Congress, IOA, Washington, D.C.

_____. 1985. Removal of Phenolic Compounds by Polymerization in Water Treatment: Use of Ozone or Enzymes. Proc. IOA Symp., Wasser Berlin.

_____. 1987. Development in Ozonation Techniques Using Combination of Ozone and Ultraviolet Light. Proc. 8th Ozone World Congress, IOA, Zürich, Switzerland.

Du Ron, B. 1982. Ozone Generation with Ultraviolet Radiation. In *Handbook of Ozone Technology and Applications*. Edited by R.G. Rice and A. Netzer. Ann Arbor Science Publishers, Ann Arbor, Mich.

Eisenhauer, H.R. 1968. The Ozonation of Phenolic Wastes. *Jour. WPCF*, 40:1887.

_____. 1971. Dephenolization by Ozonolysis. *Wtr. Res.*, 5:467.

Elmghari-Tabib, M. 1981. Analyse de la Micropollution Azotée des Eaux en cours de Potabilisation. Action de l'Ozone sur Quelques Constituants. Thèse Rennes no. 697/329, France.

Elmghari-Tabib, M. et al. 1981. Proc. 5th Ozone World Congress, Wasser Berlin, 582.

_____. 1982. Ozonation of Amines in Water. *Wtr. Res.*, 16:223.

Erickson, R.E. et al. 1977. The Reaction of Sulfur Dioxide with Ozone in Water and Its Possible Atmospheric Significance. *Atmos. Envir.*, 11:813.

European IOA. 1989a. Standard Iodometric Method for Ozone in the Gas Phase. International Ozone Association, Zürich, Switzerland.

_____. 1989b. Standard Photometric Method for Ozone in Gas. International Ozone Association, Zürich, Switzerland.

Fabjan, Ch. 1977. Electrochemical Method for Production and Analytical Determination of Ozone. 3rd Ozone World Congress, Paris.

Farhataziz, & Ross, A.B. 1977. Selected Specific Rates of Reactions of Transients from Water in Aqueous Solution. National Bureau of Standards, Washington, D.C. (no. 59).

Fauris, C. et al. 1985. Toxicological Evaluation of Water Treatments. *Envir. Technol. Letters*, 6:279–288.

Fielding, M. & Horth, H. 1986. Formation of Mutagens and Chemicals During Water Treatment Chlorination. *Wtr. Supply*, 4:103–126.

_____. 1987. The Formation and Removal of Chemical Mutagens During Drinking Water Treatment. In *Organic Micropollutants in the Aquatic Environment*. Edited by G. Angeletti and A. Bjorseth. Kluwer Academic Publishers, Dordrecht, The Netherlands.

Flogstad, H. & Odegaard, H. 1985. Treatment of Humic Waters by Ozone. *Ozone Sci. Engrg.*, 7:2:121.

Foller, P. & Goodwin, M.L. 1984. The Electrochemical Generation of High Concentration Ozone for Small-Scale Applications. *Ozone Sci. Engrg.*, 6:29.

Fronk, C.A. 1987. Destruction of Volatile Organic Contaminants in Drinking Water by Ozone Treatment. *Ozone Sci. Engrg.*, 9:3:265.

Garland, J.A. et al. 1980. The Mechanism for Dry Deposition of Ozone to Seawater Surfaces. *Jour. Geophys. Res.*, 85:7488.

Gauthier, L. et al. 1989a. Evaluation of the Clastogenicity of Water Treated With Sodium Hypochlorite or Monochloramine Using a Micronucleus Test in Newt Larvae. *Mutagenesis*, 4:3:170–173.

_____. 1989b. Application du Test Micronoyau Triton à l'Étude Directe de la Génotoxicité des Procédés de Désinfection des Eaux. *Jour. Français d'Hydrologie* (in press).

GEHRINGER, P. ET AL. 1988. Decomposition of Trichloroethylene and Tetrachloroethylene in Drinking Water by Combined Radiation/Ozone Treatment. *Wtr. Res.*, 22:8:645.

GIANNISSIS, D. ET AL. 1985. Influence des Substances Humiques sur l'Elimination du Manganèse se dans une Filière de Potabilisation. *Revue Française des Sciences de l'Eau*, 4:149.

GILBERT, E. 1974. Elimination of Organic Compounds in Water by Ozone. *Vom Wasser*, 43:275.

_____. 1983. Investigations on the Changes of Biological Degradability of Single Substances Induced by Ozonation. *Ozone Sci. Engrg.*, 5: 137.

GILBERT, E. & ZINECKER, H. 1980. Ozonation of Aromatic Amines in Water. *Ozone Sci. Engrg.*, 2:1:65.

GINOCCHIO, J.C. 1979. Ozone and Catalysts in the Treatment of Drinking Water. *Jour. Eur. Appl. Ozone Trait. Eaux*. 371–394. Com. Eur. Ass. Int. Ozone, Paris.

GLAZE, W.H. ET AL. 1986. Oxidation of Water Supply Refractory Species by Ozone with U.V. Radiation. U.S. EPA, Res. Div., EPA 600/2-86-110.

GOLDSTEIN, F. 1903. Ueber Ozonbildung. *Chemische Berichte*, 36:3042.

GOOCH, P.C. ET AL. 1976. The Cytogenetic Effects of Ozone: Inhalation and In Vitro Exposures. *Envir. Res.*, 12:188–195.

GORDON, G. 1987. The Very Slow Decomposition of Aqueous Ozone in Highly Basic Solutions. Proc. 8th Ozone World Congress, IOA. Zürich, Switzerland.

GORDON, G. & GRUNWELL, J. 1983. Comparison of Analytical Methods for Residual Ozone. Proc. 2nd Natl. Symp.—Municipal Wastewater Disinfection, EPA Report, EPA 600/9-83-009.

_____. 1984. Improved Techniques for Residual Ozone. U.S. EPA Project Summary, EPA 600/S2-84-036.

GORDON, G. ET AL. 1987a. A Survey of the Current Status of Residual Disinfectant Measurement Methods for All Chlorine Species and Ozone. American Water Works Association Research Foundation, Denver, Colo.

_____. 1987b. Disinfectant Residual Measurement Methods. American Water Works Association Research Foundation, Denver, Colo.

GOULD, J.P. 1987. Correlation Between Chemical Structure and Ozonation Kinetics: Preliminary Observations. *Ozone Sci. Engrg.*, 9:207.

GOULD, J.P. & GROFF, K.A. 1987. The Kinetics of Ozonolysis of Synthetic Dyes. *Ozone Sci. Engrg.*, 9:2:153.

GOULD, J.P. & WEBER, W.J.H. 1976. Oxidation of Phenols by Ozone. *Jour. WPCF*, 48:1:47.

GRASSO, D. 1987. Ozonation Dynamics in Water Treatment: Autocatalytic Decomposition, Mass Transfer and Impact on Particle Stability. PhD Dissertation, The University of Michigan, Ann Arbor, Mich.

GREEN, M.H.L. ET AL. 1976. Use of Simplified Fluctuation Test to Detect Low Levels of Mutagens. *Mut. Res.*, 38:33–42.

GRINFELD, S. ET AL. 1986. Mononuclei in Red Blood Cells of the Newt *Pleurodeles Waltl* After Treatment With Benzo(a)pyrene: Dependence on Dose, Length of Exposure, Post-Treatment Time, and Uptake of the Drug. *Envir. Mutagen.*, 8:41.

GUITTONNEAU, S. ET AL. 1988a. Etude de la Dégradation de Quelques Composés Organochlorés Volatils par Photolyse du Peroxyde d'Hydrogène. *Revue des Sciences de l'Eau*, 1:1/2:35.

_____. 1988b. Comparative Study of the Photodegradation of Aromatic Compounds in Water by UV and H_2O_2/UV. *Envir. Technol. Letters*, 9:1115.

GUROL, M.D. & BREMEN, W. 1983. Reactions of Ozone with Free Cyanide Species in Aqueous Solution. Extended Abstract, Div. Envir. Chem., American Chemical Society, Washington, D.C.

GUROL, M.D. & NEKOUINAINI, S. 1984. Kinetic Behavior of Ozone in Aqueous Solution of Substituted Phenols. *Indus. Engrg. Chem. Fund.*, 23:1:54.

GUROL, M.D. & SINGER, P.C. 1982. Kinetics of Ozone Decomposition. A Dynamic Approach. *Envir. Sci. Technol.*, 16:7:377.

_____. 1983. Dynamics of the Ozonation of Phenol II Mathematical Simulation. *Wtr. Res.*, 17:9:1173.

GUROL, M.D. & VATISTAS, R. 1987. Oxidation of Phenolic Compounds by Ozone and Ozone + UV Radiation: A Comparative Study. *Wtr. Res.*, 21:8:895.

GUYON, S. 1984. Contribution à l'Etude de l'Action Directe de l'Ozone sur les Composés Organiques en Milieu Aqueux Dilué. Thèse de 3ème cycle, Université de Poitiers, France.

HAAG, W.R. & HOIGNÉ, J. 1983. Ozonation of Bromide-Containing Waters: Kinetics of Formation of Hypobromous Acid and Bromate. *Envir. Sci. Technol.*, 17:261.

_____. 1984. Kinetics and Products of the Reactions of Ozone with Various Forms of Chlorine and Bromine in Water. *Ozone Sci. Engrg.*, 6:103.

HAAG, W.R. ET AL. 1984. Improved Ammonia Oxidation by Ozone in the Presence of Bromide Ion During Water Treatment. *Wtr. Res.*, 18:9: 1125.

HALL, L.W. ET AL. 1981a. Comparison of Ozone and Chlorine Toxicity to the Developmental Stages of Striped Bass, *Morone saxatilia. Can. Jour. Fish Aquat. Sci.*, 38:752–757.

_____. 1981b. *Power-Plant Chlorination. A Biological and Chemical Assessment.* Ann Arbor Science Publishers, Ann Arbor, Mich.

HARRINGTON, T.R. ET AL. 1983. Suitability of the Modified Fluctuation Assay for Evaluating the Mutagenicity of Unconcentrated Drinking Water. *Mut. Res.*, 120:97–103.

HART, E.J. ET AL. 1983. Molar Absorptivities of Ultraviolet and Visible Bands of Ozone in Aqueous Solution. *Anal. Chem.*, 55:46.

HARTEMANN, P. ET AL. 1987. Cellular Toxicity and Mutagenicity Assays From On-Site Sampling of Drinking Water Treatment Plants Using Multistage Ozonation. *Ozone Sci. Engrg.*, 9:179–194.

HARUTA, K. & TAKEYAMA, T. 1981. The Kinetics of Oxidation of Aqueous Bromide Ion by Ozone. *Jour. Phys. Chem.*, 85:2383.

HATANAKA, J. ET AL. 1982. Usefulness and Rapidity of Screening for Toxicity and Carcinogenicity of Chemicals in Medaka. *Oryzias latipes. Japan Jour. Exp. Med.*, 52:243–253.

HELLAWELK, J.M. 1988. Toxic Substances in Rivers and Streams. *Envir. Pollut.*, 50:1–2:61.

HIGBIE, R. 1935. *Trans. AICLE*, 35:365.

HOIGNÉ, J. 1982. Mechanisms, Rates and Selectivities of Oxidations of Organic Compounds Initiated by Ozonation in Water. In *Handbook of Ozone Technology and Applications.* Edited by R.G. Rice and A. Netzer. Ann Arbor Science Publishers, Ann Arbor, Mich.

_____. 1988. The Chemistry of Ozone in Water. In *Process Technologies for Water Treatment.* Edited by S. Stucki. Plenum Press, New York.

HOIGNÉ, J. & BADER, H. 1976. The Role of Hydroxyl Radical Reactions in Ozonation Processes in Aqueous Solutions. *Wtr. Res.*, 10:377.

_____. 1977a. Ozonation of Water: Selectivity and Rate of Oxidation of Solutes. Proc. 3rd IOA Congress, Paris, France.

_____. 1977b. Rate Constants for Reactions of Ozone With Organic Pollutants and Ammonia in Water. IOA Symp., Toronto, Canada.

_____. 1978a. Ozone Initiated Oxidations of Solutes in Wastewater: A Reaction Kinetic Approach. *Prog. Wtr. Technol.*, 10:657.

_____. 1978b. Ozonation of Water: Kinetics of Oxidation of Ammonia by Ozone and Hydroxyl Radicals. *Envir. Sci. Technol.*, 12:79.

_____. 1979. Ozonation of Water: Selectivity and Rate of Oxidation of Solutes. *Ozone Sci. Engrg.*, 1:73.

_____. 1983a. Rate Constants of Reactions of Ozone with Organic and Inorganic Compounds in Water. I. Non-Dissociating Organic Compounds. *Wtr. Res.*, 17:2:173.

_____. 1983b. Rate Constants of Reactions of Ozone with Organic and Inorganic Compounds in Water. II. Dissociating Organic Compounds. *Wtr. Res.*, 17:2:185.

_____. 1985. Rate Constants of Reactions of Ozone with Organic and Inorganic Compounds in Water. III: Inorganic Compounds and Radicals. *Wtr. Res.*, 19:8:993.

HOLLMAN, P.J. 1868. *Mémoire sur l'Equivalent Calorifique de l'Ozone.* Edited by Van der Post Ir, Utrecht.

HUCK, P.M. & DAIGNAULT, S.A. 1987. Assessment of the Products of Reaction of Drinking Water Disinfectants With Humic Substances and Trace Organic Pollutants. *Sci. Total Envir.*, 62:315–328.

HUNT, T.S. 1848. On the Anomalies Presented in the Atomic Volume of Sulphur and Nitrogen. *Amer. Jour. Sci.*, 6:171.

INOUE, E. & SUGINO, K. 1959. Inhibiting Action of Hydrocarbons on Ozone Formation by Silent Electrical Discharge. *Ozone Chem. Technol.* (Advances in Chemistry Series), 21:313.

ISHIZAKI, K. ET AL. 1981. Degradation of Nucleic Acids with Ozone. I. Degradation of Nucleobases, Ribonucleosides and Ribonucleoside-5′-Monophosphates. *Chem. Pharm. Bull.*, 29:3:868.

_____. 1984. Degradation of Nucleic Acids with Ozone. V. Mechanism of Action of Ozone on Deoxyribonucleoside-5′-Monophosphates. *Chem. Pharm. Bull.*, 32:9:3601.

_____. 1987. Effect of Ozone on Plasmid DNA of *Escherichia coli* in Situ. *Wtr. Res.*, 21:7:823.

JAFFEE, L.S. 1967. *Jour. Amer. Indus. Hyg. Assoc.*, 37:329–334.

JAHN, J. 1908. Beiträge zur Kenntnis des Ozons. IV. *Zeit. Anorg. Chem.*, 60:337.

JAYLET, A. ET AL. 1986. A New Micronucleus Test Using Peripheral Blood Erythrocytes of the Newt *Pleurodeles Waltl* to Detect Mutagens in Fresh-Water Pollution. *Mut. Res.*, 164:245–257.

_____. 1987. Detection of Mutagenicity in Drinking Water Using a Micronucleus Test in Newt Larvae. *Mutagenesis*, 2:3:211–214.

JOHNSON, D. ET AL. 1986. Extent of Ozone's Reaction with Isolated Aquatic Fulvic Acid. *Envir. Sci. Technol.,* 20:739–742.

JOHNSON, G.R.A. & WARMAN, J.M. 1964. Formation of Ozone from Oxygen by the Reaction of Ionizing Radiations. *Discussions Far. Soc.*, 37:87.

JOSHI, M.G. & SHAMBAUGH, R.L. 1982. The Kinetics of Ozone Phenol Reaction in Aqueous Solutions. *Wtr. Res.*, 16:6:933.

Joy, P. et al. 1980. A Quantitative Investigation of the Reaction of Ozone with *p*-Toluene Sulfonic Acid in Aqueous Solution as a Model Compound for Anionic Detergents. *Wtr. Res.*, 14:10:1509.

Kavlock, R. et al. 1980. Studies on the Developmental Toxicity of Ozone. I. Prenatal Effects. *Toxicol. Appl. Pharmacol.*, 48:19–28.

Kawamura, S. 1932. The Study of Ozone. *Jour. Chem. Soc. (Japan)*, 53:783.

Kerzhner, B.K. et al. 1985. Oxydation of Hydroxyozobenzene by Ozone in Aqueous Solutions. *Sorret Jour. Wtr. Chem. Technol.*, 7:4:19.

Khadraoui, A.B. et al. 1988. Détermination de la Solubilité de l'Ozone dans l'Eau. Mémoire présenté à l'I.S.I.B., Bruxelles, Belgique.

Killops, S.D. 1986. Volatile Ozonization Products of Aqueous Humic Material. *Wtr. Res.*, 20: 2:153.

Kilpatrick, M.L. et al. 1956. The Decomposition of Ozone Aqueous Solution. *Jour. Amer. Chem. Soc.*, 78:784.

Kircher, J.F. et al. 1960. Irradiation of Gaseous and Liquid Oxygen. *Radiation Res.*, 13:452.

Kobylinski, E.A. & Peterman, B.W. 1979. Evaluation of Ozone Oxidation and UV Degradation of Dimethylnitrosamine. *C.A.*, 93:26: 244743 f.

Kogelschatz, U. 1987. UV Absorption of Ozone in the Gas Phase, Proc. 8th Ozone World Congress, IOA, Zürich, Switzerland.

Kogelschatz, U. et al. 1987. Ozone Generation from Oxygen and Air: Discharge Physics and Reaction Mechanisms. Proc. 8th Ozone World Congress, IOA, Zürich, Switzerland.

Kosak-Channing, L.F. & Helz, G.R. 1983. Solubility of Ozone in Aqueous Solutions of 0–0.6 M Ionic Strength at 5–30°C. *Envir. Sci. Technol.*, 17:3:145.

Kowbel, D.J. et al. 1982. Determination of Mutagenic Activity in *Salmonella* of Residual Fulvic Acids After Ozonation. *Wtr. Res.*, 16: 1537–1538.

______. 1984. Chlorination of Ozonated Soil Fulvic Acid: Mutagenicity Studies in *Salmonella*. *Sci. Total Envir.*, 37:171–176.

______. 1986. Mutagenicity Studies in *Salmonella:* Residues of Ozonated and/or Chlorinated Water Fulvic Acids. *Envir. Mutagen.*, 8:253–262.

Kroyer, G. et al. 1986. Studies on the Influence of Ozone on Selected Organic Contaminants in Water. I. *Wien. Tiennegth Monatssch.*, 73:4:156.

Kruithof, J.C. 1978. Treatment of Rhine Water by Coagulation and Ozonation. PhD Thesis, Delft Technical Highschool.

Kuo, C.H. 1984. Reaction of Dissolved Pollutants with Ozone in Aqueous Solutions. EPA Project Summary, EPA 600/3-84/072.

______. 1985. Reaction of Ozone with Organics in Aqueous Solutions. EPA Project Summary, EPA 600/3-85/031.

Ladenburg, A. 1898. Ueber das Ozon. *Berichte*, 31:2508.

Laforge, P. et al. 1982. Photolyse-UV de l'Ozone Résiduel dans le Gaz d'Echappement lors de l'Ozonisation de l'Eau. *Rev. Français Sci. de l'Eau*, 1:3:255.

Langlais, B. 1985. Ozone Dosage in The Gasous Phase. *Proc. Int'l Conf.: The Role of Ozone in Water and Wastewater Treatment* (R. Perry and R.E. McIntyre, Eds.), Selper Ltd., London, p. 33.

Laplanche, A. 1982. Approche des Schémas d'Ozonation de Molécules Organiques: Etude Particulière des Composés Organophosphorés. PhD Thèse, Université de Rennes, France.

Laplanche, A. et al. 1976. Contribution to the Control of Organophosphorus Pollutants. Application to Free Water. *Jour. Français d'Hydrologie*, 7:1:11.

______. 1984. Ozonation Schemes of Organophosphorus Pesticides: Application in Drinking Water Treatment, *Ozone Sci. Engrg.*, 6:197.

______. 1985. Ozonation of Various Organic Compounds: Reaction of Different Oxidants on Phenylalamine. Wasser Berlin, IOA, 69.

______. 1989. Simulation of Ozone Transfer in Water: Comparison with a Pilot Unit. Proc. 9th Ozone World Congress, IOA, New York.

Larose, J. et al. 1982. Elimination of Humic Materials. *Ozone Sci. Engrg.*, 4:2:79.

Lawrence, J. et al. 1980. The Ozonation of Natural Waters: Product Identification. *Ozone Sci. Engrg.*, 2:1:55.

Layton, R.F. & Kinman, R.H. 1970. A New Analytical Method for Ozone. Proc. Natl. Specialty Conf. on Disinfection, Amer. Soc. Civil Engrs., New York.

Le Cloirec Renaud, C. 1984. Analyse et Evolution de la Micropollution Organique Azotée dans les Stations de Potabilisation. Effet de la Chloration sur les Acides Aminés. Thèse Rennes no. 174/121, France.

Le Cloirec Renaud, C. et al. 1984. Action Comparée du Chlore et de l'Ozone sur des Acides Aminés en Solution. Symp. Ozone & Biologie, Rennes, France.

Lecnher, J.A. & Huffman, E.W.D. Jr. 1976. Classification of Organic Solutes in Water by Using Macroreticular Resins. *Jour. Res. US Geol. Survey*, 4:737.

Lefebvre, E. et al. 1990. The Effect of Ozonation on the Removal of Organics by Coagulation-Flocculation. *Ozone Sci. Engrg.* (in press).

Legube, B. 1983. Contribution à l'Etude de l'Ozonation de Composés Aromatiques en Solution Aqueuse. PhD Thèse, Université de Poitiers, France.

LEGUBE, B. ET AL. 1981. Identification of Ozonation Products of Aromatic Hydrocarbon Micropollutants: Effect on Chlorination and Biological Filtration. *Ozone Sci. Engrg.*, 3:1:33.

_____. 1985. Identification of a Few Organics and Attempted Quantification upon Disinfection with Ozone of a Biologically Treated Wastewater. *Proc. Intl. Conf., The Role of Ozone in Water and Wastewater Treatment.* Edited by R. Perry and R.E. McIntyre. Selper Ltd., London.

_____. 1986. Ozonation of Naphthalene in Aqueous Solution a) Part 1: Ozone Consumption and Ozonation Products; b) Part 2: Kinetics Studies of the Initial Reaction Step. *Wtr. Res.*, 20:2:197.

_____. 1987a. Ozonation des Bases Puriques en Milieu Aqueux: Etudes Cinétiques de la Réaction. *Wtr. Res.*, 21:9:1101.

_____. 1987b. Ozonation of Aqueous Solution of Nitrogen Heterocycle Compounds: Benzotriazoles, Atrazine and Amitrole. *Ozone Sci. Engrg.*, 9:3:233.

_____. 1988. Removal of Humic Substances and Manganese in Slightly Mineralized Water by Means of Iron Salts and Ozone. Proc. AWWA Ann. Conf., Orlando, Fla.

_____. 1989a. Etude sur les Acides Fulviques Extraits d'Eaux Superficielles Françaises: Extraction, Caractérisation et Réactivité avec le Chlore. *Revue Français Sci. de l'Eau*, à paraître.

_____. 1989b. Ozonation of an Extracted Aquatic Fulvic Acid: Theoretical and Practical Aspects. *Ozone Sci. Engrg.*, 11:1:69.

_____. 1989c. Enhancement of Radical Chain Reactions of Ozone in Water in Presence of Aquatic Fulvic Acids. 9th Ozone World Congress, IOA, New York.

LEHNINGER. A.L. 1973. *Biochimie*, Flammarion Science Publishers, Paris.

LEITZKE, O. 1977. Instrumental Ozone Analysis in Water and Gas Phases. Intl. Symp. Ozon und Wasser (Berlin, Federal Republic of Germany: Colloquium Verlag Otto H. Hess), p. 164.

LENARD, P. 1900. Ueber Wirkungen des Ultravioletten Lichtes auf Gasförmige Körper. *Annalen für Physik*, 1:486.

LI, K.Y. 1977. PhD Dissertation, Mississippi State University, Mississippi State, Miss.

LIAO, W. ET AL. 1982. Structural Characterization of Aquatic Humic Material. *Envir. Sci. Technol.*, 16:403.

LISSI, E. & HEICKLEN, J. 1973. The Photolysis of Ozone. *Jour. Photochem.*, 1:39.

MAIER, D. 1982. Methods for the Determination of Ozone in Concentrated Gas Mixtures. In *Ozonization Manual for Water and Wastewater Treatment*, Edited by W.J. Masschelein. John Wiley & Sons, Inc., New York.

MAIER, D. & KURZMANN, G.E. 1977. The Determination of Higher Ozone Concentrations. *Wasser, Luft und Betrieb*, 21:125.

MAILFERT, H. 1894. Sur la Solubilité de l'Ozone. *Comptes Rendus*, 119:951.

MALAVEILLE, C. & BARTSCH, H. 1989. Genotoxicity of Carcinogenic Substances in Humans and Animals. *Jour. Toxicol. Clin. Exp.*, 9:15–25.

MALCOLM, R.L. & MCCARTHY, P. 1986. Limitations in the Use of Commercial Humic Acids in Water and Soil Research. *Envir. Sci. Technol.*, 20:904.

MALLEVIALLE, J. 1982. Stability of Ozone in Water. In *Ozonization Manual for Water and Wastewater.* Edited by W.J. Masschelein. John Wiley & Sons, Inc., New York.

MALLEVIALLE, J. ET AL. 1975. Détermination Expérimentale des Coefficients de Transfert de l'Ozone dans l'Eau. *Cebedeau*, 28:377:175.

MARCHENKO, P.V. & OBOLONCHIR, V.V. 1979. *Chim. Technol. Vody* 1, (1), 32. *C.A.* 93 (26): 244732 b.

MARLEY, K.A. ET AL. 1987. Ozonolysis of Naphthalene Derivatives in Water and in Kerosene Films. *Ozone Sci. Engrg.*, 9:1:23.

MARTIN, G. 1966. Relations entre le Déplacement Chimique du Proton de Dérivés Mono et Désubstituts du Benzène et diverses Propriétés Physicochimiques. Thèse Rennes, France.

MARZIN, D. 1984. Intérêt et Limites des Tests de Mutagénèse Pour la Recherche des Agents Cancérogènes Dans Nos Aliments. *Cah. Nutr. Diét.*, 19:2:81–87.

MASSCHELEIN, W.J. 1977. L'optimalisation de la Capacité de Production d'Ozone dans le Traitement des Eaux. *TSM l'Eau*, 72:177.

_____. 1982a. Contacting of Ozone with Water and Contactor Offgas Treatment. In *Handbook of Ozone Technology and Applications*, edited by R.G. Rice and A. Netzer. Ann Arbor Science Publishers, Ann Arbor, Mich.

_____. 1982b. Continuous Amperometric Residual Ozone Analysis in the Tailfer (Brussels, Belgium) Plant. In *Ozonization Manual for Water and Wastewater Treatment.* Edited by W.J. Masschelein. John Wiley & Sons, Inc., New York.

_____. 1982c. The Use of an Oxygen-Enriched Process Gas. In *Ozonization Manual for Water and Wastewater Treatment.* Edited by W.J. Masschelein. John Wiley & Sons, Inc., New York.

MASSCHELEIN, W.J. & GOOSSENS, R. 1984. Nitrophenols as Model Compounds in the Design of Ozone Contacting and Reacting Systems. *Ozone Sci. Engrg.*, 6:2:143.

MASSCHELEIN, W.J. ET AL. 1979. Mesure Electrochimique en Continu de Traces d'Ozone dans l'Eau. *Electrochemical Analysis*, 7:432.

_____. 1980. L'ozonation des Eaux. Manuel Pratique Technique et Documentation. Lavoisier, Paris.

_____. 1987. Competitive Inhibition in Ozone Gas to Liquid Transfer Dependent Reactions. In *Role of Ozone in Water and Wastewater Treatment.* Edited by D.W. Smith and G.R. Finch. Tektron International Ltd., Kitchener, Ontario.

MASSCHELEIN, W.J. ET AL. 1989. Determination of Residual Ozone or Chlorine Dioxide in Water with ACVK—An Updated Version. *Ozone Sci. Engrg.*, 11:2:209.

MATROZOV, V.I. ET AL. 1975. Solubility of Ozone in Water. *Zhur. Prikl. Khim.*, 48:8:1838.

MATSUI, M. ET AL. 1981. Ozonation of Dyes. *Jour. Soc. Dyers Colour.*, 97:5:210.

MAUERSBERGER, K. ET AL. 1986. Measurements of the Ozone Absorption Cross-Section at the 253.7 nm Mercury Line. *Geophys. Res. Letters*, 13:671.

MCCANN, J. ET AL . 1975. Detection of Carcinogens and Mutagens in the *Salmonella*/ Microsome Test: Assay of 300 Chemicals. *Proc. Nat. Acad. Sci.*, 72:5135–39; 73:950.

_____. 1988. Statistical Analysis of *Salmonella* Test Data and Comparison to Results of Animal Tests. *Mut. Res.*, 205:183–195.

MCCARTHY, P. ET AL. 1986. Establishment of a Collection of Standard Humic Substances. 3rd Intl. Mtg. IHSS, Oslo, Norway.

MCGRATH, W.D. & NORRISH, R.G.W. 1960. Studies of the Reactions of Excited Oxygen Atoms and Molecules Produced in the Flash Photolysis of Ozone. *Prog. Royal Soc., Series A.*, 254:317, London, England.

MEIER, J.R. 1988. Genotoxic Activity of Organic Chemicals in Drinking Water. *Mut. Res.*, 196:211–245.

MERLET, N. ET AL. 1980. Role of Ozone in Trihalomethane Formation. *Envir. Technol. Letters*, 1:8:384.

MOLINA, L.T. & MOLINA, M.J. 1986. Absolute Absorption Cross Section of Ozone in the 185 to 350 nm Wavelength Range. *Jour. Geophys. Res.*, 91:501.

MORIMOTO, K. & KOIZUMI, A. 1983. Trihalomethanes Induce Sister Chromatid Exchanges in Human Lymphocytes In Vitro and Mouse Bone Marrow Cells In Vivo. *Envir. Res.*, 32:72–79.

MOROZZI, G. ET AL. 1986. The Fate of Some Organic Pollutants in Wastewater During Ozone and Ozone GAC Adsorption Treatment. *Jour. Envir. Sci.* Health, A21:6:523 (Part A).

MORRIS, J.C. 1988. The Aqueous Solubility of Ozone: A Review. *Ozonews*, 16:1:14.

MUDD, J.R. ET AL. 1969. Reaction of Ozone with Amino Acids and Proteins. *Atmos. Envir.*, 3:669.

NAKAZAWA, T. ET AL. 1985. Histochemistry of Liver Tumors Induced by Diethyl-Nitrosamine and Differential Sex Susceptibility to Carcinogenesis in *Oryzias latipes. Jour. Natl. Cancer Inst.*, 75:567–573.

NARKIS, N. ET AL. 1977. Ozone Effect on Nitrogenous Matter in Effluents. *Jour. Envir. Engrg. Div., ASCE*, 10.

_____. 1980. Ozone Induced Biodegradability of a Non-Ionic Surfactant. *Wtr. Res.*, 14:9:1225.

NAZAROV. V.I. ET AL. 1979. Use of Ozone for Removing Phenols and 3,4 Benzopyrene from Technical Grade Condensates from Slow Cooking Apparatus. *Khim. Technol. Topl. Hasel Hasel*, 10, 54. *C.A.* 92:14:115813 j.

NISHINA, R. & MOTOYUKI, S. 1978. Treatment Technology of Dying Waste Water. Effect of Azogroup on Ozone Oxidations of Azo Dyes. Okayama-Ken Kogyo. *Gijutsu Santa Hokoku*, 3: 193.

OLIVER, B.G. & THURMAN, E.M. 1983. Influence of Aquatic Humic Substances in Trihalomethane Potential. In *Water Chlorination: Environmental Impact and Health Effects*, vol. 4. Edited by R.L. Jolley et al. Ann Arbor Science Publishers, Ann Arbor, Mich.

ONDA, K. ET AL. 1968. Mass Transfer Coefficients Between Gas and Liquid Phases in Packed Columns. *Jour. Chem. Engrg.* (Japan), 1:56.

ORLOV, V. ET AL. 1983. Ozonation of Urea in Water. *Envir. Technol. Letters*, 4:12:491.

OUEDERNI, A. ET AL. 1987. Ozone Absorption in Water: Mass Transfer and Solubility. *Ozone Sci. Engrg.*, 9:1.

PAILLARD, H. ET AL. 1987. Application of Oxidation by a Combined Ozone/UV Radiation Systems to the Treatment of Natural Water. *Ozone Sci. Engrg.*, 9:4:391.

_____. 1988. Conditions Optimales d'Application du Système Oxydant Ozone—Peroxyde d'Hydrogène. *Wtr. Res.*, 22:1:91.

_____. 1989a. Iron and Manganese Removal With Ozonation in Presence of Humic Substances. *Ozone Sci. Engrg.*, 11:93.

_____. 1989b. The Effect of Ozonation on Organics Removal by Coagulation-Flocculation. Proc. 9th Ozone World Congress, IOA, New York.

PALLER, M.H. & HEIDINGER, R.C. 1979. The Toxicity of Ozone to the Bluegill. *Jour. Envir. Sci. Health*, A14 (3):169–193.

_____. 1980. Mechanisms of Delayed Ozone Toxicity to Bluegill *Lepomis machrochirus rafinesque. Envir. Poll.* (Series A), 22:229–239.

PARÉ, M. 1982. Analytical Instrumentation for Control of Ozonation Systems. In *Handbook of Ozone Technology and Applications* (Vol. 1). Edited by R.G. Rice and A. Netzer. Ann Arbor Science Publishers, Ann Arbor, Mich.

Park, Y.K. & Lee, C.H. 1986. Optimum Ozone Demand for Removal of Micropollutants Contained in Gallery Infiltered Water. *Wtr. Supply*, 4:1:461.

Penkett, S.A. 1972. Oxidation of SO_2 and Other Atmospheric Bases by Ozone in Aqueous Solution. *Nature Phys. Sci.*, 240:105.

Perez, R. et al. 1987. Ozone Reactions with Carbohydrates in Aqueous Medium. Proc. 8th Ozone World Congress, IOA, Zürich, Switzerland.

Peyton, G.R. & Glaze, W.H. 1983. New Developments in the Ozone/UV Process. Proc. 6th Ozone World Congress, Washington, D.C.

_____. 1988. Destruction of Pollutants in Water with Ozone in Combination with Ultraviolet Radiation. 3. Photolysis of Aqueous Ozone. *Envir. Sci. Technol.*, 22:761.

Pommery, J. et al. 1989. SOS Chromotest Study Concerning Some Appreciation Criteria of Humic Substances' Genotoxic Potency. *Mut. Res.*, 223:183–189.

Pottenger, L.Y. 1983. Etude de l'Effet Mutagène, Transformant et Chromatidien des Concentrats d'Eaux. PhD Thesis, University of Nancy, France.

Pouvreau, P. 1984. Elimination Specifique du Fer et du Manganèse. *Jour. Français d'Hydrologie*, 15:169.

Prengle, H.W. & Mauk, C.E. 1978. Ozone/UV Oxidation of Pesticides in Aqueous Solutions. In *Ozone/Chlorine Dioxide Oxidation Products of Organic Materials*. Edited by R.G. Rice and J.A. Cotruvo, Intl. Ozone Assoc., Norwalk, Conn.

Puranik, S.S. et al. 1974. Effective Interfacial Area Irrigated Packed Columns. *Chem. Engrg. Sci.*, 29:501.

Quentin, K.E. et al. 1978. Elimination of Phenoxy Acetic Acid Pesticides in Water by Ozone. *Z. Wass. Abwass. Forsch*, 11:118.

Quillardet, P. et al. 1985. The SOS Chromotest, A Colorimetric Bacterial Assay for Genotoxins: Validation Study With 83 Compounds. *Mut. Res.*, 147:79–95.

Rao, M. & Reddy, J. 1987. Peroxisome Proliferation and Hepatocarcinogenesis. *Carcinogenesis*, 8:631–636.

Rasmussen, R.E. 1986. Inhibition of DNA Replication by Ozone in Chinese Hamster V79 Cells. *Jour. Toxicol. Envir. Health*, 17:119–128.

Rawson, A.E. 1953. Studies in Ozonisation. Part 2. Solubility of Ozone in Water. *Wtr. and Wtr. Engrg.*, 57:3:102.

Razumovskii, S.D. & Zaikov, G.E. 1971. The Solubility of Ozone in Various Solvents. *Isvestia Akad. Nauk. SSSR. Ser. Khim.*, 4:689.

Reckhow, D.A. 1984. Organic Halide Formation and the Use of Pre-Ozonation and Alum Coagulation to Control Organic Halide Precursors. PhD Dissertation, Univ. of North Carolina, Chapel Hill, N.C.

Reckhow, D.A. et al. 1986. The Ozonation of Organic Halide Precursors: Effect of Bicarbonate. *Wtr. Res.*, 20:8:987.

Reynolds, G. 1989. Aqueous Ozonation of Pesticides, A Review. *Ozone Sci. Engrg.*, 11:6:339.

Reynolds, S.I. 1955. Test Methods for Measuring Energy in a Gas Discharge. Symp. on Corona, Special Tech. Publ. No. 198, American Society for Testing and Materials, Philadelphia, Pa.

Rice, R.G. 1980. The Use of Ozone to Control Trihalomethanes in Drinking Water Treatment. *Ozone Sci. Engrg.*, 2:1:75.

Richard, Y. & Brener, L. 1984. Removal of Detergents from Drinking Water with Ozone. In *Handbook of Ozone Technology*. Edited by R.G. Rice. Butterworth Publishers, Stoneham, Mass.

Richardson, L.B. et al. 1983. Lethal and Sublethal Exposure and Recovery Effects of Ozone-Produced Oxidants on Adult White Perch (*Morone americana* Gmelin). *Wtr. Res.*, 17:205–213.

Rideal, E.K. 1920. *Ozone*. Constable Co. Ltd., London, England.

Rogozhkin, G.I. 1970. Kinetics of Ozone Absorption by Alkaline Solutions. *Tr., Vses, Nauch.-Issled. Inst. Vodosnabzh., Kanaliz., Gidrotekh. Sooruzhenii Inzh. Gidrogeol.* 1970, 25–76. USSR *C.A.* 73:81108 y.

Roselund, B. 1975. Disinfection of Hatchery Influent by Ozonation and the Effects of Ozonated Water on Rainbow Trout. In *Aquatic Applications of Ozone*. Edited by W. Blogoslawski and R.G. Rice. Intl. Ozone Inst., Cleveland, Ohio.

Rosen, H.M. 1972. Ozone Generation and Its Economical Application in Wastewater Treatment. *Wtr. Sewage Works*, 119:9:114.

Roth, J.A. & Sullivan, D.E. 1981. Solubility of Ozone in Water. *Indus. Engrg. Chem. Fund.*, 20:137.

_____. 1983. Kinetics of Ozone Decomposition in Water. *Ozone Sci. Engrg.*, 5:1:37.

Rothmund, V. & Burgstaller, A. 1912, 1913. Über die Geschwindigkeit der Zersetsung des Ozons in wässeriger Lösung. *Nernst Festschrift* p. 391 (1912) *Monatsh. Chemie.*, 34:665 (1913).

_____. 1913. Über die Geschwindigkeit der Zersetsung des Ozons in wässeriger Lösung. *Monatsh. Chem.*, 34:47.

Roustan, M. & Mallevialle, J. 1982. Theoretical Aspects of Ozone Transfer into Water. In *Ozonization Manual for Water and Wastewater Treatment*. Edited by W.J. Masschelein. John Wiley & Sons, Inc., New York.

Roustan M. et al. 1987. Mass Balance Analysis of Ozone in Conventional Bubble Contactors. *Ozone Sci. Engrg.*, 9:3:289.

Roustan, M. et al. 1989. Proc. 9th World Ozone Congress. IOA, New York.

Sayato, Y. 1987. Mutagenicity of Products Formed by Ozonation of Naphtoresorcinol in Aqueous Solutions. *Mut. Res.*, 189:217.

Sayato, Y. et al. 1989. Mutagenicity on Chlorination of Products Formed by Ozonation of Naphtoresorcinol in Water. *Mut. Res.*, 226:151–155.

Schönbein, C.F. 1840. Recherches sur la Nature de l'Odeur qui se Manifeste dans Certaines Actions Chimiques. *Compt. Rend. Dcad. Sci.*, Paris 10:706.

_____. 1849. Beobachtungen über den bei der Elektrolysation des Wassers und dem Ausströmen der gewöhnlichen Elektricität aus Spitzen sich entwikkelnden Geruch. *Pogg. Ann. Physik*, 50:616.

Schöne, E. 1873. Ueber das Verhalten von Ozon und Wasser zu einander. *Berichte Chem. Ges.*, 6:1224.

Schultze, H. & Schultze-Frohlinde, D. 1975. OH Radical Induced Oxidation of Ethanol. *Trans. Faraday Soc.*, 71:5:1099.

Shah, J. & Maxie, E.C. 1966. Gamma-Ray Radiosynthesis of Ozone from Air. *Intl. Jour. Appl. Radiation Isotopes*, 17:155.

Sheffer, S. & Esterson, G.L. 1982. Mass Transfer and Reaction Kinetics in the Ozone/Tap Water System. *Wtr. Res.*, 16:383.

Sheng der Chang & Singer, P.C. 1988. The Impact of Ozone on the Removal of Particles, Total Organic Carbon and Trihalomethane Precursors. Proc. AWWA Ann. Conf., Orlando, Fla.

Shinriki, N. et al. 1981. Degradation of Nucleic Acids with Ozone. II. Degradation of Yeast RNA, Yeast Phenylalanine tRNA and Tobacco Mosaic Virus RNA. *Biochimica & Biophysica Acta*, 665:323.

Shinriki, N. et al. 1983. Degradation of Nucleic Acids with Ozone. III. Mode of Ozone—Degradation of Mouse Proline Transfer Ribonucleic Acid (tRNA) and Isoleucine tRNA. *Chem. Pharm. Bull.*, 31:10:3601.

Shinriki, N. et al. 1984. Degradation of Nucleic Acids with Ozone. VI. Labilization of the Double-Helical Structure of Calf Thymus Deoxyribonucleic Acid. *Chem. Pharm. Bull.*, 32:9:3636.

Shorter, J. 1972. The Separation of Polar Steric and Resonance Effects by the Rise of Linear Free Energy Relationships In *Advances in Linear Free Energy Relationships.* Edited by N.B. Chapman. Plenum Publishing Co., Ltd., London.

Siboulet, R. 1984. Micronuclei in Red Blood Cells of the Newt *Pleurodeles Waltl Michah*: Induction With X Rays and Chemicals. *Mut. Res.*, 125:275.

Sierka, R.A. & Cowen, W.F. 1978. The Ozone Oxydation of Hydrazine Fuels. Civil & Envir. Engrg. Devel. Office, Tyndall Air Force Base, Fla. CEEDO—TR 78–43:117.

_____. 1981. The Catalytic Ozone Oxidations of Aqueous Solutions of Hydrogen, Monomethylhydrogen and Unsymmetrical Dimethyl Hydrogen. Proc. Indus. Waste Conf., 35:406.

Singer, P.C. & Gurol, M.D. 1983. Dynamics of the Ozonation of Phenol. I. Experimental Observations. *Wtr. Res.*, 17:9:1163.

Singer, P.C. & Zilli, W.B. 1975. Ozonation of Ammonia in Wastewater. *Wtr. Res.*, 9:127.

Somiya, I. et al. 1986. Biodegradability and GAC Adsorbability of Micropollutants by Preozonation. *Ozone Sci. Engrg.*, 8:1:11.

Spanggord & McClurg. 1978. Ozone Methods and Ozone Chemistry of Selected Organics in Water. In "Basic Chemistry," in *Ozone/Chlorine Dioxide Oxidation Products of Organic Materials.* Edited by R.G. Rice and J.A. Cotruvo, IOA, Cleveland, Ohio.

Staehelin, J. & Hoigné, J. 1982. Decomposition of Ozone in Water: Rate of Initiation by Hydroxide Ions and Hydrogen Peroxide. *Envir. Sci. Technol.*, 16:10:676.

_____. 1983. Mechanism and Kinetics of Decomposition of Ozone in Water in the Presence of Organic Solutes. *Vom Wasser*, 61:337.

_____. 1985. Decomposition of Ozone in Water in the Presence of Organic Solutes Acting as Promotors and Inhibitors of Radical Chain Reactions. *Envir. Sci. Technol.*, 19:120–126.

Staehelin, J. et al. 1984. Ozone Decomposition in Water Studied by Pulse Radiolysis. 2. OH and HO_4 as Chain Intermediates. *Jour. Phys. Chem.*, 88:5999.

Standard Methods for the Examination of Water and Wastewater 1985. Edited by A.E. Greenburg, R.R. Trussell, L.S. Clesceri, and M.A.H. Franson. American Public Health Association, American Water Works Association, and Water Pollution Control Federation (16th ed.).

Standard Methods for the Examination of Water and Wastewater. 1989. Edited by L.S. Clesceri, A.E. Greenburg, R.R. Trussell, and M.A.H. Franson. American Public Health Association, American Water Works Association, and Water Pollution Control Federation (17th ed.).

Stanley, J.H. & Johnson, J.D. 1979. Amperometric Membrane Electrode for Measurement of Ozone in Water. *Anal. Chem.*, 51:2144.

Steinberg, M. & Beller, M. 1970. Ozone Synthesis for Water Treatment by High Energy Radiation. *Chem. Engrg. Progress* (Symp. Series), 66:205.

Stoebner, R.A. & Rollag, D.B. 1981. Ozonation of a Municipal Groundwater Supply to Reduce Iron, Manganese, and Trihalomethane Formation, *Aqua*, 4:291.

STRAKA, M.R. ET AL. 1984. Residual Ozone Determination by Flow Injection Analysis. *Anal. Chem.*, 56:1973.

STRAKA, M.R. ET AL. 1985. Residual Aqueous Ozone Determination by Gas Diffusion Flow Injection Analysis. *Anal. Chem.*, 57:1799.

STUCKI, S. ET AL. 1985. In Situ Production of Ozone in Water Using a MEMBREL Electrolyzer. *Jour. Electrochem. Soc.*, 132:2.

STUMM, W. 1956. Einige chemische Gesichtspunkte zur Wasserozonisierung. *Schweiz. Zeits. für Hydrologie*, 18:201.

SUZUKI, J. ET AL. 1983. UV Irradiation Effect on the Ozonation of Polyethylene Glycol. *Suishitsu Odaku Kenkyn*, 6:1:59.

TAUBE, H. 1956. Photochemical Reactions of Ozone in Solution. *Trans. Faraday Soc.*, 53:656.

TAUBE, H. & BRAY, W.C. 1940. Chain Reactions in Aqueous Solutions Containing Ozone, Hydrogen Peroxide and Acid. *Jour. Amer. Chem. Soc.*, 62:3357.

THOMPSON, E.D. 1986. Comparison of In Vivo Cytogenic Assay Results. *Envir. Mutagen.*, 8: 753–767.

THURMAN, E.M. 1985. *Developments in Biogeochemistry: Organic Geochemistry of Natural Waters.* N. Nijhoff and Dr. W. Junk Publishers, Dordrecht, The Netherlands.

THURMAN, E.M. & MALCOLM, R.L. 1981. Preparative Isolation of Aquatic Humic Substances. *Envir. Sci. Technol.*, 15:599.

_____. 1983. Structural Study of Humic Substances: New Approaches and Methods. In *Aquatic and Terrestrial Humic Materials.* Edited by R.F. Christman and E.T. Gjessing. Ann Arbor Science Publishers, Ann Arbor, Mich.

TICE, R.R. ET AL. 1978. Cytogenetic Effects of Inhaled Ozone. *Mut. Res.*, 58:293.

TOMIYASU, H. ET AL. 1985. Kinetics and Mechanisms of Ozone Decomposition in Basic Aqueous Solution. *Inorg. Chem.*, 24:2962.

TYE AND WAITE, 1981.

URIEN, A.F. 1986. Substances Chimiques Naturelles et de Synthèse—Réactivité Avec l'Ozone, le Chlore—Implication Toxicologiques. PhD Thesis, Université de Poitiers, France.

URIEN, A.F. ET AL. 1986. Synthetic Humic Acid Model Complexing Ability and Toxic Effects. In *Chemicals in the Environment.* Edited by J.N. Lester, R. Perry, and S. Sterrit. Selper Ltd., London, England.

VAN DER KOOIJ, D. ET AL. 1987. The Effects of Ozonation, Biological Filtration and Distribution on the Concentration of Easily Assimilable Organic Carbon (AOC) in Drinking Water. Proc. 8th Ozone World Congress, IOA, Zürich, Switzerland.

VAN HOFF, F. ET AL. 1985. Formation of Linear Aldehydes During Surface Water Preozonation and Their Removal in Water Treatment in Relation to Mutagenic Activity and Some Parameters. *Sci. Total Envir.*, 47:187.

VEENSTRA, J.N. ET AL. 1983. Ozonation: Its Effects on the Apparent Molecular Weight of Naturally Occurring Organics and Trihalomethane Production. *Ozone Sci. Engrg.*, 5:4:225.

VENOSA, A.D. ET AL. 1985. Reliable Ozone Disinfection Using Off-Gas Control. *Jour. WPCF*, 57:929.

Verein Deutscher Ingenieure. 1978. VDI Method 2468, Blatt 1.

VERWEIJ, H. ET AL. 1982. Different Pathways of Tyrosine Oxidation by Ozone. *Chemosphere*, 11: 8:721–725.

VON SIEMENS, W. 1857. Über die electrostatische Induktion und die Verzögerung des Stroms in Flaschendrähten. *Poggendorff's Ann.*, 102:120.

WABNER, D. & GRAMBOW. C. 1983. Erzeugung von reaktivem Sauerstoff durch Elektrolyse. *Vom Wasser*, 60:182.

WALLER, J.G. & MCTURK, G. 1965. Storage of Compressed Gaseous Ozone. *Jour. Appl. Chem.*, 15:363.

WATERS, R. 1984. DNA Repair Test in Cultured Mammalian Cells. In *Mutagenicity Testing.* Edited by S. Venitt and J.M. Parry. IRL Press, Oxford, Wash.

WATTS, C.D. 1985. Organic By-Products of Ozonization of Humic and Fulvic Acid. *Proc. Intl. Conf., Role of Ozone in Water and Wastewater Treatments.* Edited by R. Perry and R.E. McIntyre. Selper Ltd., London, England.

WAYNE, R.P. 1972. Reactions of Excited Species in the Photolysis of Ozone. *Faraday Discussions of the Chemical Society*, 53:172.

WEDEMEYER, G.A. ET AL. 1979. Physiological and Biochemical Aspects of Ozone Toxicity to Rainbow Trout (*Salmo gairdneri*). *Jour. Fish. Res. Board Can.*, 36:605–614.

WEISS, J. 1935. Investigation on the Radical HO_2 in Solution. *Trans. Faraday Soc.*, 31:668.

WOOD, D. ET AL. 1989. The Factors Influencing the Potassium Iodide and the Neutral Buffered Potassium Iodide Methods for the Determination of Ozone. *Jour. AWWA*, 81:6:72.

YAMADA, H. & SOMIYA, I. 1980. Identification of Products Resulting from Ozonation of Organic Compounds in Water. *Ozone Sci. Engrg.*, 2:251.

YANG, T.C. & NEELY, W.C. 1986. Relative Stoichiometry of the Oxidation of Ferrous Ion by Ozone in Aqueous Solution. *Anal. Chem.*, 58: 7:1551.

YURTERI, C. & GUROL, M.D. 1988. Ozone Consumption in Natural Water: Effects of Background Organic Matter, pH and Carbonate Species. *Ozone Sci. Engrg.*, 10:3:277.

ZELAC, R.E. ET AL. 1971. Inhaled Ozone as a Mutagen: I. Chromosome Aberrations Induced in Chinese Hamster Lymphocytes. *Envir. Res.*, 4:262.

ZURER, P.S. 1987. Complex Mission Set to Probe Origins of Antarctic Ozone Hole. *Chem. Engrg. News*, 7.

III

Practical Application of Ozone: Principles and Case Studies

Guy Bablon
William D. Bellamy
Gilles Billen
Marie-Marguerite Bourbigot
F. Bernard Daniel
Françoise Erb
Cyril Gomella
Gilbert Gordon
Phillippe Hartemann
Jean-Claude Joret
William R. Knocke
Bruno Langlais
Alain Laplanche
Bernard Legube
Benjamin Lykins, Jr.
Guy Martin
Nathalie Martin
Antoine Montiel
Marie Françoise Morin
Richard S. Miltner
Daniel Perrine
Michele Prévost
David A. Reckhow
Pierre Servais
Philip C. Singer
Otis J. Sproul
Claire Ventresque

III.A INTRODUCTION

III.A.1 Uses of Ozone

Ozonation is used in drinking water treatment to achieve a variety of treatment goals. The purposes of ozonation include the following:

1. Disinfection and algae control
2. Oxidation of inorganic pollutants
   a. Iron and manganese

Table III–1 Summary of Ozone Applications

Control of:	Point of Application	Ozone Dose	Best Pathway*	Notes
Fe/Mn	Pre, Inter	Med	Molec.	Inter may be best with high–DOC waters.
Color	Inter	Med–High	Molec.	Two-step stoichiometry.
Taste and odor	Inter	High	Rad.	T&O may be produced by low ozone doses.
SOCs	Inter	Med–High	Rad.	Molec. may be best for some compounds.
Particles	Pre	Low	Unknown	May require high calcium concentration.
Algae	Pre, Inter	Low–Med	Unknown	Can be used with flotation.
Pathogens	Pre, Post	Med–High	Molec.	Pre in U.S.; post in Europe.
Cl_2 by-products	Inter, Pre	Low–High	Molec.	High levels of removal require Rad.
Biodegradables	Inter	Med	Unknown	Design of downstream filtration process is important.

*Choice of molecular (Molec.) or radical (Rad.) pathway.

3. Oxidation of organic micropollutants
   a. Taste and odor compounds
   b. Phenolic pollutants
   c. Pesticides
4. Oxidation of organic macropollutants (that is, nonspecific organic targets)
   a. Bleaching of color
   b. Increasing the biodegradability of organics
   c. Destruction of trihalomethane formation potential (THMFP), total organic halide formation potential (TOXFP), and chlorine demand
5. Improvement of coagulation

Ozone is used in three ways: as a biocide (1), as a classical oxidant (2, 3, and 4a), and as a pretreatment for improving the performance of subsequent processes (4b, 4c, and 5). The ability of ozone to act as a disinfectant and classical oxidant has long been known. Research in this area has revealed important mechanisms and engineering perspectives (for example, optimal treatment sequence, ozone dose) to help achieve these goals. However, there is less fundamental understanding of the effects of ozone on subsequent treatment processes. It is likely that applications in this latter category can be attributed to ozone-induced oxidation of macropollutants (e.g., humic materials). For example, the use of ozone in minimizing chlorination by-products relies on ozone's effectiveness as a general alternative oxidant/disinfectant as well as its ability to specifically oxidize some by-product precursors. Also, the benefits of ozone prior to granular activated carbon (GAC) and sand filtration are due, in part, to the increase in biodegradability of the macropollutants as a result of oxidative reactions. The carbon adsorbent or sand then serves as support for active biota. When ozonation is used prior to carbon adsorption, there is evidence that significant physicochemical interactions may also play a role. The third combined process, ozonation prior to coagulation, is exclusively a physicochemical treatment. At present, the mechanisms responsible for ozone's coagulation effects are poorly understood. It is classified above in a separate category, because not enough is currently known about this process to properly categorize these reactions.

In most water treatment plants, ozone is used for multiple applications. When oxidation of free iron and manganese or improved coagulation is the objective, ozone is usually applied at the head of the treatment plant (Table III–1). The use of ozone to improve GAC or slow sand filtration requires that it be added shortly before adsorption or filtration. This is usually midway through the process train and is referred to as intermediate ozonation. Taste and odor control may also be best practiced through intermediate ozonation. For other uses, ozone may be applied at almost any point in the treatment process. For example, color removal by ozone can be accomplished at any treatment stage, although concerns over minimizing ozone dose requirements and regrowth problems make intermediate application very attractive (see sec. III.C). In the United States, application of ozone for primary disinfection (i.e., major disinfection steps within the treatment plant) is most common at the head of the plant. However, in Europe, primary disinfection with ozone is more commonly practiced at the end of treatment. Ozone can also be used at any intermediate point if it is needed to control growths within the plant. The addition of ozone at multiple locations within water treatment plants is thus a natural outgrowth of its diverse applications. In the United States and Europe, it is not uncommon to use ozone in two stages (preozonation and intermediate ozonation). Several French plants are now using three-stage ozonation (preozonation, intermediate ozonation, and postozonation).

Another important consideration in the use of ozone is the relative effectiveness of molecular O_3 versus its radical decomposition products in water. As discussed in chapter II, the molecular attack is more selective and, therefore, less hindered by competitive reactions (i.e., oxidant-demanding substances). On the other hand, certain reactions with molecular ozone are too slow to be economically feasible in water treatment, and, in these cases, the radical pathway may be preferable. Table III–1 lists the particular pathway that is believed to be the most feasible (although in some cases this may be site specific). Recall from chapter II that low pH and high inorganic carbon concentration encourage the direct molecular attack. Indirect radical attack is favored by high pH, low concentration of carbonates, and presence of activating substances, such as hydrogen peroxide or ultraviolet (UV) light (these are advanced oxidation processes).

III.A.2 Current Practice

The increase in the use of ozone worldwide is partly due to the pioneering work of a small number of engineers and scientists. These individuals have devoted much of their time to the study and application of ozone, and have communicated their knowledge at professional meetings and through scientific publications. For example, Gomella (1967, 1972) and Guillerd (1968) in France, and later Rosen (1980), Rice, and Miller (Miller et al. 1978) in the United States were instrumental in advancing the science and promoting the use of ozone. Current ozone practice owes much to these individuals and to the many others who have laid the groundwork for our present understanding of ozone technology.

North America. Current applications in the United States include nearly all of those listed at the beginning of this chapter. The most common are oxidation of taste and odor compounds, bleaching of color, and control of chlorination by-products (Fujikawa et al. 1989). Figure III–1 is a frequency histogram of the stated purposes of approximately 55 US ozone plants. This list includes plants that are currently operating, some that are under construction, and a few that are under design. Most of these plants credit their choice of ozone to more than one purpose (for example, control of disinfection by-products [DBPs] along with some disinfection).

At least seven U.S. plants are now using ozone at two application points (Robson et al. 1990). Many of these are plants treating water with high concentrations of natural organic matter. In order to control both color and chlorination by-products without creating regrowth problems, these plants generally add ozone prior to coagulation and again prior to rapid-rate filtration.

With the new U.S. Safe Drinking Water Act Amendments, more attention will be paid to ozone application for disinfection and DBP control. This means that ozone will be used as a primary disinfectant and that more two-stage applications will be considered.

In Canada, the application of ozone in water treatment has been heavily influenced by the French experience and philosophy. Most (approximately 80 percent) of the

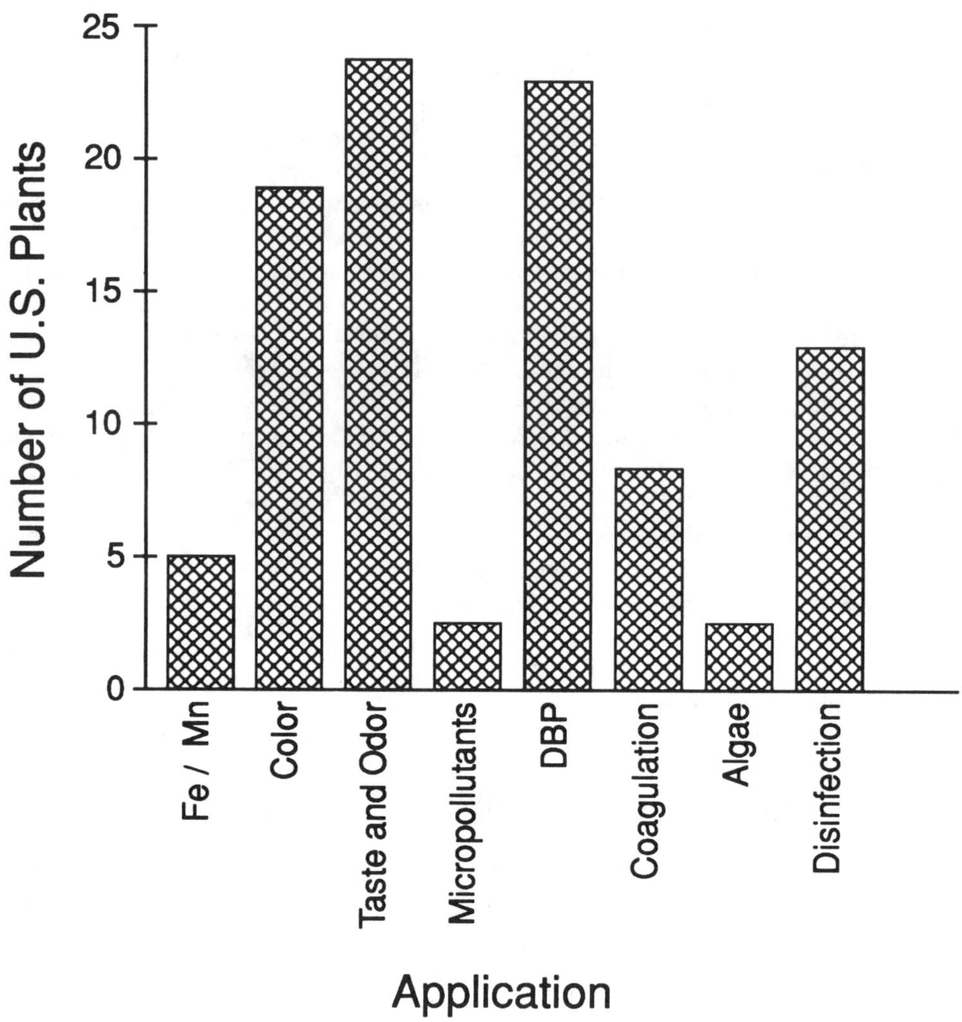

Figure III–1 Applications of Ozone in U.S. Drinking Water Treatment Plants

Canadian ozone drinking water plants are located in the province of Quebec (Bell 1987; Huck 1990). The majority of these employ ozone as a primary disinfectant at the end of treatment. More recently, ozone has been used in conjunction with GAC adsorption (Topley 1987). One of the first applications of biological activated carbon (for biological stabilization) in Canada is the full-scale plant at Laval, Quebec (Prévost et al. 1989b). Canadian plants outside of Quebec conform more to the U.S. model, in which ozone is used as a primary disinfectant in order to minimize the formation of chlorination by-products. It is also commonly used for controlling color, taste, and odor.

Europe. In Europe, and particularly in France, ozone has been primarily used for disinfection at the end of the water treatment train (see chapter I, sec. I.A.2). Based on studies of virus inactivation, the maintenance of a residual of 0.4 mg/L for 4 min has been a de facto standard (see sec. III.H). Among the 700 French water utilities using ozone, a very large majority (more than 80 percent) are equipped with postozonation. In other European countries that have been using ozone for a long period of time (e.g., Switzerland, Germany, USSR, Belgium), postozonation has been and still is widely used. For example, there are currently more than 20 European plants with an ozone capacity of 20 kg O_3/h (1060 lb/day) or greater that employ postozonation (Bourbigot 1988).

More recently, the use of intermediate ozonation (ozone prior to filtration) has become common in France, Switzerland, and Germany. It is becoming widespread as more is learned regarding biodegradation of natural organic matter (see sec. III.K) and the ability of biological processes to remove ammonia. When activated carbon is used as a filtration media, this process can also effectively remove tastes and odors (at the Choisy-le-Roi and Morsang-sur-Seine plants in France; see sec. III.D.5). Intermediate ozonation can also be used with sand filtration or dual-media filtration (anthracite/sand or activated carbon/sand). This type of application is being used in countries that did not previously have extensive experience with ozone (for example, Spain and Italy). Intermediate ozonation is also beginning to be used in the United Kingdom for taste and odor control or color reduction. A more unique use of intermediate ozonation in the United Kingdom is for delaying head loss buildup in slow sand filtration (at the Fobney plant, England; see sec. III.G.4). Intermediate ozonation can also be applied for the purpose of inactivating viruses (at the Choisy-le-Roi plant—France; see sec. III.D.5). Lastly, intermediate ozonation has recently been used for iron and manganese removal from eutrophic waters in western and central France (Paillard et al. 1990a). This application is especially effective when it follows remineralization (see sec. III.B.2) and it represents a practical application of fundamental research on aqueous ozone chemistry. It is expected that this use of ozone will be more widely applied in Europe in the future.

The third point of ozone application, preozonation, was first used for iron and manganese removal in groundwaters from France (see sec. III.B.2) and Germany (for example, the Düsseldorf process [Masschelein 1990]). Now iron and manganese removal by preozonation is also being developed in eastern European countries such as Hungary, Yugoslavia, and Bulgaria. Preozonation for improvement of flocculation and sedimentation has also been widely used, even though little is known of the fundamental basis for this effect. Extensive pilot studies are, therefore, generally performed before ozonation is installed for this purpose (for example, at the Choisy-le-Roi plant [Paillard et al. 1990b]). Preozonation is used as a coagulant aid in Germany (Mulheim, Essen), France (Choisy-le-Roi), Switzerland (Lengg, Lauzanne), and the USSR (Moscow). Between these two principal applications, about 15 percent of the 700 French ozone plants are equipped with a preozonation facility. Preozonation is also used for color removal in a few plants in the United Kingdom (see sec. III.C.3). Lastly, the use of preozonation combined with flotation has been examined in France (see sec. III.G.4). This process may also be appropriate for European countries having to treat eutrophic water (for example, the United Kingdom, the USSR, and Bulgaria).

At many locations, multistage application of ozone has proven to be most appropriate and cost-effective. Masschelein (1990) estimates that about 300 plants in Europe use a two-stage ozonation treatment. In France, about 10 percent of the 700 plants using ozone have a two-stage ozonation facility, with all the possible arrangements being represented (pre + inter, inter + post, pre + post). Approximately the same percentage of two-stage plants can be found in Germany, Switzerland, Italy, and Spain. In the United Kingdom, two-stage ozonation is beginning with the plants at Bristol and Campion Hills. About 70 percent of the new ozone plants in eastern Europe will include two ozonation steps. In France, three plants are equipped with three stages of ozonation. These include one large plant (Mery-sur-Oise, modified in 1980; see sec. III.H.2) and two smaller plants (Guy Robin treatment plant at St. Coulitz, Finistère, installed in 1985, with a capacity of 417 m^3/h [2.6 mgd] and 1.5 kg O_3/h [79 lb/day]; and Le Marillet, Vendée, installed in 1986, with a capacity of 1000 m^3/h [6.3 mgd] and 6 kg O_3/h [318 lb/day].

Based on trends in ozone use in Europe, and especially in France, one can make the following observations:

- Many plants that originally used postozonation for disinfection are looking to intermediate ozonation with activated carbon. This is especially true of plants treating high–dissolved organic carbon (DOC) waters.

- Plants using preozonation for iron and manganese removal from high-DOC waters (for example, lakes or reservoirs) are looking to intermediate ozonation so that more of the organic matter can be removed before addition of ozone.

The result of these trends will be that intermediate ozonation will be used more frequently in the future due to increases in the use of high-DOC surface sources. In the case of groundwaters, preozonation for iron and manganese control and postozonation for disinfection will continue to be applied. The use of preozonation for enhancing coagulation will continue to be developed, and it will likely be coupled with an intermediate stage of ozonation for biological stabilization. Lastly, advanced oxidation processes will be used at some full-scale plants. The development of these systems is based on extensive laboratory research performed in Europe and the United States. Most promising are the processes using ozone with hydrogen peroxide and possibly ozone plus metallic oxides. These processes have special applications for the removal of specific micropollutants such as chlorinated solvents, pesticides, and herbicides.

Eastern and southern hemispheres. As mentioned in chapter I, ozone is generally not used for the treatment of drinking water in the Middle East, Far East, Africa, and South America. A few older plants have used ozone for disinfection, based on the French philosophy. The small number of plants that have been installed in the past few years and those that are under design incorporate some of the more modern ozone applications. For example, in Japan, several future plants will use intermediate ozonation followed by activated carbon filtration. In Malaysia, the first ozone plant will be equipped with both preozonation and intermediate ozonation stages.

III.B IRON AND MANGANESE REMOVAL

III.B.1 Principles of Removal

The origin of the soluble forms of iron and manganese found in water and the oxidation–reduction reactions appropriate to the oxidation of Fe^{2+} and Mn^{2+} by ozone were described in chapter II, sec. II.A.3. This section will review the literature that deals with reduced iron and manganese removal, emphasizing studies utilizing ozone. Examples of pilot- and full-scale applications of ozone for removal of these metals will also be presented.

It is important to remember that the removal of Fe^{2+} and Mn^{2+} from drinking water is not related to a specific health problem. Rather, the problems associated with elevated concentrations of either species are aesthetic in nature. Water discoloration is a frequent consumer complaint. Also, growth of iron- and manganese-oxidizing bacteria on water mains can lead to a general deterioration of the quality of the water distributed. For these reasons, standards have been established for maximum concentrations of both soluble species in drinking water. The current French standards list fixed concentrations at a maximum of 50 μg/L for manganese and 200 μg/L for iron (Decree no. 89–3 1989). In the United States, recommended secondary maximum contaminant levels (SMCLs) are 50 μg/L for manganese and 300 μg/L for iron.

Two primary types of treatment for removing these metals exist. Treatment includes:

- physicochemical treatments that involve an initial oxidation step followed by a solid–liquid separation step for capture of the resulting insoluble metal oxide species (settling and/or filtration); and
- biological treatments that use bacteria, specific to these metals, on fixed-bed filters (Mouchet et al. 1985; Philipot 1985).

The physicochemical treatment is most commonly used; the oxidation stage, which enables insoluble varieties of these metals to be formed, can be achieved by all oxidants

Table III–2 Reactions of Fe(II) and Mn(II) with Alternative Oxidants and the Theoretical Reaction Stoichiometry

Metal/Oxidant	Reaction	Stoichiometry	Eq. No.
1. Fe(II)			
O_2 (aq)	$2Fe^{2+} + \frac{1}{2}O_2 (aq) + 5H_2O \rightarrow 2Fe(OH)_3 (s) + 4H^+$	0.14 mg O_2/mg Fe	[1]
O_3 (aq) → O_2 (aq)	$2Fe^{2+} + O_3 (aq) + 5H_2O \rightarrow 2Fe(OH)_3 (s) + O_2 (aq) + 4H^+$	0.43 mg O_3/mg Fe	[2]
HOCl	$2Fe^{2+} + HOCl + 5H_2O \rightarrow 2Fe(OH)_3 (s) + Cl^- + 5H^+$	0.64 mg HOCl*/mg Fe	[3]
$ClO_2 \rightarrow ClO_2^-$	$Fe^{2+} + ClO_2 + 3H_2O \rightarrow Fe(OH)_3 (s) + ClO_2^- + 3H^+$	1.20 mg ClO_2/mg Fe	[4]
$KMnO_4$	$3Fe^{2+} + MnO_4^- + 2H_2O \rightarrow 3Fe(OH)_3 (s) + MnO_2 (s) + 5H^+$	0.94 mg $KMnO_4$/mg Fe	[5]
2. Mn(II)			
O_2(aq)	$Mn^{2+} + \frac{1}{2}O_2 (aq) + H_2O \rightarrow MnO_2 (s) + 2H^+$	0.29 mg O_2/mg Mn	[6]
O_3(aq) → O_2(aq)	$Mn^{2+} + O_3 (aq) + H_2O \rightarrow MnO_2 (s) + O_2 (aq) + 2H^+$	0.88 mg O_3/mg Mn	[7]
HOCl	$Mn^{2+} + HOCl + H_2O \rightarrow MnO_2 (s) + Cl^- + 3H^+$	1.30 mg HOCl*/mg Mn	[8]
$ClO_2 \rightarrow ClO_2^-$	$Mn^{2+} + 2ClO_2 + 2H_2O \rightarrow MnO_2 (s) + 2ClO_2^- + 4H^+$	2.45 mg ClO_2/mg Mn	[9]
$KMnO_4$	$3Mn^{2+} + 2MnO_4^- + 2H_2O \rightarrow 5MnO_2 (s) + 4H^+$	1.92 mg $KMnO_4$/mg Mn	[10]

Source: Reprinted with permission from Pouvreau, P., *Jour. François d'Hydrologie,* 2:169. © 1984 *Journal François d'Hydrologie.*
*HOCl expressed as Cl_2 by convention.

normally used in water treatment (chlorine, chlorine dioxide, potassium permanganate, ozone) (Pouvreau 1984; Knocke et al. 1987). Dissolved oxygen is also effective for Fe^{2+} oxidation; Mn^{2+} oxidation by O_2(aq) is effective only under fairly alkaline pH conditions (Furgason and Dayor 1975; Morgan and Stumm 1964a, 1964b; Stoebner and Rollag 1981). Table III–2 lists appropriate chemical reactions between the various oxidants and Fe^{2+} and Mn^{2+}.

The choice of a reagent will depend principally on several parameters: efficiency of oxidation observed, required dosage and associated chemical cost, and competitive oxidant demands in solution (for example, sulfide, nitrite, ammonia, DOC). Often, the required oxidant dosages are significantly above those listed in Table III–2 due to competing oxidation reactions that occur during treatment (Stoebner and Rollag 1981; Vigreux and Richard 1979; Knocke et al. 1987).

Oxidation by dissolved oxygen. In the case of iron-containing well waters, aeration can be an effective form of treatment (see Eq. 1, Table III–2). The rate of this reaction increases significantly as a function of both O_2(aq) and solution pH (Stumm and Lee 1960). Although oxidation of manganese by air is thermodynamically possible, it is a slow process and requires alkaline pH conditions (pH above 9.2–9.5) to achieve acceptable rates of Mn^{2+} oxidation by O_2(aq) during water treatment (Morgan and Stumm 1964b). Few treatment facilities employ dissolved oxygen specifically for manganese removal; instead, such removal is often a by-product during water softening operations. Morgan and Stumm (1964a, 1964b) further demonstrated that Mn^{2+} removal in the presence of MnO_x(s) was best described by an autocatalytic model.

Complexation of dissolved iron by organic matter (DOC) greatly decreases the Fe^{2+} oxidation efficiency of O_2(aq) (Stumm and Singer 1966; Jobin and Ghosh 1972). Theis and Singer (1974) presented a mechanistic description that considered the relative roles of complexation and dissolved oxygen in the stability of iron in natural waters.

Oxidation by free chlorine. Oxidation of Fe^{2+} and Mn^{2+} by free chlorine is accomplished by the chemical reactions listed as Eq. 3 and Eq. 8 in Table III–2. The reaction between uncomplexed Fe^{2+} and HOCl is rapid. Knocke et al. (1990) reported that the required HOCl contact time to oxidize 2.0 mg/L Fe^{2+} varied from 3–5 min at pH 5.0 to less than 5 s at pH 7.0. Reduced manganese is oxidized much less efficiently than iron when considering free chlorine addition. Alkaline pH conditions and dosages well above the stoichiometric requirement are required to produce effective rates of oxidation (Knocke et al. 1987, 1990). The efficiency of chlorine for Mn^{2+} removal is significantly enhanced when chlorine is used in conjunction with oxide-coated filter media (Edwards and McCall 1946; Weng et al. 1986; Knocke et al. 1988).

Free chlorine is relatively ineffective for the oxidation and subsequent removal of complexed iron (Knocke et al. 1990). The presence of DOC does not reduce the efficiency of free chlorine for Mn^{2+} oxidation since soluble manganese is not readily complexed by organic matter (Knocke et al. 1990).

In the case of well water containing sulfides, the use of chlorine can lead to the formation of polysulfides, which give off odors that are much stronger than those produced by sulfides alone.

With regard to impounded water, chlorine also reacts with the organics present in the water, leading to compounds that are suspected of having adverse health effects (see chapter II, sec. II.A.6, and sec. III.I).

Oxidation by chlorine dioxide. Application of ClO_2(aq) to waters containing uncomplexed iron and manganese results in rapid oxidation of either metal for solution pH conditions above 5.0–5.5 (see Eq. 4, Eq. 9, and Table III–2). However, elevated DOC concentrations can significantly reduce the efficiency of ClO_2(aq) to oxidize either Fe^{2+} or Mn^{2+}. Reduced iron that had been complexed by humic and fulvic acids was found to be relatively stable in the presence of ClO_2 concentrations of 3–5 mg/L as Cl_2, even though the ClO_2 persisted in solution for over 30 min. Manganese oxidation by ClO_2 was impeded by the competitive oxidant demand exerted by the DOC (Knocke et al. 1990).

Oxidation by potassium permanganate. The application of potassium permanganate for Fe^{2+} and Mn^{2+} oxidation has been practiced by water treatment facilities for several decades (Ficek 1978; also see Eq. 5, Eq. 10, and Table III–2). Uncomplexed iron is oxidized almost instantaneously by $KMnO_4$ for pH conditions above 5.0. Likewise, the oxidation of Mn^{2+} by $KMnO_4$ is rapid for pH values above 5.5 (Knocke et al. 1990).

The presence of significant amounts of DOC will decrease the effectiveness of potassium permanganate for iron and manganese removal. Knocke et al. (1990) evaluated the application of $KMnO_4$ for the oxidation of soluble iron that had been previously complexed by humic and fulvic acids from natural sources. Results indicated that permanganate was ineffective when applied to iron–humate complexes; removal was more efficient when considering iron complexed by fulvic acids, although long detention times (greater than 60 min) and $KMnO_4$ dosages well above stoichiometric requirements were necessary. With respect to Mn^{2+} oxidation, the presence of DOC may produce a significant competitive oxidant demand, requiring higher dosages of $KMnO_4$ to accomplish the desired degree of manganese removal (Knocke et al. 1987).

Oxidation by ozone. In this section, the basic kinetics and stoichiometry of iron and manganese oxidation by ozone are applied to drinking water treatment. In particular, the interference from organic matter is discussed. For a discussion of the mechanisms and kinetics of iron and manganese oxidation, the reader is referred to chapter II, sec. II.A.3.

The standard oxidation–reduction potential and reaction rate of ozone are such that it can readily oxidize iron and manganese in groundwater and water of low organic content. Due to the extremely fast kinetics of iron oxidation, uncomplexed Fe(II) can easily outcompete Mn(II) for ozone. Therefore, in groundwater containing iron, little manganese oxidation occurs at low doses, such that only after the Fe(II) is nearly gone does the oxidation of Mn(II) begin (Stoebner and Rollag 1981). With groundwater and water of low organic content, the doses of ozone required to completely oxidize iron and manganese are close to the theoretical stoichiometric doses—0.43 mg O_3/mg Fe and 0.88 mg O_3/mg Mn (Eq. 2 and Eq. 7, Table III–2) providing the treated water contains no other ozone-demanding substances, such as nitrites or sulfide.

However, excessive doses will lead to the formation of permanganate, which gives the water a pinkish color. This soluble form of manganese corresponds to a theoretical stoichiometry of 2.20 mg O_3/mg Mn. The oxidation of manganese with ozone is less dependent on pH than for other oxidants. As a result, ozone is more likely to offer kinetic advantages at low pH's than at high pH's. With regard to groundwater, the use of ozone has the additional advantage of reoxygenating the water at the same time.

The presence of organic material in colored surface water inhibits the removal of iron and manganese. Organic matter (for example, as radical scavengers or promotors), may change the main pathway of ozone reactions (direct or indirect), change the speciation of the metal (for example, through complexation), or simply compete with the metal for oxidant. It is well known that humic materials and model organic compounds can complex iron and reduce its susceptibility to oxidation (Hem 1960; Jobin and Ghosh 1972; Theis and Singer 1974). The degree to which complexed iron is oxidized by ozone is likely to be dependent on the pH and the nature and concentration of the organic matter. While low concentrations of humic substances can prevent the oxidation of Fe(II) by oxygen, oxidation by ozone is generally more successful. Cromley and O'Connor (1976) found that high ozone doses caused the oxidation of most of the iron present in a colored water, while at the same time resulting in the formation of some very stable iron-organic complexes that could not be subsequently oxidized. These stable complexes were not formed when the water was simply oxygenated.

In the case of impounded water, ozone also reacts on the organic matter contained in the water. Humic substances can successfully compete with manganese for ozone, thereby requiring doses far in excess of the stoichiometric values (Giannissis et al. 1985). For this reason, it is best to first remove organics before applying ozone for the purpose of removing manganese. This can be accomplished by raw water coagulation and sedimentation, preferably at an acid pH (around pH 5) using either aluminum (Dempsey et al. 1984; Reckhow and Sibony 1986) or iron salts ($FeCl_3$ or $FeSO_4Cl$) (Bourbigot et al. 1986). In addition, coagulation itself may result in modest removals of reduced manganese. This is because a small fraction of the Mn(II) is organic-bound. Still, there will always be some residual interfering organic matter present in coagulated water. The amount of excess ozone needed to oxidize all of the manganese present will depend on the residual organic carbon concentration and on the reactivity of that organic carbon.

Laboratory studies on both synthetic, reconstituted water and natural water high in iron, manganese, fulvic acid, and humic acids (Paillard et al. 1989a, 1989b; Lefebvre 1987) have shown that for a fixed alkalinity (150 mg/L of $CaCO_3$) and at a constant ozone/DOC application rate, the unoxidized manganese residual after treatment increases as the organic concentration increases. For the water under study (pH = 8.5, manganese concentration in the region of 250 μg/L), a dose of 0.3 mg O_3/mg C resulted in a manganese residual, after sand filtration, of less than 60 μg/L; an ozonation dose of 0.5 mg O_3/mg C resulted in a filtered-water manganese residual of under 30 μg/L. On the other hand, higher levels of ozone ultimately led to the formation of permanganate, which persisted through filtration. Similar results were obtained with actual impounded water (Table III–3).

In addition to effective manganese removal, intermediate recarbonation and ozonation can lead to a reduction in the UV absorbance of the filtered water and in the concentration of trihalomethane (THM) precursors (see sec. III.I).

Another recent study (Knocke et al. 1989) using synthetic water (pH 6.3, 1 mg/L

Table III–3 Efficiency of Recarbonation and Ozonation*

Type of water	pH	TOC (mg/L)	UV (254 nm)	Fe (μg/L)	Mn (μg/L)	THM (μg/L) after 24 h chlorination†		
						$CHCl_3$	$CHCl_2Br$	$CHClB_2$
RW‡–October 1986	8.3	12.5	0.262	540	270			
Settled water (Clairtan,§ 130 g/L)	5.7	4.4	0.067	500	500			
Settled water (+ $NaHCO_3$, 140 mg/L $CaCO_3$) and filtered	8.4	4	0.062	100	410	39	16.5	8.5
Settled water (+ $NaHCO_3$, 140 mg/L $CaCO_3$ + O_3: 1 mg/L) and filtered	8.3	3.8	0.033	none detected	45	31	13	7

Source: Paillard et al. (1989a).
*Moulin Papon dam water, pH adjusted at 8.4, total alkalinity maintained at 140 mg $CaCO_3$/L, ozonation rate 1 mg/L.
†Chlorine dose = 12.5 mg/L.
‡RW = Raw water.
§Ferric chlorosulfate.

Mn(II), 2–5 mg/L total organic carbon [TOC], 50–200 mg $CaCO_3$/L) showed that 75 percent manganese removal was achieved after application of an ozone dose of 0.2–0.7 mg O_3/mg C in excess of that required for manganese oxidation alone (according to the theoretical stoichiometry). The precise excess ozone dose depended on the bicarbonate concentration. The reasons for this are discussed below.

The presence of high levels of bicarbonate can modify consumption of oxidant such that less ozone is required (Figure III–2). This is presumably due to the stabilizing effect of bicarbonate on ozone. In the absence of bicarbonate and the presence of naturally occurring initiator species (for example, hydroxide, humic substances), much of the transferred ozone will decompose to hydroxyl radicals (see chapter II, sec. II.A.2 and II.A.6). Although these species are potent oxidants, they are less selective than molecular ozone. Thus, more ozone will be required to achieve the same degree of Mn oxidation. For example, the excess ozone dose required to oxidize 75 percent of the manganese as reported by Knocke et al. (1989) was inversely proportional to the alkalinity. At an alkalinity of 50 mg $CaCO_3$/L, about 0.7 mg O_3/mg C was needed, whereas at 200 mg $CaCO_3$/L only 0.2 mg O_3/mg C was needed. For this reason, intermediate recarbonation has been used to improve the efficiency of manganese oxidation by ozone. The application of CO_2 has the additional advantage of providing supplementary buffer capacity against sudden changes in pH, which could cause the filter to release manganese.

Once the manganese is oxidized to Mn(III) or Mn(IV), it can be removed by flocculation and sedimentation, rapid sand filtration, or multi-media filtration (Collienne and Cornet 1978). Filtration through "aged" sand is especially efficient as it also permits the removal of small amounts of reduced manganese (Griffin 1960; Cleasby 1975; Weng et al. 1986). With time, the sand grains become coated with oxides of manganese and iron, and exhibit the "greensand" effect. This is a catalytic phenomenon that includes adsorption of Mn(II) on the oxide surface (Morgan and Stumm 1964a) and subsequent oxidation. For this latter process to occur over long periods of time, the sand surface must be periodically exposed to a chemical oxidant such as chlorine or permanganate. The oxidants undergo a redox reaction at the oxide surface, thereby renewing its oxidative power. As a result, small amounts of permanganate can be removed through filtration; however, excessive amounts will result in breakthrough.

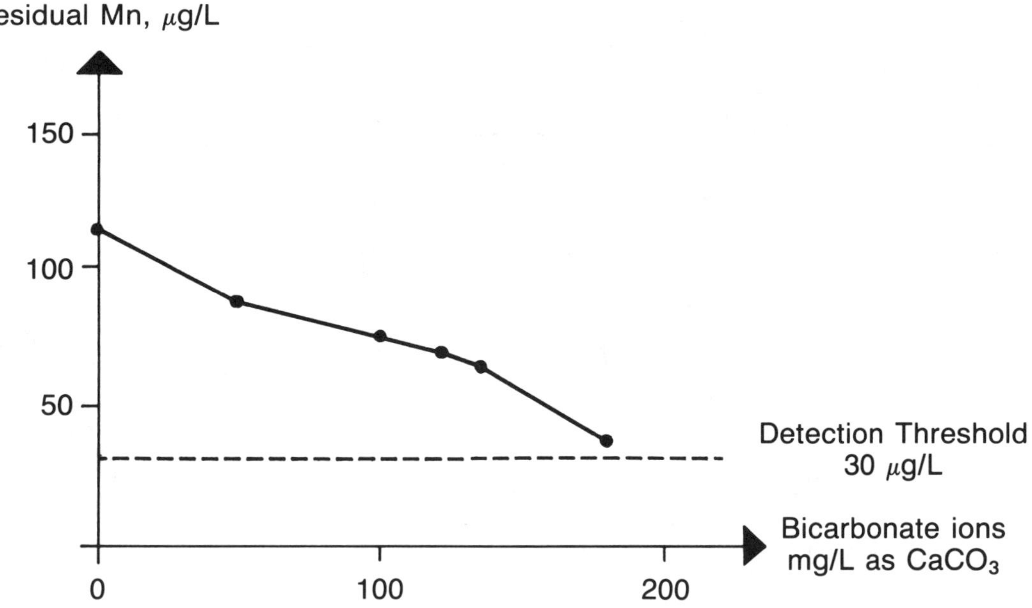

Source: Reprinted with permission from Paillard, H. et al., *Ozone Sci. Engrg.*, 11:93–114. © 1989 IOA.

Case of reconstituted water with fulvic acid from Pinail (France) (TOC: 5.4 mg C/L).

Figure III–2 Influence of Bicarbonate Concentration on Manganese Removal at a Constant Ozone Dose of 1 mg/L

III.B.2 Case Studies

Treatment of groundwater.

Villefranche-sur-Saône water works (Rhône, France). For several years, the quantities of iron and manganese have been increasing in the waters of the Beauregard catchment area, through which the alluvial layers of the Saône River are tapped for use as raw water. In 1984, a new treatment facility with a capacity of 1250 m^3/h (about 8 mgd) was commissioned. Figure III–3 shows the treatment line managed by the Lyon regional office of Compagnie Générale des Eaux (CGE).

Eleven wells drilled to a depth of 13.50 m (44 ft 3 in.) are used to supply a 500-m^3 (1.32×10^5–gal) raw water reservoir. This raw water is divided between two identical and parallel treatment trains, each treating 625 m^3/h (about 4 mgd) of water. On each of these trains, iron and manganese removal is performed by ozonation; the contact column consists of a series of three compartments. In the first chamber, where the water remains for 2 min, offgases from the postozonation chamber (disinfection stage) are injected by an immersed self-suction turbine. If necessary, this turbine can also be fed ozonized air from the ozone generators. In the second compartment, which is designed for a contact time of 2 min, ozone from the generators is diffused by an immersed pressure turbine. No ozone is introduced into the third compartment, which also has a 2-min contact time. Upstream of this preozonation column, it is possible to add potassium permanganate. The ozone dosage averages 0.1 mg/L, for average iron and manganese concentrations in the influent of 50 and 75 µg/L, respectively. Before filtration, aluminum sulfate ($Al_2(SO_4)_3 \cdot 18H_2O$) is added directly into the pipeline at a rate of 5 mg/L.

Two sand filters with a filtering surface of 43 m^2 (463 ft^2) each (grain size of sand—0.8–1.2 mm [0.03–0.05 in.]; height of layer—1.3 m [4 ft 3 in.]), operating at a velocity of 7.2 m/h (2.97 gpm/ft^2), arrest the oxidized forms of these metals. After

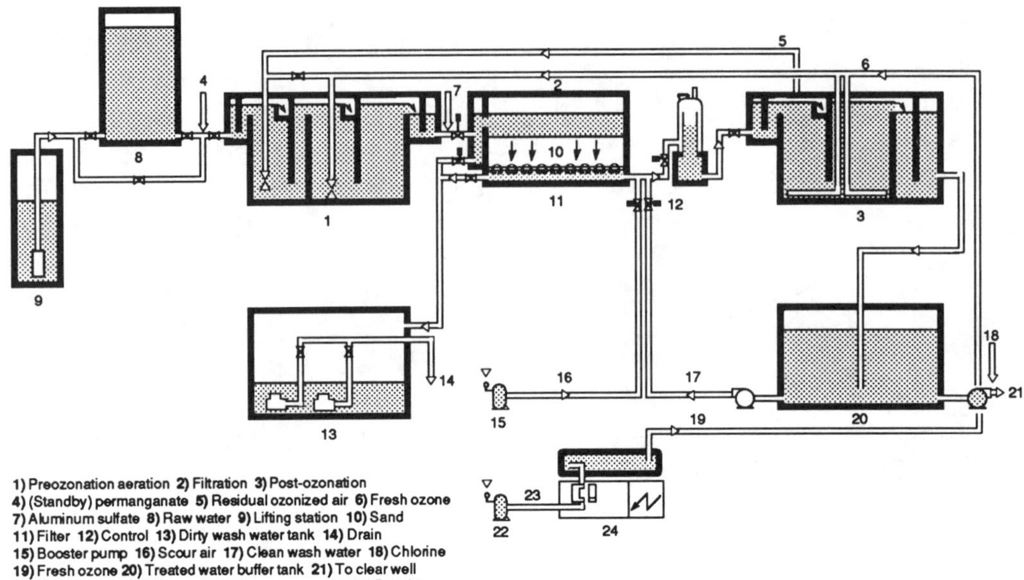

Source: Lyon's CGE brochure.

1250 m^3/h (8 mgd).

Figure III–3 Treatment Line at Water Works Supplying the Greater Villefranche-sur-Saône District

filtration, disinfection by ozone is performed by the usual methods (see sec. III.H and sec. IV.D), the water remains 9 min in the contact column where the ozonized air is diffused via porous plates. The average ozonation dosage applied for disinfection is between 0.4 and 0.5 mg/L. The ozone produced from air is supplied by two generators with a unit capacity of 2 kg O_3/h (105 lb/day). The postozonated water is stored in a 400-m^3 (1.06×10^5-gal) buffer tank, then chlorinated at 0.15 mg Cl_2/L before flowing to a 1000-m^3 (2.6×10^5-gal) reservoir from which drinking water is distributed to the town of Villefranche-sur-Saône.

Table III–4 gives the minimum, average, and maximum concentrations of total iron and manganese in the raw and treated water for 1988.

Table III–5 shows concentrations of total and dissolved manganese in random samples collected during visits to the treatment facility. The ozone dose used to eliminate manganese is small since the water contains low concentrations of manganese. This explains why the abatement is not very good. However, the treated water meets current standards. Iron concentrations along the treatment train are below the specific detection limits (less than 20 μg/L) for the two sampling days.

Saumur water works (Maine et Loire, France). This facility uses raw water from the alluvial layers of the Loire River. This water is drawn from the water table via five large-diameter wells and three deep drillings. The latter are used only in cases of emergency, because the water they provide contains more manganese than the well water, as well as more hydrogen sulfide. The treatment train installed in this plant, commissioned in 1988, is able to treat 1100 m^3/h (6.97 mgd). The plant is illustrated in Figure III–4.

The preozonation column consists of two chambers. In the first chamber, which provides a contact time close to 3 min, offgases are reinjected from the second chamber by an immersed self-suction turbine. In the second chamber, where the water undergoes a 5-min contact stage, ozone from the generator (capacity is 2.5 kg/h [132.3 lb/day]) is diffused by porous plates. The average ozone dosage is 0.5 mg/L, although the

Table III–4 Total Iron and Manganese in Raw Water and Treated Water at Villefranche-sur-Saône Water Works in 1988*

Type of Water	Total Mn (μg/L)			Total Fe (μg/L)		
	Min.	Av.	Max.	Min.	Av.	Max.
Raw water	30	75	250	<10†	50	110
Treated water	10	30	50	<10†	10	50

*Results obtained by the authors from CGE's plant operators.
†The detection limit of total Fe given by the laboratory's analytical method is <10 μg/L.

Table III–5 Total and Dissolved Manganese Contents Measured in Samples from Villefranche-sur-Saône Facility

	Raw Water	Filtered Water	Postozonated Water
	Sampling Performed 3/6/89		
Dissolved Mn, μg/L	90	35	35
Total Mn, μg/L	90	35	38
	Sampling Performed 8/22/89		
Dissolved Mn, μg/L	38	18	26
Total Mn, μg/L	40	—	30

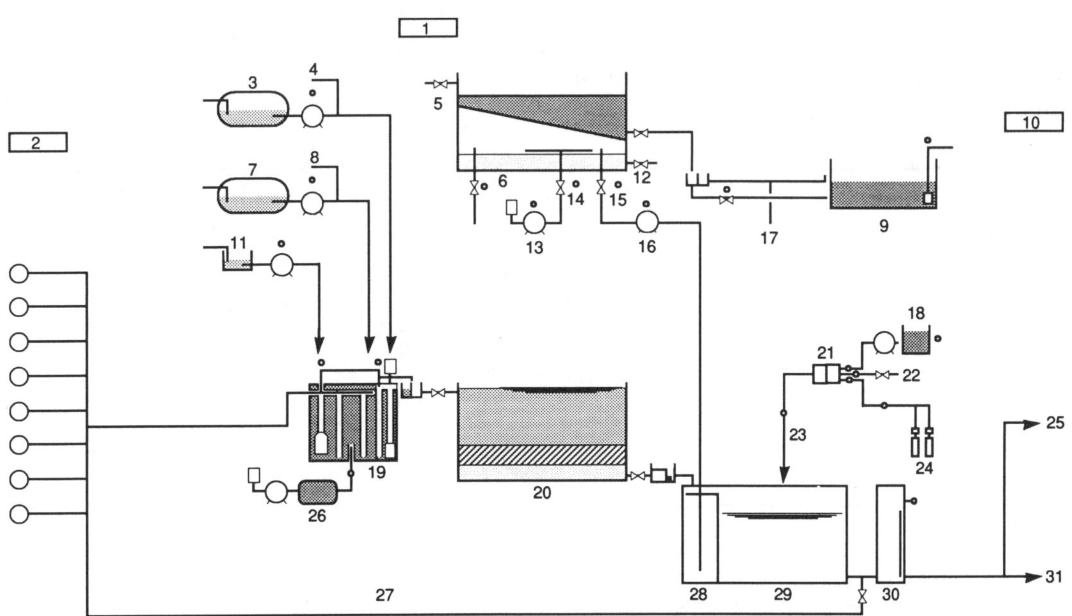

1) Treatment train 2) Pumping station 3) Caustic 4) Dilution 5) Inlet 6) Filter washing 7) Ferric chloride 8) Dilution 9) Waste water tank 10) Retaking pumps 11) Potassium permanganate 12) Outlet 13) Air booster 14) Air 15) Water 16) Wash water pump 17) Drain 18) Sodium chlorite 19) Aeration-ozonation-flash mix 20) Single-layer sand filters 21) Reactor 22) Process water 23) Chloride dioxide 24) Chlorine 25) Distribution to upper network 26) Ozone generator 27) Water works by-pass 28) Wash water tank 29) Clear well 30) Recovery tank 31) Distribution to lower network

Source: Saumur technical service department brochure.

Nominal flow rate = 1100 m^3/h (6.97 mgd).

Figure III–4 Treatment Train at Saumur Urban Water Works

Table III–6 On-Stream Development of Iron and Manganese at Saumur Treatment Facility

Date	Iron and Manganese Concentration, µg/L	Raw Water	Filtered Water	Treated Water
12/13/88*	Dissolved Fe	263	<20	<20
	Dissolved Mn	230	<20	<20
1/11/89*	Dissolved Fe	23	–	48
	Dissolved Mn	107	–	<20
1/26/89*	Dissolved Fe	130	–	40
	Dissolved Mn	290	–	27
2/8/89*	Dissolved Fe	38	–	<20
	Dissolved Mn	300	–	<20
3/16/89†	Dissolved Fe	30	<20	–
	Total Fe	40	30	–
	Dissolved Mn	20	<10	–
	Total Mn	20	10	–
8/16/89†	Dissolved Fe	32	<20	<20
	Total Fe	420	<20	44
	Dissolved Mn	212	<10	<10
	Total Mn	920	<10	21

Data noted * and † were not analyzed by the same laboratory, so the detection limits are not the same.
*Results from the Laboratoire Départmental d'Hydrologie de Maine-et-Loire, Angers, France.
†Results from the CGE Research Center Laboratory, Maisons Laffitte, France.

Table III–7 Moulin Papon Impounded Raw Water Quality and EEC Standards on Water for Consumption

Parameters	High/Low Values over 10 Years 1976–1986		High/Low Values, Summer 1987, on Treated Water		EEC Standards, 7/15/80 Directive	
	Min.	Max.	Min.	Max.	GL*	MAC†
pH at 20°C (68°F)	6.7	9.8	6.7	9.8	6.5–8.5	9.5
Color, mg Pt-Co/L	35	160	–	–	1	20
Turbidity, NTU	4	40	9	20	0.4	4
Hardness, mg/L Ca^{2+}	18	20	20.8	23.2	–	–
Alkalinity, mg/L $CaCO_3$	20	40	36	46	–	–
TOC, mg C/L	–	–	6.8	12	–	–
COD (H^+, $KMnO_4$), mg O_2/L	6	15	7.8	11.3	2	5
NH_4^+, mg/L	0.05	0.6	0.08	0.45	0.05	0.5
NO_2^-, mg/L	0.02	0.14	<0.005	<0.005	0.0	0.1
NO_3^-, mg/L	0.25	10.00	–	–	25	50
Total Fe, µg/L	300	1500	170	1000	50	200
Total Mn, µg/L	30	550	150	450	20	50

Source: Paillard et al. (1988).
*GL: Guidelines.
†MAC: Maximum Admissible Concentration.

maximum application, justified to allow for the poor quality of water from the deep drillings, is 1.9 mg O_3/L. Downstream of the ozonation stage, a flash-mixing tank provides a dose of 3–5 mg/L of ferric chloride, the pH being adjusted between 7.6 and 7.8 using caustic (8 to 10 mg/L of caustic). If necessary, permanganate and chlorine dioxide can also be added via the flash-mixing tank. The water is then distributed over four sand filters with a filtering area of 48 m^2 (516 ft^2) (grain-size distribution—0.8–1.2 mm [0.03–0.05 in.]; bed height—1 m [39 in.]); filtration rate—5.7 m/h [2.35 gpm/ft^2]). After this treatment, the water is disinfected with chlorine dioxide (average dosage is 0.2 mg/L).

Table III–6 gives a summary of the results obtained for this plant. The quality of treated water with regard to dissolved Fe and Mn is very good. For the most part, concentrations are below the detection limits of the analytical methods. These results are more significant than those for the previous case study since Saumur's raw water contains more Fe and Mn.

Treatment of impounded water.

Moulin Papon water works, La Roche sur Yon (Vendée, France). This 840-m^3/h (5.33-mgd) plant uses poor-quality raw water supplied by a reservoir. Table III–7 shows the main characteristics of the water to be treated, compared with European Economic Community (EEC) standards on drinking water quality.

The water utility, which originally consisted of a chlorine or potassium permanganate preoxidation stage, flocculation using aluminum sulfate, settling, interozonation, sand filtration, recarbonation and disinfection with ozone, followed by final chlorination with sodium hypochlorite (to protect the bacteriological quality of the water in the distribution system), had to be altered. Results were, in fact, poor with regard to the residual concentration of organics, color, and turbidity, as well as iron, manganese, and ammonia concentrations. After laboratory studies (see sec. III.B.1), which enabled an appropriate treatment train to be designed in view of the problems to be solved, tests were made directly onsite (Paillard et al. 1988a). Allowances were made during these tests for certain conditions imposed by the existing facility. Table III–8 depicts the three treatment lines tested on site. Table III–9 shows the average water quality obtained on each of the tested systems. These data indicate the high level of iron and manganese removal on train 3, despite the fact that the concentrations of these metals in the raw influent were greater than at the time the other two systems were tested. During some of the tests, recarbonation (pH = 8.2; alkalinity = 100 mg/L of $CaCO_3$) and ozonation (0.8 mg/L) plus filtration helped to reach residual manganese concentrations of under 30 μg/L, as opposed to settled water concentrations of 550 μg/L.

Additionally, the table shows that the results are favorable to train 3 with respect to removal of organic matter and control of THMs.

This example of a modified treatment train shows that the ozonation stage must be incorporated in the treatment train in a way that optimizes the potential of ozone as an oxidant.

In view of the results obtained by the use of treatment train 3, the Moulin Papon utility was altered accordingly, i.e., elimination of prechlorination, use of coagulation with ferric chloride (120 mg/L [~70 mg/L as $FeCl_3$]) at an acid pH (5.4–5.5), and addition of a flocculating aid (starch being replaced by anionic polymer); a flash-mixing tank will be installed upstream of the settling tank. After recarbonation, the ozonation dose varies between 0.8 and 1.3 mg O_3/L. Furthermore, existing sand filters will be replaced by dual-media sand-plus-activated carbon beds in order to encourage biological nitrification and the removal of biodegradable carbon (Bablon et al. 1987a) (see sec. III.K). Figure III–5 is a flow diagram of the new treatment train.

A sampling program performed in February and August 1989 for the purpose of monitoring iron and manganese concentrations through the treatment train led to the results given in Table III–10.

Hackensack pilot-plant studies. The Hackensack Water Company, in Hackensack, N.J., undertook a five-year pilot-plant investigation during the early 1980s to investigate the use of preozonation for treatment of water from the Oradell Reservoir. Representative water quality data for this reservoir for a three-year period are given in Table III–11. Principal treatment objectives were the removal of color, iron, and manganese as well as the minimization of THM concentrations in the finished water.

Experimental variables that related to ozone application during treatment included variations in ozone dosage and contact time and the use of preozonation versus both pre- and postozonation. More detailed information on the experimental protocol used has been presented by Weng et al. (1986).

Table III–8 Treatment Trains Tested Full Scale at Moulin Papon Plant

	Raw Water	Preoxidation + Lime Slurry	Flocculation, Settling	Remineralization or pH Adjustment	Interozonation	Filtration	Remineralization or pH Adjustment	Primary and Final Disinfection
Train 1: Original treatment, May '87	pH = 7.0 Alk.* = 37.5 mg/L $CaCO_3$	$KMnO_4$ = 0.5 mg/L	$Al_2(SO_4)_3$ = 90 mg/L	Lime slurry = 9 mg/L CaO	Ozone = 1–1.5 mg/L	Sand	Lime slurry = 18 mg/L CaO CO_2 = 20 mg/L	Ozone = 1.2 mg/L NaOCl = 0.5–2.5 mg/L Cl_2
	TH† = 53 mg/L $CaCO_3$	NaOCl = 12 mg/L Cl_2	Starch = 3.5 mg/L				pH = 8.0	
	18°C (64.4°F)		pH = 6.3	pH = 7.1				
		Lime slurry = 9–15 mg/L CaO	Alk. = 22 mg/L $CaCO_3$	Alk. = 39 mg/L $CaCO_3$			Alk. = 65 mg/L $CaCO_3$	
Train 2: Tested in June '87	pH = 7.1 Alk. = 39 mg/L $CaCO_3$	$KMnO_4$ = 0.5 mg/L	$Al_2(SO_4)_3$ = 90 mg/L	Lime slurry = 8–10 mg/L CaO	Ozone = 1 mg/L	Sand	Lime slurry = 18 mg/L CaO CO_2 = 20 mg/L	Ozone = 1.2 mg/L NaOCl = 4.5 mg/L Cl_2
	TH = 56 mg/L $CaCO_3$	NaOCl = 0	Starch = 3.5 mg/L					
	20°C (68°F)		pH = 6.3	pH = 7.1			pH = 8.0	
		Lime slurry = 9.15 mg/L CaO	Alk. = 21 mg/L $CaCO_3$	Alk. = 38 mg/L $CaCO_3$			Alk. = 64 mg/L $CaCO_3$	
Train 3: Experi-mental treatment, tested July–Oct. '87	pH = 7.0	$KMnO_4$ = 0–0.65 mg/L	$FeCl_3$ = 50 mg/L	Lime slurry = 18 mg/L CaO				Ozone = 1 mg/L
	Alk. = 40 mg/L $CaCO_3$				Ozone = 0.5–1 mg/L	Sand	pH = 7.9–8.0	
	TH = 56 mg/L $CaCO_3$	NaOCl = 0	Starch = 3.5 mg/L	$NaHCO_3$ = 114 mg/L $CaCO_3$			Alk. = 100–110 mg/L $CaCO_3$	NaOCl = 1.5–2 mg/L Cl_2
	20°C (68°F)		pH = 5.6–5.9	pH = 7.9–8.2				
		Lime slurry = 11.2 mg/L CaO	Alk. = 10 mg/L $CaCO_3$	Alk. = 100–110 mg/L $CaCO_3$				

Source: Paillard et al. (1988a).
*Alk. = total alkalinity.
†TH = total hardness.

Table III–9 Quality of Water Treated at Moulin Papon Water Works on Tested Treatment Trains*

	Train 1: May 1987 Original Treatment With Prechlorination		Train 2: June 1987 Original Treatment Without Prechlorination		Train 3: July–October 1987 Experimental Treatment	
	Raw	Treated	Raw	Treated	Raw	Treated
COD (H^+, $KMnO_4$) (mg O_2/L)	10.4	4	11	4.6	10.5	2.3
TOC (mg C/L)	6.7	4.1	8.2	4.9	7.5	2.65
NH_4^+ (mg/L)	0.10–0.42	<0.05	0.10–0.15	<0.05	0.10–0.50	<0.05–0.07
Color (mg Pt-Co/L)	–	–	–	–	12–24	1.5–2.5
Turbidity (NTU)	–	0.2–0.4	–	0.4	40–70	0.2–0.3
Total Mn (μg/L)	150	36	150	30–40	200	≤30
Total Fe (μg/L)	600	<10	450	10	700	≤10
Total Al (μg/L)	–	20–160	–	45	–	30
Free chlorine (mg Cl_2/L)	–	0.45–0.75	–	0.3–1	–	0.8
Total chlorine (mg Cl_2/L)	–	0.55–1	–	0.5–1.3	–	0.9
$CHCl_3$ (μg/L)	<2	168.4	–	10	–	<5
$CHCl_2Br$ (μg/L)	<1	54.4	–	9	–	5
$CHClBr_2$ (μg/L)	<1	6.5	–	6	–	6
$CHBr_3$ (μg/L)	<2	Trace	–	Trace	–	7
Total THMs (μg/L)	<2	230	–	25	–	20

Source: Paillard et al. (1988a).
*See Table III–8 for treatment details.

Representative experimental results from the pilot-plant operations are contained in Table III–12. In general, the use of preozonation in conjunction with alum coagulation and direct filtration was found to result in successful treatment of the Oradell Reservoir water. Efficient removal of turbidity, color, and iron was achieved on a daily basis. In most instances, residual manganese concentrations were well below 50 μg/L in the filtered water; however, there were several days when effluent manganese concentration exceeded the 50-μg/L level. Studies indicated that the most efficient manganese removal was accomplished when oxide-coated filter media were utilized. The data indicated that the media functioned best for manganese capture when free chlorine was added to the filter-applied water in sufficient quantities to obtain free chlorine in the filter effluent. Trihalomethane formation was minimized by the application of ammonia to the filter effluent water, resulting in the conversion of any residual chlorine to chloramines.

Certain pilot-plant trials involved the use of GAC following filtration. Results indicated that GAC produced enhanced removal of DOC, color, iron, and manganese; a reduction in THM levels was also observed when GAC influent and effluent samples were considered.

The treatment scheme chosen for the full-scale facility (34,700 m^3/h [220 mgd]) is shown in Figure III–6; design criteria are contained in Table III–13. Construction of the treatment facility was completed in 1989, and included four treatment trains that treat 8675 m^3/h (55 mgd) with four ozone generators with a capacity of 12.8 kg O_3/h (675 lb/day). Initial measurements of treatment performance indicate that the facility produced finished water of desired quality with respect to color, DOC removal, iron and manganese removal, and residual THM concentrations.

III.B.3 Design Considerations

Treatment of groundwater. It has been noted in sec. III.B.1 that for groundwater in which iron and manganese are uncomplexed, the oxidation by ozone is fast;

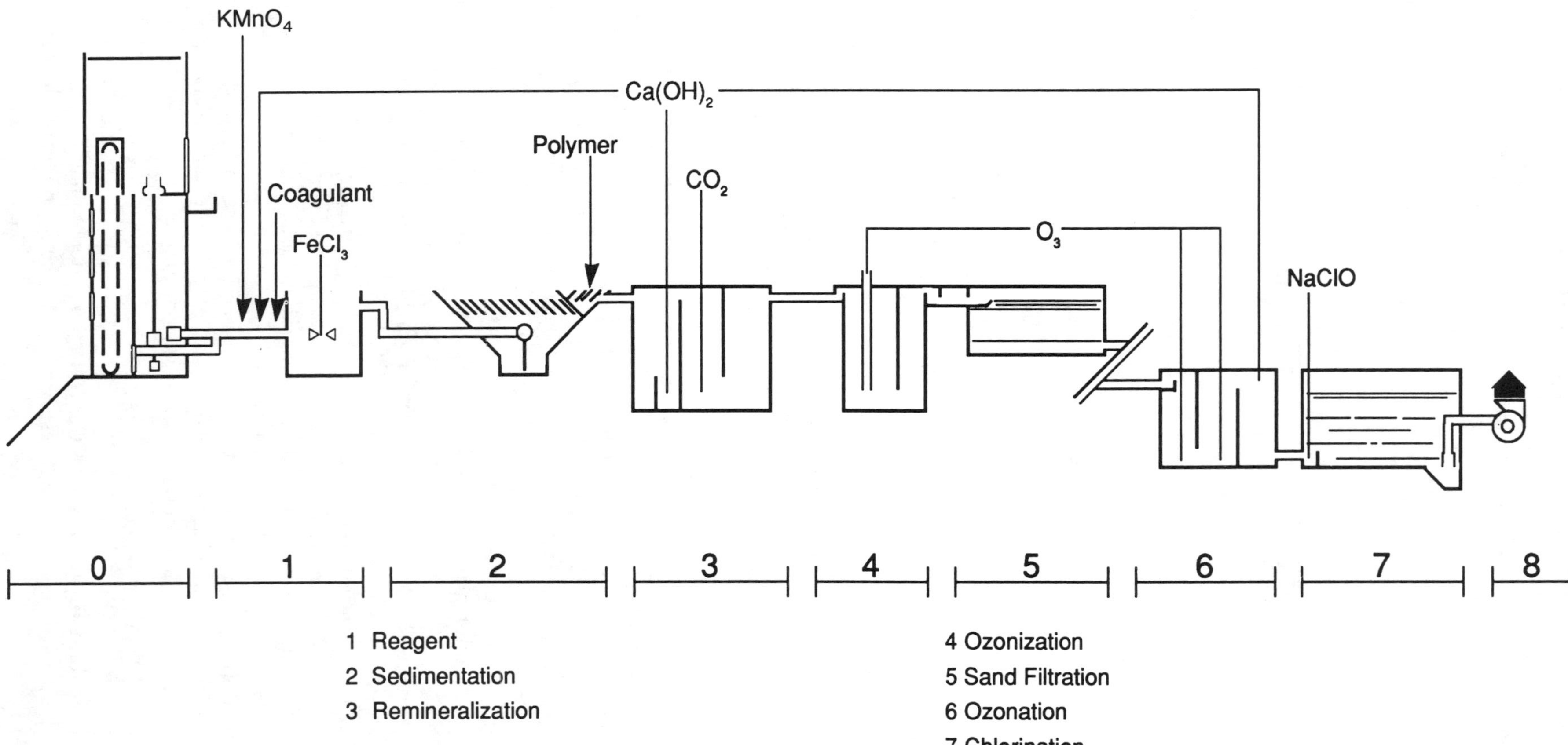

Source: Reprinted with permission from Paillard, M. et al., Symposium: "Traitements Curatifs en Vue de la Potabilisation des Eaux Chargées en Microphytes Dulçaquicoles." © 1988 Compagnie Générale des Eaux.

840 m^3/h (5.33 mgd) at maximum operating capacity.

Figure III–5 New Treatment Train at Moulin Papon Water Works, La Roche sur Yon, France

Table III–10 On-Stream Development of Iron and Manganese Concentrations in New Treatment Train at Moulin Papon Water Works

Iron/Manganese Concentration (μg/L)	Raw Water	Settled Water	Filtered Water	Discharged Water
		Sampling Performed 2/7/89		
Dissolved Fe	264	<20*	<20*	<20*
Total Fe	918	94	<20*	25
Dissolved Mn	87	60	<10*	<10*
Total Mn	152	69	13	13
		Sampling Performed 8/21/89		
Dissolved Fe	760	120	14	14
Total Fe	1370	16	34	28
Dissolved Mn	300	80	24	32
Total Mn	440	85	36	51

Source: Results from the CGE Research Center Laboratory, Maisons Laffitte, France.
*The detection limits of Fe and Mn given by the laboratory's analytical method are Fe < 20 μg/L and Mn < 10 μg/L.

Table III–11 Raw Water Characteristics for Oradell Reservoir (October 1977–July 1980)*

Parameter	Average	Maximum	Minimum
Temperature, °C (°F)	12.3 (54.1)	28 (82.4)	1 (33.8)
pH	7.7	9.1	6.9
Apparent Color, mg Pt-Co/L	36	94	12
Turbidity, NTU	5.0	19	0.2
Iron, mg/L	0.32	2.26	0.03
Manganese, mg/L	0.15	0.51	0.01
Alkalinity, mg/L $CaCO_3$	78	108	38
Hardness, mg/L $CaCO_3$	104	148	56

*Data provided in consultant report to Hackensack Water Company.

moreover, the ozone consumption is very close to the theoretical stoichiometry. This means that preozonation can be used and the contact time chosen for the transit of water through the contact column can be short (2 or 3 min). Often this contact time is slightly higher to allow for hydraulic short circuits in the column. After checking that the water to be treated does not contain other reducing agents, the ozone dosage will be slightly higher than required by the stoichiometry.

When the water contains high concentrations of iron and a low concentration of manganese, it is best to reuse the offgas of the contact chamber in a prechamber to allow a partial oxidation of the iron by the ozone residual and the oxygen remaining in the offgas. The reinjection can be performed by a self-suction turbine or by an emulsifier (see chapter IV, sec. IV.D).

It is necessary to avoid an overdosage of ozone, which leads to the production of permanganate (a soluble species) going through the filters, and pink water. Otherwise, it may be useful to include a second compartment for the reduction of permanganate to insoluble manganese species.

Treatment of impounded water. In sec. III.B.1, it was shown that kinetic competition occurs when iron and manganese are present in water along with organics. To remove these metals from water, it is advisable to optimize the elimination of organics prior to oxidation with ozone. In this case, ozonation will be used as an intermediate ozonation step (between settling and filtration) to remove the manganese. The diffusion can be performed through porous diffusers.

Table III–12 Performance Data for Pilot-Scale Preozonation Water Treatment Facility in Hackensack, New Jersey*

Date	Sample†	Ozone Dose (mg/L)	Chlorine‡ (mg/L)	Alum§ (mg/L)	Polymer CFT-1§ (mg/L)	$Na_2S_2O_3$** (mg/L)	Filter Head Loss (ft,m)	pH	Color (mg Pt-Co/L)	Turbidity (NTU)	Fe (mg/L)	Mn (mg/L)
6/21–22	Raw							7.6	32	3.4	0.46	0.15
	Filter influent***	1.7	2.6	10.7	1.7	5.3		7.2	18	3.1	0.17	0.01
	Filter effluent						7.7,	7.5	4	0.3	0.04	0.01
	GAC effluent						2.35	7.3	3	0.3	0.03	0.02
6/29–30	Raw							7.7	32	3.9	0.39	0.20
	Filter influent***	1.6	4.4	7.4	1.4	5.4		7.3	22	3.0	0.17	0.23
	Filter effluent						5.8,	7.6	3	0.3	0.01	0.03
	GAC effluent						1.77	7.2	2	0.2	<0.01	0.01
7/7–8	Raw***							7.2	80	9.5	0.81	0.44
	Filter influent***	1.7	4.4	9.8	2.3	5.1		7.1	37	5.5	0.38	0.30
	Filter effluent						7.6,	7.1	13	0.6	0.04	0.05
	GAC effluent						2.32	7.1	3	0.4	0.03	0.03
10/13–14	Raw							7.9	50	10.0	0.73	0.37
	Filter influent***	1.6	0.0	6.3	1.5	0.0		7.5	28	8.9	0.27	0.26
	Filter effluent						9.3,	7.6	4	0.4	0.06	0.05
	GAC effluent						2.83	7.5	2	0.3	0.04	0.02
10/18–19	Raw							7.9	47	7.0	0.46	0.26
	Filter influent***	1.6	2.7	6.8	1.6	4.9		7.6	22	5.6	0.28	0.25
	Filter effluent						5.9,	7.5	5	0.4	0.01	0.03
	GAC effluent						1.80	7.4	2	0.3	<0.01	<0.01

*O_3 contact time = 92 min; filtration rate = 12.1 m/h (5.0 gpm/ft^2); average filter run length = 22 h.
†Composite sample.
‡Applied to filter influent.
§Applied to raw water stream before ozonation (as $Al_2(SO_4)_3 \cdot 18\ H_2O$).
**Applied to filter outlet.
***Grab sample.

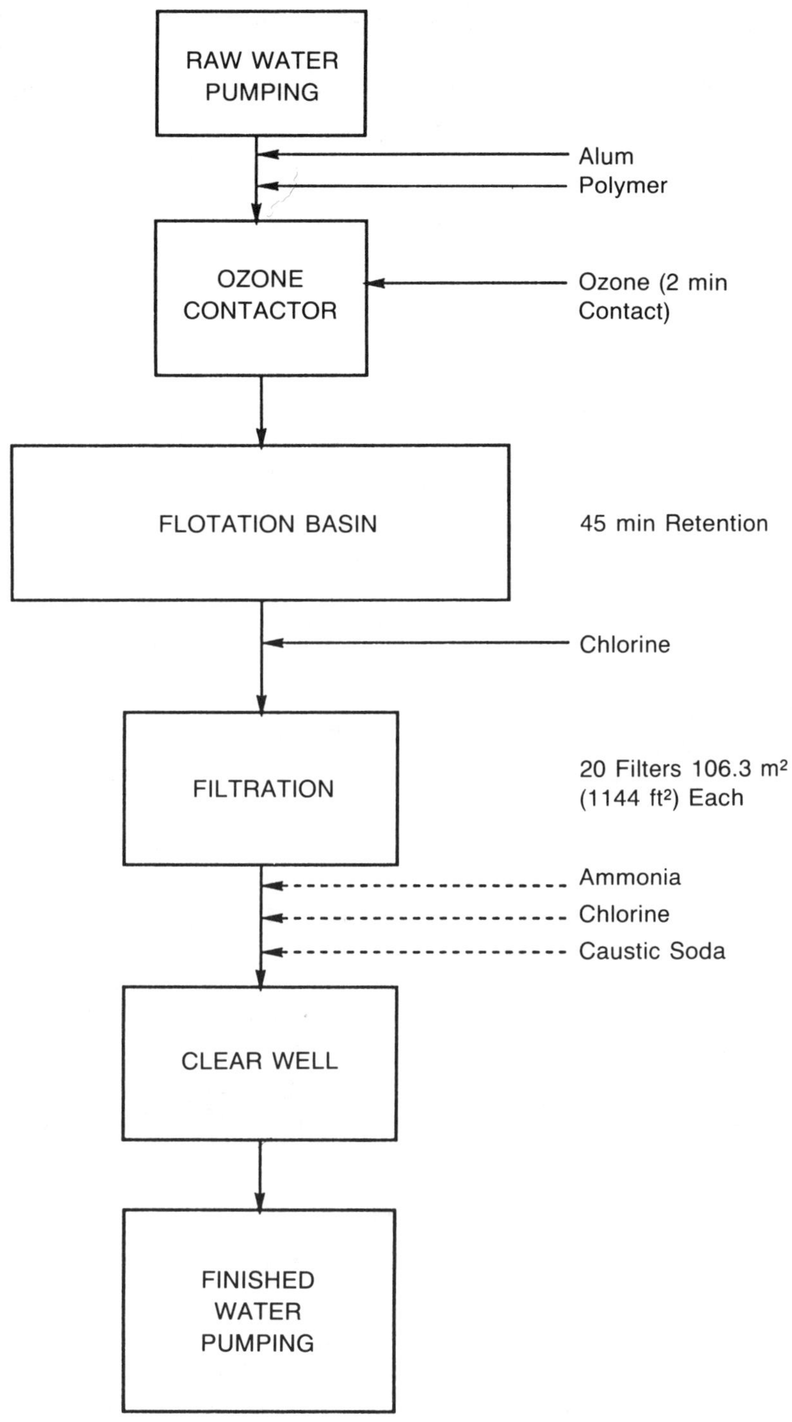

34,700 m^3/h (220 mgd).

Figure III–6 Schematic of Full-Scale Water Treatment Facility for Hackensack, New Jersey

Manganese removal can be facilitated by recarbonation of the water upstream of the oxidation step (addition of $NaHCO_3$ or CO_2 plus lime), as demonstrated at Moulin Papon (see sec. II.B.2). The ozone dosage to apply is linked to the residual TOC concentration before the ozonation step (see sec. III.B.1). In the particular case of Moulin Papon, the necessary ozone dosage for achieving a residual manganese concentration

Table III–13 Design Criteria for Full-Scale Treatment Plant (34,700 m^3/h—220 mgd) Based on Pilot-Plant Results

Coagulation
Alum dose = 7.5 mg/L as $Al_2(SO_4)_3$, 18 H_2O
Cationic polymer dose = 1.25 mg/L
Ozonation
Dose = 1.3 mg/L
Contact time = 2 min
Filtration
Dual-media filters (sand and anthracite)
Effective size of sand = 0.6 mm (0.024 in.)
Effective size of anthracite = 1.5 mm (0.059 in.)
Depth of new sand = 0.30 m (12 in.)
Depth of oxide-coated sand = 0.15–0.23 m (6–9 in.)
Depth of anthracite = 0.68 m (27 in.)
Filtration rate = 14.5 m/h (6 gpm/ft^2)
Chlorination
Chlorine dose = 2.5 mg/L
Ammonia dose = 0.4 mg/L

less than or equal to 50 μg/L varies between 0.8 and 1.3 mg/L for a manganese concentration between 30 and 550 μg/L and a TOC concentration between 3.8 and 4.0 mg C/L in the settled water. These dosages are very far from the theoretical stoichiometry due to the partial oxidation of organics that occurs simultaneously with the manganese removal.

Due to the variable nature and reactivity of TOC in waters of different origins, pilot tests must be performed to determine the necessary ozone dosage and contact time (generally between 2 and 10 min).

Special considerations for post-ozone plants. When a treatment line includes a post-ozonation step at the end of the train for disinfection purposes without subsequent filtration (see sec. III.K), manganese removal achieved during pre-ozonation or interozonation must be very well controlled. Otherwise, further oxidation of residual manganese may occur during postozonation, which could produce pinkish water.

III.C COLOR ABATEMENT

III.C.1 Definition and Origin of Color

Water appears to be colored when dissolved matter absorbs visible light or when suspended particles scatter light (Rayleigh scattering). These two sources of color are the basis for distinction between apparent color and true color. Apparent color is due to both light absorption and scattering. True color depends exclusively on the dissolved species in the water. Since light scattering is perceptible only when the size of the particles is within the field of visible wavelengths (400–800 nm), it is possible to eliminate the effect of light scattering by filtering on a 0.45–μm membrane. It should be noted that the difference between apparent color and true color is related to the water's turbidity.*

Compounds responsible for color. True color implies the presence of substances that absorb light of wavelength 400–800 nm or possibly those that fluoresce in

*In this section, color units will be defined as Pt-Co units (U.S. convention) or mg Pt-Co/L (European convention). These units should be considered equivalent.

Table III–14 Standards and Recommendations Concerning Color in Drinking Water

French Standard	EEC		WHO Recommendations
MAC:* 15 mg Pt-Co/L	Guidelines: 1 mg/L Pt-Co	MAC:* 10 mg/L Pt-Co	Guidelines: 15 mg/L Pt-Co

*Maximum admissible concentration.

the range 200–400 nm. This would suggest compounds with a polyaromatic structure, a substituted aromatic structure (azoic, urenes, etc.), polyenes, condensed heterocyclic molecules, or complex ions (for example, carotenoids, humic acids, permanganate, manganese or iron (II) humates, or azo-dyes). It should be noted that the π bonds absorb in the near UV spectrum (less than 200 nm) and that conjugated structures (polyenes) are necessary for absorption to occur in the visible spectrum.

Most molecules responsible for colored water contain one to several aromatic rings and begin to absorb light at 250 nm. Many researchers have established correlations between color measured on a spectrophotometer in the range of 400–500 nm and UV absorption at 254 nm—a measure being used more frequently in France and the United States to characterize water quality. Although not all aromatic rings produce color, there is almost certainly a correlation between color and aromaticity when the color is largely attributable to humic-type compounds. This is generally the case in natural waters. The origin and structure of these compounds is described in chapter II, sec. II.A.6.

It is important to remember that fulvic acids, which are more soluble in water than humic acids, play a major role in the origin of color in water, particularly in swamps and bogs. In addition, the color of soft water is even more pronounced due to the fact that soft water has low levels of calcium, which is capable of precipitating humic substances. Depending on the concentration of these humic substances, water will look yellow to almost black.

Under certain conditions, the color of water will be affected by compounds other than humic substances. In eutrophic water, color is accentuated by the presence of chlorophyll and xanthic coloring matter (like spirilloxanthine). In water containing a great deal of sulfur-based products, red and green colors develop according to the season. In some cases, synthetic coloring agents generated by certain industries (textile, paint), metal compounds (metal plating), or natural pigments (slaughterhouses) that are not completely removed in wastewater treatment plants are found in surface water.

III.C.2 Treatments for Color Removal

Why remove color? Colored water is unattractive and is responsible for the majority of complaints from water utility customers, along with taste and odor. Unpleasant aesthetic qualities may induce consumers to seek other sources of water, which may be of poorer bacteriological quality or more expensive.

Water with a true color that is linked to the presence of fulvic or humic acids will consume large amounts of chlorine at the final residual disinfection stage, leading to formation of organohalides (see chapter II, sec. II.A.6 and sec. III.I). It is for these reasons that there are standards limiting the amount of color in water (Table III–14).

General treatment techniques for color removal. Seventy percent or more of true color can be removed through direct filtration or conventional treatment (coagulation with hydrolyzing metals, e.g., alum or iron salts, followed by flocculation and filtration or sedimentation plus filtration). Even greater removals of humic substances and color are possible with activated carbon adsorption (Lemarchand et al. 1981). However, activated carbon can have a very short life, depending on the initial color level and

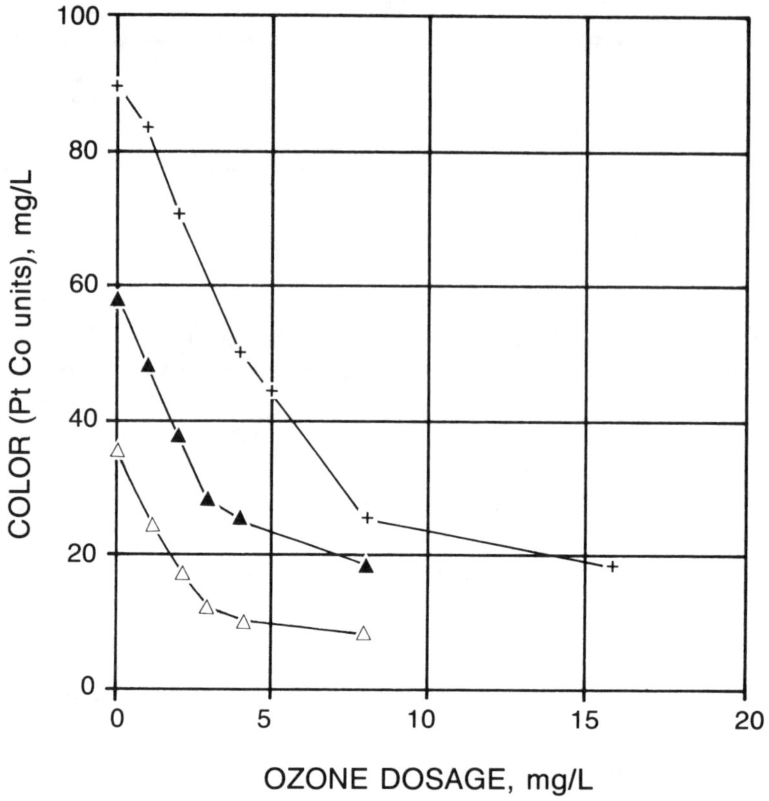

Source: Reprinted with permission from Flogstad, H. and Odegaard, H., *Ozone Sci. Engrg.* 7:121. © 1985 IOA.

Contact time: 4 min; concentration of ozone in air: 4–20 mg/L.

Figure III–7 Effect of Raw Water Color and Ozone Dose on Color Removal Versus the Initial Color

the nature of the organics in the water. Other treatments such as reverse osmosis or microfiltration are also effective (Odegaard et al. 1986). Oxidation with chlorine and chlorine dioxide can lead to satisfactory color abatement, but ozone remains the most efficient oxidizer, and ozonation is the treatment most often mentioned in the literature for oxidative color removal.

For highly colored water, color removal will usually be performed by combining several types of treatment (Meijers 1977; Constantine 1982). It should be remembered that poor control of oxidation treatment can, in itself, be the initial cause of color. A case in point is manganese removal whereby manganese ions are oxidized until permanganate is formed (see sec. III.B), causing the water to turn pink.

Direct effects of an ozonation step on color. Because humic substances are the primary cause of color in natural waters, it is useful to review the reactions of ozone with humic and fulvic acids (see chapter II, sec. II.A.6). According to different authors, ozone doses of 1–3 mg O_3/mg C (Watts 1985; Killops 1986) lead to almost complete color removal. Considering that these compounds account for 30–50 percent of the DOC in the water (Black and Christman 1963), the ozone dosages to be applied in order to reach treatment goals for color can be very high. It is interesting to note that when the ozone dosage is sufficient the modification of the organic structure is such that the final chlorine demand can decrease.

Studies carried out by Flogstad and Odegaard (1985), from which Figure III–7 is taken, confirm that the consumption rates of ozone for color removal can be high and that beyond a certain threshold the residual color is difficult to remove. Many studies show that for water with a color between 20 and 50 Pt–Co units, the application of 1 mg O_3/L will lead to a color abatement of about 10 Pt–Co units (Britton and McFadzean 1984; Thureson 1962; Flogstad and Odegaard 1985; Wallentin and Nyberg 1962).

The contribution to overall color from specific compounds, such as chlorophyll or certain pigments, is small (less than 0.25 Pt–Co units), and their ozonation has not been specifically studied. Ozonation does, however, seem to reduce the concentration of chlorophyll in algae-laden water (Hyde et al. 1984). Colored synthetic organic compounds are often highly oxidizable (see chapter II, sec. II.A.4). They are also adsorbable. The choice of a treatment process for water containing these types of compounds must take into account all problems connected with such water before the best treatment train is chosen. The removal of color caused by complex inorganic ions and organo-metallic compounds does not appear to have been studied in the field of drinking water.

Little is known regarding the relative effectiveness of ozone versus an advanced oxidation process for the removal of color. Kato et al. (1988) showed that a combination of ultrasound and ozonation can improve the color removal achievable by ozone alone by 20 percent.

Effect of ozonation on subsequent removal of color. As stated earlier, when treating highly colored water, color removal will require a treatment sequence containing several steps. Klimkina et al. (1987), for instance, used ozonation (ozone dose: 8–13 mg/L) followed by filtration on activated carbon (filtration rate: 5–10 m/h [2.0–4.1 gpm/ft^2]). Color was removed by 20–60 percent during the ozonation step. Then, following activated carbon filtration, most of the remaining color was removed, resulting in an overall removal of 90–95 percent.

A number of authors (Gould 1985; Gould et al. 1984; Greaves et al. 1988) have suggested an ozonation stage before slow filtration. The results obtained using this type of treatment are interesting not only from the point of view of color, as shown in Figure III–8, but also in relation to TOC removal (see chapter II, sec. II.A.6). Other authors have tested ozonation followed by diatomaceous earth filtration and have achieved satisfactory results (Fulton 1980; Bryani and Yapijakij 1977).

In some cases, multistage ozonation may be an attractive solution (Britton and McFadzean 1984). A treatment train of this type will be described in the following case study.

III.C.3 Case Study: Tullich Water Treatment Works, Oban, Scotland

This 480-m^3/h (3-mgd) treatment facility, controlled by the Strathclyde Regional Authority, supplies the town of Oban from two raw water resources—Loch na Gleann and Loch Nell. The water from these two lochs is mixed in the same reservoir (Bhearraidh Reservoir). This water quality is typical of that usually found in Scotland. It is colored, soft, and slightly turbid, with a pH of over 7.0. Table III–15 lists the principal characteristics of the waters from these two lochs. The analyses were performed by the Counties Public Health Laboratories of London, before the period of pilot tests. These tests were designed to determine the best treatment for color removal. The treatment train that was ultimately used has been in service since 1978 (Gowans and McFadzean 1982).

Brief summary of pilot studies performed. In the late 1960s, the firm of Glenfield & Kennedy engineered a large mobile pilot unit to enable testing of various treatment trains directly on water from Loch Nell and on the mixed waters from the two above-mentioned lochs. The choice of a treatment system was heavily influenced by seasonal growth of green algae and the need to remove color. The process adopted was as follows: microscreening, ozonation, sand filtration, and finally ozonation.

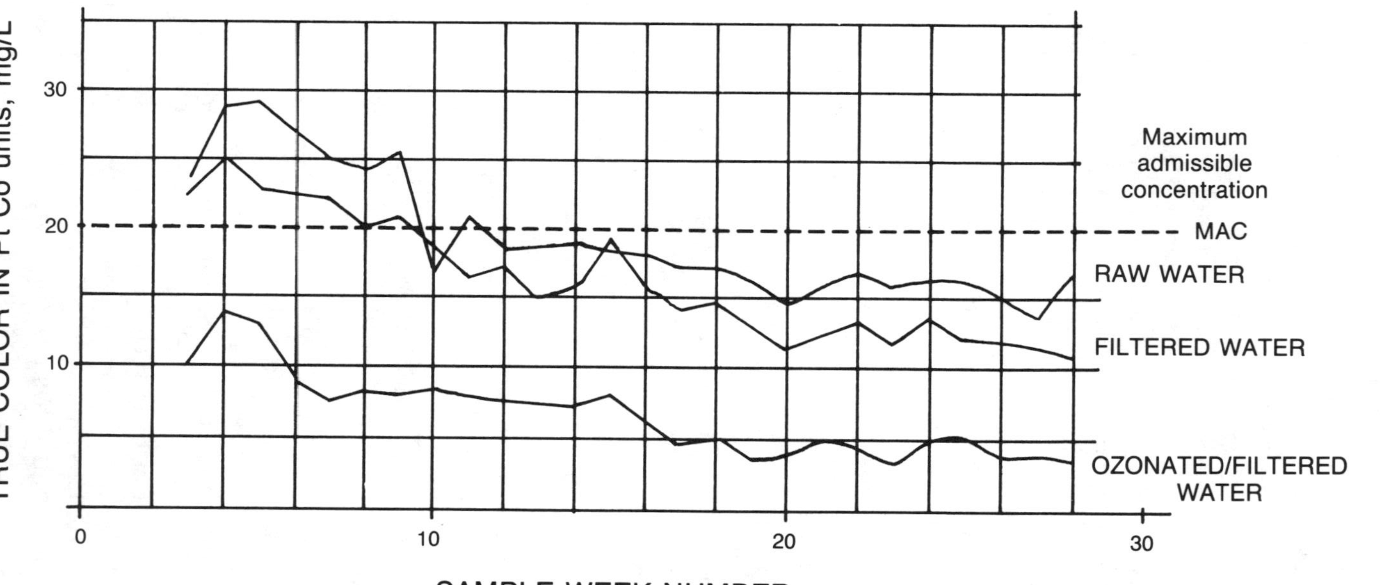

Source: Reprinted with permission from Greaves, G.F. et al., in "Proc.: Advances in Slow Sand Filtration." © 1988 Ozotech and North West Water Authority.

Figure III–8 Graph of True Color Removal Through Slow Sand Filters Alone and with Ozone Pretreatment

Table III–15 Summary of Loch na Gleann and Loch Nell Raw Water Quality

Parameter	Loch na Gleann at Bhearraidh Reservoir (20 samples)			Loch Nell (12 samples)		
	Max.	Min.	Mean	Max.	Min.	Mean
Turbidity (NTU)	4	1	3	5	1	1.8
Color (mg Pt-Co/L)	45	25	37	70	15	45
pH	7.7	7.1	7.3	8.1	6.8	7.3
Free CO_2 (mg/L)	5.0	Trace	2.0	7.0	Trace	2.0
Conductivity (μmho/cm)	120	85	100	120	70	90
Dissolved solids at 180°C 356°F (mg/L)	85	65	75	90	50	65
Alkalinity (mg/L $CaCO_3$)	35	10	30	30	15	20
Hardness—total (mg/L $CaCO_3$)	40	30	35	30	20	25
• Nitrate-N (mg/L)	1.7	0.0	0.3	1.2	0.0	0.3
• Nitrite-N (mg/L)	0.01	–	–	0.01	ND*	–
• Ammonia-N (mg/L)	0.07	0.00	0.01	0.10	0.00	0.03
Dissolved oxygen (mg/L)	4.5	2.7	3.1	5.5	3.1	–
Iron (mg/L)	0.20	0.03	0.10	0.50	ND*	–
Manganese (mg/L)	0.50	ND*	–	0.10	ND*	–

Source: Gowans and McFadzean (1982).
*ND: Below detection limit.

Figure III–9 shows the color abatement obtained using this system when testing was performed in September 1969 on Loch Nell water. Total ozone dose was between 1 and 6 mg/L. The ozone was either applied at one point (upstream of filtration) or split between two contact stages (upstream and downstream of filtration). These tests showed that two ozonation steps separated by sand filtration gave greater color removal efficiency for the same overall dose. Moreover, during tests when ozone was applied upstream of the filters only, the pH adjustment, performed later by adding lime until the equilibrium pH was attained, led to a reformation of color. This phenomenon was never observed when tests were performed on a system where ozonation was applied at two different points in the treatment train.

Description of the existing treatment train. The system that was finally adopted is composed of the following stages:

- Microscreening that is designed to remove algae and some other constituents, such as iron and manganese, some of which may be found in the form of precipitates;
- An initial ozonation step that removes the majority of color by partial oxidation of the chromophoric groups of dissolved organics, plus oxidation of dissolved iron and manganese, when these metals are periodically present in the water;
- A sand filtration step to remove the suspended solids remaining at this stage of treatment (substances that are not removed by microscreening and metal hydroxides formed during preozonation);
- A second ozonation stage that ensures disinfection and additional color abatement when used in accordance with effective viricidal and bactericidal conditions;
- A chlorination step that is designed to maintain the bacteriological quality of the water in the distribution system (1.5–1.6 mg Cl_2/L); and
- Readjustment of pH to equilibrium level by lime addition.

Figure III–10 shows a flow diagram of this treatment system.

Ozonation stages. Preozonation takes place in a contact column divided into two compartments. The first compartment receives both the offgases from the second compartment of the preozonation column and those of the postozonation column. These offgases are transported through a liquid ring compressor and diffused through ceramic

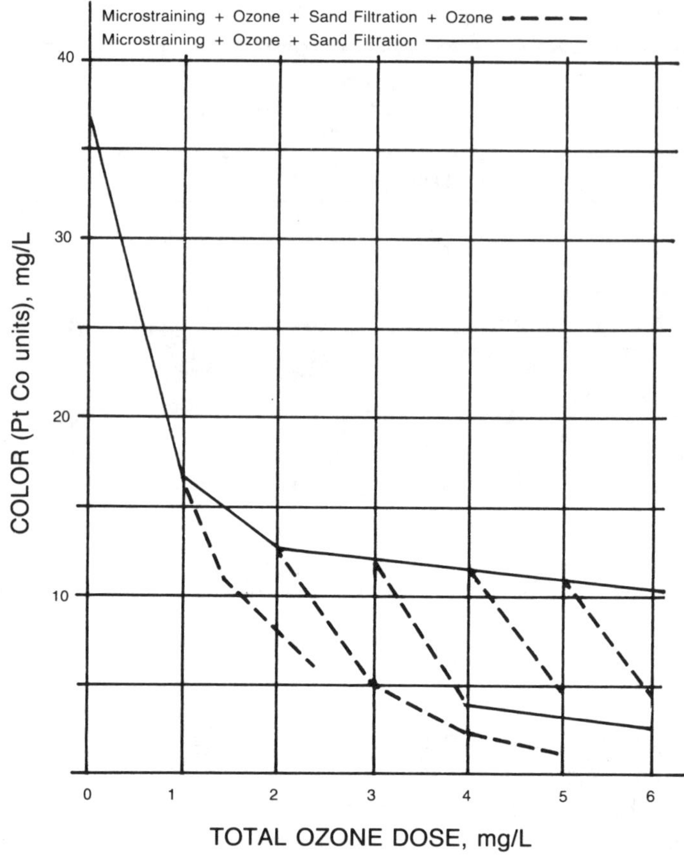

Source: Reprinted with permission from Gowans, I.A. and McFadzean, C.J., *Jour. Inst. Wtr. Eng. Sci.*, 37:5:437–459. © 1982 Institution of Water Engineers and Scientists.

Figure III–9 Removal of Color from Loch Nell Water as a Function of Ozone Dose and Point of Application

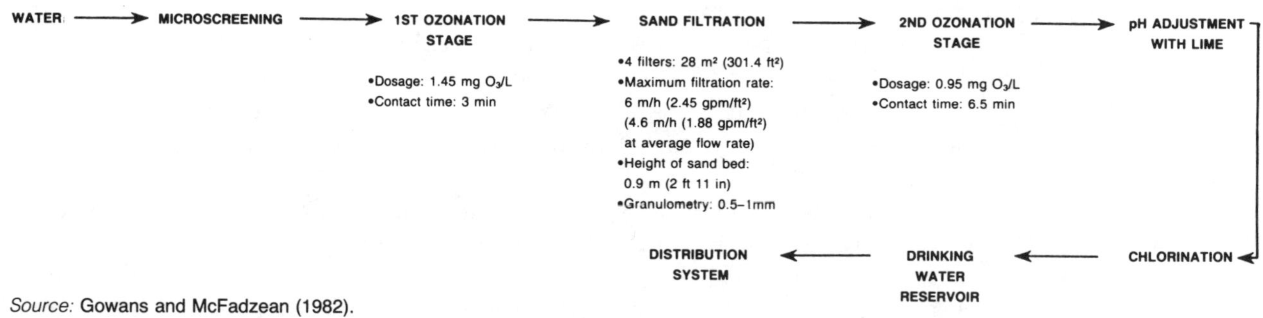

Source: Gowans and McFadzean (1982).

480 m^3/h (3 mgd).

Figure III–10 Flow Diagram of Treatment Train at Tullich Water Works (Oban, Scotland)

rod type diffusers. In the second compartment, fresh ozone issuing from the ozone generators is injected via porous ceramic rod type diffusers. Thus, the overall system has a countercurrent flow regime. The contact time during preozonation is 3 min. The ozone dosage applied in preoxidation is 1.45 mg/L.

The second ozonation stage is performed in a chamber with two compartments in which ozone, dosed at 0.95 mg/L, is diffused through ceramic rod type diffusers. The hydraulic retention time, first in countercurrent then in cocurrent conditions, is 6.5 min. Overall transfer efficiency in the ozonation chambers is above 90 percent, providing the

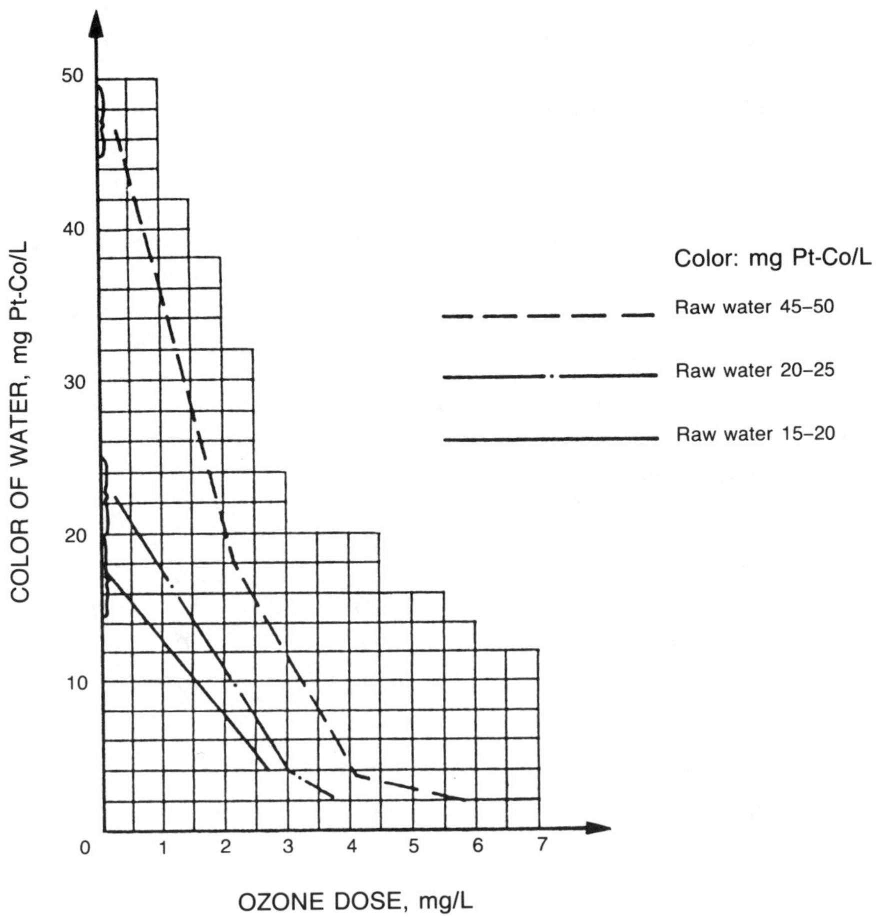

Source: Reprinted with permission from Gowans, I.A. and McFadzean, C.J., *Jour. Inst. Wtr. Eng. Sci.*, 37:5:437–459. © 1982 Institution of Water Engineers and Scientists.

Figure III–11 Tullich Water Treatment Works (Oban, Scotland): Ozone Dose and Treated Water Color Reduction

offgases are reinjected. Ozone is generated from air by two ozone generators with a unit capacity of 1500 g O_3/h (73.4 lb/day).

Full-scale performance. During commissioning of this facility in 1978, the treatment train was tested at a maximum flow rate of 480 m^3/h (3 mgd) at a time of year when the color of the raw water varied between 15 and 50 Pt–Co units. Figure III–11 shows the results obtained during these tests according to the total ozone dosages applied. The ozone distribution system was the one described above. The results achieved are very similar to those observed during the test run (see Figure III–9). During 1988, the color of the raw water varied between 13 and 32 Pt–Co units, and those of the treated water varied between 1 and 10 Pt–Co units. This gave an average color removal of approximately 77 percent (Strathclyde Regional Council Water Department 1989).

III.C.4 Design Considerations

Color can be removed by preozonation, intermediate ozonation, or both. The use of ozonation for color removal at the end of the treatment line must be avoided, due to increased biodegradability of organic carbon (see sec. III.K). Recall that the Tullich water treatment plant was designed during the 1970s. At that time, the notion of biodegradable organic carbon and the importance of biological stabilization was not known. If treatment includes GAC filtration at the end of the train, it may be best to place ozonation upstream of the GAC filter. Moreover, the combination ozonation–GAC filtration is certainly the best train for treating highly colored water (see sec. III.C.2). When the treatment line includes only direct filtration, ozone should be used as a pretreatment. Lastly, if the process corresponds to conventional treatment (settling then filtration), it is necessary to test the three possibilities—ozonation before settling, ozonation before filtration, and two-step ozonation—simultaneously to optimize color removal for the water being treated.

Regarding the kinetics and stoichiometry of color oxidation, two phases are observable (see Figure III–7). To decrease the ozone consumption, it may be judicious to choose the ozone dosage to satisfy the first ozone demand period, during which the rate of oxidation (i.e., color removal) is fast. The residual color should then be removed by a second treatment step. In practice, this means that the contact time does not need to be very high—3–6 min is acceptable. Recent studies (Bellamy et al. 1990) seem to demonstrate that the use of an in-line ozone dissolution system is more efficient than a conventional bubble contactor. Regarding the ozone dosage, many studies show that for water with a color level between 20 and 50 Pt–Co units, the application of 1 mg O_3/L will lead to a color abatement of about 10 Pt–Co units (Britton and McFadzean 1984; Thureson 1962; Flogstad and Odegaard 1985; Wallentin and Nyberg 1962).

Lastly, it is important that the ozonation tests for color removal be performed in a dynamic pilot system. There are indications that the ozone concentration in gas influences the rate of color removal. This effect cannot be easily simulated using laboratory-scale tests.

III.D CONTROL OF TASTES AND ODORS

III.D.1 Introduction

The development of tastes and odors in treatment plants and distribution networks is the origin of many customer complaints. Although tastes and odors can be very difficult to control, it is the objective of every treatment plant operator to distribute odorless water with good taste.

In France, Decree no. 89.3 (1989), in connection with the Council of the European Communities directive 80778, sets a limit on tastes and odors. Expressed as dilution to a perceptible level, this limit is 2 at 12°C (53.6°F) and 3 at 25°C (77°F). In the United States, the National Secondary Drinking Water Regulations recommend that the threshold odor number (TON) be 3 or less in finished waters (U.S. EPA 1977).

III.D.2 Definition of Tastes and Odors and Evaluation Techniques

Taste can be defined as follows:

- the sensations perceived by the odor- or taste-receiving organs following stimulation by certain soluble substances; and
- the quality of the particular sensations provoked by these soluble substances (Vade Mecum 1987).

Describing the taste of water is very subjective and depends, above all, on the individual. Four fundamental tastes can be distinguished: sweet, bitter, salty, and sour (*Standard Methods* 1985). About 30 to 40 basic odors exist (Amoore 1986). The simultaneous detection of different tastes and odors produces a flavor associated with a certain food or drink. For purposes of simplicity, the terms "odor," "taste," and "flavor" will be used interchangeably in this text.

Different techniques for sensory analysis of water exist. There are two principal

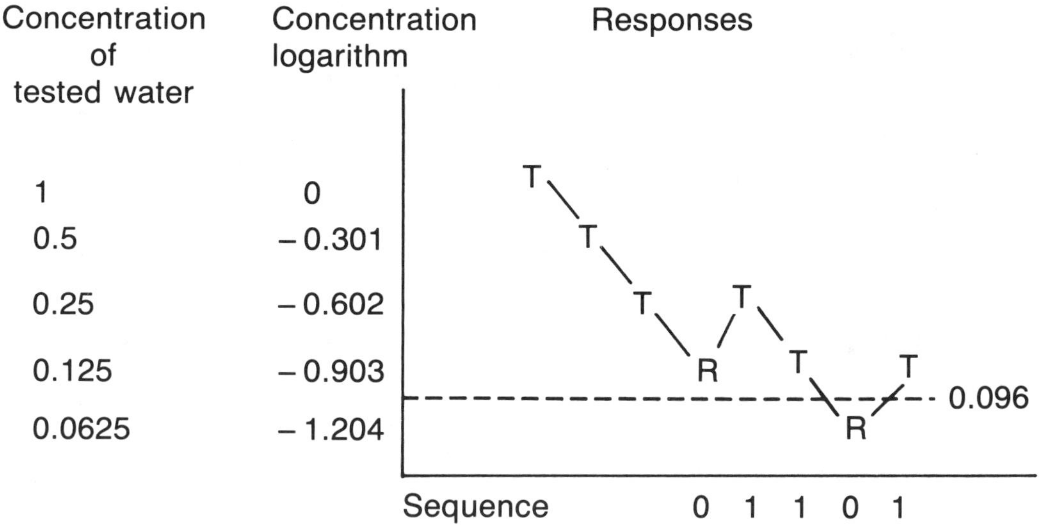

Log (estimated threshold) = −0.903 + [−0.381 × 0.301] = −1.018

Concentration estimated threshold: 0.096

where: −0.381 = Correction Factor (according to Dixon's tables)

0.301 = Log 0.5 = Log (dilution factor)

Source: Reprinted with permission from Jestin, J.M. et al., *TSM L'Eau,* June, pp. 281–289. © 1987 Techniques, Sciences, Méthodes.

T: identification of tested water ("good" response); R: reference water identified as the most odorous sample ("error").

Figure III–12 Typical Scheme of Up-and-Down Session

methods for measuring a water's taste: absolute measurements or comparative measurements (classification by rank). These methods are discussed below.

Absolute measurements. These methods, according to French standards (AFNOR 1984) or U.S. standard methods (*Standard Methods* 1985), define a detection limit by dilution. The disappearance of taste is reached when subsequent dilution results in a water without detectable taste.

The technique of flavor-profile analysis (FPA) was developed in the U.S. food industry (Little 1970) and modified for analysis of tastes and odors in drinking water (Krasner et al. 1983). At least four panelists participate in the determination of a water sample's organoleptic qualities. This is done by determining the presence or absence of each odor or taste encountered and its intensity. A rating system of 0 to 12 is used (Anselme et al. 1987a).

The up-and-down method was developed at the research center of Generale des Eaux (Jestin et al. 1987a; Levi and Jestin 1988). Members of a tasting panel are presented with two glasses of water: one glass is the sample to be tested and the other glass is a reference sample. Panel members then determine whether or not there is a perceptible taste in the water contained in either of the glasses. If the water being analyzed is found to have the stronger taste, it is diluted by one half and once again presented to the tasters along with the reference sample. This allows constant recalibration. The typical progression of a tasting session is shown in Figure III–12. The test continues until the panel members have gone through four additional comparisons beyond the point at which the first error was made (that is, when the sample under analysis was not recognized as having the stronger taste). A statistical calculation based on Dixon's tables (Dixon and

Massey 1960) is then used to determine the exact threshold for each individual taster, allowing for the fact that responses at low levels may be random. The results are analyzed individually, as an arithmetic and as a geometric average.

Comparative measurements (classification by rank). Comparative measurements rank different waters from the best (rank 1) to the worst (rank n). This process, using a much simpler methodology than the previous procedures, does not give an absolute quantification of the water taste. Nevertheless, it allows valid comparisons of different waters on a repetitive basis and provides for a relative comparison on diverse water sources or different treatment processes (Chédal 1987).

III.D.3 Odor- and Taste-Producing Compounds and Their Origin

Taste-producing compounds in drinking water can have different origins. They may be present in the raw water, formed during treatment (especially with oxidation processes), or generated in the distribution network. These compounds may stem from industrial activity, or they may be the result of activity of living organisms whose metabolisms release malodorous products. Also, the decomposition of natural substances in the environment leads to the generation of tastes and odors. Mallevialle and Suffet (1987) have recently prepared an extensive summary of current knowledge on taste and odor compounds in drinking water.

Inorganic compounds. Certain inorganic compounds present in water may generate tastes. Iron, copper, and zinc can be tasted in distilled water at concentrations of 0.05, 2.5, and 5 mg/L, respectively (Cohen et al. 1960). Nevertheless, only iron is likely to be present at such concentrations in potable water (Water Research Center 1981). Hydrogen sulfide may also cause characteristic tastes, and it is used as an odor descriptor (rotten egg) with an odor threshold concentration of 0.0011 mg/L (Mallevialle and Suffet 1987). Reduced sulfur compounds and iron (II) may be readily eliminated by an appropriate treatment (oxidation, aeration).

Organic metabolites. In the majority of cases, tastes are attributable to organic compounds present in very small concentrations, on the order of nanograms per liter. They may stem from the decomposition of plant matter, but normally they are produced by the activity of specific organisms present in water (Sigworth 1957; Dice 1975). The organisms responsible for secretion of compounds causing tastes and odors are principally actinomycetes and blue-green algae. These odors are particularly familiar to water distributors who rely entirely on raw water from impoundments. Tastes are often more pronounced in summer when the water demand is high and the level of the reservoirs is low (Burlingame et al. 1986).

The compounds most frequently associated with tastes and odors from these microorganisms are the alicyclic alcohols, geosmin and 2-methylisoborneol (MIB). Numerous studies that identify these metabolites have been reported in the literature (Gerber 1979; Wood et al. 1983; Medsker et al. 1968). However, geosmin and MIB are not the only products that cause tastes; numerous compounds are formed and released by different types of algae (Hayes and Burch 1989). These products are varied (Jüttner 1983), and include phenols, aliphatic alcohols (saturated or nonsaturated), aldehydes, aromatics, ketones, alkanes, esters, thioesters, and sulfides. They are responsible for tastes either in their natural form (Palmer 1962) or after being exposed to oxidative treatment such as chlorination (Burttschell et al. 1959). Two compounds that have strong characteristic odors in their natural form are cadine-4-ene-1-ol (Gerber 1971) and 2-isopropyl-3-methoxy-pyrazine (Buttery and Ling 1973).

Tables III–16 and III–17 give examples of the most typical odor-producing algae (Mouchet 1982) and the minimal concentrations of certain species likely to produce tastes and odors (Seppovaara 1971). Also, other organisms such as zooplankton or

Table III–16 Examples of the Most Typical Odor-Producing Algae

Group	Type	Odor Produced, Quantity: Moderate	Odor Produced, Quantity: Abundant
Cyanophyceae	*Anabaena*, *Aphanizomenon*, *Microcystis**, *Nostoc*	Herbal, Earthy, Musty	Rotten, Sewer, Pharmaceutical
	Oscillatora	Herbal	Earthy, Spicy
Chlorophyceae	*Closterium*, *Cosmarium*, *Pediastrum*, *Scenedesmus*, *Spirogyra*, *Ulothrix*		Herbal
	Chlamydomonas, *Dictyosphaerium*, *Eudorina*, *Gonium*, *Pandorina*, *Tribonema*, *Volvox*	Herbal	Fishy
Diatomaceae	*Asterionella*, *Cyclotella*, *Stephanodiscus*, *Tabellaria*	Herbal, Spicy	Fishy
	Fragillaria, *Mellosira*	Geranium	Musty
	Synedra	Herbal	Musty, Fishy
Chrysophyceae	*Dynobryon*, *Mallomonas*	Violet, Fishy	Fishy
	Synura, *Uroglenopsis*	Cucumber, Rotten	Fishy
Euglenophyceae	*Euglena**		Fishy
Dinophyceae	*Ceratium*	Fishy	Rotten, Sewer, Pharmaceutical
	Peridinium, *Glenodinium*	Cucumber	Fishy

Source: Mouchet (1982).
*Also, appearance of a sweet taste.

nematodes (Lin 1976; Chang et al. 1960) may occasionally pose problems of taste and odor formation.

Compounds attributed to industrial pollution. Tastes and odors present in polluted water from a river or underground aquifer may not be removed by treatment. The taste can originate from chlorinated solvents, hydrocarbons, pesticides, or other very diverse industrial pollutants (Van Gemerts and Nettenbreijer 1977).

Treatment by-products. Different compounds present in raw water may be transformed during treatment to produce tastes and odors. These changes are caused primarily by oxidation processes. In particular, the action of chlorine on certain

Table III–17 Critical Concentrations of Certain Algae Producing Odorous Metabolites

Kind of Algae	Critical Limits	
	Cells/100 mL	Colonies/100 mL
Cyanophyceae		
Anabaena	530,000	20,000
Aphanizomenon	660,000	20,000
Gomphosphaeria	–	17,000
Microcystis (or *Anacystis*)	3,500,000	–
Oscillatoria	5,300,000	300,000
Chlorophyceae		
Ankistrodesmus	400,000	–
Chlamydomonas	360,000	–
Closterium	20,000	–
Eudorina	–	8,000
Pandorina	20,000	–
Scenedesmus	–	150,000
Euglenophyceae		
Euglena	80,000	–
Chrysophyceae		
Dinobryon	300,000	–
Mallomonas	45,000	–
Synura	–	1,000
Diatomaceae		
Asterionella	300,000	–
Cyclotella	220,000	–
Melosira	250,000	–
Synedra	300,000	–
Tabellaria	75,000	–
Pyrrophyceae		
Ceratium	20,000	–
Cryptomonas	120,000	–

Source: Reprinted with permission from Seppovaara, O., *Aqua Fennica,* pp. 118–129 © 1971 *Aqua Fennica.*

compounds can lead to the development of tastes and may be accompanied by undesirable secondary effects, such as the production of THMs (Rook 1974). The chlorination process, in turn, may lead to pharmaceutical tastes or to the appearance of chlorine-tasting compounds (Jestin et al. 1987b). After oxidation with chlorine, phenols and their homologues will lead to malodorous chlorine-substituted compounds. Table III–18 shows that the detection limits of some chlorinated by-products are significantly lower than those of the unchlorinated forms (Van Gemerts and Nettenbreijer 1977).

Also, ammonia or organo-nitrogen compounds susceptible to forming chloramines after chlorination may cause odors (Krasner and Barrett 1984).

The aldehydes formed after the oxidation of numerous organic compounds by ozone (see sec. II.A.4 through sec. II.A.6) lead to fruity tastes (Suffet et al. 1986). Certain aldehydes are also formed from other oxidants (chlorine and chlorine dioxide) (Hrudey et al. 1987, 1988; Stevens 1982).

Compounds formed in distribution networks. Treated water may not have any particular taste as it leaves the treatment plant, but it may have a pronounced taste at the consumers' tap. This phenomenon can be attributed to the following factors:

- Under certain conditions (increase in temperature, long detention time in the distribution reservoirs and networks), regrowth of microorganisms can occur, which leads to release of malodorous compounds (Silvey and Roach 1953).
- Large concentrations of residual oxidant in the treated water lead to slow

Table III–18 Comparison Between the Detection Odor Threshold Values of Chlorinated and Nonchlorinated Compounds

Compound	Threshold Value (mg/L)
Phenol	1.0–5.9
4-Chlorophenol	0.0005–1.2
2,4-Dichlorophenol	0.002–0.21
Anisole	0.05
2,3,6-Trichloroanisole	3×10^{-10}
2,4,6-Trichloroanisole	3×10^{-8}

Source: Reprinted with permission from Van Gemerts, J.L. and Nettenbreijer, A.H., Compilation of Odor Threshold Values in Air and Water.

secondary reactions between the residual oxidant and the organic compounds that persist after treatment. These reactions may bring about additional odor development (Montiel et al. 1987).

- The oxidants used to protect the bacteriological quality of the network (Cl_2, ClO_2) may themselves be causing the taste.
- Some products used for lining pipes and reservoirs contain compounds that may be released into the distributed water and form tastes (Zoeteman 1980).

III.D.4 Odor and Taste Elimination Processes

Depending on the nature of the compounds creating tastes and odors, different treatment processes can be used for their elimination. Nevertheless, in most cases, only multiple treatment stages will lead to the production of potable water in accordance with the taste and odor levels required by legislation.

Nonoxidative processes. Aeration eliminates most dissolved gases, such as hydrogen sulfide, by stripping. The efficiency of the process depends on the pH of the water. It also depends on the stripping condition. Stripping also leads to the removal of certain volatile solvents responsible for odor formation (Lalezary et al. 1983). It should be noted that these processes usually result in a simple transfer of the compounds from the water to the atmosphere.

Because of its adsorption capacity, powdered activated carbon (PAC) can also be used for taste and odor control in conjunction with settling and filtration. However, high concentrations are often necessary to eliminate odor compounds (Montiel 1977). Pilot studies at Choisy-le-Roi (Compagnie Générale des Eaux 1977) have shown that the addition of 60 mg/L of PAC during conventional treatment of water enriched with geosmin or MIB gives a product with acceptable taste. However, Fiessinger and Richard (1975) contend that GAC is a more economical alternative when PAC dosages exceed 20 mg/L.

Granular activated carbon filters retain certain taste and odor compounds (Hattori 1987), but the efficiency of GAC adsorption for taste and odor is very dependent on the organic compounds present in the water. Because of competitive adsorption of different dissolved products, GAC filtration will be more or less effective depending on the specific compounds that are to be retained. In addition, a risk of release by desorption of these compounds exists (De Laat 1988). It may, therefore, be necessary to replace the used GAC rather often (Richard and Conan 1978).

Slow sand filtration can also result in the elimination of certain odors, such as those characterized as earthy and musty (Yagi et al. 1986; Ando et al. 1987).

Oxidative processes.

Aeration. Aeration is used for the oxidation of iron in groundwater and for stripping and oxidation of H_2S, both of which can be responsible for tastes and odors.

Aeration is probably not effective for the removal of geosmin and MIB (Mallevialle and Suffet 1987).

Chlorine. Chlorine has been used in some circumstances to remove tastes and odors. However, chlorine can form odors by itself or through its by-products (Hrudley et al. 1987, 1988; Glaze et al. 1990). In these cases, much more intense odors are formed (Jestin et al. 1987b; Burttschell 1959).

It has been shown that chlorine has no effect on musty tastes (Dougherty and Morris 1967). In particular, it does not react with geosmin and MIB, both of which are saturated compounds (Gauntlett and Packham 1973; Glaze et al. 1990).

If the water has a high concentration of organic pollutants, treatment by chlorine leads to a significant degradation in the flavor of the water, with an impact proportional to its dose. In this case, prechlorination should be avoided to prevent the formation of tastes, odors (Bablon et al. 1987b), and other undesirable compounds (see sec. III.I).

Chlorine dioxide. The efficiency of chlorine dioxide in taste and odor elimination remains very controversial (Gauntlett and Packham 1973; Lin 1976). In addition, some ClO_2 oxidation by-products show toxicity at certain levels (Ben Amor 1988; Stevens 1982).

Potassium permanganate. At high concentrations, potassium permanganate is effective at oxidizing metabolites of actinomycetes (Cherry 1962; Lalezary et al. 1986). However, Glaze et al. (1990) have shown that it is relatively ineffective at oxidizing geosmin or MIB.

Ozone. Used by itself, ozone can sometimes solve taste and odor problems. However, if the taste- and odor-causing compounds are saturated, ozone may have little effect (see sec. II.A.2 through II.A.5). Unsaturated aldehydes, which are important taste and odor by-products of *Synura* species (Juttner 1981; Hayes and Burch 1989), should be readily oxidizable by ozone. The effectiveness of ozone has recently been substantiated for one of these compounds, 2,4-decadienal (Glaze et al. 1990). A study covering five compounds responsible for musty odors (2,3,6-trichloroanisole [TCA], 2-isopropyl-3-methoxy-pyrazine [IPMP], 2-isobutyl-3-methoxy-pyrazine [IBMP], geosmin, and MIB) has shown the relative efficiency of the different cited oxidants in organic-free water (Lalezary et al. 1986). This study showed that MIB was the most difficult to oxidize followed by geosmin, trichloroanisole, and the two pyrazines. The authors also showed that ozone and permanganate were the most efficient of the oxidants at destroying the three unsaturated compounds considered (Lalezary et al. 1986). Despite their relative unreactivity, most studies using natural waters have shown ozone to be quite effective at removing geosmin and MIB (Terashima 1988; Daniel and Meyerhofer 1989; Duguet et al. 1989). It is likely that ozonation is most effective in waters that support the hydroxyl radical pathway (Glaze et al. 1990).

With a water containing large quantities of organic matter, it was observed that the action of ozone on the odor level was variable and dependent on the treatment conditions (Figure III–13) (Richard and Fiessinger 1977). Further studies have shown that an increase in the odor level for some treatment rates at intermediate doses was generally linked to formation of aldehydes, giving the water a fruity taste (Anselme et al. 1987a, 1987b). Tables III–19 and III–20 show the effect of ozonation at the plant in Morsang sur Seine, France (described in sec. III.D.5) and list the compounds responsible for the fruity flavor after ozonation.

Advanced oxidation processes. Combined oxidation has been used in Japan by Kato et al. (1983) to remove musty odors from water intended for human consumption. These authors used the combination of ozone and UV light to eliminate organic compounds such as geosmin and MIB. The experimental conditions were as follows:

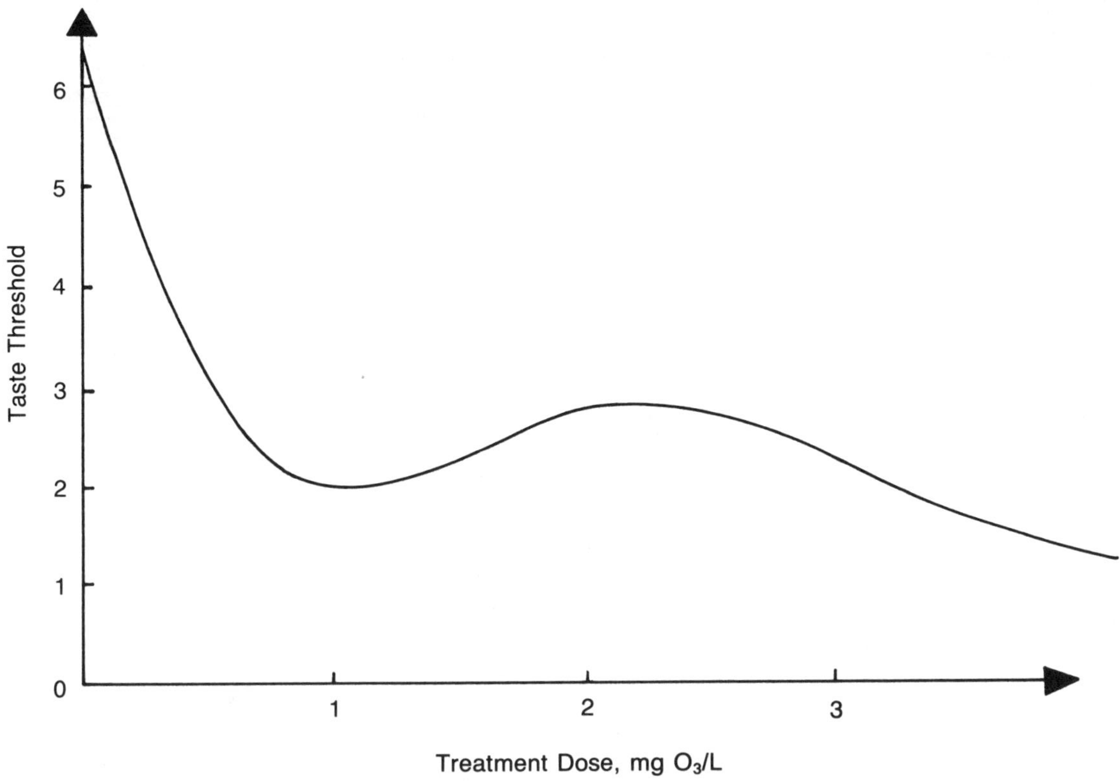

Source: Reprinted with permission from Richard, Y. and Fiessinger, F., in "International Ozone Institute Congress." © 1977 IOA.

Figure III–13 Influence of Ozone Dose on Taste Threshold

Table III–19 Taste and Odor Changes Caused by Ozonation at Morsang sur Seine (1984)

Description of Taste and Odor	Taste or Odor Intensity
Fruity	Both increase
Musty	Both decrease
Muddy	Both decrease
Earthy	Both decrease
Fishy	Both decrease
Astringent	No change
Plastic	No change

Source: Reprinted with permission from Anselme, C. et al. Proc. 8th Ozone World Congress, pp. C102–C127. © 1987 IOA.

Water flow: 2.0 m^3/h (0.013 mgd)
O_3 contact time: 10–20 min
O_3/UV contact time: 3–30 s
Applied O_3 dose: 0–5 mg/L
UV intensity: 0.54 W/cm^2 s

Figures III–14 and III–15 report the experimental results. Elimination of 100 percent of the geosmin (initial concentration 22 ng/L) and 90 percent of the MIB (initial concentration 130 ng/L) was achieved with an ozone dose of 5 mg/L alone or 4 mg/L if combined with UV. Thus, with this water the necessary ozone dose for odor treatment may be reduced by 20–40 percent by using the combined oxidant O_3/UV. Later studies

Table III–20 Taste and Odor Correlations During Treatment at Morsang sur Seine

Description of Taste and Odor	Compounds—CLSA*
Orange-like odor	Heptanal Hexane Methyl ethyl benzene
Fruity odor Sweet taste	Aldehydes C8, C11, C14 Alkanes C15 to C22
Fragrant odor Oxidant (T & O) Sour taste	Nonanal Heptanal Trimethylcyclohexane

Source: Reprinted with permission from Anselme, C. et al., Proc. 8th Ozone World Congress, pp. C102–C107. © 1987 IOA.
*Compounds analyzed by closed loop stripping analysis.

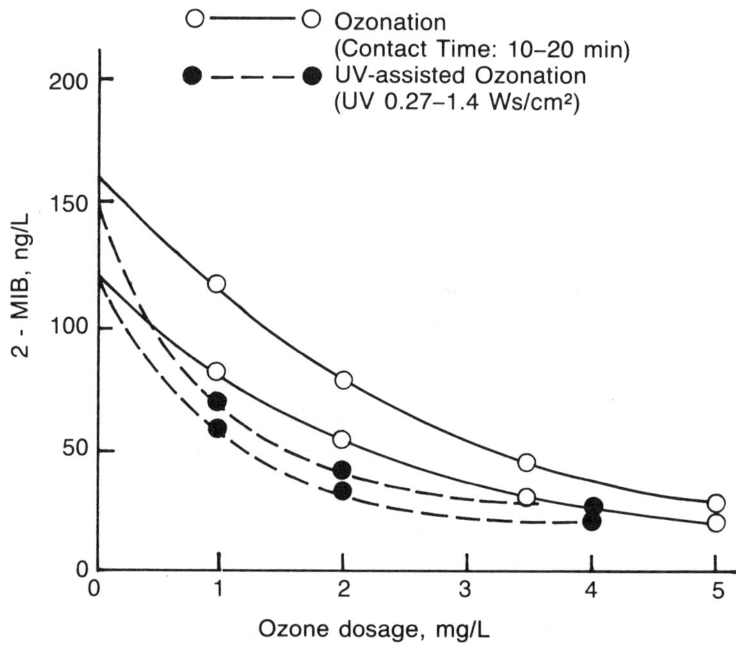

Source: Reprinted with permission from Kato, M. et al., *Wtr. Supply,* 6:349–353. © 1983 *Water Supply.*

Figure III–14 Influence of Combined Oxidation (UV and O_3) on MIB Elimination

with Colorado River water also showed a modest improvement with O_3/UV compared to O_3 alone (Glaze et al. 1990).

Recent work concerning the combination of ozone with hydrogen peroxide (McGuire et al. 1989) was performed on natural raw water containing 100 ng/L of both geosmin and MIB. The reductions obtained with ozone alone or with ozone coupled with hydrogen peroxide are shown in Figures III–16 and III–17. The combination increased geosmin and MIB elimination by 35 percent compared to ozone alone.

The improvement in removal rates for taste and odor compounds obtained by advanced oxidation processes (O_3/UV and O_3/H_2O_2) can be explained by the action of initiators (hydrogen peroxide, UV rays) to induce the decomposition of ozone in water, thus generating hydroxyl radicals that are very reactive when the water has a low concentration of scavengers (HCO_3^-, CO_3^{2-}, see sec. II.A.1). The reader is referred to sec. III.E.2 for a discussion of advanced oxidation process kinetics and stoichiometry.

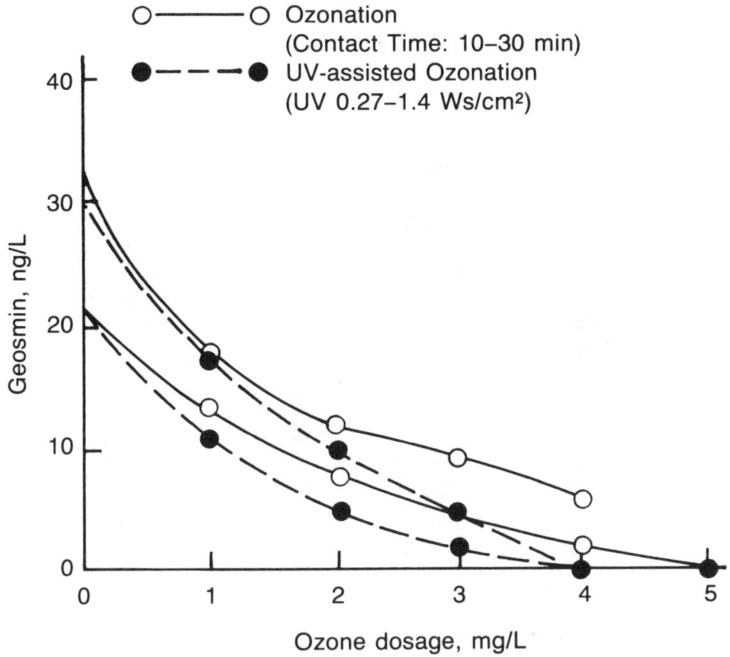

Source: Reprinted with permission from Kato, M. et al., *Wtr. Supply*, 6:349–353. © 1983 *Water Supply*.

Figure III–15 Influence of Combined Oxidation (UV and O_3) on Geosmin Elimination

Ozonation followed by filtration and adsorption. As discussed in the introduction, every treatment stage may have an influence on odor levels. The pilot plant at St. Maur, France, treating Marne River water, has been used to study the different combinations of chemical and biological treatment available for taste and odor control. The results are presented in Table III–21 and show the advantage of ozonation in combination with filtration (Montiel 1983).

Lundgren et al. (1986) have evaluated (by gas chromatography and sniffing columns) the destruction and disappearance of tastes and odors during ozonation and biological sand filtration. With an ozone dose of 7 mg/L, the compounds with high molecular weight as well as geosmin and MIB were eliminated in a satisfactory manner, whereas other odorous compounds such as aldehydes persisted.

Vik et al. (1988) conducted pilot studies on the removal of geosmin and MIB, comparing GAC filtration to ozonation followed by GAC. In this case, the GAC filters alone were shown to be more effective than the O_3/GAC process. This is because ozone by-products are formed that can more effectively compete with geosmin and MIB for the adsorption sites of GAC (Vik et al. 1988; see sec. III.D.4). The degree to which this occurs is doubtlessly related to the dose of ozone applied before activated carbon filtration, the nature of the organics in the raw water, and the degree of biological activity on the GAC (see sec. III.K).

Generally, the process of oxidation followed by filtration seems to be the most efficient taste and odor treatment process. The elimination of tastes and odors from Rhine River water at Düsseldorf, West Germany, was carried out by ozone treatment followed by filtration through two layers of activated carbon (Schenk 1962). Anselme et al. (1985) showed a significant reduction in tastes and odors from water after ozonation and GAC filtration (sec. III.D.5). In the same way, Richard and Fiessinger (1977) showed the evolution of tastes during different stages of treatment. When ozonation was followed by filtration on GAC, lower tastes and odors were obtained. In addition, the carbon's useful life was extended (Figure III–18 and sec. III.K).

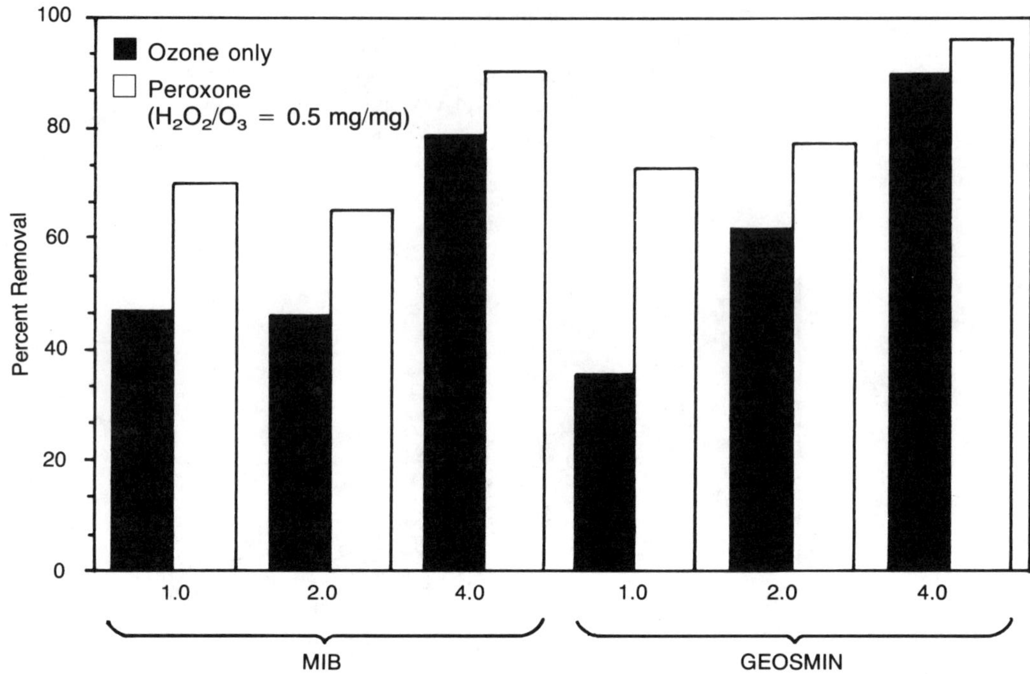

Source: Reprinted with permission from McGuire, M.J. et al., in "Wasser Berlin: 1989 Conference of the IOA." © 1989 IOA.

Figure III–16 Percent Removal of MIB and Geosmin in Colorado River Water—Bench-Scale Results

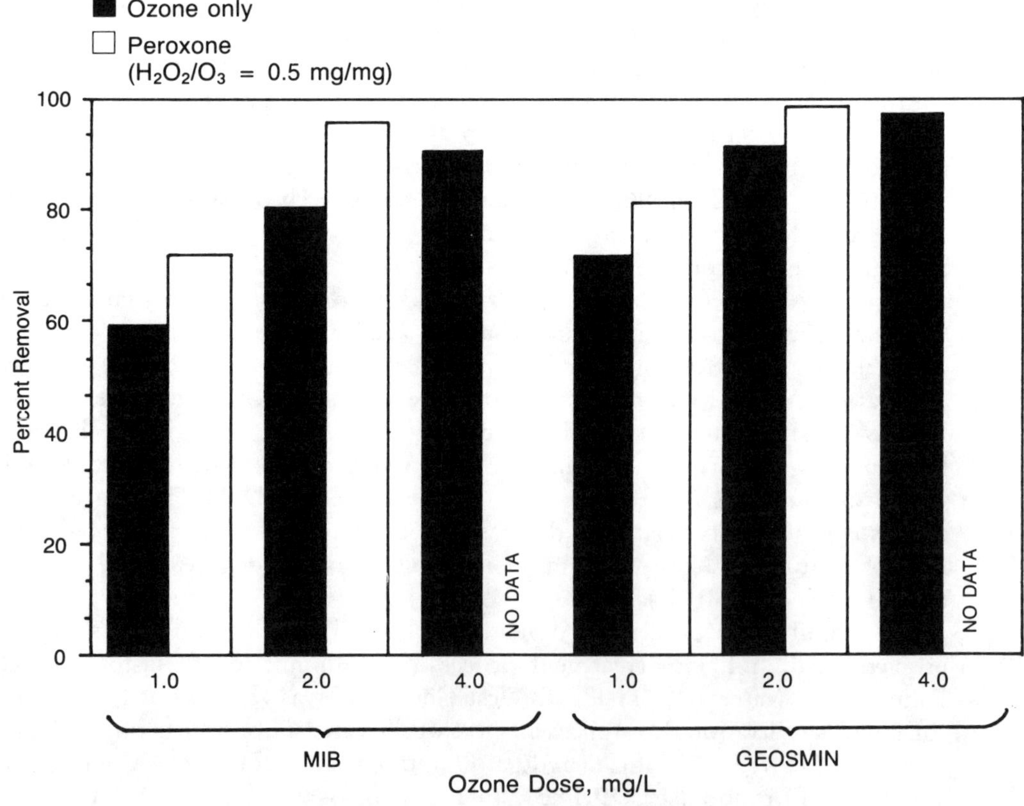

Source: Reprinted with permission from McGuire, M.J. et al., in "Wasser Berlin: 1989 Conference of the IOA." © 1989 IOA.

Figure III–17 Percent Removal of MIB and Geosmin in State Project Water—Bench-Scale Results

Table III–21 Effect of Ozonation and Subsequent Treatment on the Taste Threshold Number at the Saint-Maur Pilot Plant (France)

Treatment	Average Taste Threshold	Flavor Reduction (%)
Raw water	10	–
Settled water	7	30
Filtered water	5.3	47
Filtered water with ozone	3.3	67
Sand without ozone	3	70
Sand with ozone	3	70
Sand and activated carbon without ozone	2.8	72
Sand and activated carbon with ozone	1.8	82
Activated carbon without ozone	2.3	77
Activated carbon with ozone	2.8	72

Source: Reprinted with permission from Montiel, A.J., *Wtr. Sci. Tech.,* 15:6/7:279. © 1983 IAWPRC.

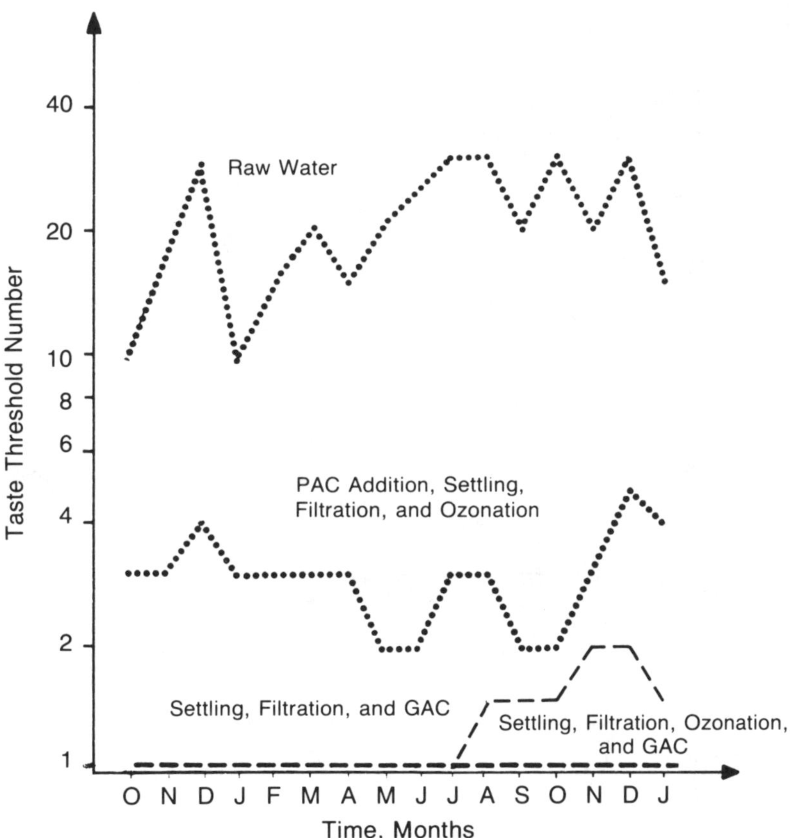

Source: Richard and Fiessinger (1977).

Figure III–18 Removal of Taste by Adsorption and Oxidation Treatment

It should be noted that taste and odor problems of algal origin (in reservoirs, lakes and eutrophic rivers) can be partially resolved by preventive treatment, such as protection against eutrophication (Bernhardt 1988). However, preventive treatments (mixing of water in artificial lakes, construction of preliminary retention basins), curative treatments (elimination of sediments, withdrawal of bottom waters), and temporary treatments (distribution of copper sulfate [Means and McGuire 1986] or lime) are expensive and have not really been proven reliable. The ultimate barrier is still the drinking water treatment plant (Auvray and Dupuy 1988).

III.D.5 Case Studies: Choisy-le-Roi and Morsang sur Seine

The results obtained at the Choisy-le-Roi and Morsang sur Seine treatment plants in France are particularly interesting since they show the evolution of tastes during different treatments and the impact of ozonation on flavors. These two plants in the Paris area treat Seine River water. The treatment lines are shown in Figures III–19 and III–20.

The results obtained in Morsang sur Seine are expressed using the FPA (sec. III.D.2) and presented in Figures III–21 and III–22. Before the ozonation and activated carbon filtration stages, the odors with the highest frequency of occurrence and the most intensity were earthy, fishy, and musty. After ozonation and carbon filtration, the frequency and intensity of odors were greatly reduced. The same beneficial effect of ozone and GAC was observed at Choisy-le-Roi (Jestin et al. 1987b). The median level measured by the up-and-down method (sec. III.D.2) was 1.4, which is a particularly low level for this test (Figure III–23).

At Morsang sur Seine (Anselme et al. 1985), the frequency of appearance of flavors after ozonation and filtration was below 20 percent for the majority of cases (see Figure III–21) and the intensity never exceeded 4 (Figure III–22). Nevertheless, it is important to control the ozone dosage so that the aldehydes (ozone by-products) are adsorbed during GAC treatment; this means that it is necessary to avoid an overdosage, which could lead to excessive production of short-chain aldehydes that are not well adsorbed. (This probably explains the poor removal by GAC of fruity odors noted in Figure III–21).

At both plants, it is evident that final chlorination increased the taste and odor of the water, as shown in the previously cited figures. Comparison of the taste and odor data

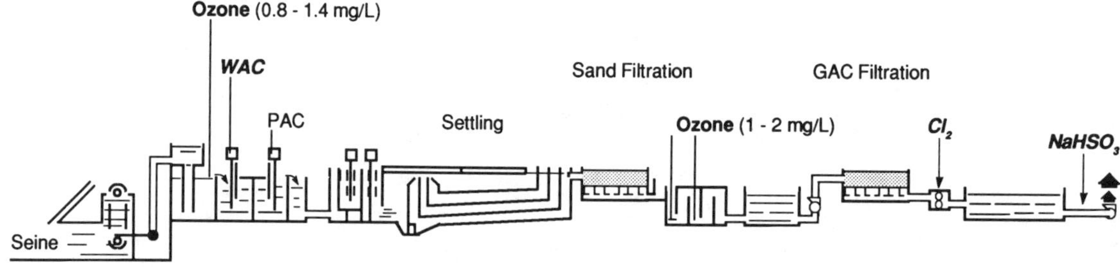

Source: Reprinted with permission from Jestin, J.M. et al., *Trib. Cébédeau,* 522:40:17–26. © 1987 La Tribune du Cébédeau.

Figure III–19 Treatment Line of the Plant at Choisy-le-Roi

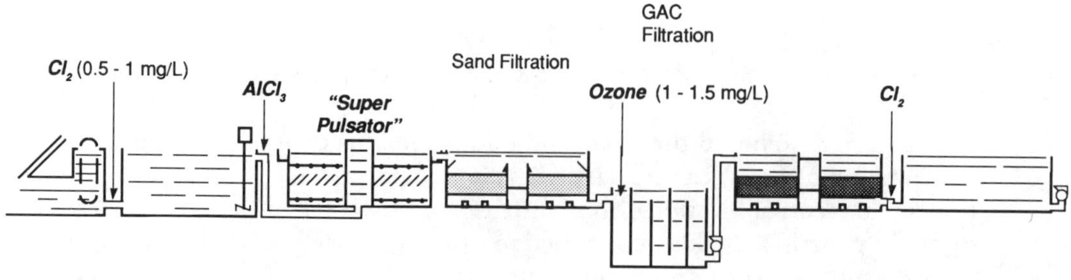

Source: Reprinted with permission from Anselme, C. et al., in "38eme Journée Internationale du Cébédeau," 12:233–259. © 1985 La Tribune du Cébédeau.

Figure III–20 Treatment Line of Plant No. 2 at Morsang sur Seine

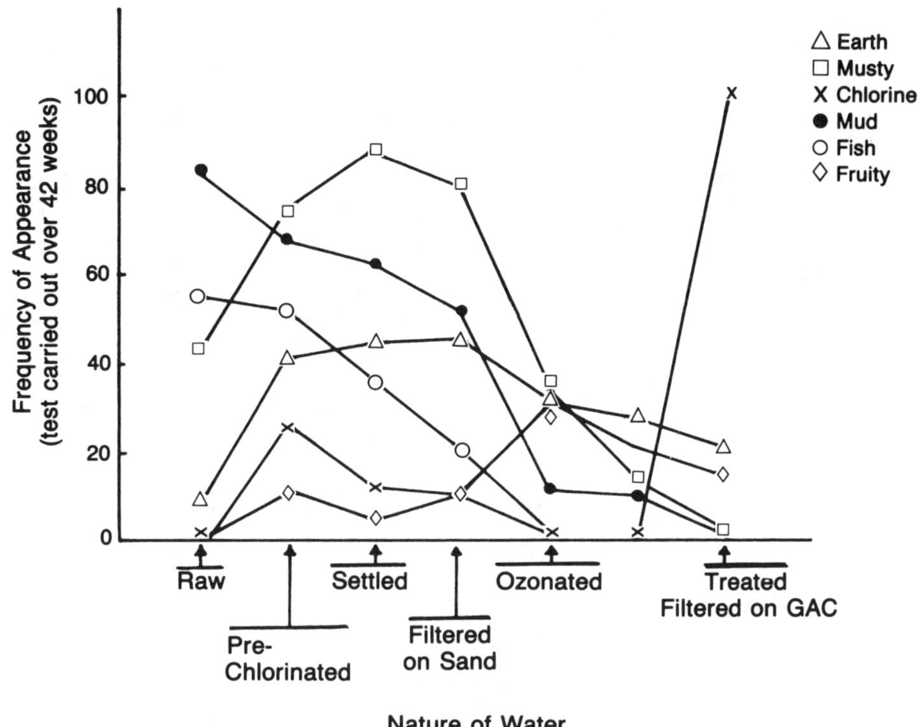

Source: Reprinted with permission from Anselme, C. et al., in "38eme Journée Internationale du Cébédeau," 12:233–259. © 1985 La Tribune du Cébédeau.

Figure III–21 Frequency of Appearance of Odors at Morsang Plant No. 2

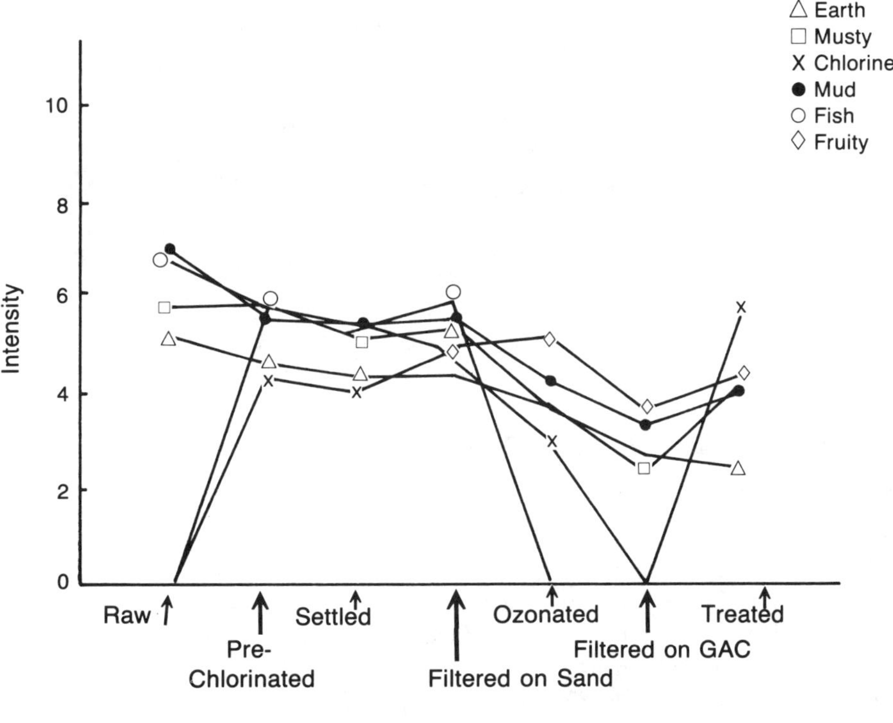

Source: Anselme et al. (1985).

Figure III–22 Average Intensity of Odors at Morsang Plant No. 2

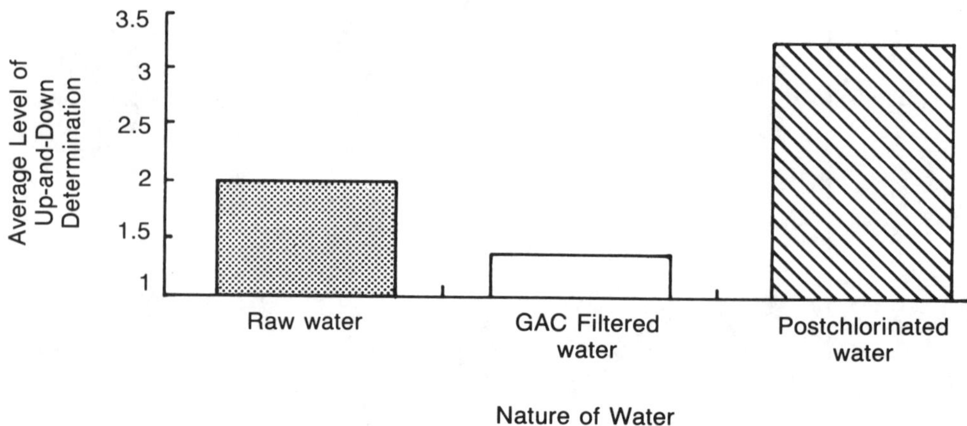

Source: Jestin et al. (1987b).

Figure III–23 Variation of the Odor Perception Level at the Choisy-le-Roi Plant

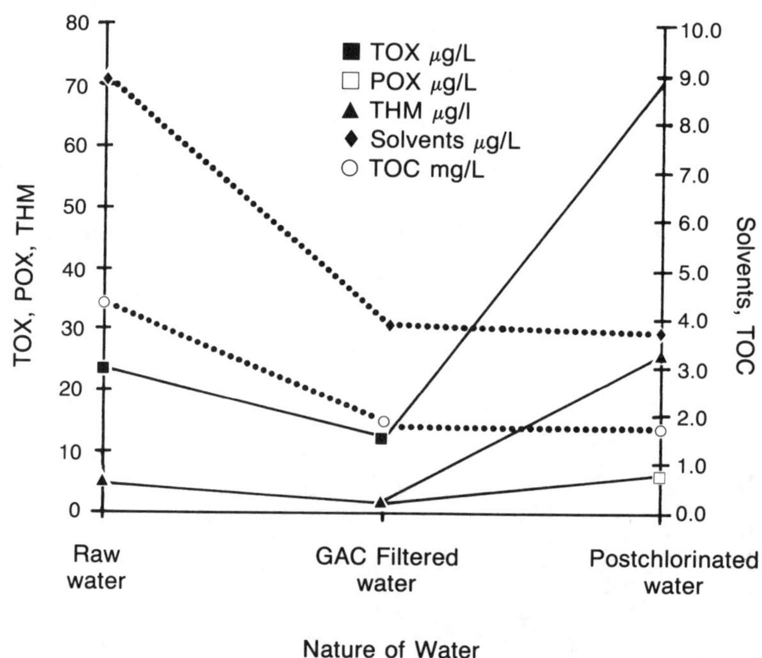

Source: Jestin et al. (1987b).

Figure III–24 Variation of Global Organic Pollution Parameters on the Choisy-le-Roi Treatment Line

in Figure III–23 for Choisy-le-Roi with the organic parameters in Figure III–24 leads to the supposition that the tastes increase in parallel with the formation of chlorination by-products (for example, THMs or total organic halides [TOX]). A series of variance analyses (Jestin et al. 1987b) confirmed this hypothesis and showed that flavors perceived after posttreatment are due to organics oxidizable by chlorine present at very low concentrations in the activated carbon–filtered water. This study at Choisy-le-Roi confirmed the influence of chlorine dose and chlorine contact time on the tastes and odors produced during terminal chlorination (Jestin et al. 1987b).

III.D.6 Design Considerations

It has been shown that ozone can be effective at treating water for taste and odor problems, especially when the water is relatively free from radical scavengers. The effectiveness of the process will be different for different waters and different taste- and odor-causing compounds. In some cases, use of ozone may actually increase the taste of the water. Site-specific analyses and testing will be required in order to properly test for use of ozone for taste and odor control, and to determine ozone dosages and contact times.

It has also been observed that ozone, in combination with other downstream treatment processes, especially GAC filtration, can greatly increase taste and odor treatment efficiency and reliability. Again, careful consideration of the cause of taste and odor compounds and the waters to be treated should be made prior to embarking on a treatment system design. Analysis and possibly pilot-scale experimentation may be required to determine the optimum choice of ozone and downstream treatment.

III.E ELIMINATION OF SYNTHETIC ORGANIC CHEMICALS

Raw drinking water often contains small quantities of organic matter that must be removed or reduced in concentration. This organic matter is usually characterized by nonspecific measurements (ultraviolet [UV] absorbance at 254 nm or 270 nm, TOC, chemical oxygen demand [COD], etc.).

When more detailed analyses are conducted, it becomes apparent that these organics belong to one of two groups:

- large molecules produced by natural processes: humic acids, fulvic acids, or proteins (see chapter II, sec. II.A.5 and sec. II.A.6); or
- small molecules, either monomers of the above molecules or synthetic organic chemicals (SOCs).

The latter group is included in the general class of compounds called micropollutants, because they are always present in small quantities (often on the order of micrograms per liter or nanograms per liter) and because they are sometimes toxic. Structure, reactivity, and concentration of these compounds (including biocides, hydrocarbons, phenols, plasticizers, dyes, amines, and solvents) are described in chapter II, sec. II.A.4.

Ozone or advanced ozonation processes can remove many SOCs. This removal leads to the chemical transformation of these molecules into toxic or nontoxic by-products. Such transformation can theoretically lead to complete oxidation to CO_2; however, this is rarely the case in water treatment. Any observable reduction in TOC is due either to a small degree of CO_2 formation (for example, decarboxylation of amino acids) or the formation and loss of volatile compounds through stripping.

Some of these oxidation by-products are easily removed in subsequent stages of treatment, especially when GAC is used. The case study at Rouen La Chapelle, France, presented at the end of this section, will highlight the efficiency of the combined ozone plus activated carbon process at removing SOCs.

III.E.1 Uncatalyzed Ozonation

Reaction kinetics. Removals through ozonation of easily oxidized micropollutants take place primarily as the result of direct reaction (see sec. II.A.1 and sec. II.A.4). The contribution of indirect radical reactions for these compounds is small, owing to the presence of free radical scavengers as carbonates and bicarbonates. Under such conditions, micropollutants reacting strongly with ozone are most likely to be removed (for example, alkenes, activated aromatics such as phenols, compounds possessing a reactive hetero-atom [N or S]). In chapter II, the kinetics of molecular O_3 reactions with SOCs were discussed (sec. II.A.4). A general second-order rate law was proposed:

$$- d[M]/dt = k [O_3] [M]$$

In subsequent tables second-order rate constants were presented for selected hydrocarbons (Table II–5), phenol (Table II–8), carbohydrates (Table II–12), amines, amino acids and aliphatic acids (Table II–13), and bases and nucleotides (Table II–16). Many of these constants were determined by measuring the loss of ozone as a function of time in distilled water containing the compound of interest and free-radical scavengers. Constants determined in this way are represented as k_{O_3} rather than k. They are equal to k when the O_3:SOC stiochiometry is 1 mol/mol. In some cases two or even three moles of ozone will be rapidly consumed with each mole of SOC oxidized. In this case k will be ½ or ⅓ of k_{O_3}. In Table III–22, the names and reactivity of a few additional micropollutants of interest are summarized. In Table III–23, the rate constant necessary to eliminate 10, 50, or 90 percent of a given micropollutant is calculated by assuming that the reaction rate is second order and that a constant dissolved ozone concentration of 0.4 mg/L is maintained in the reactor.

The values are theoretical and must be used with caution. In fact, most contactors are intermediate between plug flow and completely mixed flow. Also, one must take into account competition between the micropollutant, the by-products, and the other chemicals in the water. Moreover, these reactions are not always simple direct attack, and different compounds can play the role of initiators, promotors, or scavengers to render the principal pathway dependent on species other than the micropollutant of interest (see Chapter II, sec. II.A.2). In sec. II.A.4, a mathematical model was presented for the oxidation of SOCs via both the direct and indirect pathways. Unfortunately, not enough is known about the values of some of these rate constants. As a result, this model is of limited practical use.

The survey carried out by Fronk (1987) on behalf of the U.S. Environmental Protection Agency (EPA) on 29 toxic substances highlights the selective nature of ozone. An overview of the results obtained is given in Table III–24. The conclusions of this research were as follows:

- Ozone is capable of destroying some volatile compounds, in particular alkenes and aromatics, under the conditions of treatment applied to drinking water.
- The destruction of alkanes is very slow except if a radical mechanism is encouraged through increasing the pH (pH greater than 9). However, compared with distilled water, the high concentrations of free radical scavengers found in rivers or groundwaters strongly limit the degree to which alkanes can be oxidized.

Required ozone dosage. To calculate required ozone dosage, it is necessary, in theory, to know not only the concentration of micropollutant but the concentration of all compounds consuming ozone. In practice, this is not possible. However, Yurteri and Gurol (1988) have proposed the following equation for the overall consumption of ozone, requiring only data on the TOC, pH, and alkalinity:

$$d[O_3]/dt = - w [O_3]$$

with: $\log w = -3.98 + 0.66 \text{ pH} + 0.61 \log(\text{TOC}) - 0.42 \log(\text{alkalinity}/10)$

The success of this type of model underscores the importance of bulk organic matter (as measured by TOC) in determining the loss of ozone. Rarely will micropollutants be present at high enough concentrations to significantly affect the ozone residual. However, in some waters the TOC may not accurately represent the reactivity of organic matter with ozone. In this case, it may be necessary to use another parameter, which may change over the course of the reaction, unlike the TOC. For the purpose of disinfection, applied ozone doses are determined experimentally so as to secure a sufficient dissolved residual ozone concentration during a sufficient length of time (see sec. III.H). In general, the ozone dosages applied along the treatment train will be on the order of a few milligrams per liter. An illustration of the ozone demand (mol/mol) of a few compounds is given in Figure II–21 (see chapter II, sec. II.A.4).

Table III–22 Reactivity and Rate Constants of Some Micropollutants Toward Ozone: Direct Attack (k_{O_3}) and Radical Mechanism (k_{OH})

Product or Family	k_{O_3} ($M^{-1}s^{-1}$) Dissociated or Protonated Form (B^- or BH^+)	k_{O_3} ($M^{-1}s^{-1}$) Nondissociated Form (HB)	k_{OH} ($M^{-1}s^{-1}$)	References*
Alkanes	–	10^{-3} to 1	10^{6} to 10^{9}	a
Dibromochloropropane	–		1.47×10^{8}	b
Chloroform	–	0.1	8.5×10^{6}	c
Carbon tetrachloride	–	<0.005	–	d
Olefins		1 to 10^{5}	10^{8} to 10^{11}	a
Linoleic acid		10^{6}	10^{9}	e
Styrene	–	3×10^{5}	–	d
1,1-Dichloroethylene	–	1.1×10^{2}	4×10^{9}	e
cis-1,2-Dichloroethylene		3.1×10^{2}		i
Trichloroethylene	–	17	10^{8}	d
Tetrachloroethylene	–	0.1		d c
Aromatics		1 to 10^{3}	10^{8} to 10^{10}	a
Benzene	–	2	5×10^{9}	d
Chlorobenzene	–	0.75	1.5 to 2.10^{9}	f
Naphthalene	–	3×10^{3}	–	d
Nitrobenzene	–	9×10^{-2}	3×10^{9}	e
Phenols	10^{5} to 10^{9}	10^{3} to 10^{6}	10^{9} to 10^{10}	a
Phenol	1.4×10^{9}	1.3×10^{3}	5×10^{9}	f
4-Chlorophenol	0.2×10^{9}	0.6×10^{3}		d
2-Chlorophenol	2×10^{8}	1.1×10^{3}		d
2,4-Dichlorophenol	8×10^{9}	$<1.5 \times 10^{3}$		d
2,3-Dichlorophenol	–	$<2 \times 10^{3}$		d
2,4,5-Trichlorophenol	$>10^{9}$	$<3 \times 10^{3}$		d
2,4,6-Trichlorophenol	$>10^{8}$	$<10^{4}$		d
2-Cresol	–	1.2×10^{4}		d
3-Cresol	–	1.3×10^{4}		d
4-Cresol	–	3.0×10^{4}		d
4-Nitrophenol	1.6×10^{7}	50		d
Pentachlorophenol	–	3×10^{5}		d
Resorcinol	3×10^{5}	3×10^{5}		d
Aldehydes		1 to 10	10^{9}	a
Acetaldehyde	–	1.5	–	d
Benzaldehyde	–	2.5	–	d
Ketones		10^{-2} to 1	10^{9} to 10^{10}	a
Acetone	–	0.03	–	d
Alcohols		10^{-2} to 1	10^{8} to 19^{9}	a
Methanol	–	2.4×10^{-2}	10^{9}	e
Ethanol	–	0.37	–	d
Cyclopentanol	–	2	–	d
Glucose	–	0.5	2×10^{9}	e
Methyl α-D glucoside	–	1.60	–	a
Carboxylic acids		10^{-5} to 10	10^{7} to 10^{9}	a
Formic acid	100	5	–	d
Acetic acid	3×10^{-5}	3×10^{-5}	2×10^{8}	d
Succinic acid	3×10^{-2}	–	–	d
Glyoxylic acid	1.9	0.17	–	d
N-Containing organics	0 to 1	10^{2} to 10^{7}	10^{9} to 10^{10}	a
Methylamine	0	1.4×10^{5}	–	d
Dimethylamine	–	1.9×10^{7}	–	d
Trimethylamine	0.13	4.1×10^{6}	10^{10}	d
Pyridine	0.01	3	–	d
Alanine	0.01	6.4×10^{3}	–	d
Aniline	3×10^{-3}	9×10^{7}	9×10^{9}	e
S-Containing organics		10 to 10^{5}	10^{8} to 10^{10}	a
Ethylmercaptan		$>2 \times 10^{5}$	–	d
Dipropylsulfide		$>2 \times 10^{5}$	–	d
Pesticides				
Methoxychlor		2.7×10^{2}		i
Atrazine		13†		i
Parathion		70	–	h

*a: Barker and Jones (1988); b: Glaze and Lay (1989); c: Francis (1987); d: Hoigné and Bader (1983); e: Sonntag (1989); f: Anbar and Neta (1967); g: Rey et al. (1987); h: Laplanche et al. (1976); i: Yao and Haag (1990).

†This stoichiometry is 2 mol O_3/mol atrazine; therefore, $k \approx 6\ M^{-1}\ s^{-1}$.

Table III–23 Theoretical Calculations of the Rate Constant *k* Necessary to Obtain 10, 50, or 90 Percent Loss of SOC by Direct O_3 Pathway*

Degradation Rate (percent)	Hydraulics	Contact Time: 180 s	Contact Time: 360 s	Contact Time 600 s	Contact Time 1200 s
10	Plug flow	7.03×10^1	3.5×10^1	2.1×10^1	1.05×10^1
	Completely mixed flow	7.4×10^1	3.7×10^1	2.22×10^1	1.11×10^1
50	Plug flow	4.62×10^2	2.31×10^2	1.39×10^2	6.93×10^1
	Completely mixed flow	6.67×10^2	3.33×10^2	2.00×10^2	1.00×10^2
90	Plug flow	1.53×10^3	9.21×10^2	4.61×10^2	2.30×10^2
	Completely mixed flow	6.00×10^3	4.39×10^3	2.63×10^3	1.32×10^3

*Conditions: Plug flow or completely mixed flow. $[O_3]_{ss} = 0.4$ mg/L (8.33×10^{-6} mol/L).
Rate equation: $-d[M]/dt = k[O_3][M]$.

III.E.2 Advanced Oxidation Processes

There are two principal ways of forcing ozone to react via a free radical pathway and thereby increasing its reactivity:

- activation by hydrogen peroxide ($O_3 + H_2O_2$), and
- activation by UV radiation (O_3 + UV).

The radical mechanisms that come into play have been described in detail in chapter II, sec. II.A.2.

Many molecules that are not reactive via the direct ozonation pathway can be degraded by this free radical process. Pilot-plant experiments have been conducted on alcohols and aliphatics acids (Nakayama et al. 1979; Namba and Nakayama 1982; Comyns and Redhayne 1983), chlorinated solvents (Brunet et al. 1982; Glaze and Lay 1989), tributylphosphate (Bollyky and Siler 1989), geosmin and methylisoborneol (Duguet et al. 1989), and pesticides (Prengle and Mauk 1978). Examples reflecting their efficiency and conditions for use are described in the following paragraphs.

Ozone/hydrogen peroxide processes. Figure III–25 shows the concentration of trichloroethane as a function of ozonation dose, during oxidation with ozone only and with a mixture of ozone and hydrogen peroxide. As long as there remained a residual of peroxide, the O_3/H_2O_2 system was far more effective than the O_3 system at oxidizing this saturated halocarbon.

Paillard et al. (1990) studied the elimination of atrazine in filtered Seine River water (Figure III–26). Results showed a more complete degradation of this pesticide with the O_3/H_2O_2 combined system as compared to ozone alone. The optimal H_2O_2/O_3 ratio was 0.35 to 0.45 g/g. The percent of elimination was a function of ozone dose, contact time, and alkalinity.

Brunet et al. (1984) studied the action of O_3 and H_2O_2 on oxalic acid, a product often formed by the primary ozonation of aromatic molecules. The rate of oxidation by ozone alone or by the combination of $O_3 + H_2O_2$ is shown in Figure III–27. The radical system is far more effective at pH's found in natural water. However, under conditions of high pH, the addition of H_2O_2 greatly inhibits the destruction of oxalic acid. This may be due to rapid formation of HO_2 and OH radicals from the reaction of O_3 with HO_2^-. When formed quickly, these radicals may react to give O_2 and water. The UV absorbance of water is commonly attributed to the presence of aromatic structures. Although the majority of these structures probably are not hazardous to human health, their destruction may parallel the destruction of toxic aromatic micropollutants.

Duguet et al. (1985a) examined the efficiency of an O_3/H_2O_2 system for the disappearance of UV absorbance at 254 nm in a lake water (Cholet, France). The initial absorbance of 0.7 prior to ozonation was reduced to 0.5, 0.4, and 0 with H_2O_2

Table III–24 Effect of Water Quality on the Oxidation of Volatile Organic Contaminants by Ozone

Applied ozone dose (mg/L):	2		6		4		20							
Conditions*	A		A		B		A		B		C		D	
Compounds	PR[†]	T[‡]	PR[†]	T[‡]	PR[†]	T[‡]	PR[†]	T[‡]	PR[†]	T[‡]	PR[†]	T[‡]	PR[†]	T[‡]
Alkanes														
1,1-Dichloroethane	0	0.4	0	0.5	0[§]	0.9, 2	0[§]	6, 10	4	6, 9	75	11, 19	25	17
1,2-Dichloroethane	0	0.4	0	0.5	–	–	9[§]	6, 7	–	–	80	11, 19	26	17
1,1,2-Trichloroethane	0	0.4	0	0.5	2[§]	0.9, 2	0[§]	6, 10	3	6, 9	65	11, 19	17	17
Carbon tetrachloride	11	0.4	29	1	–	–	12[§]	7, 6	–	–	11	9.4, 19	3	17
1,2-Dichloropropane	0	0.4	0	0.5	8[§]	0.9, 2	15[§]	6	21	6, 9	91	19	42	17
Average percent removal	**2**		**6**		**3**		**5**		**9**		**64**		**23**	
Aromatics														
Benzene	49	1.5	81[§]	3.6	71	2	91[§]	12, 10, 11	94	9	99	12	–	–
Toluene	49	1.5	84[§]	3.6	77[§]	2, 2	91[§]	12, 10, 11	96[§]	10, 9	98	12	–	–
o-Xylene	54	1.5	87[§]	3.6	86[§]	2, 2	97[§]	12, 10, 11	95[§]	10, 9	97	12	89	19
p-Xylene	–	–	95[§]	2.5	87	2	96[§]	10, 9	97[§]	10, 9	98	14	–	–
m-Xylene	55	1.5	91[§]	3.6, 2.5	88[§]	2, 2	98[§]	9, 10, 11	97[§]	10, 9	98[§]	12, 14	–	–
Ethylbenzene	47	1.5	86[§]	6.2, 5	77[§]	2, 2	92[§]	10, 9, 12	95[§]	10, 9	92[§]	12, 14	–	–
Chlorobenzene	86	0.4	89[§]	3.1	47	3	95[§]	12, 7, 6	84	6	99[§]	8, 19	91	17
o-Dichlorobenzene	–	–	79	2.5	67[§]	2, 2	92[§]	10, 9	81[§]	10, 9	90	14	74	19
p-Dichlorobenzene	32	1.5	65	6	75	2	94[§]	10, 12	100	9	87	12	–	–
m-Dichlorobenzene	–	–	84	2, 5	40[§]	2, 2	91[§]	6, 9	60[§]	10, 9	94	14	61	19
Average percent removal	**53**		**72**		**72**		**94**		**90**		**95**		**79**	
Alkenes														
cis-1,2-Dichloroethylene	82	0.4	99	0.5	93	2	94[§]	7, 6	87	10	94	11	98	19
trans-1,2-Dichloroethylene	100	0.4	100	1	100	0.9, 2	100[§]	10, 6	100[§]	6, 9	98	8	–	–
1,1-Dichloroethylene	76	0.4	98	0.5	84	3	98[§]	7, 6	–	–	100[§]	11, 19	90	17
Trichloroethylene	39	1.5	76	25, 6	45	3	95	9, 12	–	–	96	14, 12	–	–
Tetrachloroethylene	40	0.4	60[§]	1, 2.5	13	3	72[§]	10, 9, 6	–	–	96[§]	8, 14, 19	65	17
Average percent removal	**67**		**97**		**67**		**92**		**94**		**97**		**84**	

Source: Fronk (1987).
*A: Distilled water; pH not adjusted.
B: Groundwater; pH not adjusted.
C: Distilled water; pH 9–10.
D: Groundwater; pH 8–10.
– Test not conducted.
All removals blank corrected for stripping effects.
[†]PR: Percent removal.
[‡]T: Transferred ozone dose (mg ozone/L water): applied ozone dose—CFF gas.
[§]Average of two or more tests.

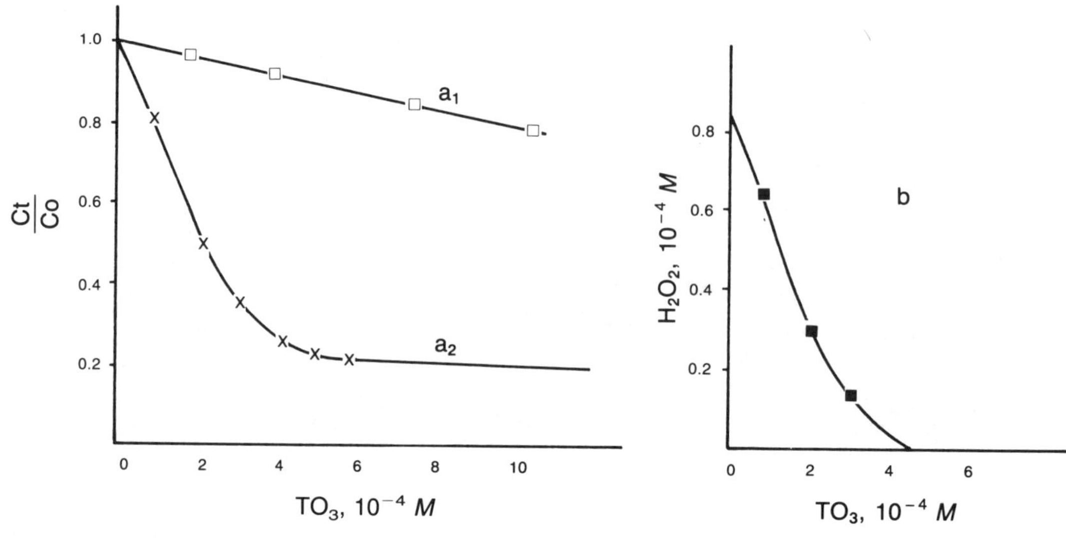

Source: Brunet (1985).

Figure III–25 Trichloroethane Oxidation by O_3 and O_3/H_2O_2 Systems

concentrations of 0, 0.24, and 0.47 mg/L, respectively, and an ozone dose of about 20 mg/L (Figure III–28).

Parallel experiments have shown that the point of H_2O_2 addition is an important parameter (Table III–25). The optimum moment appears to be the point at which the ozone alone has already oxidized the highly reactive molecules. The implementation of the radical system makes it possible to oxidize the refractory molecules. This allows one to take full advantage of the selectivity of molecular ozone before converting the process to one of nonselective free radical attack.

Ozone/UV processes. With ozone/UV processes, performances have also been measured either on group parameters (TOC, THMFP) or on specific micropollutants, such as volatile organic chlorinated compounds or pesticides.

Although organohalides can be destroyed by the O_3/UV combination, the reactivity will depend on the type of halogen in the molecule ($CHBr_3$ > $CHBr_2Cl$ > $CHCl_3$). According to Paillard et al. (1988b), CCl_4 and $Cl_3C–CCl_3$ are refractory to O_3 and O_3/UV in natural water. Their removal is affected only by stripping (Figure III–29). On the other hand, Francis (1987) was able to remove CCl_4 when conducting experiments in deionized water (Figure III–30). The impact of carbonate and bicarbonate can be seen when these two experiments are compared. These are free radical scavengers, which decrease the efficiency of the oxidation, possibly to the extent of total inhibition in the case of the most refractory products.

Although phenolic compounds (phenol, *p*-cresols, 2,3-xylenol, and 3,4-xylenol) may be easily oxidized by ozone alone, complete oxidation to CO_2 is uncommon. However, the use of a radical system encourages oxidation to CO_2. In Figure III–31, Gurol and Vatistas (1987) show a much greater decrease of TOC with the O_3/UV system as compared to ozone alone.

In summary, a certain number of factors influence the course of these reactions:

- pH (Glaze et al. 1980),
- relative concentration of oxidants (O_3/H_2O_2) (Paillard et al. 1988),
- photon flux in the O_3/UV system (Glaze et al. 1980), and
- radical scavenger concentration (Peyton et al. 1982; Doré et al. 1987; Guittonneau et al. 1988).

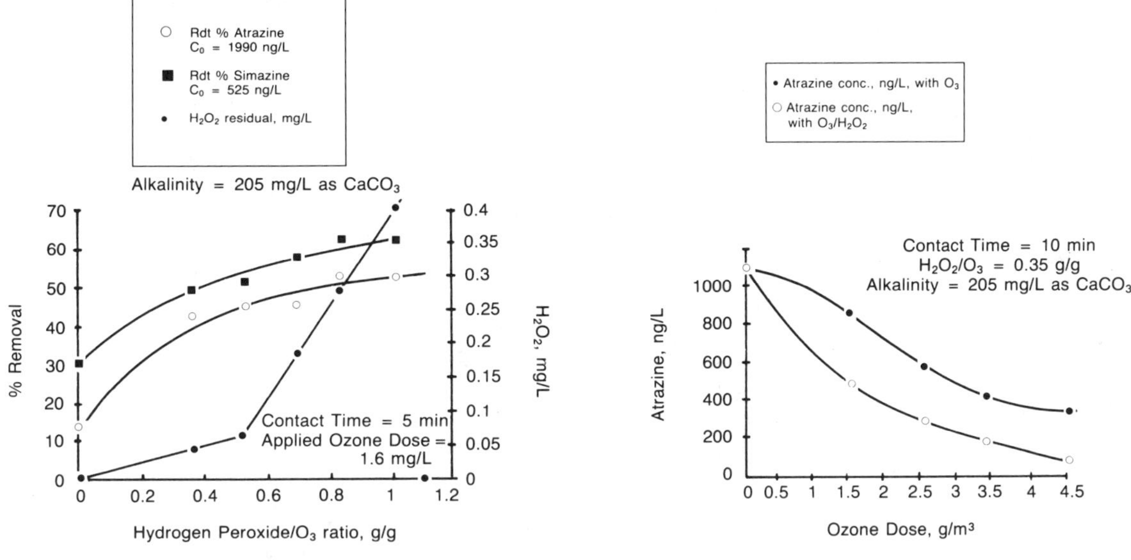

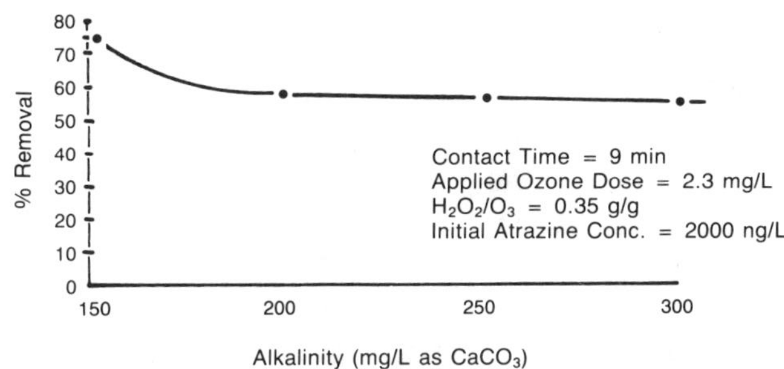

Source: Paillard et al. (1990).

pH = 7.95; TOC = 2 mg/L; T = 23°C (73.4°F).

Figure III–26 Elimination of Atrazine on the Seine River

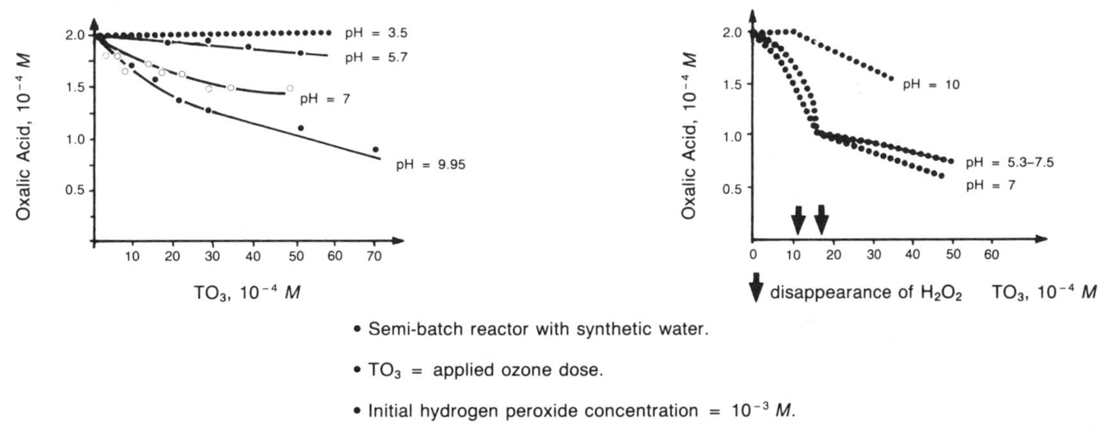

- Semi-batch reactor with synthetic water.
- TO_3 = applied ozone dose.
- Initial hydrogen peroxide concentration = 10^{-3} *M*.

Source: Brunet et al. (1984).

Semi-batch reactor, oxalic acid in distilled water.

Figure III–27 Action of O_3 and O_3 + UV on Oxalic Acid

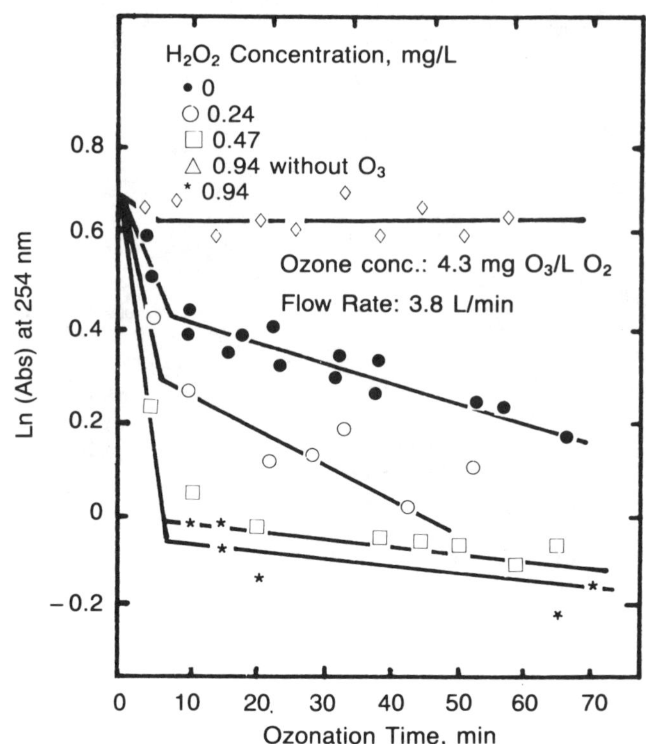

• 3–L semicontinous reactor.
• Experiments on lake water: pH = 7.5; alkalinity = 80 mg $CaCO_3$/L

Source: Duguet et al. (1985a).

Figure III–28 Effect of O_3/H_2O_2 on UV Absorbance

Table III–25 Change in the Removal of UV Absorption After 15 min of Semicontinuous Ozonation for Two Types of Water. Effect of Addition of Hydrogen Peroxide and of the Optimum Injection Time

Type of Water	Initial UV Absorbance (254 nm)	Percent Removal UV Absorbance		
		O_3 Alone	$O_3 + H_2O_2$ (T = 0★)	$O_3 + H_2O_2$† (Optimum T‡)
Clarified Seine River water	3.6	28	58	75
Groundwater	2.4	36	39	67

Source: Duguet et al. (1985a).
★Hydrogen peroxide added at the commencement of ozone addition.
†Applied ozone dose: 3 mg/L; hydrogen peroxide dose: 0.6–0.7 mg/L.
‡Hydrogen peroxide added 4–8 min after the start of ozonation. This was found to be the optimum delay time from laboratory experiments.

Modeling of advanced oxidation processes. Table III–26 (Glaze et al. 1987) gives the theoretical quantities of oxidant required to obtain a certain quantity of OH radical. As a result of their studies, Glaze and co-workers conclude that the O_3/H_2O_2 combination gives a good yield of OH, is easily adaptable to existing facilities, and would have the lowest relative process cost based on OH. The use of ozone at a high pH is of limited use. Although the O_3/UV combination seems difficult to implement on a large scale, it can be used for low flow rates, especially when the substrates to be oxidized are strongly UV absorbing.

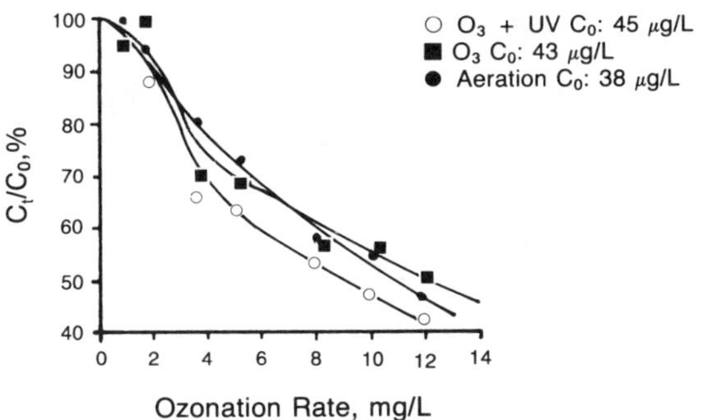

- C_0 = initial concentration.
- Surface water with halogenated organic compounds —
  pH = 7.9; TOC = 2 mg/L; alkalinity = 235 mg/L $CaCO_3$; T = 20°C.

Source: Paillard et al. (1987).

UV: 100 mW · s/cm^2; ozonation rate: 12 mg/L/h.

Figure III–29 Oxidation of CCl_4, Semicontinuous Experiment

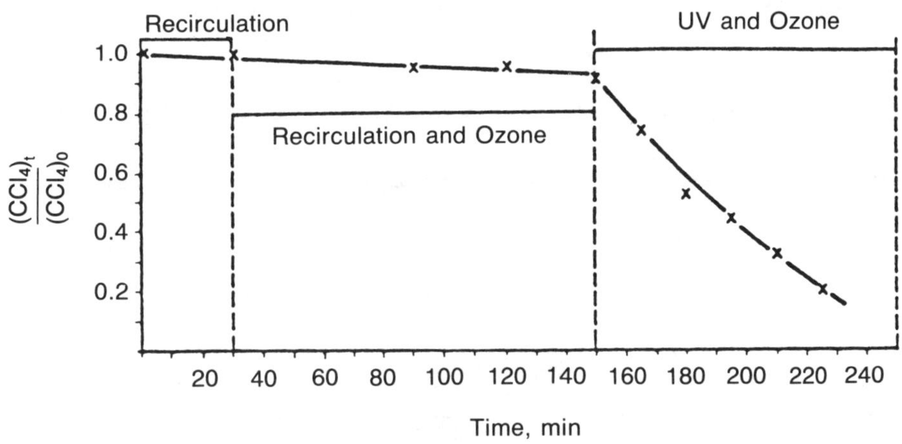

- 300-L semi-batch reactor.
- Applied ozone dose: 3.45 10^{-3} *M*/minute.
- Axial 2-kw antimony halide lamp.

Source: Francis (1987).

Figure III–30 Oxidation of 1×10^{-4} mol/L Carbon Tetrachloride Dissolved in Deionized Water

Recently, based on experiments using tetrachloroethylene as a model compound, Glaze and Kang (1989) described a kinetic model of the O_3/H_2O_2 process. This model takes into account doses of ozone and peroxide, mass transfer of ozone, and radical scavenging reactions. Figure III–32 shows the evolution of rate constant k_0 versus bicarbonate concentration and H_2O_2/O_3 ratio.

In the case of free-radical oxidation through the photolysis of ozone by UV irradiation, the superposition of different free-radical generating mechanisms is likely, but the O_3/UV system can also act in the following ways on the aqueous solution:

- direct excitation of certain compounds by UV irridation (k_{UV}),
- direct ozonation due to the presence of excess molecular ozone (k_{O_3}),
- free-radical oxidation (k_{UV,O_3}), and
- stripping of volatile compounds (k_e).

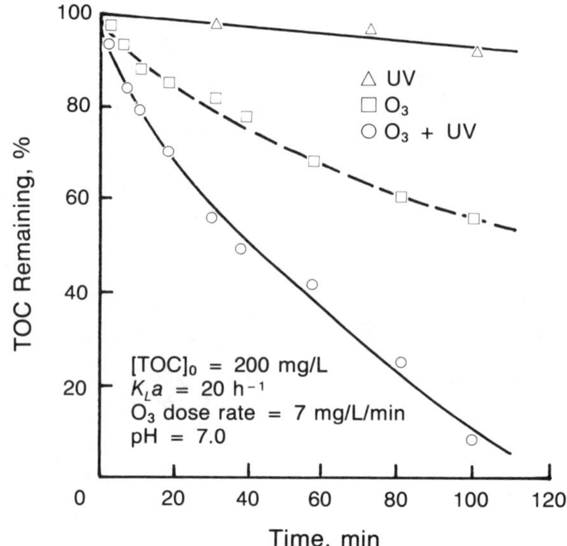

Source: Gurol and Vatistas (1987).

Figure III–31 Comparison of TOC Removal by UV Radiation, Ozone and Ozone + UV at pH 7.0

Table III–26 Theoretical Amounts of Oxidants and UV Required for Formation of Hydroxyl Radical in Ozone-Peroxide-UV System

System	Moles of Oxidant Consumed per Mole of OH Formed		
	O_3	UV★	H_2O_2
Ozone-hydroxide ion†	1.5	–	–
Ozone-UV	1.5	0.5	(0.5)‡
Ozone-hydrogen peroxide†	1.0	–	0.5
Hydrogen peroxide-UV	–	0.5	0.5

Source: Glaze et al. (1987).
★Moles of photons (Einsteins) required for each mole of OH formed.
†Assumes that superoxide formed in the primary step yields one OH radical per O_2, which may not be the case in certain waters.
‡Hydrogen peroxide formed in situ.

Also, the overall equation for the rate disappearance of M can be represented by a general expression of the following type (Peyton et al. 1982):

$$-\frac{dM}{dt} = k_{UV}\, I^a\, (M)^b + k_{O_3}\, (O_3)^c\, (M)^d + k_{UV,O_3}\, I^e\, D^f\, (M)^g + k_e\, (M)$$

where:

I = radiation flux transferred inside the reactor
(O_3) = ozone concentration in the liquid phase
D = ozonation rate.

III.E.3 Case Study on Ozone–GAC Treatment: Rouen La Chapelle Water Utility (Seine Maritime, France)

The water utility at Rouen La Chapelle, France, supplies drinking water to four boroughs belonging to the Rouen South Suburban Water Board, representing a

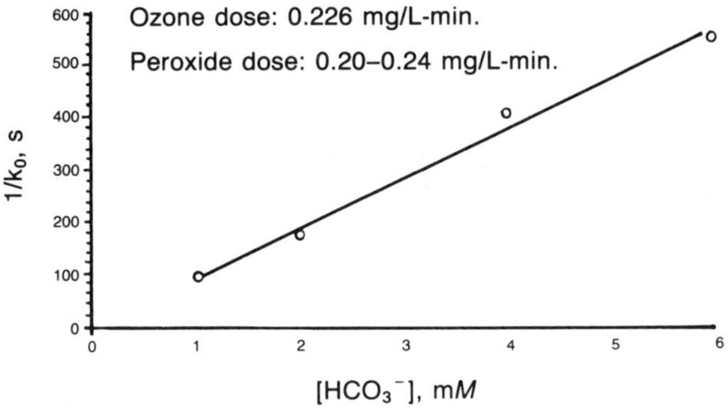

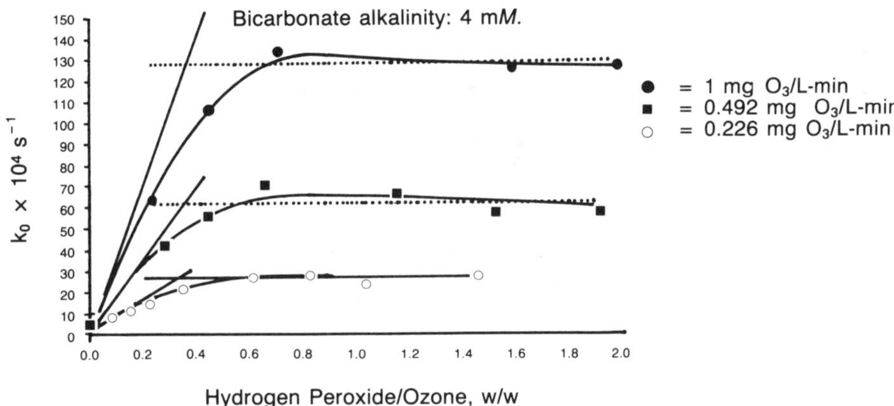

Source: Glaze and Kang (1989).

k_O: pseudo first-order rate constant for the oxidation of tetrachloroethylene (pH = 7.6–7.8). 70-L stainless steel sparged, stirred reactor.

Figure III–32 Ozone-Hydrogen Peroxide Process: Effects of Peroxide-Ozone Ratio and Bicarbonate Concentration

population of over 100,000. This plant obtains raw water from three wells drilled in the calcotic aquifer (Seine Valley). Initially, heavy use of this water table, combined with pollution related to industrial development, led to a deterioration in water quality (high concentrations of ammonia, iron, manganese, and various organic micropollutants). In order to solve these problems, an advanced treatment train, treating 2080 m^3/h (13.2 mgd), was installed in 1976. It includes two-stage ozonation on either side of dual filtration (on sand followed by activated carbon) (Figure III–33). At the present time, iron, manganese, and ammonia concentrations have sharply decreased; the main problem remaining is the removal of micropollutants.

The first stage of treatment begins with introduction of ozonated air into two preozonation tanks that are positioned in series. The plant contains three independent ozonation lines, each one associated with a well pumping station. The ozone is injected into the first compartment by an immersed self-suction turbine fed by offgases coming from the postozonation contactor (disinfectant step) and into a second preozonation compartment. In the second compartment, ozone coming from the generators is diffused via porous rod type diffusers. The overall contact time during this preoxidation step is 4 min. The average ozone applied dosage is 0.5 mg/L.

The preozonated water then undergoes dual filtration in a specially designed

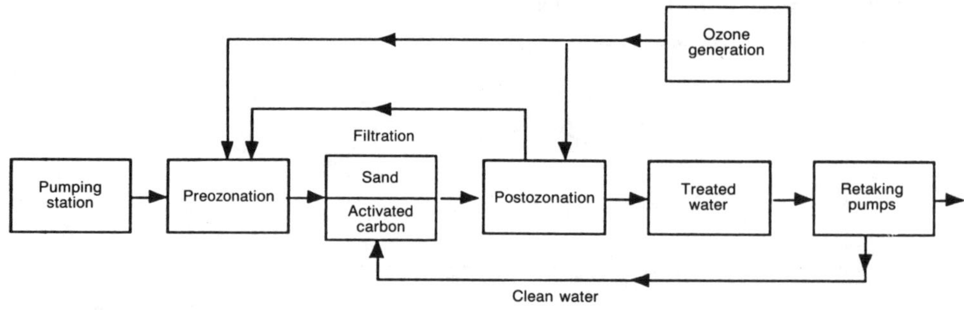

Source: CGE/CTE brochure.

2080 m^3/h (13.2 mgd).

Figure III–33 Schematic of the Treatment Line at the Rouen la Chapelle Water Works

Table III–27 Bulk Removal Tests on Organic Micropollutants

	Well Water	Treated Water
UV Absorbance, 254 nm		
Min.	0.018	0.005
Average	0.026	0.014
Max.	0.035	0.021
COD (mg/L O_2)		
Min.	0.9	0.3
Average	1.7	0.7
Max.	3.0	1.4
Anionic detergents (μg/L DBS)		
Min.	8	5
Average	41	15
Max.	63	23

Source: Reprinted with permission from Gomella, C. and Versanne, D., *TSM L'Eau,* 5 (May). © 1980 TSM L'Eau.

Table III–28 Removal of Volatile Chlorinated Hydrocarbons

	Groundwater	Treated Water
Chloroform (μg/L)		
Max.	<1	<1
Average	<1	<1
Min.	<1	<1
Trichlorethane (μg/L)		
Max.	6	2.9
Average	3.0	1.7
Min.	0.9	0.9
Carbon Tetrachloride (μg/L)		
Max.	0.02	0.02
Average	0.01	0.01
Min.	<0.01	<0.01
Trichlorethylene (μg/L)		
Max.	10.7	2.1
Average	9.7	1.2
Min.	8.2	0.3
Tetrachloroethylene (μg/L)		
Max.	15.6	0.27
Average	12.3	0.12
Min.	4.7	<0.1

Source: Reprinted with permission from Gomella, C. and Versanne, D., *TSM L'Eau,* 5 (May). © 1980 TSM L'Eau.

Table III–29 Removal of Organic Micropollutants: Determination by Gas Chromatography

Type of Compound Identified	Number of Analyses	Approximate Concentration (ppb)	
		Raw Water	Treated Water
Isoalkane > C 8	1	80	2
Methylcyclohexane	3	13	4
Isoalkane > C 8	3	16	5
Unidentified	2	12	4
Isoalkane > C 8	3	27	2
"	2	13	4
"	1	4	Trace
n-Nonane	3	8	4
Isoalkane > C 9	3	5	3
"	2	4	3
"	2	8	4
n-Decane	3	3	1
Isoalkane > C 10	3	2	Trace
"	3	2	Trace
n-Undecane	3	1	Trace
	1	18	1
	2	3	Trace
Isomers of 4-tertio-butylcresol	3	11	Trace
tertiobutyl-anisole	3	5	Trace
	3	6	Trace
	3	3	Trace
Bicyclohexyl	3	30	13
Unidentified	3	5	2
n-Tetradecane	3	2	Trace
Unidentified	1	Trace	Trace
n-Pentadecane	3	2	Trace
Dimethylphthalate	2	7	3
Isoalkane > C 15	3	Trace	Trace
n-Hexadecane	3	2	Trace
n-Heptadecane	3	2	Trace
Diallylphthalate	3	Trace	Trace
n-Octadecane	1	1	Trace
Diisobutylphthalate	3	6	3
Butyl isobutylphthalate	3	3	2
Dibutylphthalate (DBP)	3	400	190
n-Cicosane	3	3	Trace
Isoalkane ≥ C 20	3	2	Trace
"	3	2	Trace
"	3	2	Trace
Unidentified	2	3	Trace
"	3	Trace	Trace
Isoalkane ≥ C 20	3	3	Trace
Diethylhexylphthalate	3	195	95

Source: Reprinted with permission from Gomella, C. and Versanne, D., *TSM L'Eau,* 5 (May). © 1980 TSM L'Eau.

structure containing two stacked stages that work in series. The sand layer (thickness: 1 m [39.37 in], effective size: 0.5 mm [0.02 in]) is located above a layer of activated carbon (Picactif 15–40, thickness: 0.75 m [29.53 in]). These two layers of filtration media are supported and separated by floors; the floor supporting the sand is equipped with nozzles. The floor supporting the carbon is constructed of porous concrete; the space between the sand filter bottom and the layer of activated carbon is 1.5 m (4.92 ft). The filter bank consists of six filters, each with a surface area of 75 m^2 (807.3 ft^2) and working at a filtration rate of 3 m/h (1.24 gpm/ft^2).

Treatment ends with postozonation in two parallel contactors, with a hydraulic residence time of 10 min. This final treatment also helps to improve the taste and odor characteristics of the water. The ozone is injected via porous rod type diffusers at a dose rate of 0.6 mg/L. The ozone generated from air is supplied by three generators, each with a unit capacity of 1.9 kg O_3/h (100.5 lb/day). The treated water is stored in two 1250-m^3 (3.3×10^5-gal) reservoirs before being postchlorinated at 0.2 mg/L and sent to a 500-m^3 (1.32×10^5-gal) clear well.

Analyses of the raw and treated waters for various organic parameters were carried out over a period of 40 months (Gomella and Versanne 1980). Results of gross organic analyses show overall removals of 50% or more (Table III–27). Likewise, the action of ozone plus dual-stage filtration was effective at removing volatile organochlorine compounds (Table III–28). More sophisticated analyses employing cyclohexane extraction and gas chromatography–mass spectrometry (GC–MS) show that this treatment train successfully removes a variety of nonhalogenated hydrocarbons and phthalate derivatives (Table III–29).

III.E.4 Design Considerations

The elimination of a specific micropollutant can be a problem for the designer, who must choose the most appropriate system and the principal design parameters. In the case where the rate constant is greater than 10^5 or 10^6 $M^{-1}s^{-1}$, the micropollutant will certainly be eliminated if a measurable ozone residual is maintained. In the case where the rate constant is below 10^2 $M^{-1}s^{-1}$, an oxidation system involving a radical pathway (O_3/H_2O_2 or O_3/UV) or another process should be investigated. However, some waters may encourage the decomposition of ozone to the point where relatively refractory SOCs will degrade (see discussion of geosmin and MIB in section III.D). For intermediate values, it would be necessary to conduct pilot-plant tests under real conditions to obtain the necessary information.

III.F PARTICULATE REMOVAL

III.F.1 Principles

Coagulating effects of ozone—introduction. This section deals with the effects of ozone on removal of clay turbidity through coagulation. The phrase "coagulating effects of ozone" will be used in reference to all physicochemical phenomena involving the improvement of subsequent flocculation–settling, flocculation–flotation, direct filtration, or in-line filtration as a result of preozonation. These phenomena may take the form of a wide range of secondary effects induced by preozonation, such as a shift in particle size distribution towards larger sizes; the formation of colloidal particles from "dissolved" organic matter; the improved removal of TOC or turbidity during subsequent settling, flotation, or filtration; a decrease in coagulant dose necessary for achieving desired effluent turbidity or TOC; an increase in floc settling velocities; and extended filter run length due to slower head loss buildup or delayed breakthrough. These are called secondary effects because they do not occur instantly at the moment of ozone reaction. Rather, they begin to manifest themselves only after a period of mixing or after the addition of a classical coagulant. Although it is considered unlikely that a single mechanism is responsible for all of the above secondary phenomena, several of them may be attributed to the same primary effect. Furthermore, all of these phenomena are related in some way to ozone-induced changes in the properties of dissolved and particulate matter.

It is important to understand that the coagulating effects of ozone go beyond any direct oxidative effects on organic macropollutants as presented elsewhere in this book (for example, the bleaching of color, the increase in biodegradability, and the direct oxidation of THM precursors). For this reason, one must be wary of studies claiming improved removal of organic matter when the data are based solely on removal of color or UV absorbance. Also, when studying the removal of disinfection by-products (for example, THMs), one must be careful to incorporate controls permitting the separate evaluation of ozone's direct effects. Finally, the coagulating effects of ozone may not be observed with all waters. Whenever considering the use of ozone as a coagulant aid, the effects of preozonation should be critically evaluated in pilot studies incorporating the proper controls.

History of ozone as a coagulant aid. The observed coagulating effects of ozone were first commercially employed on moderately colored lake waters in the form of the "Micellisation–Demicellisation" (MD) process (Guillerd 1968; Diaper 1970; Meyer–Konig and Carl 1972). Water was pretreated by filtration through microscreens to remove ozone-consuming particles. Then, following preozonation, the resulting newly formed particles were removed by rapid sand filtration, perhaps with the aid of a polymer or aluminum coagulant. In cases where the colloidal turbidity was not appreciable following ozonation, final filtration might be omitted, and this was termed the "Microzon" process. The elimination of final filtration was not always successful, however, as exmplified by the regrowth problems encountered at Loch Turret, Scotland (Thorburn et al. 1975).

In the past decade, the coagulating effects of ozone have been successfully used in modern conventional flocculation-sedimentation-filtration plants and direct-filtration plants employing a variety of filtration media. While in most cases, ozonation prior to conventional treatment has little effect on ultimate particle removal (that is, the removal that can be achieved at optimal coagulant doses), ozone often facilitates the removal of readily coagulatable material (Richard 1978; Lienhard et al. 1980; Gerval 1980; Jekel 1983). This results in savings in coagulant, sludge treatment, and sludge disposal costs. In the case of direct filtration, there are indications that preozonation may encourage the removal of particles that could not otherwise be removed, regardless of coagulant dose. The particles formed by preozonation have been observed to be more efficiently enmeshed by precipitating $Al(OH)_3$ than those originally present (Jekel and Reicherter 1982). Preozonation has also been observed to cause an increase in aluminum floc size (Hodges et al. 1979; Chedal and Schulhof 1979) and in the rate of settling (Saunier et al. 1983). The effect of preozonation on the coagulation of DOC has not been well documented, and firm conclusions regarding DOC removal cannot as yet be made.

The coagulating effects of ozone have sometimes been referred to as the "microflocculation" phenomenon (Maier 1979). The use of this term implies that a certain fundamental knowledge of this process exists; for this reason, it is not used here. In fact, the mechanisms behind ozone's effects on coagulation are still unknown. Even the raw water characteristics that might indicate a water's susceptibility to the coagulating effects of ozone have proven elusive (Fiessinger et al. 1979; Hodges et al. 1979; Saunier et al. 1983; Reckhow et al. 1986a). As a result, the job of assessing the suitability and optimal operating conditions of an ozonation–coagulation process is, at present, mired in empiricism.

Mechanisms for the coagulating effects of ozone. Laboratory-scale experiments on the coagulating effects of ozone have often failed to show the same beneficial effects that have been reported in the field. Many have found preozonation to hinder subsequent bench-scale coagulation (Albert 1976; Van Breeman et al. 1979; Scrivner et al. 1980; Jekel 1983; Reckhow and Singer 1984a; Lefebvre et al. 1989), and others have been unable to demonstrate a coagulating effect (Alizadeh Astari 1978; Fiessinger et al. 1979; Larose et al. 1982). There are many potentially significant differences in the operation and design of laboratory-scale systems as compared to full-scale systems. It is important to understand these differences and bridge them where possible. However, before discounting the possibility of serious scale-up problems, the underlying mechanisms for the coagulating effects of ozone must be understood.

Ozone can react with both organics and inorganics in raw water to form a diverse

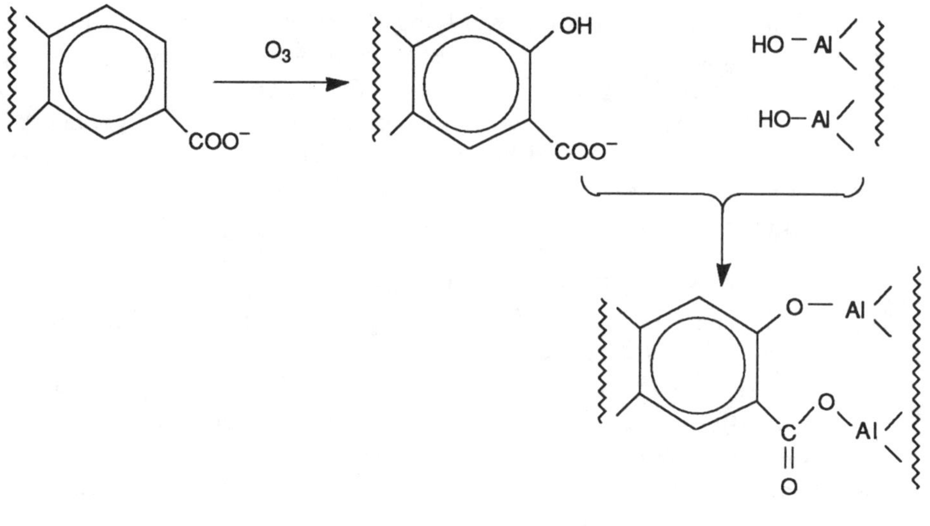

Figure III–34 Mechanisms for the Coagulating Effects of Ozone: Increase in Organic-Aluminum Associations

group of reaction products (see chapter II, sec. II.A.6). The formation of these products is instantaneous or nearly so. As indicated earlier in this section, these reactions will be referred to as primary effects. It is proposed that these primary effects then give rise to a family of secondary effects, which manifest themselves as some type of coagulating behavior. What follows is a discussion of both types of effects including possible mechanisms for the coagulating behavior induced by ozone. Although the removal mechanisms refer explicitly to coagulation and flocculation processes, they should also be considered important with respect to subsequent filtration. For example, formation of larger particles during flocculation should lead to a lower rate of head loss accumulation (Habibian and O'Melia 1975), and if the particles are larger than 1–2 μm, better clean-bed removal should also occur. Even in the absence of improved flocculation, particle destabilization should lead to better overall removal during filtration (Tobiason and O'Melia 1988). The six mechanisms presented below are not intended to comprise an exhaustive list, but rather are intended to indicate the nature by which ozone may impact conventional treatment or direct filtration. A more extensive discussion of these proposed mechanisms may be found elsewhere (Reckhow et al. 1986a).

1. Increased association of aluminum with ozonated organic matter. In chapter II, sec. II.A.6, the primary effects of ozone on natural organic matter were discussed. One of these effects is an increase in oxygenated functional groups following ozonation, especially the carboxylic acids. Based on several model compound studies (Parfitt et al. 1977; Davis and Leckie 1978; Kummert and Stumm 1978), it is reasonable to assume that the carboxylic and phenolic functional groups in natural organic matter form complexes with aluminum oxide and clay surfaces, and that these groups are, therefore, responsible for organic matter–aluminum surface associations (Davis and Gloor 1981). Accordingly, research has shown that preozonation can increase the degree of adsorption of natural organic material to activated alumina (Chen et al. 1987) and aluminum hydroxide (Kuhn et al. 1978). A commonly espoused corollary to these observations is that increases in carboxyl and phenolic content should lead to improved removal of DOC or DOC-coated clays by alum coagulation due to (a) the increased formation of "insoluble aluminum humates;" (b) charge neutralization from Al complexation; or (c) the increased adsorption of humic materials to aluminum hydroxide floc.

Figure III–34 shows, in a simplified form, one way this process might proceed. Both the organic ligand and the metal coagulant could be present initially in a dissolved state or bound to a colloidal or precipitating surface. Note that while this mechanism renders more of the DOC susceptible to coagulation, it may also lead to a higher coagulant demand. Furthermore, if the new acidic functional groups do not form strong complexes with aluminum, the additional charge and polarity of the compounds may render them more soluble and poorly removed by coagulation regardless of the coagulant dose.

Jekel (1983) examined the specific complexation chemistry of aluminum as it relates to ozonated humic materials. Contrary to the argument made above, he was unable to find an increase in specific aluminum complexation despite significant increases in total carboxylic acid groups. In the same experiment, he noted that alum coagulation of TOC was actually less effective following preozonation, especially at low coagulant doses. Jekel attributed this to increased formation of soluble Al–organic complexes (as opposed to insoluble Al–organic precipitates) due to the greater density of free acidic functional groups. Note that this phenomenon should also lead to greater coagulant demand and delay the precipitation of aluminum hydroxide (Snodgrass et al. 1984). Later work by Lefebvre and co-workers (1989) suggested that ozone has a similar effect when using ferric chloride as a coagulant. Increasing doses of ozone, resulting in higher carboxyl acidities, caused a shift in the removal of DOC to higher coagulant doses.

2. Improvement in calcium complexation by ozonated organic matter. An increase in the number of carboxylic acid groups may also lead to an increase in the degree of calcium complexation. This could result in direct precipitation of the organic matter (Guillerd 1968; Maier 1979; Liao and Randtke 1985) or increased particle destabilization by polymeric bridging, as illustrated in Figure III–35. In the latter case, absorbed strands of natural organic matter from two particles may form a bridge in binding with the same calcium atom(s).

Organics in ozonated lake waters and humic acid solutions have been reported to adsorb more completely to calcium carbonate (Kuhn et al. 1978; Fiessinger et al. 1979) and to precipitate more readily in the form of "calcium humate" (Jekel and Ernst 1981) as compared to the untreated controls. An increase in the complexation of calcium may also lead to improved adsorption of organics to both alum floc, as suggested by the results of Randtke and Jepsen (1981), and to other precipitating metal oxide surfaces such as MnO_2 (Colthurst and Singer 1982; Dempsey and O'Melia 1983). Recently, Dowbiggin and Singer (1989) found that high concentrations of calcium could induce the coagulating effects of ozone on precoated alumina particles. Later work on actual raw waters substantiated these findings (Chang and Singer 1988). These authors also found that there was a clear optimum ozone dose range with waters containing moderate to high levels of calcium. Chang and Singer noted that, in many cases where the rate of flocculation increased upon ozonation, the particle charges became more negative. This implies that simple electrostatic effects alone cannot explain this phenomenon.

3. Loss of organic matter from the surface of clay particles. One might also expect any decreases in molecular weight (MW) (see chapter II, sec. II.A.6) or increases in hydrophilicity of the organic matter following ozonation to lead to desorption from clay particle surfaces. This alteration of the adsorbed organic layer is important, because it is well established that small amounts of adsorbed humic materials stabilize particles in water (Tipping and Higgins 1982; Gibbs 1983; Jekel 1983; Ali et al. 1984; Felix–Filho 1985). Furthermore, laboratory studies with natural organic matter have shown that it is the highest-MW fraction that both adsorbs most strongly to Al_2O_3 particles (Davis and Gloor 1981) and is most capable of stabilizing alumina (Felix–Filho 1985) and silica suspensions (Jekel 1983). Limited data suggest that many types of algal extracellular matter may also stabilize natural particles. After alteration of the organic coatings, particle destabilization may occur via a minimization of the negative surface charges or by a shrinking of the adsorbed organic layers, which cause a steric hindrance to particle–particle contacts. This latter effect is illustrated in Figure III–36. It should be noted

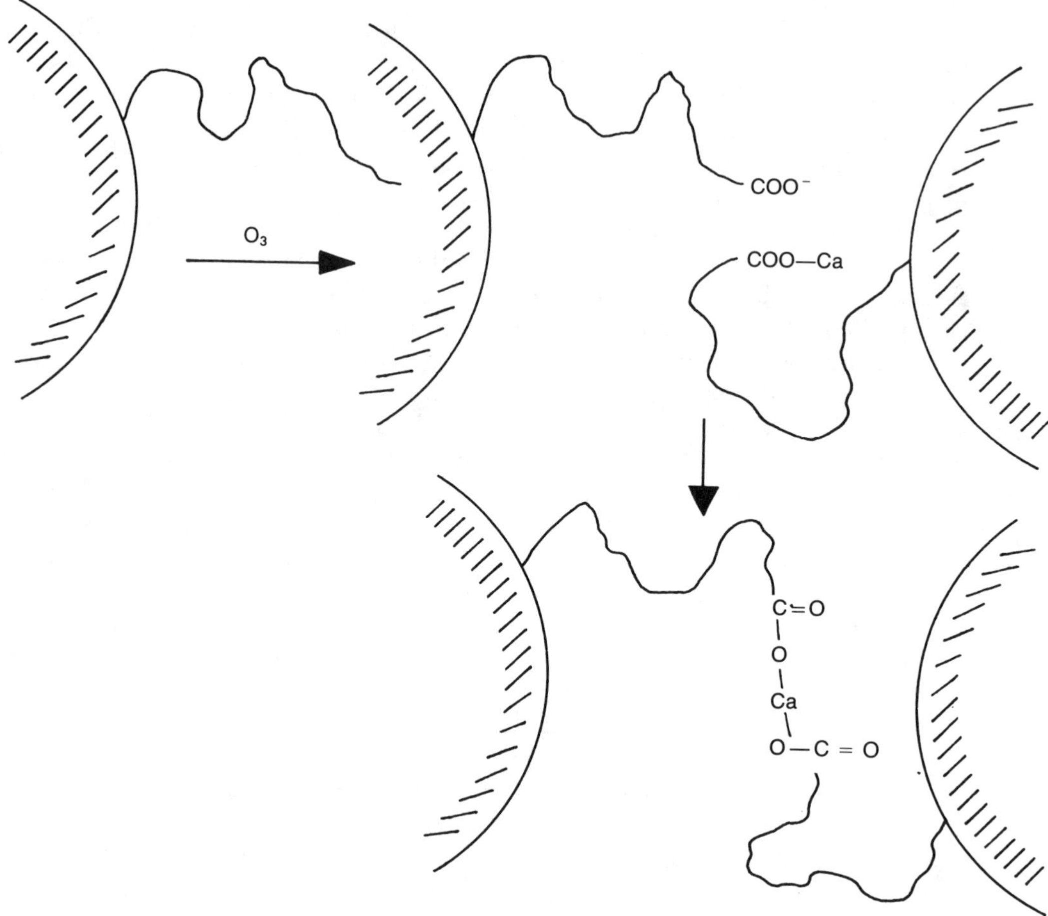

Figure III–35 Mechanisms for the Coagulating Effects of Ozone: Change in Calcium Complexation

that this mechanism pertains strictly to turbidity removal, because only a minor fraction of the organic content in most natural waters is associated with clay particles (Lienhard et al. 1980; Jekel 1983; Felix–Filho 1985; Dowbiggin and Singer 1989). Also, this mechanism may not explain the deterioration of particle removal seen at high ozone doses.

Jekel (1983) has shown, using solutions of Ruhr River humic acid and silica particles, that low doses of ozone can negate the particle-stabilizing influence of humic acid. He has accordingly interpreted this as being due to the reduced adsorption of organics leading to a reduction in steric stabilization. However, Felix–Filho (1985) was unable to demonstrate an analogous destabilization with mixtures of fulvic acid and α-Al_2O_3. Later, Hoyer and co-workers (1987) presented data suggesting that ozone can also reverse the inhibitory effect caused by certain types of algal extracellular matter.

4. Polymerization of organic matter. It is believed that ozonation of natural organic material leads to the formation of metastable organics, such as ozonides, organic peroxides, and organic-free radicals, which continue to react long after the ozone has disappeared (Buydens 1970; Gilbert 1983a). Given enough time and sufficiently high organic concentrations, these compounds might encounter other stable or metastable organics and undergo condensation or polymerization reactions (Figure III–37). For waters less concentrated in organic matter, this process could be catalyzed through locally enhanced organic concentrations at aluminum–polymer surfaces, clay particle surfaces, or gas–liquid interfaces created during ozone contact. Such reactions may lead

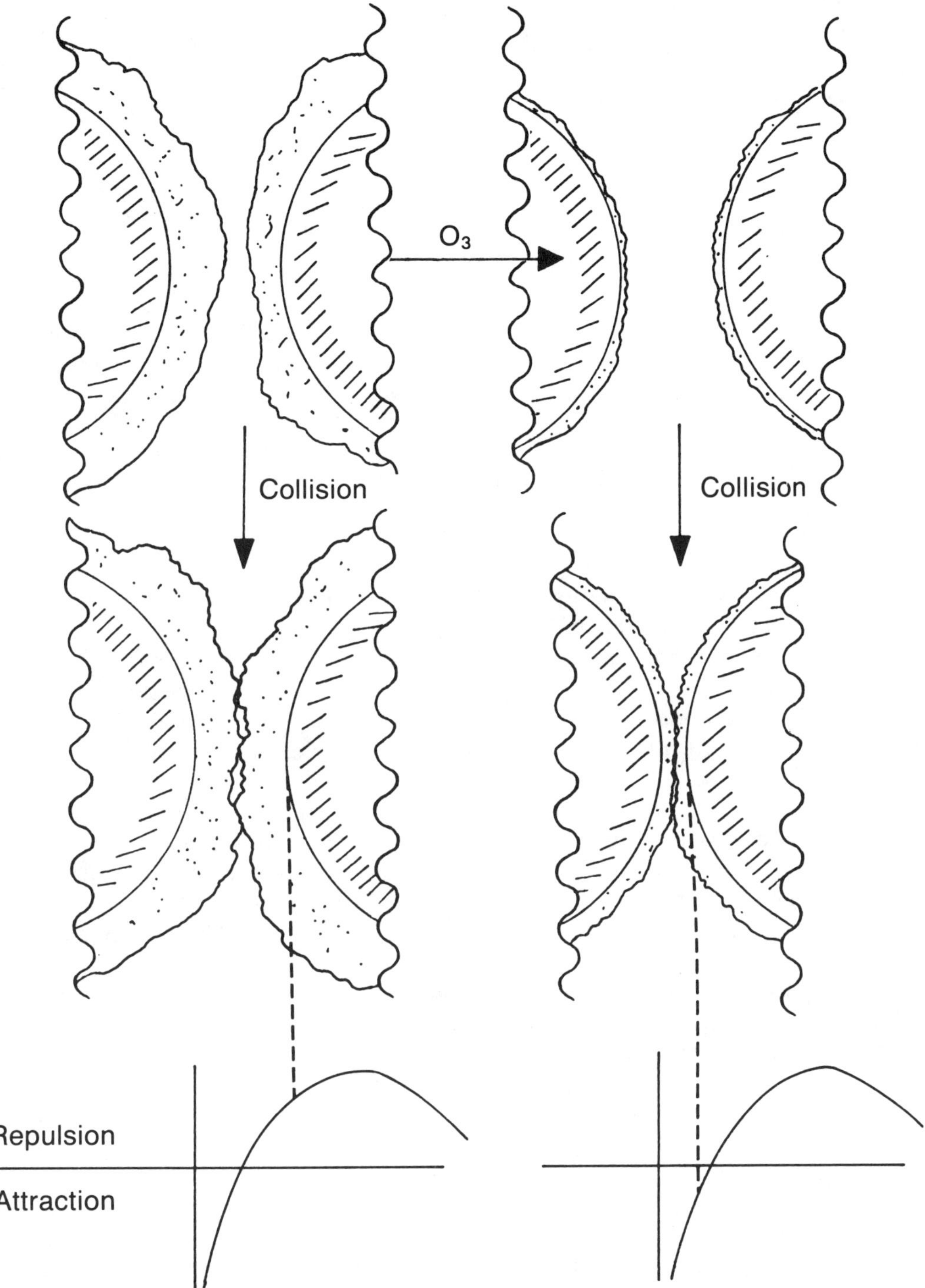

Figure III–36 Mechanisms for the Coagulating Effects of Ozone: Reduction in Organic Coating on Clay Particle Surfaces

to the formation of a polymer that can (a) precipitate spontaneously; (b) become more easily enmeshed by $Al(OH)_3$; (c) associate more strongly with solid surfaces; (d) act as a "bridging" polyelectrolyte; or (e) just by virtue of its size, "consume" less aluminum coagulant per carbon atom removed. Regardless of the exact removal mechanism, there is significant empirical evidence for the enhanced removal by Al or Fe species of

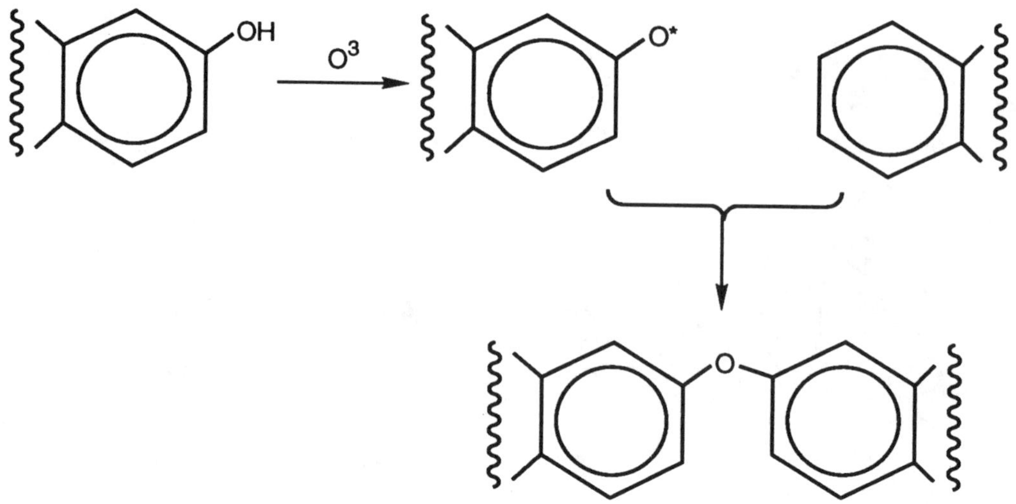

Figure III–37 Mechanisms for the Coagulating Effects of Ozone: Oxidative Coupling of Organic Matter

high-MW organics over lower-MW compounds (Davis and Gloor 1981; Thebault et al. 1981; Sinsabaugh et al. 1986a, 1986b).

Studies using size-exclusion chromatography have suggested that under certain circumstances, increases in apparent molecular size can be observed (Bruchet et al. 1987). Furthermore, research with dilute solutions of phenolic and N-containing organics have unequivocally demonstrated the presence of specific polymerization products following ozonation (Li and Kuo 1980; Elmghari–Tabib et al. 1982; Doré and Legube 1983; Kuo 1984; Duguet et al. 1985b). It should be recognized that the complexity of humic materials hinders the detection of similar polymerization reactions for natural waters. Recently, Grasso and Weber (1988) reported increases in high-MW organics following ozonation of filtered river water. In the same experiment, they reported improved removal of turbidity by bench-scale flocculation/settling (preceded by ozonation) of this water.

5. Breakup of metal–organic complexes. Under the proper conditions, ozone can break up metal–humic complexes (Cromley and O'Connor 1976; Mallevialle et al. 1978; Shambaugh and Melnyk 1978). The subsequent release of oxidized metals such as Fe(III) and Mn(IV) can be considered an *in situ* production of metal coagulant. Ferric salts are, of course, widely used as primary coagulants, and hydrated Mn(IV) species have been found to remove significant amounts of natural organic material by adsorption, especially in the presence of calcium (Colthurst and Singer 1982; Dempsey and O'Melia 1983). Also, the loss of metal cations should increase the net negative charge of de-complexed organic matter. This may render the organics more hydrophilic and less likely to stabilize particles via adsorption to their surfaces. Alternatively, they may be more susceptible to adsorption on polar aluminum hydroxide precipitates (Figure III–38). This group of mechanisms has particular relevance to the use of preozonation with groundwaters and certain lake waters rich in Fe and Mn. Felix–Filho (1985) found low levels of Fe(II) (0.5–0.8 mg/L) to be quite effective at catalyzing the coagulating effects of ozone in mixtures of alumina and fulvic acid.

6. Reactions with algae. Ozone readily kills or lyses many types of algae (Pak et al. 1981; see sec. III.G). The subsequent liberation of biopolymers (for example, nucleic acids, proteins, polysaccharides) may be equivalent to the release of natural organic coagulants (Figure III–39). Alternatively, some naturally released algal biopolymers can actually hinder coagulation, and the use of ozone may minimize this adverse effect.

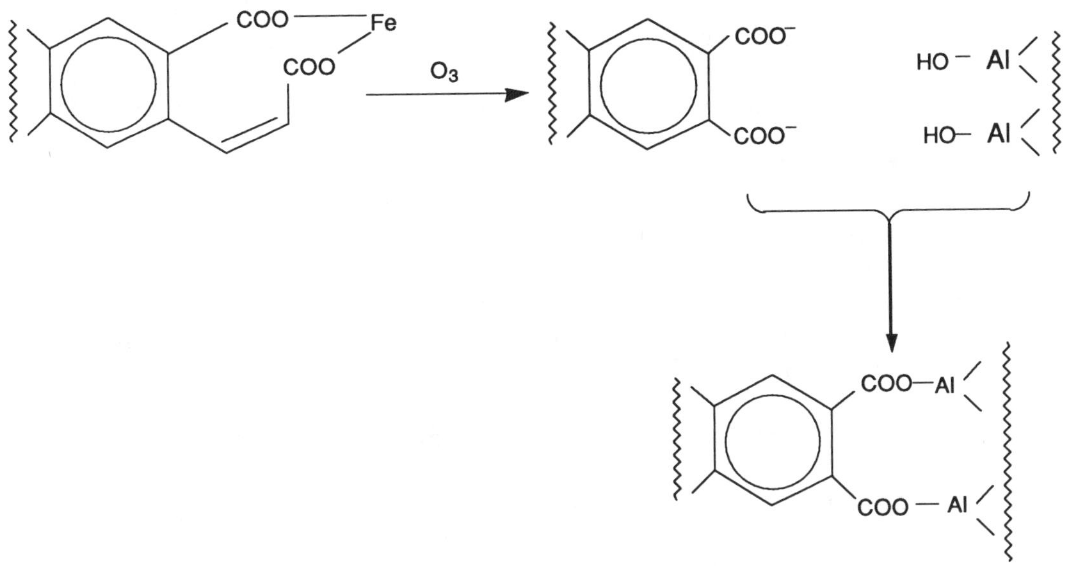

Figure III–38 Mechanisms for the Coagulating Effects of Ozone: Breakup of Metal-Organic Complexes

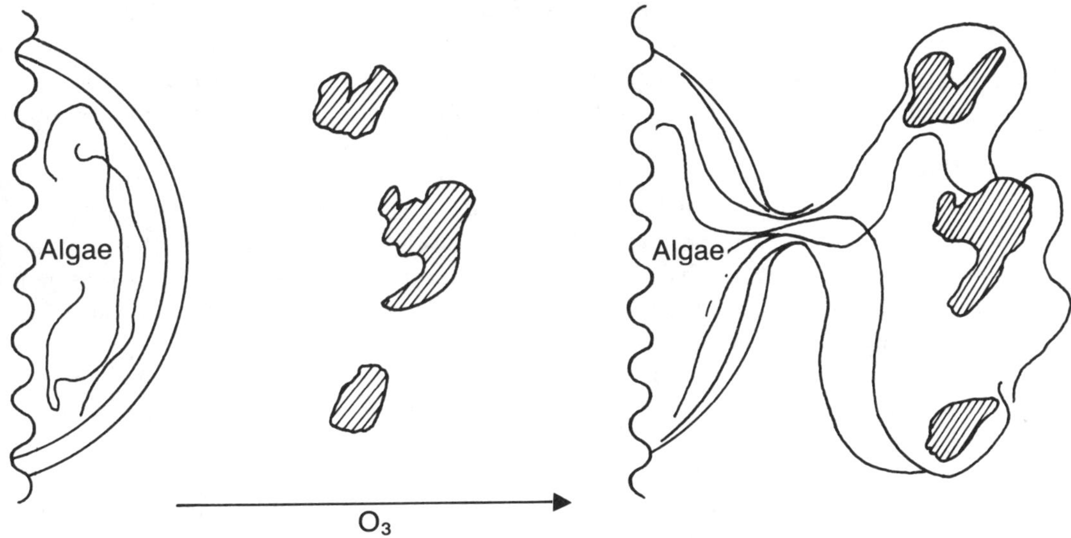

Figure III–39 Mechanisms for the Coagulating Effects of Ozone: Reactions with Algae

Regardless of the mechanism, ozonation has been observed to improve the removal of algae by coagulation–settling (Ginnochio 1981) and flotation (Betzer et al. 1980). In addition, improved coagulation of particles and TOC following preozonation has been noted to correspond quite strongly with phytoplankton concentrations (Volga River [Damez and Dernaucourt 1979]; Lake Constance [Maier 1979]) or to occur during the time of year when high algal concentrations are common (Seine River [Pascal et al. 1984]).

Many types of high-MW compounds released by algae could potentially affect subsequent treatment. For example, it has been proposed that deoxyribonucleic acid (DNA), once liberated during cell lysis, becomes a major factor in the flocculation of activated sludge (Tenney and Stumm 1965; Vallom and McLoughlin 1984). As for the polysaccharides, the flocculating effects of one particularly prominent group of algal exudates, the alginates, are well documented (Bontoux et al. 1976). It should be noted that this class of polysaccharides is known to be quite surface-active and able to complex calcium very strongly. The liberation of surface-active polymers could also explain the presence of the thick organic foam noted in preozone contact chambers at several plants (Campbell and Pescod 1966; Trussell et al. 1975; Maier 1979; Afanasyev et al. 1981; Gerval and Bablon 1983). This foam is strongly correlated with algal concentration (Maier 1979; Afanasyev et al. 1981), and its formation is apparently catalyzed by the presence of ozone (Gerval and Bablon 1983). On the other hand, certain types of extracellular algal matter are known to interfere with coagulation processes, perhaps via formation of a steric layer on clay particles. Recent studies have shown that ozone can readily break up this polymeric material and give lower filtered water turbidities (Hoyer et al. 1987).

Effect of raw water quality on its susceptibility to the coagulating effects of ozone.

Sensitivity to raw water quality. Based on theoretical considerations of the various proposed mechanisms discussed above, one would expect many aspects of water quality to potentially play a role in determining the efficacy of ozone as a coagulant aid. From empirical observations, it is clear that differences in water quality can have a profound effect on this process, as demonstrated by the early literature on electronic particle size analysis of ozonated water. At low ozone doses (less than 2.5 mg/L), some researchers have observed an immediate flocculating behavior (an increase in the number of large particles at the expense of the smaller ones) (Schalekamp 1977; Alizadeh Astari 1978; Jekel 1983), an anti-flocculating behavior (a reduction of the number of large particles with an accompanying increase in the small particle concentration) (Alizadeh Astari 1978; Hodges et al. 1979; Mathonnet et al. 1985), an ambiguous mixture of the two (Alizadeh Astari 1978; Richard 1982), or no effect at all (Colin et al. 1982). The results were equally variable when turbidity or total particle numbers were measured following ozonation (Maier 1979). Some have reported increases (Guillerd 1968; Somerville and Rempel 1972; Gerval 1978; Chedal and Schulhof 1979), some decreases (O'Donovan 1965; Schalekamp 1977; Richard 1982; Jekel 1983), and some a mixture of the two (Alizadeh Astari 1978). In one particular study that demonstrated improved flocculation with ozone, particle zeta potentials were found to be unchanged by preozone treatment (Richard 1982), whereas another researcher observed either a less negative potential or no change at all (Chedal 1982).

Even studies using a single source of raw water have sometimes revealed complex and conflicting effects of preozonation on subsequent coagulation. This is exemplified by the series of reports on the treatability of the Seine River near Paris. The quality of this water source is subject to wide seasonal variations as well as changes along its length due to point and nonpoint discharges. Richard (1978) found improved coagulation of organics and turbidity following preozonation of water collected downstream of Paris, whereas Seine River water collected upstream, containing less organic material, failed to show this behavior. At about the same time, a similar study on water downstream of Paris failed to show improved coagulation due to preozonation (Alizadeh Astari 1978). Later studies at Choisy-le-Roi (upstream) showed improved coagulation of particles and turbidity as a result of the ozone treatment, but not of TOC (Bablon 1981; Colin et al. 1982; Saunier et al. 1983). However, during a more recent phase of this same study, the coagulation of TOC was also reported to have benefitted from preozonation (Bablon and Saunier 1983).

Specific aspects of raw water quality necessary for the use of ozone as a coagulant aid. Very little is known with any degree of certainty about the relationship between water quality and the coagulating effects of ozone. Based on limited data, it has been suggested that a certain critical concentration of organic material must be present to

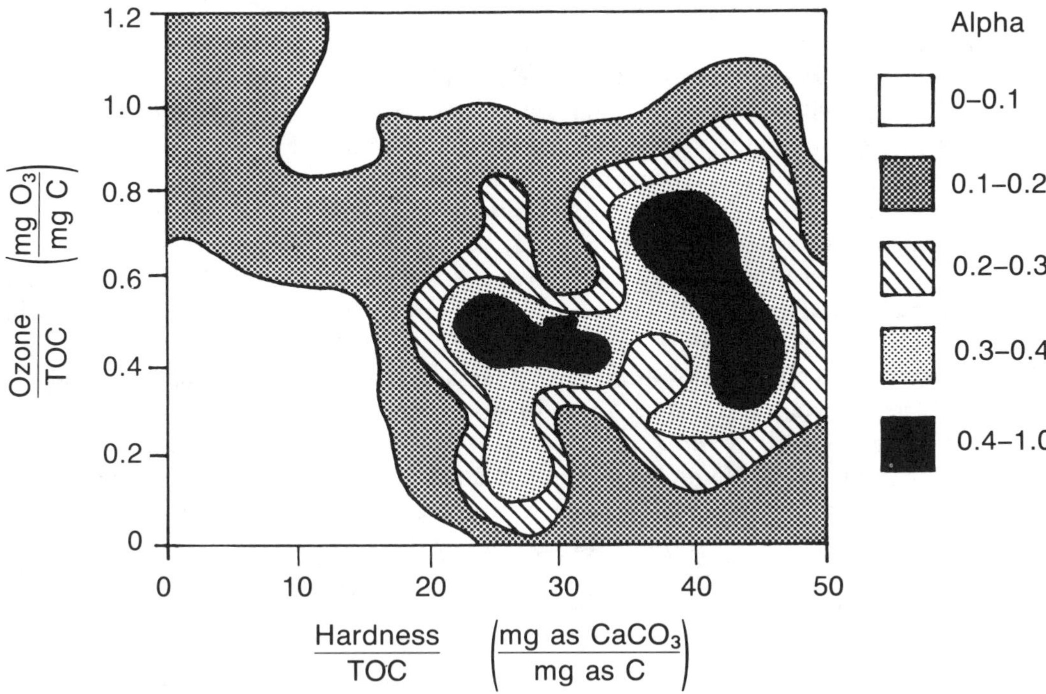

Source: Chang and Singer (1988).

Figure III–40 Effect of Hardness and Ozone Dose on Particle Stability in Ozonated Waters

observe coagulating effects (Alizadeh Astari 1978; Richard 1982), and that these effects often show seasonal patterns, increasing in autumn (Buydens 1970) and at times of high algal concentrations (Damez and Dernaucourt 1979; Maier 1979; Pascal et al. 1984). Others hold that the coagulating behavior of ozone occurs most often with waters of high turbidity as well as with highly colored waters (Damez and Dernaucourt 1979). Furthermore, the benefits of preozonation have been observed to increase with raw water particle numbers (Chedal 1982).

Perhaps the most convincing link between raw water quality and coagulating effects of ozone concern calcium hardness. Singer and Chang (1988) examined the impact of ozone on particle stability in 12 raw waters from 7 locations. They found that ozone's effect was dependent on calcium concentration and ozone dose when expressed on a per-carbon basis (Figure III–40). In studies of this type, particle stability is typically characterized by α, the ratio of particle collisions that result in attachment to the total number of particle collisions (also referred to as the collision efficiency or "stickiness" factor). Thus, the magnitude of α is directly related to the rate of flocculation. Figure III–40 indicates that the ratio of hardness to TOC has to be at least 20 mg/mg before preozonation will have a significant beneficial effect on particle coagulation.

III.F.2 Case Studies

Two case studies will be presented here—a conventional plant (Choisy-le-Roi, near Paris) and a direct filtration plant (Los Angeles, Calif.). In addition to these two plants, many other installations in North America and in Europe are currently using ozone as a coagulant aid (Diaper 1970; Meyer–Konig and Carl 1972; Barker and Palmer 1977; Sontheimer et al. 1978; Miller et al. 1978; Damez and Dernaucourt 1979; Maier 1979; Lienhard et al. 1980; D'Elia et al. 1981; IOA 1982; Gowens and McFadzean 1983; Zabel

1984; Stankovic and Tucovic 1985; Le Pauloue 1985a, 1985b; IOA 1986; Cryer 1986). However, most of these have not been studied as extensively as the two examples given.

Ozone–coagulation–settling: the Choisy-le-Roi plant. The treatment plant at Choisy-le-Roi (800,000 m^3/day [210 mgd]) serves a large portion of the Paris suburbs. It uses water from the Seine River, a moderately polluted source subject to seasonal algal blooms (Table III–30). Preozonation was incorporated into this plant in the early 1980s as a result of extensive pilot studies (Saunier et al. 1983). The current treatment scheme includes preozonation, addition of WAC (partially neutralized aluminum chloride), flocculation, settling, rapid sand filtration, intermediate ozonation, GAC filtration, chlorination, storage, and partial dechlorination (Figure III–41). Early work with a 12-m^3/day (2.2 gpm) pilot plant (Saunier et al. 1983) and corroborative studies with a 800-m^3/day (150 gpm) "simulator" (Pascal et al. 1984) generally showed improved particle removals as a result of preozonation. Figure III–42 shows that ozone improved turbidity removals over a wide range of coagulant doses. In addition, formation of larger and faster-settling floc was noted. Based on these studies, the optimal ozone dose was assessed to be 0.8 mg/L. Analysis of TOC removal, however, gave mixed results. The effects of preozonation on any given day were unpredictable and apparently related to changes in raw water quality. Early pilot work suggested that increases in particle numbers and pH enhanced the beneficial effects of ozone.

Later, Gerval and Bablon (1983) concluded that preozonation would permit a 30 percent coagulant savings, reduced sludge volume, and reduced filter wash water volume while maintaining the same filtered water quality. These authors also presented new pilot data that suggested an improved removal of TOC as a result of preozonation. More recently, the effects of preozonation in the full-scale plant were investigated by operating the preozonation facility in a periodic, on–off mode for several months (Gerval et al. 1985). The removal of turbidity by coagulation–settling and by coagulation–settling–filtration was found to be greater during periods of preozonation (average dose = 0.7 mg/L). However, these authors were unable to demonstrate a significant effect on TOC removal. In summary, there is no doubt that preozonation improves the removal of turbidity by conventional treatment at Choisy-le-Roi. However, its effect on the removal of organic matter has not been adequately demonstrated.

Ozone–direct filtration: the Los Angeles Aqueduct Filtration Plant. In contrast to experiences in Europe, much of the early interest in preozonation in the United States has centered on its beneficial effects with direct filtration. This rapidly growing interest is partly due to the success of the Los Angeles Aqueduct Filtraton Plant. The establishment of stricter drinking water quality standards in the state of California prompted the Los Angeles Department of Water & Power to investigate treatment alternatives for its Owens River source (see Table III–30). Preozonation was one of the alternatives the department's consultants chose to investigate. Early pilot studies with the Owens River water gave conflicting results with regard to the impact of ozonation on the removal of particles during subsequent filtration (McBride et al. 1977). Preozonation was, however, recommended based on its superior disinfecting power and its ability to control tastes and odors and destroy THM precursors.

Later work at Los Angeles demonstrated a clear improvement in filtered water turbidity when preozonation was used in place of prechlorination or as an alternative to no oxidative pretreatment (Hodges et al. 1979). This change in behavior was attributed to changes in raw water quality when the source was switched from groundwater to surface water. The authors hypothesized that the resulting reduction in total dissolved solids (TDS), hardness, and alkalinity, or the increased importance of organic matter associated with the particles, may have been responsible. During this later phase, improvements in filtered water turbidity were noted as soon as 20 min from the time the ozonator was turned on (Figure III–43). This small lag period corresponded precisely to the detention time of the pilot system. Beneficial effects of ozone were noted for sand,

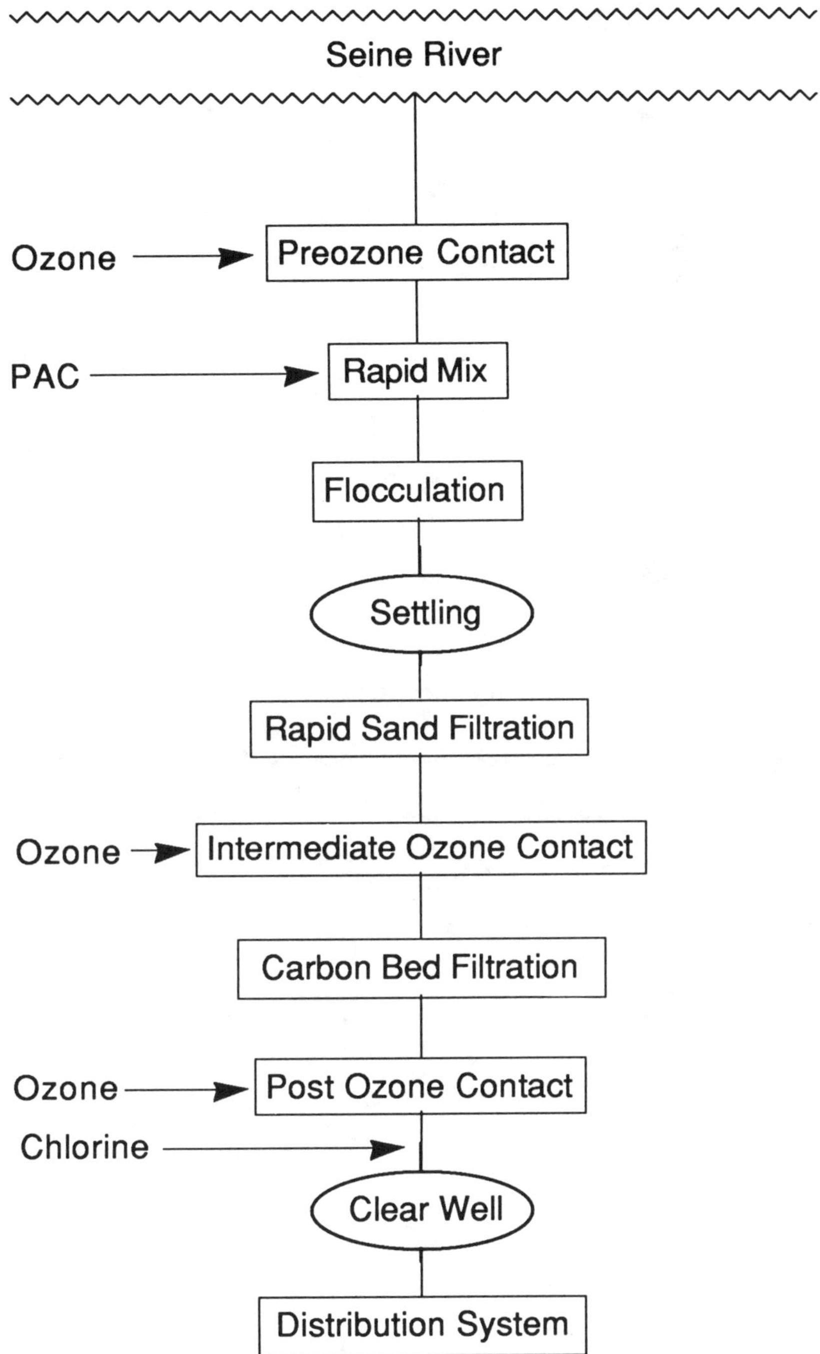

Figure III–41 Choisy-le-Roi Water Treatment Plant Schematic

anthracite, and dual-media filters at a variety of coagulant doses, and the improvement in effluent turbidities increased with increasing ozone doses up to 1.9 mg/L. Also, it was noted that at a given alum dose, preozonation resulted in lower optimum polymer doses and greater ultimate turbidity removals as compared to controls using prechlorination. Improved turbidity removals were verified when mass balances on parallel backwashed filters showed a significantly greater mass of particles in the backwash water coming from the filter that received preozonated water. The authors reported that fewer and more voluminous particles were visually apparent in the flocculated waters that had

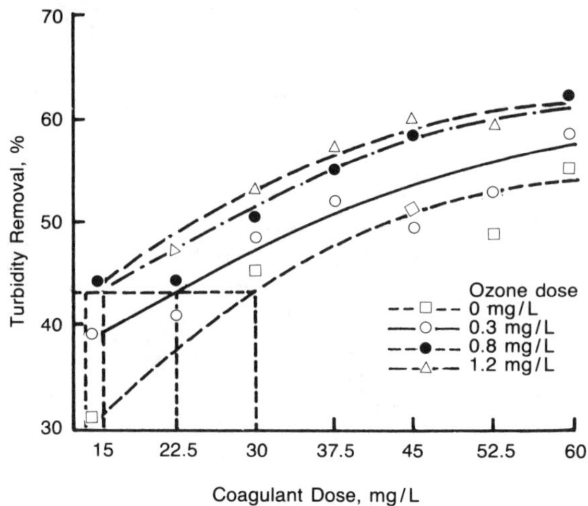

Source: Reprinted with permission from Saunier, B.M. et al., *Jour. AWWA,* 75:5:239–245. © 1983 AWWA.

Figure III–42 Effects of Preozonation on Settled Water Turbidity at Choisy-le-Roi (from Saunier et al., 1983)

Table III–30 Raw Water Quality Characteristics for the Plants at Choisy-le-Roi and Los Angeles

Parameter	Choisy-le-Roi	Los Angeles
Organic		
TOC (mg/L)	4	3
COD (mg/L)	4	5
Color (Pt-Co Units)		8
UV absorbance (per cm)	0.090	
Inorganic		
TDS (mg/L)	250	200
Hardness (mg/L)	240	80
Alkalinity (mg/L as $CaCO_3$)	200	110
Ca (mg/L)	95	25
Mg (mg/L)		5
Fe (mg/L)	0.25	0.06
Particulate		
Turbidity (NTU)	15	6
Particles (per μL)	120	20
Chlorophyll (mg/L)	0.030	
pH	8.2	7.5

been preozonated as compared to those that had been prechlorinated. These observations were substantiated with an electronic particle size analyzer (Hodges et al. 1980). Preozonation also greatly reduced filter conditioning time; however, these authors were unable to detect a significant effect of preozonation on head loss. Several years of additional pilot study indicated that preozonation at a 1-mg/L dose permitted an increase in filtration rate from 9 gpm/ft^2 (with prechlorination at a 2-mg/L dose) to 13.5 gpm/ft^2 (preozonation only [Prendiville and McBride 1983]). Accompanying this increased rate was a decrease in optimum coagulant dose of 50 percent for $FeCl_3$ and 25 percent for the cationic polymer (CDM–BC 1982).

A small number of THMFP measurements made during a later phase of the study indicated that preozonation–filtration removed about 50 percent of the THM precursors, while prechlorination–filtration failed to show any precursor removal. Such an improved removal of THMFP may be due to a change in the point of chlorination, the

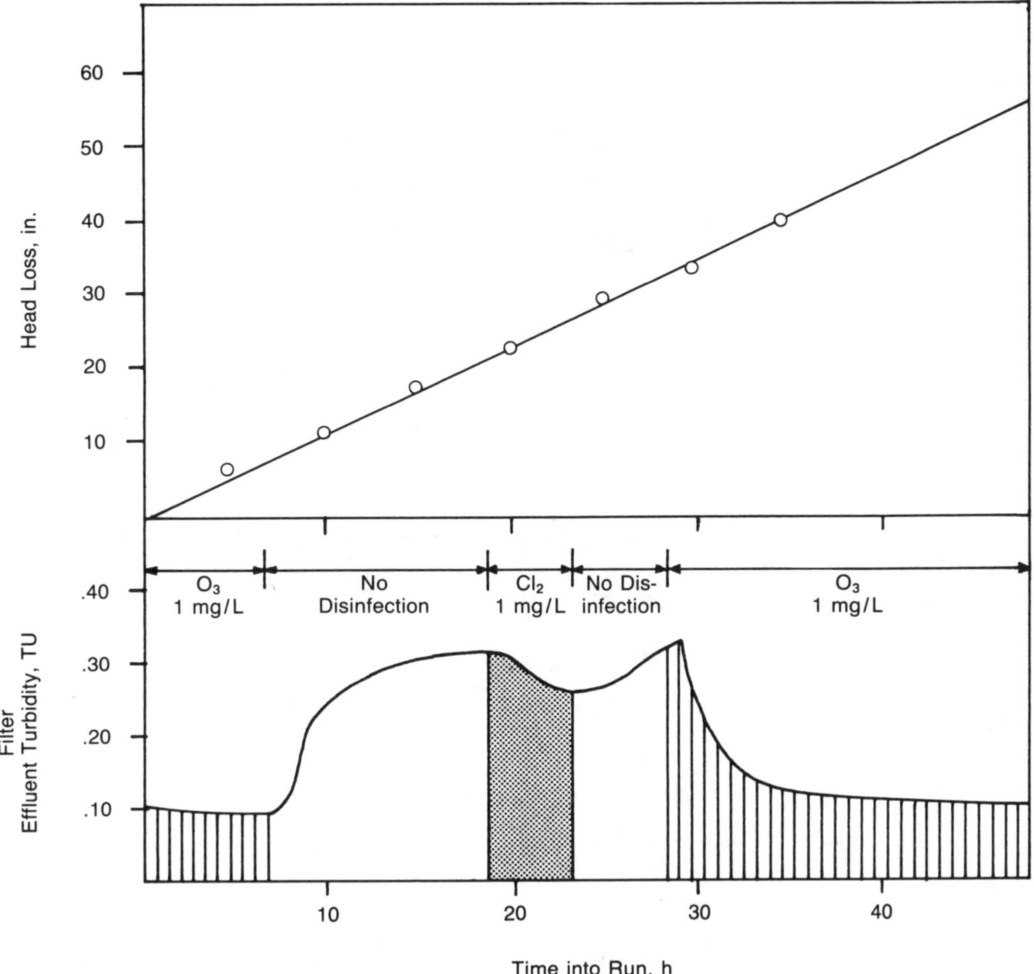

Source: Hodges et al. (1980).

Figure III–43 Effect of Preoxidants on Filtered Water Turbidity and Head Loss at Los Angeles

enhanced removal of organic carbon with preozonation, or a combination of the two. Unfortunately, little or no data were collected regarding organic carbon removal in these pilot studies.

The 2,200,000-m^3/day (600-mgd) preozonation–direct filtration plant was completed and placed in operation in December of 1986 (Figure III–44). A series of full-scale tests performed in the first seven months of operation verified much of the pilot-plant data (Georgeson and Straub 1987). An ozone dose of 0.6 mg/L (deemed optimum by the authors) resulted in a reduction in filtered water turbidity from 0.14 NTU without ozone to about 0.08 NTU with preozonation. Turbidity removal increased slightly up to an ozone dose of 1.35 mg/L, above which no additional benefit was seen (Georgeson and Karimi 1988). Other tests indicated that preozonation was far superior to prechlorination as evaluated by filtered water turbidity. In summary, there is no doubt that preozonation improves the removal of turbidity by direct filtration at the Los Angeles plant. Its effect on the removal of organic matter has not been determined.

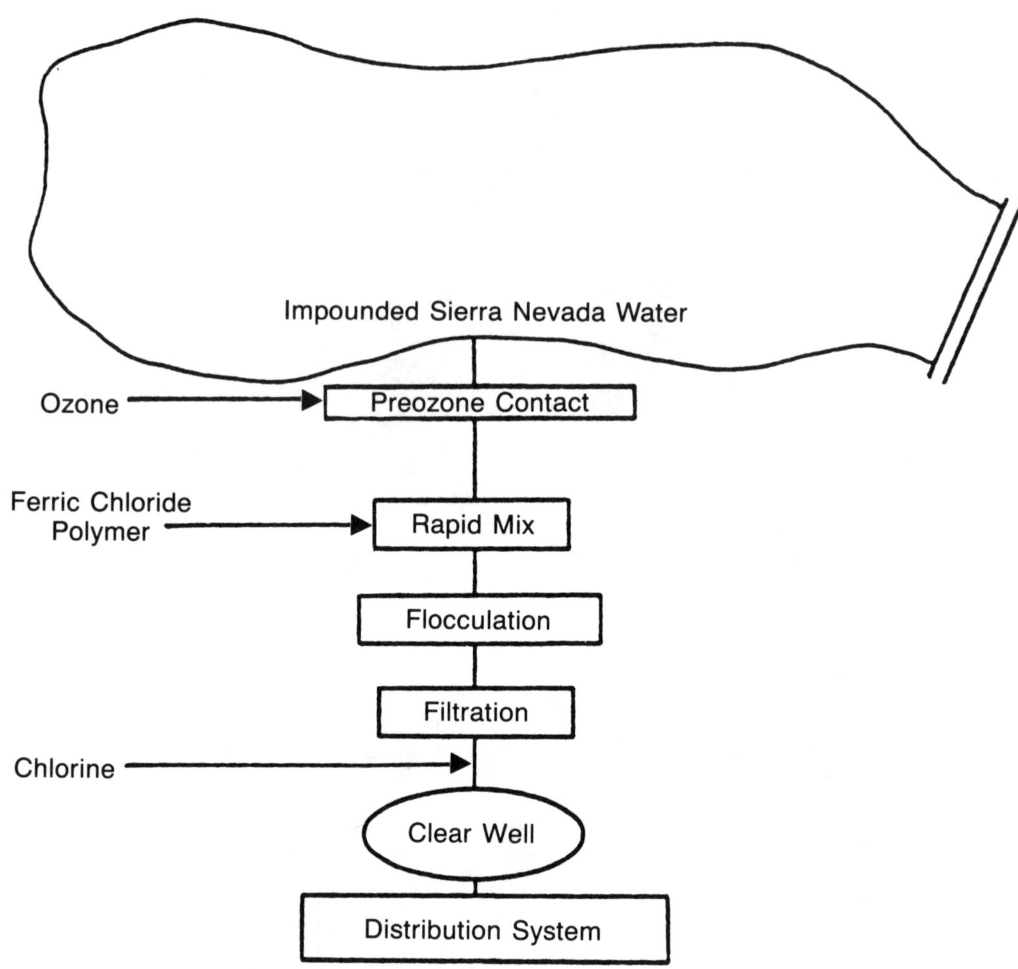

Figure III–44 Los Angeles Department of Water and Power Aquaduct Filtration Plant Schematic

III.F.3 Special Considerations for Process Evaluation, Design, and Control

Process evaluation. Preozonation has been used successfully as a coagulant aid with conventional treatment, direct filtration, and in-line filtration. For conventional treatment, the benefits of preozonation have been seen with ferric chloride, polyaluminum chloride, and organic cationic polyelectrolytes as well as with alum. When used with direct filtration, pilot studies have shown that ozone improves process performance with single- and dual-media filters, ranging in depths from about 1.5 m (60 in.) to about 2.3 m (90 in.) (Hodges et al. 1980; Montgomery 1986). In addition, the beneficial effects of ozone have been seen with filtration rates from 14 m/h (6 gpm/ft^2) to as high as 43 m/h (18 gpm/ft^3). Although prechlorination can improve filter performance, preozonation generally does a better job when both preoxidants are optimized with respect to dose. The improved performance with ozone has been seen in lower filtered water turbidities, slower head loss buildup, and faster filter ripening (Montgomery 1986). Other studies have shown a comparable rate of head loss buildup but a delayed turbidity breakthrough (Hodges et al. 1980). As with direct filtration, the use of preozonation prior to in-line filtration has been shown to result in significantly lower filtered water turbidities as compared to prechlorination or no preoxidation. This has been observed with both dual- and single-media (sand or anthracite) filters (Hodges et al. 1980).

The use of ozone on raw surface waters subject to algal blooms has resulted in the formation of a thick organic foam in the contactors. This may be due to the formation or release of surface-active biopolymers by reaction of ozone with algae. Given the natural tendency of algae to float, it is likely that preozonation combined with a flotation process would be a viable alternative to conventional treatment in some cases (see sec. III.G). Even with waters not especially high in algae, the use of flotation in conjunction with ozonation may be cost-effective. For example, engineers at Hackensack, N.J., noticed that the particles produced upon ozonating raw water treated with alum and polymer had a natural tendency to float. As a result, they designed a full-scale system employing turbine diffusers and post-contact flotation skimmers. This was used as a pretreatment to dual-media filtration (Fung et al. 1988).

As discussed in sec. III.F.1, some waters do not show a coagulating effect of ozone. In fact, it is likely that all waters will exhibit a deterioration in the performance of settling, flotation, or filtration at some elevated ozone dose. A negative result with ozone presents two possibilities: (1) the dose or conditions chosen are suboptimal, and further treatment studies will show a beneficial effect of ozone; or (2) the raw water quality is such that ozone simply will not help the removal of turbidity or DOC regardless of the conditions. Although it is difficult to prove the latter, there are waters that have been extensively studied that fail to show any coagulating effects of ozone. This is especially true for organics removal. In fact, there is very little irrefutable evidence for enhanced DOC removal by physicochemical processes following ozonation in any water.

Finally, when making conclusions on physicochemical mechanisms in ozonation–coagulation processes, one must be careful to recognize any possible biological effects. These could include increased suspended biomass activity in aluminum hydroxide sludge during settling, as has been observed in pilot-plant studies at Choisy-le-Roi (Chedal and Schulhof 1979), or an increase in the activity of fixed-film biomass during direct or in-line filtration of preozonated–coagulated water. For guidance on the use of ozone in conjunction with biological treatment, the reader is referred to sec. III.K.

Process design and control.

Ozone dose. It is clear from the growing literature on the coagulating effects of ozone that low ozone doses (most often 0.5–1.5 mg/L) are most effective, and excessive doses can lead to deterioration of subsequent coagulation (see Schalekamp 1977). Rather than citing absolute ozone doses, it may be more appropriate to present them as ozone-to-carbon ratios. This is because organic carbon generally controls the lifetime of ozone in water and thereby controls its effectiveness. Likewise, the ozone-to-carbon ratio influences the degree to which organic matter is altered by the applied ozone. Figure III–45 shows some preozone doses used at several U.S. plants that employ ozone for the purpose of improving subsequent coagulation. Note that there is a trend toward higher doses when treating waters higher in TOC. The fact that there is not a simple stoichiometric relationship between ozone dose and TOC may be attributed to the complex nature of this process, the multipurpose application of ozone, and the many nontechnical constraints that must be considered in establishing full-scale plant doses. Note that the median carbon-specific dose from this figure is about 0.3–0.4 mg/mg C, which is within the optimum dose range based on the work of Singer and Chang (1988) (see Figure III–40). Pilot-scale studies have indicated that higher ozone doses may be needed to effectively treat surface water by direct filtration during times of high algal populations (Montgomery 1986).

Point of ozone application and contact time. Some have suggested, directly or indirectly, that ozone should be added at the point of addition of coagulant or just before in order to maximize its effect on conventional treatment (Sontheimer et al. 1978; Jekel and Reicherter 1982; Richard 1984). It is possible that the synergistic effects between ozone and aluminum coagulants are catalyzed by transient surfaces (for example, ozone

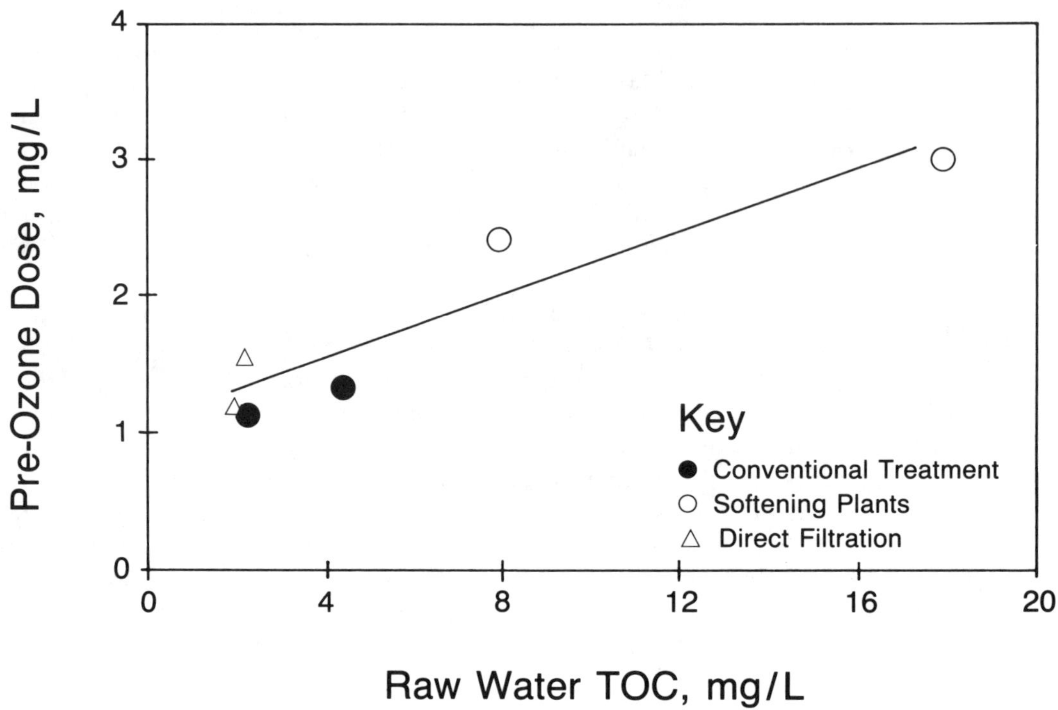

Source: Singer and Chang (1988).

Figure III–45 Selected U.S. Plants Using Ozone as a Coagulant Aid: Preozone Doses Employed vs. TOC

bubble, fresh aluminum hydroxide surfaces) or rely on reactions of transient species (for example, organic peroxides). If this were the case, one would expect that the benefits of preozonation would be lost when contacting and coagulant addition are too widely separated in time. In fact, the addition of coagulants or coagulant aids prior to ozone contact may allow one to use the turbulence of the contactor for improving flocculation. Studies at Hackensack, N.J., have indicated that this scheme provides the additional slow mixing needed for optimal flocculation prior to direct filtration (Weng et al. 1986).

Bicarbonate alkalinity and pH. Current knowledge of aqueous ozone chemistry indicates that a water's pH and bicarbonate alkalinity must play a role in the use of ozone as a coagulant aid. Changes in pH and bicarbonate concentration are known to alter the rate of ozone decomposition (see chapter II, sec. II.A). This affects the relative degree of direct oxidation by molecular ozone as compared to that by hydroxyl radicals. While with the first three mechanisms proposed here, it is not possible to predict with certainty which oxidant will be more effective, some speculations can be made with respect to the last three. For example, polymerization reactions (mechanism 4) should occur more readily as a result of rapidly decomposing ozone (for example, high pH, low bicarbonate) via radical mechanisms (Nonhebel et al. 1979). This has been observed with the laboratory ozonation of phenolic compounds (Chrostowski et al. 1983). The oxidation of Fe(II) to Fe(III) (mechanism 5), however, should occur more efficiently via the direct mechanism (Hoigné et al. 1985). In addition, iron itself may be able to change the nature of the organic oxidation products by accelerating the decomposition of ozone (Hoigné et al. 1985) and by acting as an electron-transfer catalyst between ozone and organics as it may between oxygen and organics (Theis and Singer 1974). Finally, the direct molecular ozone attack should be more effective at killing or stressing algal cells (mechanism 6). This is because the short-lived hydroxyl radicals are statistically less likely to encounter a large particle.

Despite these numerous arguments for the effect of pH and alkalinity on ozone's role in coagulation, there is very little empirical evidence indicating which conditions are best. Saunier and co-workers (1983) report that the coagulating effects of ozone improve with increasing pH of the raw water (that is, under conditions of rapid ozone decomposition). However, this could be due to concomitant increases in calcium hardness, which happen to be correlated with pH. In addition, changes in alkalinity or pH may also affect ozone transfer efficiency. Before conclusions can be drawn regarding effects of pH and alkalinity, carefully controlled studies must be conducted with many types of raw waters, and the fundamental mechanisms must be better understood.

III.G ALGAE REMOVAL

III.G.1 Problems Connected with the Presence of Plankton in the Water

People involved in water treatment speak generally of an "algae problem." In fact, it is more accurate to use the term "plankton." This term includes all the micro-algae that under favorable conditions (the presence of ideal amounts of nutrients, heat, and sunlight in the environment) can undergo periods of explosive growth. It also includes animal plankton (zooplankton), which belong to a higher level in the food chain, as well as actinomycetes. All these organisms are sized within a range of a few microns to a few millimeters.

Two important reasons can be cited for the elimination of these organisms. First is their potential for development into an algal bloom at a moment's notice, a phenomenon that occurs particularly in lakes and reservoirs and can cause significant disturbances during the various stages of the conventional treatment train (Godart 1988; Mouchet 1982), such as:

- Poor coagulation and flocculation due to sudden variations in raw water pH, related to the photosynthetic activity of the algae. Flocculation also becomes difficult because a number of these organisms behave like negatively charged biocolloids. Moreover, some species are also likely to release substances into the environment that complex the metallic cations generally used in coagulation.
- Poor settling characteristics connected with poor flocculation and the low apparent density of the algae.
- Short filtration runs due to clogging in the bottom of the filters or blanketing on the surface, depending on the size and morphology of the different species. Filter malfunction may also be due to gas release in the filters from water that is supersaturated in oxygen. This supersaturation is caused, as is the case with pH fluctuations, by the photosynthetic activity of the algae.
- Various odors and flavors (depending on plankton species and concentration) linked to metabolites excreted by plankton in the water (see sec. III.D).

Secondly, because of their size, some planktonic species are liable to get through the treatment train. Their presence in the distribution networks will then further degrade utility water quality (Trancart et al. 1988). The decay of these organisms causes an increase in the concentration of organic matter. They are the origin of taste and odors; in some cases, they are toxic. Finally, these microorganisms, whether in a live or decayed state, can be responsible for bacterial regrowth or can provide favorable conditions for the development of other forms of microbes in distribution systems. Hence, European guidelines on potable water quality stipulate that plankton must be absent (EEC Council Directive 1980).

Table III–31 presents a list of plankton species generally predominant in water and the effects they produce (Krauter 1974; Palmer 1980).

Table III–31 List of Species Generally Predominant in Water and the Effects They Produce*

	Form Properties				Specific Properties		Dimension	Interference in Water Treatment: Interference of Biochemical Origin				Interference in Water Treatment: Interference of Morphological Origin			
Algae	Circular/oval	Filamentous	Rectangular, Lancet-Needle-Bar-shaped	Irregular, Including Star-shaped	Flagellated	Slime Capsule	Average Dimension (μm)	Taste and Odor Problems	Coagulation-Interference	Slime-Producing	Causing Natural Softening of Water	Filter Clogging	Filter Breakthrough	Algae Persistent in Distribution Systems	Effect Indefinite
Cyanophyceae															
Aphanizomenon	0	1	0	0	0	0	5	1	0	0	0	0	0	0	1
Microcystis	1	0	0	0	0	1	5	1	0	0	0	0	0	0	1
Oscillatoria	0	1	0	0	0	0	10	1	0	1	0	1	9	0	0
Chorophyceae															
Chlamydomonas	1	0	0	0	0	1	20	1	0	0	0	0	1	0	0
Chlorella	1	0	0	0	0	0	8	1	0	0	0	1	9	1	0
Coelastrum	1	0	0	0	0	1	20	0	0	0	0	0	0	0	1
Mougeotia	0	1	0	0	0	0	35	0	0	0	0	1	9	0	0
Pandorina	1	0	0	0	1	0	10 (30)	1	0	0	0	0	0	0	1
Pediastrum	0	0	0	1	0	0	30	1	0	0	0	1	9	0	0
Scenedesmus	1	0	0	0	0	0	15	1	0	0	1	0	0	1	1
Ulothrix	0	1	0	0	0	0	10	0	0	0	0	1	9	0	0
Volvox	1	0	0	0	1	0	5 (500)	1	0	0	0	0	0	0	1
Cryptomonas	1	0	0	0	1	0	20 (40)	0	0	0	0	0	0	0	1
Euglenophyceae															
Euglena	1	0	0	0	1	0	70	1	1	1	0	0	1	1	0
Crysophyceae															
Dinobryon	0	0	0	1	1	0	40	1	0	0	0	1	9	1	0
Bacillariophyceae															
Asterionella	0	0	0	1	0	0	80	1	1	0	0	1	9	1	0
Coscinodiscus	1	0	0	0	0	0	200	0	0	0	0	0	0	0	1
Cyclotella	1	0	0	0	0	0	30	1	0	0	0	1	9	1	0
Diatoma sp.	0	0	0	1	0	0	60	0	0	0	0	1	9	0	0
Melosira	0	1	0	0	0	0	25	1	0	0	0	1	9	0	0
Nitzschia	0	0	1	0	0	0	40	0	0	0	0	1	9	0	0
Synedra	0	0	1	0	0	0	200	1	1	0	1	1	9	1	0

Sources: Krauter (1974); Palmer (1980). Reprinted with permission from Janssens, J.G. et al., *Wtr. Supply,* 4:347–366. © 1986 IAWPRC.

*0 = no; 1 = yes; 9 = no information available.

III.G.2 Review of the Principal Treatments Currently Used for Plankton Removal

Traditional treatment trains do not usually insure adequate water quality from disturbances connected with the presence of plankton. Several methods can be employed to solve these problems, including removal by microscreening, filtration without adding a flocculant, flotation, flocculation in a filter, or oxidation. Although described separately, these different treatment stages can be used sequentially, or even simultaneously, since no single process is 100 percent effective. In some cases, problems caused by secondary effects of plankton (tastes and odors, high concentrations of dissolved organics) must also be addressed.

Microscreening. This purely mechanical process physically removes the organisms and has no other action than eliminating plankton and other types of suspended solids likely to be encountered in raw water. The mesh size of the filter cloths can be chosen according to the size of the algal species (20 to 40 μm mesh). The filtering efficiency of these microscreens is greater than is apparent from their mesh size, owing to the formation of a film of plankton on the cloth. Depending on the species of plankton present, the average removal efficiencies given in the literature are between 60 and 100 percent (Klassen et al. 1970; Lynch et al. 1963; Lehn 1969). For very minute species, when effective removal depends on the formation of a plankton film, the performance of microscreening becomes very poor if there are no filamentous species to increase the filtering efficiency.

Flotation. In waters that are low in alkalinity, high in organics, and subject to algal blooms, flotation usually leads to better results than settling. Richard et al. (1983) described the Moule plant in northern France, where flotation offers a better alternative than sedimentation. This plant has a capacity of 1200 m^3/h (7.6 mgd) and treats water with a TOC of 9 to 14 mg C/L, a COD (permanganate test in acid medium) of 12 to 16 mg O_2/L, and an algae concentration of 15,000 to 50,000 organisms/mL.

The flotation stage (overflow rate: 6.3 m/h [2.6 gpm/ft^2], percentage of recirculated water for pressurization: 7 to 15 percent) and the original settling stage (overflow rate: 3.6 m/h [1.49 gpm/ft^2]) provides similar water quality (turbidity in the region of 1 NTU). Flotation required 27 percent less flocculant (210 instead of 290 mg/L ferric chlorosulfate) than gravity settling. Moreover, the sludge generated by flotation was far more concentrated, reaching a level of 20 g/L or more of total suspended solids (TSS).

Likewise, the Water Research Center (Rees et al. 1979) performed a study on the clarification of water by flotation and concluded that for algae-laden water, this process is more effective than sedimentation. Removal of 90 percent of the algae is achieved, which results in longer filtering runs and a better filtered water quality.

Burrows and Woodward (1988) compared three treatment trains: flocculation-sedimentation, flocculation-sedimentation-flotation, and flocculation-flotation, for the removal of *Aphanizomenon,* a type of algae particularly difficult to remove (2824 filaments/mL). These authors showed that a flocculation-flotation step removed 95 percent, while flocculation-sedimentation removed only 44 percent. The redundancy of a sedimentation-flotation step brought no extra benefit. Burrows and Woodward (1988) also tested a process in which flotation was performed on top of filtration and downstream of flocculation-settling facilities. When the algae concentration was low, the flotation process made it possible to extend the filtration run length by 78 percent. In periods of high *Aphanizomenon* concentration (4200 filaments/mL), only 53.9 percent removal efficiency was achieved by this process.

Filtration with and without the use of a flocculant. Another means to improve the removal of algae is to optimize the type of filter used. Ginocchio (1981) tested different kinds of filters, and Figure III–46 shows some of the results he obtained. Sand filters with a large grain size yielded results similar to those obtained through microscreening with a filter mesh size of 25 μm. However, a sand filter with a finer particle size was more effective against small algae.

Flocculation on a dual-media filter is much more effective (about 90 percent for most algae). This was confirmed by other authors (such as Bernhardt 1988). Janssens et al. (1986) compared the results obtained by flocculation-filtration on dual-media (hydroanthracite-sand) with results obtained using triple-media (pumice stone-hydroanthracite-sand) sand filters. They concluded that triple-media filters were still more effective in removing very small plankton (nanoplankton).

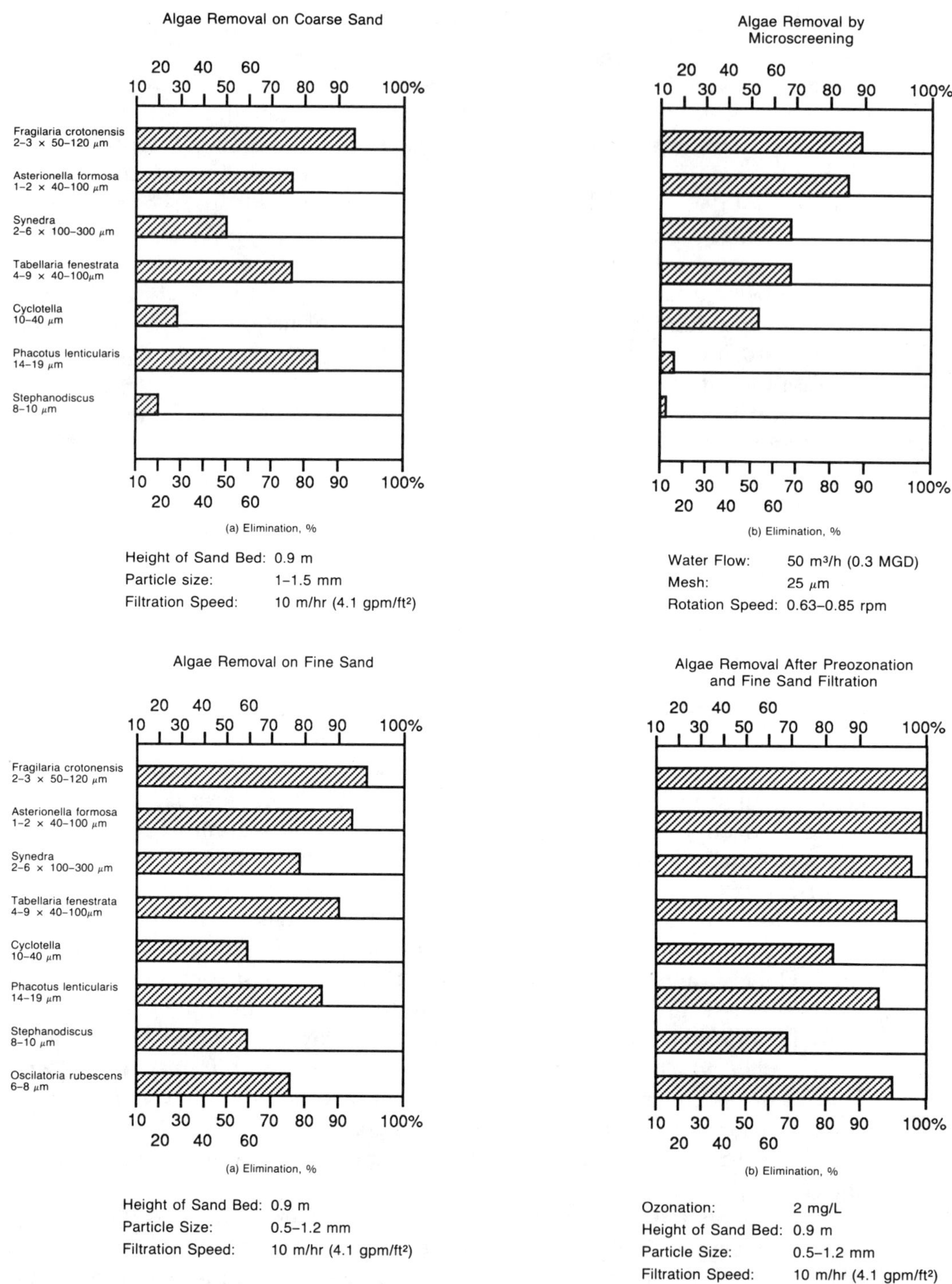

Source: Reprinted with permission from Ginocchio, J.C., *L'Eau et L'Industrie,* 56:19–24. © 1981 L'Eau, L'Industrie et les Nuisances.

Figure III–46 Removal of Different Types of Algae According to Type of Treatment

III.G.3 Advantages of Ozonation in Treating Plankton-Laden Water

Destruction of plankton or limitation of its growth. Ozone, like any other oxidant such as chlorine or chlorine dioxide, has a lethal effect on some algae (Pak et al. 1981) or limits their growth (Hyde et al. 1984) (see sec. G.4). Some species of plankton

Table III–32 Inactivation of *Notholca caudata* by Oxidation: Laboratory Tests

Oxidant Used	Oxidation Conditions		Inactivation Percentage
	Dose (mg/L)	Reaction Time (min)	
Potassium Permanganate	1	30	9.5
	2	30	68
	3	30	89
	1	60	68
	2	60	98
	3	60	100
Hydrogen Peroxide	100	30	50
	100	60	75
Ozone	1	1	14
	2	1	64
	3	1	70
	1	5	26
	2	5	72
	3	5	76

Source: Bernhardt and Lüsse (1989).

are particularly sensitive to a specific oxidant (Sukenik et al. 1987). Although chlorine and chlorine dioxide are most effective for the elimination of algae in raw water, use of these oxidants is declining for the reasons explained in sec. III.D.

Ozone is also capable of inactivating certain zooplankton, e.g., mobile organisms like *Notholca caudata*. Bernhardt and Lüsse (1989) showed that in order to remove such organisms by flocculation and filtration, they must first be inactivated. Table III–32 shows the efficiency of ozone, compared with other oxidants, in inactivation of this species. Table III–33 illustrates the advantage of inactivation as a preliminary step in the removal of certain types of plankton.

Improvement of plankton removal methods by the use of flocculants. Several authors have demonstrated ozonation's ability to improve flocculation efficiency in pilot units or industrial treatment plants, especially during periods of algal bloom (Damez and Dernaucourt 1979; Janssens et al. 1986; Pascal et al. 1984).* For example, Richard et al. (1983) showed that preozonation at the Moule water works (dose rate: 0.6–1.4 mg O_3/L) led to an approximate 40 percent reduction in the amount of coagulant required to obtain a turbidity level of about 1 NTU in the treated water (i.e., 130 mg/L of ferric chlorosulfate with preozonation instead of 210 mg/L without). Likewise, tests performed by Ginocchio (1982) and summarized in Figure III–47 show how preozonation improves the effectiveness of flocculation plus dual-media filtration for the removal of various types of algae (dosage: 2 mg O_3/L; dissolved residual ozone after a contact time of 10 min: 0.25 mg O_3/L).

Ozoflotation. Ozoflotation is a new process combining the physical phenomenon of flotation with the oxidizing properties of ozone (see sec. D.4, chapter IV). The ozoflotation process (Bourbigot and Faivre 1986) can be considered as a pretreatment stage where the main objective is to reduce the load on the remainder of the treatment train. The structure for this process requires little floor space. However, its prime advantage is that it permits the use of a simplified ozoflotation–dual media filtration train for water; hitherto, it was impractical to use a single step of dual media filtration

*The mechanisms that make these improvements possible are still relatively unknown, despite the efforts of quite a few authors to find an explanation (Betzer et al. 1980; Bernhardt et al. 1985; Sukenik et al. 1987; Hoyer et al. 1987). For further information, readers are encouraged to consult the bibliographic synopsis by Reckhow (1985), Reckhow et al. (1986), and sec. III.F.1.

Table III–33 Removal Efficiency of Flocculation-Flotation on Zooplankton Species, With or Without Prior Inactivation: Laboratory Tests

Zooplankton Species Tested	Removal Percent Without Prior Inactivation	Removal Percent With Prior Inactivation by $KMnO_4$
Notholca candata	77	90.4
Artemia salina	82	99.7

Sources: Bernhardt (1988); Bernhardt and Lüsse (1989).

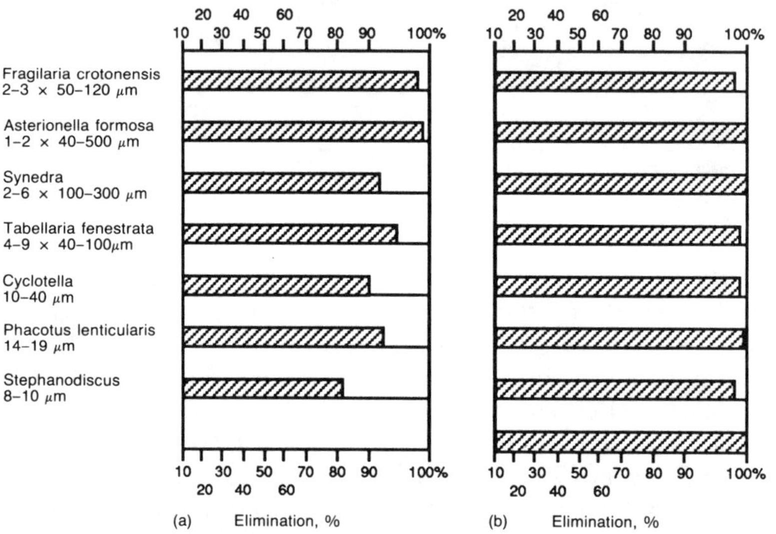

(a): Flocculation (2 mg/L $Al_2(SO_4)_3$), dual media filtration (pumice: depth = 30 cm, size range = 1.5–2 mm; sand: depth = 60 cm, size range = 0.5–1.2 mm; filtration speed = 10 m/h [4.1 gpm/ft²]).

(b): Ononation (2 mg/L), flocculation, dual media filtration under the same conditions as in (a).

Source: Reprinted with permission from Ginocchio, J.C., in *Ozonization Manual for Water and Wastewater Treatment* (W.J. Masschelein, ed.). © 1982 John Wiley & Sons.

Figure III–47 Influence of Ozonation on the Removal of Several Species of Algae Through Subsequent Flocculation and Dual Media Filtration

during certain months of the year owing to excessive turbidity and algal growth (Faivre et al. 1988).

The process consists of a flash-mix tank and a structure suitable for ozoflotation. The coagulant is injected at concentrations that can reach twice the dosage normally used for colloidal destabilization with small quantities of coagulant on a filter. Examples of the characteristics of raw water treatable with this process are given in the case studies (sec. G.4).

The structure, shown in Figure III–48, is in two compartments. In compartment A, the ozonation compartment, a maximum quantity of fine bubbles of ozonized gas is obtained by sweeping the porous diffusers with an additional water stream. The bigger bubbles rise to the surface of the compartment as in a conventional ozonation column. The speed of the water crossing this compartment from top to bottom entrains the fine bubbles (diameter between 200 and 500 μm) into compartment B, the flotation compartment (Adler et al. 1985). In this second compartment, the fine bubbles thus selected

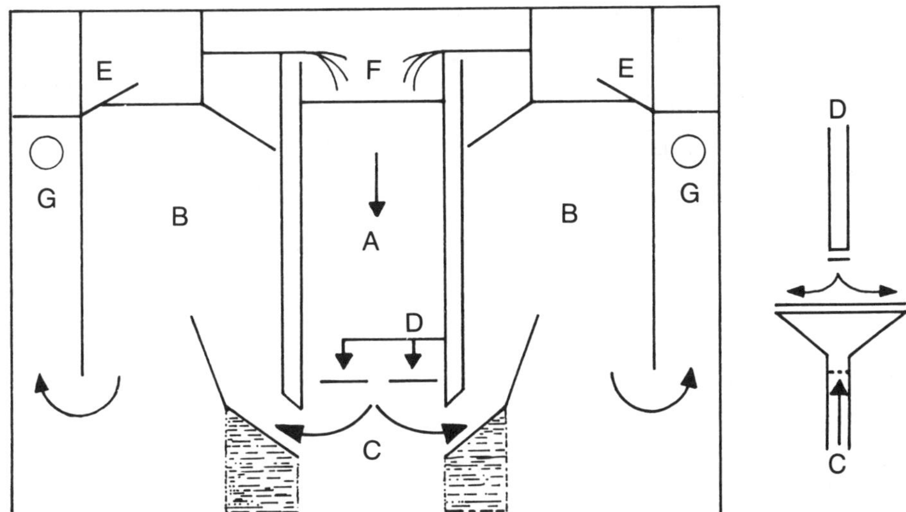

A: Ozonation Compartment
B: Flotation Compartment
C: Porous Disc
D: Sweeping Water Stream
E: Scum Collector
F: Raw Water Inlet
G: Floated Water Collector

Source: Bourbigot and Faivre (1986).

Figure III–48 Schematic of an Ozoflotation Unit

will cause flotation of most of the clogging substances (mainly algae and other elements caught in the floc). This preozonated and floated water is then collected from the bottom of the flotation unit and discharged through the water collector (G). The floating particles are periodically removed by raising the water level so that the scum runs off into the appropriate collectors (E). Scum removal is usually made easier by the installation of a spraying manifold.

It is important to note that this process represents a compromise between conventional preozonation, which results in inefficient flotation, and a classical dissolved air flotation stage in which the flowthrough speed of the water in the structure must be limited in such a way as to avoid carry-over of the fine bubbles (diameter: 50 μm) produced as part of the process.

In order to obtain optimum fine bubble production, it is necessary to apply a water stream, which sweeps the porous plates. (This water stream needs to be less than 10 percent of treated water flow for economical reasons.) Lower gas flows lower the amount of sweeping water necessary to produce small bubbles. This makes the production of ozone from oxygen more attractive than from air because a smaller gas flow results from the same ozone dosages. This is especially true when the ozone demand in the water is high.

The results obtained at a semi-industrial pilot plant and with new applications of this process are presented in the case studies.

III.G.4 Case Studies

The Fobney water treatment plant in England. This water works, located in Reading, England, has a treatment capacity of 1875 m^3/h (11.9 mgd) and a maximum

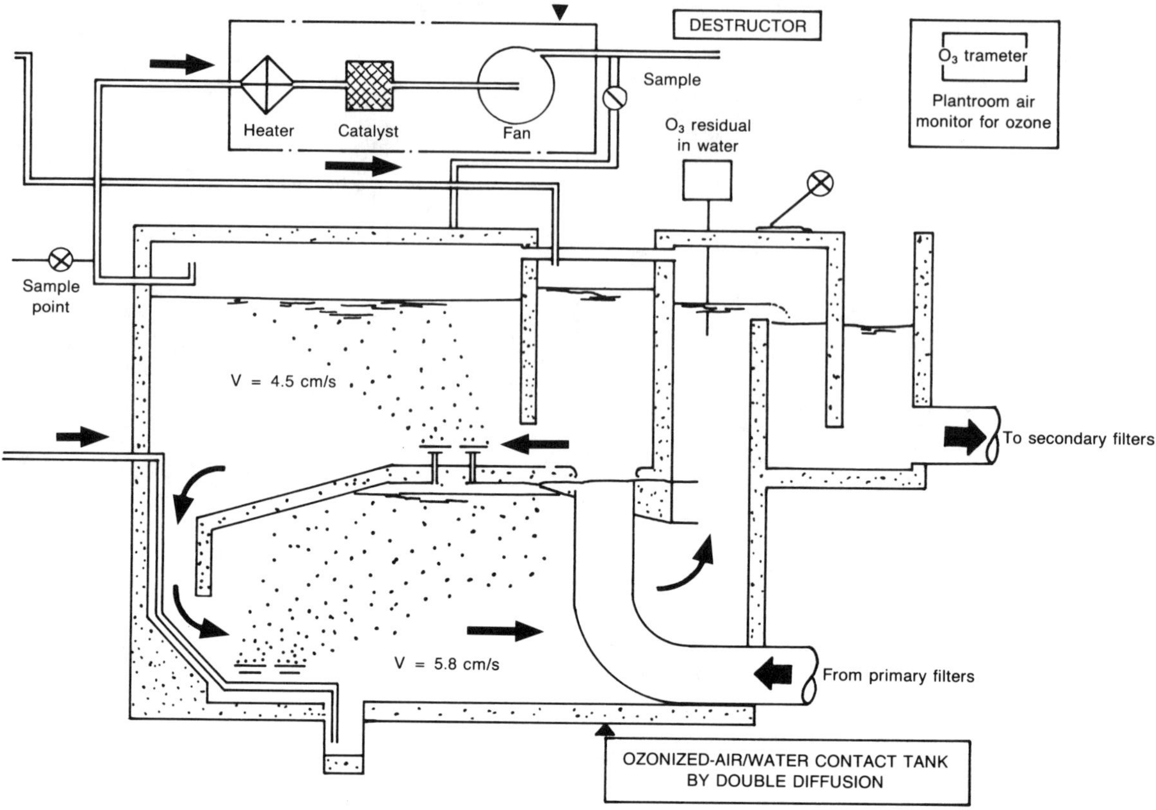

Source: Greaves (1989).

Figure III–49 Ozonation Chamber at Fobney Water Works, England

design flow of about 2290 m^3/h (14.5 mgd). It was used by the Thames Water Authority for full-scale ozonation trials before slow-rate filtration on sand. The raw water is pumped from the Kennet River; the initial treatment train consisted of coarse bar and cup screening, ferric sulfate dosing when necessary, primary high-rate upflow filtration at 12 m/h (4.9 gpm/ft^2), secondary slow sand filtration at 0.25 m/h (0.1 gpm/ft^2), and disinfection with chlorine.

In 1982, the eight slow filters were divided into two banks of four each (i.e., a reference bank that continued receiving water straight from high-rate filtration and one receiving water that had been ozonated after high-rate filtration). The original ozone generator was replaced in 1984 by a new 2-kg O_3/h (about 106 lb/day) ozone generator, representing a dosage of 2 mg/L. This dose was enough to maintain a residual ozone concentration of 0.5 mg/L in the contact chamber, but it was noted that the residual was nil in the top layer of the water above the slow filters. The specially designed contactor is shown in Figure III–49. Plant performance was monitored jointly by the Thames Water Authority and the Water Research Center's Stevenage laboratory (Hyde et al. 1984).

In summer when the effects of algae are more pronounced, preozonation was instrumental in increasing the abatement of TOC in the filtered water by 20 percent compared with water from the original treatment train. Slow filtration runs on the ozonated train were always far longer than on the reference trains, as shown in Figure III–50. The ozonation step enabled overall filter runs to be lengthened by 60 percent during the summer. On the other hand, in the fall and winter, when algae growth is slight, the influence of ozonation on run length remained less than 10 percent.

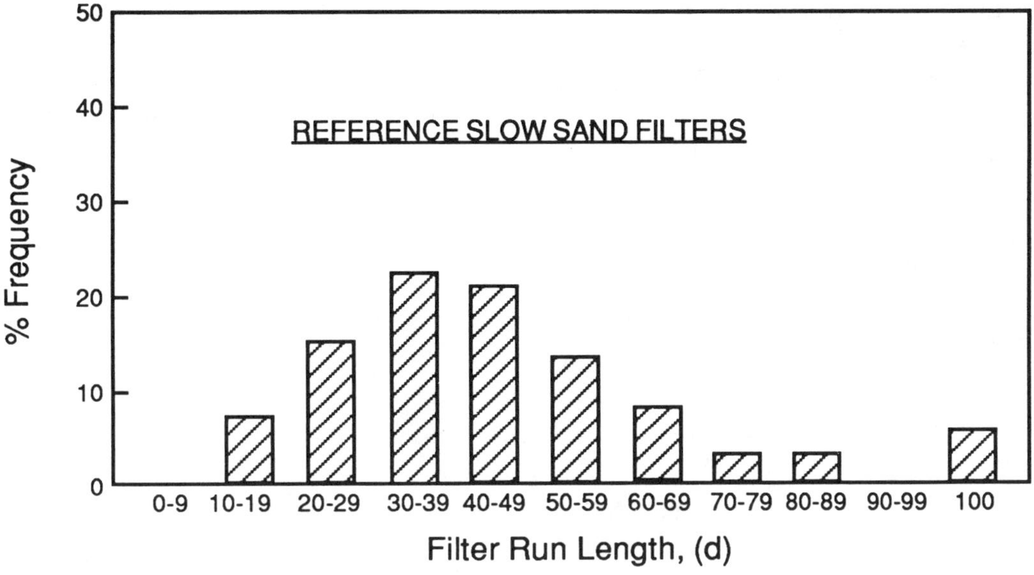

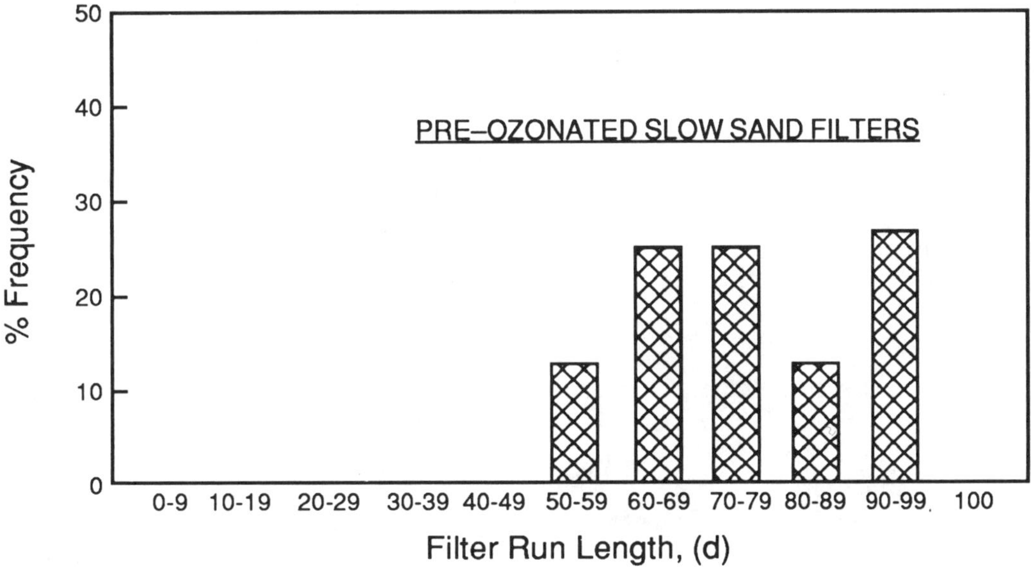

Source: Reprinted with permission from Hyde, R. et al., in "37th International Conference Cébédeau." © 1984 La Tribune du Cébédeau.

Figure III–50 Effect of an Upstream Ozonation Step on Slow Sand Filter Run Length

Ozonation at Fobney has reduced the problems associated with algae clogging of the slow sand filters. This has been confirmed by observing the various types of algae found on filters, as is shown in Table III–34. The predominant algal species found on the reference filters, i.e., filamentous varieties well-known for their clogging effects, were not present on filters receiving ozonated water. Certain species were capable of penetrating inside the filters, but they did not seem to have a harmful effect on filter run length. Analysis of the chlorophyll in the water on the slow filters revealed a reduction of about 88 percent between ozonated and unozonated water.

Ozoflotation plants in France. Ozoflotation plants have been constructed at Autun, Le Creusot, and Lyon using the treatment train shown in Figure III–48 (Faivre

Table III–34 Evolution of the Algal Species Found on Slow Sand Filters Receiving Ozonated or Nonozonated Water

Reference Slow Sand Filter	Slow Sand Filter Receiving Ozonated Water
Cladophora (dominant)	*Scenedesmus*
Melosira (dominant)	*Synedra*
Cymbella	*Pediastrum*
Phoicosphenia	*Cymbella*
Cocconeiss	*Pinnularia*
Mycrocystis	

Source: Reprinted with permission from Hyde, R. et al., Proc. 37th Intl. Conf. CEBEDEAU. © 1984 La Tribune du Cebedeau.

Table III–35 Characteristics of Raw Water at the Autun Water Works (Summer Season, 1987 and 1988)

	Summer 1987		Summer 1988	
Parameters	Range	Average	Range	Average
Color (mg Pt-Co/L)	40–80	65	25–60	43
Turbidity (NTU)	2–7	3.5	2.6–3.6	3.0
COD ($KMnO_4$ test in acid medium) (mg O_2/L)	4–12	8	6.2–8.1	7.3
TOC (mg C/L)	–	–	5.5–6.4	5.9
pH	–	–	6.3–6.6	6.4
Chlorophyll A (μg/L)	2.05–16.3	5.10	4.0–21	9.36
Beaudrey Clogging Index*	2–80	25	–	–

Source: Bourbigot (1990).
*AFNOR, 1986a; NFT 90.030.

et al. 1989). Following several tests on a 2-m^3/h (about 8.8 gpm) pilot plant, the Autun water works (220 m^3/h—1.4 mgd) was used to test the ozoflotation process in a full-scale application. The main characteristics of the raw water during the 1987 and 1988 summer test periods are shown in Table III–35. This raw water, from a small forest lake, is low in dissolved solids (conductivity approximately 40 μmho/cm; dry residual solids about 30 mg/L; total alkalinity about 10 mg/L $CaCO_3$). A significant strong algal growth was observed during the summers of 1987 and 1988.

The treatment train was as follows (refer to Table III–36 for ozoflotation conditions):

Saint Blaise Water Works (Autun, France): 220 m^3/h (1.4 mgd)

ozoflotation on half the flow → flocculation-sedimentation (aluminum sulfate: 10–25 mg/L; Hazen velocity: 2 m/h [0.8 gpm/ft^2]) → sand filtration (velocity: 2.3 m/h [0.9 gpm/ft^2]) → disinfection with sodium hypochlorite

A significant drop in the clogging index of the water at the outlet of the ozoflotation step accompanied algae removal. As shown in Figure III–51, this drop increased in proportion to the ozone dosage (which is increased by elevating the ozone concentration in the oxygen without altering the flow rate of gas entering the ozoflotation structure).

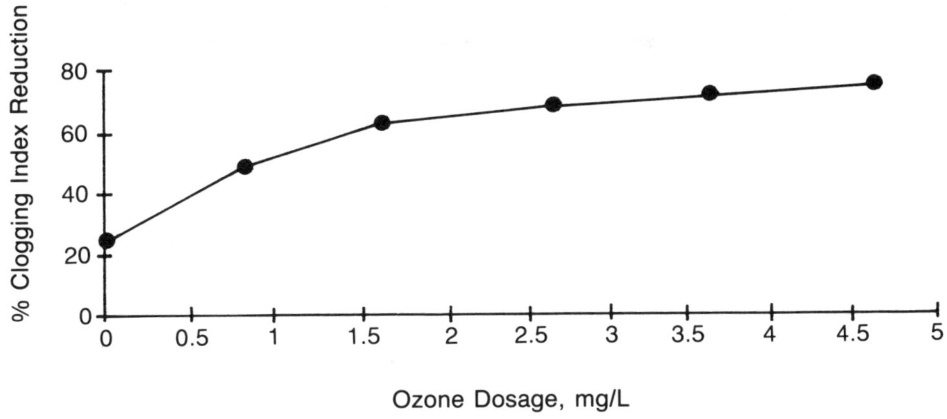

Source: Faivre et al. (1989).

Each sample represents the average of three tests.

Figure III–51 Decrease of the Clogging Index (AFNOR 1986a; NFT 90.030) During the Ozoflotation Step: Autun

No fundamental explanation regarding the observed correspondence between clogging index and ozone dosage is available.

The removal of turbidity in filtered water compared to raw water reached 90 percent. Color and COD in the ozoflotated water (permanganate test in acidic medium) decreased 30 to 50 percent and 10 to 20 percent, respectively (compared to raw water) under the ozoflotation conditions described in Table III–36. At the same time, a reduction of 5 to 10 percent in the TOC content was also obtained (Faivre et al. 1989). Color removal was another by-product of the process; at the end of treatment train, the color of the water was below 5 Pt-Co units.

In view of the results achieved in this first plant, two other French water works have started to use an ozoflotation step. Their treatment trains are as follows:

La Couronne Water Works (Le Creusot, France): 600 m^3/h (3.8 mgd)

ozoflotation, recarbonation	→ lamellar clarification (Hazen velocity = 0.75 m/h [0.3 gpm/ft^2])	→ sand filtration (filtration rate = 6.4 m/h [2.6 gpm/ft^2])	→ neutralization with lime	→ ozone disinfection	→ chlorination

In this water works the ozoflotation structure installed in the old flocculator is also used for recarbonation: lime water (40 to 45 mg CaO/L) is injected at the top of the compartment while the CO_2 gas is injected (at a dose of 60 to 70 mg/L) in the form of soda water sweeping the porous discs.

Miribel-Jonage Water Works: (Lyon, France): 6750 m^3/h (42.8 mgd)

Ozoflotation	→ dual-media filtration (anthracite/sand; filtration rate = 7.5 m/h [3.1 gpm/ft^2])	→ ozone disinfection

Table III–37 gives details of the ozoflotation stage operating conditions for these new facilities as compared with the Autun plant.

Table III–36 Ozoflotation Efficiency in Summer at Autun Water Works (1987 and 1988)

Test Period and Operating Conditions During Ozoflotation Stage	Algal Species	Concentration in Raw Water*	Average Removal (Percent)
1987			
Ozonation dosage: 3 mg O_3/L	*Onglena* sp. *(Euglenophyceae)*	10.1×10^5	98
Flocculant dosage: 20–25 mg/L			
$(Al_2(SO_4)_3 \cdot 18\ H_2O)$	*Melosira granulaea (Bacillariophyceae)*	$0.6\text{–}11.8 \times 10^5$	39
1988			
Ozonation dosage: 2 mg O_3/L	*Melosira granulaea (Bacillariophyceae)*	$2.8\text{–}3.1 \times 10^5$	51
Flocculant dosage: 5–20 mg	*Symra urella (crysophyceae)*	$0.5\text{–}1.9 \times 10^4$	60
$(Al_2(SO_4)_3 \cdot 18\ H_2O)$	*Dinobrion mallomonas (chrysophyceae)*	$0.5\text{–}1 \times 10^4$	55
	Staurasinum megacanthum (chlorophyceae)	$1.0\text{–}1.5 \times 10^4$	79
	Scenedesmus armatus (chlorophyceae)	$0.5\text{–}1 \times 10^4$	87
	Aphanizomenon flos aquae (cynophyceae)	2×10^4	42
	Total algal population	$3.2\text{–}33.5 \times 10^5$	69

*Number of colonies or cells per liter.

Table III–37 Summary of Operating Conditions for Three French Ozoflotation Systems in Three Water Works

	Name of Plant (Location)		
	Saint Blaise (Autun)	La Couronne (Le Creusot)	Miribel-Jonage (Lyon)
Nominal operating flow rate	220 m^3/h (1.4 mgd)	600 m^3/h (3.8 mgd)	6750 m^3/h (42.8 mgd)
Ozone feed gas	Oxygen	Air	Oxygen
Ozone concentration in gas (NTP)	45–50 g O_3/m^3	15–20 g O_3/m^3	45–50 g O_3/m^3
Applied ozonation dosage	1–3 mg O_3/L	0.5 mg O_3/L	0.5–1 mg O_3/L
Flocculant dosage	10–25 mg/L of $Al_2(SO_4)_3 \cdot 18\ H_2O$	50–60 mg/L of WAC* Density = 1.2	3–5 mg/L of $FeCl_3$ (41%)
Overflow rate	16.7 m/h (6.8 gpm/ft^2)	21.6 m/h (8.8 gpm/ft^2)	20.7 m/h (8.5 gpm/ft^2)

Source: Villessot (1989).
*A commercial polyaluminum chloride coagulant.

III.H DISINFECTION

III.H.1 Principles

General concepts.

Introduction. Disinfection (the elimination or inactivation of bacteria, viral particles, and parasites) is one of the major objectives of drinking water suppliers. There is no point in distributing an agreeable, clear water free of any traces of toxic, carcinogenic, or mutagenic products if it contains vectors of traditional or new (pathologies due to opportunist organisms) diseases.

In France in the early '60s, the requirement for a disinfectant at least as efficient as chlorine and without tastes and odors was the driving force for substantial accomplishments made in the application of ozonation of drinking water. At the same time that highly reliable equipment was being developed to produce large quantities of ozone, new engineering concepts were also being developed for its application to water. The main feature of these concepts was the need to maintain a certain minimum concentration for a given time (e.g., 0.4 mg/L × 4 min under ideal laboratory conditions) (Coin et al. 1964, 1967). The full scale contact time, at a constant residual, was increased over that used in the laboratory (e.g., from 4 min to 10–12 min) as a safety factor designed to offset the effects of hydraulic short-circuiting.

Bourbigot (1988) lists the 23 largest facilities using at least one ozonation stage that have been designed to meet the above guidelines. Studies have not revealed any problems in the disinfection process at these plants. It is important to note that to produce the recommended 0.4 mg/L residual concentration at these 23 plants, an average dose of 3–4 mg/L, or nearly 10 times more than the minimum, is required. This is an important point that shows why it is necessary to express experimental results in terms of the residual concentration obtained rather than the doses applied to the water.

Disinfection kinetics. Chick's law can be represented in the mathematical form (Chick 1908):

$$N_t = N_0 \exp(-kt)$$

Where:

N_t = number of microorganisms surviving after time t
N_0 = initial number of microorganisms
k = rate constant, which depends on the type of microorganism and disinfectant
t = time the water is in contact with the disinfectant.

If the reaction conforms to Chick's law, a straight line results when the data are plotted on a log scale. In practice, experimental results do not always show a linear relationship (Block et al. 1981). These deviations are due to different factors, especially microorganism aggregation, protection offered by extraneous matter "binding" microorganisms, and nonsteady disinfectant concentration. Age and physical condition frequently varies among the population, and they may possess extracellular matter that restricts the diffusion of the disinfectant. Experimental conditions can also have a considerable effect. These considerations led to the use of the C · t concept based on the empirical equation developed by Watson (1908). The models developed by Chick and Watson may be summarized as follows:

$$k = \Lambda\, C^n$$

and:

$$k' = C^n \cdot t = -\frac{\ln(N_t/N_0)}{\Lambda}$$

Where:

Λ = coeffient of specific lethality
C = disinfectant concentration
t = time required to achieve a given inactivation level of a microorganism exposed under defined conditions
k and k' = constants
n = constant known as the "coefficient of dilution."

The underlying hypotheses for Watson's law are:

- The disinfectant concentration, C, is supposed to be constant during reaction time, t.

Table III–38 Proposed U.S. EPA C · t Values (mg · min/L) for the Inactivation of *Giardia* Cysts with Ozone at Different Temperatures and pHs from 6 to 9

	Temperature (°C)					
Inactivation	0.5	5	10	15	20	25
0.5 log	0.48	0.32	0.23	0.16	0.12	0.08
1 log	0.97	0.63	0.48	0.32	0.24	0.16
1.5 log	1.5	0.95	0.72	0.48	0.36	0.24
2 log	1.9	1.3	0.95	0.63	0.48	0.32
2.5 log	2.4	1.6	1.2	0.79	0.60	0.40
3 log	2.9	1.9	1.4	0.95	0.72	0.46

Source: U.S. EPA (1989a).

Table III–39 Values of Specific Coefficients of Lethality for the Main Disinfectants (L/mg/min)

Disinfectant	Enterobacteria	Viruses	Bacterial Spores	Amoebic Cysts
O_3	500	5	2	0.5
HOCl	20	1 up	0.05	0.05
OCl^-	0.2	<0.02	<0.0005	0.0005
NH_2Cl	0.1	0.005	0.001	0.02

Source: Morris (1975).

- The microorganisms must be a single strain of synchronous development.
- Genetically similar microorganism populations are assumed.
- The killing action must be of the single-hit and single-site type.

When inactivation reactions obeying this law are plotted on log-log paper, the result is a straight line with slope n (Rubin et al. 1983). If n = 1, C · t remains constant regardless of the disinfectant concentration. This strict linearity in Watson's plots is not often observed with a wild mixed population of microorganisms. If n > 1 (for example, n = 2), the value of C · t increases very sharply when the disinfectant concentration increases. (The disinfectant concentration is the predominant factor and contact time is less important.) However, if n < 1 (for example, n = 0.5), the value of C · t increases when the disinfectant concentration increases, but the rate of increase in the C · t product is less. Under this condition, contact time is the predominant factor. According to Haas and Karra (1984) most values of n obtained from this type of test are between 0.7 and 1.3 for chlorine. In the United States, the Safe Drinking Water Committee (1980), and later Hoff (1987) at the U.S. EPA, have discussed the utility of the C · t concept.

Wickramanayake et al. (1984a) studied the inactivation of *Giardia* and *Naegleria* cysts by ozone and showed that the inactivation mechanisms observed did not follow Chick's law due to an initial latent phase; however, the results for 99 percent inactivation were accurately described by Watson's law with a coefficient of dilution close to 1. They concluded that the C · t product could be confidently used to compare the efficiency of disinfectants against cysts, which, it must be remembered, show higher resistance than other microbial forms. Based on this work, the U.S. EPA (1989a) proposed C · t values for *Giardia* inactivation, suitable at pH 6–9, at different temperatures, shown in Table III–38. Thus, for example, a C · t of 0.7 provides 3 logs of inactivation of *Giardia* cysts at 20°C. The 1.6 C · t product given when 0.4 mg/L of ozone residual is applied during a 4-min contact time for virus inactivation provides a 99.9 percent inactivation level at 10°C.

Overview of inactivation kinetics. Morris (1975) demonstrated the high inactivating capacity of ozone compared to chlorine-based disinfectants. Table III–39, drawn

Table III–40 C · t Values (mg · min/L) for 99 Percent Inactivation of Microorganisms with Disinfectants at 5°C

	Disinfectant			
Microorganism	Free Chlorine (pH 6 to 7)	Preformed Chloramine (pH 8 to 9)	Chloride Dioxide (pH 6 to 7)	Ozone (pH 6 to 7)
E. coli	0.034–0.05	95–180	0.4–0.75	0.02
Polio 1	1.1–2.5	770–3740	0.2–6.7	0.1–0.2
Rotavirus	0.01–0.05	3810–6480	0.2–2.1	0.006–0.06
Phage f2	0.08–0.18	–	–	–
G. lamblia cysts	47–>150	–	–	0.5–0.6
G. muris cysts	30–630	1400	7.2–18.5	1.8–2.0

Source: Hoff (1987).

Table III–41 Results of Comparative Studies on the Inactivation of Various Microorganisms by Ozone (Microorganisms Classified by Decreasing Resistance to Ozone from Top to Bottom)

Bacteria			Viruses			Protozoan Cysts		
Species	Estimated C · t Reduction 99 Percent (mg · min/L)	Reference	Species	Estimated C · t Reduction 99 Percent (mg · min/L)	Reference	Species	Estimated C · t Reduction 99 Percent (mg · min/L)	Reference
M. fortuitum	0.53 (24°C)	Farooq et al. (1977a)	Polio 1	0.2 (5°C)	Roy et al. (1980)	*N. gruberi*	4.23 (5°C)	Wickramanayake et al. (1984a)
Bacillus sp.	ND*	Haufele and Sprockhoff (1973) Broadwater et al. (1973)	Coxsackie B3	ND	Snyder and Chang (1974) Evison (1977)	*G. muris*	1.94 (5°C)	Wickramanayake et al. (1984a)
S. typhimurium	ND	Burleson et al. (1975)	Human Rota	0.006–0.06 (4°C)	Vaughn et al. (1987)			
E. coli	0.006–0.02 (1°C)	Katzenelson et al. (1974)	Rota SA-11	0.02–0.06 (4°C)	Vaughn et al. (1987)	*G. lamblia*	0.53 (5°C)	Wickramanayake et al. (1984b)
S. marcescens	ND	Haufele and Sprockhoff (1973)						

Source: Block (1978) and Sobsey (1989) and original references.
*Not determined. Classification based on relative resistance of different microorganisms to ozone according to authors.

from this author's work, shows specific coefficients of lethality that comply with Chick's law. A more recent comparison by Hoff (1987) also illustrates the greater effectiveness of ozone as compared to other water treatment disinfectants (Table III–40).

Ozone is very effective against bacteria. Wuhrmann and Meyrath (1955) studied the effect of very small concentrations of dissolved ozone, i.e., 0.6 μg/L on *E. coli.* They showed that at 12°C an *E. coli* strain added to pure water was reduced by 4 logs (99.99 percent) in less than 1 min with 9 μg/L of dissolved ozone. Very similar results were obtained with *Staphylococcus* sp. and *Pseudomonas fluorescens; Streptococcus faecalis* required a contact time twice as long with the same dissolved ozone concentration, and *Mycobacterium tuberculosis* required six times as long, for the same reduction observed with *E. coli*.

Block (1978) and Sobsey (1989) have summarized much of the available literature on ozonation of various microorganisms (Table III–41). The results can be summarized as follows:

Table III–42 C · t Values (mg · min/L) for 99 Percent Inactivation of *Giardia* Cysts with Ozone, Free Chlorine, and Chlorine Dioxide

		Giardia lamblia		*Giardia* sp.				*Giardia muris*		
Experimental Conditions	Oxidant pH	7*	7*		7–9†			7‡	7‡	9‡
	Temp.	5°C	25°C	<5°C	5°C–15°C	16°C–25°C	<25°C	5°C	25°C	25°C
C · t products (mg · min/L)	Ozone	0.53	0.17	–	–	–	–	0.27	1.94	0.11
	Cl_2	–	–	300	240	180	90	–	–	–
	ClO_2	–	–	–	–	–	–	11.2	5.3	2.1

*According to Wickramanayake et al. (1984b).
†According to Jarroll et al. (1981); Rice et al. (1982); Leahy (1985); Hibler et al. (1987).
‡According to Wickramanayake et al. (1984a).

Bacteria: Of the vegetative bacteria, *E. coli* is one of the most sensitive, while the Gram-positive cocci (*Staphylococcus* and *Streptococcus*), the Gram-positive bacillae (*Bacillus*), and the *Mycobacteria* are most resistant. However, no significant difference is found among the Gram-negative bacillae, which include *E. coli* and pathogens such as *Salmonella* that are all very sensitive to inactivation by ozone. Sporular forms are always far more resistant to ozone than vegetative forms.

E. coli cannot be considered as the best indicator of the microbiological quality of water disinfected by ozone. A bacterial indicator of ozone disinfection efficiency might be more appropriately chosen from *Mycobacteria* (for example, *M. forfuitum*) or sporular bacillae (for example, *Bacillus cereus*).

Viruses: Viruses are generally more resistant to ozone than vegetative bacteria but not more than sporular forms of *Mycobacteria*. Phages are the most sensitive, while little difference seems to exist between the polio- and coxsackie viruses. Human rotavirus (Vaughn et al. 1987) showed no significant difference in sensitivity from those mentioned above.

Parasites: Protozoan cysts are much more resistant than vegetative forms of bacteria and viruses. *Giardia lamblia* cysts have a sensitivity to ozone equivalent to sporular forms of *Mycobacteria*. *Naegleria* and *Acanthamoeba* cysts are much more resistant than *Giardia* cysts. Wickramanayake et al. (1984a, 1984b) found that C · t products for 99 percent inactivation of *G. lamblia* and *N. gruberi* at 5°C were 0.53 and 4.23 mg · min/L, respectively. Available data on *Cryptosporidium* oocyst inactivation (Peeters et al. 1989; Langlais et al. 1990) suggest that among waterborne pathogenic protozoans, this microorganism stands out as being especially resistant even to chlorine dioxide and to ozone treatment (see later discussion). More data are needed on *Cryptosporidium* inactivation.

Ozone is very effective against free-living amoebae trophozoites (reported C · t values for 99 percent inactivation at 20°C are less than 0.1 mg · min/L [Langlais et al. 1990]).

The effects of various parameters on disinfection efficiency.

Temperature and pH effects. Microorganism destruction rate is generally believed to increase with increasing temperature. According to the van't Hoff–Arrhenius theory (Fair et al. 1968), temperature partly determines the rate at which the disinfectant diffuses through the microorganism surface and its rate of reaction with the substrate. It is generally accepted that increasing the temperature by 10°C increases the reaction rate by a factor of 2 or 3. When the temperature increases, ozone is less soluble and less stable in water (Katzenelson et al. 1974) (see chapter II, sec. E), but the ozone reaction rate with the substrate increases. U.S. EPA (1989a) and Wickramanayake et al. (1984a) cite increasing inactivation rates for cysts with increasing temperatures (see Tables III–38 and III–42). A number of experiments have shown that although increasing the temperature from 0 to 30°C significantly affects the solubility of ozone and its rate of decomposition, it had virtually no effect on the disinfection rate evaluated by the total bacteria count (data cited by Kinman, 1975). Thus the increase of ozone reaction rates exceeds all other effects of ozone instability. Numerous authors have found similar results (Farooq et al. 1977b; Roy et al. 1980). The only limitation imposed by these phenomena is that the C · t values for ozone are less accurate than those for other disinfectants (Hoff, 1987) due to its volatility and reactivity and the difficulty of keeping the residual concentration constant.

The impact of pH on the bactericidal or virucidal effects of ozone seems minor (Diaper 1972) or at least, is far less important than it is for the chlorine data cited by Kinman (1975). The relative unimportance of pH has been noted over pH ranges of 5.8 to 8 (Suchkov 1964), 6 to 8 (Evison 1977), and 5.7 to 10.1 (Farooq et al. 1977b). The changes in disinfecting efficiency with variations in pH seem to be due to changes in the ozone decomposition rate. Ozone decomposes faster in a high-pH aqueous solution and forms various types of oxidants with differing reactivities (see chapter II, sec. A.1). Tests carried out at different pH values but constant residual ozone concentrations showed that the degree of microorganism inactivation by ozone remained virtually unchanged (Farooq et al. 1977b).

Morris (1975) found that the variations in pH that occur in water treatment did not significantly affect disinfecting efficiency by ozone. The work of Wickramanayake and colleagues (1984a) showed that the inactivation of *Giardia muris* cysts improved when the pH increased from 7 to 9. This was attributed to possible changes in the cyst chemistry making it easier for ozone to react with cell constituents. But they noted somewhat slower inactivation of *Naegleria gruberi* cysts at pH 9 than at lower pH levels, indicating that pH effects are organism-specific.

More recent studies reviewed by Sobsey (1989) indicated decreased virus inactivation by ozone at alkaline pH (pH 8–9) for poliovirus type 1 (Harakeh and Butler 1984) and rotaviruses SA-11 (simian) and Wa (human) (Vaughn et al. 1987).

Effects of turbidity (suspended material) and ozone-consuming compounds. Microorganisms in a water to be treated are not generally in the free state; they may be in the form of aggregates, fixed on the surface of mineral or organic matter, incorporated with human, animal, or cell debris, or associated with flocs, such as aluminum hydroxide. The form in which microorganisms are present during the disinfection stage is important. Aluminum hydroxide (Walsh et al. 1980) or bentonite (Boyce et al. 1981) do not seem to protect poliovirus and *E. coli* against inactivation by ozone. Turbidity would not, therefore, seem to be valid as a parameter to determine the residuals of ozone required for disinfection. The inhibiting effect depends on the types of turbidity rather than the turbidity level. For example, the work of Walsh et al. (1980) on ozone inactivation of aluminum floc–associated viruses and bacteria (size range comparable to that which could escape filtration of a potable water supply) demonstrated that these microorganisms were afforded no protection at floc turbidity levels of 1 and 5 NTU. Similar results have been obtained for poliovirus 1, coxsackie virus A9 and *E. coli* associated with bentonite clay, but adsorption of the f2 bacteriophage at 1 and 5 NTU of bentonite clay retarded the rate of inactivation by ozone (Boyce et al., 1981).

Although turbidity due to unoxidizable mineral substances seems to have little effect, the same cannot be said for organic substances, which consume large quantities of ozone. Thus, viruses associated with cells or cell fragments are protected from inactivation by ozone doses that readily inactivate purified virus (Emerson et al. 1982). Similar results have been obtained with fecal-associated coliform bacteria and poliovirus 1 distributed in water to produce a 5-NTU suspension at initial ozone concentrations of 0.028 and 0.013 mg/L, respectively, with 30 s of contact time (Foster et al. 1980).

However, it is still true that water must be of the best possible quality, both in terms of turbidity and dissolved organic matter, to ensure optimum disinfection. Dissolved organics interfere with ozone disinfection by consuming ozone to produce compounds with little or no microbiocidal activity and reducing the concentration of active species available to react with microorganisms. Harakeh and Butler (1984) observed that poliovirus 1 inactivation by 0.2 mg/L ozone in sewage effluent was reduced in the presence of increased levels of organics. Majumdar et al. (1973) reported that the degree of inactivation of this virus in secondary effluents by 1 mg/L ozone residual was greater in preozonated than in nonozonated wastewater. Preozonation satisfied the ozone demand of the wastewater organics as indicated by reduced concentrations of suspended solids and chemical oxygen demand (Sobsey 1989).

Ozone inactivation mechanism and comparison of C · t values for selected organisms.

Effects on bacteria. The inactivation of bacteria by ozone can be considered as an oxidation reaction (Bringmann 1954). Oxidation is caused by the high oxidation-reduction potential of ozone in water (Chang 1971; Kurzmann 1975), by ozone's ability to diffuse to sites to be deactivated, and by chemical species such as free radicals, which are reactive by-products of ozonation (see chapter II, sec. A.1).

The bacterial membrane seems to be the first site of attack (Giese and Christensen 1954; Christensen and Giese 1954), either by way of the glycoproteins or glycolipids (Scott and Lesher 1963) or through certain amino acids, for example, tryptophan (Goldstein and McDonagh 1975).

Ozone also disrupts enzymatic activity of bacteria by acting on the sulfhydryl groups in certain enzymes. Vrochinskii (1963) noted that ozonated bacteria lost their ability to degrade sugars and produce gases. Glucose 6-phosphate dehydrogenase is affected by ozone in the same way as all enzyme systems (Chang 1971). Bacterial death could be due directly to changes in cellular permeability (Murray et al. 1965), possibly followed by cellular lysis. However, lysis probably is not the primary inactivation mechanism (Perrich 1976) but is the result of an extremely high oxidant concentration (at the gas-liquid phase interface) or prolonged ozonation.

The oxidation and bacterial inactivation reactions always occur very quickly, which led Bringmann (1954) to think that ozone acted differently than chlorine. He believed chlorine selectively oxidized certain enzyme systems while ozone acted as a "general oxidizing agent."

Beyond the membrane and cell wall, ozone may act on the nuclear material. Christensen and Giese (1954) and Scott and Lesher (1963) showed that ozone affects both purines and pyrimidines in nucleic acids. More recently, Prat et al. (1968) showed that the pyrimidine bases of nucleic acids in *E. coli* were modified by ozone treatment, thymine being more sensitive than cytosine and uracil. This was confirmed by Scott (1975). The action of ozone on this kind of compound is described in chapter II, sec. A.5.

Ozone is also extremely effective against *Legionella pneumophila,* where a mean ozone concentration of 0.21 mg/L (average of 6 trials with a range from 0.10 to 0.30 mg/L) within a minimum contact time of 5 min gave an inactivation greater than 99 percent. This reaction was not markedly affected by pH or temperature and caused a lowering of the bacteria's unsaturated fatty acid content (Domingue et al. 1988).

Effects on viruses. The first site of action on viruses is doubtless the virion capsid, particularly the proteins that form it (Cronholm et al. 1976; Riesser et al. 1976). The ozone seemed to modify the viral capsid sites that the virion uses to fix on the cell surfaces. Higher concentrations of ozone dissociated the capsid completely.

The mechanism of ozone inactivation of bacteriophage f2 ribonucleic acid (RNA) studied by Kim et al. (1980) included releasing of RNA from the phage particles after the phage coat is broken into many protein subunit pieces. These findings suggest that ozone breaks the protein capsid, liberating RNA and disrupting adsorption to the host pili, and that the naked RNA may be secondarily inactivated by ozone but at a rate less than that for RNA within the intact phage. The mechanism of ozone inactivation of the deoxyribonucleic acid (DNA) bacteriophage T4 is quite similar: ozone attack of the protein capsid, liberation of the nucleic acid, and inactivation of the DNA by ozone (Sproul et al. 1982).

However, recent work by Shinriki et al. (1988) on the Tobacco Mosaic Virus (TMV), using molecular biological techniques, seems to show that ozone has a specific effect on RNA. It was shown that the quantity of RNA decreased as the ozone concentration increased and that the drop in TMV infectiousness correlated closely to the reduction in RNA recovery. The authors concluded by posing an inactivation hypothesis. Ozone attacks both protein coat and RNA. The damaged RNA crosslinks with amino acids of the coat protein subunits. TMV loses its infectivity because of its inability to uncoat.

Effects on spores. Rickloff (1987) reported on spore inactivation with ozone under hospital sterilization conditions. It would seem that, in a liquid phase at ambient temperature with an ozone concentration of about 10 mg/L, *Bacillus* and *Clostridium* spores were inactivated within 10 min, although no details were given on the residual ozone level maintained.

Effects on amoebae. Microscopic observation carried out during assays of inactivation of trophozoites of *Naegleria* and *Acanthamoeba* (Perrine et al., 1984a) showed that they were rapidly destroyed and the cell membrane was ruptured. Perrine and Langlais (1989) showed that in the case of *Naegleria gruberi* cysts, ozone seemed to affect the plugs that are the weak point in the structure of these cysts. These plugs at least partially consist of mucopolysaccharide compounds that can be stained and viewed under a microscope. After ozonation, these cystic plugs disappear or are partially destroyed, depending on the ozonation conditions. In addition, the number of inactivated cysts correlates well with the number of cysts that undergo morphological modification (the cystic plugs are attacked and become permeable to the stain).

Based on electron microscopic studies, Wickramanayake (1984) speculated that ozone initially affects the *Giardia muris* cyst wall and makes it more permeable. Aqueous ozone subsequently penetrates into the cyst and damages the plasma membranes. Additional penetration of ozone into the cyst eventually affects the nucleus, ribosomes, and other ultrastructural components.

***The special case of free amoebae,* Giardia, *and* Cryptosporidium.** *Giardia*, amoebae, and *Cryptosporidium* are protozoa. Their life cycles comprise two differentiated forms: fragile trophozoites for multiplication within the host and resistant cysts for conservation and dissemination outside the host.

For parasites such as *Giardia*, the dysenteric amoeba *Entamoeba histolytica*, and *Cryptosporidium*, the trophozoites are too fragile to live in the environment. They stay in the infected host intestine, where they multiply and encyst. Cysts, the only form found in the environment, are discharged with feces and disseminate the species. Thus, they can enter waterways through the fecal route or in sewage discharges from infected hosts. If they are not inactivated or physically removed, they can be recovered in drinking waters and their ingestion may lead to an infection.

There are other free-living protozoa that spend their life cycles entirely in the environment, where trophozoites and cysts can be simultaneously recovered. These protozoa are generally common soil and fresh-water microorganisms. Nevertheless, some of them merit particular attention in water treatment, especially free-living amoebae in the genera *Naegleria* and *Acanthamoeba*, because several strains are opportunistic pathogens for humans.

Giardia *cysts:* The evidence of waterborne giardiasis has been well documented (Craun 1988). *Giardia* cysts (*G. lamblia, G. muris*) leading to diarrheal syndromes are disseminated in the environment from a large variety of animals: domestic cats and dogs, wild rodents (rats, beavers), and humans. They are continuously introduced into

Table III–43 C · t Values for 99 Percent Inactivation of *Naegleria* and *Acanthamoeba* Trophozoites (Pathogenic and Nonpathogenic Strains) with Ozone and Free Chlorine

Oxidant	Microorganism	T (°C)	pH	C · t (mg.min/L)	Reference
Ozone	*Naegleria gruberi* *Acanthamoeba* *Naegleria gruberi* *fowleri* *Acanthamoeba*	25	7.4	<0.1 to >0.4*	Perrine et al. (1989)
Free chlorine	*Naegleria gruberi* *fowleri*	25	7.4	0.69–1.39	Perrine et al. (1989)
	Acanthamoeba	25	7.4	1.31–1.51	
	Naegleria/Acanthamoeba	28	7.4–7.5	<3.5	Derreumaux et al. (1974)

*Inactivation kinetics of *Naegleria* and *Acanthamoeba* trophozoites with ozone did not conform to Watson's law with $n = 1$:

$C \cdot t < 0.1$ at $C = 0.2$ mg/L;
$C \cdot t > 0.4$ at $C = 0.1$ mg/L.

surface waters where they are very resistant: at 5°C, more than 90 percent of *G. muris* survived after 25 days (Wickramanayake et al., 1985). According to the same authors, cysts aged by storage in water for 17 days at 5°C required C · t values for 99 percent inactivation by ozone similar to those for fresh cysts.

C · t values for ozone, chlorine, and chlorine dioxide obtained during inactivation tests of *Giardia lamblia, Giardia muris,* and undefined *Giardia* strains are presented in Table III–42. These results indicate that at either 5° or 25°C, ozone is a much more effective *Giardia* cysticide than chlorine or chlorine dioxide. On the other hand, C · t values for ozone recorded for *G. muris* cysts are greater than those for *G. lamblia* cysts (1.94 versus 0.53 mg · min/L at pH 7 and 5°C, 0.27 versus 0.17 mg.min/L at pH 7 and 25°C).

Entamoeba histolytica *cysts: E. histolytica* cysts are the causative agent of human amoebiasis and are transmitted similar to the scheme for *Giardia*. C · t values for inactivation of *E. histolytica* cysts by ozone are not well known. For chlorine, the C · t values reported by Chang and Fair (1941) and Stringer and Kruse (1970) range from 15 to 75 mg · min/L at pH 7–9 (10°C to 30°C).

*Free-living amoebae—*Naegleria *and* Acanthamoeba: *Naegleria* was the causative agent of fatal primary amoebic meningoencephalitis cases contracted during swimming and diving in artificially heated pools and also through the use of domestic waters (Anderson and Jamieson 1972). *Acanthamoeba* has been a causative agent of opportunistic diseases associated with AIDS (Martinez and Janistschke 1985; Gonzalez et al. 1986). Some strains of these amoebae have been associated with severe keratitis in contact lens wearers (Moore et al. 1985) whose lenses may have been contaminated with amoebae from tap water used to rinse the lenses or their cases (Georges and Perrine 1989). *Acanthamoeba* is also thought to facilitate the dissemination of *Legionella* (Perrine and Langlais 1986).

Elimination of trophozoites: Trophozoites of *Naegleria* and *Acanthamoeba* strains have a high sensitivity to ozone. A residual of 0.2 mg/L was sufficient to reduce active trophozoites by 2–4 log units in 30 s (Perrine et al. 1984a). These observations were confirmed by inactivation kinetic studies (Perrine et al. 1989) with ozone and free chlorine. Low C · t values were recorded for both disinfectants (Table III–43). For ozone, the low C · t value (0.1 mg · min/L) was valid only with at least a 0.2-mg/L residual (little inactivation of trophozoites occurred after 4 min contact time with 0.1-mg/L residual).

Table III–44 C · t Values for 99 Percent Inactivation of Free-Living Amoeba Cysts with Ozone, Free Chlorine, and Chlorine Dioxide

Oxidant	Microorganism	T (°C)	pH	C · t (mg · min/L)	Reference
Ozone	*N. gruberi* 1518	25	7.4	0.70	Perrine et al. (1989)
	Paris	25	7.4	0.68	
	MO5	25	7.4	0.73	
	NEG	5	7	4.23	Wickramanayake et al. (1984a)
		15	7	2.04	
		25	7	1.29	
		30	7	1.21	
		25	5	1.33	
		25	6	1.55	
		25	8	1.36	
		25	9	2.12	
	N. fowleri 0359	25	7.4	0.30	Perrine et al. (1989)
	A. castellanii	25	7.4	1.00	
	A. species MR4	25	7.4	1.10	
	Brest	25	7.4	1.12	
Free Chlorine	*N. gruberi* 1518	25	7.4	27	Perrine et al. (1989)
	Paris	25	7.4	31	
	MO5	25	7.4	34	
	NEG	25	5	7.72	Rubin et al. (1983)
		25	7	12.1	
		25	9	197	
	N. fowleri 0359	25	7.4	34	Perrine et al. (1989)
	A. strains	25	7.4	>160	
Chlorine Dioxide	*N. gruberi* NEG	5	7	15.47	Chen et al. (1985)
		15	7	9.58	
		25	7	5.51	
		30	7	3.77	
		25	5	6.35	
		25	6	5.25	
		25	8	3.92	
		25	9	2.91	

Inactivation of cysts: Ozone C · t values for inactivation of free-living amoeba cysts, summarized in Table III–44, show that at 20 to 25°C and pH 6 to 8, a 0.4-mg/L ozone residual for 4 min would give an inactivation of some pathogenic and nonpathogenic cysts of *Naegleria* and *Acanthamoeba* in excess of 99 percent (Perrine et al. 1989; Wickramanayake et al. 1984a). The best efficiency of ozone for *N. gruberi* NEG was found at 30°C and pH 7. At 25°C, ozone efficiency decreased at pH 9 (Wickramanayake et al. 1984a). At 25°C and pH 7.4, *Naegleria* cysts were more sensitive to ozone than those of *Acanthamoeba* (Perrine et al. 1989).

As shown in Table III–44, the C · t values for free chlorine are considerably less favorable than those for ozone. Even for the more sensitive pathogenic cysts, they are much higher than those found with ozone. Alkaline pH sharply reduced the chlorine efficiency (Rubin et al. 1983). For 99 percent inactivation of *Acanthamoeba* strains, C · t values greater than 160 were recorded (Perrine et al. 1989).

Finally, as shown in Table III–44, chlorine dioxide has an efficiency intermediate to ozone and chlorine that is increased by alkalinity and decreased by lower temperature (Chen et al. 1985). No C · t values are available for inactivation of *Acanthamoeba* by chlorine dioxide. Since *Acanthamoeba* is the predominant free-living amoebae in fresh waters, and since they are opportunistic pathogens, they have to be removed or inactivated for protection of public health. Because of their high resistance to free chlorine, ozone is an appropriate means to inactivate *Acanthamoeba* cysts.

Table III–45 Effect of Ozone on *Cryptosporidium* Oocysts Infectivity for Rats at 20°C and pH 7–7.1

No. of Oocysts by Oral Route	O_3 (mg/L)		Time (min)	No. of Infected Rats
	Initial	Final		
10^3	0	–	–	0/8
10^4	0	–	–	9/9
10^7	0.63	0.59	4	1/3
	0.83	0.78	4	0/12
10^8	0.61	0.58	4	3/3
	0.61	0.58	8	0/12
	0.80	0.75	4	2/3
	1.10	1.02	4	0/12

Source: Langlais et al. (1990).

Cryptosporidium *oocysts:* *Cryptosporidium* is a parasite that infects a wide variety of host species including humans. It is found in both domestic animals (cattle, sheep, goats, dogs, and cats) and wild animals (beavers, muskrats, rabbits, etc). In humans, *Cryptosporidium* is the causative agent of cryptosporidiosis. This infection may produce, in immunocompetent humans, a self-limited flulike gastrointestinal illness that resolves spontaneously in 1 to 4 weeks. In immunocompromised patients, particularly those with acquired immunodeficiency syndrome, cryptosporidiosis may produce severe and prolonged diarrhea for which there is no satisfactory chemotherapy and which contributes to mortality. *Cryptosporidium,* like many other important waterborne pathogens, completes its life cycle in the gastrointestinal tract of the infected hosts in which oocysts are formed. These oocysts are fully sporulated, infective, and environmentally stable when passed in feces (Current et al. 1983). Their transmission from host to host by the fecal-oral route suggests the possibility of waterborne transmission. Many mammals may serve as reservoirs of infection for humans (Current 1987). The cross-species transmission (Fayer and Ungar 1986) increases the potential for waterborne transmission, because animals, in addition to humans, contaminate water sources. The contamination of many surface waters by oocysts and several waterborne outbreaks caused by *Cryptosporidium* contamination of drinking waters have been documented (Rose et al. 1986; Madore et al. 1987; Ongerth and Stribbs 1987; Logsdon et al. 1988; and Colbourne et al. 1990).

Water treatment plays a significant role in the prevention of waterborne diseases, but exclusion of *Cryptosporidium* from public water supplies is complicated. The small size of oocysts (4 to 5 μm) makes removal by filtration difficult: sand filtration reduces oocyst concentrations, but it does not eliminate them completely (Madore et al. 1987). Moreover, oocysts are apparently resistant to a number of disinfectants as well as to water chlorination (Campbell et al. 1982). Nevertheless, preliminary results of a recent study (Langlais et al. 1990) using a rat model to ascertain oocyst viability showed that ingestion of 10^4 oocysts was the infection threshold and that ozonation inactivated *Cryptosporidium* oocysts. These results, shown in Table III–45, indicate that ozonation corresponding to a $C \cdot t$ of about 3.3 mg · min/L inactivated at least 10^3 oocysts, while a $C \cdot t$ of 4.4 to 4.9 mg · min/L was required for an inactivation of at least 10^4 oocysts. Peeters et al. (1989), using neonatal mice, found an ozone concentration of 1.1 mg/L for 6 min sufficient to disinfect water containing 10^4 oocysts/mL while some oocysts remained viable after a 30-min contact time with 0.43 mg/L of chlorine dioxide. Korich et al. (1990) also used mouse infectivity to show that 90 percent inactivation could be achieved after 5 min contact with 1 mg/L ozone. This gave a $C \cdot t$ value of 5 mg · min/L as compared to 78 mg · min/L for chlorine dioxide and 7200 mg · min/L for chlorine and monochloramine.

Limitations in applying C · t values. The sensitivity of microorganisms to different disinfectants varies widely depending on the type of microorganisms considered and the disinfectant used. It must also be remembered that experimental conditions (in particular, the physicochemical composition of the water) are major factors influencing the disinfection action of ozone. C · t values should be applied judiciously, with consideration of the local environmental conditions. Also, the total amount of inactivation needed, i.e., 2, 3, or 4 logs or more of disinfection capacity, must be determined. This latter factor has not been considered in this review.

III.H.2 Case studies

The use of ozone as a disinfectant in large-scale plants will be illustrated by three case studies from France: the Neuilly-sur-Marne and Méry-sur-Oise water works (Langlais et al. 1989; Schulhof 1987), and a pilot plant located in Choisy-le-Roi (Joret et al. 1986). A fourth case study will be presented based on the Los Angeles Aqueduct Filtration Plant (California) where C · t calculations from full-scale disinfection studies were compared with expected inactivation from C · t values cited in the literature (Stolarik and Christie 1988).

Bacteriological control. Water quality in the Paris-area water treatment plants is regularly monitored by a series of analyses at different points in each treatment train. These measurements will be used to demonstrate the influence of ozone on bacteriological control during 1988 at the Neuilly-sur-Marne plant.

Description of Neuilly-sur-Marne treatment plant. The Neuilly-sur-Marne plant produces 600,000 m^3/day (158 mgd) (a nominal flow rate) using water from the Marne River about 30 km upstream of Paris and supplies a population of 1,500,000 (Langlais et al. 1989). To insure safety of the water supply, river water is examined upstream of the plant and the intake is shut down in the event of a major pollution incident in the river. Raw water may also be transferred between two of the water works in the Paris region, Neuilly-sur-Marne and Choisy-le-Roi, at flow rates of 500,000 m^3/day (132 mgd) in the event of serious pollution. The Neuilly-sur-Marne treatment train consists of the following stages: pretreatment, flocculation, settling, sand filtration, ozonation, and chlorination/dechlorination.

The pretreatment stage consists of a specially engineered tank designed to avoid cross-reaction between chemicals. Pretreatment includes an injection of coagulant: aluminum polysulfate chloride (a commercial product) at a dose rate of 70 mg/L, sulfuric acid during algal growth periods, and powdered activated carbon in the event of accidental pollution. After flocculation, sedimentation takes place in a rectangular settler with continuous sludge removal. The settled water then flows through 42 filters offering 5040 m^2 (54,250 ft^2) of filtering area containing a 1.4-m (4.6-ft) layer of sand with a grain-size distribution of 1.2 to 1.4 mm (Uniformity Coefficient: 1.2–1.4). The filters operate at a flow rate of 4 m/h.

The filtered water is then disinfected with ozone at an average dosage of 1.5 mg/L, but this rate can be varied from 0.6 to 3 mg/L. Ozonation takes place in four contact chambers where ozone is diffused via porous plates. Based on flowthrough calculations, time in the contactors is 9 to 15 min (12 min on average). The ozone is generated by four ozonators, each with a capacity of 30 kg/h. The ozone concentration at the end of the contactors is 0.4 mg/L. Variations from the set point of 0.4 mg/L of ozone at the end of the contactors are automatically detected and the ozone dosage is varied through a feedback control loop.

Postdisinfection with chlorine is performed at a dose of 1.2 to 1.5 mg/L to maintain a residual concentration of the oxidizer in the distribution system. Dechlorination

Table III-46 Bactericidal Efficiency of Ozone at Neuilly-sur-Marne Plant (1988; Monthly Determinations; Simultaneous Sampling)

Parameter		Postfiltration	Postozonation
Heterotrophic Plate Count*/mL 20°C—72 h	No. samples	10	10
	Positive	10	5
	Min.	120	0
	Max.	7700	19
	Mean	1580	2.7
	St. dev.	2275	5.8
Heterotrophic Plate Count*/mL 37°C—24 h	No. samples	10	10
	Positive	10	3
	Min.	7	0
	Max.	520	30
	Mean	186	5.4
	St. dev.	155	10.7
Total Coliform†/100 mL	No. samples	10	10
	Positive	10	2
	Min.	300	0
	Max.	7600	1
	Mean	3580	0.2
	St. dev.	2308	0.4
Fecal Coliform†/100 mL	No. samples	9	10
	Positive	9	2
	Min.	4	0
	Max.	460	1
	Mean	198	0.2
	St. dev.	167	0.4
Fecal Streptococci‡/100 mL	No. samples	10	10
	Positive	9	0
	Min.	0	0
	Max.	145	0
	Mean	32	0
	St. dev.	46	0
Sulfate-Reducing Anaerobes‡/20 mL	No. samples	9	10
	Positive	2	2
	Min.	0	0
	Max.	5	1
	Mean	0.7	0.2
	St. dev.	1.7	0.4

Source: Langlais and Perrine (1990).

*Pour plate method on plate count agar (Difco) according to French Standards (AFNOR 1986b, NFT 90.402 [20°C—72 h] and NFT 90.401 [37°—24 h]).

†Membrane filtration method according to French Standards NFT 90.414 (AFNOR 1986b).

‡Membrane filtration method according to French Standards NFT 90.416 (AFNOR 1986b).

is used to obtain a maximum residual chlorine concentration of 0.4 mg/L leaving the plant.

Bacteriological results. The sampling sequences carried out over the year at Neuilly-sur-Marne monitored bacteria levels before and after ozonation. Table III–46 reports the bacterial counts (according to French Standard; AFNOR 1986b) found in the water after the sand filtration and postozonation steps.

Heterotrophic plate count reductions at 20°C (72-h incubation period) and at 37°C (24-h incubation period) were 99.8 and 97.1 percent, respectively. Total and fecal coliform inactivation were 99.994 and 99.9 percent, respectively. Fecal streptococci were never detected in 100-mL samples after postozonation. Ozonation achieves an almost total reduction of bacteria.

Control of protozoa. A free amoebae count was carried out along the treatment train at Méry-sur-Oise in order to show the impact of each unit process on these protozoa.

Description of Méry-sur-Oise treatment plant. The Méry-sur-Oise water treatment plant has an output capacity of 270,000 m^3/day (71 mgd). It treats raw water from the Oise River 15 km upstream of its confluence with the Seine and provides water for a population of some 630,000. Commissioned in 1980, the Méry-sur-Oise facility uses biological treatment to treat this poor quality source water. It includes a preozonation stage followed by conventional flocculation-sedimentation treatment, sand filtration, intermediate ozonation followed by filtration on granular activated carbon, postozonation, and postchlorination.

Preozonation may be used in the initial stage when the raw water is especially contaminated. These conditions were not present during the two days when the plant was monitored. Depending on the raw water quality, dosage can be varied between 0.8 and 1.7 mg/L of ozone with a contact time of about 1 min. The water is then stored for 2 to 3 days in a reservoir covering 5.3 ha (depth: 9 m [30 ft]). In the event of pollution in the Oise, this reservoir provides the necessary reserve to make it possible to stop pumping water from the river until the pollution front has passed the intake structure.

Chemicals are injected and mechanically stirred in a pretreatment tank. These include flocculants (basic aluminum sulfate [Aqualenc®] and ferric chloride), caustic for pH adjustment and, if necessary, powdered activated carbon. After flocculation, the water settles in multi-stage settlers. Filtration then takes place on 12 sand filters with a total filtering area of 1300 m^2 (14,000 ft^2) operating at a rate of 6 m/h. The filtering bed is 1.40 m (4.6 ft) thick and the effective size of particles is 1 to 1.2 mm (Uniformity Coefficient: 1.2–1.4).

The ozonation step midway through the treatment train has two contact chambers in which the ozonized air is diffused through porous plates. The water moves at a flow rate of 6000 m^3/h (26,000 gpm) through the contactors in 10 to 20 min (theoretical time) where it is given an average ozone dose of 1.5 mg/L (min. 0.9 mg/L, max. 2.8 mg/L). Ozonated water then flows through 12 granular activated carbon filters (effective size 1 to 1.4 mm) affording a total filtering surface of 1300 m^2 (14,000 ft^2) by 1 m (3.3 ft) deep. The filtration rate is 6 m/h.

The final disinfection stage is performed in two ozonation columns where an average ozone dose of 1.5 mg/L (min. 0.7 mg/L, max. 3.7 mg/L) is diffused via porous plates. The theoretical contact time is in the range of 10 to 20 min. Postdisinfection, using an average dosage of 1.3 mg/L of chlorine, provides a residual of chlorine at the outlet of the water works that is reduced if required by dechlorination with sodium bisulfate until a chlorine concentration of about 0.3 mg/L is reached.

The ozone for the preozonation stage is generated by two ozonators with a production capacity of 3.6 kg O_3/h per unit.

Free amoeba count along the Méry-sur-Oise treatment train. Free amoebae were counted by an MPN method adapted by Perrine et al. (1984b). However, it should be noted that the microscope detection system adopted for observing the amoebae could only be used within the range of 20 to 24,000 organisms per liter.

Water was sampled at different treatment stages. Tables III–47 and III–48 report the operating conditions of the facility (oxidizing rates and residuals) on the two sampling days and the results of the physicochemical analyses and free amoeba counts.

Each of the treatment units reduced the amoeba concentrations on the two sampling days. The reductions each day from the raw water to the effluent were about 99 percent. After ozonation with a dose rate of 1.4 mg/L or 2.4 mg/L and an average residual of about 0.4 mg/L, no amoebae were detected by the method used.

Virus control. The virucidal effects of ozone used as a pretreatment or as an intermediate treatment were studied in two types of plants: an experimental pilot

Table III–47 Operating Conditions of the Méry-sur-Oise Plant (Case Study: Control of Protozoa)

		Sampling Dates			
		1/16/89		9/14/89	
Treatment Step	Chemicals	Dosage (mg/L)	Oxidant residual (mg/L)	Dosage (mg/L)	Oxidant residual (mg/L)
Preozonation	O_3	0	0	0	0
Coagulation-Flocculation	Basic aluminum sulfate	16.0	–	8.0	–
	$FeCl_3$	6.6	–	3.2	–
Interozonation	O_3	1.53	0.11	0.34	0
Postozonation	O_3	1.39	0.43	2.42	0.42
Postchlorination	Gaseous chloride	0.60		1.28	
	Sodium hypochlorite	0.54	0.72		0.57
Dechlorination	Sodium bisulfite	0.37	–	0.19	–
	Total chlorine	–	0.33	–	0.28
	Free chlorine	–	0.25	–	0.23

Source: Langlais and Perrine (1990).

Table III–48 Physicochemical and Biological Characteristics of Water at Different Points of Méry-sur-Oise Treatment Plant

Type of Water	Date	Temperature (°C)	Turbidity (NTU)	Baudrey Clogging Indicator*	pH	Dissolved O_2 (mg/L)	Ammonia (mg/L)	Nitrites (mg/L)	COD† (mg/L)	Free Amoebae (MPN/L)
Raw	1/16/89	7.3	–	6.84	7.9	9.7	0.55	–	4.5	11,000
	9/14/89	19.5	–	–	–	–	–	–	3.8	40,000
Stored	1/16/89	–	51	0.37	7.9	6.9	0.31	–	3.6	930
	9/14/89	–	197	–	7.9	–	0.12	0.46	2.9	24,000
Settled	1/16/89	–	1.42	0.15	7.8	7.4	–	–	2.6	430
	9/14/89	–	1.54	–	7.8	–	–	–	2.4	2,400
Sand-filtered	1/16/89	–	0.17	0.05	7.8	6.6	0	0	2.4	210
	9/14/89	–	0.18	–	7.7	4.2	0	0	2.2	930
Interozonated	1/16/89	–	–	–	–	10.7	–	–	–	150
	9/14/89	–	–	–	–	8.2	–	–	1.4	750
GAC-filtered	1/16/89	–	0.10	–	7.8	8.9	–	0	1.9	40
	9/14/89	–	–	–	7.7	4.8	0	0	1.2	430
Postozonated	1/16/89	–	–	–	–	–	–	–	–	<20
	9/14/89	–	–	–	–	–	–	–	–	<20
Postchlorinated	1/16/89	–	0.05	–	7.7	11.2	0	–	1.4	<20
	9/14/89	–	0.05	–	7.7	8.2	–	–	1	<20

Source: Langlais and Perrine (1990).
*According to French Standards NFT 90.030 (AFNOR 1986b).
†Permanganate oxidation in acidic media according to French standards NFT 90.018 (AFNOR 1986b).

plant processing water from the Seine River (Joret et al. 1986) and the Méry-sur-Oise plant described earlier. Methods used for the concentration, enumeration, and identification of viruses have been previously described (Joret and Block 1981; Joret et al. 1986).

Virucidal action of river water preozonation (pilot-plant study). The experimental pilot plant had two processing lines, operating in parallel on water from the Seine. Figure III–52 summarizes the operating conditions and the different water treatment

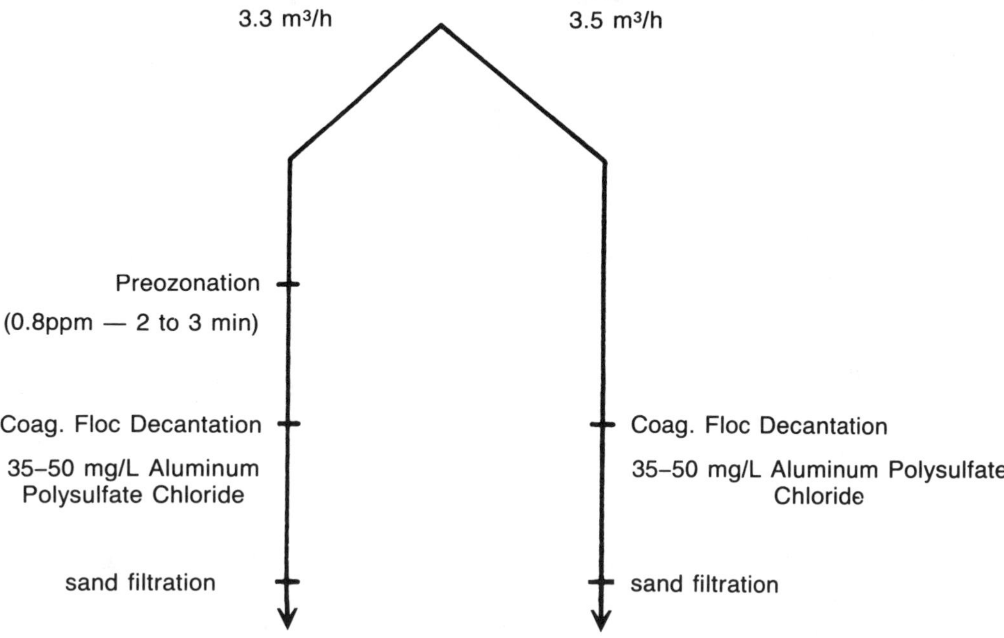

Figure III–52 Characteristics of the Seine Pilot Plant

Table III–49 Effect of Preozonation of Seine River Water on Inactivation and Removal of Viruses (Choisy-le-Roi Pilot Plant)

Virus Concentrations (PFU/1000 L)		Virus Removal (percent)				
			Coag. Floc Settling		After Sand Filtration	
Raw Water	Preozonated Water	Preozonation Alone	Without Preozonation	With Preozonation	Without Preozonation	With Preozonation
950	<27	>97	84	>99.8	99	>99.8
1420	20	98	90	97	99	>99.9
190	20	64	31	>95	>99.3	>99.2
380	<10	>97	37	>95	92	>99.6
1150	120	90	66	>99	98	>99.9
910	300	67	59	>99	91	>99.8

Source: Joret et al. (1986).

steps in this plant during the experimental period. The properties of the raw water were as follows: temperature (°C) = 6.4–13.1; dissolved O_2 (mg/L) = 10–11.4, turbidity (NTU) = 36–55; and pH = 7.8–8.1. Results of the virological study are summarized in Table III–49. Viruses were isolated in all samples of river water (35 to 58 L) with contamination levels ranging from 190 to 1420 PFU/1000 L. River water was ozonated at a dosage of 0.8 mg/L for 2–3 min, but measured residuals were not observed. Virus inactivation equalled or exceeded 97 percent in three of the six samples.

Comparison of the virus removal results at each stage of the two lines working in parallel highlights the advantages of preozonation. When no preozonation occurred, viruses were isolated in all of the 75- to 100-L samples of decanted water (virus removal:

Table III–50 Virus Concentrations (PFU/1000 L)* at Various Stages in the Méry-sur-Oise Plant

Date	Stored Water	Settled Water	Sand-Filtered Water	Intermediate Ozonated Water	GAC-Filtered Water	Postozonated Water	Postchlorinated Water
6/15/82	10.4	<25	7.1	<1.2	<1.2	<1.2	ND[†]
6/16/82	4.6	13.2	<1.1	<1.1	0	<1.1	0
6/17/82	100	75	<2	<1.6	<1.2	<2	<1.2
10/12/82	20	5	0	<1.1	0	0	0
10/13/82	10	20	3	<1.1	0	<1.1	0
10/14/82	30.7	10	5	<1.1	0	0	0
3/8/83	<17	20	<1.3	<2.3	<1.4	<1.3	0
3/9/83	20	<10	2.3	<3	<1.2	<1.3	0
3/10/83	10	10	1.2	<1.4	<1.2	<1	0

Source: Joret et al. (1986).
*Calculated from PFU observed in volume sampled (30-100 L for stored water, 550-1000 L for other sampling points).
[†]Not determined.

31 to 90 percent) and five of six 1000-L sand-filtered samples (overall virus removal: 78 to 95 percent for positive samples). Along the preozonated lines, viruses were isolated in only one of the six 100-L samples of decanted water; the overall reduction was then 97 percent. After rapid sand filtration, no viruses were isolated in any of the six 800-L samples, and the overall viral reduction obtained at this point in the process was 2 to 3 logarithms (compared to 1 to 2 logarithms without ozonation).

Virucidal effect of ozone as an intermediate treatment: studies on the Méry-sur-Oise plant. Nine series of analyses were carried out at each stage in the Méry-sur-Oise plant described earlier; the results are shown in Table III–50. Viruses were found eight times in raw water from the reservoir, seven times in water after coagulation, flocculation, and decantation, and five times after rapid sand filtration. These results confirmed those found during the pilot-plant study: virus removal was not complete after clarification nor after rapid sand filtration.

The critical stage for the elimination of viruses in this plant was the intermediate ozonation (ozone concentrations of 1.5 mg/L for 10 min, residual ozone = 0.05 to 0.2 mg/L). In fact, no viruses were isolated during the series of tests carried out on water samples after intermediate ozonation, GAC filtration, postozonation, and postchlorination.

Overall plant removal of viruses. Data are presented in Table III–51 for overall plant virus removal for the five-year period 1984–1988. These data are for the Méry-sur-Oise and Neuilly-sur-Marne plants described earlier and for the Choisy-le-Roi plant. The latter plant is identical with Méry-sur-Oise except that there is no post-ozonation. Monthly monitoring of the plants, before and after treatment, confirmed the absence of viruses in the water produced.

Type of virus isolated. Among the 580 PFU isolated during the pilot-plant and the Méry-sur-Oise full-scale plant studies, 96 were randomly identified and sereotyped. The viruses isolated were mainly echoviruses (49 and 41 percent, respectively) and Coxsackie B viruses (41 and 48 percent, respectively). Poliomyelitis viruses were isolated in only 8 and 11 percent of the cases, respectively. No selective effect for any of the types of virus isolated was detected at any of the treatment stages studied (Joret et al. 1986).

Conclusions. In a conventional pilot plant treating river water, without prechlorination or preozonation, the reduction in virus content obtained by clarification with coagulation, flocculation, decantation, and rapid sand filtration ranged from 1 to 2 logarithms. If the raw water was heavily contaminated, these processes could not

Table III–51 Virus Concentrations Measured in Three Plants Before and After Treatment (Monthly Samples Taken from 1984 to 1988)

Plant	Type of Water	Samples Analyzed	Positive Samples*	Percent	Virus concentrations (PFU/1000 L)	
					Min.-Max.	Average
Méry-Sur-Oise	Raw	58	49	84	<50–3,450	510
	Stored	53	41	77	<50–2,400	320
	Finished	57	0	0	0	0
Choisy-le-Roi	Raw	59	56	95	<50–20,200	2,695
	Finished	59	0	0	0	0
Neuilly-sur-Marne	Raw	59	50	85	<50–6,650	910
	Finished	59	0	0	0	0

Source: Joret (1990).
*Volume of concentrated samples: raw and stored water, 20 L; finished water, 1000 L.

eliminate all viruses. However, pretreating river water at low ozone concentrations (0.8 mg/L for 2 to 3 min) significantly improved the reduction in virus content (64 to 98 percent in this study). It then became possible to completely eliminate viruses after rapid sand filtration. If, for any reason, the final disinfection stage suffers a brief malfunction, this type of pretreatment provides an additional safety factor to ensure that the water produced will be free of any virus contamination.

If the virus contamination level of the water is low, intermediate ozonation at low doses (1.5 mg/L for 10 min, 0.05 to 0.2 mg/L residual) eliminates all viruses that have survived the clarification and rapid sand filtration stages (based on 1000-L volume sampled, Méry-sur-Oise). If more heavily contaminated water is subjected to the same preliminary treatment, the virucidal ozonation rate recommended by Coin et al. (1964) (0.4 mg/L residual for 4 min), currently applied in many plants, is quite sufficient to eliminate viruses.

The strongly germicidal effects (inactivation of bacteria, viruses, and protozoa) of ozone under laboratory conditions have been extensively explored by other authors (see sec. H.1 of this chapter). However, more extensive field studies are required to check whether these doses could be reduced (while still maintaining a safety factor) to minimize the energy costs incurred in this type of treatment.

Projection of ozone C · t values at the Los Angeles Aqueduct Filtration Plant. The Los Angeles Department of Water and Power (LADWP), which treats drinking water for the city, studied the disinfection treatment requirements of the pending United States Surface Water Treatment Rule (Stolarik and Christie 1988). The rule's disinfection requirement proposed determining the effectiveness of disinfection from application of C · t values under known conditions of water temperature and pH. The study was conducted at the Los Angeles Aqueduct Filtration Plant (LAAFP).

The effectiveness of disinfection was measured by monitoring coliform, heterotrophic plate counts, and *Giardia* cysts before and after ozone addition.

The LAAFP receives water from two major California surface water supplies, the Los Angeles Aqueduct and the State Water Project. Ozone is used for predisinfection ahead of coagulation, flocculation, and filtration, and is followed by postdisinfection with chlorine. The 7900-lb/day (3600-kg/day) ozonation system at LAAFP was started up in late November, 1986. The facilities used for ozone contacting were configured for two-stage ozone disinfection and chemical oxidation reactions.

Control of the ozonation system is accomplished by a programmable controller; ozone residuals are sampled by process instrumentation at the end of each stage and checked daily by grab samples analyzed by wet chemistry methods (KI and Indigo trisulphonate).

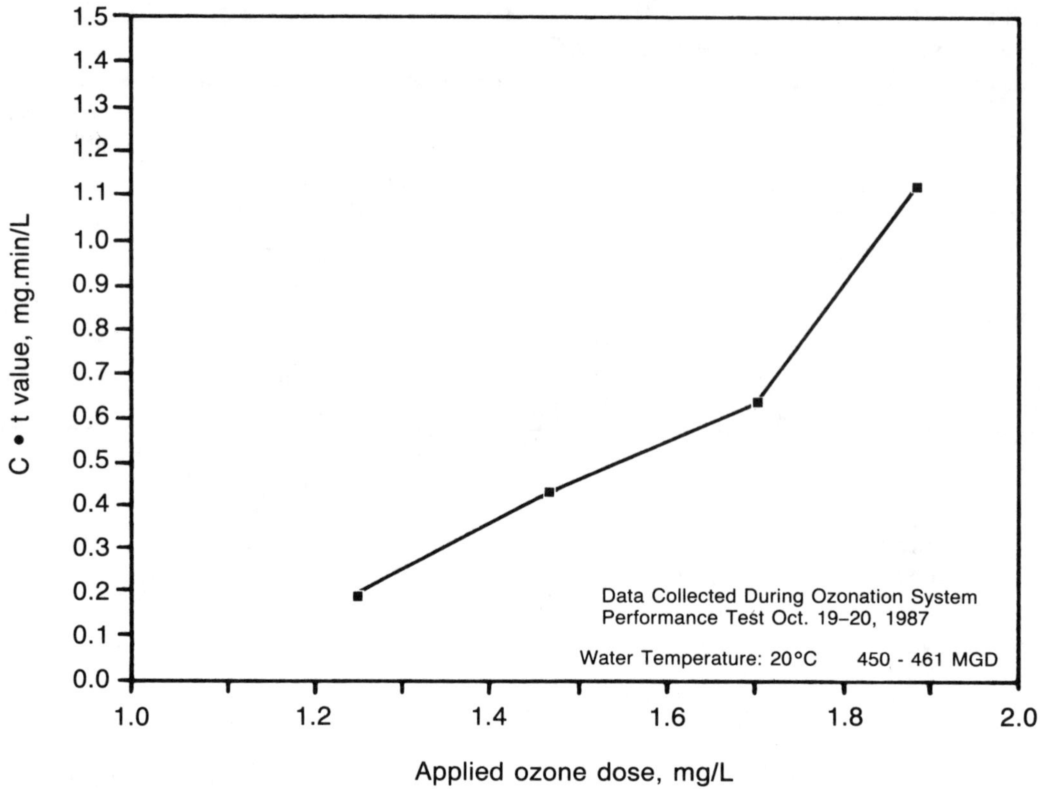

Source: Stolarik and Christie (1988).

Figure III–53 C · t Values Versus Applied Ozone Dose—Los Angeles Aqueduct Filtration Plant

Table III–52 Ozone Power Cost Versus Applied Ozone Dose and C · t Values (Los Angeles Aqueduct Filtration Plant Data Collected During Ozonation System Performance, October 19–20, 1987)

Dose (mg/L)	C · t Value (mg · min/L)	Power Cost* ($/MG)
1.24	0.19	4.84
1.47	0.44	5.62
1.70	0.64	6.35
1.88	1.12	7.41

Source: Stolarik and Christie (1988).
*At 0.065 $/kWh.

Figure III–53 and Table III–52 show the C · t value and ozone treatment cost for a given applied ozone dose during ozonation system power consumption testing.

The bacteriological study showed that ozone is a highly effective disinfectant (see Tables III–53 and III–54). Even at an absorbed ozone concentration of 1.0 mg/L (but without measurable ozone residual), influent coliforms were completely inactivated and bacteria counts were reduced by 1 to 2 logs in raw water from the State Water Project (Table III–53).

On Los Angeles Aqueduct (LAA) raw water, increasing the ozone residual from 0.08 to 0.46 mg/L had no further impact on coliform and heterotrophic plate count reductions, which were at minimum detection levels (see Table III–54). A minimum

Table III–53 Bacteriological Results on State Project Water, Los Angeles Aqueduct Filtration Plant, October 27–28, 1987

Ozone Dose* (mg/L)	C · t Value (mg · min/L)	Sample Location†	Total Coliform MPN/ 100 mL	Total Coliform CFU‡/ 100 mL	Fecal Coliform (FC/100 mL)	Heterotrophic Plate Count 48 h—37°C PCA (CFU/mL)§	Heterotrophic Plate Count 7 days—28°C R2A (CFU/mL)§
1.00	N/A**	Raw	23	–	0	1,000	1,600
		After O_3	<2	–	0	42	25
		Clear well	–	0/0	0	–	22
1.63	N/A	Raw	33	–	0	1,300	24,000
		After O_3	<2	–	0	<1	6
		Clear well	–	0/0	0	–	13
2.02	0.97	Raw	170	–	0	1,100	61,000
		After O_3	<2	–	0	<1	<1
		Clear well	–	0/0	0	<1	14
3.98	12.3	Raw	120	–	2	>300	50,000
		After O_3	<2	–	0	2	<1
		Clear well	–	0/0	0	<1	17

Source: Stolarik and Christie (1988), Stewart (1990).
*Absorbed ozone dose based on transfer efficiency; ozone residual and C · t based on iodometric method.
†Raw—before ozone; after O_3—exit contact basin; clear well—after post-Cl_2 at 1.0 mg/L.
‡Coliform/non-coliform.
§Heterotrophic plate count (CFU/mL) with PC agar, 48 h, 37°C and R2A agar, 7 day, 28°C.
**N/A (not applicable): either residual not measured or no residual.

Table III–54 Bacteriological Results on Los Angeles Aqueduct Water, Los Angeles Aqueduct Filtration Plant, October 19–20, 1987

Ozone Dose	Ozone Residual (mg/L)	After Ozone C · t (mg · min/L)	After Ozone Total Coliform (TC/100 mL)	After Ozone HPC* (CFU/mL)
1.19	0.08	0.19	0	1
1.41	0.20	0.44	0	2
1.63	0.27	0.64	0	<1
1.80	0.46	1.12	0	1

Source: Stolarik and Christie (1988), Christie (1990).
*Heterotrophic plate count with plate count agar, 48 h, 37°C.

ozone residual of about 0.10 mg/L exiting the ozone contact basin was necessary to achieve effective predisinfection of LAA water.

The rapid decay of ozone residual after the water leaves the ozone contacting system may allow regrowth of a microbiological population in the filters; postdisinfection with chlorine returned bacteriological parameters to low levels at the plant's clear well (data not shown).

Results of *Giardia lamblia* testing were negative in both plant influent and effluent samples. However, past quarterly sampling of plant influent had indicated the presence of *Giardia* cysts in the LAA surface water supply. C · t calculations are sensitive to sampling techniques and to short holding times (seconds). Tracer-determined detention times have shown that contact basin hydraulic residence times are about one half of the theoretical values. The resulting C · t values projected for ozone show that LAAFP could meet EPA's proposed values (see Table III–38), but with significantly reduced operating margins and higher treatment costs to sustain the ozone production necessary to achieve the pretreatment goal of one log reduction of *Giardia*.

Bacteriological monitoring for coliform and heterotrophic bacteria indicated that ozone was effective at doses yielding C · t values less than the proposed U.S. EPA values for *Giardia*. This is not unexpected, however, since it is known that bacteria are much more sensitive to ozone than are *Giardia* cysts.

III.I DISINFECTION BY-PRODUCT CONTROL

III.I.I Introduction

The use of ozone may cause changes in the concentration of disinfection by-products (DBPs) in finished water. These changes may come about because ozone can either (1) immediately destroy or form by-products, (2) immediately destroy or create new by-product precursors, (3) alter the nature of the water so that subsequent processes are better or worse at removing the by-products, (4) alter the nature of the water so that subsequent processes are better or worse at removing the by-product precursors, or (5) permit use of lower doses of other disinfectants or changes in their point of addition. These first two effects are easily studied in the laboratory. As a result, we know a great deal more about them than the latter three. The first two mechanisms are termed primary effects. The latter three are then the secondary effects of ozone on disinfection by-product formation.

The current knowledge of chlorination by-products from natural organic matter is much greater than the knowledge of by-products from other disinfectants. As a result, chlorination by-product formation has been more intensively studied in water treatment. This section will present a summary of much of the knowledge gained in chlorination by-product studies. Unfortunately, this information cannot be readily extrapolated to other disinfectants, because the effects of treatment may be highly by-product–specific.

III.I.2 Formation of Ozonation By-products

The direct, molecular ozone pathway and the free radical pathway, resulting from the complex decomposition of ozone, likely both play a role in the formation of ozone by-products in natural waters. The nature of natural organic matter, which serves as precursor material for ozone by-products, can vary widely. Other important parameters defining natural waters, like pH and the concentration of free radical scavengers, can also vary widely. Ozonation conditions within water treatment plants vary with treatment objectives. For all of these reasons, the number and relative amounts of ozone by-products can be expected to vary from plant to plant.

In water treatment plants, smaller amounts of ozone are used than would be required to fully mineralize complex humic and fulvic precursor materials. From information presented at two ozone conferences (IOA 1988a, 1988b), the typical ozone dose–to–carbon ratio is below 1.2 mg O_3/mg C. For purposes of enhancing flocculation the ratio may be nearer to 0.4 mg O_3/mg C (chapter III.F); whereas for purposes of oxidation, the ratio can be higher. At these levels, the ozonation of humic and fulvic materials is not complete. Ozonated humic material is different from the raw material, and is characterized by a mixture of lower-molecular-weight, more polar products and high-molecular-weight products remaining closer in identity to the raw material. Because it is difficult to identify and quantify the lower-molecular-weight materials with analytical methods historically employed in drinking water analyses, the database for understanding ozone by-products is relatively small.

The U.S. EPA (1988) cited ozone by-products as a generic class of compounds for regulatory consideration because of their potential health effects. This action prompted a number of studies to identify ozone by-products under drinking water treatment plant

Table III–55 Aldehyde Formation at the Lengg Plant in Zurich, Switzerland

	Concentration (ng/L)							
	1 mg/L O_3 Applied		2.5 mg/L O_3 Applied		5 mg/L O_3 Applied		7 mg/L O_3 Applied	
Aldehydes	Pre	Post	Pre	Post	Pre	Post	Pre	Post
Hexanal	ND*	40	ND	78	ND	74	ND	24
Heptanal	8	82	30	140	18	145	20	68
Octanal	6	74	22	190	16	320	26	110
Nonanal	7	160	34	340	26	680	55	164
Decanal	12	260	36	240	38	920	80	134
Undecanal	ND	40	ND	64	ND	82	26	16
Dodecanal	ND	28	16	24	ND	58	24	12
Tridecanal	ND	ND	ND	12	ND	24	20	8

Source: Schalekamp (1978).
*ND = not detected.

conditions. Prior to that time, studies addressing this subject largely dealt with ozonation of concentrated solutions of humic and fulvic materials or employed atypical ozonation conditions, i.e., high ozone/carbon ratios or long ozone contact times (Lawrence et al. 1980; Killops et al. 1985; Anderson et al. 1985; Killops 1986). These studies identified a number of aliphatic aldehydes, mono- and dicarboxylic acids, mono- and diketones, alkanes, and phthalates. Additionally, Singer (1988a), in an assessment of research needs involving ozonation of drinking waters, noted organic peroxides, epoxides, and unsaturated aldehydes as products that may be formed and may be of interest because of their potential health effects. Glaze (1988) summarized the status of both ozone by-products and requisite analytic methods.

Formation of aldehydes. Ozone has long been known to produce aldehydes as by-products from a wide variety of unsaturated organic compounds via the Creigee mechanism (see chapter II, sec. A.1). It is therefore not surprising that aldehydes should be formed from the ozonation of natural organic matter like humics (see chapter II, sec. A.6) or natural cell material like amino compounds (see chapter II, sec. A.5). In addition to their possible effects on human health, higher-molecular-weight aldehydes produced by ozonation have been linked to fruity, fragrant, and orange-like odors in treated waters (Anselme et al. 1988). In general, concentrations of C_1 to C_{14} aldehydes have been reported to increase upon ozonation. The degree to which this increase occurs depends on ozone dose, with some intermediate dose giving a maximum formation (Schalekamp 1978; Glaze et al. 1988). It is clear that the nature of the background organic matter, not just the DOC level, is critical in determining the distribution and concentration of aldehydes formed (Glaze et al. 1989a, 1989b). Precursor removal has been shown to result in reduced aldehyde formation (Van Hoof et al. 1985). In some plants secondary reactions between aldehydes and other consistuents may lead to the disappearance or transformation of some of these ozonation by-products (Glaze et al. 1989b).

Schalekamp (1978) reported the presence of a number of aldehydes in ozonated waters in Swiss drinking water treatment plants. Using closed loop stripping analysis (CLSA), he measured increases in the concentrations of C_6 through C_{13} compounds in the ng/L range (Table III–55). The homologous series of heptanal through decanal was also seen in source waters at concentrations below 100 ng/L, likely present there as the result of natural evironmental oxidation processes. Schalekamp investigated ozone dose dependency and found that aldehyde formation peaked near 5 mg/L ozone dose, after which aldehyde concentration fell off.

Glaze et al. (1989a) also used CLSA to investigate the presence of aldehydes. The results were similar to Schalekamp's but demonstrated the effect of varying background

Table III–56 Aldehyde Formation at the Los Angeles Aqueduct Filtration Plant

	Estimated Concentration (ng/L)							
	State Project Water★						LAAW†	
	O_3 = 1 mg/L		O_3 = 3 mg/L		O_3 = 4 mg/L		O_3 = 1 mg/L	
Aldehydes	Raw	O_3	Raw	O_3	Raw	O_3	Raw	O_3
Hexanal	ND‡	130	ND	135	ND	82	8	578
Heptanal	ND	195	ND	168	ND	121	ND	2,900
Octanal	ND	48	ND	ND	ND	ND	47	214
Nonanal	55	107	30	83	ND	92	139	618
Decanal	199	270	250	220	199	130	615	1,130
Undecanal	ND	ND	ND	ND	ND	ND	75	495
Dodecanal	ND	ND	ND	ND	ND	ND	14	155
Tridecanal	ND	ND	ND	ND	ND	ND	4	233
Tetradecanal	ND	ND	ND	ND	ND	ND	16	258

Source: Glaze et al. (1989a).
★pH = 7.7–7.8, TOC = 2.6 mg/L, turbidity <1 NTU, temperature = 18°C.
†Los Angeles Aqueduct Water, pH = 8, TOC = 1.6 mg/L, turbidity = 3.4 NTU.
‡ND = not detected.

water quality on by-product formation. Two different supplies were ozonated under similar conditions at the Los Angeles Aqueduct Filtration Plant. Results are given in Table III–56. When treating Los Angeles Aqueduct water, the concentrations of C_6 through C_{14} aldehydes increased significantly beyond their source water levels. When treating State Project water, however, fewer aldehydes were detected in both source and ozonated waters. Ozone dose dependency was also suggested in State Project water where aldehyde formation appeared to peak near a 1-mg/L dose.

Lykins et al. (1986) tracked octanal and nonanal through a pilot plant treating Mississippi River water. The sum of these two raw water constituents averaged 14 ng/L. Upon ozonation, their sum increased an average of 144 percent. Of interest was the fact that in a parallel pilot plant, the sum increased an average of 56 percent upon chlorination, indicating that aldehyde formation is not limited to ozonation.

Using a derivatization/pentane extraction/HPLC/UV method, Van Hoof et al. (1985) examined formation of C_1 through C_7 aldehydes in ozonated Belgian surface waters. Lower detection levels approached 1 μg/L. They found formaldehyde, acetaldehyde, propanal, and pentanal at concentrations below 8 μg/L as the principal products. Aldehyde concentrations increased with transferred ozone dose. The treated water, from which 50 percent of the TOC had been removed, yielded generally lower aldehyde formation than its companion ozonated raw water, supporting the premise that the humic and fulvic materials present in source water are the precursors for aldehyde formation.

Glaze et al. (1989b) modified a method presented by Yamada and Somiya (1989) that employed aqueous derivatization, extraction, and gas chromatography allowing for the quantitation of simple aldehydes at microgram-per-liter levels. Samples were taken from three water treatment plants and one pilot plant. Formaldehyde was found in all ozonated waters and in some raw waters as well. In raw waters, concentrations were below 4 μg/L; in ozonated waters, they reached 7 to 28 μg/L. At only one of these four locations were C_2 and higher aldehydes detected (see Table III–57). Because of the differences in by-product formation at this plant compared to the other three, these data support the contention that the make-up of the natural organic material in the source water plays a significant role in by-product formation. The plant producing the family of C_1 through C_7 aldehydes treated a relatively high-TOC surface water (near 8 mg/L). However, another plant treating a groundwater with a TOC also near 8 mg/L produced

Table III–57 Aldehyde Formation in a Full-Scale Ozonated Surface Water

Aldehydes	Concentration (μg/L) Raw*	Ozonated†
Formaldehyde	3.2	28.3
Acetaldehyde	ND‡	9.7
Propanal	ND	2.7
Butanal	ND	4.1
Pentanal	ND	15.2
Heptanal	4.6	3.7
Glyoxal	ND	13.0
Methyl Glyoxal	ND	28.3

Source: Glaze et al. (1989b).
*TOC = 8.4 mg/L.
†5.5 mg/L pre-O_3, followed by 1.5 mg/L mid-point O_3.
‡ND = not detected.

only formaldehyde near 4 μg/L. Based on TOC, comparable formation of aldehydes might have been expected. Glaze et al. hypothesized that high levels of hydrogen sulfide in the groundwater depleted the applied ozone, or more significantly, that sulfur compounds degraded the ozone by-products to produce unidentified secondary by-products. The issue of secondary by-products may be as important as that of primary by-products. In laboratory studies, Glaze et al. demonstrated the loss of formaldehyde with the addition of hydrogen sulfide or with sulfite, an oxidation product of sulfide. Samples were also taken in the distribution system of the plant treating a lower-TOC surface water. Formaldehyde concentrations found there were comparable to those leaving the plant, i.e., near 7 μg/L. Formaldehyde was also found in parts of the distribution system carrying water from plants that did not employ ozonation. Whether it resulted from chlorination (as did nonanal and octanal in Mississippi River water) or was a raw water constituent passing through the plant is unknown.

Jacangelo et al. (1989) reported similar results. In plants where ozonation was employed, summed concentrations of formaldehyde and acetaldehyde increased to as high as 57 μg/L. Free chlorine alone also produced these two aldehydes with concentrations reaching 13 and 11 μg/L, respectively. It could not be determined, however, if these concentrations were significantly different than those in the raw water.

McGuire (1989), like Jacangelo et al., utilized the aldehyde method of Glaze et al. (1989b). He surveyed 35 U.S. water plants, many of which do not treat with ozone. Based on sampling done over nine months, they reported formaldehyde and acetaldehyde medians ranging from 2 to 5 μg/L for formaldehyde and from 2 to 3 μg/L for acetaldehyde in finished waters.

In pilot studies treating Ohio River water, Miltner et al. (1990) also employed the Glaze et al. aldehyde method. Three parallel plants were studied: one employed preozonation with postchloramination, another preozonation with postchlorination, and the third postchlorination only. Data are given in Figures III–54 and III–55 and represent mean results based on daily sampling for a week. Figure III–54 clearly illustrates elevated levels of formaldehyde as a result of ozonation (ozone/TOC = 0.8), but subsequent loss of the compound through sedimentation and filtration processes. The loss was attributed to biodegradation. (See sec. K.3 in this chapter.) Clear well chlorination boosted formaldehyde levels in water that had or had not been ozonated. It was interesting to note that clear well chloramination also elevated formaldehyde levels, although the statistical confidence in this increase was lower than the confidence for the increase resulting from clear well chlorination. Figure III–55 illustrates similar formation of acetone and glyoxal resulting from ozonation or chlorination and similar biological control.

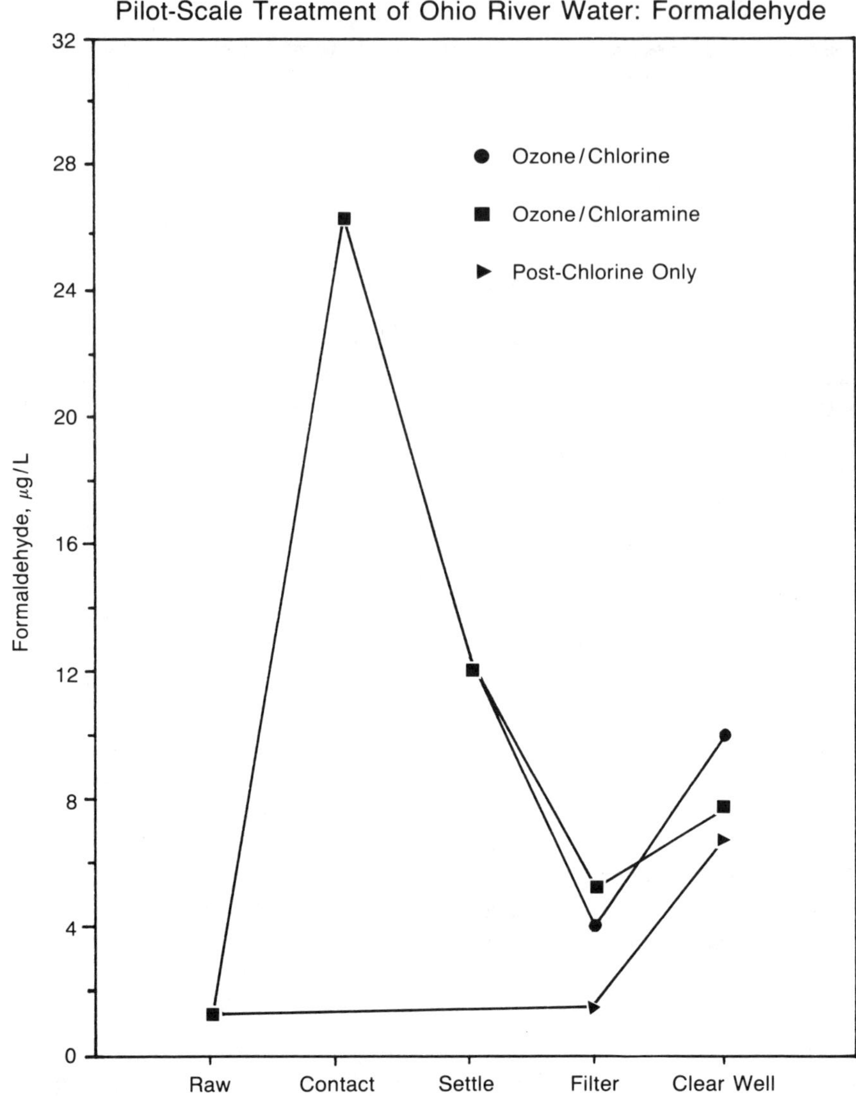

Source: Miltner et al. (1990).

Ozone at contactor = 1.93 mg/L applied; O_3/TOC = 0.8; contact time = 7.5 min; chlorine = 2.82 mg/L; chloramine = 1.52 mg/L; postchlorine only = 3.04 mg/L; clear well contact time = 8.5 h.

Figure III–54 Formation and Control of Formaldehyde

Glaze and co-workers (1989a) compared formaldehyde levels in ozonated State Project water and Los Angeles Aqueduct water. As with the C_6 and higher aldehydes, higher concentrations were found treating the latter source.

Formation of carboxylic acids. Concentrations of low-molecular-weight monocarboxylic acids ($\leq C_{13}$) (Glaze et al. 1989) and dicarboxylic acids (Anderson et al. 1985) have been shown to increase upon ozonation. However, this may not be true for many of the higher-molecular-weight acids, as this group contains many naturally occurring unsaturated fatty acids that are susceptible to ozone attack. Some of the low-molecular-weight acids may come directly from ozonation of the unsaturated higher-molecular-weight acids (Glaze et al. 1988; Reynolds et al. 1989); some come from the oxidation of aldehydes; and some are likely to be derived from humic materials or other natural organic matter of unknown composition. The formation of acids from aldehydes is supported by the observation that at high ozone doses, aldehyde

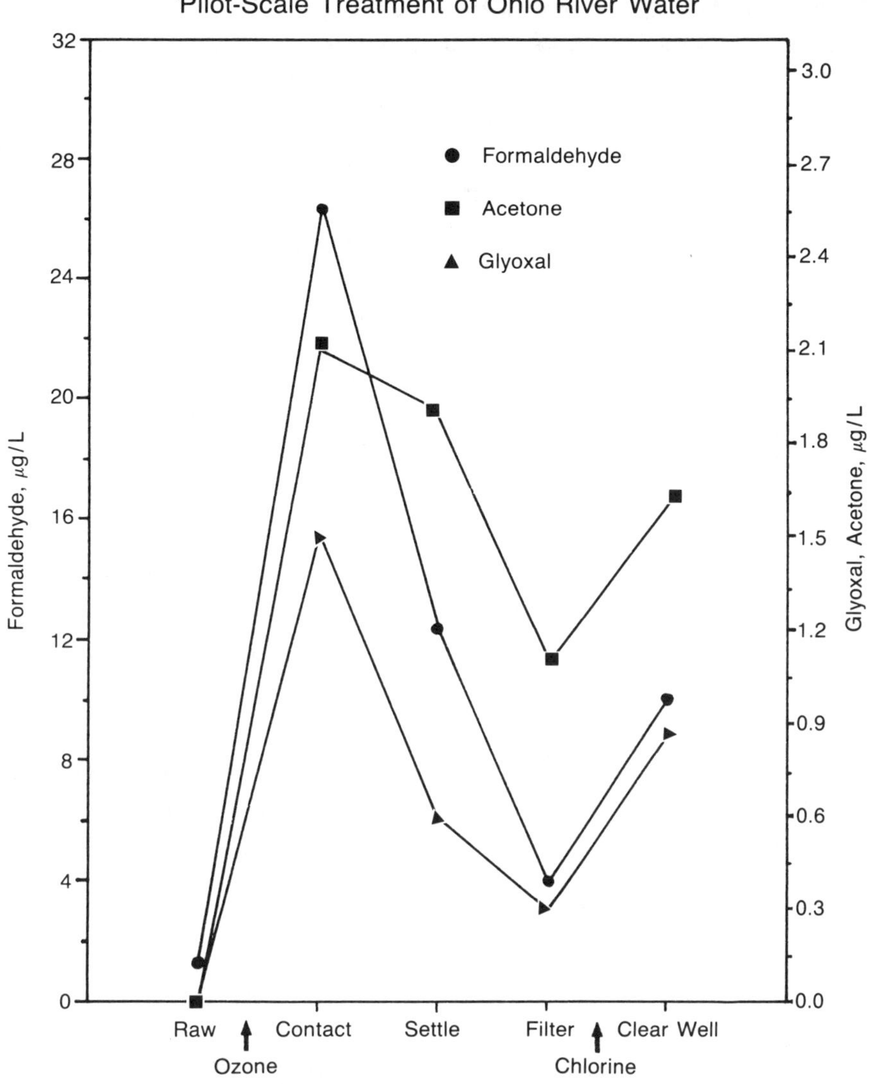

Source: **Miltner et al. (1990).**

Ozone = 1.93 mg/L applied, O_3/TOC = 0.8, contact time = 7.5 min; chlorine = 2.82 mg/L applied, contact time = 8.5 h.

Figure III–55 Formation and Control of Aldehydes

concentrations decrease (Schalekamp 1978; Glaze et al. 1988). As with the aldehydes, the formation of acids is highly dependent on the nature of the organic matter present in the water or the concentration of free radical scavengers (Glaze et al. 1989a).

Monocarboxylic acids were detected by Glaze et al. (1989a) using a derivatization and XAD resin accumulation method. C_9 through C_{24} compounds were found in both raw and ozonated Los Angeles Aqueduct and State Project waters. Because of the uncertainty in extraction efficiencies, only estimated concentrations were reported (Table III–58). The data suggest that concentrations of C_{13} and lower carboxylic acids may increase upon ozonation. Trends in changes in concentrations of naturally occurring fatty acids, i.e., C_{14}, C_{16}, C_{18}, and C_{20}, and of other higher-molecular-weight acids were less clear, however, given their extraction uncertainty.

One naturally occurring fatty acid, 9-hexadecenoic acid, was found to be completely destroyed by ozonation in the Los Angeles Aqueduct water. Because it contains

Table III–58 Carboxylic Acid Concentration in Ozonated Surface Waters

Carboxylic Acids	Estimated Concentration (ng/L)			
	LAAW*		SPW†	
	Raw	O_3	Raw	O_3
Nonanoic	ND‡	42	ND	ND
Decanoic	4	30	ND	46
Dodecanoic	48	73	180	180
Tridecanoic	6	34	ND	46
Tetradecanoic	314	235	90	770
Pentadecanoic	86	71	490	460
Hexadecanoic	793	341	1900	1600
Heptadecanoic	42	42	75	90
Octadecanoic	561	153	280	280
Nonadecanoic	8	11	ND	ND
Eicosanoic	112	35	ND	ND
Heneicosanoic	1	21	ND	ND
Docosanoic	29	37	ND	ND
Tetracosanoic	12	28	ND	ND
9-Hexadecenoic	433	ND	1200	690

Source: Glaze et al. (1989a).
*Los Angeles Aqueduct Water, pH = 8, TOC = 1.6 mg/L, turbidity = 3.4 NTU.
†State Project Water, pH = 7.7–7.8, TOC = 2.6 mg/L, turbidity <1 NTU, temperature = 18°C.
‡ND = not detected.

a double bond, Glaze et al. (1988) suspected it was a precursor for some of the aldehydes and carboxylic acids formed. In laboratory studies by Glaze et al. (1988), spiked solutions of this fatty acid were ozonated and yielded heptanal and heptanoic acid, presumably through the formation of an intermediate as part of the Criegee mechanism (Bailey 1978). While heptanal and heptanoic acid were the result of the reaction on one side of the double bond, 9-oxononanoic acid and nonanedioic acid were the result of the other side. In other laboratory studies, Glaze et al. (1988) ozonated raw water at varying doses. The naturally occurring C_9 through C_{11} aldehydes all increased in concentration up to a given dose (near 1 mg/L), after which their concentrations fell off. This is consistent with the trend in Schalekamp's (1978) data. Lower ozone doses produced higher levels of heptanal; higher ozone doses brought about lower levels of heptanal, presumably through conversion to heptanoic acid. These types of results have been substantiated in model systems using pure solutions of unsaturated fatty acids (Reynolds et al. 1989).

The above data suggest the presence of a corresponding acid for each aldehyde. Van Hoof et al. (1986) analyzed pilot plant samples for formic and acetic acid. They employed derivatization with HPLC and UV detection. The two acids were detected in both raw and ozonated waters. Ozone doses ranged from 0.2 to 1.5 mg O_3/mg C, and resulting acids in ozonated waters were as high as 23 and 46 μg/L, respectively. No comparisons were made between these acids and their companion aldehydes (Van Hoof et al. 1985). Further, while the concentrations of these two acids increased with ozonation in some source waters, they decreased in others.

The reason for differences in by-product formation when ozonating two different source waters under similar conditions at the same plant is not well understood. Formation of aldehydes was higher in Los Angeles Aqueduct water than in State Project water. Again, a relationship with TOC was not apparent, as the TOC of the State Project water (2.6 mg/L) was higher than that of the Los Angeles source (1.6 mg/L). While the precursors for these aldehydes are largely unknown, at least one, 9-hexadecenoic acid, is known. Table III–58 indicated that this unsaturated fatty acid was a constituent of both source waters. In the Los Angeles source, while it was of lower estimated concentration, its loss was complete upon ozonation. Conversely, in the State

Project source, its estimated concentation was higher, but it did not appear to be fully oxidized. The Creigee mechanism, wherein this unsaturated acid produces aldehydes, is one involving the direct ozone pathway. It is interesting to note that the State Project water contained relatively less total alkalinity, i.e., near 86 mg/L, as $CaCO_3$. Glaze et al. (1989a) hypothesized that the higher concentration of hydroxyl radical scavengers in the Los Angeles source, i.e., 120 mg/L total alkalinity as $CaCO_3$, favored the presence of more molecular ozone in its treated water than would have resulted in treated State Project water; therefore, a Creigee-type reaction with compounds like 9-hexadecenoic acid was favored in the Los Angeles source and resulted in relatively higher aldehyde production.

Formation of other ozonation by-products. With the Creigee mechanism, organic peroxides are anticipated intermediates. Glaze and Wang (1989) utilized a derivatization method coupled with HPLC and fluorescence detection to search for these intermediates following ozonation of 9-hexadecenoic acid. They found an unidentified organic peroxide that was moderately stable in distilled water but unstable in Los Angeles Aqueduct water. With the loss of the compound in distilled water, hydrogen peroxide formed. When the Glaze and Wang method (1989) was employed at the Los Angeles Aqueduct Filtration Plant, no organic peroxides were detected above 0.15 μg/L in ozonated or finished waters (Liang et al. 1989).

Liang et al. searched for hydrogen peroxide (H_2O_2) downstream of the ozone contactor at Los Angeles. In the contactor effluent, H_2O_2 was at 34 μg/L when the applied ozone was 1.05 mg/L. Only 6 μg/L H_2O_2 was detected when applied ozone was 4.4 mg/L. The H_2O_2 was short-lived, however, as it was not detected in the effluent from the downstream unchlorinated direct filtration process under either dose condition.

Using an XAD method, Glaze et al. (1988, 1989a) identified other by-products in ozonated drinking waters. These included mono- and diketones, phthalates, alkanes, and phenols. These were also present at trace or estimated nanogram-per-liter concentrations in source waters.

Analytical methods for ozonation by-products. As noted, the database for understanding the formation of ozone by-products and their control in downstream processes is relatively small and is limited by the availability of analytical methods to quantitate the compounds at microgram or nanogram levels that might be expected based on fundamental organic chemistry. In the studies presented in sec. III.I.2, several methods were employed for aldehydes (Schalekamp 1978; Van Hoof et al. 1985; Lykins et al. 1986; Yamada and Somiya 1989; Glaze et al. 1989a, 1989b), for organic peroxides (Glaze and Wang 1989), and for other by-products such as ketones (Glaze et al. 1988, 1989b). Depending on their application, these methods were targeted at a few specific compounds or at a broad spectrum of compounds. The choice of methods for future studies of ozonation by-products may not be limited to those noted here, as method development studies are underway for epoxides, organic peroxides, organic bromides, and other theoretical by-products of interest (AWWA Research Foundation 1989).

III.I.3 Primary Effects on General Disinfection By-product Formation

Reactions of ozone with pre-formed disinfection by-products. It is well established that ozone will not react directly with trihalomethanes (THMs) within a reasonable period of time. However, the process of ozone contacting may result in small losses of THMs due to volatilization. Other chlorination by-products besides the THMs may be more susceptible to ozone attack. From pilot studies on Seine River water, Bruchet et al. (1985) found that low doses of ozone can directly reduce instantaneous TOX concentrations in partially treated water by as much as 50 percent.

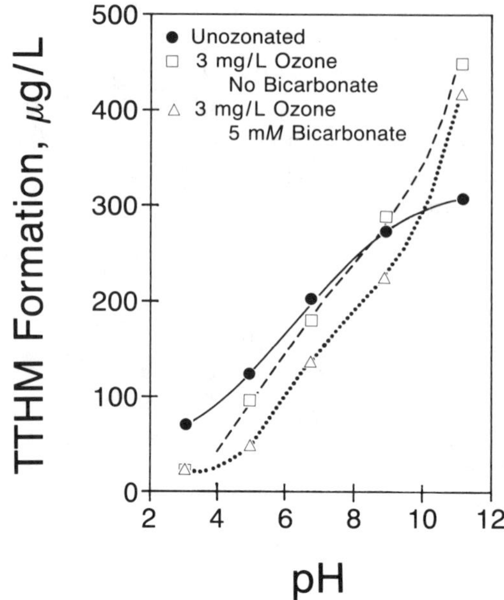

Source: Reprinted with permission from Reckhow, D.A. et al., *Wtr. Res.*, 20:8:987–998. © 1986 Pergamon Press PLC.

Black Lake fulvic acid; chlorination conditions: 20 mg/L dose, 3 days.

Figure III–56 Impacts of Ozonation on THM Precursors: Importance of pH and Bicarbonate

Primary effects on disinfection by-product precursors.

Reactions of ozone with THM precursors. Early work on oxidation of THM precursors seemed to indicate that the effects of preozonation were highly site-specific and unpredictable (for example, see Umphres et al. 1979). As more controlled laboratory studies were conducted, a clearer picture of ozone's effects on THM precursors emerged (Riley et al. 1978; Brunet et al. 1980; Reckhow and Singer 1984). The key variables that seem to determine ozone's effect are the ozone dose, pH, alkalinity, and nature of the organic matter. Figure III–56 illustrates the importance of pH and alkalinity. Note the characteristic base-catalyzed formation of THMs in the absence of preozonation. When the water is preozonated, the THM formation curve is significantly altered. At low pH's of chlorination, precursor destruction by ozone is quite effective; however, above some critical pH, net precursor enhancement is observed. The exact value of this critical pH depends on the alkalinity and on the nature of the organic material. Most aquatic humic substances do not show precursor enhancement at pH's of 7.5 or below. However, one notable exception is the Les Landes fulvic acid extracted from waters of the Bordeaux region of France (Legube et al. 1985; see sec. II.A.6).

The presence of HCO_3/CO_3 is expected to induce a shift in the ozonation mechanism from the nonselective OH radical attack to the highly specific molecular ozone attack (refer to chapter II, sec. A). The fact that bicarbonate has a beneficial effect on THM formation potential (THMFP) suggests that direct molecular ozone attack is either more selective for THM precursors than is the indirect, radical pathway, or it forms fewer precursors as by-products. Figure III–57 summarizes data from three independent studies on the ozonation of natural aquatic fulvic acids. Malley and co-workers (1986) used pH 7–buffered (1 m*M* bicarbonate) waters from the highly colored Provencial Brook in Princeton, Massachusetts. Reckhow and co-workers (1986b) used pH 7–buffered solutions (at various bicarbonate concentrations) of extracted fulvic acid

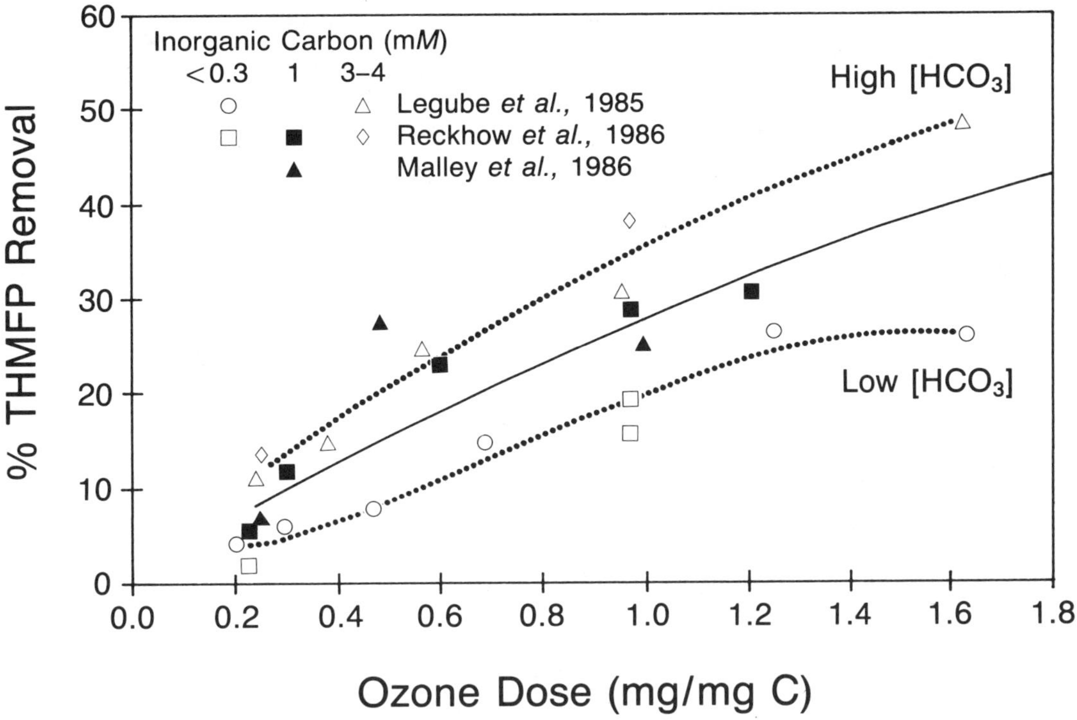

Source: Reckhow and Sibony (1986).

Figure III–57 Summary of Laboratory Results on Direct Destruction of THM Precursors by Ozone

from Black Lake near Elizabethtown, North Carolina. Legube et al. (1985) used similar extracts obtained from a pond near Poitiers, France and buffered at pH 7.5. Note from Figure III–57 that these data compare quite well when plotted as a function of ozone/carbon dose and bicarbonate concentration. This suggests that natural aquatic fulvic acids from these geographically distant areas behave in a similar fashion. Given neutral pH's and moderate levels of bicarbonate, one might expect to observe THMFP reductions of 3 to 20 percent at the ozone doses most commonly applied in drinking water treatment. Data from six full-scale U.S. plants suggests that these laboratory data may accurately reflect the actual precursor reductions (Figure III–58).

Ozone also seems to change the relative rate at which a water forms THMs (see Riley et al. 1978) and total organic halides (TOX) (Hubbs and Holdren 1986). This may mean that ozone can reduce THMs and TOX at the consumer's tap to a greater extent than would be predicted based on THMFP and TOX formation potential (TOXFP) reductions. Formation potential tests are often conducted under extreme conditions (high chlorine dose, long contact time); as a result, they may be insensitive to changes in the rate of rapid by-product formation.

Although it is well known that ozone itself will not produce chlorinated organic by-products, small amounts of bromoform may be produced in the presence of bromide (Figure III–59). This is a result of ozone's ability to oxidize bromide to bromine. This probably does not occur to a significant extent at low O_3/TOC ratios (e.g., <0.5 mg/mg) where organic matter can effectively compete for the ozone (Doré et al. 1988). However, when higher doses are used, and the highly reactive organics become exhausted, oxidation of bromide may occur, leading to the formation of instantaneous

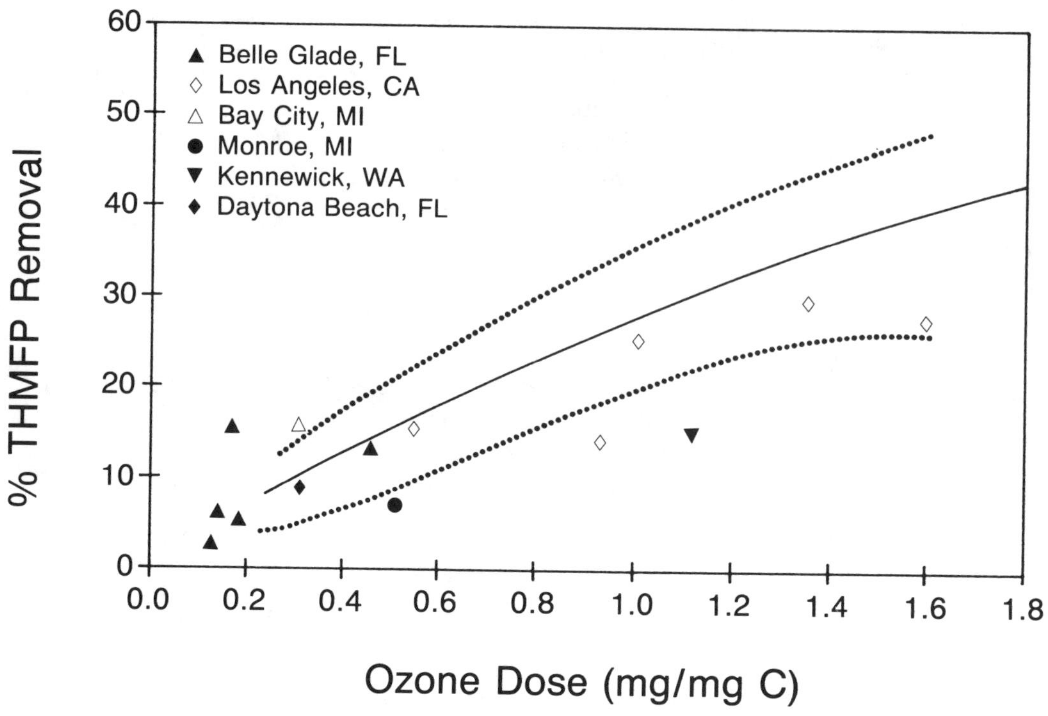

Sources: Singer (1989); Georgeson and Karimi (1988).

Figure III–58 Summary of Full-Scale Results on Direct Destruction of THM Precursors by Ozone

bromoform. For example, California State Project water containing 0.13 mg/L bromide gave 4 μg/L of bromoform after ozonation at doses of 1–4 mg/L (Glaze et al. 1989a). When this is the case, the addition of ammonia prior to ozonation may effectively control bromoform formation (James et al. 1988). As suggested above, many waters having a significant ozone demand will consume all of the transferred ozone before oxidation of bromide can occur. However, in the process, ozone may be satisfying some of the future rapid chlorine demand. Later, when chlorine is applied, it will react more slowly with the organic matter, and the oxidation of bromide will occur to a greater extent. Thus, ozonation may indirectly lead to a shift in THMs toward the brominated forms (Kruithof et al. 1985; Singer 1988b).

This enhancement of brominated THMs upon ozonation was exhibited in pilot studies of Ohio River water conducted by Miltner et al. (1990). Figure III–60 illustrates the means of daily samples collected over a week's time. For waters where chlorine was the only oxidant, there was no significant change in dibromochloromethane precursor levels as a result of conventional treatment. Three-day storage of finished water resulted in instantaneous dibromochloromethane levels near to the raw water formation potential level. With preozonation, however, precursor levels increased 31 percent and with clear well chlorination and 3-day storage, instantaneous levels exceeded the raw water formation potential level. For chloroform, however, conventional treatment lowered precursor levels 33 percent. The reaction of ozone with chloroform precursors is demonstrated in Figure III–61 where precursor removal improved to 39 percent. This resulted in lower instantaneous levels upon clear well chlorination in finished and 3-day stored waters. Because THMs are regulated as a group, finished water instantaneous TTHMs were lower as a result of preozonation. In high-bromide waters, however, this would likely not be the case. (In these studies, terminal levels resulted from exhausting

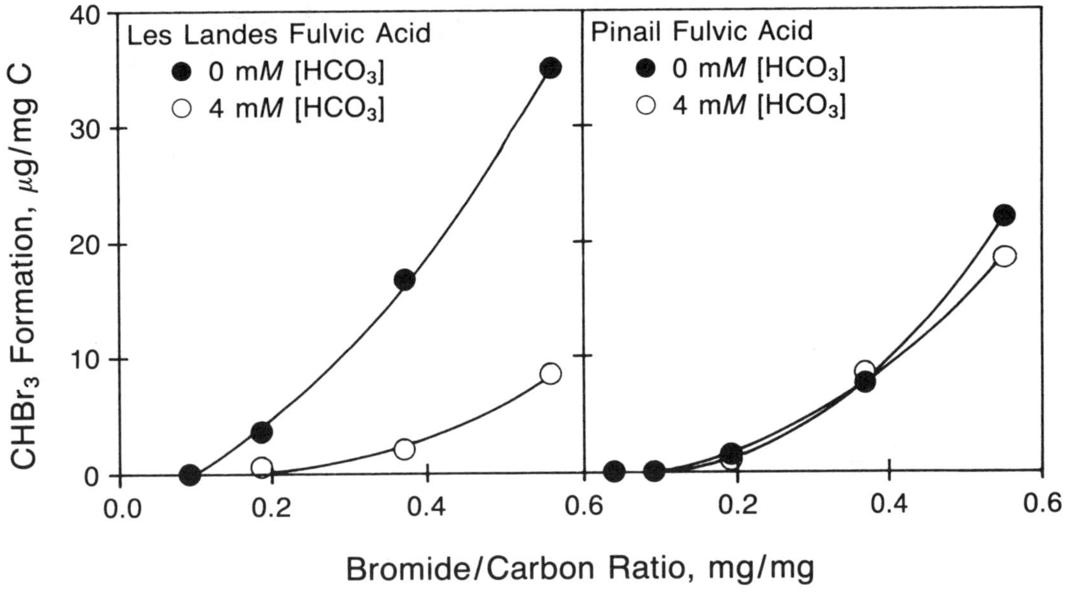

Source: Croué (1987).

Ozone dose equals 1 mg/mg C.

Figure III–59 The Ozone-Induced Formation of Bromoform (Les Landes Fulvic Acid)

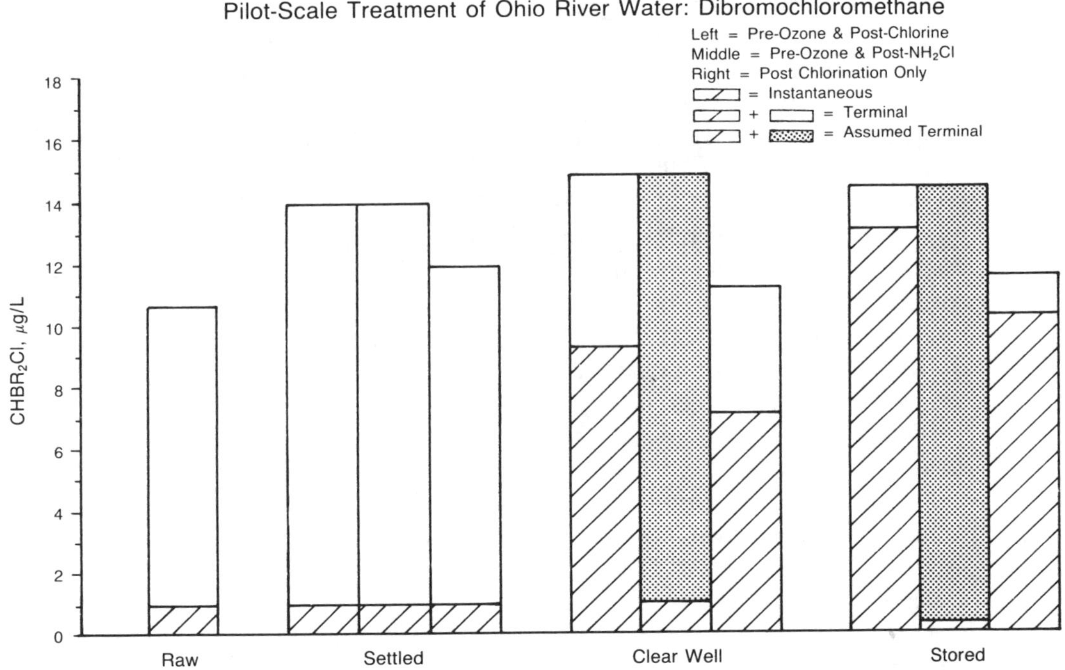

Source: Miltner et al. (1990).

For oxidant conditions, see Figure III–54; terminal conditions = 12 mg/L chlorine dose, 7 days, pH 8.6–9.0, 20°C; storage conditions = 3 days, no chlorine added, 20°C.

Figure III–60 Formation and Control of Dibromochloromethane

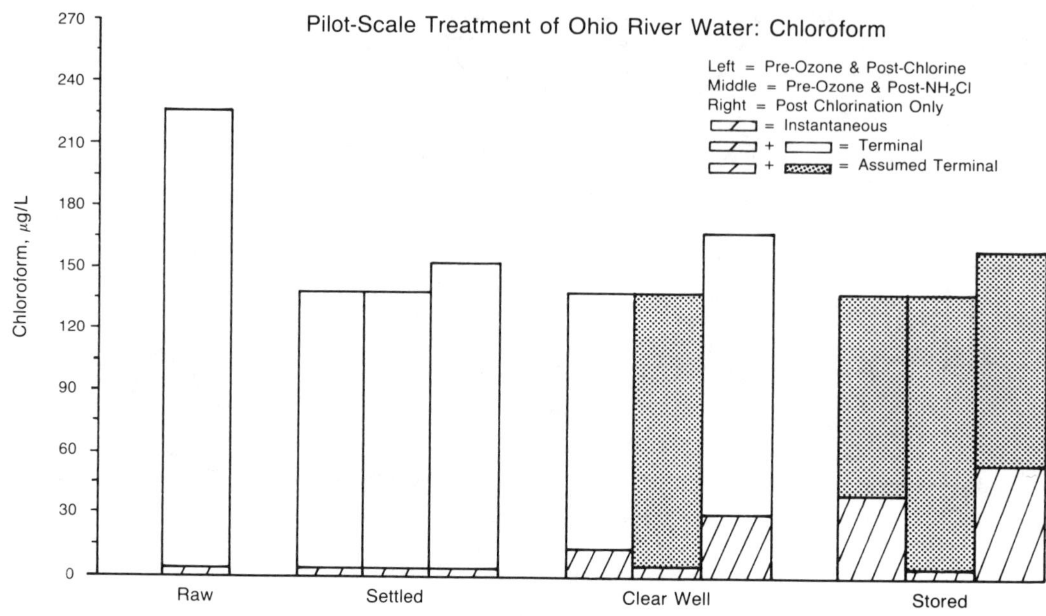

Source: Miltner et al. (1990).

For oxidant conditions, see Figure III–54; for terminal and storage conditions, see Figure III–60.

Figure III–61 Formation and Control of Chloroform

precursors by driving the reactions to completion over 7 days with 12 mg/L free chlorine dose. Assumed terminal levels are given where no terminal samples were collected.)

Reactions of ozone with other chlorination by-product precursors. Ozone has been found to readily destroy chlorine-produced TOXFP and precursors to some other specific organic halides (e.g., trichloroacetic acid, or TCAA) in the laboratory. The relative precursor removals for some of these are similar to those observed for THMFP (Reckhow and Singer 1984; Legube et al. 1985; see Figure III–62). However, several types of precursors do not follow the general trend set by THMFP. Highly reactive precursors such as dichloroacetonitrile formation potential (DCANFP) may be more easily destroyed (Reckhow and Singer 1984). The dichloroacetic acid precursors (dichloroacetic acid formation potential, or DCAAFP) appear to be relatively unaffected by preozonation. Finally, at least one haloketone precursor (i.e., 1,1,1-trichloroacetone formation potential, or TCACFP) is generally enhanced by preozonation (Reckhow and Singer 1984; Jacangelo et al. 1989)

The effects of pH and bicarbonate on the destruction of other precursors by ozone are qualitatively similar to the observations for THMFP. That is, ozone is generally most effective at destroying precursors at low pH, and high levels of bicarbonate seem to result in greater overall precursor destruction. Consider, for example, the TOX data in Figure III–63. Here TOX formation in the unozonated control goes down as pH increases, a behavior that is quite different from the THM case. Nevertheless, the greatest degree of precursor destruction occurs at low pH, and some enhancement may be seen at high pH. Recent pilot-scale and full-scale data substantiate the neutral-pH laboratory results on destruction of TOXFP by ozone (Lykins et al. 1986; Singer 1989; U.S. EPA 1989b).

In contrast to most of the precursors discussed above, chloropicrin precursor concentrations were thought to be enhanced by preozonation at neutral pH (Becke et al.

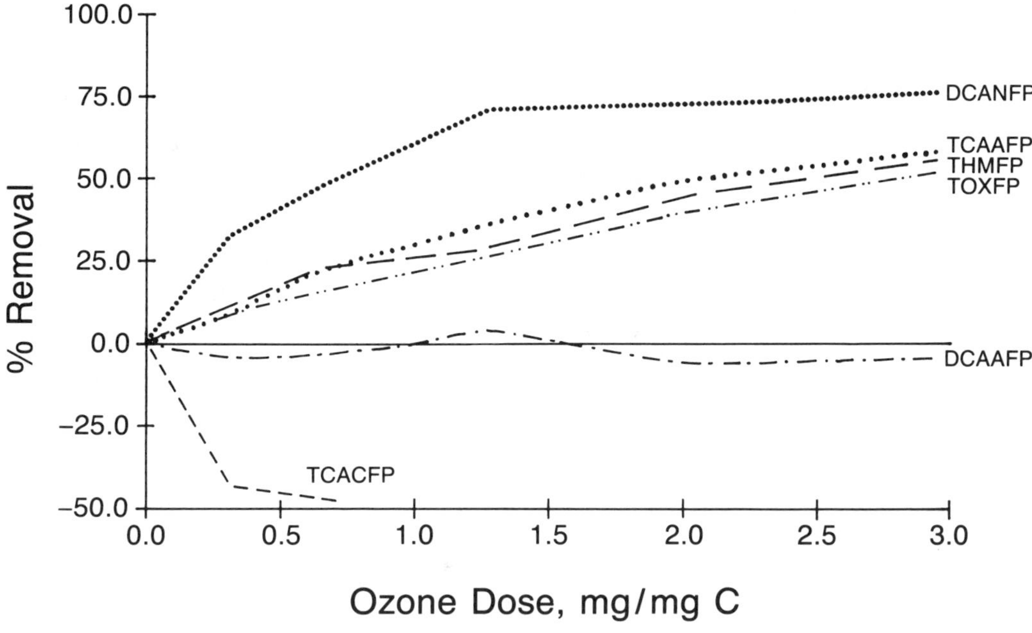

Source: Reprinted with permission from Reckhow, D.A. et al., *Wtr. Res.*, 20:8:987–998. © 1986 Pergamon Press.

Black Lake fulvic acid; pH 7, 20 mg/L HOCl dose, 3 days. DCANFP, dichloroacetonitrile formation potential; TCAAFP, trichloroacetic acid formation potential; THMFP, trihalomethane formation potential; TOXFP, total organic halide formation potential; DCAAFP, dichloroacetic acid formation potential; TCACFP, trichloroacetone formation potential.

Figure III–62 The Effect of Ozone Dose on Various Chlorination By-product Precursors

1984; Hoigne and Bader 1988). Although the precise structure of these precursors is unknown, they are believed to contain nitrogen. While amines have been shown to form chloropicrin, nitroorganics give much higher yields (Merlet et al. 1985). Preozonation may enhance chloropicrin formation by (1) oxidizing amines, forming nitroorganics; or by (2) formation of highly reactive nitrogen pentoxide in the ozone generator, which may lead to nitration of organic matter at the bubble interface. The formation of nitrogen pentoxide from nitrogen gas has been observed in corona discharge generators using air. Thus, the use of pure oxygen may help reduce the formation of chloropicrin precursors. In pilot studies employing pure oxygen as a source to a corona-discharge-type generator, chloropicrin precursors increased significantly, suggesting oxidation of amines as the mechanism (Miltner et al. 1990). Figure III–64 illustrates the effect of preozonation (0.8 mg O_3/mg TOC) on precursors by comparing raw and settled water terminal levels. Also interesting to note is the continued increase in chloropicrin precursors occurring downstream of sedimentation, hours after the ozone had dissipated. This suggests reactions initiated by ozone but subsequently driven slowly by intermediates. As was the case for dibromochloromethane, instantaneous 3-day stored levels exceeded raw water terminal levels. Without ozonation, chloropicrin precursors were reduced.

Reactions of AOPs with THM precursors. Various combinations of ozone, hydrogen peroxide, and ultraviolet light are sometimes used for the in situ production of hydroxyl radicals. These combined oxidants are often referred to as advanced oxidation processes (AOPs). Factors such as the intensity of UV light, the peroxide-to-ozone ratio, the pH, and the presence of radical scavengers strongly affect such processes (Glaze et al. 1987; Paillard et al. 1987; see sec. II.A.2 and III.E.2). Earlier it was

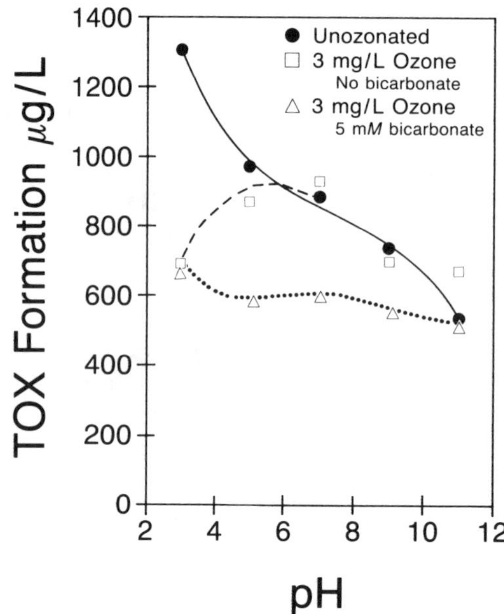

Source: Reprinted with permission from Reckhow, D.A. et al., *Wtr. Res.*, 987–998. © 1986 Pergamon Press.

Black Lake fulvic acid; chlorination conditions: 20 mg/L dose, 3 days.

Figure III–63 Impacts of Ozonation on TOX Precursors: Importance of pH and Bicarbonate

established that molecular ozone is more selective than hydroxyl radicals for highly reactive humic-associated THM precursors. However, it may be desirable to remove some of the slow-reacting THM precursors (i.e., certain ketonic structures). In order to oxidize these structures within a reasonable contact time, it might be necessary to employ hydroxyl radicals. Thus, AOPs could be useful for THM control, especially following the removal of reactive precursors (e.g., after uncatalyzed ozonation or coagulation). Pilot-scale work has indicated that UV light can significantly enhance the ability of ozone to remove THM precursors, especially at high ozone doses (Glaze et al. 1982; Sierka and Amy 1985). Sierka and Amy (1985) report that at high ozone doses, THMFP destruction may be more efficient when concentrations of bicarbonate are lower due to an improvement in ozone contactor efficiency or by accelerating the chemical kinetics. The ability of bicarbonate to retard the ozone/UV oxidation of simple organic molecules is well documented (Bourbigot et al. 1985a). In a fashion analogous to UV, the addition of hydrogen peroxide can also enhance the ability of ozone to remove THM precursors at high ozone doses (Duguet et al. 1985a; Wallace et al. 1988). Results with AOPs can sometimes be ambiguous due to improved ozone transfer under conditions of rapid decomposition of aqueous-phase ozone. The effect of ozone decomposition on gas transfer may be especially pronounced in small bench-scale systems where ozone transfer is commonly less than 80 percent. In full-scale systems, where transfer efficiency is already very good, any increase in transfer kinetics can do comparatively little to improve the overall efficiency. For this reason, lab-scale transfer rates should be carefully documented.

Reactions of ozone with chloramine by-product precursors. Jacangelo et al. (1989) have shown that ozone can partially destroy precursors that react with monochloramines to form trihalomethanes, haloacetonitriles, haloacetic acids, and TOX. On the other hand, ozone may enhance the formation of haloketones and chloropicrin upon subsequent chloramination. These effects are qualitatively similar to those reported for chlorination by-products.

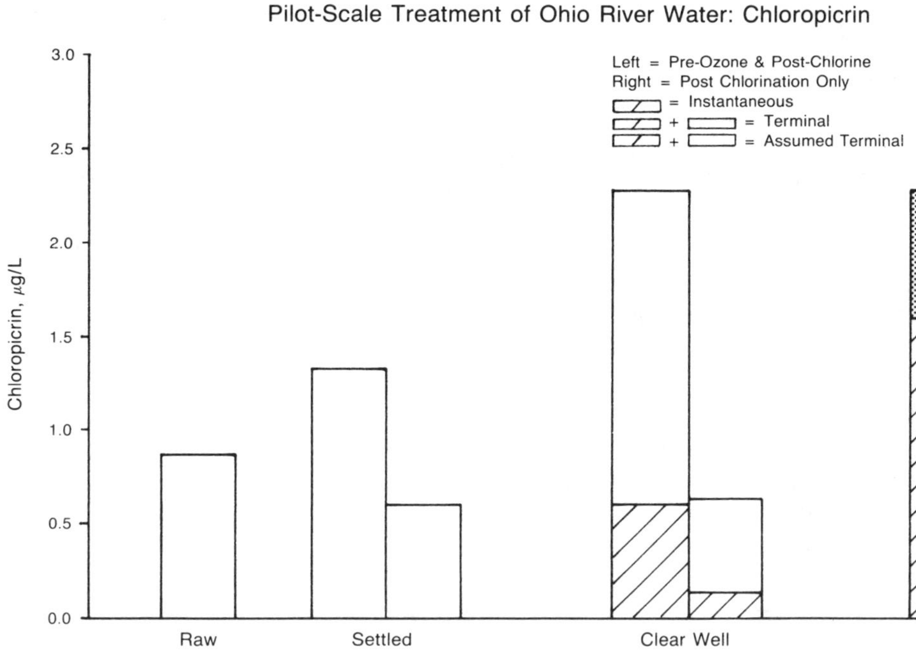

Pilot-Scale Treatment of Ohio River Water: Chloropicrin

Source: Miltner et al. (1990).

For oxidant conditions, see Figure III–54; for terminal and storage conditions, see Figure III–60.

Figure III–64 Formation and Control of Chloropicrin

III.I.4 Secondary Effects of Preozonation on Disinfection By-products

Effect of ozone on subsequent processes.

Ozonation prior to coagulation. In addition to the ability of ozone to immediately oxidize some organohalide precursors, it may also improve the removal of organic precursors in subsequent coagulation (Chedal 1985). The reasons for this have not yet been established; however, there exist many plausible explanations (see sec. III.F.1). A recent study at the Choisy-le-Roi treatment plant in the Paris suburbs suggests that preozonation can improve the coagulation of THM precursors to the same extent as it does that of DOC, or to a greater extent (Gerval et al. 1985; see sec. III.F.2). Some pilot plant data from the United States indicate that preozonation may also significantly improve the removal of THM precursors by subsequent precipitative softening and direct filtration (Wagner and Elefritz 1983; Prendiville and McBride 1983). Despite widespread success with ozone as a coagulant aid for the removal of turbidity, the effects of ozone on organics removal are still quite controversial. As a result, one should look critically at ozone/coagulation pilot data on organics removal. In particular, it is important not to presume that improvements in organics removal will naturally follow improvements in particle removal.

Ozonation prior to GAC. Ozonation prior to granular activated carbon (GAC) adsorption sometimes leads to lower steady-state TOC concentrations following breakthrough. This applies to disinfection by-product precursors as well. Research has shown that the improved TOC removal is a result of increased biological activity on the GAC that occurs when influent water is ozonated and the organics are rendered more biodegradable (Brunet et al. 1982; Bourbigot et al. 1985b; see sec. III.K). Others suggest that preozonation acts by reducing the molecular size of the bulk organic matter. This would help accelerate the slow diffusion of solutes into the micropores. A recent study suggests that preozonation does improve the adsorption of high-molecular-weight organics, but hinders the adsorption of the low-molecular-weight fraction (Maloney et al. 1985). Since the higher-molecular-weight organics tend to be richer in THM and TOX precursors per carbon atom, one might presume that ozonation improves the removal of THMFP and TOXFP by GAC to a greater extent than it improves the removal of TOC.

A few researchers have reported data on precursor removals by ozone-GAC. One pilot study showed improved removal of TOX and mutagenicity by ozone-GAC as compared to GAC alone (Van der Gaag et al. 1985). Other pilot studies in the Netherlands have shown that preozonation significantly reduced THMFP in GAC-treated waters (Kruithof et al. 1985). However, this additional precursor removal may have been due solely to immediate oxidation by ozone. Similar effects were observed for Extractable Organic Halide (EOX) and Adsorbable Organic Halide (AOX). A recent laboratory study using GAC mini-columns demonstrated a significant improvement in DOC, THMFP, and TOXFP adsorption following preozonation (Malley et al. 1988). This beneficial effect of preozonation was found to increase with increasing ozone dose, and it was attributed to the accelerated adsorption kinetics that would be expected from a reduction in the molecular weight of the organics. One must be careful, however, that such effects do not improve the competitive adsorption of the bulk organics and reduce the capacity of the activated carbon for target micropollutants (e.g., pesticides).

Ozonation prior to slow sand filtration. Slow sand filters generally have a thick layer of microbiota, called the schmutzdecke, on the upper sand surface. This contains a diverse community of microorganisms including bacteria, protozoa, rotifers, algae, and other groups. Filter performance is closely linked to the activity of the schmutzdecke, and for this reason slow sand filtration is considered to be a biological as well as a physical process. Ozonation prior to slow sand filtration can lead to significant THMFP and TOC removals even when sand filtration or ozonation alone does not have an important effect. This occurs presumably as a result of increased biological activity on the sand when the influent water is ozonated and the organics are transformed into a more biodegradable form (Symons et al. 1984). For a more complete discussion of this process refer to sec. K of this chapter.

Reactions of ozone to lower chlorine demand. At the same time ozone reacts with precursor material, it can react with those naturally occurring materials that exhibit chlorine demand. The combination of lower concentrations of both precursors and chlorine demand can result in lower formation of disinfection by-products, depending on the ozone conditions. It can also result in lower formation of taste and odor compounds, which may not have the health implications of the by-products but are certainly undesirable.

The efficiency of treatment processes for the control of materials that react with chlorine can be evaluated by different means: (1) by measuring the disinfection by-products formed, i.e., instantaneous THMs, as already discussed, (2) by measuring the formation potential of the by-product, i.e., THMFP, that could be formed, as already discussed, and (3) by measuring the chlorine demand, which will be discussed here. This last measurement is more suitable when the chlorine doses applied are small, i.e., when there is a low chlorine-to-TOC ratio. This method is best performed on water in which the concentrations of ammonia and reducing agents are minimal. Typical chlorine consumption curves are of the shape shown in Figure III–65 and exhibit (1) a very rapid chlorine consumption phase immediately after the disinfectant is added and (2) a decrease in the rate of the reaction after a contact time of about four hours. This follows, irrespective of the water studied, second-order kinetics, i.e., one mole of chlorine per mole of reaction site. The kinetics of these reactions are described elsewhere

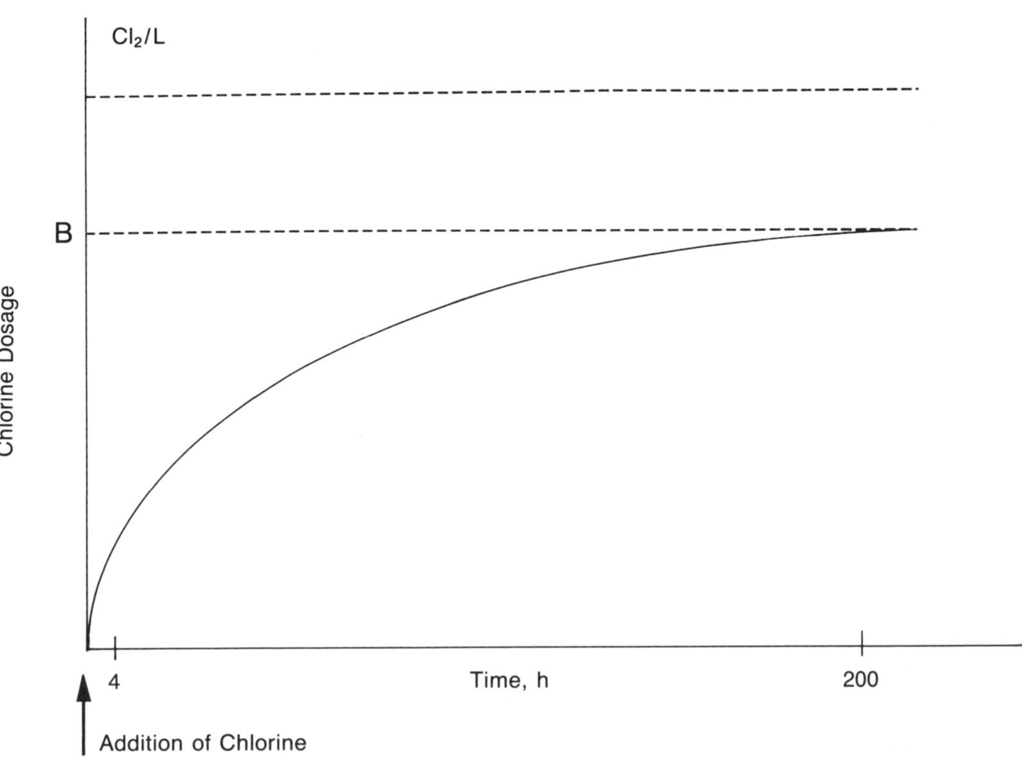

Source: Reprinted with permission from Ventresque, C. and Bablon, G., Proc. 9th Ozone World Congress, edited by J. Bollyky, pp. 207–222. © 1989 IOA.

Figure III–65 Typical Figure for Chlorine Demand Kinetics

(Jadas-Hecart et al. 1988; Ventresque et al. 1987), from which it is possible to define a simple maximum value called "B" that indirectly represents the concentration of sites susceptible to chlorine oxidation in a given water.

Ventresque and Bablon (1989) and Jadas-Hecart (1989) attempted to identify the products responsible for the chlorine demand in a water treatment plant in the Paris area. They found that chlorine-reactive sites could be roughly separated into two distinct groups (Figure III–66). One group consisted of low-molecular-weight molecules (<1000 daltons), the other of high-molecular-weight molecules (>1000 daltons). Thirty percent of the chlorine demand was due to humic substances and at least 30 percent appeared to be attributable to free or combined amino acids, the latter being highly reactive. These represented only a small fraction of DOC, but their chlorine demand was important. The remaining substances causing chlorine demand could not be identified. In order to reduce chlorine demand, then, treatment that lowers the concentration of the humic substances and amino acid–type compounds seems to be attractive. While these compounds are not readily adsorbed, amino acids and humic substances altered by ozonation become more biodegradable.

It is important to note that chlorine demand can be lowered by ozonation alone or by ozonation in combination with filtration on either inert media or GAC, wherein ozone-enhanced biodegradation can take place. Biodegradation of organic matter is discussed in sec. K of this chapter.

Control of chlorine demand by ozonation and biodegradation was studied at the Neuilly-Sur-Marne plant in suburban Paris. (The plant is described in Figure III–77). After ozonation, the water passed through two parallel GAC columns. One column was

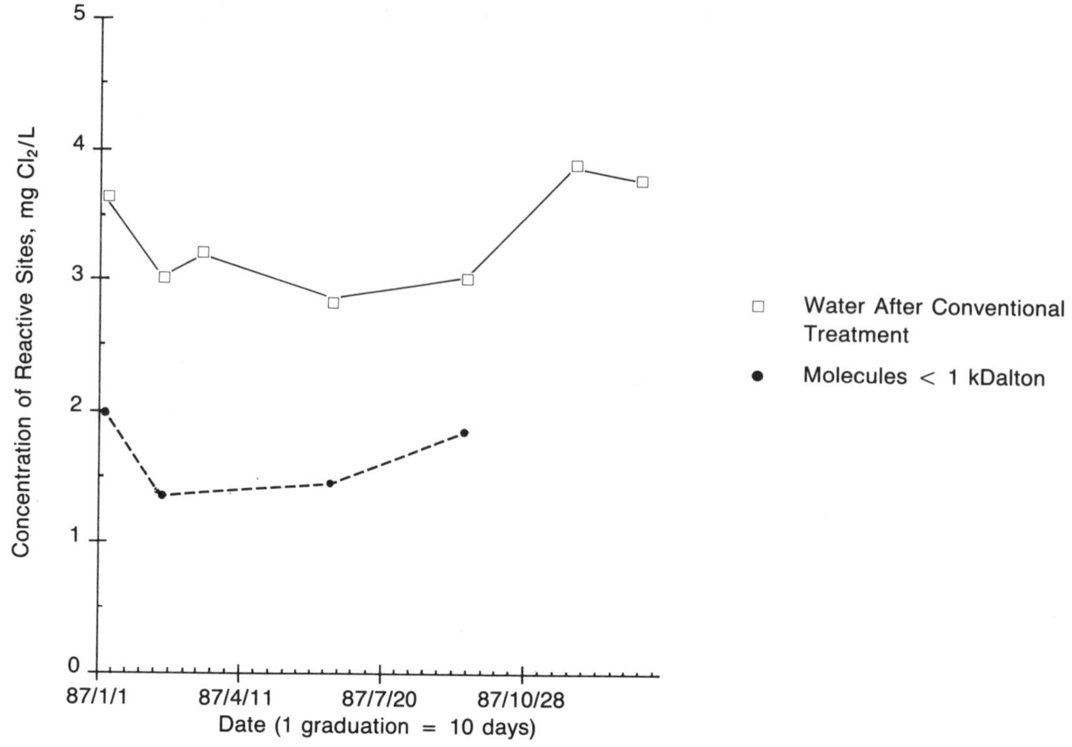

Source: Reprinted with permission from Ventresque, C. and Bablon, G., Proc. 9th Ozone World Congress, edited by J. Bollyky, pp. 207–222. © 1989 IOA.

Figure III–66 Effect of Ultrafiltration (1,000 daltons) on Chlorine-Reactive Sites Following Rapid-Rate Filtration

filled with fresh carbon, the other with saturated, five-year-old carbon that supported an active population of bacteria. At the beginning of the study, the fresh carbon was free from bacteria and the removal of chlorine-reactive substances was predominantly achieved by adsorption. Figure III–67 shows that the fresh carbon had little effect on the chlorine-reactive sites. The efficiency of this adsorption process initially reached a level of about 20 percent at a contact time of 10 min. Given the same contact time, the old carbon had a much more pronounced effect on chlorine-reactive sites. Then, as the fixed bacterial population in the fresh carbon increased, the chlorine demand was more effectively removed. The ozone/GAC process, sometimes termed biological activated carbon or BAC, appeared to be effective for the removal of chlorine-reactive substances.

The effectiveness of ozonation alone or of ozonation plus BAC was observed in studies at the Choisy-le-Roi plant in the Paris area. Until 1984, the plant employed conventional treatment with ozonation of filtered water. During 1985, BAC was added downstream of ozonation. Several benefits of BAC treatment are seen in Table III–59. First, chlorine consumption in the contact tank was reduced. The savings on chlorine were comparable to the energy costs required for ozone production. The dose of chlorine was cut by a factor of 2.5. The cost and risk reductions associated with the storage of less chlorine were considered important. Second, the stability of residual chlorine in the distribution system increased. Table III–59 shows the improved residuals recorded at different locations in the system. As a result, several of the booster chlorination stations in the system were shut down. Finally, the THM, TOX, purgeable organic halogen (POX), and chloroform in the water leaving the treatment plant (Table III–60) were greatly decreased, i.e., from 94 μg/L to 25 μg/L chloroform for ozonation alone and from 94 μg/L to 5 μg/L chloroform for ozonation and BAC. The formation potential of these same parameters was also decreased.

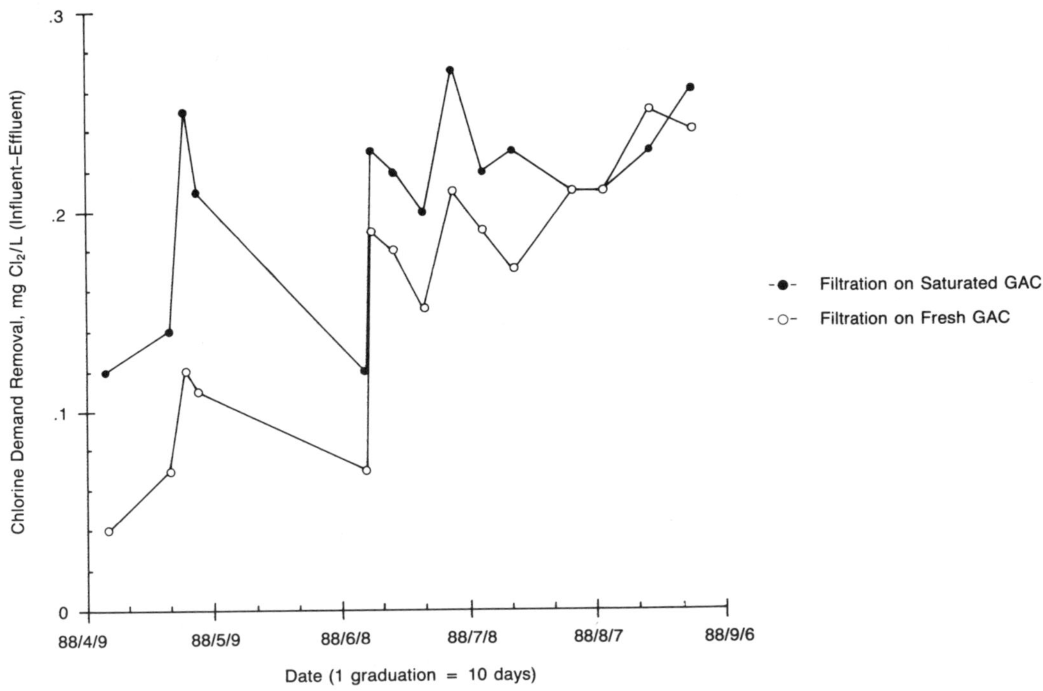

Contact time = 10 min, filtration velocity = 6 m/h.

Figure III–67 Comparison Between Adsorption and Biodegradation for the Removal of Chlorine Demand

Table III–59 Impact of Biological Filtration on Chlorination at Choisy-le-Roi

	Cl_2 (mg/L)	
Location	1984★ Conventional Treatment Plus Ozonation	1985† Additional BAC Treatment
Chlorine Consumption in Contact Tank‡	1.42	0.35
Chlorine Residual at Plant Effluent (After Dechlorination with Bisulfite)	0.45	0.42
Chlorine Residual at Massy-Anthony§	0.11	0.27
Chlorine Residual at Athis-Mons§	0.08	0.23

★Coagulation + settling + sand filtration + ozonation + chlorination + dechlorination.
†BAC placed between ozonation and chlorination.
‡2- to 4-h contact time before dechlorination.
§Massy-Anthony and Athis-Mons are distribution system reservoirs.

Chlorine demand was also investigated at the St. Rose plant in Laval, Canada. Both short-term (4-h) and long-term (7-day) demands were studied. While the reduction in chlorine demand by ozonation alone was not always significant, the reduction with both ozone and BAC was. (See Figure III–68.)

The effectiveness of ozonation alone to control chlorine demand may be site-specific. In drinking water treatment where dosages are rarely greater than 2 mg ozone per mg TOC, ozone may have only a limited effect on the concentration of chlorine reactive sites. Ventresque and Bablon (1989), using the approach described in Figure III–65,

Table III–60 Evaluation of Trihalomethanes at Choisy-le-Roi

	Concentration, (μg/L)*					
	In-plant Formation Test†			Formation Potential‡		
	SF§	SF + Ozone	SF + Ozone + BAC	SF	SF + Ozone	SF + Ozone +BAC
TTHM	135	54	20	310	237	127
Chloroform	94	25	5	186	126	60

*Mean values for 1988, 25 samples.
†Chlorinated at 1.5 mg/L for 4 h, followed by dechlorination, leaving a 0.4-mg/L residual.
‡6 mg/L added chlorine.
§SF = sand filtration.

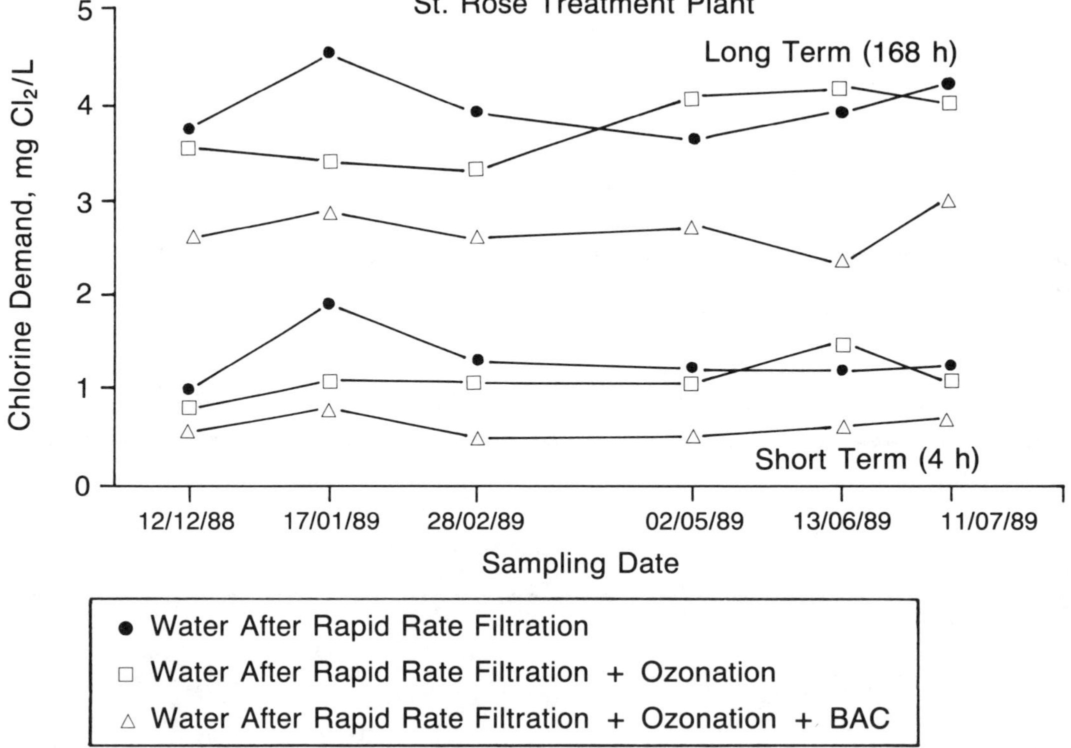

Figure III–68 Control of Chlorine Demand at Laval, Canada

examined ozone's effect on chlorine demand. Figure III–69 presents data that demonstrate, for the Choisy-le-Roi plant, the limited efficiency of ozone alone for reducing chlorine-reactive sites. Laboratory tests (Figure III–70) with the same water from the same plant demonstrated that increasing ozone dosage did not significantly reduce the concentration of chlorine-reactive sites. Similar results were achieved with fulvic acids extracted from the Seine River. In fact, when the extracts were low in bicarbonate, the concentration of chlorine-reactive sites slightly increased. This is in agreement with the results reported by Reckhow et al. (1986b) for ozonation of fulvic acid extracted from Black Lake, NC. In these studies, it appears that natural waters or extracted fulvic acids had similar chlorine reactivity before and after reaction with low doses of ozone.

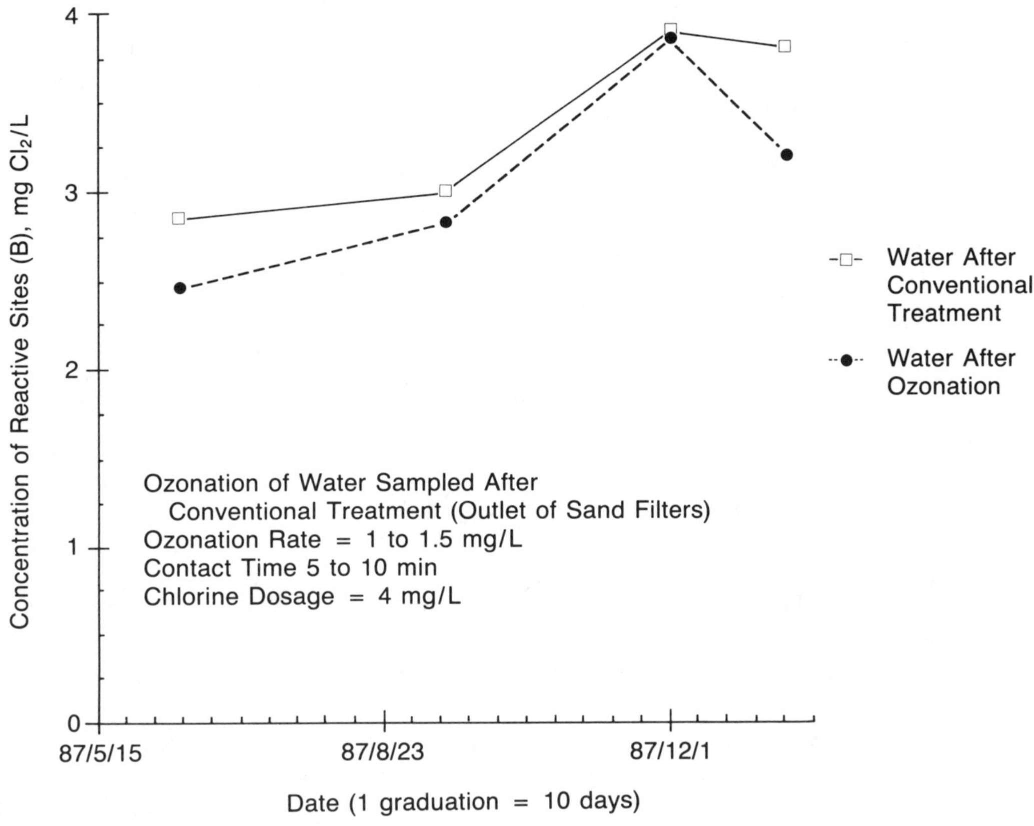

Source: Reprinted with permission from Ventresque, C. and Bablon, G. Proc. 9th Ozone World Congress, edited by J. Bollyky, pp. 207–222. © 1989 IOA.

Figure III–69 Effect of Ozonation on Chlorine-Reactive Sites Following Rapid-Rate Filtration

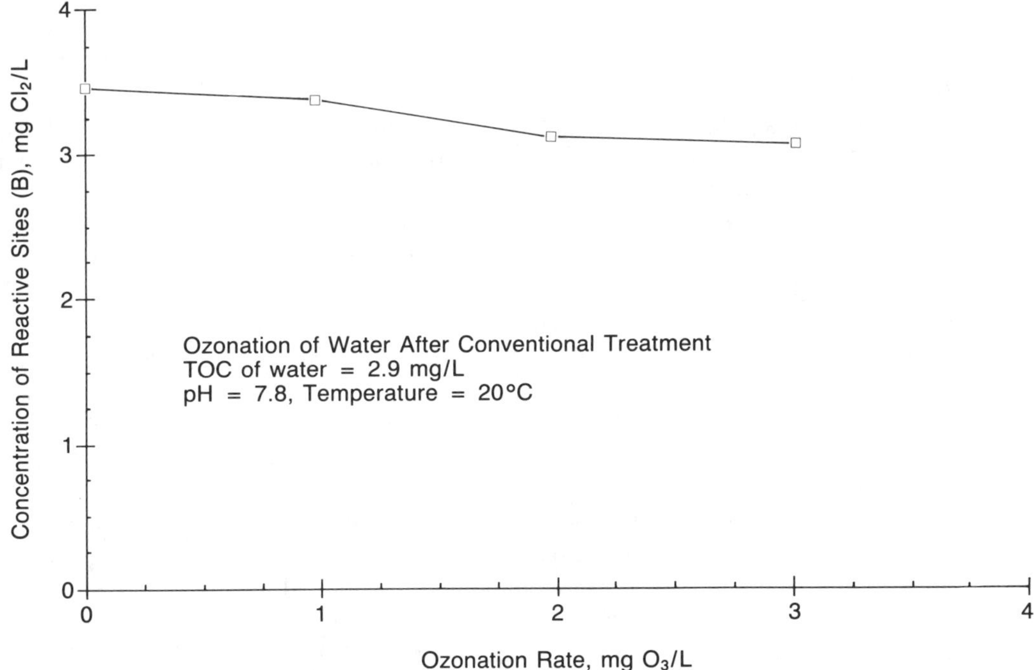

Source: Reprinted with permission from Ventresque, C. and Bablon, G., Proc. 9th Ozone World Congress, edited by J. Bollyky, pp. 207–222. © 1989 IOA.

Figure III–70 Effect of Ozone Rate on Chlorine-Reactive Sites Following Rapid-Rate Filtration

In other studies, the change in chlorine demand was more apparent. Miltner et al. (1990) studied conventional treatment of Ohio River water on parallel pilot plants. Both applied chlorine following filtration; one plant also ozonated the raw water. At the plant where only chlorine was applied, a mean dose of 3.04 mg/L was required to carry a 3-day mean free residual of 0.2 mg/L. At the plant where ozone was applied, a mean chlorine dose of only 2.84 mg/L was required to carry the same residual. Lower formation of TOX and many halogenated disinfection by-products resulted. (Representative data for chloroform are given in Figure III–61.) The lower formation of by-products at the ozonated plant may have been due to the lower level of chlorine that was required and/or to ozone's effect on precursors.

Preozonation in place of prechlorination. In many plants preozonation can be used in place of prechlorination as a primary disinfectant/oxidant. This allows one to (1) delay the application of chlorine, (2) reduce the chlorine dose, or (3) eliminate chlorination entirely in favor of an alternative secondary disinfectant. Shifting the point of chlorination can lead to reductions in by-product concentrations for three reasons. First, the in-plant chlorine contact time is reduced. Second, the lower TOC concentrations in the partially treated water permit the use of lower doses of chlorine. Third, disinfection by-product precursors are generally better removed in treatment than the by-products themselves. This is due to the higher molecular weight and lower oxidation state (lower acidity and hydrophilicity) of the precursors. Thus, by shifting the point of chlorination downstream, one is trading inefficient by-product removal for efficient precursor removal.

The overall effect of using preozonation in place of prechlorination should be thought of as the sum of the immediate effects of ozone on precursors plus the effects of changing chlorination practices. The effects of ozone on precursors have been discussed previously. The following sections, while they do not involve ozonation studies, discuss the effects of changing chlorination practices.

Reduction in prechlorine dose. A reduction in prechlorine dose can lead to reduced concentrations of most disinfection by-products in the finished water. This is due to slower by-product formation and the opportunity for improved precursor removal concurrent with the reaction between precursors and reduced chlorine levels. For example, full-scale results from the Walnut Creek, California water treatment plant suggested that by reducing the dose of prechlorine from 2.0 mg/L to 1.87 mg/L, the total THM concentration in the distribution system dropped from 116 to 94 μg/L. The plant's treatment train included alum coagulation followed by direct sand filtration (Carns and Stinson 1978).

Shift of chlorination from before prestorage to after. If chlorination is practiced prior to a long period of raw water storage, one would expect that much of the terminal THMs would be converted to instantaneous THMs. Thus, subsequent processes such as coagulation, filtration, and oxidation would have little impact on THM concentrations at the consumer's tap (because these processes will remove THMFP, but not preformed [i.e., instantaneous] THMs). Moving the point of chlorination to immediately after prestorage would allow for potentially significant precursor removal by subsequent processes as well as benefits from any precursor removal that might occur during prestorage. A pilot-plant study using Ohio River water has indicated that a reduction of 25 to 35 percent in product water THMFP can be obtained by making this shift (prestorage residence time was 2 days). These results prompted an analogous shift in the point of chlorine addition at the accompanying full-scale plant. The observed response was a 70 percent decrease in instantaneous THMs in the distribution system. Unfortunately, THMFP data were not reported for the full-scale change (Symons et al. 1981).

Shift of chlorination from before coagulation/settling to after. Shifting the point of chlorination from before coagulation/settling to after coagulation/settling will show a number of effects. Postchlorination allows one to benefit from the precursor removals

Table III–61 Effect of Shift in Chlorination on THMs

Plant	Percent Term* TTHM Reduction During Clarification	Influence of Moving Chlorination: Percent Inst† TTHM Reduction	Influence of Moving Chlorination: Percent Improved Term TTHM Reduction	Move Before/After
Cincinnati, OH	>33	39	0	Clarification
Pittsburgh, PA	26	54	0	Clarification
Wheeling, WV	16	32	0	Clarification
Contra Costa Water District, CA	–	23–37	–	Clarification
Bristol, RI	–	75	–	Clarification
Met. Water District of Southern Cal., CA	18–33	–	0	Prestorage/filtration
U.S. EPA pilot plant	–	–	40	Clarification
Cincinnati, OH	19	–	86	Prestorage
Durham, NC	–	28–40	–	Clarification
Daytona Beach, FL	41	10	0	Clarification
Daytona Beach, FL	41	33	23	Filtration
East Bay Municipal Utility District, CA	13	2	0	Filtration

Source: Symons et al. (1981).
*Term = terminal.
†Inst = instantaneous.

associated with coagulation processes as well as allowing for reduced chlorine contact time for those precursors not removed by coagulation. However, prechlorination followed by coagulation will also lead to reductions in precursors as compared to prechlorination alone, because (1) many precursors will not have been given enough time to react with the aqueous chlorine at this point, and a certain fraction of these will be removed during this process, thereby being rendered inaccessible to the chlorine; and (2) some of the higher-molecular-weight by-products may be directly removed by coagulation. Prechlorination may also affect the coagulation process itself in the same way other forms of preoxidation have been observed to impact this process (e.g., preozonation). Consequently, the effects of moving the point of chlorination can be complex and difficult to predict. The most reliable data on this process modification come from the full-scale plant studies. In some cases, marked reductions of instantaneous THM and THMFP in plant effluents have been noted (28 percent instantaneous THM reduction [Lange and Kawczynski 1978]; 40 percent instantaneous THM reduction [Young and Singer 1979]; and 40 percent and 85 percent instantaneous THM reduction [Gummerman and Heim 1983]), whereas other studies have failed to show a significant net effect (Cohen et al. 1978; ORSANCO 1980; Gummerman and Heim 1983). A similar change in the point of chlorination at the precipitative softening plant at Daytona Beach, Fla. also failed to show a significant change in THMFP (Symons et al. 1981). In softening plants, moving the point of chlorination to one following sedimentation typically involves a higher pH, which enhances THM formation. These types of full-scale studies are difficult to perform due to the many uncontrollable changes in the raw water and plant operation that can confound the interpretation of results.

Shift of chlorination from before sand filtration to after. In a pilot-plant study using Ohio River water, reductions in THMFP of 15 percent were noted when chlorination was shifted from pre–sand filtration to post–sand filtration (Symons et al. 1982). Reductions of 23 percent in THMFP were noted in the Daytona Beach plant, which also practices precipitative softening (Symons et al. 1982). Note that shifting the point of chlorination to postfiltration is necessary in order to induce fixed-film biodegradation in the filter (e.g., nitrification, carbon removal).

Table III–61 demonstrates the effect on THM levels as a result of shifting chlorination to a point further into the treatment train where the water is of a better quality. It summarizes both the studies described above, as well as others. These results, together with the reaction of ozone on precursors, suggest the overall effect on by-product control. This was demonstrated at the Belle Glade, Fla. water treatment plant.

III.I.5 Case Study: The Belle Glade Water Treatment Plant

The city of Belle Glade in south central Florida uses Lake Okeechobee as a source for its drinking water supply. This water has an extremely high and variable TOC (up to 75 mg/L) and a high level of hardness, and it is subject to periodic algal blooms. Through the early '80s, Belle Glade was treating this water with very high prechlorine doses, followed by lime softening (with alum addition), filtration, and postchlorination. Trihalomethane concentrations as high as 1000 μg/L were produced, taste and odor problems existed, and color removal was not complete.

The maximum contaminant level (MCL) for THMs forced the city to reconsider its water treatment practices. The city's consultant operated an onsite pilot plant for 12 weeks and examined the following process changes: (1) use of ozone (3-4 mg/L) prior to softening, (2) use of ozone immediately prior to filtration, and (3) shifting of the point of chlorination to postfiltration (Wagner and Elefritz 1983). This pilot treatment train gave effluent instantaneous THM concentrations that were about 20 percent of those observed for the full-scale plant during the same period. Total organic carbon removals observed for the pilot ozonation/softening process were about 60 percent. Based on these results, the city decided to modify the plant for 2-stage ozonation while at the same time making several other process modifications (e.g., terminate prechlorination, re-carbonate, improve clarification and filtration). The modified plant went on-line in August 1984, and its operation reached steady state in October 1984. The full-scale ozone contact chambers had detention times of 4 min (1st stage) and 10.7 min (2nd stage) at a plant flow of 950 m^3/h (6 mgd). Despite combined doses of as much as 15 mg/L, aqueous ozone concentrations in the contactor effluents were always below the detection limit (Singer et al. 1989).

Samples collected from the full-scale plant showed about 40 percent TOC removal through softening when preceded by ozonation (0.45 mg/mg TOC [Singer 1988b]). Removals of TOC throughout the entire plant averaged 49 percent after the modifications as compared to 44 percent before (Singer et al. 1989). The effect of ozone on precursors was not evaluated, but the combined benefits of adding ozone, shifting chlorination, and making other changes resulted in instantaneous THMs being reduced significantly. Compliance monitoring samples, however, indicated that concentrations in the distribution system were still above the MCL (see Figure III–71). In addition, THM speciation shifted toward the brominated forms (see inset, Figure III–71). This could have been due to partial oxidation of the organic material by ozone, resulting in a lower chlorine demand and more opportunity for bromide oxidation. Figure III–71 also shows the effects of these process changes on TOX concentrations. In general, the TOX concentrations followed the same pattern as the THMs. The mass ratio of TOX to TTHM was always about 2, which is within the range expected for alkaline waters.

Despite the above changes, the Belle Glade plant was still out of compliance. As a result, in September 1987 the plant operators started adding chlorine and ammonia (3:1 weight ratio) to the effluent of both ozone contractors (Singer et al. 1989). The absence of a free residual of chlorine brought the treatment system into compliance with the MCL. At the same time, the average ozone dose was reduced from about 8 mg/L to about 6 mg/L (split evenly between 1st and 2nd stages). Following this last set of changes, the plant operators observed that better-settling floc were formed. Figure III–72 shows the new plant's treatment train.

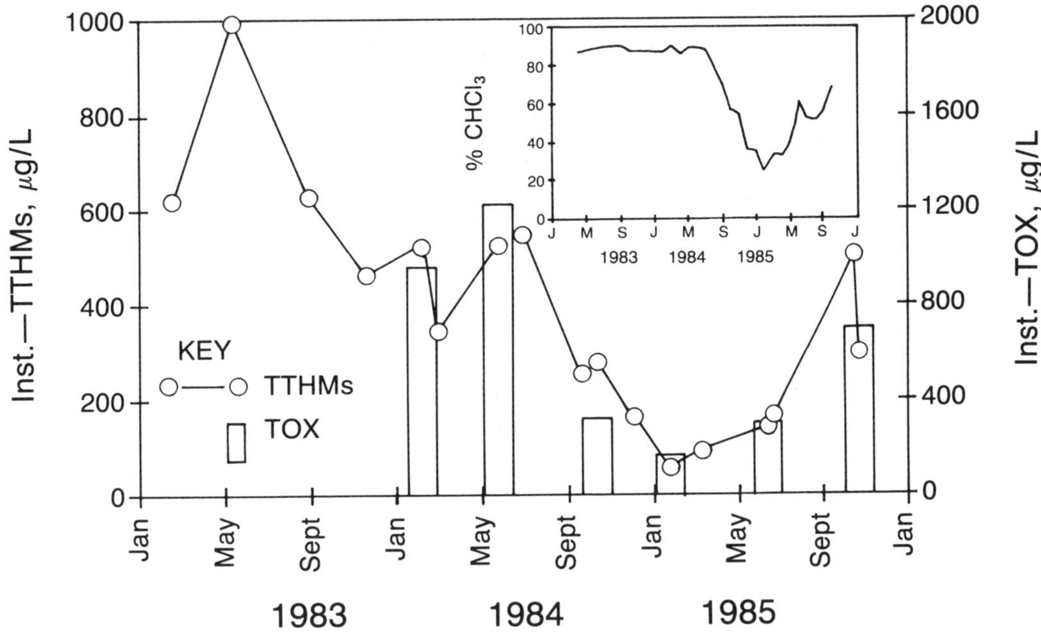

Source: Singer (1988b).

Figure III–71 Chlorination By-product Formation at the Belle Glade Water Treatment Plant

III.I.6 Design of Pilot Studies for DBP Control

The determination of design criteria for control of disinfection by-products is not as straightforward as for control of many other targets. For disinfection, the components of the C · t term need to be defined; for oxidation of SOCs, the rate constant *k* needs to be defined; similarly, for oxidation of color, kinetics can be defined. But the situation with disinfection by-products is different because of the complex nature of the mechanisms of their formation and the concern about numerous by-products with health implications, rather than a single target. When designing treatability studies and determining design criteria, several questions must be addressed. Is the objective to control disinfection by-products themselves, or is the objective to use ozone for some other purpose, like disinfection, and to then determine its effect on by-products? If the objective is to control disinfection by-products, are the targets halogenated by-products like THMs or haloacetic acids, or are they oxygenated by-products like aldehydes?

Treatability studies for control of halogenated by-products. The rate constants for ozone's reactions with already-formed halogenated by-products (see sec. E of this chapter) indicate slow reactions under water-treatment-plant conditions. Further, if free chlorine is used to maintain a distribution system residual, by-product levels will again increase downstream of its application point. Consequently, direct oxidation of halogenated by-products is generally not practical. Treatability studies are typically directed at ozone's reaction with the precursor material.

Which of the halogenated by-products is the target? Generally, if the targets are chlorinated by-products, ozone can be used to improve their control. An example is given in Figure III–61 for chloroform. If brominated by-products are the targets, however, ozone may not be the best option. An example is given in Figure III–60 for dibromochloromethane. For grouped targets like TTHMs encompassing mixed halogenated compounds, ozone's role may depend on the level of bromide in source waters. Although chloropicrin is a chlorinated by-product, its precursors involve nitrogenous material, which can be enhanced by ozonation. For this compound, ozone may not be the best option, as demonstrated in Figure III–64.

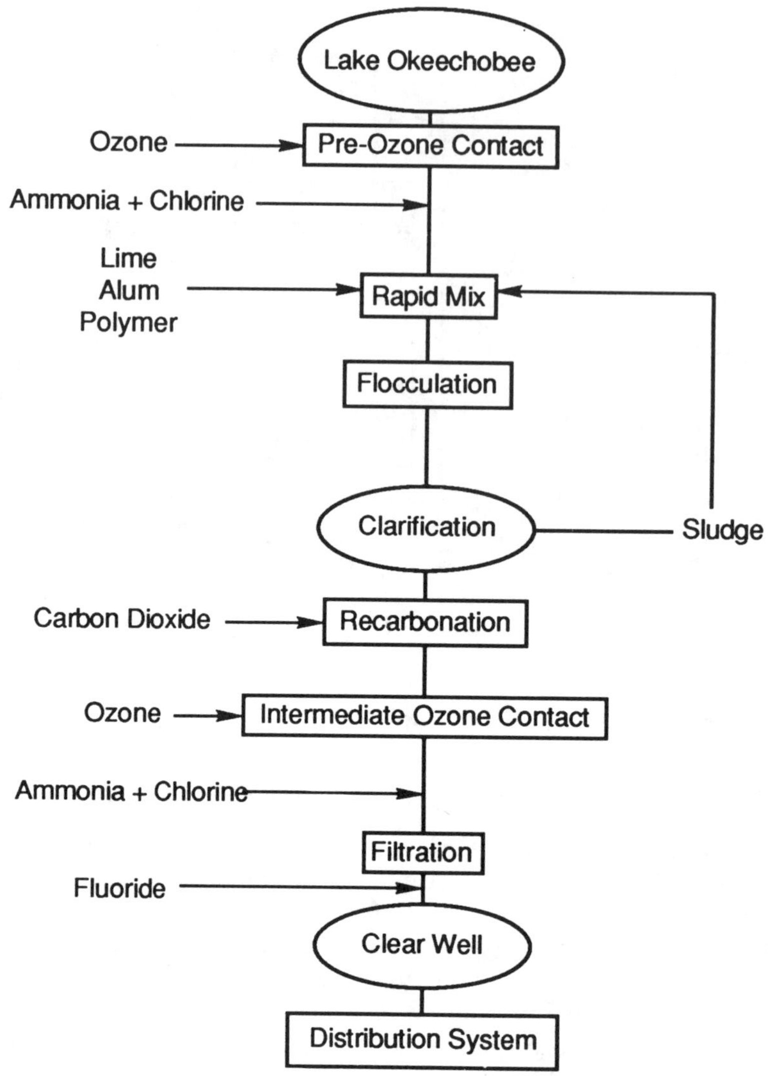

Source: Singer et al. (1989).

Figure III–72 Belle Glade Drinking Water Treatment Plant Schematic

If consideration is given to adding ozone for halogenated by-product control along with making other treatment modifications at the same time, the study must be carefully designed and the ozonation and other modifications must be sequenced in a treatability matrix so that the benefits of each can be properly defined. Using a situation like Belle Glade's as an example, one should ascertain what portion of the TTHM control was attributed to moving the point of chlorination, to improving clarification (which may have enhanced precursor removal), and to adding ozone. Belle Glade's precursor levels were so high that postfiltration chlorination could not produce compliance THM levels. Consequently, ozone/chloramine treatment was implemented. In situations where precursor levels are not as extreme, shifting the point of chlorination and improving precursor control through clarification may sufficiently lower by-product levels such that the cost of ozonation for additional precursor control may not be warranted. A treatability study designed to separate the effects of these types of modifications can define the trade-offs. On the other hand, if ozonation provides adequate disinfection as well as improving precursor control, its cost may be warranted.

In assessing ozone's role in precursor control, the conditions should be defined that optimize the lowering of formation potential. Formation potential is an indirect measure of the unreached precursors at a sampling point and is defined as the difference between the instantaneous or background by-products and the terminal level of by-products at that point. The terminal level results when the reaction is driven to completion in bench studies. Symons et al. (1981) discuss this approach at length for THMs, and an example describing formation of chloroform and reduction in its precursor concentration is given in Figure III–61. The reaction conditions for other by-products, while at an early stage of understanding, have also been addressed (Krasner et al. 1989; Stevens et al. 1989; Jacangelo et al. 1989). In optimizing the ozone/precursor reaction, variation in ozone concentration, contact time, and the ratio of ozone to target formation potential should be studied. In addition to an ozone/THMFP ratio (for example, an ozone/TOC ratio) might be studied as a surrogate. Further, the many factors affecting both ozone concentration and precursor concentration should be studied, as will be discussed.

Treatability studies for control of ozone by-products and AOC. As noted, the basis of knowledge on formation and control of ozone by-products like aldehydes is in its infancy. Even less is known about carboxylic acids and other by-products. For whatever purpose ozone is employed in water treatment, it appears that aldehyde concentrations will be enhanced. Further, aldehydes may be formed by chlorination or chloramination (see sec. III.I.2), or both aldehydes and ketones may be enhanced by chlorine dioxide treatment (Stevens 1982; U.S. EPA 1989b). It also appears (see sec. K.3 in this chapter) that enhancement of several of the aldehydes correlates with enhancement of assimilable organic carbon (AOC) as a result of ozonation. Consequently, AOC should be considered an ozone by-product. While the health implications of aldehydes may at present be uncertain (see sec. III.J), the detrimental impact of high levels of AOC on biological stability and distribution system regrowth is better understood. Therefore, treatability studies should examine control of AOC. As Figure III-82 indicates, biodegradation can reduce AOC levels in-plant.

If control of aldehydes and ketones or control of AOC is an objective, then treatability studies should attempt to define those conditions that minimize their formation or should attempt to optimize bioactivity in downstream processes. In attempting to minimize formation, variation in ozone concentration and contact time should be studied. Sec. III.I.2 suggested an important ozone dose dependency. Figure III-82 indicates that formation can be rapid; in that case contactor residence time was 7.5 minutes. In attempting to optimize biological activity, the control offered by each downstream process should be studied. Sec. K.3 of this chapter indicates that heterotrophic plate count (HPC) bacteria surviving the ozonation process feed on aldehydes and AOC, thereby lowering by-product concentrations before leaving the plant. Treatment studies should, therefore, (1) define the relationship between AOC and distribution system regrowth, and (2) balance the advantage of bioactivity in in-plant processes to degrade aldehydes and AOC with the required control of bacteria in finished waters. While the data presented in Figure III-82 represent only five consecutive days of operation at the pilot scale, they indicate sufficient buildup of HPC bacteria to reduce AOC to its source water level through clarification and filtration. In this case, filter backwash with chlorinated water did not destroy bioactivity on the filter (0.76 m [30 in.] of sand and anthracite operated at 6.1 m/h [2.5 gpm/ft^2]). Note that while AOC increased upon clear well chlorination, coliform growth response (CGR) did not. In this study, AOC was determined by the van der Kooij et al. (1982) method; therefore, the AOC increase upon chlorination suggests regrowth potential for species like *Pseudomonas fluorescens*, P-17. However, the CGR (Rice 1989) decrease upon chlorination suggests no coliform regrowth potential.

General by-product treatability studies. Several aspects of treatability studies are common to the control of either halogenated by-products or ozone by-products. One is the question of where to apply ozone in the treatment train. Where preozonation is employed, ozone demand will be higher. Consequently, a portion of the transferred ozone will be utilized on nontarget compounds. Further, precursor levels will be higher and ozone will be competing with other processes (coagulation and clarification), which can also lower precursor levels. Delaying ozonation until coagulation and clarification have taken place will result in ozonating a water lower in both demand and precursors and will likely result in more efficient use of ozone and improved precursor control, but may not result in sufficient control of AOC, depending on the ability to establish and maintain bioactivity in the remaining downstream processes. If use of multiple ozone application points is an option, piloting will allow an examination of single or multiple points to determine the optimum use of ozone in combination with other processes. If GAC is employed to adsorb or if BAC is employed to adsorb or biodegrade these materials, treatability studies will be further complicated in attempting to simultaneously optimize ozonation, adsorption, and biodegradation.

Another important question is how long to pilot. Seasonal variation in water quality parameters such as TOC, the formation potential of the target compound(s), pH, alkalinity, temperature, bromide levels, and demand can all affect the ozone dose required to accomplish the objective, as discussed elsewhere. Based on historical water quality data, or on sampling to establish those data, piloting at different times of the year may be prudent. Additionally, seasonal demand for water may affect contact times in ozonation chambers. Finally, after treatability studies have determined the ozone doses, contact times, application points, ozone/TOC ratios, etc., that will best control by-products or AOC, a matrix study should be conducted to establish operational ranges to meet these criteria. For example, water depth, water flow rate, gas-to-liquid ratio, gas-phase ozone concentration, bubble size, and demand all affect transfer efficiency and ozone utilization. Chapter IV addresses these factors in their relation to ozone contactor design.

III.J MINIMIZATION OF ADVERSE HEALTH EFFECTS

III.J.1 Introduction

Ozonation has been used as a method of drinking water disinfection in France since 1905 (and, to a lesser extent, in the rest of Europe) (see sec. K, this chapter). In contrast, ozonation has been used only sparingly as a drinking water disinfectant in North America, where public water works have predominantly employed chlorination for that purpose. It has been known for some time that the process of chlorine disinfection produces large numbers of (mostly chlorinated) by-products, and that some of these by-products, although present at very low levels, are mutagenic and carcinogenic (Meier 1988). The observation that chlorination produces chloroform and other trihalomethanes and that these compounds are carcinogenic in 2-year rodent bioassays has been a matter of concern in the United States for some time. For this reason, there is currently great interest in alternative methods of drinking water disinfection, with ozonation being one of the most frequently mentioned choices.

The chemistry of ozone in aqueous solutions is complex (see chapter II, sec. A); however, it is clear that ozone reacts with natural products in the water supply (e.g., humic acids) to form numerous disinfection by-products. The common opinion holds that the reaction by-products of ozonation appear to be less toxic than those produced by chlorination. For example, ozonation does not produce THMs in water and in fact can destroy THM precursors (Symons et al. 1981). Likewise, ozonation of humic acids does not produce chlorohydroxyfuranones, highly mutagenic reaction by-products of chlorination (Meier 1988). However, there is insufficient knowledge regarding the toxicity of ozonated water to be certain of its safety under a variety of circumstances. Moreoever, it has been suggested that a small amount of chlorination will be required after ozonation to provide residual protection for the distribution system. The reaction by-products of this combined process are even less well understood. In this section the literature dealing with the toxicology of ozone disinfection (or reaction) by-products will be discussed.

III.J.2 Case Studies: Subchronic, Carcinogenic, and Mutagenic Effects of Ozonated Drinking Water

***In vivo* subchronic assays for carcinogenicity.** Since the utilization of ozone as a public drinking water disinfectant is not likely to expose the public to ozone itself but rather to ozone reaction by-products, the U.S. EPA has conducted several short-term *in vivo* bioassays designed to detect carcinogenic potential in ozonated water residues (concentrates). However, to date no standard 2-year rodent bioassays or other chronic toxicity tests on aqueous ozone or ozone disinfection by-products have been reported.

A 2-year rodent bioassay would be very useful for risk assessment since it should be noted here that, while short-term *in vivo* bioassays have some utility in predicting carcinogenic potential of pure chemicals and/or complex mixtures, they do not by themselves provide definitive evidence of carcinogenicity. In most countries, only the two-year rodent bioassay is used in quantitative risk assessments. On the other hand, the finding of a tumorigenic response for a chemical and/or environmental mixture via an *in vivo* short-term carcinogenesis bioassay is evidence of cancer-causing potential, and such results should be treated with due concern.

In fact, the correlation between some short-term *in vivo* bioassays for tumorigenesis and the 2-year rodent cancer bioassay is reasonably good (Pereira and Stoner 1985). Unfortunately the converse observation, that is, the finding that a short-term *in vivo* cancer bioassay is negative, provides only modest assurance of safety, since many of the requirements of the 2-year bioassay (e.g., administration of the maximally tolerated dose) are generally not achievable or applicable to the short-term *in vivo* bioassay.

An important study on the subchronic effects of drinking water disinfection by-products on mice was conducted by Miller et al. (1986). The mice were treated for 30 days with 100- to 400-fold reverse osmosis concentrates of drinking water samples taken from the Mississippi River and disinfected in a pilot treatment plant (Jefferson Parish, Louisiana). The influent stream from the Mississippi River (3.4 mg/L TOC) was split into four sub-streams; one of these was disinfected with ozone (0.5 mg/L, pH 6–7), and the other three were individually treated with either chlorine, chlorine dioxide, or chloramine. The concentrates from each disinfectant type were administered in drinking water to 10 male mice and 10 female mice. After the 30-day treatment, a necropsy was performed on each animal to determine gross pathological changes and organ-to-body weight ratios. No apparent toxicity was observed in any of the treated groups. However, because no microscopic histopathological examination was performed, this study cannot be considered sensitive to subtle toxicologic changes. Nevertheless, this study on mice agrees with the more detailed study of Daniel et al. (1989), which was conducted on rats.

Miller et al. (1986) also reported the results of three short-term *in vivo* bioassays designed to measure the carcinogenic potential of disinfection by-products in water samples from the same pilot water treatment plant. The disinfected water was concentrated (4000-fold) via XAD and the resulting organic residues were used for toxicological testing. The various concentrates (of the disinfectant reaction by-products) were evaluated in the rat liver enzyme altered focus assay (designed to measure early preneoplastic events in the liver) using a histochemical staining method for γ-glutamyl-transpeptidase (GGTase)–positive foci. None of the water concentrates increased the number of focal areas (preneoplastic lesions) per liver. Oral administration of the concentrate from the ozone-disinfected water resulted in significantly lower total focal volume than the concentrate from the nondisinfected control water (i.e., 0 versus 21 ± 21 GGTase-positive foci/cm^3 liver, respectively). By comparison, the positive control, a single oral administration of 50 mg/kg diethylnitrosamine, a potent liver carcinogen, resulted in 383 ± 97 GGTase-positive foci/cm^3.

In like manner, the testing of the 4000-fold concentrated ozone-treated water via oral gavage in the Strain A mouse lung adenoma assay gave a response not significantly different from the concentrate prepared from the nondisinfected water (i.e., 0.40 versus 0.28 adenomas/mouse, respectively) for the male animals. The positive control, a single oral administration of urethane (10 mg/mouse), induced 11.6 adenoma/mouse (Miller et al. 1986). The results obtained in the female animals were similar.

Finally, the 4000-fold concentrates were also tested in the mouse skin initiation-promotion bioassay by administration of three oral doses of the aqueous concentrate and subsequent topical application of the potent mouse skin tumor promoter 12-O-tetradecanoylphorbol-13-acetate (TPA). This assay is designed to detect tumor initiators via the induction of benign tumors (papillomas) and malignant tumors (carcinomas) on the skin of the SENCAR mouse. The water concentrates from the ozone-disinfected water induced a response equal to that induced by concentrates from the nondisinfected water (both 0.16 tumors/mouse), whereas the positive control, a single oral administration of urethane (500 mg/kg), produced 0.9 tumors/mouse.

Taken together, these results indicate that the reaction by-products (recoverable via reverse osmosis) from ozone disinfection do not appear to have strong carcinogenic potential. However, it bears repeating that the failure of these short-term *in vivo* bioassays to respond can not be considered to be definitive evidence for the absence of carcinogenic potential. In a comparative sense, it should be mentioned that the aqueous concentrates prepared from the other treatments (i.e., chlorine, chloramine, and chlorine dioxide) were also negative. Finally, it must be noted that these three short-term bioassays were designed primarily to detect tumor initiators (i.e., genotoxic carcinogens); thus, the possible presence of a nongenotoxic carcinogen can not be excluded on the basis of the data reported by Miller et al. (1986). For these reasons, it could be argued that these negative cancer test results are inadequate evidence of noncarcinogenicity or that ozone is better or worse, from a toxicological viewpoint, than the other disinfectants. Nevertheless, the lack of a response using these 4000-fold concentrates is encouraging from the standpoint of public health.

In an earlier study, Bull et al. (1982) reported mixed results from a SENCAR mouse tumor initiation bioassay (also using topical TPA promotion) on the organic residue from (100- to 190-fold) concentrates from Ohio River water that had been disinfected by various methods. In one sample, the ozonated (1–3 mg/L) water concentrate produced a skin papilloma response (0.28 tumors/mouse) that was comparable to the response from chlorinated (0.16 tumors/mouse) and chloraminated (0.2 tumors/mouse) water concentrates. There were no increased tumor incidences produced by the concentrates from undisinfected water or from water disinfected with chlorine dioxide. However, in two subsequent experiments using water obtained from the Ohio River at other times of the year, no evidence of increased tumor-initiating activity was observed in mice treated with concentrates resulting from any of the disinfectant processes, including ozonation.

These two studies point out the difficulties of attempting to evaluate in experimental animals, even using short-term tests, the potential carcinogenicity of drinking water disinfection by-products. In summary, the preponderance of the evidence collected up to this point indicates that the reaction by-products resulting from ozone disinfection do not produce carcinogenic responses in several of the widely accepted short-term *in vivo* bioassays.

***In vitro* short-term bioassays for genotoxicity and cytotoxicity.** Short-term bioassays for genotoxicity on ozonated drinking waters and ozone disinfection byproduct mixtures have produced varied results. A few studies have addressed the toxicological potential associated with reaction by-products produced by the combined processes of ozonation and chlorination.

Ozonation of aqueous solutions of both humic and fulvic acids, as well as ozone disinfection of drinking water, have been shown both to enhance (Dolara et al. 1981; Gruener 1978; Carlberg et al. 1981; Cognet et al. 1986; Van der Gaag et al. 1982; Van Hoof 1983) and to reduce (Zoeteman et al. 1982; Denkhaus et al. 1980; Kowbel et al. 1986; Miller et al. 1986) the mutagenic activity detected in various strains of *Salmonella typhimurium* (Ames test). For example, Zoeteman et al. (1982) found that ozonation of Rhine River water resulted in a reduction of the mutagenicity detected in strain TA98 (frame-shift sensitive) with and without metabolic activation produced by a mammalian liver S9 homogenate. In contrast, Cognet et al. (1986) reported that ozonation of the raw water from the Moulle Water Treatment Plant in northern France significantly increased the level of mutagenicity detected in the extractable organics. The water entering this plant from the Houlle River is high in organic content (5–15 mg/L DOC). Chlorination of the water from this source prior to extraction produced levels of mutagenicity comparable to that of ozonation (Cognet et al. 1986).

These disparate results appear to be due to differences in the ratio of ozone to organic in the particular samples. Several studies have demonstrated increased mutagenicity after low-level ozonation, yet decreased mutagenicity after more extensive ozonation. For example, Kool et al. (1985) found that XAD-4 7500-fold concentrated ozonated water (0.8–2 mg/L) from the Meuse River in the Netherlands exhibited decreased mutagenic activity in *Salmonella* strains TA98 and TA100. This effect was dependent on the water quality and treatment conditions, as a slight increase in mutagenic activity occurred after light ozonation.

Similarly, Backlund et al. (1988) showed that ozone treatment (10 mg/L) prior to chlorination (20 mg/L) of humic water (21 mg/L TOC) from Lake Savojarvi in western Finland did not reduce the mutagenic activity toward *Salmonella* strain TA100 (without metabolic activation) compared to chlorination alone. Earlier, these investigators had shown that when extremely high doses of ozone (33 mg/L) were utilized, the mutagenicity induced by subsequent chlorination could be eliminated (Backlund et al. 1985). No mutagenic activity was detected after ozone treatment alone.

Similar results have been obtained in the United States by Meier (1989), who investigated the effect of ozonation and of ozonation followed by chlorination of humic acid solutions (2.5-3 mg/L TOC). In another work, Bourbigot et al. (1986a) found that if a sufficient dose of ozone ($\geq$3 mg/L) was used to disinfect drinking water at the Méry-sur-Oise plant (in suburban Paris), the resulting 4000-fold concentrate (XAD-2/7 derived) did not increase the frequency of forward mutations in Chinese hamster V79 cells. Similarly, using the SOS-Chromotest, an assay based on the induction of the SOS DNA repair system in *E. coli* (see chapter II, sec. B.2), Bourbigot et al. (1986b) demonstrated that low-level ozonation of the water (0.75 mg/L) resulted in the production of extractable mutagens, while disinfection with 1.5 mg/L ozone did not. Further, the use of an even higher dose of ozone (3 mg/L) resulted in a concentrate that had less genotoxic activity than the concentrate from nondisinfected water. Presumably, the mutagenic ozone by-products formed by the 0.75-mg/L dose of ozone are themselves destroyed by the higher 3-mg/L treatment. These investigators also observed that the mutagenic compounds produced by low-level ozonation could be removed by granular activated carbon (GAC) filtration.

Based on the Bourbigot et al. (1986b) study, it could be reasoned that the increase in mutagenic activity following ozonation of river water reported by Dolara et al. (1981) may well be ascribed to the low concentration of ozone (0.4 mg/L) employed for disinfection.

In a later study, at a pilot treatment plant on the Oise River (France), Hartemann et al. (1987) demonstrated that concentrates (1000-fold, via XAD-8/XAD-4) of raw river water exhibited both mutagenicity (2 to 2.5 times background) and tumor promoter potential as assessed by the Chinese hamster V79 cell-TPA assay (see chapter II, sec. B.2). Treatment of this water with 2.5 mg/L ozone was effective in destroying the tumor-promoting activity, but the mutagenic activity was refractory to this treatment. However, filtration of the ozone-disinfected sample through GAC resulted in the removal of both activities. Likewise, the GAC apparently removed precursors of these mutagens, since chlorination (0.9 mg/L) or ozonation (0.9 mg/L) of the GAC-treated water resulted in no activity either as a mutagen or a tumor promoter.

It is clear, however, that the definition of light versus extensive ozonation is a relative concept, and the effectiveness of various levels of ozonation will depend on the water sample in question. For example, in a related study, Kool and Hrubec (1986) found that application of 3 mg/L ozone to a series of surface waters from various sources resulted in the generation of a low level of mutagenic activity in the Ames test. When these same water samples were treated with 10 mg/L ozone, the activity either disappeared or was reduced. The waters being used in this study ranged from 0.8 to 7.5 mg/L TOC. Thus, while 3 mg/L ozone destroyed the mutagenic activity in the French drinking water study (Bourbigot et al. 1986b) this level was not sufficient to achieve the same degree of destruction for the samples used in the Dutch drinking water study (Kool and Hrubec 1986).

It is quite possible that differences in pH of the reaction medium during the ozonation may also account for some of the discrepancies in the reports of disinfectant-induced mutagenicity observed by different laboratories. For example, in one study (Kowbel et al. 1984), water obtained from Mer Blue Lake in Canada was filtered and acidified (pH 2), and the organics were concentrated on XAD-8 resin. The organic fraction (mostly fulvic acids) was adjusted to 50 mg/L TOC and oxidized with 50 mg/L ozone at several pH values (pH 4, 6, and 8). The resulting ozonated-only samples were either nonmutagenic or only weakly mutagenic toward *Salmonella* strain TA100, regardless of the pH at ozonation. On the other hand, direct chlorination of these fulvic acid solutions or chlorination of previously ozonated solutions resulted in the formation of highly mutagenic residues. The mutagenic activity of the ozonated/chlorinated samples, however, decreased as the pH of the water during ozonation increased. Very similar results had been obtained by these same workers when they studied the ozonation of soil-derived fulvic acids (Kowbel et al. 1984).

Taken together, these studies all suggest that weak ozonation (i.e., levels of ozone [in milligrams per liter] less than or equal to the TOC [in milligrams per liter]) results in the modest production of mutagens and/or fails to destroy the organic substrate that serves as the precursor to the production of chlorine disinfection by-products. In contrast, the use of higher doses of ozone destroys the ozone-induced mutagens as well as some of the precursors to the chlorine-induced mutagens.

These observations are analogous to those seen by Hartemann et al. (1987) who evaluated the cytotoxic effects of aqueous ozonation by-products via the assessment of RNA synthesis in HeLa cells (derived from a human cervical carcinoma) using water from the Seine River (France) at the Vigneux pilot treatment plant. These studies demonstrate that the raw, unconcentrated river water did induce measurable reductions in the HeLa cell RNA synthesis (i.e., exhibited cytotoxicity) but that the prior disinfection of the water with 5 mg/L ozone (10 min) was sufficient to destroy the cytotoxic compounds. These compounds were apparently not destroyed by lesser degrees of ozonation (1 mg/L, 10 min). In addition, chlorination (0.2 mg/L) of the lightly ozonated water (1 mg/L, 10 min) resulted in a cytotoxic concentrate, whereas chlorination of the more extensively ozonated water (5 mg/L, 10 min) did not. These studies would indicate that under some conditions, ozone can destroy the precursors of the toxic chlorine disinfection by-products.

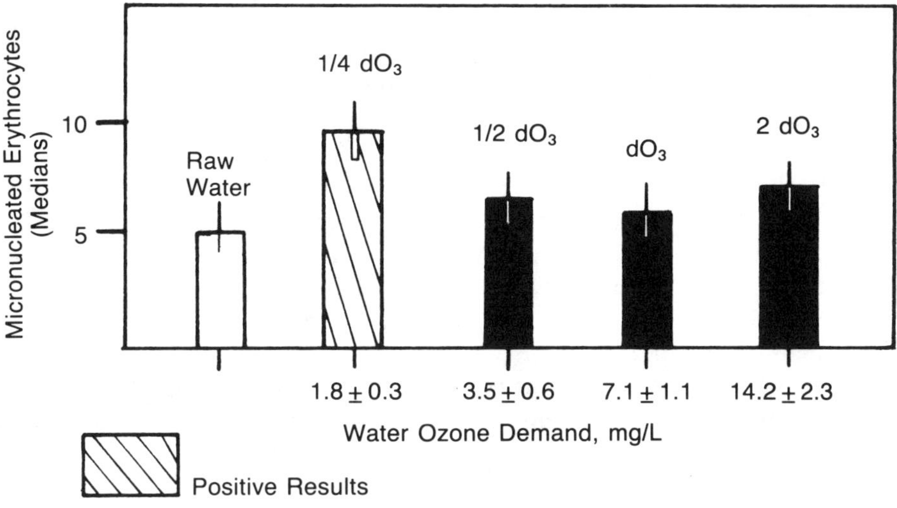

Figure III–73 Genotoxicity of the Seine River Depending on Ozone Dose, February, 1988

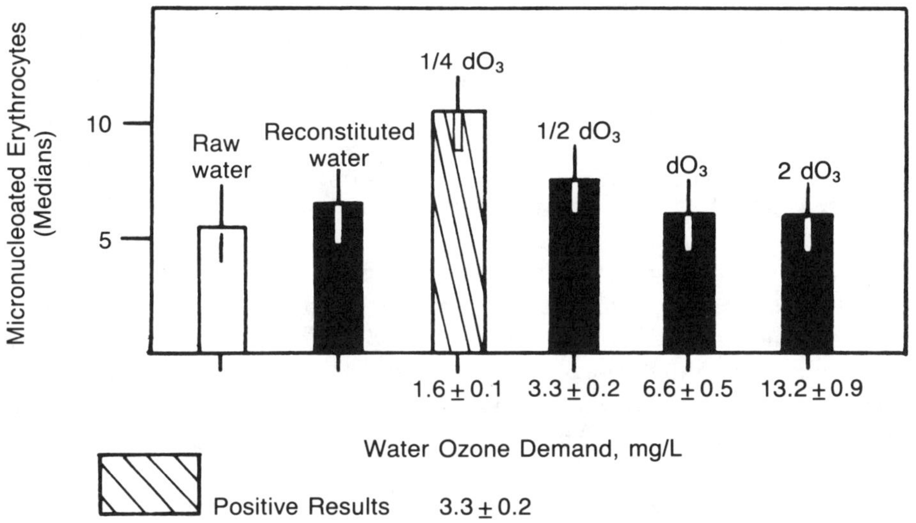

Figure III–74 Genotoxicity of the Seine River Depending on Ozone Dose, June, 1988

In addition, Gauthier et al. (1989) used the micronucleus test on batrachians to show that the use of insufficient amounts of ozone in relation to the organic load (dosage corresponding to 1/4 the ozone water demand) may induce increased genotoxicity, and that this latter is eradicated when ozone levels are augmented (see Figures III–73 and III–74). In this study, the test has been applied on water samples from the Seine River. The micronucleus test allows the direct detection of genotoxicity in water samples without prior extraction or concentration of micropollutants (see chapter II, sec. B.2).

III.J.3 Unresolved Questions Concerning the Toxicology of Ozonation By-products

The reaction of ozone with organic material is extremely complex and can vary greatly, depending on the substrate and other conditions. As mentioned, ozone is an extremely powerful oxidant that can break down long-chain organic molecules (see chapter II, sec. A). These reactions can be hindered, however, whereby a double bond is only partially cleaved and epoxides are formed. In addition, free radicals (molecules with single unpaired electrons) may result from the ozonation process whereby a covalent bond is dissociated, leaving two free radicals. These radicals are highly reactive with biological materials.

Relatively little is known about ozonation reactions with natural products, pesticides, and other organics present in the water during disinfection (see chapter II, sec. B.3). Certainly, one concern is the fact that these various contaminants serve as the substrates for reactions with ozone leading to the production of epoxides, aldehydes, carboxylic acids, and other disinfection by-products in drinking water. Some are potential alkylating or arylating agents and, as such, may be highly mutagenic (Carmichael et al. 1982). The carcinogenicity of epoxides, however, is highly variable and is strongly dependent on the structure and on the substituents. The range of disinfection by-products is likely to depend on the chemical characteristics of the untreated water, which in turn will affect the toxicological properties of the finished water. Thus, the detailed assessment of the toxicological consequences of ozone disinfection will require further research on the chemistry of the ozone reaction and of the distribution of the reaction by-products.

Another area of uncertainty in the interpretation of these toxicological studies is the effect that the concentration processes might have on the validity of the sample generated for the toxicological testing. For example, differential adsorption characteristics of the mutagens on XAD could result from altered properties of the chemical mixture (Zoeteman et al. 1982). Likewise, it is not known whether observed mutagenic activity in a concentrated sample is linearly related to the mutagenicity of the source water, or if it is influenced by the concentration process itself. Van Hoof et al. (1986) also showed that XAD-8 effectively removed mutagens formed during the ozonation (1–4 mg/L) of surface waters under neutral and acidic conditions.

To circumvent this problem, some investigations have reacted high concentrations of disinfectant with humic acid solutions to facilitate the identification and isolation of disinfectant by-products as well as to conduct toxicological studies (see chapter II.B). However, while this approach clearly seems to be an improvement for some disinfectants (e.g., chlorine: cf. Coleman et al., 1984), it is not clear whether such an approach could be used for ozone. Thus, it is not certain whether toxicological studies conducted on ozonated-humic models are relevant to the drinking water situation. For example, because ozone can react by indirect mechanisms involving transient radical intermediates, compounds might be formed by reactions within the concentrate that might not occur under realistic water treatment conditions. Such anomalous reactions are likely to be more common for ozone disinfectant by-products than for those from chlorination.

As more knowledge of ozone disinfection by-products becomes available, toxicological testing can be conducted on individual contaminants (see chapter II, sec. B.3). The priority for these toxicology studies on single by-product contaminants will be determined by considerations of the occurrence and concentration of these by-products as well as by their known or predicted toxic properties. Such an approach has been successfully applied to the disinfection by-products of chlorination.

Another approach is to use a very sensitive toxicological test such as the micronucleus test on *pleurodele* larvae to allow the direct detection of genotoxicity in water samples without prior extraction and concentration of micropollutant. Some results are presented above.

III.J.4 Conclusions and Recommendations

The 90-day subchronic toxicity study recently completed in rats (Daniel et al. 1989) is by itself an insufficient basis for judging the potential health effects of ozonation by-products. Likewise, the toxicity study in mice by Miller et al. (1986) does not adequately confirm the negative findings of the 90-day study in rats because test animals were screened only for gross abnormalities. On the other hand, both of these studies, which used aqueous humic acids or water concentrates several orders of magnitude higher in ozone disinfection by-products than could be expected in drinking water, are encouraging in that they were essentially devoid of significant toxicological effects. More subchronic studies would be useful, and research must be extended to chronic studies.

Further, a comparison of the study on chlorinated humic acids (Condie et al. 1985) with the very similar study on ozonated humic acids (Daniel et al. 1989) suggests that ozonation by-products may be relatively less toxic than chlorination by-products. In agreement, the preponderance of data collected on various short-term tests both *in vivo* and *in vitro* indicate that ozonation produces a less genotoxic assay of disinfectant by-products than chlorination. Many studies show that the results are pH- and dose-dependent.

In the short-term bioassays, a substantial effort should be required to identify appropriate preconcentration methods that minimally alter the activity of compounds of concern. An alternative is to use a technique that allows the direct detection of toxicity in water samples without prior extraction and concentration of the micropollutant (e.g., the micronucleus test on *pleurodele* larvae). Concerning the humic acids, screening should not be limited to commercially available products, as these may not reflect real water conditions.

In future studies, biological species and markers should be selected that are likely to be sensitive to the types of compounds produced by ozonation. A variety of bioassays should be used that cover a broad range of toxicological endpoints. Beyond the *Salmonella*/microsome assay, the research should include bioassays based on mammalian cells for both cloestogenesis and mutagenesis, as well as for nongenotoxic endpoints such as general cytotoxicity and possible specific target organ effects.

III.K BIOLOGICAL STABILIZATION

III.K.1 Introduction

In previous sections concerning the applications of ozone, the efficiency of ozonation, used as a single treatment or combined with others, was described in terms of removal of color, taste and odor, micropollutants, and disinfection by-products. These substances or parameters are all attributed to the presence of certain organic compounds in the water to be treated. There is a considerable amount of practical experience, especially in Europe, indicating that both natural organic matter and industrial pollutants can be effectively removed in drinking water treatment plants by biological processes. Biological processes have been found to remove color, taste and odor, disinfection by-products, and disinfection by-product precursors. In addition, the promotion of biological activity in a treatment plant can help prevent uncontrolled growth in distribution systems. This is especially important for plants using oxidative treatments such as ozonation, which can enhance the biodegradability of natural organic matter. Greater biological stability may partially eliminate the need for maintaining a disinfection residual in the distribution system and thereby further reduce the presence of disinfection by-products. This section describes the implementation and advantages of such biological stabilization processes in drinking water treatment and how they are affected by ozonation.

Historical background of biological filtration. The first efforts at surface water purification in Europe consisted of engineered processes that imitated the natural processes observed in the environment. For this reason, slow sand filtration was one of the first engineered processes. However, the advantages of the sand system that made it attractive in the beginning (e.g., mechanical simplicity) have been partially negated by its limited ability to ensure the effective treatment of increasingly contaminated waters.

Some of the limits of slow sand filtration are:

1. high sensitivity to variations in raw water turbidity; and
2. poor performance with regard to complexed minerals and organic micropollutants, especially those producing color.

Therefore, slow sand filtration technology has had to be improved in Europe in order to adapt to the changing quality of raw waters. As a result, slow sand filter applications today often incorporate:

1. pretreatment to eliminate algae and turbidity (microfiltration, flocculation);
2. pretreatment to improve micropollutant removal (ozonation, coagulation); and
3. posttreatment with chlorine to guarantee disinfection.

In many cases, water treatment engineers, aware of the improvements they have to bring to water treatment, have preferred to implement different technologies, especially for large systems where requirements of personnel and the size of the filters render slow sand filtration prohibitively expensive. Thus, rapid-rate filtration with prechlorination and chemical treatment has been replacing slow sand filtration.

The presence of high concentrations of ammonia in many raw waters has led to a reevaluation of biological treatment and filtration practices. This reevaluation was accelerated by the trihalomethane regulations and the resulting limitations on breakpoint chlorination as an ammonia control strategy. In the 1970s, treatment plant designers tried to incorporate biological treatment while retaining the obvious advantages of high-rate filtration (e.g., the Mulheim Process [Jekel 1979]). This could be easily justified based on the risks associated with chlorination by-products and the desire to improve taste, odor, and aesthetic qualities of the treated water (Rittman and Snoeyink 1984). It was observed that, under certain conditions, ammonia could be eliminated by the existing rapid-rate filters using media such as anthracite or granular activated carbon. As a result, the dose of chlorine to achieve adequate disinfection (and the THM formation) could be reduced.

New approaches to biological stabilization. The objective of any drinking water system is to supply a quality product with respect to bacteriology, toxicology, and aesthetics. The type and degree of treatment necessary depends on the quality of the available raw water. Contamination of raw water sources by organic and inorganic compounds is a predominant concern of water treatment engineers. Many of the dissolved organic compounds are not significantly reduced by conventional treatment (i.e., flocculation, settling, rapid-rate filtration). Biological treatment can remove biodegradable compounds that would otherwise be a food source for bacteria present in the distribution network. The growth of bacteria ultimately leads to the development of higher organisms such as *Asellae*. Moreover, biological treatment permits lower doses of chlorine to be added in order to maintain free chlorine residual at the end of the distribution network, thus decreasing the risks of THM and odor formation.

III.K.2 Fundamental Aspects

Biodegradability tests. Biodegradable or assimilable organic carbon (depending on the author and the methods) is defined as either (1) the fraction of dissolved organic carbon (DOC) that can be used by the bacteria for growth and cell maintenance or (2) the degree to which microbial growth is stimulated by this DOC. In view of the large number and complexity of the dissolved organic compounds in natural waters, it is presently impossible to obtain an identification of each molecule constituting the dissolved organic matter. The characterization of biological availability of dissolved organic matter requires the use of bioassay techniques.

Table III–62 Biodegradability Tests

Test	Microorganisms Utilized	Variable Measured*	Classification†	Required Time (days)
van der Kooij et al. (1982, 1987)	P17, NOX	CFU	AOC	4–8
Jago and Stanfield (1984)	Natural‡	ATP	AOC	§
Reasoner and Rice (1989) Rice et al. (1990)	*E. cloacae*	CFU	AOC	5
Joret and Levi (1986)	Natural**	DOC	BDOC	3–4
Servais et al. (1987) Hascoet et al. (1986)	Natural‡	Bacterial Growth	AOC	~20
Servais et al. (1987, 1989)	Natural‡	DOC	BDOC	21–28
Werner and Hambsch (1986)	Natural‡	Turbidity*** or DOC	AOC*** or BDOC	1–2

*Variable measured: CFU, colony forming units; ATP, adenosine triphosphate; DOC, dissolved organic carbon; bacterial growth; turbidity.
†Classification: AOC, assimilable organic carbon; BDOC, biodegradable organic carbon.
‡Bacterial population from raw, in-plant, or treated waters.
§Time until ATP is maximized.
**Bacterial population is attached to filter sand.
***Werner and Hambsch (1986) also propose the use of DOC analysis, which would make this an "absolute" (i.e., BDOC) method as well.

In heavily polluted waters (e.g., municipal wastewater), the familiar biochemical oxygen demand (BOD) or chemical oxygen demand (COD) tests have been used for years as a means of assessing the effectiveness of biological treatment. However, these methods are not sufficiently sensitive for use with treated drinking waters (Rittmann and Huck 1989). In recent years, several methods have been proposed to determine the easily assimilable organic carbon (AOC) or biodegradable dissolved organic carbon (BDOC) in dilute systems. Several of these methods are summarized in Table III–62 and presented in greater detail below.

The first method proposed for measuring assimilable organic carbon was established by van der Kooij et al. (1982). It calls for one to follow the growth of a pure culture by a plate count. The bacteria are inoculated in the water sample after heating to 60°C for pasteurization. The bacteria commonly used are *Pseudomonas fluorescens* P17 and *Spirillum* NOX. The maximum plate count generally achieved after 4 to 8 days of incubation (15°C) is related to the maximum growth of a strain on a specific substrate. It is converted into units of sodium acetate by referring to a calibration line. This calibration curve is established by measuring the growth of P17 or NOX on a series of simple substrate dilutions, namely acetate and oxalate (van der Kooij et al. 1982; Van der Kooij 1987). The results may be expressed as micrograms of equivalent carbon, based on either acetate or oxalate.

Some possible disadvantages of van der Kooij's pure culture approach are:

1. A pure culture of microorganisms has always a more limited biodegradation capacity than a more heterogenous bacteria population with a wider range of biodegradation possibilities. This may lead to an underestimation of the AOC value.
2. Maximum growth counts may be influenced by the physiological state of the *Pseudomonas fluorescens* culture.
3. Growth of the P17 strain may be inhibited by aluminum salts and some ozonation by-products (Huck et al. 1990).
4. Use of pure cultures requires special care.

Some laboratories, such as the Water Research Center in England, have modified the method of van der Kooij et al. (1982) by using an inoculum of an autochthonous bacterial population and by following the growth rate through measurements of adenosine triphosphate (ATP) concentration (Jago and Stanfield 1984). Thus, they eliminate the problems arising from use of a pure culture, as cited previously.

Rice and co-workers (1990) and Reasoner and Rice (1989) have developed another pure culture procedure that attempts to measure the ability of coliforms to proliferate as a result of nutrient enhancement by ozonation or other processes. In this method, a pure culture of coliform (*Enterobacter cloacae*) is seeded and the ratio of log growth after 5 days (20°C) and time zero is determined. A higher ratio indicates a higher coliform growth response (CGR) and cautions that distribution system regrowth of coliforms may be possible. Rice's data generally indicate a correlation between increases in CGR and van der Kooij et al.'s AOC with ozonation.

The methods of van der Kooij et al. (1982), Jago and Stanfield (1984), Rice et al. (1990), and Reasoner and Rice (1989) yield a biodegradability index that is determined from the amount of bacterial growth that has occurred after a certain incubation time. Methods of this type, based on bacterial growth, will be referred to as AOC assays. The AOC, usually presented in terms of carbon equivalents of some standard electron donor (e.g., acetate), is probably a function of the amount of biodegradable electron donor (organic and inorganic), the change in oxidation state of the electron donor, the yield factor, and bacterial decay. Other methods have been developed with the objective of directly measuring the DOC that can be oxidized completely to CO_2 or incorporated into the cell biomass. These will be referred to as BDOC tests. The determination of BDOC and nonbiodegradable dissolved organic carbon (NBDOC = DOC − BDOC) is convenient with respect to assessing dissolved organic carbon mass balance.

Servais and co-workers (1987, 1989) and Hascoet et al. (1986) proposed a bioassay technique that incorporates some aspects of the van der Kooij and Werner methods (discussed below). The water sample containing dissolved organic matter is sterilized by 0.2-μm membrane filtration. A second sample of the same water or a similar water downstream of sand filtration is used as an inoculum (following removal of protozoans by 2-μm filtration). After approximately 3 weeks of incubation at 20°C in the dark, the following are measured:

1. the reduction of the DOC concentration; and
2. the development of bacterial biomass (measured by epifluorescence microscopy after staining of the bacterial DNA with acridine orange).

The BDOC can, therefore, be determined by two independent methods:

1. from the difference of DOC in the sample at the beginning and the end of incubation; and
2. from a mass balance on the bacterial biomass knowing a yield coefficient and a decay coefficient.

Servais et al. (1987) have shown a good correlation between the two methods of BDOC determination. The method that measures the difference between initial and final DOC (#1 above) seems to be simple and reliable, the only disadvantage being the 3-week incubation period and the sensitivity of the procedures to small concentrations of BDOC. The monitoring of biomass (#2 above) offers additional information on the biodegradation kinetics of organic matter in the sample (Servais and Billen 1989) but obviously requires more elaborate analysis.

In order to minimize the response time of this type of experiment, Joret and co-workers (Joret and Levi 1986; Joret et al. 1988) proposed that water samples could be tested in the presence of an inoculum of bacterial biomass attached to sand, and the decrease of DOC could be followed in accordance with Servais et al. (1987). Thorough washing of the sand is necessary in order to avoid contamination of the water sample. These authors found that with the use of sand the response time could be reduced to 3 to 4 days. Furthermore, an interlaboratory study in Europe (unpublished) showed a good correlation between BDOC obtained with and without sand. These two methodologies, the rapid test with sand and the slow test without sand, allow one to obtain precise values for BDOC and NBDOC.

It may also be useful to characterize the rate of biodegradation, and not just the

Table III–63 Biodegradability Determined by Various Methods on Surface Waters

		Test Results				
Method	Sample	DOC (mg C/L)	AOC (mg C/L)	BDOC (mg C/L)	Percentage of DOC	Ref.
AOC*	Lek River (Netherlands)	6.8	0.062–0.085		0.9–1.25	*
AOC*	Meuse River (Netherlands)	4.7	0.118–0.128		2.5–2.7	*
AOC*	Brabantse Bierbosch (Netherlands)	4.0	0.08–0.103		2–2.6	*
AOC*	Mille Iles River (Canada)	6.9	0.1		1.4	†
BDOC‡	Seine River (France)	2.9		0.9	30	‡
BDOC‡	Oise River (France)	3.9		1.4	35	‡
BDOC‡	Marne River (France)	2.7		0.8	28	‡
BDOC§	Meuse River (Belgium)	3.5		1.2	34	**
BDOC§	Scheldt River (Belgium)	9.8		3.4	35	**
BDOC§	Rupel River (Belgium)	7.5		2.0	26	**
BDOC§	Mille Iles River (Canada)	6.9		0.65	9.4	†

*van der Kooij et al. (1982); AOC values in mg C/L as acetate.
†Prevost et al. (1989a).
‡Joret et al. (1986).
§Servais et al. (1987).
**Servais and Billen (1989), unpublished results.

ultimate extent of biodegradation. Werner proposed a test which uses autochthonous preconcentrated bacteria as an initial inoculum (Werner 1985; Werner and Hambsch 1986). At the end of the test (20–50 h), BDOC is calculated by the difference of the initial and final DOC, much like Servais et al. (1987). Werner admits that even 50 h is too short a time to measure all of the BDOC. Also, Werner follows the evolution of bacterial biomass by turbidimetry. This is used to determine the specific growth rate, μ, and the total cell yield. By running several tests at different degrees of dilution, the Monod coefficients, μ_{max} (maximum growth rate) and K_s (half-saturation constant), can be estimated. The advantages of having this type of kinetic information must be weighed against the expense of obtaining a set of computer-controlled turbidimeters as recommended by Werner.

Taking a different approach, Servais and Billen (1989) define two classes of assimilable organic matter: H1 (rapidly biodegradable organic matter) and H2 (slowly biodegradable organic matter). The difference in these two classes is determined through monitoring of the bacterial biomass by epifluorescence microscopy accompanied by simulation with a mathematical model.

Thus, there exist two basic types of methods for measuring biodegradability (1) the "index" methods based on the comparison of bacterial growth in the water sample and on various concentrations of a known substrate, and (2) the "absolute value" methods, essentially based on the difference in concentration of DOC after and before incubation in the presence of heterotrophic bacteria. As mentioned earlier, methods of the former type will still be referred to as AOC tests, and methods of the latter type will be called BDOC tests. The index methods may be superior for waters with very low concentrations of biodegradable organics, because of the inability of modern DOC analyzers to distinguish differences of 0.1 mg/L or less. However, the index methods are complicated by variable yield and decay coefficients that may depend on the seed and nature of the electron donor. Within each of these two groups are methods that use pure cultures and those that use mixed cultures. One possible disadvantage of the mixed culture methods is that they may only predict biodegradability in the systems from which the seed bacteria were taken; therefore, applicability will be site-specific. The pure culture method, on the other hand, may not predict biodegradability in a system if that system is not accurately represented by the particular strains used (e.g., P17 and NOX).

Table III–63 presents a comparison of the value of BDOC measured on raw waters

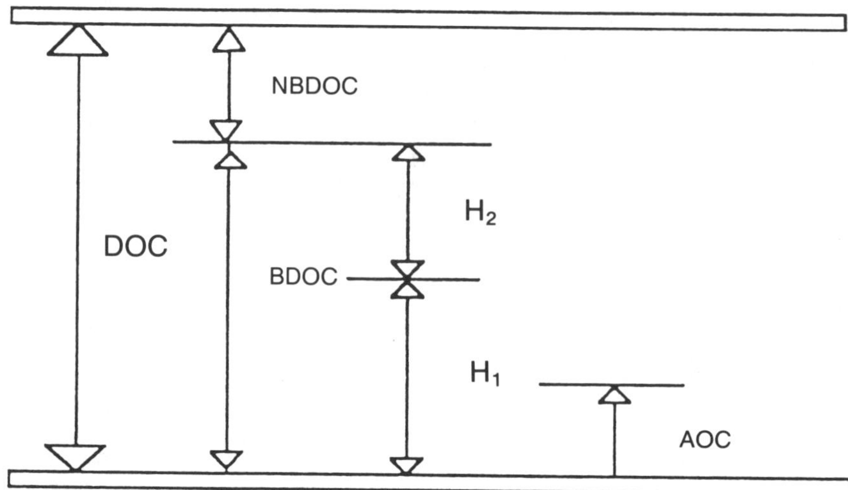

Figure III–75 Hypothetical Fractionation of DOC According to Biodegradability

by two types of methods. From a comparison of this type, it is clear that the "index" methods, such as that of van der Kooij using P17 (van der Kooij et al. 1982), largely underestimate the proportion of DOC that can be used by the bacteria. Also, the methods based on the measurement of BDOC give a measure of the nonbiodegradable or refractory organic carbon in addition to the measurement of AOC. This concept of classifying DOC according to its biodegradability can be very useful in the study of a water treatment system. The qualitative relationship between DOC, BDOC, NBDOC, AOC, and Servais and Billen's kinetic fractions (H1 and H2) are summarized in Figure III–75.

Effect of ozone on biodegradability. It has long been well established that ozone reacts with humic substances, reducing their color and ultraviolet (UV) absorbance (see sec. III.C). In all likelihood, it is the aromatic rings in the humic substances that are responsible for most of the UV absorbance. Ozonation of these structures gives rise to lower-molecular-weight carboxylic acids that are more easily biodegradable. The lowering of net molecular size has been established by analyzing the molecular weight of natural and synthetic waters before and after ozonation (see sec. II.A.6).

Numerous authors have shown evidence of the increase in the biodegradability of dissolved natural organic matter following ozonation (Sontheimer 1979; Janssens et al. 1985; Somiya et al. 1986; see sec. II.A.6 for a discussion of the effects of ozone on biodegradability of extracted humic substances). Despite methodological controversies concerning the determination of biodegradability (e.g., AOC versus BDOC), this tendency is now accepted whichever method is used. Hubele (1986) used an entirely different method to demonstrate the effect of ozone on the biodegradability of natural waters. He showed that the removal of organic matter by GAC significantly increased with the ozone dose. This increased removal was attributed to biodegradation rather than increased adsorption, because the GAC's adsorption capacity was already saturated. Table III–64 summarizes a few results on the biodegradability (measured by different methods) of water sampled upstream and downstream of ozonation in several drinking water treatment plants. All these results show a significant increase in biodegradability across the ozonation stage. Although the doses of ozone used are different, the AOC methods seem to give a much larger relative increase as compared to the BDOC methods.

Some authors have tried to relate the increase in BDOC to the ozone dose applied. Werner and Hambsch (1986) showed, on humic substances extracted from the Fuhrberg water treatment plant (near Hanover, Germany), that BDOC increased sharply at doses

Table III–64 The Effect of Ozonation on Biodegradability in Several Drinking Water Treatment Plants

Method	Treatment Plant	Biodegradability Before Ozonation	Biodegradability After Ozonation*	Percent Increase
AOC†	Kralingen (Netherlands)	0.025	0.173	692
AOC†	Weesperkarspel (Netherlands)	0.012	0.108	900
BDOC‡	Choisy-le-Roi (France)	0.41	0.71	173
BDOC‡	Neuilly-sur-Marne (France)	0.32	0.56	175
BDOC§	Choisy-le-Roi (France)	0.46	0.70	152

*Dose sufficient to achieve 0.4 mg/L residual ozone for 4 min.
†van der Kooij et al. (1982); units: mg C/L as acetate.
‡Joret and Levi (1986); units: mg C/L.
§Servais et al. (1987); units: mg C/L.

Table III–65 Effect of Ozone Dose on BDOC and Bacterial Growth Kinetics*

Ozone Dose (mg O_3/mg DOC)	BDOC/DOC	μ_{max} (h^{-1})	K_s (mg/L)†
0	0.10	0.20	4.2
0.5	0.25	0.38	6.2
1	0.32	0.61	7.7
1.5	0.37	0.61	5.5

Source: Werner and Hambsch (1986).
*Natural organic matter extracted from the Fuhrberg water treatment plant.
†"Apparent" K_s, because this is not a single organic electron donor, but rather a mixture of many organic compounds.

up to 1 mg O_3/mg DOC, and continued to increase a small amount when the ozone dose was raised to 1.5 mg O_3/mg DOC (Table III–65). Figure III–76 shows the effect ofthe ozone dose on waters sampled after rapid-rate filtration in three plants not using prechlorination. In England, Jago and Stanfield (1984) tested ozone doses between 1 and 3.5 mg O_3/L and showed a maximum AOC production at a dose of 2 mg O_3/L. Other authors have suggested a limit to the ozone dose beyond which no more increase of AOC occurs (Brunet et al. 1982). van der Kooij and co-workers (1989) also noted an attenuation of the ozone effect at values higher than 1.5 to 2 mg O_3/mg DOC. Huck and co-workers (1989) studied the effect of ozone on raw water from the Rossdale Water Treatment Plant (Edmonton, Alberta, Canada) at various times of the year. They found that ozone doses of 0.5 mg O_3/mg DOC always resulted in an increase in AOC (method of van der Kooij et al. [1982]). However, during the winter months, a dose of 1.0 mg O_3/mg DOC generally led to a net decrease in AOC as measured with P17. This may be due to inhibition of the growth of P17 by some ozonation by-products and unrelated to the concentration of biodegradable organic substrate (Huck et al. 1990). The NOX strain always showed an increase upon ozonation.

It seems, therefore, that certain molecules are either resistant to the action of ozone or are transformed into products that remain nonbiodegradable. However, many natural organic compounds show substantial improvements in biodegradability at doses of 0.5 mg O_3/mg DOC. Studies of the ozonation of simple aromatic compounds have shown that little change occurs in biodegradability with partial oxidation, whereas at higher ozone doses, ring cleavage occurs and biodegradability increases greatly (Doré et al. 1980; Gilbert 1983b). The same appears to be true for much of the organic matter in natural waters. Beyond 1 mg O_3/mg DOC, in most cases, all the biodegradable organic carbon likely to be formed has already been transformed. This ozonation dosage of 1 mg

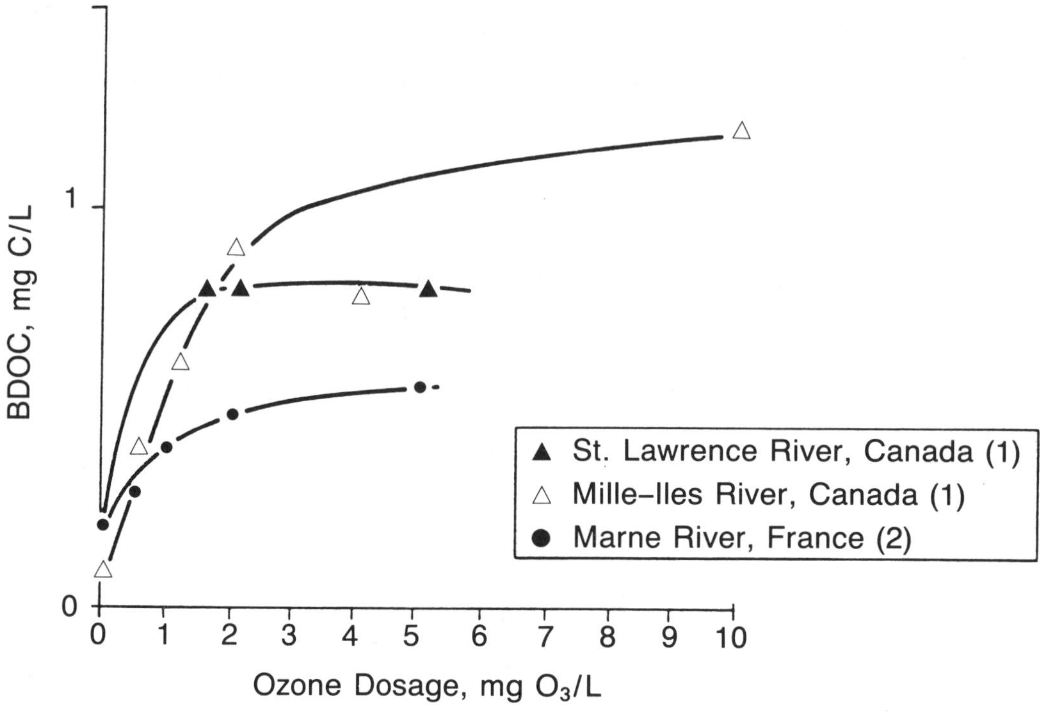

Sources: Prevost et al. 1989b (method of Servais et al. 1987); unpublished results (method of Joret and Levi 1986).

Figure III–76 Effect of Ozone Dose on the Biodegradability of Three River Waters

O_3/mg DOC falls within the range of doses currently applied in full-scale water treatment plants.

In addition, Servais and Billen (1989) have shown that ozonation primarily increases the easily biodegradable fraction. At the Choisy-le-Roi plant (France), the rapidly biodegradable organic carbon, H1, increased from 0.35 mg C/L to 0.6 mg C/L during the ozonation stage, whereas the slowly biodegradable organic carbon, H2, was not modified (0.1 mg C/L). These observations correspond to those of Werner and Hambsch (1986) who showed a change in the maximum growth rate of the bacteria (μ_{max}). This rate, considered as representative of the rate of biodegradability, increased across the ozonation stage (see Table III–65).

In conclusion, all the methods proposed for analysis of biodegradability are unanimous in showing that an increase in the biodegradable fraction of organic carbon occurs as a result of moderate levels of ozonation. In addition, the rate at which the ozone-modified compounds are used by the bacteria increases. Furthermore, there seems to be a limiting value, depending on the type and concentration of the organic matter present in water, above which additional ozonation will not increase the biodegradable fraction of the DOC. This value is typical of the doses commonly applied for the purpose of disinfection. Therefore, good disinfection may be accompanied by optimal stimulation of biodegradation. However, in contrast to disinfection, long contact times may not be necessary for the conversion of refractory DOC to biodegradable DOC.

Biological activity on granular media: special emphasis on GAC. The encouragement of biological activity in water treatment plants is becoming widespread in Europe as more is learned of its potential benefits (Rittmann and Huck 1989). However, in the United States, biological processes have not been extensively used due to a perception that encouragement of bacterial growth contradicts the principal objective of water treatment, which is to eliminate human pathogenic organisms. However, due to the low substrate concentrations found in drinking waters, bacterial populations on granular media are dominated by species that can use a wide range of organic compounds and have low K_s values (i.e., oligotrophs [Bouwer and Crowe 1988; Rittmann and Huck 1989]). This efficiency of substrate utilization allows them to obtain enough energy to support normal cell functions under conditions where other organisms would perish. It is unlikely that pathogenic organisms could compete and survive under these conditions.

Due to the important economic benefits that biological activity may impart to GAC adsorption processes (see sec. III.K.3), biodegradation phenomena on GAC media have been extensively studied. As a result, this section will focus on results with GAC as the bacterial support. One research question that has attracted attention is the possibility of bioregeneration of the GAC, an actual loss of GAC surface coverage of adsorbed organic matter due to biodegradation. Studies of DOC removal using humic substances and CO_2 production seem to substantiate this phenomenon (Topalian 1982). However, the issue is greatly complicated by the tendency of GAC itself to take up oxygen and give off CO_2 (Hubele 1985), and firm conclusions on bioregeneration of natural organic matter cannot yet be drawn (Sontheimer et al. 1988). Nevertheless, several researchers have shown that in simple model systems, bioregeneration does occur (Speitel and DiGiano 1987; Chang and Rittmann 1987; Speitel et al. 1989). This bioregeneration may be practically limited to compounds adsorbed on the outer carbon surfaces, and it likely depends on the adsorption and biodegradation properties of the organic matter.

Sontheimer et al. (1988) presented a summary of several recent mathematical models aimed at describing the behavior of biologically active GAC columns. They concluded that due to the large number of site-specific model parameters, the lack of a fundamental understanding of biomass displacement within a filter, and the diverse nature of aquatic organic matter, these models cannot be used as quantitative predictors of GAC column performance. Rather, they should still be considered as research tools.

The full-scale ozonation/GAC systems at the Choisy-le-Roi and Méry-sur-Oise water treatment plants (suburbs of Paris) and the pilot system at Neuilly-sur-Marne (France) have been studied extensively (Billen et al. 1985, 1989). This work provides some valuable insights into biological activity in full-scale BAC processes; therefore, its principal findings will be summarized below. In all three plants studied by Billen and co-workers, ozone/GAC follows conventional treatment without prechlorination: i.e., flocculation, settling, rapid-rate filtration (see Figure III–77). Biological removal of ammonia is primarily accomplished in the rapid sand filters. In these two plants, a total of 30 contactors measuring 117 m^2 (1260 ft^2) are filled with PICABIOL★ activated carbon, which is specially designed to promote bacterial attachment. These filters (bed depth: 2.5 m [8.2 ft]) have been operated at a filtration rate of 15 m/h (6.2 gpm/ft^2) for over 5 years. The ozone dose used prior to GAC contacting is typically 1–2 mg/L.

Through extensive monitoring of these two plants, Billen and co-workers (1985, 1989) have characterized the nature of the microbial coverage on GAC. In cases where the amount of BDOC is less than 2 mg C/L, and in all situations at the end of a potable water treatment train, the biomass consists of bacterial colonies, fixed in protected areas such as in the macrostructure of the activated carbon (Figure III–78). This is in contrast to the biomass found in wastewater systems, which is in the form of a fixed biofilm. Hence, with such a low microbial density, the release of bacteria by fermentative conditions is not likely to occur as it does in wastewater systems. Others have noted the isolated nature of bacterial growth in activated carbon (e.g., Werner 1981).

Billen and co-workers (1985, 1989) have also proposed a conceptual model for biodegradation on GAC (Figure III–79). They assert that the nature of biodegradable

★PICA; Levallois, France.

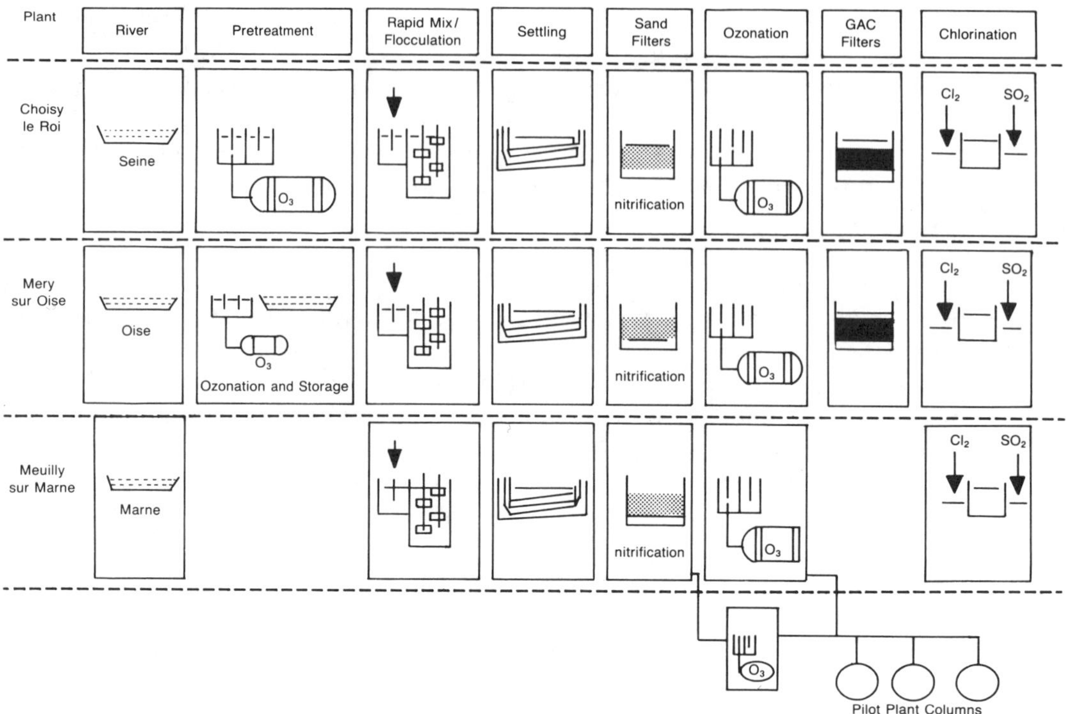

Source: Reprinted with permission from Ventresque, C. and Bablon, G., Proc. 9th Ozone World Congress, edited by J. Bollyky, pp. 207–222. © 1989 IOA.

Figure III–77 Treatment Train for the Water Treatment Plants at Choisy-le-Roi, Méry-sur-Oise, and Neuilly-sur-Marne, France

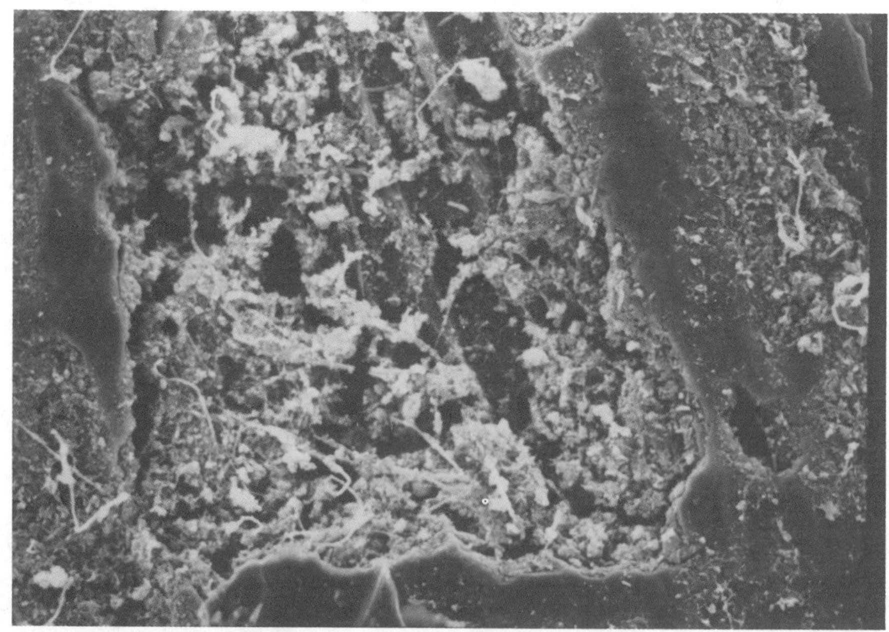

From a GAC column of 1 m height, 6 m/h flow velocity, 1 mg/L BDOC in the influent.

Figure III–78 Heterotrophic Biomass Attached to a Grain of PICABIOL® Activated Carbon

organic material is an essential factor for its treatability. Three classes of assimilable compounds can be distinguished using the BDOC method of Servais et al. (1987). A small fraction of the total biodegradable organics consists of monomeric substances of low molecular weight directly assimilable by bacteria (direct substrates designated

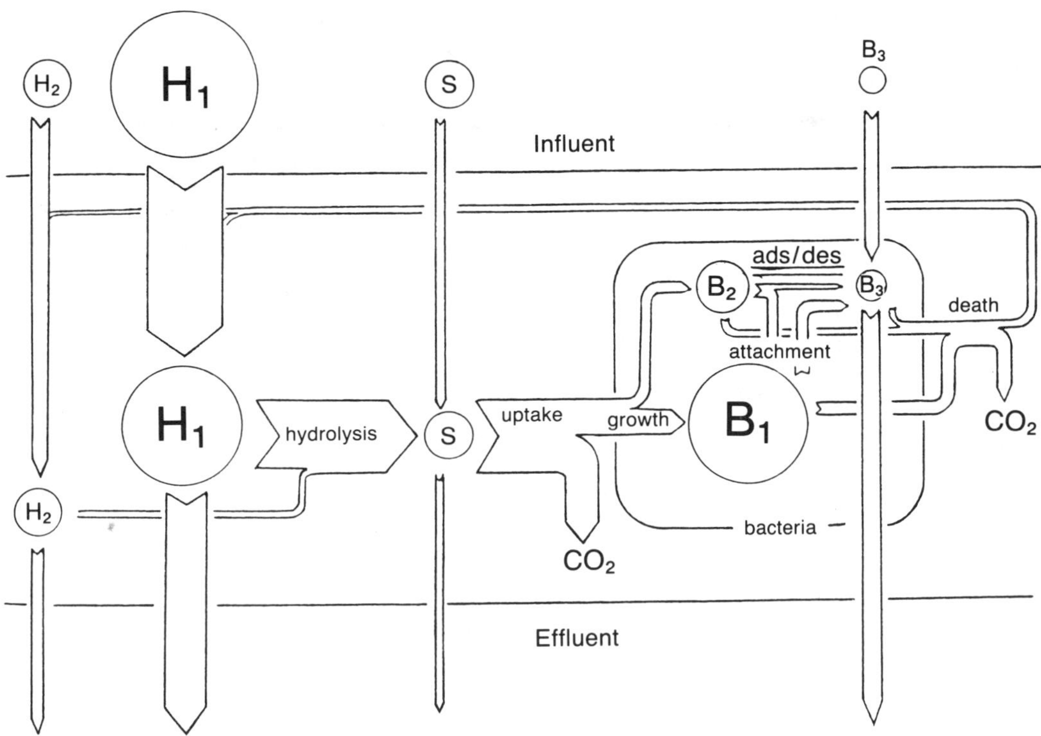

Source: Billen et al. (1989).

H1 = Fraction of BDOC that is rapidly biodegradable
H2 = Fraction of BDOC that is slowly biodegradable (H1 + H2 = BDOC)
S = Substrate directly available for bacterial growth
B1 = Bacteria bound irreversibly to the GAC
B2 = Bacteria bound reversibly to the GAC
B3 = Free bacteria
Death = Loss of bacteria through decay and predation
ads/des = adsorption and desorption of bacteria on the GAC

Figure III–79 Conceptual Model for the Utilization of BDOC in a GAC Contactor

"S" in Figure III–79). The two other fractions are formed by macromolecules "H," of higher molecular weight, which have to undergo exoenzymatic hydrolysis into direct substrates before being available to bacteria. This hydrolysis is the rate-limiting step of the biodegradation process. Depending on the ease of hydrolysis, the high-molecular-weight fractions can be separated into a fraction H1, rapidly hydrolyzable, and a fraction H2, more slowly available.

In rapid-rate biological filtration, it is primarily the directly assimilable "S" and the fraction H1 that are eliminated (Table III–66). The reduction of the fraction H2 would require contact times much longer than 10 min. The fraction "S" is the most readily assimilable organic matter. As long as a significant fraction of "H" is present, its hydrolysis leads to additional "S" and the concentration of "S" remains at an approximately steady-state concentration. The complete removal of "S" thus depends on the elimination of the "H" fractions.

This removal phenomenon is illustrated in Figure III–80, which shows the removal of organic materials by rapid-rate filtration on biological activated carbon as a function of molecular size. The refractory fraction that cannot be eliminated by biological filtration is composed of approximately 80 percent of organic molecules of a size greater than 500 dalton. The biodegradable fraction of the organics in the feed water is about 60 percent macromolecular (H1 and H2) and about 40 percent are small molecules (<500 dalton ["S"]). The elimination of this biodegradable fraction is mainly associated with the macromolecular fraction H1; the monomeric fraction "S" is hardly reduced.

Table III–66 Effect of Biological Stabilization Through GAC on Rapidly Biodegradable (H1) and Slowly Biodegradable (H2) BDOC (Choisy-le-Roi Treatment Plant, 10 min GAC Contact Time)

Date	Water Sample	BDOC (mg C/L)	H1 (mg C/L)	H2 (mg C/L)
9/18/84	Influent	0.69	0.35	0.34
	Effluent	0.54	0.21	0.35
5/22/85	Influent	0.70	0.60	0.10
	Effluent	0.58	0.48	0.10
10/28/85	Influent	0.46	0.40	0.06
	Effluent	0.34	0.28	0.05
10/30/85	Influent	0.52	0.48	0.04
	Effluent	0.22	0.18	0.04
10/31/85	Influent	0.50	0.40	0.10
	Depth 20 cm	0.29	0.20	0.09
	Depth 40 cm	0.30	0.21	0.09
	Effluent	0.18	0.10	0.08

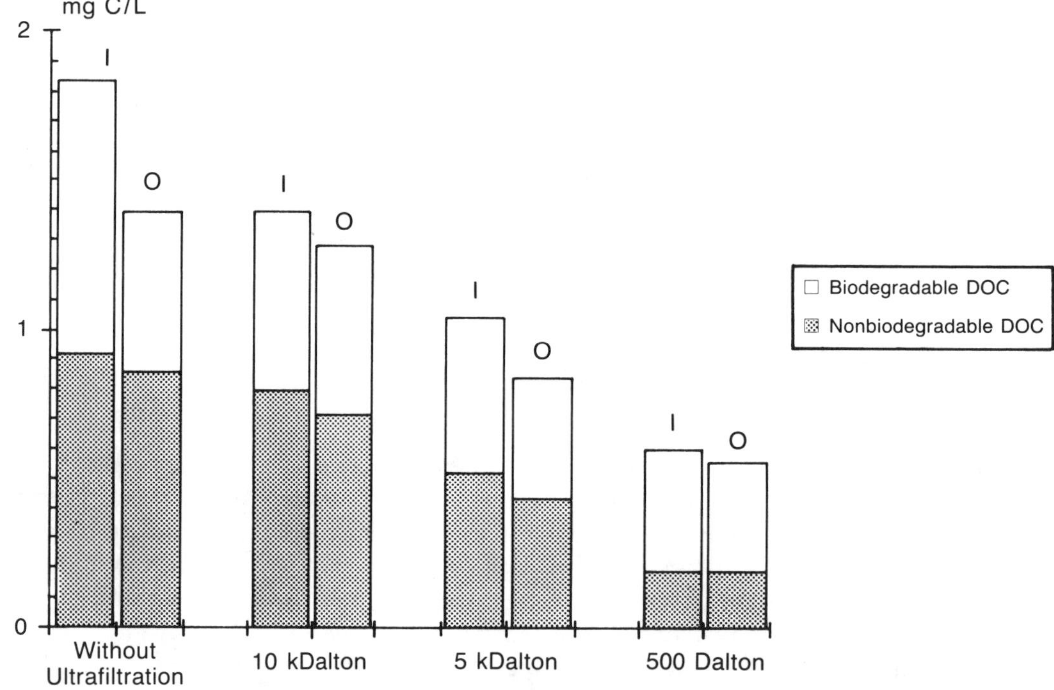

Source: Billen (1989).

Figure III–80 Characterization of DOC Based on Size and Biodegradability in the Influent (I) and the Effluent (O) of a GAC Filter

Billen et al. (1985, 1989) have also determined DOC and particulate organic carbon mass balances on the plants at Choisy-le-Roi and Méry-sur-Oise. The overall BDOC balance of the water flowing through the GAC contactors is shown schematically in Figure III–79. The amount of bacterial biomass fixed to the GAC is at steady state. A part of the BDOC entering the biological reactor is biodegraded. Of this biodegraded carbon, 70 percent is released in the form of CO_2, and 30 percent serves to produce new bacterial cells. It has been found that only 5 percent is removed during backwashing. In fact, 85 percent of the bacteria are subject to decay due to lysis or grazing by protozoans. The protozoans also produce CO_2, but their growth corresponds only to a small percentage of the initial carbon content, since the majority are removed during backwashing.

BDOC removal by BAC filtration occurs in very cold waters (1–2°C), even if these low temperatures last 4–5 months (Prevost et al. 1989a). Experience shows that temperature has two effects on the use of organic material:

1. a direct effect on the enzymatic kinetics of a given population of microorganisms; and
2. an effect due to the adaptation of the microorganisms to a change of temperature (at lower temperatures, different colonies of bacteria are present than at higher temperatures).

The response to temperature of such populations has been modeled by Lehman and co-workers (1975) and field tested by Billen et al. (1989). Figure III–81 illustrates why long-term operation at low temperatures is more efficient than one would predict based on laboratory data. Curve B was obtained from a GAC column originally acclimated at 20°C. These data suggest that a 10°C drop in temperature would result in a drop in bacterial activity of a factor (Q_{10}) of 2.3. On the other hand, a second column that was acclimated at 10°C (curve A) gave a different type of biomass. This system was less active than the 20°C system by only a factor of 1.6. It is not surprising, then, that successful biodegradation occurred at Laval, Canada when temperatures were below 10°C for several months (Prevost et al. 1989a). The work of Billen et al. suggests that unique colonies of bacteria, especially adapted to low temperatures, became acclimated to Laval's GAC.

III.K.3 Stabilization Processes

Considering the distribution network, the bacteriological quality of the water at the consumer's tap depends on several factors, such as:

- the time taken for the water to reach the consumer from the water treatment plant (the problems of bacterial regrowth are greater in the remote branches and/or in small-diameter pipes); and
- the nature and concentration of the residual disinfectant at the sampling point.

Water quality will deteriorate faster in the distribution network if residual chlorine is rapidly consumed and if the fixed and free bacteria can easily find nutrients to support growth. Because of the gradual quality deterioration of many raw waters, it is becoming increasingly difficult to produce water that will remain biologically stable throughout the distribution system. While application of chlorine at high doses will limit biological activity, it will also result in the formation of chlorination by-products and tastes. However, in the absence of free residual chlorine, bacteria will develop, utilizing the biodegradable carbon (van der Kooij et al. 1982; Huck 1989).

In some cases, the installation of an extra treatment process is required to reduce the biodegradability and to avoid excessive formation of chlorination by-products. Biological filtration, whether slow-rate or rapid-rate, can meet these two objectives. When ozonation is placed upstream of filtration, and other conditions are favorable (e.g., adequate micronutrients, no inhibitors), microbiological activity is increased (e.g., Werner 1981), and the removal of biodegradable organic matter is enhanced.

Biological stabilization also leads to lower dissolved organic carbon concentrations in the finished water. The net removal of dissolved organic carbon by biological stabilization processes occurs through two steps:

1. During ozonation, a portion of the nonbiodegradable dissolved organic carbon is transformed into BDOC (refer to earlier discussion on the effects of ozone on biodegradability).

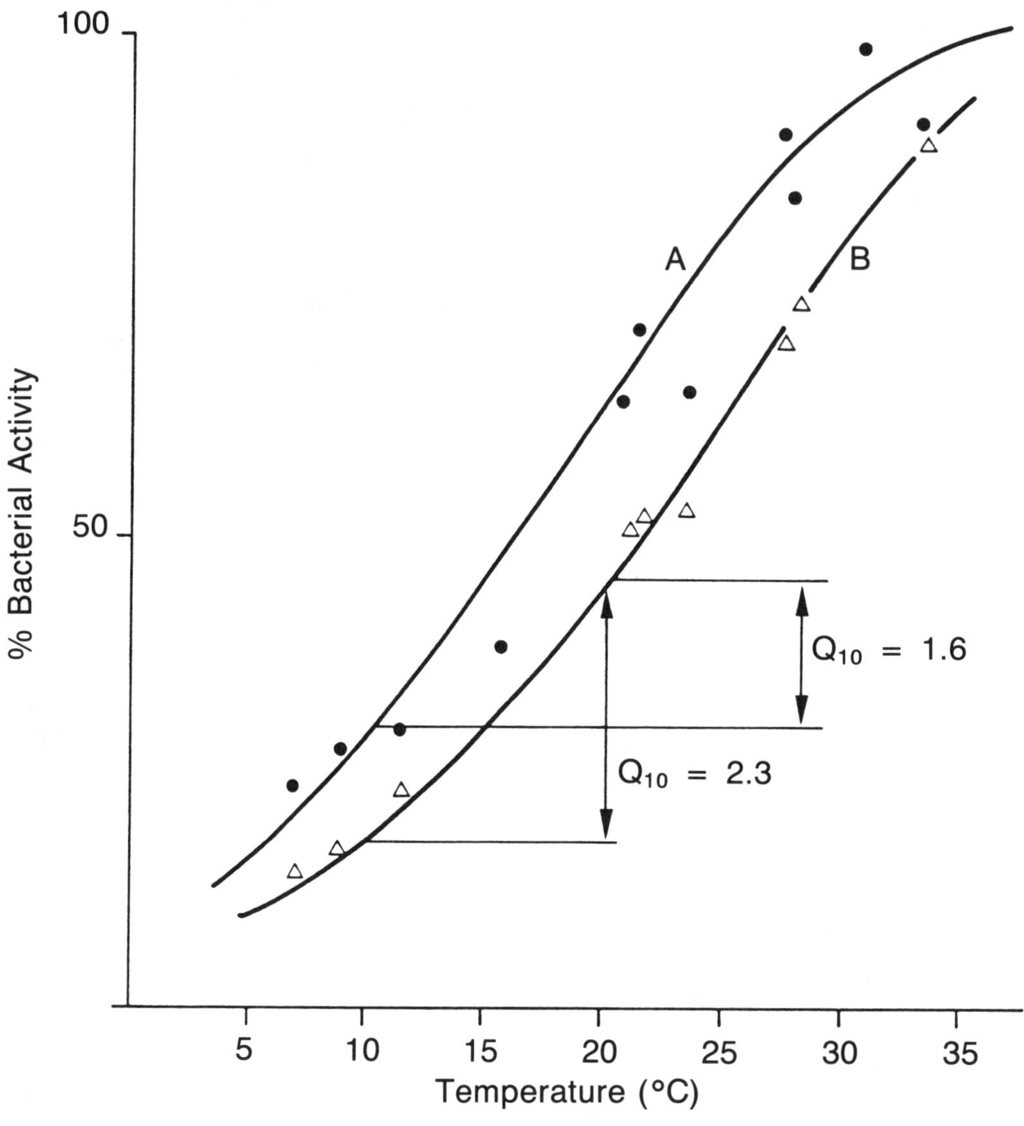

Source: Billen et al. (1989).

The solid circles and curve A represent data and model predictions (based on Lehman et al. 1975) from GAC operated at 10°C (50°F). The open triangles and curve B represent data and model predictions for GAC operated at 20°C (68°F).

Figure III–81 Effect of Temperature on Relative Bacterial Activity

2. The BDOC is converted to carbon dioxide and biomass by the microorganisms attached to the downstream filter.

Slow sand filtration. Although slow sand filtration was the first biological treatment process for potable water, it is limited in its ability to remove color, some taste and odor compounds, and contaminants such as synthetic organic compounds (SOCs) and volatile organic chemicals (VOCs). In addition, slow sand filtration is very sensitive to high water turbidities; therefore, most treatment systems using this process are preceded by a pretreatment process to reduce and stabilize the turbidity of the influent water (Bellamy et al. 1985). Examples of pretreatment include water storage basins (Huisman and Wood 1974) microstraining, roughing treatment, prefiltration (Wegelin 1988); contact coagulation or coagulation on the filter (Welte 1988); and complete pretreatment with clarification (Montiel et al. 1983; Montiel and Ouvrard 1985a, 1985b).

Table III–67 Effect of Ozone on the Performance of Slow Sand Filtration

	Percent Removal	
Parameter	Slow Sand Filtration Without Preozonation	Slow Sand Filtration With Ozonation
Color	30	74
Turbidity	70	70
Iron	53	61
TOC	8–15	25–35

Source: Reprinted with permission from Greaves, G.F. et al., in *Slow Sand Filtration: Recent Developments in Water Treatment Technology* (N.J.D. Graham, ed.). © 1988 Ellis Horwood Ltd.

Table III–68 Effect of Ozone Dose on the Removal of Dissolved Organic Matter by Slow Sand Filtration

		Percent Removal Across Slow Sand Filter	
Ozone Dose (mg/L)	Residual Ozone (mg/L)	Absorbance at 254 nm	Absorbance at 400 nm
0	0	12	24
1	0	47	71
3	0.2	65	86
3	0.5	66	86

Source: Reprinted with permission from Rachwal, A.J. et al., in *Slow Sand Filtration: Recent Developments in Water Treatment Technology* (N.J.D. Graham, ed.). © Ellis Horwood Ltd.

Ozone prior to slow sand filtration can be used for somewhat different purposes than those for which it is used with conventional systems. The purpose of ozone in this treatment configuration includes color removal, removal of iron and manganese, and breakdown of refractory organic molecules to increase biodegradability. Table III–67 shows the effect of ozone on the removal of color, turbidity, iron, and TOC (Greaves et al. 1988). Others have shown that the application of ozone before slow sand filtration can increase the TOC removal efficiency of the filters by about 35 percent (Rachwal et al. 1988; Zebel 1985) with a higher performance in the summer season. Rachwal et al. (1988) also report an improvement in removal of substances absorbing at 254 nm and 400 nm when ozone is placed before the slow sand filters (Table III–68).

Rapid-rate filtration. Sand or anthracite offers significantly less surface area for bacterial attachment than does GAC. Therefore, for an identical ozone application ahead of either one of these inert media or GAC, more complete biodegradation would be expected on GAC (e.g., Hubele 1985; this is also discussed in sec. III.K.2 and illustrated in Figure III–88). However, rapid-rate filtration, employing either sand or dual media, has been reported to lower AOC levels following ozonation. This has been demonstrated in Dutch water treatment plants both with direct filtration and conventional treatment (van der Kooij et al. 1982; van der Kooij 1987). These plants employed biologically active, i.e., nonchlorinated, filters.

Similar results have been observed in the United States. Table III–69 describes four pilot-scale tests and demonstrates the increase in AOC following ozonation and subsequent decreases with sedimentation and rapid-rate filtration (LADWP 1988). Although total AOC is given, the behavior was similar for its components of the P17 strain of *Pseudomonas* (measured as acetate equivalents) and its NOX strain of *Spirillum* (measured as oxalate equivalents). This suggests (according to van der Kooij [1987]) that ozone produces carboxylic acids like oxalate, formate, and glyoxalate, i.e., those utilized by NOX, as well as a wider spectrum of by-products, i.e., those utilized by P17. Sec. I.2 of this chapter demonstrated the increase in higher-molecular-weight carboxylic

Table III–69 Pilot-Scale Control of AOC*

Study	#1	#2	#3	#4
Influent	418	390	263	237
Ozonated†	807	578	918	588
Settled‡	752	370	579	410
Filtered§	>340	229	159	212

Source: LADWP (1988).
*μg acetate and oxalate C eq/L using P17 and NOX (van der Kooij 1987).
†Ozone dose: #1, #2 = 7 mg/L; #3, #4 = 5.2 mg/L; all contact times = 4 min.
‡30 min.
§5.3 m/h (2.2 gpm/ft²), dual media.

Table III–70 Full-Scale* Control of Nutrients: Case #1

	AOC†	CGR‡
Raw	78	0.38
Ozonated	134	1.06
Filtered§	95	0.49
Raw	78	0.38
Ozonated	107	0.77
Filtered**	107	1.5

Source: LADWP (1988).
*U.S. high-rate direct filtration plant.
†ug C/L as acetate (van der Kooij et al. 1982).
‡Coliform growth response, $\log_{10} N_5/N_0$ (Reasoner and Rice 1989; Rice et al. 1990).
§Biologically active filter; no chlorination.
**Growth-controlled filter; chlorination 3 times/week.

acids in drinking waters upon ozonation and discussed the conversion of aldehydes to carboxylic acids resulting from ozonation. It is presumed in these pilot studies that some plate count bacteria like *Pseudomonas* and *Spirillium* survived ozonation and subsequently caused the biodegradation of these carboxylic acids in downstream sedimentation and nonchlorinated, rapid-rate filtration processes.

van der Kooij (1987) discussed the importance of a free chlorine residual, and its relationship to the bacteria in the filter and their ability to biodegrade AOC. This was confirmed in a U.S. study of parallel, direct filters (LADWP 1988; see Table III–70). In a nonchlorinated, biologically active filter, both AOC and coliform growth response decreased. However, in a chlorinated filter, the biodegradable organic matter was not attenuated as AOC levels were unchanged and CGR increased to a level indicative of potential coliform regrowth, i.e., greater than 1.0.

In pilot-scale treatment of Ohio River water, AOC and CGR levels that resulted from ozonation also fell during subsequent sedimentation and dual-media, rapid-rate filtration (Miltner et al. 1990). Heterotrophic plate count (HPC) was also determined (see Figure III–82). The data indicate that bacteria surviving ozonation flourished in the nutrient-rich downstream waters and biodegraded organic matter as measured by AOC (as per van der Kooij et al. [1982]) and CGR (as per Reasoner and Rice [1989]; Rice et al. [1990]) to levels below that of the source water. These data represent five days of operation. The filter only experienced chlorination during backwash, and this level of disinfection (1.2 mg/L free chlorine) was apparently not sufficient to destroy bioactivity. Postchlorination elevated AOC levels (although the magnitude of the increase was atypical of chlorination), slightly decreased CGR, and lowered the HPC to acceptably low densities in finished water. This tradeoff between chlorine's ability to raise AOC in

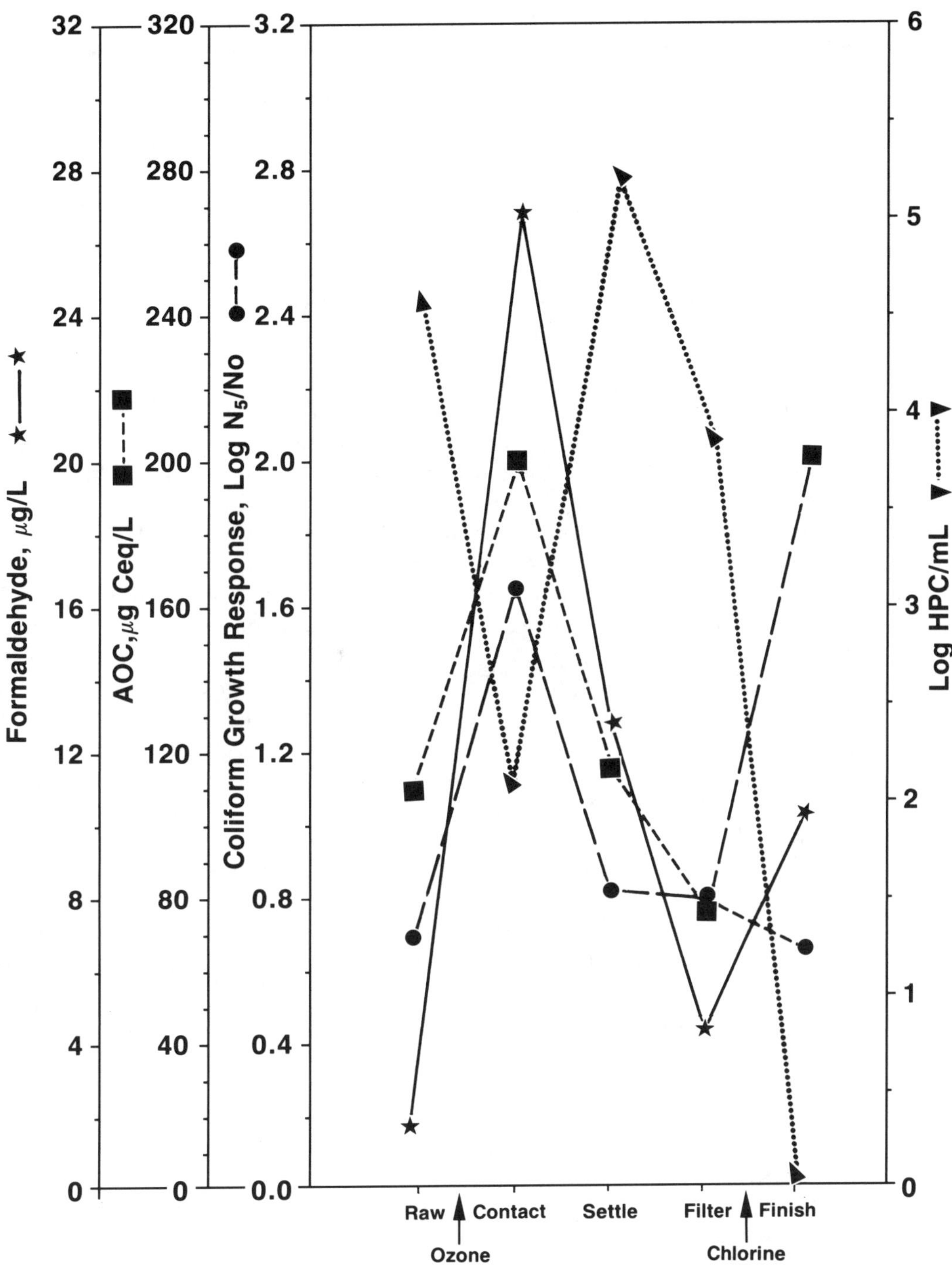

Source: Miltner et al. (1990).

1.93 mg/L applied ozone dose (0.8 mg O_3/mgC); 7.5 min ozone contact time; 9 h sedimentation; 0.76 m filter media depth; 6.1 m/h filtration rate; 2.82 mg/L chlorine dose; 8.5 h chlorine contact time.

Figure III–82 Formation and Control of Formaldehyde, CGR, HPC, and AOC in the Ohio River Pilot Plant

the clear well in some waters (van der Kooij 1987) and provide final disinfection is important. Also important is the tradeoff between promoting biodegradation of AOC in rapid-rate filters and keeping microbiological conditions in the filter below nuisance levels.

Table III–71 Full-Scale Control of Nutrients: Case #2

	AOC*	CGR†
Raw	231	0.08
Ozonated‡	243	1.64
Filtered§	216	1.22

Source: Rice (1989).
*μg acetate C eq/L (van der Kooij et al. 1982).
†Coliform growth response, $\log_{10} N_5/N_0$.
‡1.3 mg/L, 2 min, followed by 19 min detention.
§Flotation skimmer, followed by chlorination, dual media filtration at 14.6 m/h (6 gpm/ft^2), and chloramination.

Concentrations of formaldehyde, an ozonation by-product, are also given in Figure III–82 and are seen to correlate well with AOC. These formaldehyde data are also given in Figures III–54 and III–55 along with two other ozonation by-products, acetone and glyoxal. Together, these data indicate that such simple aldehydes and ketones are well correlated with AOC, and they may themselves be substrates for HPC bacteria like the *Pseudomonas* strain used in the AOC determination.

While data from these pilot- and full-scale studies indicate significant AOC increase upon ozonation followed by degradation during sedimentation and rapid-rate filtration, this is not always the case. An exception was noted at a full-scale water treatment plant in the United States (Rice 1989). Table III–71 gives mean data of ten tests over time showing statistically insignificant changes in AOC through the plant, which employed disinfected filters. Changes in CGR, however, were significant.

As discussed in sec. K.2 of this chapter, AOC does not measure all the BDOC, but measures only that portion of the BDOC that is more easily assimilable or more easily biodegradable. By using AOC, the fate of the remaining biodegradable materials may not be known. The data presented here suggest that biodegradation of materials can occur in rapid-rate filters. These data should be viewed with caution, however, as the more slowly biodegradable material not measured by AOC may be passing rapid rate filters. This has been the French experience which led to their employment of GAC with its increased surface area.

Biological activated carbon. A very important unit process for the removal of dissolved organic matter and micropollutants is granular activated carbon adsorption. Perhaps the major economic concern with GAC is when to regenerate or replace the carbon. A practical approach used by some German water works is simply to reactivate when a given quantity of water has been filtered through the carbon. The criterion used, when TOC removal is the objective, is in the range of 50 m^3 of water treated per kg of GAC. In many of those cases, this translates into a six-month bed life. However, when GAC is used to remove disinfection by-products, the bed life may only be a few weeks (Symons et al. 1981; Graese et al. 1987).

These short run times are not often applied in practice because of the high costs of carbon regeneration or replacement. Depending on the way in which the filters are operated, a significant amount of biological activity may develop in the carbon bed. When the influent water is ozonated, this biological activity is enhanced. For this reason the combination of ozone and GAC, often called biological activated carbon, is especially effective. However, carbon can also be biologically active when preceded by other oxidants or by no oxidant. The presence of biological activity can result in the removal of organic molecules that would have otherwise been adsorbed. This serves to increase the capacity of the carbon for removal of DOC or specific micropollutants. The time between replacements or regenerations may be extended as much as fourfold compared with simple adsorption (Jekel 1979).

Table III–72 Removal of Biodegradable Dissolved Organic Carbon (BDOC) and Nonbiodegradable Dissolved Organic Carbon (NBDOC) in Parallel GAC Contactors: Fresh GAC Versus Old GAC (Neuilly-sur-Marne)

Date of Measurement	Influentm mg/L		Effluent, mg/L		Difference, mg/L (%)	
	BDOC	NBDOC*	BDOC	NBDOC	BDOC	NBDOC
April 1988						
Activated Carbon 1†	0.38	1.16	0.22	0.61	0.16 (42)	0.55 (47)
Activated Carbon 2‡	0.35	1.14	0.20	1.10	0.15 (43)	0.04 (4)
August 1988						
Activated Carbon 1, After 5 Months	0.47	1.28	0.25	1.23	0.22 (47)	0.05 (4)
Activated Carbon 2, After 5 Months	0.47	1.28	0.27	1.27	0.20 (43)	0.01 (1)

*NBDOC = (DOC − BDOC).
†Virgin GAC.
‡5-year-old GAC.

Initiation of biological activity. When initially placed in operation, carbon possesses its maximum adsorption capacity. As the adsorption capacity becomes exhausted, the biological activity becomes more important. Table III–72 compares removal of organic matter on granular activated carbon at various stages of bed life. The characteristics of the filters are filter depth = 1 m (3.3 ft), operating at 6 m/h (2.36 gpm/ft^2), for a contact time of 10 min. The performances are almost identical in terms of total organic removal when the adsorption capacity is spent after a few months, as biological removal becomes the main mechanism for BDOC removal. These same data (Table III–72) show that the biodegradable organic carbon is greatly reduced by the old GAC, whereas the nonbiodegradable or refractory DOC (NBDOC) is only removed through adsorption on new GAC (Figure III–83).

Sontheimer and co-workers (1988) have reviewed the literature on biodegradation on GAC. They conclude that significant biological activity can be expected to begin in GAC drinking water systems after 5–20 days of operation.

BDOC reduction. Although there is an increase in BDOC during ozonation, the efficiency of subsequent biodegradation on GAC is such that the BDOC of the GAC effluent is often lower than the BDOC of the ozone influent. This is even true of systems with GAC that has been in use for a long period of time (Figure III–84). These systems will also, of course, show a moderate destruction of nonbiodegradable organic carbon, attributable primarily to the ozonation step (Figure III–85). For example, at the Méry-sur-Oise plant (empty bed contact time [EBCT] = 10 min; see Figure III–77 for a description of this plant) ozone plus GAC removes a significant additional percentage of BDOC and DOC above that removed through conventional treatment (Table III–73).

The degree to which biodegradable organics are removed by ozone/GAC will depend on the process conditions, the temperature, and the quantity of assimilable matter to be treated. For example, the data in Figure III–86 show that for an influent BDOC concentration of 0.8 mg/L and a temperature of 20°C, BDOC can be reduced by 70 percent for a contact time of 10 min (independent of filtration rate up to 20 m/h [8.3 gpm/ft^2]) and by 90 percent for a contact time of 20 min (Neuilly-sur-Marne pilot experiment). Hubele (1985) has also shown that only minor improvements in DOC biodegradation can be gained in going beyond an EBCT of 5–10 min.

Billen et al. (1985, 1989) have studied the importance of hydraulics on the BAC process. All other things being equal, the velocity of the water (up to 20 m/h or 8.2

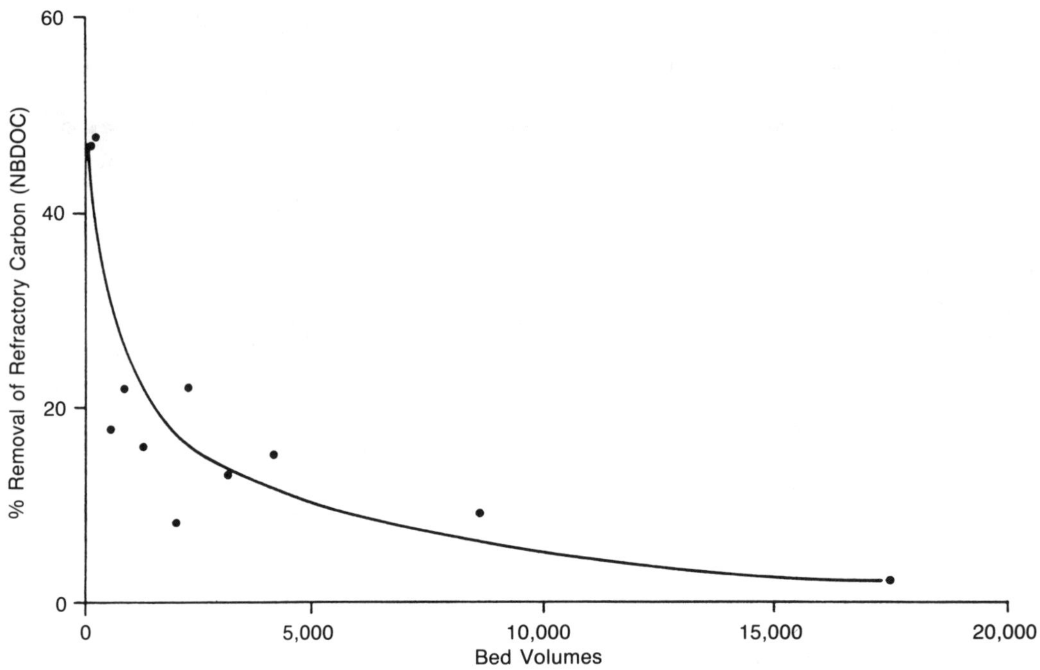

Figure III–83 Removal of Refractory DOC (NBDOC) at the Neuilly-sur-Marne Pilot Plant

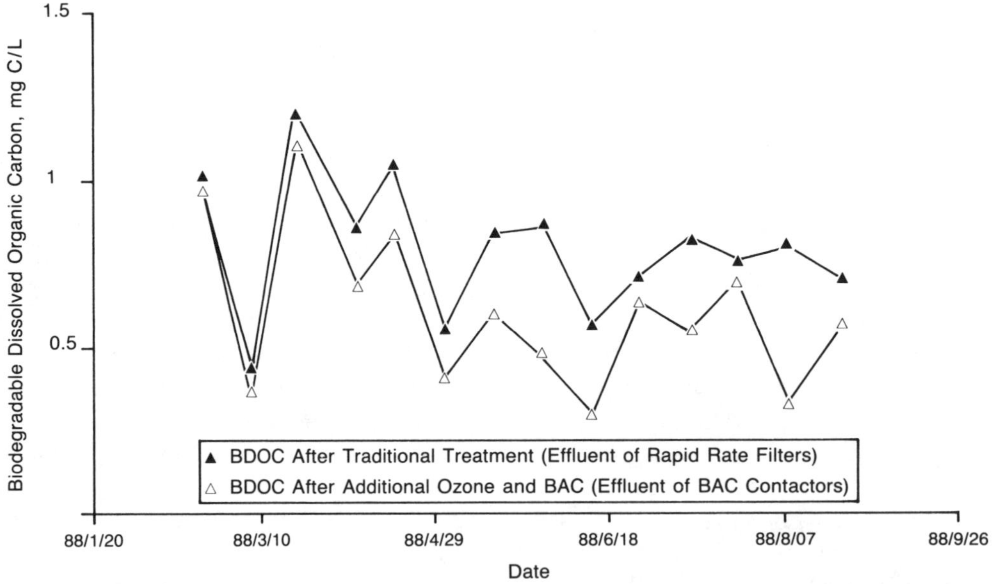

Figure III–84 Removal of BDOC by the Combination of Ozone and GAC at the Méry-sur-Oise Water Treatment Plant (Method of Joret and Levi 1986)

gpm/ft^2) appears to have no effect on the bacterial growth and activity. However, backwashing conditions can be very important for the purpose of controlling predator populations. The maintenance of a certain predator population may be desirable to provide a check on the excess proliferation of bacterial biomass.

Billen and co-workers (1989) have characterized the effect of influent BDOC and EBCT on effluent BDOC in BAC processes (Figure III–87). Knowing the value of

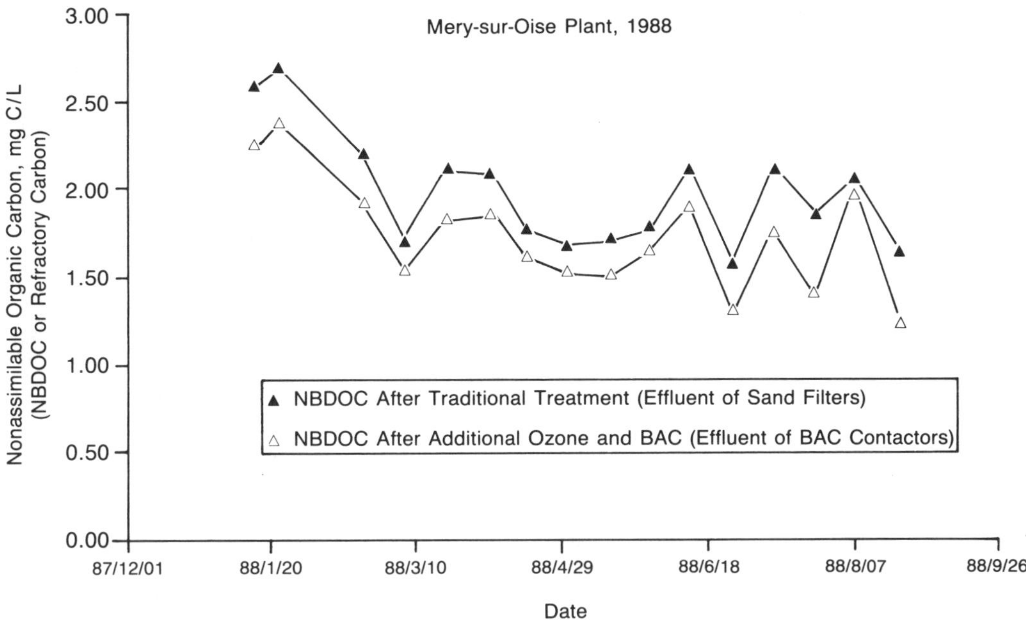

Figure III–85 Removal of NBDOC by the Combination of Ozone and GAC at the Méry-sur-Oise Water Treatment Plant (Method of Joret and Levi 1986)

Table III–73 Comparison Between Conventional Treatment and Ozone + GAC for the Removal of Dissolved Organic Carbon in the Mery-sur-Oise Treatment Plant (1988). BDOC Measured According to Method of Joret and Levi (1986)

	Percent Removal*	
	Conventional Treatment†	O_3 + GAC
Dissolved Organic Carbon	15	15
Biodegradable Dissolved Organic Carbon	30	20

*Percent removal across conventional treatment or across subsequent O_3/GAC treatment.
†Coagulation + gravity settling + rapid sand filtration.

BDOC at the influent (BDOC produced by the conventional treatment train), and based on a desired value in the distribution system, it may be possible to design the ozonation/GAC filtration step by establishing curves similar to the ones shown in Figure III–87. For example, with an influent BDOC of approximately 0.65 mg C/L and a 10-min EBCT, one would expect an effluent BDOC of about 0.25 mg/L. The effluent BDOC could then be lowered by either (1) adding ozone, which would increase the influent BDOC and therefore lower the effluent BDOC, or (2) adding more GAC or decreasing loading rate, which would give a 30-min EBCT and therefore lower the effluent BDOC.

Fixed bed processes that have been in use for a long time in wastewater treatment are heavily loaded with assimilable products. These processes are not influenced by the type of media (sand, anthracite, polysytrene, activated carbon, etc.). In drinking water treatment applications, where concentrations of assimilable substrate are very low, the nature of the bacterial media support has been shown to be of primary importance (Duchesne et al. 1988; Nakamura et al. 1989). For example, the number of bacteria found in PICABIOL granular carbon, with its macroporous structure, was three to four

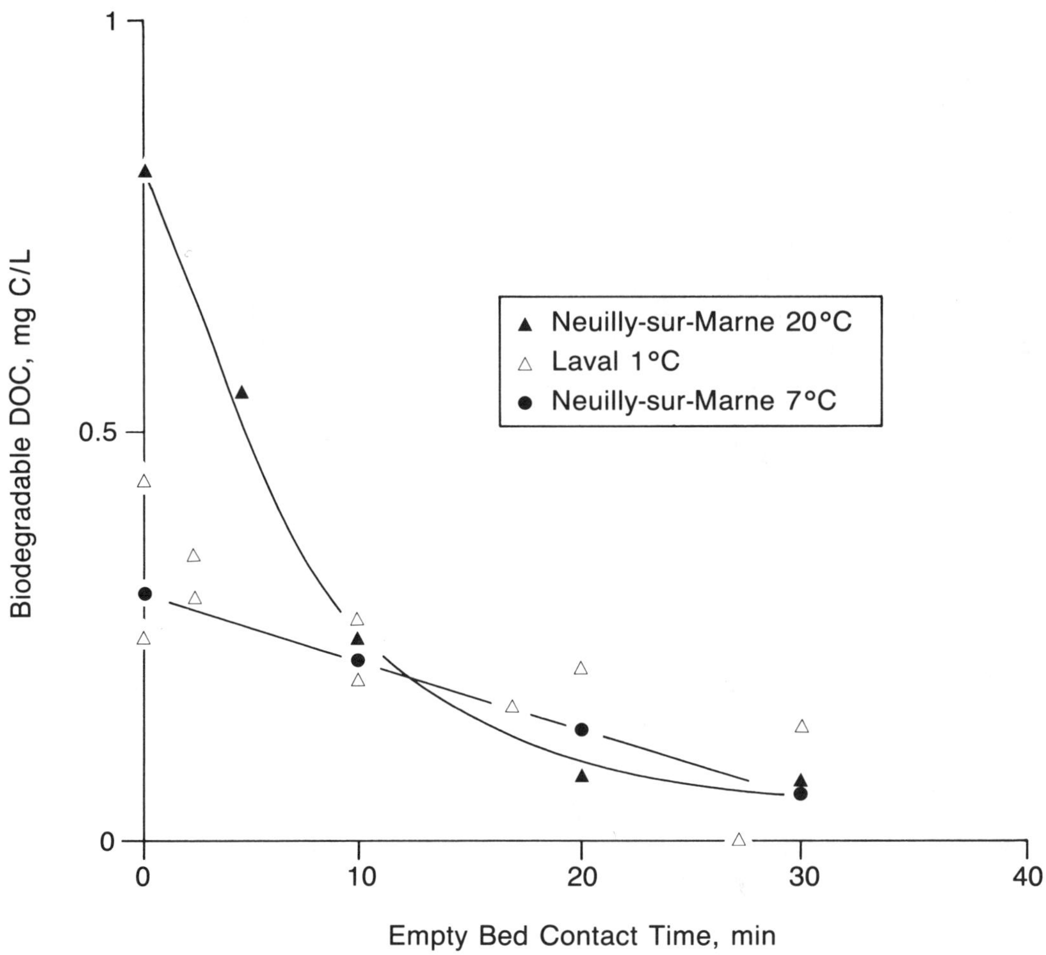

Source: Neuilly-sur-Marne: unpublished results; Laval: from Prevost et al. (1989).

Figure III–86 Effect of EBCT on the Removal of BDOC in GAC Filters Receiving Ozonated Water

times greater than that found on sand and other types of carbon placed in the same conditions. Although the large cavities (10–100 μm) in the PICABIOL® carbon grains are an important factor in their superior performance (LeChevallier et al. 1984), there are other aspects, such as grain size distribution and chemical or charge properties of the surface, that may also have an influence on colonization by bacteria. The influence of the media on BDOC removal is shown in Figure III–88 for a contactor operating at 20°C. In this case, an EBCT over 1 h would be necessary when using sand in order to reach the performance levels achieved in 10 min with PICABIOL®.

Abiotic considerations of GAC following ozonation. In addition to effects of biodegradation, ozone may also alter the abiotic adsorption of organic molecules onto carbon. In chapter II, sec. A.6, the ability of ozone to reduce molecular size, increase charge density, and increase hydrophilicity of natural organic matter was presented. Increases in hydrophilicity may tend to reduce equilibrium adsorptive capacity of the carbon (Kuhn et al. 1978; Topalian 1982; Hubele 1985; Chen et al. 1987). However, the accompanying decrease in molecular size may render more of the molecules accessible to the carbon micropores. The net effect of these two opposing phenomena cannot be predicted with certainty. Furthermore, smaller molecules have higher rates of diffusion; therefore, they should move more quickly into these micropores. These latter effects have been used to explain improved removals of DOC by GAC under conditions where

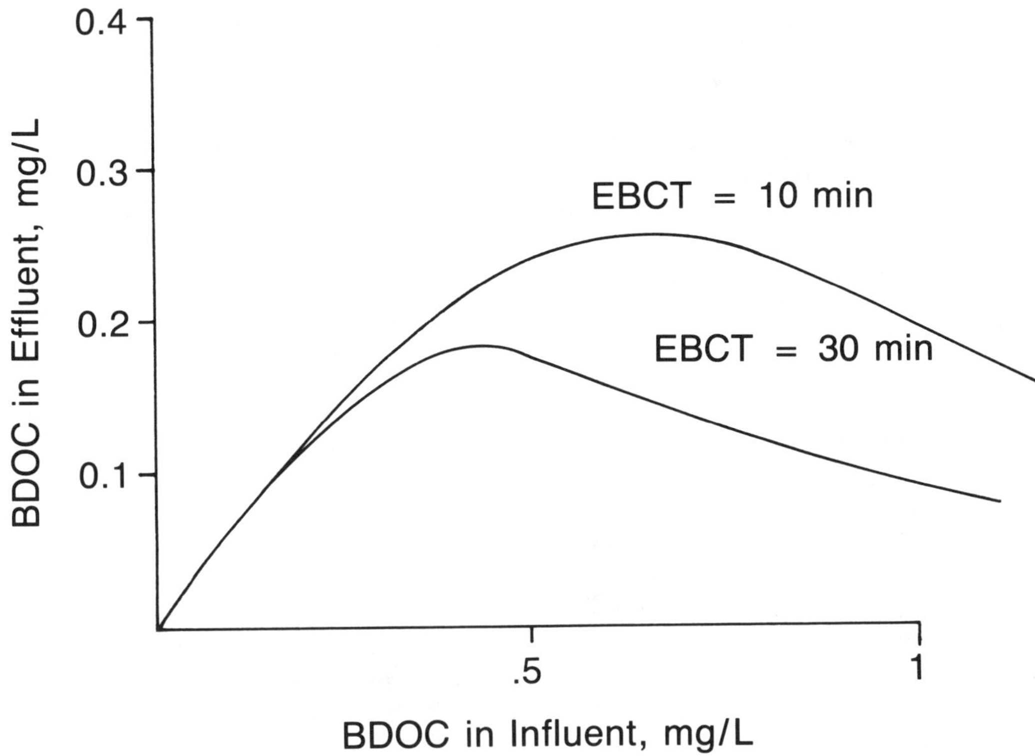

Source: Data from the Neuilly-sur-Marne pilot plant (from Billen et al. 1989).

Figure III–87 Effect of Influent BDOC on the GAC Effluent BDOC Concentration at Two Empty Bed Contact Times and 20°C (68°F)

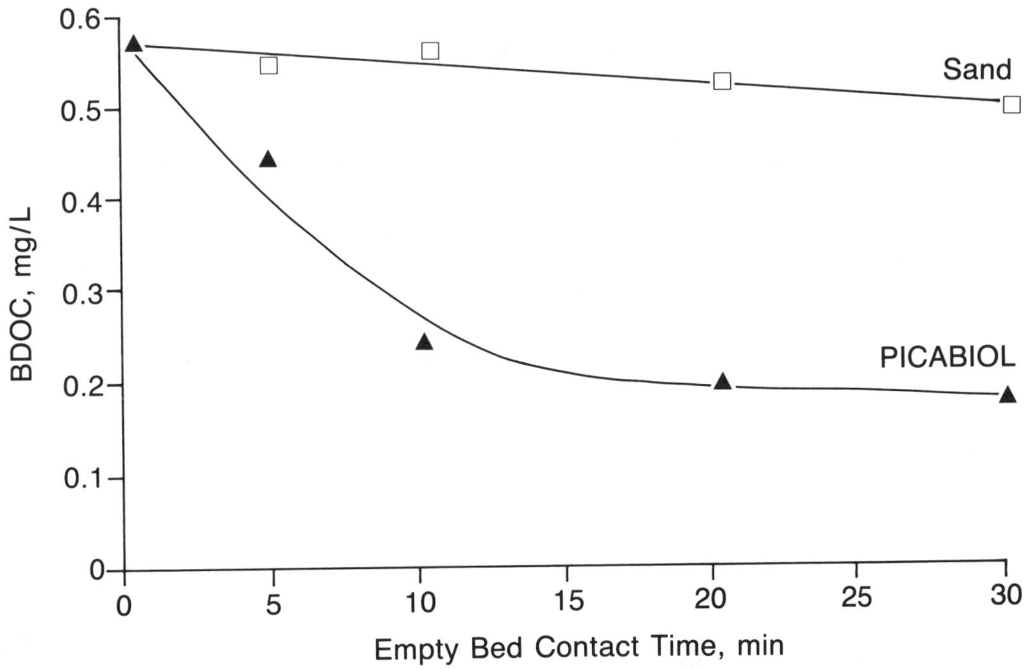

Source: Data from the Neuilly-sur-Marne pilot plant (method of Joret and Levi 1986).

Figure III–88 Effect of Filter Media and EBCT on the Removal of BDOC

biological activity is thought to be unimportant (Kaastrup et al. 1986; Malley et al. 1988). It is possible that with high-molecular-weight organic matter, low doses of ozone will result in improved adsorption characteristics, whereas higher doses should lead to a deterioration (e.g., Benedek et al. 1979). For additional information on abiotic adsorption onto GAC, the reader is referred to Sontheimer et al. (1988).

Secondary benefits of BAC processes. Experience at the Choisy-le-Roi plant indicates that BAC is especially effective at controlling THM concentrations in the finished water (Jadas-Hecart et al. 1988). Although ozone by itself did not significantly change the chlorine demand of the partially treated water, it did increase the biodegradability of the chlorine-demanding organic compounds. This led to a significant reduction in the chlorine demand of the BAC-treated water. As a result, lower chlorine doses could be applied to the finished water, while the desired chlorine residual is still maintained at the extremities of the distribution system. These lower chlorine doses resulted in significantly lower terminal THM concentrations. Tastes and odors associated with chlorination were also reduced (Bablon et al. 1986). The reader is referred to sec. III.I.4 for more information on the use of ozone alone and with BAC to lower chlorine demand and control disinfection by-products.

III.K.4 Design Considerations

Combined ozonation-GAC processes should be placed after settling and either before or after filtration in conventional treatment plants (i.e., as an additional treatment step) or in place of sand filtration in direct filtration plants. It is therefore often necessary with retrofitted systems to provide additional head (i.e., pumping) in order to work within the constraints of the existing plant hydraulics. In cases where some of the existing high-rate filters are available for conversion to BAC, the backwashing procedure and equipment must be reevaluated.

Based on the work at Choisy-le-Roi and Mery-sur-Oise (Billen et al. 1985, 1989), the important design parameters for biological stabilization were found to be:

1. the type of the media,
2. the empty bed contact time, and
3. the ozonation dose.

These must be evaluated for different temperatures and different concentrations of the water to be treated.

Design of the ozone contactors. The major role of ozonation upstream of the biological filtration stage is to increase the quantity of biodegradable dissolved organic carbon in the water. Numerous experiments (see sec. III.K.2) have shown that there exists an optimum ozone dose beyond which the biodegradable carbon no longer increases. It seems that the maximum production of BDOC is not affected by contact time, but rather (contrary to disinfection [see sec. III.H]) by the quantity of ozone transferred. In the case of new installations, the effect of ozone dosage on BDOC formation must be examined for each type of water. Even then, a verification of the influence of contact time on BDOC production is recommended. However, as a guideline, Sontheimer et al. (1988) recommend 0.5 mg O_3/mg C. This represents a compromise between maximum BDOC formation and excessive oxidation of the organic matter (i.e., poor adsorbability).

An example of an ozone contactor designed for BAC in the Paris suburban plants is given by Gerval et al. (1989; Figure III–89). Experience with these systems has led to the following general rules:

- Since ozone is typically used for biological stabilization downstream of clarification, ozone contacting can be done using porous plates, a system that is inexpensive and resistant to wear.
- The contactors must be designed for a plug-flow hydraulic regime (in water). This will reduce the size needed for adequate reaction.

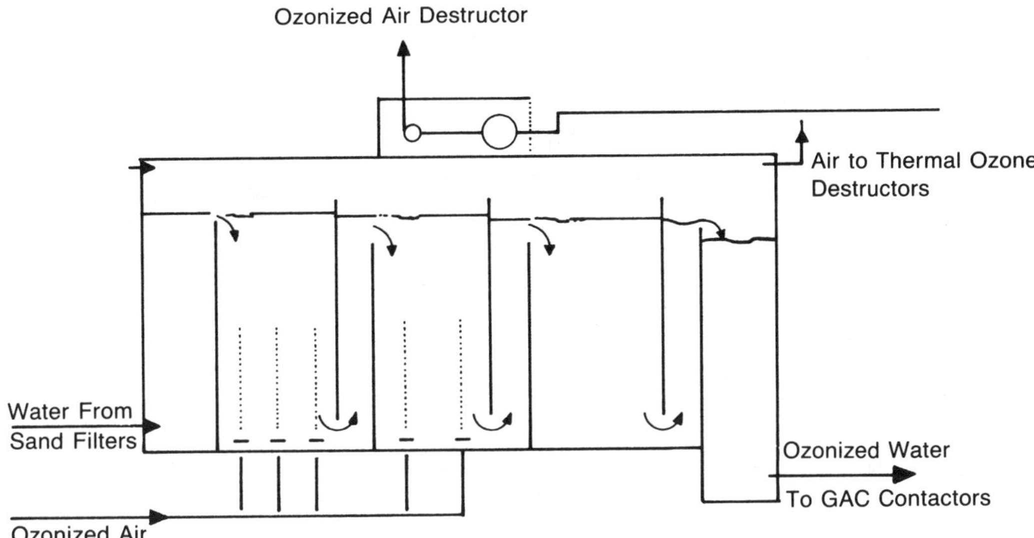

Source: Reprinted with permission from Gerval, R. et al., Proc. 9th Ozone World Congress, edited by J. Bollyky, pp. 486–495. © 1989 IOA.

Figure III–89 Ozone Contactor Design

- Counter-current contactors should be used to further minimize their size.
- The height of the water column, as well as the parameters assuring good contact in general, are in accordance with those generally recommended for ozone contactors (see sec. IV.D).

Design of GAC contactors. It is well established that the most important parameter for biological treatment in filters is the contact time. For a given water (specific BDOC concentration), and temperature, the use of activated carbon optimized for biomass support (e.g., PICABIOL®) makes it possible to minimize the contact time. For economic considerations, it is often preferable to use high-rate, deep-bed filters.

Again, the BAC system described by Gerval and co-workers (1989) provides one example of how a GAC contactor might be designed for biological treatment. For this particular plant, the filters were filled with 2.5 m (8.2 ft) of PICABIOL® and were operated at 18 m/h (7.4 gpm/ft^2). A schematic of this biological GAC contactor is shown in Figure III–90. It consists of:

- A concrete structure fitted with a special type of filter bottom with nozzles. The originality lies in the equal distribution of the air through the media during backwashing and air scouring.
- A system for introducing the treated water, consisting of a siphon and a depressurized air generator.
- A water flow rate regulator (POLYHYDRA® [Blanchard 1967]) that serves both to distribute the influent equally between the contactors and to regulate the output of each one. At the Choisy-le-Roi water treatment plant, this system is controlled by a computer.
- Backwashing control center common to all the contactors. Backwashing is indispensable to maintain the biological balance and to avoid the growth of higher organisms. When determining the frequency of backwashing runs, particular attention was given to the removal of protozoans and algae that provide food for predator organisms like *Nais*. In short, the backwashing operation is triggered by means of two pressure probes adequately positioned in the contactor. For the sequence especially devised for PICABIOL® GAC, the following schedule was adopted (Figure III–91):

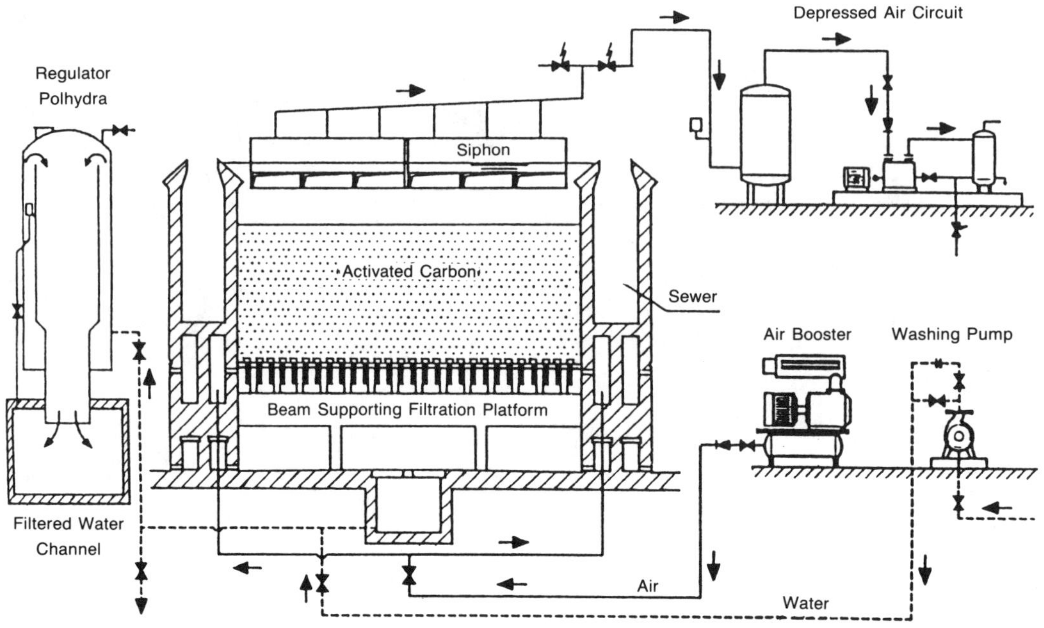

Figure III–90 Flow Diagram of a GAC Biological Contactor

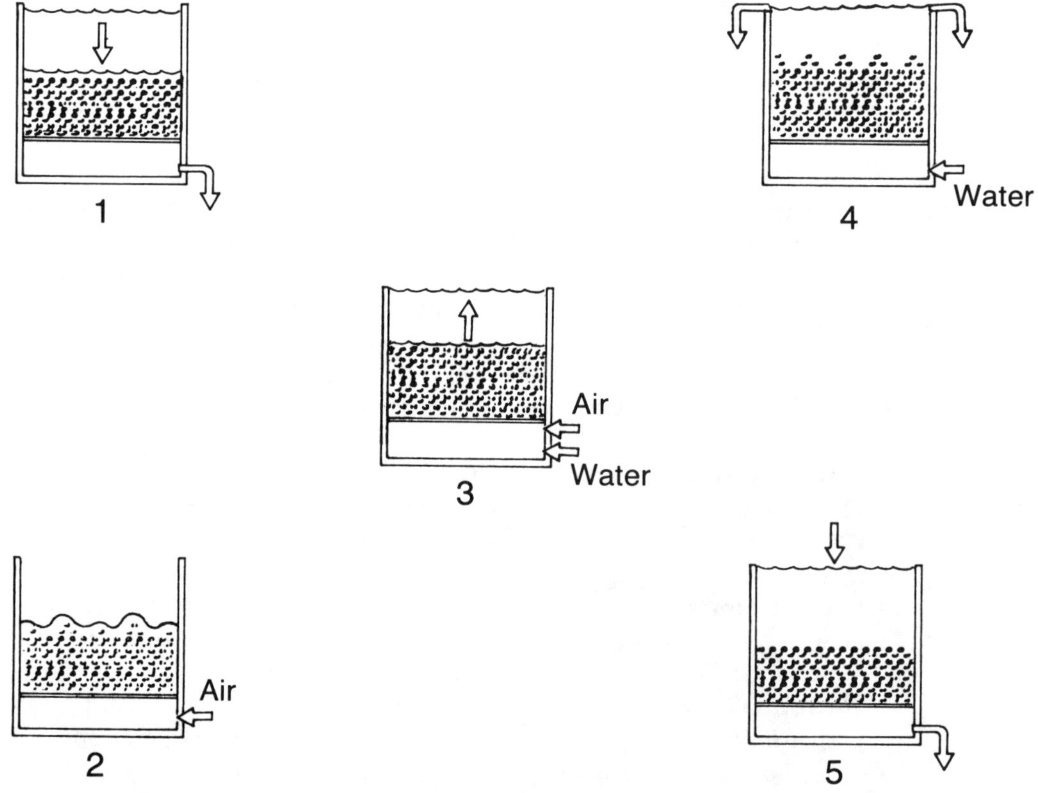

Figure III–91 Backwashing of a GAC Biological Contactor

Step 1: The water level on the contactors is lowered to a level sufficient for the simultaneous backwash and air-scour sequence.

Step 2: When the level of the water reaches the top of the GAC, scour air is blown in to loosen the grains.

Step 3: Air and water are injected simultaneously without causing a loss of carbon, since the operation is designed to keep the water level below the weir.

Step 4: The air flow is stopped and a rinsing with water is performed for about a quarter of an hour.

Step 5: Filtration to waste. The objective of this phase is to discharge the interstitial water to the drain before resuming contactor services. The effluent produced by the contactor at the beginning of the new run is of poor quality, a situation that will last about 10 min, depending on the circumstances. This stage, in fact, gives the carbon time to recover its normal state of permeability after disturbance by fluidization, when the bed expansion can reach as much as 14 percent of its normal expansion.

References

ADLER, P.M. ET AL. 1985. Ozone Transfer Enhancement in Bubble Columns: New Injection Device. Proc. Intl. Conf.: The Role of Ozone in Water and Wastewater Treatment, edited by R. Perry and A.E. McIntyre. Selper Ltd., London (pp. 18–32).

AFNOR. 1984. Recueil des Normes Françaises (NFT 90-038). Paris Association de Normalisation.

_____ . 1986a. Recueil des Normes Françaises NFT 90-030. Paris Association de Normalisation.

_____ . 1986b. Recueil de Normes Françaises (eaux, méthodes d'essai). 3éme édition. Paris Association de Normalisation (edit).

AFANASYEV, V.A. ET AL. 1981. Ozonation of Potable Water in Moscow. Proc. Wasser Berlin '81, pp. 362–366.

ALBERT, G. 1976. The Influence of Dissolved Organic Compounds in Flocculation. In *Special Problems of Water Technology—Adsorption,* EPA-600/9-76-030, pp. 355–379.

ALI, W. ET AL. 1984. Colloidal Stability of Particles in Lakes: Measurement and Significance. *Wat. Sci. Tech.,* 17:701–712.

ALIZADEH ASTARI, A. 1978. Influence de la Préozonation sur l'Efficacité de la Coagulation des Eaux. Report, D.E.P.S. Hydrologic, Université de Paris, XI.

AMOORE, J.E. 1986. The Chemistry and Physiology of Odor Sensitivity *Jour. AWWA,* 78:3:70.

ANBAR, M. & NETA, P. 1967. A Compilation of Specific Bimolecular Rate Constants for the Reaction of Hydrated Electrons, Hydrogen Atoms and Hydroxyl Radicals with Inorganic and Organic Compounds in Aqueous Solution. *Intl. Jour. Appl. Rad. Iso.,* 18:493.

ANDERSON, K. & JAMIESON, A. 1972. Primary Amoebic Meningoencephalitis. *Lancet,* 1:902.

ANDERSON, L.J. ET AL. 1985. The Reaction of Ozone with Isolated Aquatic Fulvic Acid. *Org. Geochem.,* 8:1:65.

ANDO, M. ET AL. 1987. Deodorization by Two Step, High Rate Biological Filtration. Second International Symposium on Off-Flavors in the Aquatic Environment, Kagoshima, Japan.

ANSELME, C. ET AL. 1985. Influence des Traitements de Désinfection et d'Oxydation sur les Qualités Organoleptiques de l'Eau: Cas de l'Usine de Morsang sur Seine. 38ème Journée Internationale du Cebedeau.

_____ . 1987a. Les Problèmes de Mauvais Goûts dans l'Eau Potable. Construction d'un Modèle Prédictif. *TSM l'Eau,* Juin:261.

_____ . 1987b. Removal of Tastes and Odors by Ozonation. Proc. of the 8th Ozone World Congress, IOA, Zürich, Switzerland (pp. C102–C127).

_____ . 1988. Effects of Ozone on Tastes and Odors. *Jour. AWWA,* 80:10:45.

AUVRAY, J. & DUPUY, S. 1988. Traitements Possibles pour les Retenues. Colloque, Traitement Curatifs en Vue de la Potabilisation des Eaux Chargées en Microphytes Dulsaquicoles, Rennes, France (Nov. 15–16).

AWWA RESEARCH FOUNDATION, 1989. Identification and Occurrence of Ozone By-products in Drinking Water. J. Oxenford, project officer, Denver, CO.

BABLON, G. 1981. L'Analyse Global Multicritère Multivariable: Une Démarche Nouvelle pour le Traitement des Eaux de Surface: Application au Couplage Ozonation-Coagulation. *La Tech. de l'Eau et de l'Assain.,* 416/417:9–22.

BABLON, G. & SAUNIER, B.M. 1983. *Ozonation: Environmental Impact and Benefit,* IOA.

BABLON, G. ET AL. 1986. Combined Use of Ozone and BAC in a Large Treatment Plant. *Wtr. Supply,* 4:3:35.

_____ . 1987a. Evolution des Techniques de Filtration Rapide: Le Bicouche Sable-Charbon Actif. *TSM L'Eau*, 82:4:153.

_____ . 1987b. Evolution of Organics in Potable Water Treatment System. *Aqua*. 2:110.

BACKLUND, P. ET AL. 1985. Mutagenic Activity in Humic Water and Alum Flocculated Humic Water Treated with Alternative Disinfectants. *Sci. Total Environ.*, 47:257.

_____ . 1988. Formation of Ames Mutagenicity and of the Strong Bacterial Mutagen 3-Chloro-4-(dichloromethyl)-5-hydroxy-2 (5H)-furanone and other Halogenated Compounds During Disinfection of Humic Water. *Chemosphere*, 17: 1329–36.

BAILEY, P.S. 1978. *Ozonation in Organic Chemistry, Vol. 1, Olefinic Compounds*. Academic Press, New York.

BARKER, M.R. & PALMER, D.J. 1977. *Water SA*, 3:2:58–65.

BARKER, R. & JONES, A.R. 1988. Treatment of Malodorants in Air by the UV/Ozone Technique. *Ozone Sci. Engrg.*, 10:4:405.

BECKE, CH. ET AL. 1984. Herkunft von Trichlornitromethane in Trinkwasser. *Vom Wasser*, 62: 125.

BELL, J. 1987. Ozone Production and Application Technology. Proc. 2nd Intl. Conf.: The Role of Ozone in Water and Wastewater Treatment, edited by D.W. Smith and G.R. Finch (pp. 105–124). Tektran Int. Ltd., Kitchener, Ontario.

BELLAMY, W.D. 1989. Unpublished. CH2M Hill, Inc., Denver, CO.

BELLAMY, W.D. ET AL. 1985. Slow Sand Filtration: Influence of Selected Process Variables. *Jour. AWWA*, 77:12:62.

_____ . 1990. In-line Ozone Dissolution Demonstration-Scale Evaluation. Presented at the IOA PAC. Congress, Shreveport, Louisiana, March 27–29.

BEN AMOR, H. 1988. Le Bioxyde de Chlore dans la Filière de Production des Eaux de Consommation: Contribution à l'Etude de son Mécanisme d'Action sur des Molécules Modèles et sur des Substances Humiques Aquatiques. Thèse de l'Université de Poitiers, France.

BENEDEK, A. ET AL. 1979. The Effect of Ozone on the Biological Degradation and Activated Carbon Adsorption of Natural and Synthetic Organics in Water. Part II. Adsorption. *Ozone Sci. Engrg.*, 1:4:347–356.

BERNHARDT, H. 1988. Studies on the Treatment of Eutrophic Water. Presented at the 17th Congress IWSA, Special Subject 12, Rio de Janeiro, Brazil (Sept. 12–16).

BERNHARDT, H. & LÜSSE, B. 1989. Elimination of Zooplankton by Flocculation and Filtration. *Aqua*, 38:1:23.

BERNHARDT, H. ET AL. 1985. Reaction Mechanisms Involved in the Influence of Algogenic Organic Matter on Flocculation. *Z. Wasser-Abwasser-Forsch*, 18:18.

BETZER, N. ET AL. 1980. Effluent Treatment and Algae Recovery by Ozone Induced Flotation. *Wtr Res.*, 14:1003–1009.

BILLEN, G. ET AL. 1985. Action des Populations Bacteriennes Vis-a-Vis des Matieres Organiques dans les Filtres Biologiques. Report to Compagnie Générale Des Eaux, Paris.

_____ . 1989. Etude des Conditions de Fonctionnement des Contacteurs Biologiques dans le Traitement des Eaux. Report to Compagnie Générale Des Eaux, Paris.

BLACK, A.P. & CHRISTMAN, A.F. 1963. Chemical Characteristics of Fulvic Acids. *Jour. AWWA*, 55:897.

BLANCHARD, P. 1967. Equal Division and Regulation of Treatment Plant Flows. *Houille Blanche*, 4:431.

BLOCK, J.C. 1978. Revue Bibliographique de l'Action de l'Ozone sur les Microorganismes. Séminaire Ozone du GRUTTEE, Paris.

BLOCK, J.C. ET AL. 1981. Désinfection des Eaux par l'Ozone. *L'Eau et L'Industrie*, 58:69–76 (Oct.).

BOLLYKY, L.J. & SILER, J. 1989. Removal of Tributylphosphate from Water by Advanced Oxidation Methods. Proc. 9th Ozone World Congress, New York, 1:634.

BONTOUX, J. ET AL. 1976. Mecanisme d'Action des Alginates dans le Traitement des Eaux de Boissons. *J. Français d'Hydrol.*, 7:2:71–84.

BOURBIGOT, M.M. 1988. Ozone Disinfection in Drinking Water. NEWWA 107th Ann. Conf., Dixville Notch, N.H.

BOURBIGOT, M.M. & FAIVRE, M. 1986. Patent No. 86.08780: June 18, 1986. Société OTV. "Dispositif de Transfert de Gaz et Flottation dans le Traitement d'Eau à Epurer."

BOURBIGOT, M.M. ET AL. 1985a. The Simultaneous Use of Ozone and Ultraviolet Rays in Water Treatment. Presented at the International Ozone Association Congress, Berlin.

_____ . 1985b. The Role of Ozone and Granular Activated Carbon in the Removal of Mutagenic Compounds. Presented at the 2nd Intl. Symp. on Health Effects of Drinking Water Disinfectants and Disinfection By-products, Cincinnati, OH.

_____ . 1986a. Traitement des Eaux de Barrage. Cas de l'Usine de Moulin-Papon. Séminaire GRUTTEE: "Substances Humiques" (Rennes, France).

_____ . 1986b. Role of Ozone and Granular Activated Carbon in the Removal of Mutagenic Compounds. *Envir. Health Persp.*, 69:159–163.

_____ . 1990. L'Ozoflottation. Presented at the AGHTM Workshop (Upper and Lower Normandy section), Mont St. Michel, France (April 20).

Bouwer, E.J. & Crowe, P.B. 1988. Biological Processes in Drinking Water. *Jour. AWWA,* 80: 9:82–93.

Boyce, D.S. et al. 1981. The Effect of Bentonite Clay on Ozone Disinfection of Bacteria and Viruses in Water. *Wtr. Res.,* 15:759–767.

Bringmann, G. 1954. Determination of the Lethal Activity of Chlorine and Ozone on *E. Coli. Z. f. Hygiene,* 139:130–139.

Britton, A. & McFadzean, C.J. 1984. The Effect of Ozone on the Performance of Rapid Gravity Filters. Presented at the IWES and WRC Ozone Seminar.

Broadwater, W.T. et al. 1973. Sensitivity of Three Selected Bacterial Species to Ozone. *Appl. Microbiol.,* 26:3:391–393.

Bruchet, A. et al. 1985. Characterization of Total Halogenated Compounds During Various Water Treatment Processes. In *Water Chlorination: Chemistry, Environmental Impact and Health Effects, vol. 5,* edited by R.L. Jolley et al. Lewis Publishers, Inc., Chelsea, MI, pp. 1165–1184.

_____ . 1987. Use of Gel Permeation Chromatography to Study Water Treatment Processes. In *Organic Pollutants in Water,* edited by I.H. Suffet and M. Malaiyandi. Advances in Chemistry series 214, American Chemical Society, Washington, D.C. (pp. 381–399).

Brunet, R. 1985. Rapport d'Etudes: "Oxydants Combinés," Convention de Recherche. Compagnie Générale des Eaux. Laboratoire de Chimie de l'Eau et des Nuisances. Université de Poitiers (France).

Brunet, R. et al. 1980. Evolution Chimique des Matieres Organiques au Cours de l'Ozonation d'une Eau de Surface Filtrée. *Aqua,* 4:76.

_____ . 1982. The Influence of the Ozonation Dosage on the Structure and Biodegradation of Pollutants in Water and Its Effect on Activated Carbon Filtration. *Ozone Sci. Engrg.,* 4:1:15.

_____ . 1984. Oxidation of Organic Compounds Through the Combination of Ozone and Hydrogen Peroxide. *Ozone Sci. Engrg.,* 6: 3:175.

Bryani, E.A. & Yapijakij, C. 1977. Ozonation Diatomite Filtration: Removes Color and Turbidity. *Wtr. Sew. Works,* 24:9:96.

Bull, R.J. et al. 1982. Use of Biological Assay Systems to Assess the Relative Carcinogenic Hazards of Disinfection By-products. *Envir. Health Persp.,* 46:215–227.

Burleson, G.R. et al. 1975. Inactivation of Viruses and Bacteria by Ozone, With and Without Sonication. *Appl. Microbiol.* 29:340–344.

Burlingame, G.A. et al. 1986. A Case of Geosmin in Philadelphia's Water. *Jour. AWWA,* March, pp. 56.

Burrows, G.O. & Woodward, A.J. 1988. Practical Approaches to Minimizing Algal Effects on Water Quality. Presented at the workshop "Traitements Curatifs en Vue de la Potabilisation des Eaux Chargées en Microphytes Dulçaquicoles" (Rennes, France), November 15–16.

Burttschell, R.H. et al. 1959. Chlorine Derivatives of Phenol Causing Taste and Odor. *Jour. AWWA,* 51:2:205.

Buttery, R.G. & Ling, L.C. 1973. Earthy Aroma of Potatoes. *Jour. Agric. Food Chem.* 21: 745.

Buydens, R. 1970. L'ozonation et ses Répercussions sur le Mode d'Epuration des Eaux de Rivières. *Trib. Cebedeau* 319–320:286–291.

Campbell, I. et al. 1982. Effect of Disinfectants on Survival of *Cryptosporidium* Oocysts. *Vet. Rep.,* 111:414–415.

Campbell, R.M. & Pescod, M.B. 1966. *Wat. Sew. Works,* July, pp. 268–272.

Carlberg, G.E. et al. 1981. Bleaching of Dissolving Sulfite Pulps. *Das Papier,* 35:257–264.

Carmichael, N.G. et al. 1982. The Health Implications of Water Treatment with Ozone. *Life Sci.,* 30:117–129.

Carns, K.E. & Stinson, K.B. 1978. Controlling Organics: The East Bay Municipal Utility District Experience. *Jour. AWWA,* 70:11:637.

CDM–BC. 1982. Los Angeles Aqueduct Water Filtration Plant Final Design Report, January. Camp Dresser & McKee, and Brown and Caldwell, Consulting Engineers, Boston, MA.

Chang, E.H. et al. 1960. Survey of Free-Living Nematodes and Amoebas in Municipal Supplies. *Jour. AWWA,* 52:5:613.

Chang, H.T. & Rittmann, B.E. 1987. Verification of the Model of Biofilm on Activated Carbon. *Envir. Sci. Technol.,* 21:3:280–288.

Chang, S.D. & Singer, P.C. 1988. The Impact of Ozonation on the Removal of Particles, TOC, and THM Precursors. Proc. AWWA Ann. Conf., Orlando, Fla., pp. 1325–1342.

Chang, S.L. 1971. Modern Concept of Disinfection. *J. Sanit. Engin. Division,* 97:689–707.

Chang, S.L. & Fair, G.M. 1941. Viability and Destruction of the Cysts of *Entamoeba histolytica. Jour. AWWA,* 33:1705–1715.

Chedal, J. 1982. In *Ozone Technology and its Practical Applications,* Ann Arbor Science Publishers, Inc., Ann Arbor, MI.

_____. 1985. "Raw Water Ozonation." In *Handbook of Ozone Technology and Applications,*

Vol. II: Ozone for Drinking Water Treatment, edited by R.G. Rice & A. Netzer, Butterworth Publishers, Stoneham, MA.

______. 1987. La Saveur de l'Eau depuis la Ressource jusqu'au Robinet du Consommateur. *TSM l'Eau,* 10:419.

CHEDAL, J. & SCHULHOF, P. 1979. Reduction of the Content of Chlorine Compounds by a Treatment Combining Physico-chemical and Biological Process. In *Oxidation Techniques in Drinking Water Treatment,* pp. 176–194, EPA-570/9-79-020, U.S. Government Printing Office.

CHEN, A.S.C. ET AL. 1987. Activated Alumina Adsorption of Dissolved Organic Compounds Before and After Ozonation. *Envir. Sci. Technol.,* 21:1:83–90.

CHEN, Y.S.R. ET AL. 1985. Inactivation of *Naegleria gruberi* Cysts by Chlorine Dioxide. *Wtr. Res.,* 19:6:783–789.

CHERRY, A.K. 1962. Use of Potassium Permanganate in Water Treatment. *Jour. AWWA,* 54:417.

CHICK, H. 1908. An Investigation of the Laws of Disinfection. *J. Hyg.,* 8:92–158.

CHRISTENSEN, E. & GIESE, A.C. 1954. Changes in Absorption Spectra of Nucleic Acids and their Derivatives Following Exposure to Ozone and Ultraviolet Radiations. *Arch. Biochem. Biophys.,* 51:208–216.

CHRISTIE, J.D. 1990. Personal communication. Los Angeles Dept. of Water and Power, Los Angeles, CA.

CHROSTOWSKI, P.C. ET AL. 1983. Ozone and Oxygen Induced Oxidative Coupling of Aqueous Phenolics. *Wat. Res.* 17:11:1627–1633.

CLEASBY, J.L. 1975. Iron and Manganese Removal—A Case Study. *Jour. AWWA,* 67:3:147.

COGNET, L. ET AL. 1986. Mutagenic Activity of Disinfection By-products. *Envir. Health Persp.,* 69:165–175.

COHEN, J.M. ET AL. 1960. Taste Threshold Concentrations of Metals in Drinking Water. *Jour. AWWA,* 52:5:660.

COHEN, R.S. ET AL. 1978. Controlling Organics: The Metropolitan Water District of Southern California Experience. *Jour. AWWA,* 70:647.

COIN, L. ET AL. 1964. Inactivation par l'Ozone du Virus de la Poliomyelite Présent dans les Eaux. *La Presse Méd.,* 72:37:2153–2156.

______. 1967. Inactivation par l'Ozone du Virus de la Poliomyelite Présent dans les Eaux (Nouvelle Contribution). *La Presse Méd.,* 75:38:1883–1884.

COLBOURNE, J. ET AL. 1990. Thames Water's Experiences with *Cryptosporidium.* IAWPRC. International Symposium on Health Related Water Microbiology, Tübingen, Germany, April 1–6.

COLEMAN, E. ET AL. 1984. Gas Chromatography/Mass Spectrometry of Mutagenic Extracts of Aqueous Chlorinated Humic Acid. A Comparison of the By-products to Drinking Water Contaminants. *Envir. Sci. Tech.,* 18:674–681.

COLIN, J.L. ET AL. 1982. *Trib. Cebedeau,* 459:57–70.

COLLIENNE, R. & CORNET, J.C. 1978. Etude sur Station Pilote des Possibilités de Traitement d'Eau Douce, Agressive et Manganifère. *La Tech. de l'Eau et de l'Assain.,* 380/381: 31:45.

COLTHURST, J.M. & SINGER, P.C. 1982. Removing Trihalomethane Precursors by Permanganate Oxidation and Manganese Dioxide Adsorption. *Jour. AWWA,* 74:2:78–83.

COMPAGNIE GÉNÉRALE DES EAUX. 1977. Recherche sur Pilote à Choisy-le-Roi des Traitements Permettant de Réduire les Seuils de Saveur de l'Eau de Seine au Cours de l'Été 1976–1977. Rapport d'Etude AFBSN.

COMYNS, A.E. & REDHAYNE, W.H. 1983. The Activation of Hydrogen Peroxide in Industrial Effluent Treatment. Proc. IOA Symp., Brussels, Belgium.

CONDIE, L. ET AL. 1985. Subchronic Toxicology of Humic Acid Following Chlorination in the Rat. *J. Toxicol. Envir. Health,* 15:305–314.

CONSTANTINE, T.A. 1982. Advanced Water Treatment for Color and Organic Removal. *Jour. AWWA,* 74:6:310.

CRAUN, G.F. 1988. Surface Water Supplies and Health. *Jour. AWWA,* 80:2:40.

CROMLEY, J.T. & O'CONNOR, J.T. 1976. Effect of Ozonation on the Removal of Iron from a Ground Water. *Jour. AWWA,* 68:6:315–319.

CRONHOLM, L.S. ET AL. 1976. Enteric Virus Survival in Package Plants and the Upgrading of the Small Treatment Plants Using Ozone. Research Report No. 98, Water Resources Research Institute, University of Kentucky, Lexington, KY.

CROUE, J.-P. 1987. Contribution a l'Etude de l'Oxydation par le Chlore et l'Ozone d'Acides Fulviques Naturels Extraits d'Eaux de Surface. Doctoral dissertation, University of Poitiers, France.

CRYER, E.T. 1986. Proc. of the IOA Spec. Conf. on Develop. & Oper. of Ozonation Drinking Wat. Plants, Perrysburg, Ohio.

CURRENT, W.L. 1987. *Cryptosporidium:* Its Biology and Potential for Environmental Transmission. *Crit. Rev. Envir. Control,* 17:21–33.

CURRENT, W.L. ET AL. 1983. Human Cryptosporidiosis in Immunocompetent and Immunodeficient Persons: Studies of an Outbreak and Experimental Transmission. *New Engl. J. Med.,* 308:1252–1257.

DAMEZ, F. & DERNAUCOURT, J.C. 1979. L'Ozone avant Floculation. *La Tech. de l'Eau et de l'Assain.,* 388:17–28.

DANIEL, F.B. ET AL. 1989. Subchronic Toxicity of Ozonated/Chlorinated Humics in Sprague-Dawley Rats. *Envir. Sci. Technol.* (submitted).

DANIEL, P.A. & MEYERHOFER, P.F. 1989. Ozonation of Taste and Odor Causing Compounds. 1989. Proc. 9th Ozone World Congress, vol. 1, edited by L.J. Bollyky. IOA, New York, pp. 239–242.

DAVIS, J.A. & GLOOR, R. 1981. Adsorption of Dissolved Organics in Lake Water by Aluminum Oxide. Effect of Molecular Weight. *Envir. Sci. Technol.,* 15:10:1223–1229.

DAVIS, J.A. & LECKIE, J.O. 1978. Effect of Adsorbed Complexing Ligands on Trace Metal Uptake by Hydrous Oxides. *Envir. Sci. Technol.,* 12:12:1309–1315.

Decret No. 89.3 du 3 Janvier 1989 Relatif aux Eaux Destinées à la Consommation Humaine à l'Exclusion des Eaux Minérales Naturelles. *Journal officiel de la République Française du 4 janvier 1989:*125.

DE LAAT, J. 1988. Contribution à l'Etude du Mode d'Elimination de Molécules Organiques Modèles sur Charbon Actif en Grains. Interactions entre les Processus d'Absorption et de Biodégradation. Doctorat d'Etat, Université de Poitiers, France.

D'ELIA, P. ET AL. 1981. Experiences d'Italie dans l'Emploi de l'Ozone, Proc. Wasser Berlin '81, pp. 411–420.

DEMPSEY, B.A. & O'MELIA, C.R. 1983. "Proton and Calcium Complexation of Four Fulvic Acid Fractions." In *Aquatic and Terrestrial Humic Materials,* edited by R.F. Christman and E.T. Gjessing, Ann Arbor Science Publishers, Inc., Ann Arbor, MI.

DEMPSEY, B.A. ET AL. 1984. The Coagulation of Humic Substances by Means of Aluminium Salts. *Jour. AWWA,* April:141.

DENKHAUS, R. ET AL. 1980. Removal of Mutagenic Compounds in a Wastewater Reclamation System Evaluated by Means of the Ames *Salmonella*/Microsome Assay. *Prog. Water Technol.,* 12:571–589.

DERREUMAUX, A.L. ET AL. 1974. Action du Chlore sur les Amibes de l'Eau. *Ann. Soc. Belge Med. Trop.,* 54:415–428.

DIAPER, E.W.J. 1970. *Wat. Sew. Works,* 117: 11:373–380.

_____. 1972. "Practical Aspects of Water and Waste Water Treatment by Ozone." In *Ozone in Water and Wastewater Treatment,* edited by F.L. Evans III, Ann Arbor Science Publishers, Inc., Ann Arbor, MI, 145–179.

DICE, J.C. 1975. The Challenge to the AWWA Taste and Odor Control Committee. Proc. AWWA WQTC, Atlanta, Ga.

DIXON, W.J. & MASSEY, F.J. 1960. *Introduction of Statistical Analysis.* McGraw-Hill, New York.

DOLARA, P. ET AL. 1981. Effect of Ozonation and Chlorination on the Mutagenic Potential of Drinking Water. *Bull. Environ. Contam. Toxicol.,* 27:1–6.

DOMINGUE, E.L. ET AL. 1988. Effects of Three Oxidizing Biocides on *Legionella pneumophila* Serogroup 1. *Appl. Environ. Microbiol.,* 54:741–747.

DORÉ, M. & LEGUBE, B. 1983. Mecanisme d'Action de l'Ozone sur les Composées Aromatiques Simples. *J. Français d'Hydrologie,* 40:11–30.

DORÉ, M. ET AL. 1980. Mechanism of the Reaction of Ozone with Soluble Aromatic Pollutants. *Ozone Sci. Engrg.,* 2:1:39–54.

_____. 1987. The Role of Alkalinity in the Effectiveness of Processes of Oxidation by Ozone. In The Role of Ozone in Water and Wastewater Treatment. Proc. 2nd International Conference, edited by D.W. Smith and G.R. Finch. Tektran Int. Ltd., Kitchener, Ontario.

_____. 1988. Interactions Between Ozone, Halogens, and Organic Compounds. *Ozone Sci. Engrg.,* 10:2:153–172.

DOUGHERTY, J.D. & MORRIS, R.L. 1967. Studies on the Removal of Actinomycetes, Musty Tastes, and Odors in Water Supplies. *Jour. AWWA,* 59:1320.

DOWBIGGIN, W.B. & SINGER, P.C. 1989. Effects of Natural Organic Matter and Calcium on Ozone-Induced Particle Destabilization. *Jour. AWWA,* 81:6:77–84.

DUCHESNE, D. ET AL. 1988. Quality Control of the Water Filtered on Granular Activated Carbon at Sainte-Rose Water Plant. *Sci. Tech. Eau,* 21:1:33.

DUGUET, J.P. ET AL. 1985a. Improvement in the Effectiveness of Ozonation of Drinking Water Through the Use of Hydrogen Peroxide. *Ozone Sci. Engrg.,* 7:3:241–258.

_____. 1985b. Polymerisation des Matières Organiques de l'Eau: Comparison des Effects de l'Ozone et de la Peroxydase sur le 2,4-Dichlorophenol. Proc. Wasser Berlin '85 (pp. 165–177).

_____. 1989. Geosmin and 2-Methylisoborneol Removal Using Ozone or Ozone/Hydrogen Peroxide Coupling. Proc. 9th Ozone World Congress, edited by L.J. Bollyky. IOA, New York (pp. 709–719).

EDWARDS, S.E. & MCCALL, G.B. 1946. Manganese Removal by Breakpoint Chlorination. *Wat. Sew. Works,* 93:303.

EEC Council Directive to the Quality of Water Intended for Human Consumption (80778/EEC). 1980. *Official Journal of the European Communities,* L229:11.

Elmghari-Tabib, M. et al. 1982. Ozonation Reaction Patterns of Alcohols and Aliphatic Amines. *Ozone Sci. Engrg.*, 4:4:195–205.

Emerson, M.A. et al. 1982. Ozone Inactivation of Cell Associated Viruses. *Appl. Environ. Microbiol.*, 43:603–608.

Evison, L.M. 1977. Disinfection of Water with Ozone: Comparative Studies with Entero Viruses, Phage and Bacteria. 3rd Congrès I.O.I., Paris.

Fair, G.M. et al. 1968. *Water and Wastewater Engineering. Vol. 2. Water Purification and Wastewater Treatment and Disposal.* John Wiley and Sons, Inc., New York.

Faivre, M. et al. 1988. Ozoflottation et Techniques Complémentaires. Presented at the workshop "Traitements Curatifs en Vue de la Potabilisation des Eaux Chargées en Microphytes Dulçaquicoles," Rennes, France (Nov. 15–16).

_____. 1989. Essais de Désalgage des Eaux par le Procédé "d'Ozone-Flottation" sur l'Usine d'Autun. *TSM l'Eau,* Sept., p. 485.

Farooq, S. et al. 1977a. The Effect of Ozone Bubbles on Disinfection. *Prog. Wat. Tech.*, 9: 233–247.

_____. 1977b. Basic Concepts in Disinfection with Ozone. *Jour. WPCF,* 49:1818–1831.

Fayer, R. & Ungar, B.L.P. 1986. *Cryptosporidium* spp. and Cryptosporidiosis. *Microbiol. Rev.*, 50:458–483.

Felix-Filho, J.A. 1985. The Effects of Aquatic Humic Substances and Ozonation on the Stability of Particles, Ph.D. Dissertation, University of North Carolina, Chapel Hill, NC.

Ficek, K.J. 1978. "Potassium Permanganate for Iron and Manganese Removal." In *Water Treatment Plant Design for the Practicing Engineer,* edited by R.L. Sanks, Ann Arbor Science Publishers, Inc., Ann Arbor, Mich.

Fiessinger, F. & Richard, Y. 1975. La Technologie du Traitement des Eaux Potables par le Charbon Actif Granulé: I. Le Choix du Charbon. *TSM L'Eau,* 7:271.

Fiessinger, F. et al. 1979. *Journees Europeennes sur les Applications de l'Ozone du Traitement des Eaux,* IOA, pp. 245–321.

Flogstad, H. & Odegaard, H.O. 1985. Treatment of Humic Waters by Ozone. *Ozone Sci. Engrg.*, 7:2:121.

Foster, D.M. et al. 1980. Ozone Inactivation of Cell- and Fecal-Associated Viruses and Bacteria. *Jour. WPCF,* 52:2174–2184.

Francis, P.D. 1987. Oxidation by UV and Ozone of Organic Compounds Dissolved in Deionized and Raw Mains Water. *Ozone Sci. Engrg.*, 9:4:369.

Fronk, C.A. 1987. Destruction of Volatile Organic Contaminants in Drinking Water by Ozone Treatment. *Ozone Sci. Engrg.*, 9:3: 265.

Fujikawa, E.G. et al. 1989. Ozonation in America: An Evolution of Success. *Wtr. Engrg. Mgmt.*, Oct., pp. 20–24.

Fulton, G.P. 1980. New York City's Pilot Plant Filters Studies II. *Public Works,* 11:5:85.

Fung, L.C. et al. 1988. Ozone: From Pilot to Design to Operation. In Ozonation Systems: Design, Operation, and Maintenance, Proc. of the Monroe, MI, Conf., IOA.

Furgason, R.R. & Day, O.R. 1975. Iron and Manganese Removal with Ozone. *Wat. Sew. Works,* July, 61–63.

Gauntlett, R.B. & Packham, R.F. 1973. The Removal of Organic Compounds in the Production of Potable Water. *Chem. Ind.*, 17:8–12.

Gauthier, L. et al. 1989. Application du Test Micronoyau Triton à l'Etude Directe de la Genotoxicité des Procédés de Désinfection des Eaux. *Jour. Français d'Hydrologie* (in press).

Georges, P. & Perrine, D. 1989. Possibilité de Survie d'Amibes Libres Responsables de Kératites dans les Solutions Décontaminates Proposees pour l'Entretion des Lentilles de Contact. *Bull. Acad. Natl. Med.*, 173:421–427.

Georgeson, D.L. & Karimi, A.A. 1988. Water Quality Improvements with the Use of Ozone at the Los Angeles Water Treatment Plant. *Ozone Sci. Engrg.*, 10:3:255–276.

Georgeson, D.L. & Straub, W.O. 1987. Interpretation of Turbidity and Trihalomethane Data at the Los Angeles Aqueduct Filtration Plant, Proc. Water Quality Technology Conf., Baltimore, MD., pp. 383–413.

Gerber, N.N. 1971. Sesquiterperiods from Actinomycetes. *Phytochemistry,* 11:385.

Gerber, N.M. 1979. Volatile Substances from Actinomycetes: Their Role in the Odor Pollution of Water. *CRC Critical Reviews in Microbiol.*, 7:191.

Gerval, R. 1978. Proc. 4th Intl. Ozone Congress, IOA, Los Angeles.

_____. 1980. Les Differents Emplacements de l'Ozonation dans la Chaine de Traitement d'Eau Potable. *L'Eau et l'Industrie,* 45:77–84.

Gerval, R. & Bablon, G. 1983. Optimization of the Ozonation-Flocculation Stage, Proc. 6th Ozone World Congress, IOA, Washington, D.C., pp. 115–118.

Gerval, R. et al. 1985. Combined Preozonation-Coagulation at the Choisy-le-Roi Drinking Water Treatment Plant. Proc 7th Ozone World Congress, IOA, Tokyo, pp. 239–244.

_____. 1989. Unpublished report.

Giannissis, D. et al. 1985. Influence of Humic Substances on Manganese Removal in a Drinking Water Treatment Train, *Revue Française des Sciences de l'Eau,* 4:149.

Gibbs, R.J. 1983. *Envir. Sci. Technol.,* 17:4:237–239.

Giese, A.C. & Christensen, E. 1954. Effects of Ozone on Organisms. *Physiol. Zool.,* 27:101–115.

Gilbert, E. 1983a. Proc. 6th World Ozone Congress, IOA, Washington, D.C.

_____. 1983b. Investigations on the Changes of Biological Degradability of Single Substances Induced by Ozonation. *Ozone Sci. Engrg.,* 5:3: 137–149.

Ginocchio, J.C. 1981. Action de l'Ozone sur l'Elimination de Différentes Algues lors de la Filtration. *L'Eau et l'Industrie,* 56:19–24.

_____. 1982. "Effect of Ozone on Elimination of Various Algae by Filtration." In *Ozonization Manual for Water and Wastewater Treatment,* edited by W.J. Masschelein. John Wiley and Sons, Inc., New York.

Glaze, W.H. 1988. Summary: Workshop on By-products of Ozonation. AWWA Research Foundation, Denver, CO.

Glaze, W.H. & Kang, J.W. 1989. Evaluation of the Ozone/Hydrogen Peroxide Process Using Tetrachloroethylene as a Model Compound. Proc. 9th Ozone World Congress, IOA, New York, 1:596.

Glaze, W.H. & Lay, Y. 1989. Oxidation of 1,2-Dibromo-3-chloropropane Using Advanced Oxidation Process. Proc. 9th Ozone World Congress, IOA, New York, 1:688.

Glaze, W.H. & Wang, K. 1989. A Postcolumn Derivitization Method for Simultaneous Determination of Organic Peroxides and Hydrogen Peroxide. School of Public Health, University of California, Los Angeles.

Glaze, W.H. et al. 1980. U.S. EPA Report EPA-600/2-80-110.

_____. 1982. Destruction of Pollutants in Water with Ozone in Combination with Ultraviolet Radiation. 2. Natural Trihalomethane Precursors. *Environ. Sci. Technol.,* 16:8:454–458.

_____. 1987. The Chemistry of Water Treatment Processes Involving Ozone, Hydrogen Peroxide, and Ultraviolet Radiation. *Ozone Sci. Engrg.,* 9:4:335–352.

_____. 1988. "Application of Closed Loop Stripping and XAD Resin Adsorption for the Determination of Ozone By-products from Natural Water" in *Biohazards of Drinking Water Treatment,* edited by R.A. Larson, Lewis Publishers, Inc., Chelsea, MI, p. 201.

_____. 1989a. Evaluation of Ozonation By-Products from Two California Surface Waters. *Jour. AWWA,* 81:8:66–73.

_____. 1989b. Ozonation By-products. 2. Improvement of an Aqueous-Phase Derivitization Method for the Detection of Formaldehyde and Other Carbonyl Compounds Formed by the Ozonation of Drinking Water. *Environ. Sci. Technol.,* 23:7:838–847.

_____. 1990. Evaluating Oxidants for the Removal of Model Taste and Odor Compounds from a Municipal Water Supply. *Jour. AWWA,* 87:5:79–84.

Godart, H. 1988. Les Algues. Internal Report, Compagnie Générale des Eaux. Brittany Center, Rennes, France.

Goldstein, B.D. & McDonagh, E.M. 1975. Effect of Ozone on Cell Membrane Protein Fluorescence I. In Vitro Studies Utilizing the Red Cell Membrane. *Environ. Res.,* 9:179–186.

Gomella, C. 1967. Le Traitement des Eaux par l'Ozone. *Trib. Cebedeau,* 287:397.

_____. 1972. Ozone Practices in France. *Jour. AWWA,* 64:1:39–45.

Gomella, C. & Versanne, D. 1980. Nitrification Biologique et Affinage d'une Eau de Forage. Le Point après 40 Mois d'Exploitation de l'Usine de la Chapelle (Banlieue Sud de Rouen). *TSM l'Eau,* 5.

Gonzalez, M.M. et al. 1986. Acquired Immunodeficiency Syndrome Associated with *Acanthamoeba* Infection and Other Opportunistic Organisms. *Arch. Pathol. Lab. Med.,* 110:749.

Gould, M.M. 1985. The Use of Ozone in Water Treatment. *Public Health Engineer,* 13:3:161.

Gould, M.M. et al. 1984. An Experimental Study of Ozonation Followed by Slow Sand Filtration for the Removal of Humic Color from Water. *Ozone Sci. Engrg.,* 6:3.

Gowans, I.A. & McFadzean, C.J. 1982. Development and Experience with the Use of Ozonation at Tullich Water Treatment Works, Oban. Presented at the Joint Meeting, Institution of Civil Engineers (Glasgow and West of Scotland Association), Institution of Water Engineers and Scientists (Scottish Section), Glasgow, U.K.

_____. 1983. Design, Construction and Operation of Tullich Water Treatment Works, Using Ozone Treatment. *Jour. Inst. Water Eng. Sci.,* 37:5:437–459.

Graese, L.S. et al. 1987. GAC Filter-Adsorbers. Report to American Water Works Association Research Foundation, Denver, CO.

Grasso, D. & Weber, W.J., Jr. 1988. Ozone-Induced Particle Destabilization. *Jour. AWWA,* 80:8:73–81.

Greaves, G.F. 1989. Personal communication.

Greaves, G.F. et al. 1988. Ozonation and Slow Sand Filtration for the Treatment of Coloured

Upland Waters—Pilot Plant Investigations. In *Slow Sand Filtration—Recent Developments in Water Treatment Technology,* edited by N. J. D. Graham. Ellis Horwood Ltd., Chichester, U.K.

Griffin, A.E. 1960. Significance and Removal of Manganese in Water Supplies. *Jour. AWWA,* 52:10:1326.

Gruener, N. 1978. Mutagenicity of Ozonated, Recycled Water. *Bull. Environ. Contam. Toxicol.,* 20:522.

Guillerd, J.R. 1968. L'évolution dans le Traitement des Eaux par l'Ozone au Cours des Quinze Dernières Années. *Tech. Sci. Munic.,* 63: 10:279–312.

Guittonneau et al. 1988. Etude de la Dégradation de Quelques Composés Organochlorés Volatils par Photolyse du Peroxyde d'Hydrogène. *Revue des Sciences de l'Eau* 1:(1–2):35.

Gummerman, R.C. & Heim, N. 1983. Evaluation of Treatment Effectiveness for Reducing Trihalomethanes in Drinking Water. U.S. Environmental Protection Agency Report, Washington, D.C.

Gurol, M.D. & Vatistas, R. 1987. Oxidation of Phenolic Compounds by Ozone and Ozone + UV Radiation: A Comparative Study. *Wtr. Res.,* 21:8:895.

Haas, C.N. & Karra, S.B. 1984. Kinetics of Microbial Inactivation by Chlorine. I. Review of Results in Demand-Free Systems. *Wtr. Res.,* 18:1443.

Habibian, M.T. & O'Melia, C.R. 1975. Particles, Polymers, and Performance in Filtration. *ASCE. Jour. Envir. Engrg. Div.,* 101:EE4:567–583.

Harakeh, M.S. & Butler, M. 1984. Factors Influencing the Ozone Inactivation of Enteric Viruses in Effluent. *Ozone Sci. Engrg.,* 6:235–243.

Hartemann, P. 1987. Cellular Toxicity and Mutagenicity Assays from On-site Sampling of Drinking Water Treatment Plants Using Multistage Ozonation. *Ozone Sci. Engrg.,* 9:179.

Hascoet, M.C. et al. 1986. Use of Biological Analytical Methods to Optimize Ozonation and GAC Filtration in Surface Water Treatment. Proc. AWWA Ann. Conf., pp. 205–222., AWWA, Denver, CO.

Hattori, K. 1987. Water Treatment Systems and Removal of Earthy Odor Compounds. Second International Symposium on Off-Flavors in the Aquatic Environment, Kagoshima, Japan.

Haufele, A. & Sprockhoff, H.V. 1973. Ozon als Desinfektionmittel gegen Vegetative Bakterien, Bazillensporen, Pilze und Viren in Wasser. *Bakteriol. Parasitenkde Infekt. Krankh. Hyg.,* 175: 53–70.

Hayes, K.P. & Burch, M.D. 1989. Odorous Compounds Associated with Algal Blooms in South Australian Waters. *Wat. Res.* 23:1:115–121.

Hem, J.D. 1960. Complexes of Ferrous Iron with Tannic Acid. *USGS Wat. Sup. Paper 1459-D,* Washington, D.C.

Hibler, C.P. et al. 1987. Inactivation of *Giardia* Cysts with Chlorine at 0.5°C to 5.0°C. American Water Works Research Foundation, Denver, CO.

Hodges, W.E. et al. 1979. Proc. ASCE Environmental Engineering Conference, San Francisco, CA.

_____. 1980. Los Angeles Aqueduct; Treatment Process Report, Los Angeles Dept. of Water & Power, Report #AX 410-6.

Hoff, J.C. 1987. Strengths and Weaknesses of Using C · t Values to Evaluate Disinfection Practice. Proc. AWWA Seminar, Assurance of Adequate Disinfection, or C · t or not C · t. American Water Works Assn., Denver, CO, 49–65.

Hoigné, J. & Bader, H. 1983. Rate Constants of Reactions of Ozone with Organic and Inorganic Compounds in Water. *Wtr Res.,* 17, Part I.; part II. 185.

_____. 1988. The Formation of Trichloronitromethane (Chloropicrin) and Chloroform in a Combined Ozonation/Chlorination Treatment of Drinking Water. *Wtr. Res.,* 22:3:313.

Hoigné, J. et al. 1985. Rate Constants of Reactions of Ozone with Organic and Inorganic Compounds in Water. III. Inorganic Compounds and Radicals. *Wtr. Res.* 19:8:993–1004.

Hoyer, O. et al. 1987. The Effect of Ozonation on the Impairment of Flocculation by Algogenic Organic Matter. *Z. Wasser-Abwasser-Forsch.,* 20: 123–131.

Hrudey, S.E. et al. 1987. Potent Odor Causing Chemicals Arising from Drinking Water Disinfection. Presented at the 2nd International Symposium of Off-Flavors in the Aquatic Environment. Kagoshima, Japan.

_____. 1988. Odorous Aldehydes Produced by Disinfectant Reaction with Common Amino Acids. Proc. AWWA Ann. Conf., Orlando, Fla.

Hubbs, S.A. & Holdren, G.C. 1986. Chloroorganic Water Quality Changes Resulting from Modification of Water Treatment Practices. American Water Works Association Research Foundation Report.

Hubele, C. 1985. Adsorption und biologischer Abbau von Huminstoffen in Aktivkohlefiltern. Doctoral dissertation, University of Karlsruhe, Germany.

_____. 1986. Design of Fixed Bed Adsorbers Using Mathematical Models. *Wtr. Supply,* 4: 197.

HUCK, P.M. 1989. Use of Biological Processes in Drinking Water Treatment in Western Europe. *Sci. Tech. Eau,* 22:2:141.

_____. 1990. Unpublished report.

HUCK, P.M. ET AL. 1989. Effect of Water Treatment Processes on Concentrations of Assimilable Organic Carbon. Proc. 12th Intl. Symp. on Wastewater Treatment; 1st Canadian Wrkshp. on Drinking Water, Montreal.

_____. 1990. Methods for Determining Assimilable Organic Carbon and Some Factors Affecting the van der Kooij Method. *Ozone Sci. Engrg.* (in press).

HUISMAN, K. & WOOD, W.E. 1974. Slow Sand Filtration. World Health Organization, Geneva.

HYDE, R. ET AL. 1984. Upgrading of Slow Sand Filters by Preozonation. Presented at the 37th Intl. Conf.: Use of Fixed Biomass for Water and Wastewater Treatment, Liege, Belgium.

INTERNATIONAL OZONE ASSOCIATION (IOA). 1982. Turin's New Drinking Water Plant. *Ozonews,* 10:6:8–9.

_____. 1986. Ozone Treatment at Ohumwa, Iowa, for Dealing with Agricultural Run-off. *Ozonews,* 14:2:5.

_____. 1988a. Proc. Ozonation Systems: Design, Operation and Maintenance. Conference at Monroe, MI, April 27–28.

_____. 1988b. Proc. Ozonation Systems and Drinking Water Treatment. Conference at Myrtle Beach, SC, December 6–7.

JACANGELO, J.G. ET AL. 1989. Ozonation: Assessing its Role in the Formation and Control of Disinfection By-products. *Jour. AWWA,* 81:8: 74.

JADAS-HECART, A. 1989. Contribution a l'Etude de la Demande en Chlore a Long Terme d'une Eau Potable. Modelisation et Identification of Precurseurs Organiques. Doctoral Thesis, Poitiers University, France.

JADAS-HECART, A. ET AL. 1988. Effect of Ozonation on the Chlorine Demand of a Treated Surface Water and Model Compounds. Proc. Int. Conf. on Ozone in Water Quality Management, International Ozone Assoc., Zürich, Switzerland.

JAGO, P.H. & STANFIELD, G. 1984. Application of ATP Determination to Measurement of Growth Potential in Water. Res. Report, Water Res. Centre, Medmenham, U.K.

JAMES, C.R. ET AL. 1988. Pilot and Prototype Scale Comparison of Preozonation/Deep Bed Filtration with GAC. Proc. IOA Conf. on Ozonation Systems: Design, Operation and Maintentance, Monroe, MI.

JANSSENS, J.G. ET AL. 1985. Ozone Enhanced Biological Activated Carbon Filtration and its Effect on Organic Matter Removal and in Particular on AOC Reduction. *Wtr. Sci. Tech.,* 17: 6/7:1055.

_____. 1986. Experiences with Direct Filtration: Plant-Scale Evaluation and Pilot-Scale Investigations. *Wtr. Supply,* 4:347.

JARROLL, E.L. ET AL. 1981. Effect of Chlorine on *Giardia lamblia* Cyst Viability. *Appl. Environ. Microbiol.,* 41:483–487.

JEKEL, M.R. 1979. "Experience with Biological Activated Carbon Filters." In *Oxidation Techniques in Drinking Water Treatment,* edited by W. Kuhn and H. Sontheimer. U.S. EPA-570/9-79-020.

JEKEL, M.R. 1983. The Benefits of Ozone Treatment Prior to Flocculation Processes. *Ozone Sci. Engrg.,* 5:21–35.

JEKEL, M.R. & ERNST, B.U. 1981. Einfluß von Huminstoffen und ihrer ozonten Produkte auf die Elektrolytkougulation von Polystyrollatex. *Vom Wasser,* 57:123–136.

JEKEL, M.R. & REICHERTER, U.F. 1982. Experience with a Combination of Preozonation and Floc Filtration Prior to Goundwater Recharge by Slow Sand Filtration. Symp. Proc. "Water Filtration," Antwerp, Belgium.

JESTIN, J.M. ET AL. 1987a. Goûts et Odeurs de l'Eau Potable. Une Méthode d'Evaluation, *TSM l'Eau* Juin:281.

JESTIN, J.M. ET AL. 1987b. Les Odeurs Générées par la Chloration. Identification du Problème, Conséquence pour le Traitement. *Trib. Cebedeau,* 522:40:17.

JOBIN, R. & GHOSH, M.M. 1972. Effect of Buffer Intensity and Organic Matter on the Oxygenation of Ferrous Iron. *Jour. AWWA,* 64:9: 590.

JORET, J.C. 1990. Private communication.

JORET, J.C. & BLOCK, J.C. 1981. Survie de Virus Enteriques Adsorbés sur Microfibre de Verre au cours d'un Transport Postal. *Can. J. Microbiol.,* 27:246.

JORET, J.C. & LEVI, Y. 1986. Methode Rapide d'Evaluation du Carbone Eliminable des Eaux par Voie Biologique. *Trib. Cebedeau,* 510:39:3.

JORET, J.C. ET AL. 1986. Inactivation des Virus dans l'Eau sur une Filière de Production à Ozonation Ètagée. *Wtr. Res.,* 20:871.

_____. 1988. Rapid Method for Estimating Bioeliminable Organic Carbon in Water. Proc. AWWA Ann. Conf., p. 1715, AWWA, Denver, CO.

JÜTTNER, F. 1981. Detection of Lipid Degradation Products in the Water of a Reservoir During a Bloom of *Synura uvella. Appl. Env. Microbiol.,* 41:1:100–106.

_____. 1983. Volatile Odorous Excretion Products of Algae and their Occurrence in the Natural Aquatic Environment. *Wtr. Sci. Technol.,* 15:6/7:247.

KAASTRUP, E. ET AL. 1986. Activated Carbon Adsorption of Humic Substances and the Influence of Preozonation on Such. Proc. AWWA Ann. Conf., pp. 607–625, AWWA, Denver, CO.

KATO, M. ET AL. 1983. The Further Development in Ozone Application Technology in Potable Water: Ultraviolet or Ultra-Sonic Assisted Ozonization. *Wtr. Supply,* 6:349.

_____. 1988. The Further Development in Ozone Application Technology in Potable Water; UV and WS Assisted Ozonation. *Wtr. Supply,* 6:12:349.

KATZENELSON, E. ET AL. 1974. Inactivation Kinetics of Viruses and Bacteria in Water by Use of Ozone. *Jour. AWWA,* 66:725–729.

KAVLOCK, R. ET AL. 1980. *Toxicol. Appl. Pharmacol.,* 48:19.

KILLOPS, S.D. 1986. Volatile Ozonation Products of Aqueous Humic Material. *Wtr. Res.,* 20: 2:153.

KILLOPS, S.D. ET AL. 1985. Ozonation of Humic and Fulvic Acid (Isolated from a Lowland River Water) and Upland Water: Organic By-products. Tech. Report 224, Water Research Center, Medmenham, U.K.

KIM, C.K. ET AL. 1980. Mechanism of Ozone Inactivation of Bactériophage f2. *Appl. Environ. Microbiol.,* 39:210–218.

KINMAN, R.N. 1975. Water and Wastewater Disinfection with Ozone: A Critical Review. *Crit. Rev. Environ. Contr.,* 5:141–152.

KLASSEN, E.H. ET AL. 1970. Pilot Studies for Quality Managment in Terminal Reservoir. Proc. AWWA Ann. Conf., Washington, D.C.

KLIMKINA, N.V. ET AL. 1987. An Evaluation of the Treatment of Colored Waters Using an Oxidizing Sorption Method. *Gigiena Sanitaria,* 1: 16.

KNOCKE, W.R. ET AL. 1987. Using Alternative Oxidants to Remove Dissolved Manganese From Waters Laden With Organics. *Jour. AWWA,* 79:3:75.

_____. 1988. Soluble Manganese Removal on Oxide-Coated Filter Media. *Jour. AWWA,* 80: 12:65.

_____. 1990. Kinetics of Iron and Manganese Oxidation by Alternative Oxidants. AWWA Research Foundation Final Report.

KOOL, H.J. & HRUBEC, J. 1986. The Influence of an Ozone, Chlorine and Chlorine Dioxide Treatment on Mutagenic Activity in (Drinking) Water. *Ozone Sci. Engrg.,* 8:217–234.

KOOL, H.J. ET AL. 1985. Evaluation of Different Treatment Processes with Respect to Mutagenic Activity in Drinking Water. *Sci. Total Environ.,* 47:229–256.

KORICH, D.G. ET AL. 1990. Effects of Ozone, Chlorine Dioxide, Chlorine, and Monochloramine on *Cryptosporidium parvum* Oocyst Viability. *Appl. Env. Microbiol.* 56:5:1423–1428.

KOWBEL, D.J. ET AL. 1984. Chlorination of Ozonated Soil Fulvic Acid: Mutagenicity Studies in *Salmonella. Sci. Total Envir.,* 37:171–176.

_____. 1986. Mutagenicity Studies in *Salmonella:* Residues of Ozonated and/or Chlorinated Water Fulvic Acids. *Envir. Mutagen.,* 8:253–262.

KRASNER, S.W. & BARRETT, S.E. 1984. Aroma and Flavor Characteristics of Free Chlorine and Chloramines. Proc. AWWA WQTC, Denver, Col.

KRASNER, S.W. ET AL. 1983. Application of the Flavor Profile Method for Taste and Odor Problems in Drinking Water. Water Quality Technology Conference, AWWA, Norfolk Virginia.

_____. 1989. The Occurrence of Disinfection By-products in U.S. Drinking Water. *Jour. AWWA,* 81:8:41.

KRAUTER, H.S.D. 1974. Das Leben in Wassertropfen. Mikroflora and Mikrofauna des Süsswassers. *Kosmos Publ.,* Stuttgart, Germany.

KRUITHOF, J.C. ET AL. 1985. "Influence of Water Treatment Processes on the Formation of Organic Halogens and Mutagenic Activity by Postchlorination." In *Water Chlorination: Chemistry, Environmental Impact and Health Effects, vol. 5,* edited by R.L. Jolley et al. Lewis Publishers, Inc., Chelsea, MI, pp. 1137–1163.

KUHN, W. ET AL. 1978. Use of Ozone and Chlorine in Water Utilities in the Federal Republic of Germany. *Jour. AWWA* 70:6:326–331.

KUMMERT, R. & STUMM, W. 1978. The Surface Complexation of Organic Acids on Hydrous γ-Al_2O_3. *Envir. Sci. Technol.* 12:12:1309–1815.

KUO, C.H. 1984. Reactions of Dissolved Pollutants with Ozone in Aqueous Solutions. EPA-600/S3-84-072.

KURZMANN, G.E. 1975. Der Einsatz von Ozon zur Aufbereitung von Wasser. In Proc. of the Second Intl. Symp. on Ozone Technology, edited by R.G. Rice, P. Pichet and M.A. Vincent. International Ozone Institute, Syracuse, New York, pp. 514–521.

LADWP. 1988. Unpublished report, Los Angeles Dept. of Water and Power, Los Angeles, CA.

LALEZARY, S. ET AL. 1983. Air Stripping of Taste and Odor Compounds from Water. Comm. Congress AWWA, Las Vega, Nev.

_____. 1986. Oxidation of Five Earthy-Musty Taste and Odor Compounds. *Jour. AWWA,* 78: 3:62.

LANGE, A.A. & KAWCZYNSKI, E. 1978. Controlling Organics—The Contra Costa County Water District Experience. *Jour. AWWA,* 70:11: 653.

LANGLAIS, B. & PERRINE, D. 1990. Private communication.

LANGLAIS, B. ET AL. 1989. L'Ozonation: une Technique de Pointe pour l'Amélioration de la Qualité des Eaux. Presented at the A.Q.T.E. Congress. 8–10 March, Ottawa, Canada.

_____. 1990. *New Developments: Ozone in Water and Wastewater Treatment*. The C · t Value Concept for Evaluation of Disinfection Process Efficiency; Particular Case of Ozonation for Inactivation of Some Protozoa, Free-Living Amoeba and *Cryptosporidium*. Presented at the Int. Ozone Assn. Pan-American Conference, Shreveport, Louisiana, March 27–29.

LAPLANCHE, A. ET AL. 1976. Etude de la Cinétique de l'Ozonation des Pesticides Organophosphorés: Exemple du Parathion *TSM l'Eau*, 4:169.

LAROSE, J. ET AL. 1982. Elimination of Humic Materials. *Ozone Sci. Engrg.*, 4:2:79–89.

LAWRENCE, J. ET AL. 1980. The Ozonation of Natural Waters: Product Identification. *Ozone Sci. Engrg.*, 2:55.

LE PAULOUE, J. 1985a. *Ozonews*, 13:1:14–16.

_____. 1985b. A Ninth Ozone Facility in the FRG. *Ozonews*, 13:3:13.

LEAHY, J.G. 1985. Inactivation of *Giardis muris* Cysts by Chlorine and Chlorine Dioxide. Master's Thesis, Ohio State University, Columbus, Ohio.

LECHEVALLIER, M.W. ET AL. 1984. Disinfection of Bacteria Attached to Granular Activated Carbon. *Appl. Envir. Microbiol.*, 48:5:918.

LEFEBVRE, E. 1987. Influence des Ions Bicarbonates sur l'Elimination par l'Ozone du Fer et du Manganèse des Eaux Riches en Substances Humiques. Diplôme d'Etudes Approfondies Chimie et Microbiologie de l'Eau. Universités de Metz, Paris et Poitiers, France.

LEFEBVRE, E. ET AL. 1989. The Effect of Ozonation on the Removal of Organics by Coagulation-Flocculation. *Ozone Sci. Engrg.*, in press.

LEGUBE, B. ET AL. 1985. "Ozonation of Organic Halide Precursors: Effects of Bicarbonate and Bromide." In *The Role of Ozone in Water and Wastewater Treatment,* edited by R. Perry and A.E. McIntyre, Selper Ltd., London.

LEHMAN, J. ET AL. 1975. The Assumptions and Rationals of Computer Model of Phytoplankton Dynamics. *Limnol. Oceanogr.*, 20:343.

LEHN, H. 1969. Phytoplankton. Untersuchungen and der Microfilter-Versuchsanlage in Wasserwerk Konstanz. *Wasser Abwasser,* 110:4: 24:90.

LEMARCHAND ET AL. 1981. Charbon Actif et Matières Organiques dans le Traitement de l'Eau Potable. *TSM l'Eau,* 11:561.

LEVI, Y. & JESTIN, J.M. 1988. Offensive Tastes and Odors Occuring after Chlorine Addition in Water Treatment Processes. *Wtr. Sci. Tech.*, 20: 8/9:269.

LI, K.Y. & KUO, C.H. 1980. *AIChE Symp. Ser.*, #197, 76:161–168.

LIANG, S. ET AL. 1989. The Big Switch: Los Angeles Aqueduct Filtration Plant Treatment of California State Project Water. Proc. 9th Ozone World Congress. International Ozone Assoc., June 3–9.

LIAO, M.Y. & RANDTKE, S.J. 1985. Removing Fulvic Acid by Lime Softening. *Jour. AWWA,* 77:8:78–88.

LIENHARD, H. ET AL. 1980. *Vom Wasser* 55:149–158.

LIN, S.D. 1976. Sources of Tastes and Odors in Water. Parts 1 and 2. *Wtr. & Sew. Works,* June: 101 (part 1) and July:64 (part 2).

LITTLE, A.D. 1970. Cambridge, Massachussets. The Flavor Profile Panel (intern. document).

LOGSDON, G.S. ET AL. 1988. Roundtable *Cryptosporidium, Jour. AWWA,* 80:2:14.

LUNDGREN, B.V. ET AL. 1986 Removal of Off-Flavor Compounds in Biologically Active Sand Filters: Combined Effects with Ozonation. Second International Symposium on Off-Flavors in the Aquatic Environment, Kagoshima, Japan.

LYKINS, B.W. JR. ET AL. 1986. Chemical Products and Toxicologic Effects of Disinfection. *Jour. AWWA,* 78:11:66–75.

LYNCH, W.O. ET AL. 1963. Experiences with Microstraining at Ilion. Presented at the New York section AWWA meeting, New York.

MADORE, M.S. ET AL. 1987. Occurrence of *Cryptosporidium* Oocysts in Sewage Effluents and Selected Surface Waters. *J. Parasitol.*, 73: 702–705.

MAIER, D. 1979. Microflocculation by Ozone. In *Oxidation Techniques in Drinking Water Treatment,* EPA-570/9-79-020, pp. 394–417.

MAJUMDAR, S.B. ET AL. 1973. Inactivation of Poliovirus in Water by Ozonation. *Jour. WPCF,* 45:2433–2443.

MALAVEILLE, C. & BARTSCH, H. 1989. Genotoxicity of Carcinogenic Substances in Humans and Animals. *J. Toxicol. Clin. Exp.*, 9:15–25.

MALLEVIALLE, J. & SUFFET, I.H., EDS. 1987. *Identification and Treatment of Tastes and Odors in Drinking Water.* Cooperative Research Report, AWWARF/LE. American Water Works Association, Denver, CO.

MALLEVIALLE, J. ET AL. 1978. "The Degradation of Humic Substances in Water by Various Oxidation Agents (Ozone, Chlorine, Chlorine Dioxide)." In *Ozone/Chlorine Dioxide Oxidation Products of Organic Materials,* edited by R.G. Rice & J.A. Cotruvo, IOA, pp. 189–199.

MALLEY, J.P. JR. ET AL. 1988. Preoxidant Effects on Organic Halide Formation and Removal of Organic Halide Precursors. *Envir. Tech. Let.*, 9:1089–1104.

MALONEY, S.W. ET AL. 1985. Ozone-GAC Following Conventional U.S. Drinking Water Treatment. *Jour. AWWA*, 77:8:66.

MARTINEZ, A.J. & JANISTSCHKE, K. 1985. *Acanthamoeba*, an Opportunistic Microorganism: A Review. *Infection*, 13:6:251–256.

MASSCHELEIN, W.J. 1990. Overview of Ozone Installations in Europe. Proc. AWWA Ann. Conf., Cincinnati, Ohio.

MATHONNET, S. ET AL. 1985. Impact of Preozonation on the Granulometric Distribution of Materials in Suspension. *Ozone Sci. Engrg.*, 7: 2:107–120.

MCBRIDE, D.G. ET AL. 1977. Pilot Plant Investigations for Treatment of Owens River Water. Proc. AWWA Ann. Conf., San Francisco, CA. Part I, 14A-3.

MCCANN, J. ET AL. 1988. Statistical Analysis of *Salmonella* Test Data and Comparison to Results of Animal Tests. *Mutat. Res.*, 205:183–195.

MCGUIRE, M.J. 1989. 6th Quarterly Progress Report to U.S. EPA Office of Drinking Water, Cincinnati, Ohio, March 29. Submitted by Metropolitan Water District of Southern California.

MCGUIRE, M.J. ET AL. 1989. PEROXONE for Control of Disinfection By-Products, Tastes and Odors and Microorganisms. Presented at Wasser Berlin '89 Conference, IOA, Berlin.

MEANS, E.G. & MCGUIRE, M.J. 1986. An Early Warning System for Taste and Odor Control. *Jour. AWWA*, March:77.

MEDSKER, L.L. ET AL. 1968. Odorous Compounds in Natural Waters. An Earthy Smelling Compound Associated with Blue-Green Algae and Actinomycetes. *Envir. Sci Technol.*, 2:241.

MEIER, J.R. 1988. Genotoxic Activity of Organic Chemicals in Drinking Water. *Mutat. Res.*, 196:211–245.

______. 1989. Unpublished observation.

MEIJERS, A.P. 1977. Quality Aspects of Ozonization. *Wtr. Res.*, 11:8:647.

MERLET, N. ET AL. 1985. Chloropicrin Formation During Oxidative Treatments in the Preparation of Drinking Water. *Sci. Total Environ.*, 47:223–228.

MEYER-KONIG, V.W. & CARL, E. 1972. *GWF-Wasser/Abwasser* 113:25–34.

MILLER, G.W. 1978. An Assessment of Ozone and Chlorine Dioxide Technologies for Treatment of Municipal Water Supplies. EPA-600/2-78-147, U.S. EPA, NTIS, Springfield, VA.

MILLER, R.G. ET AL. 1986. Results of Toxicological Testing of Jefferson Parish Pilot Plant Samples. *Envir. Health Persp.*, 69:129–139.

MILTNER, R.J. ET AL. 1990. Pilot-Scale Investigation of the Formation and Control of Disinfection By-products. Proc. AWWA Ann. Conf., June 17–21.

MONTGOMERY, J.M, CONSULTING ENGINEERS, INC. 1986. Preozonation/Deep Bed Filtration Pilot Plant Study. Report to the Contra Costa Water District, by James M. Montgomery, Consulting Engineers, Inc., Pasadena, CA.

MONTIEL, A.J. 1977 Les Goûts de l'Eau. Paper presented at the "Journées Internationales de Pharmacie," Paris, France.

______. 1983. Municipal Drinking Water Treatment Procedure for Taste and Odor Abatement—A Review. *Wtr. Sci. Tech.*, 15:6/7:279.

MONTIEL, A.J. & OUVRARD, J. 1985a. Ozone and Biological Treatment—Effects on the Elimination of Cyanides. *Ozone Sci. Engrg.*, 7:2:85.

______. 1985b. Specific Elimination of Cyanides on a Drinking Water Treatment Pilot Plant. *Tech. Sci. Munic.*, 80:1:51.

MONTIEL, A.J. ET AL. 1983. Specific Removal of Organic Compounds (Toluene-Xylene) by Coupling Physico-Chemical Treatment and Biological Treatment. *Jour. Fr. Hydr.*, 14:40:45.

______. 1987. Etude de l'Origine et du Mécanisme de Formation de Composés Sapides Responsables de Goûts de Moisi dans les Eaux Distribuées. *TSM l'Eau*, Février:73.

MOORE, M.B. ET AL. 1985. *Acanthamoeba* Keratitis Associated with Soft Contact Lenses. *Am. J. Opthalmol.*, 100:396–403.

MORGAN, J.J. & STUMM, W. 1964a. Colloïd-Chemical Properties of Manganese Dioxide. *Jour. Coll. Sci.* 19:347.

______. 1964b. Oxygenation of Aqueous Manganese. II. *Am. Chem. Soc.*, 145th. New York.

MORRIS, J.C. 1975. Aspects of the Quantitative Assessment of Germicidal Efficiency. In *Disinfection: Water and Wastewater*, edited by J. D. Johnson. Ann Arbor Science Publishers, Inc., Ann Arbor, MI, Chapter 1:1.

MOUCHET, P. 1982. Réflexions Complémentaires sur l'Importance des Phénomènes Biologiques dans le Traitement et la Distribution des Eaux de Consommation. *Tech. de l'Eau et l'Assain.*, 424:Avril:7:25.

MOUCHET, P. ET AL. 1985. Elimination du Fer et du Manganèse Contenus dans les Eaux Souterraines: Problèmes Classiques, Progrès Récents. *Wtr. Supply*, 3:Berlin B:137.

MURRAY, R.G.E. ET AL. 1965. The Location of the Mucopeptide of Sections of the Cell Wall of *Escherichia coli* and Other Gram-Negative Bacteria. *Can. J. Microbiol.*, 11:3:547–560.

NAKAMURA, K. ET AL. 1989. Substrate Affinity to Oligotrophic Bacteria in Biofilm Reactors. *Wtr. Sci. Tech.*, 21:8/9:779.

NAKAYAMA, S. ET AL. 1979. Improved Ozonation in Aqueous Systems. *Ozone Sci. Engrg.*, 1: 2:119.

NAMBA, K. & NAKAYAMA, S. 1982. *Bull Chem. Soc.*, 55:3339.

NONHEBEL, D.C. ET AL. 1979. *Radicals*, Cambridge University Press, New York.

O'DONOVAN, D.C. 1965. Treatment with Ozone. *Jour. AWWA*, 57:9:1167–1194.

ODEGAARD, H. ET AL. 1986. Advanced Techniques for the Removal of Humic Substances in Potable Water. *Wtr. Supply*, 4:129.

ONGERTH, J.E. & STIBBS, H.H. 1987. Identification of *Cryptosporidium* Oocysts in River Water. *Appl. Environ. Microbiol.*, 53:672–676.

ORSANCO (Ohio River Valley Water Sanitation Commission). 1980. Water Treatment Process Modifications for Trihalomethane Control and Organic Substances in the Ohio River. U.S. Environmental Protection Agency Report, EPA-600/2-80-028, U.S. Government Printing Office, Washington, D.C.

PAILLARD, H. ET AL. 1987. Application of Oxidation by a Combined Ozone/Ultraviolet Radiation System to the Treatment of Natural Water. *Ozone Sci. Engrg.*, 9:4:391–418.

_____. 1988a. Une Nouvelle Filière de Traitement pour l'Élimination Poussée de la Matière Organique, des Précurseurs de THM et du Manganèse. Presentation at the symposium "Traitements Curatifs en Vue de la Potabilisation des Eaux Chargées en Microphytes Dulçaquicoles," Rennes, France (Nov. 15–16).

_____. 1988b. Conditions Optimales d'Application du Système Oxydant Ozone-Peroxyde d'Hydrogène. *Wtr. Res.*, 22:1:91.

_____. 1989a. Iron and Manganese Removal with Ozonation in Presence of Humic Substances. *Ozone Sci. Engrg.*, 11:93.

_____. 1989b. Effects of Alkalinity on the Reactivity of Ozone Towards Humic Substances and Manganese. *Aqua*, 378:1:32.

_____. 1990a. Use of Ozone to Treat Eutrophic Water for Potabilization. Intl. Conf. on the Use of Ozone in Treatment of Water. IOA, Madrid, Spain (Apr. 3–5).

_____. 1990b. Ozonation of Water in France. In *Water Supply and Wastewater Disposal*, edited by K.A. Zoetermeer (vol. 11).

_____. 1990c. Elimination of Atrazine by a Combined Ozone-Hydrogen Peroxide Treatment on the Seine River. Proc. 6th European Symp. on Organic Micropollutants in the Aquatic Environment. Lisbon, Portugal (May 22–24).

PAK, H. ET AL. 1981. Some Observations on the Effect of Ozone Treatment on Suspended Particulate Matter in Sea Water. *Estuar. Coastal Shelf Sci.*, 13:327–336.

PALMER, C.M. 1962. Algae in Water Supplies. U.S. Public Health Service Publ. No. 657, U.S. DHEW, PHS.

_____. 1980. Algae and Water Pollution. The Identification, Significance and Control of Algae in Water Supplies and in Polluted Water. Costle House Publications, Ltd., U.K.

PARFITT, R.L. ET AL. 1977. *Jour. Soil Sci.*, 28: 40–47.

PASCAL, O. ET AL. 1984. La Chimiométrie pour Piloter un Traitement Combiné de Préozonation Coagulation. Exemple de l'Usine d'Eau Potable de Choisy-le-Roi (800 000 m^3/d). Proc. of the Workshop on Ozone in Potable Water Treatment: Installations, Operating and Experimental Advances, Montreal, IOA.

PEETERS, J.E. ET AL. 1989. Effect of Disinfection of Drinking Water with Ozone or Chlorine Dioxide on Survival of *Cryptosporidium parvum* Oocysts. *Appl. Environ. Microbiol.*, 55:6:1519–1522.

PEREIRA, M.A. & STONER, G.P. 1985. Comparison of the Rat Liver Foci Assay and Strain A Mouse Lung Tumor Assay to Detect Carcinogens: A Review. *Fund. Appl. Toxicol.*, 5:688–699.

PERRICH, J.R. 1976. Inactivation of *Escherichia coli* by Ozonation in an Aqueous Reactor System. Doctoral dissertation, University of Louisville, Louisville, KY.

PERRINE, D. & LANGLAIS, B. 1986. Legionelloses et Procédés de Désinfection des Eaux. *L'Eau, L'Industrie, Les Nuisances*, 104:55.

_____. 1989. Etude du Mecanisme de l'Action Kysticide de l'Ozone sur les Amibes Libres. *TSM l'Eau*, 84:4:214–218.

PERRINE, D. ET AL. 1984a. Action d l'Ozone sur les Trophozoites d'Amibes Libres Pathogènes ou Non. *Bull. Soc. Franc. Parasitol.*, 3:81.

_____. 1984b. Efficacité du Procede d'Ozonation des Eaux Recyclees de Piscine dans la Destruction des Kystes d'Amibes Libres Pathogens ou Non. *L'Eau, L'Industrie, Les Nuisances*, 82: 29–32.

_____. 1989. Cinétique d'Action du Chlore et de l'Ozone sur les Trophozoites et les Kystes d'Amibes Libres des Genres *Naegleria* et *Acanthamoeba*. Colloque: Désinfection des Eaux et Santé Publique, Lille, France, Sept. 26–27.

PEYTON, G.R. ET AL. 1982. Destruction of Pollutants in Water with Ozone in Combination With Ultraviolet Radiation. 1. General Principle and Oxidation of Tetrachloroethylene. *Environ. Sci. Technol.*, 16:8:448.

PHILIPOT, J.M. 1985. Biological Techniques Used for the Preparation of Drinking Water.

Presented at EPA Symposium, Cincinnati, Ohio, August 26–29.

Pouvreau, P. 1984. Elimination Spécifique du Fer et du Manganèse. *Jour. Francais d'Hydrologie,* 2:169.

Prat, R. et al. 1968. Effets de l'Hypochlorite de Sodium, de l'Ozone et des Radiations Ionisantes sur les Constituants Pyrimidiques d'*Escherichia coli. Ann. Inst. Pasteur,* 114:595–607.

Prendiville, P.W & McBride, D.G. 1983. Ozonation—Direct Filtration of Los Angeles Drinking Water, Proc. 6th World Ozone Congress, IOA, Washington, D.C.

Prengle, H.W. & Mauk, C.E. 1978. Ozone/UV Oxidation of Pesticides in Aqueous Solutions. In *Ozone/Chlorine Dioxide Oxidation Products of Organic Materials,* edited by R.G. Rice, Ozone Press International.

Prévost, M. et al. 1989a. Study on the Performance of Biologically Active Carbon Filters (BAC) in Cold Waters. Proc. 12th Symp. on Wastewater Treatment & 1st Workshop on Drinking Water (pp. 347–366), Montreal, Nov. 20–22.

_____. 1989b. Full Scale Evaluation of Biological Activated Carbon Filtration for the Treatment of Drinking Water. WAIC AWWA Annual Conference, Philadelphia, November 11–16.

Rachwal, A.J. et al. 1988. Advanced Techniques for Upgrading Large Scale Slow Sand Filters. In *Slow Sand Filtration—Recent Developments in Water Treatment Technology,* edited by N. J. D. Graham. Ellis Horwood Ltd., Chichester, U.K.

Randtke, S.J. & Jepsen, C.P. 1981. Chemical Pretreatment for Activated Carbon Adsorption. *Jour. AWWA,* 73:8:411–419.

Rao, M. & Reddy, J. 1987. Peroxisome Proliferation and Hepatocarcinogenesis. *Carcinogenesis,* 8:631–636.

Reasoner, D.J. & Rice, E.W. 1989. U.S. EPA Experience with AOC and Coliform Growth Response Assays. Presentation at the Workshop on "Measurement of AOC in the Field of Drinking Water Treatment," Karlsruhe, Germany.

Reckhow, D.A. 1985. Effects of Preozonation on Subsequent Physico-Chemical Treatment Processes. Presented at the SVW Water Symposium, Brussels, Belgium.

Reckhow, D.A. & Singer, P.C. 1984. The Removal of Organic Halide Precursors by Preozonation and Alum Coagulation. *Jour. AWWA,* 76:4:151–157.

Reckhow, D.A. & Sibony, J. 1986. Several Approaches to Minimizing Formations of THM and Other Hazardous Chlorination By-products. Gruttee Seminar: "Humic Substances" Rennes, France.

Reckhow, D.A. et al. 1986a. Ozone as a Coagulant Aid. In AWWA Semin. Proc.: Ozonation, Recent Advances and Research Needs, AWWA, Denver, CO.

_____. 1986b. The Ozonation of Organic Halide Precursors: Effect of Bicarbonate. *Wtr. Res.,* 20:8:987–998.

Rees, A.J. et al. 1979. Water Clarification by Flotation. 5 WRC Technical Report TR 114.

Rey, R.P. et al. 1987. Ozone Reactions with Carbohydrates in Aqueous Medium. Proc. 8th Ozone World Congress, New York, 2:E106.

Reynolds, G. et al. 1989. Aqueous Ozonation of Fatty Acids. *Ozone Sci. Engrg.,* 11:143.

Rice, E.W. 1989. Bioassay Procedures for Predicting Coliform Bacterial Growth in Drinking Water. Ph.D. Dissertation, University of Cincinnati, Ohio.

Rice, E.W. et al. 1982. Inactivation of *Giardia* Cysts by Chlorine. *Appl. Env. Microbiol.,* 43: 250–251.

_____. 1990. Bioassay Procedure for Prediction of Coliform Bacterial Growth in Drinking Water. *Env. Tech.* (in press).

Richard, Y. 1978. Ozone et Applications. Proc. Seminaries du GRUTTEE, Paris.

_____. 1982. Importance of Ozone on Oxidation Processes for the Treatment of Potable Water: Interference with Other Oxidants. *Ozone Sci. Engrg.,* 4:2:59–77.

_____. 1984. Proc. IOA/NIWR Intl. Conf., South Africa.

Richard, Y. & Conan, M. 1978. Etude du Charbon Actif Granulé, Influence de la Hauteur de Couche. *TSM l'Eau,* Mars:189.

Richard, Y. & Fiessinger, F. 1977. Emploi Complémentaire des Traitements Ozone et Charbon Actif. International Ozone Institute Congress, Paris, France.

Richard, Y. et al. 1983. Influence of Preozonation on Clarification by Flotation for Drinking Water Treatment. *Ozone Sci. Engrg.,* 5:3.

Rickloff, J.R. 1987. An Evaluation of the Sporicidal Activity of Ozone. *Appl. Environ. Microbiol.,* 53:683–686.

Riesser, V.W. et al. 1976. Possible Mechanisms of Poliovirus Inactivation by Ozone. In *Forum on Ozone Disinfection,* edited by E.G. Fochtman, R.G. Rice, and M.E. Browning. International Ozone Institute, Syracuse, NY, 186–192.

Riley, T.L. et al. 1978. The Effect of Preozonation on Chloroform Production in the Chlorine Disinfection Process. In *Water Chlorination: Environmental Impact and Health Effects,*

Vol. 2, edited by R.L. Jolley et al., Ann Arbor Science Publishers, Inc., Ann Arbor, MI.

RITTMANN, B.E. & HUCK, P.M. 1989. Biological Treatment of Public Water Supplies. *Crit. Rev. Envir. Contr.,* 19:2:119–184.

RITTMANN, B.E. & SNOEYINK, V.L. 1984. Achieving Biologically Stable Drinking Water. *Jour. AWWA,* 76:10:106.

ROBSON, C.M. ET AL. 1990. Status of U.S. Drinking Water Treatment Ozonation Systems. Unpublished report.

ROOK, J.J. 1974. Formation of Haloforms During Chlorination of Natural Waters. *Wtr. Treat. Exam.,* 23:2:234.

ROSE, J.B. ET AL. 1986. Detection of *Cryptosporidium* from Wastewater and Freshwater Environments. *Wtr. Sci. Technol.,* 18:10:233–239.

ROSEN, H.M. 1980. State of the Art of Ozonation for Commercial Applications in the U.S. *AICLE Symp. Ser.* 197, Vol. 76, pp. 97–116.

ROY, D. ET AL. 1980. Inactivation of Enteroviruses by Ozone. *Prog. Water Techn.,* 12:819–836.

RUBIN, A.J. ET AL. 1983. Disinfection of Amoebic Cysts in Water with Free Chlorine. *Jour. WPCF,* 55:9:1174–1182.

SAFE DRINKING WATER COMMITTEE. 1980. The Disinfection of Drinking Water. In *Drinking Water and Health,* vol. 2. National Academy Press, Washington, D.C., pp. 5–137.

SAUNIER, B.M. ET AL. 1983. Preozonation as a Coagulant Aid in Drinking Water Treatment. *Jour. AWWA,* 75:5:239–245.

SCHALEKAMP, M. 1977. Proc. AWWA Ann. Conf., Anaheim, CA.

_____. 1978. Experiences with Ozone in Switzerland with Special Respect to the Changes in Hygienically Questionable Materials." In *Water Berlin 1977,* edited by O. H. Hess, p. 31. Also given in *Ozonews,* May 1978, International Ozone Association.

SCHENK, P. 1962. The Water Treatment Plant at the Dusseldorf "Am Staad" Water Works. *Gas-U. Wasserforch.,* 103:791.

SCHULHOF, P. 1987. European Experience in the Use of Activated Carbon Treatment. *J. Environ. Pathol. Toxicol. Oncol.,* 7:718:55–75.

SCOTT, D.B.M. 1975. The Effect of Ozone on Nucleic Acids and Their Derivatives. In *Aquatic Applications of Ozone,* edited by W.J. Blogoslawski and R.G. Rice, International Ozone Institute, Syracuse, NY, pp. 1–15.

SCOTT, D.B.M. & LESHER, E.C. 1963. Effect of Ozone on Survival and Permeability of *Escherichia coli, J. Bacteriol.,* 85:567–576.

SCRIVNER, A.E. ET AL. 1980. *Aqua,* 7:141–144.

SEPPOVAARA, O. 1971. The Effect of Fish of the Mass Development of Brackish Water Plankton. *Aqua Fenica,* 118.

SERVAIS, P. & BILLEN, G. 1989. Unpublished data.

SERVAIS, P. ET AL. 1987. Determination of the Biodegradable Fraction of Dissolved Organic Matter in Water *Wtr. Res.,* 21:4:445.

_____. 1989. Simple Method for Determination of Biodegradable Dissolved Organic Carbon in Water. *Appl. Envir. Microbiol.,* 55:10:2732–2734.

SHAMBAUGH, R.L. & MELNYK, P.R. 1978. Removal of Heavy Metals via Ozonation, *Jour. WPCF,* 50:1:113–121.

SHINRIKI, N. ET AL. 1988. Mechanism of Inactivation of Tobacco Mosaic Virus with Ozone. *Wtr. Res.,* 22:933.

SIERKA, R.A. & AMY, G.L. 1985. Catalytic Effects of Ultraviolet Light and/or Ultrasound on the Ozone Oxidation of Humic Acid and Trihalomethane Precursors. *Ozone Sci. Engrg.,* 7:1:47.

SIGWORTH, E.A. 1957. Control of Odor and Taste in Water Supplies. *Jour. AWWA,* 49:12:1507.

SILVEY, J.K.G. & ROACH, A.W. 1953. Actinomycetes in the Oklahoma City Water Supply. *Jour. AWWA,* 45:4:409.

SINGER, P.C. 1988a. Assessment of Ozonation Research in Drinking Water, AWWA Research Foundation Report.

_____. 1988b. An Evaluation of Alternative Oxidant and Disinfectant Treatment Strategies for Controlling Trihalomethane Formation in Drinking Water, Report to the U.S. EPA, Drinking Water Research Division, Cincinnati, OH.

_____. 1989. Impact of Ozone on the Removal of Particles, TOC, and THM Precursors. AWWA Research Foundation Report.

SINGER, P.C. & CHANG, S.D. 1988. Impact of Ozone on the Removal of Particles, TOC, and THM Precursors, AWWA Research Foundation Report.

SINGER, P.C. ET AL. 1989. Ozonation at Belle Glade, Florida: A Case History. Proc. 9th Ozone World Congress, IOA.

SINSABAUGH, R.L. ET AL. 1986a. DOC Removal by Coagulation with Iron Sulfate. *Jour. AWWA,* 78:5:74–82.

_____. 1986b. Precursor Size and Organic Halide Formation Rates in Raw and Coagulated Surface Waters. *ASCE Jour. Envir. Engrg. Div.,* 112:EE1:139–153.

SNODGRASS, W.J. ET AL. 1984. Particle Formation and Growth in Dilute Aluminum(III) Solutions. Characterization of Particle Size Distributions at pH 5.5. *Wtr. Res.,* 18:4:479–488.

SNYDER, J.E. & CHANG, P.W. 1974. Relative Resistance of Eight Human Enteric Viruses to Ozonization in Saugatucket River Water. In *Aquatic Applications of Ozone,* edited by W.J. Blogoslawski and R.G. Rice, International Ozone Institute, Syracuse, NY, pp. 25–42.

SOBSEY, M.D. 1989. Inactivation of Health Related Microorganisms in Water by Disinfection Processes. *Wtr. Sci. Tech.,* 21:3:179–195.

SOMIYA, I. ET AL. 1986. Biodegradability and GAC Adsorbability of Micropollutants by Preozonation. *Ozone Sci. Engrg.,* 8:1:11.

SOMMERVILLE, R.C. & REMPEL, G. 1972. *Jour. AWWA,* 64:6:377–382.

SONNTAG, C.V. 1989. The Chemistry Behind the Upgrading of Water with UV Light: An Overview. Proc. Wasser Berlin 1989, V.1.1.

SONTHEIMER, H. 1979. Process Engineering Aspects in the Combination of Chemical and Biological Oxidation. In *Oxidation Techniques in Drinking Water Treatment,* edited by W. Kuhn and H. Sontheimer. EPA-570/9-79-020.

SONTHEIMER, H. ET AL. 1978. The Mülheim Process. *Jour. AWWA,* 70:7:393–396.

_____. 1988. *Activated Carbon for Water Treatment,* 2nd ed. (English). DVGW-Forschungsstelle, AWWA Research Foundation, Denver, CO.

SPEITEL, G.E., JR. & DIGIANO, F.A. 1987. The Bioregeneration of GAC Used to Treat Micropollutants. *Jour. AWWA,* 79:1:64–73.

SPEITEL, G.E., JR., ET AL. 1989. Biodegradation of Trace Concentrations of Substituted Phenols in Granular Activated Carbon Columns. *Envir. Sci. Technol.,* 23:1:68–74.

SPROUL, O.J. ET AL. 1982. The Mechanism of Ozone Inactivation of Waterborne Viruses. *Wtr. Sci. Technol.,* 14:303–314.

Standard Methods for the Examination of Water and Wastewater. 1985. APHA, AWWA, and WPCF. Washington, D.C. (16th ed.).

STANKOVIC, I. & TUCOVIC, A. 1985. Ozone Application in the New Belgrade Water Treatment Plant. Proc. 7th Ozone World Congress, Tokyo, IOA (pp. 60–66).

STEVENS, A.A. 1982. Reaction Products of Chlorine Dioxide. *Env. Health Persp.,* 46:101.

STEVENS, A.A. ET AL. 1989. Formation and Control of Non-Trihalomethane Disinfection Byproducts. *Jour. AWWA,* 81:8:54.

STEWART, M.H. 1990. Private communication. Metropolitan Water District of Southern California, Los Angeles, CA.

STOEBNER, R.A. & ROLLAG, D.A. 1981. Ozonation of a Municipal Groundwater Supply to Reduce Iron, Manganese, and Trihalomethane Formation. *Aqua,* 291–299.

STOLARIK, G.F. & CHRISTIE, J.D. 1988. Projection of Ozone C · t Values, Los Angeles Aqueduct Filtration Plant. IOA Conference, Monroe, MI, April.

STRATHCLYDE REGIONAL COUNCIL WATER DEPARTMENT, 1989. Personal communication.

STRINGER, R.P. & KRUSE, C.W. 1970. Amoebic Cysticidal Properties of Halogens in Water. Proc. Nat. Special. Conf. on Disinfection. Am. Soc. Civil Eng., New York.

STUMM, W. & LEE, G.F. 1960. Oxygenation of Ferrous Iron. *Ind. Eng. Chem.,* 53:143.

STUMM, W. & SINGER, P.C. 1966. Precipitation of Iron in Aerated Groundwater. Proc. Amer. Soc. of Civil Engineers. 92: No. SA5: 120.

SUCHKOV, B.P. 1964. Decontamination of Drinking Water Containing Agents Which Can Cause Intestinal Diseases and Enteroviruses by Ozonation. *Gig. I. Sanit.,* 29:22.

SUFFET, I.H. ET AL. 1986. Removal of Tastes and Odors by Ozone. Seminar on Ozonation and Water Treatment. Proc AWWA Ann. Conf., Denver, Colo.

SUKENIK, A. ET AL. 1987. Effect of Oxidants on Microalgal Flocculation. *Wtr. Res.,* 21:5:533.

SYMONS, J.M. ET AL. 1981. *Treatment Techniques for Controlling Trihalomethanes in Drinking Water.* U.S. Environmental Protection Agency, EPA-600/2-81-156, Washington, D.C., p. 124.

_____. 1984. "Removal of Organic Contaminants from Drinking Water Using Techniques Other Than Granular Activated Carbon Alone—A Progress Report." In *Adsorption Techniques in Drinking Water Treatment,* EPA 570/9-84-005, U.S. Environmental Protection Agency, Washington, D.C.

TENNEY, M.W. & STUMM, W. 1965. *Jour. WPCF,* 37:10:1370–1388.

TERASHIMA, K. 1988. Reduction of Musty Odor Substances in Drinking Water—A Pilot Plant Study. *Wtr. Sci. Tech.* 20:8/9:275.

THEBAULT, P. ET AL. 1981. Mechanism Underlying the Removal of Organic Micropollutants during Flocculation by an Aluminum or Iron Salt. *Wtr. Res.,* 15:183–189.

THEIS, T.L. & SINGER, P.C. 1974. Complexation of Iron(II) by Organic Matter and Its Effect on Iron(II) Oxidation. *Env. Sci. Tech.,* 8: 6:569–573.

THORBURN, G.W. ET AL. 1975. The Ozonization of Loch Turret Water. Proc. 2nd International Symposium on Ozone, IOA.

THURESON, L.E. 1962. Ozonizing Investigation at Falun Water Works. *Wattenhygien,* 18:53.

TIPPING, E. & HIGGINS, D.C. 1982. *Colloids and Surfaces,* 5:85–92.

TOBIASON, J.E. & O'MELIA, C.R. 1988. Physicochemical Aspects of Particle Removal in Depth Filtration. *Jour. AWWA,* 80:12:54–64.

TOPALIAN, P. 1982. Physikalisch-Chemische und Biologische Aufbereitung eines Huminstoffhaltigen Grundwassers in Aktivkohlefiltern. Doctoral dissertation, University of Karlsruhe, Germany.

TOPLEY, M.G. 1987. Application of Ozone/BAF Process for Organics and Colour Removal at a Water Treatment Plant at CFB Cornwallis, Nova Scotia (Canada). Proc. 2nd Intl. Conf.: The Role of Ozone in Water and Wastewater Treatment, edited by D.W. Smith and G.R. Finch (pp. 173–187). Tektran Intl. Ltd., Kitchener, Ontario.

TRANCART, J.L. ET AL. 1988. Potabilisation des Eaux Chargées en Algues et en Nutriments: Impact sur les Réseaux de Distribution d'Eau Potable. Presented at the workshop: "Traitements Curatifs en Vue de la Potabilisation des Eaux Chargées en Microphytes Dulçaquioles," Rennes, France (Nov. 15–16).

TRUSSEL, R.R. & UMPHRES, M.D. 1978. The Formation of Trihalomethanes. *Jour. AWWA,* 70:11:604.

TRUSSELL, R.R. ET AL. 1975. Proc. 2nd Intl. Symp. On Ozone Techn., pp. 586–611, IOA.

TYE & WAITE. 1981. Mutagens, Carcinogens on the Water Cycle. *Wat. Poll. Cont.,* 600–613.

U.S. EPA. 1977. National Secondary Drinking Water Regulation. 40 CFR 143.3. U.S. Government Printing Office, Washington, D.C.

_____. 1988. *Federal Register,* 53:14:1892, January 22.

_____. 1989a. Guidance Manual for Compliance with the Filtration and Disinfection Requirements for Public Water Systems Using Surface Water Supplies. U.S. Environmental Protection Agency, Washington, D.C.

_____. 1989b. Formation of Organic By-products of Disinfection, Task 19, R. J. Miltner, project manager, Drinking Water Research Division, Cincinnati, OH.

UMPHRES, M.D. ET AL. 1979. The Effects of Preozonation on the Formation of Trihalomethanes. *Ozonews,* 6:3, Part 2.

Vade Mecum du Chef d'Usine de Traitement d'Eau Destinée à la Consommation. 1987. Association Générale des Hygiènistes et Techniciens Municipaux. Paris. Techniques et Documentation, Lavoisier.

VAHIDI, B. ET AL. 1983. Hydrogen Peroxide Enhanced Ozonation for Removal of THM Precursors from Surface Water. 6th Ozone World Congress, Washington, D.C.

VALLOM, J.K. & MCLOUGHLIN, A.J. 1984. Lysis as a Factor in Sludge Flocculation. *Wtr. Res.,* 18:12:1523–1528.

VAN BREEMAN, A.N. ET AL. 1979. *Wtr. Res.,* 13:8:771–779.

VAN DER GAAG, M.A. ET AL. 1982. Presence of Mutagens in Dutch Surface Water and Effects of Water Process for Drinking Water Preparation. In *Progress in Clinical and Biological Research, Vol. 109, Mutagens in Our Environment,* edited by M. Sorse and H. Vanio. Alan R. Liss, New York, pp. 277–286.

_____. 1985. The Influence of Water Treatment Processes on the Presence of Organic Surrogates and Mutagenic Compounds in Water. *Sci. Total Env.,* 47:137–153.

VAN DER KOOIJ, D. 1987. The Effect of Treatment on Assimilable Organic Carbon in Drinking Water. In *Treatment of Drinking Water for Organic Contaminants* (Proc., Second National Conference on Drinking Water, Edmonton, Alberta, Canada), edited by P.M. Huck and P. Toft. Pergamon Press, New York, p. 317.

VAN DER KOOIJ, D. ET AL. 1982. Determining the Concentration of Easily Assimilable Organic Carbon in Drinking Water. *Jour. AWWA,* 74:10:540.

_____. 1989. The Effects of Ozonation, Biological Filtration and Distribution on the Concentration of Easily Assimilable Organic Carbon (AOC) in Drinking Water. *Ozone Sci. Engrg.,* 11:3:297–311.

VAN GEMERTS, J.L. & NETTENBREIJER, A.H., EDS. 1977. Compilation of Odor Threshold Values in Air and Water. Natl. Inst. for Wtr. Supply, Voorburg, Netherlands, and Centr. Inst. for Nutr. & Food Res. TNO, Zeist, Netherlands.

VAN HOOF, F. 1983. Influence of Ozonization on Direct-Acting Mutagens Formed During Drinking Water Chlorination. In: *Water Chlorination: Environmental Impact and Health Effects, vol. 4,* edited by R.L. Jolley et al. Ann Arbor Science Publishers, Inc., Ann Arbor, MI, pp. 1211–1220.

VAN HOOF, F., ET AL. 1985. Determination of Aliphatic Aldehydes in Waters by HPLC. *Anal. Chim. Acta,* 169–419.

_____. 1986. Formation of Oxidation By-products in Surface Water Preozonation and Their Behavior in Water Treatment. *Wtr. Supply,* 4:93–102.

VAUGHN, J.M. ET AL. 1987. Inactivation of Human and Simian Rotaviruses by Ozone. *Appl. Env. Microbiol.,* 53:2218–2221.

VENTRESQUE, C. & BABLON, G. 1989. Ozone, a Means of Stimulating Activated Carbon Reactors. Proc. 9th Ozone World Congress, I0A, New York.

VENTRESQUE, C. ET AL. 1987. Development of Chlorine Demand Kinetics in a Drinking Water Treatment Plant. In *Water Chlorination: Chemistry, Environmental Impact and Health Effects, vol.*

6, edited by R. L. Jolley et al. Lewis Publishers, Inc., Chelsea, MI, pp. 715–728.

Vigreux, B. & Richard, P. 1979. Déferrisation et Démanganisation d'Eau Potable. Compte-rendu SGN.

Vik, E.A. et al. 1988. Pilot Scale Studies for Geosmin and 2-Methylisoborneol Removal. *Wtr. Sci. Tech.,* 20:8/9:229.

Villessot, D. 1990. Le Filières de Traitement des Eaux Douce. *TSM l'Eau,* April :211.

Vrochinskii, K.K. 1963. Experimental Data on Water Decontamination with Ozone. *Hyg. Sanit.,* 28:3.

Wagner, R. & Elefritz, R.A. 1983. Ozonation for Effective THM Control. *Public Works,* 114: 4:46.

Wallace, J.L. et al. 1988. The Combination of Ozone/Hydrogen Peroxide and Ozone/UV Radiation for Reduction of Trihalomethane Formation Potential in Surface Water. *Ozone Sci. Engrg.,* 10:1:103–112.

Wallentin, A. & Nyberg, I. 1962. Ozone Treatment of Lygern Water. *Wattenhygien,* 18:38.

Walsh, D.S. et al. 1980. Ozone Inactivation of Floc Associated Viruses and Bacteria. *J. Environ. Eng. Div., ASCE,* 106:711–726.

Water Research Center. 1981. A Guide to Solving Water Quality Problems in Distribution Systems. Tech. Rept. Medmenham, U.K.

Watson, H.E. 1908. A Note on the Variation of the Rate of Disinfection with the Change in the Concentration of Disinfectant. *J. Hyg.,* 8:536–542.

Watts, C.D. 1985. Organic By-Products of Ozonization of Humic and Fulvic Acids. Proc. Intl. Conf. on the Role of Ozone in Water and Wastewater Treatment, edited by R. Perry and A.E. McIntyre. Selper Ltd., London.

Wegelin, M. 1988. Roughing Gravel Filters for Suspended Solids Removal. In *Slow Sand Filtration—Recent Developments in Water Treatment Technology,* edited by N. J. D. Graham. Ellis Horwood Ltd., Chichester, U.K.

Welte, 1988. Unpublished report.

Weng, C-N. et al. 1986. Ozonation: An Economic Choice for Water Treatment. *Jour. AWWA,* 78:11:83–89.

Werner, P. 1981. Mikrobiologische Untersuchungen zur Chemisch-Biologischen Aufbereitung eines Huminsaurehaltigen Grundwassers. *Vom Wasser,* 57:157–164.

_____. 1985. Eine Methode zur Bestimmung der Verkeimungsneigung von Trinkwasser. *Vom Wasser,* 65:5:257–270.

Werner, P. & Hambsch, B. 1986. Investigations on the Growth of Bacteria in Drinking Water. *Wtr. Supply,* 4:3:227–232.

Wickramanayake, G.B. 1984. Kinetics and Mechanism of Ozone Inactivation of Protozoan Cysts. Doctoral dissertation, Ohio State University, Columbus, OH.

Wickramanayake, G.B. et al. 1984a. Inactivation of *Naegleria* and *Giardia* Cysts in Water by Ozonation. *Jour. WPCF,* 56:8:983–988.

_____. 1984b. Inactivation of *Giardia lamblia* Cysts with Ozone. *Appl. Env. Microbiol.,* 48: 3:671–672.

_____. 1985. Effects of Ozone and Storage Temperature on *Giardia* Cysts. *Jour. AWWA,* 77:8:74–77.

Wood, S. et al. 1983. Microbes as a Source of Earthy Flavors in Potable Water—A Review. *Intl. Biodeterioration Bull.,* 19:3/4:83.

Wuhrmann, K. & Meyrath, J. 1955. The Bactericidal Action of Ozone Solution. *Schweitz. J. Allgen. Pathol. Bakteriol.,* 18:1060.

Yagi, M. et al. 1986. Musty Odor Problems in Japanese Water Supplies. *Wtr. Supply,* 4:195.

Yamada, H. & Somiya, I. 1989. The Determination of Carbonyl Compounds in Ozonated Water by the PFBOA Method. *Ozone Sci. Engrg.,* 11:127.

Yao, C.C.D. & Haag, W.R. 1990. Rate Constants for Ozonation of Pesticides, Solvents, PCBs, and Other Organics in Water. *Am. Chem. Soc. Preprints,* Div. of Envr. Chem., pp. 15–16.

Young, J.S. & Singer, P.C. 1979. Chloroform Formation in Public Water Supplies: A Case Study. *Jour. AWWA,* 71:87.

Yurteri, C. & Gurol, M.D. 1988. Ozone Consumption in Natural Water: Effect of Background Organic Matter, pH, and Carbonate Species. *Ozone Sci. Engrg.,* 10:277:7.

Zabel, T.F. 1984. Proc. IOA/NIWR Conf., Pretoria, South Africa.

_____. 1985. The Application of Ozone for Water Treatment in the United Kingdom: Current Practice and Recent Research. *Ozone Sci. Engrg.,* 7:1:11.

Zoeteman, B.C.J. 1980. *Sensory Assessment of Water Quality.* Pergamon Press, Elmsford, N.Y.

Zoeteman, B.C.J. et al. 1982. Mutagenic Activity Associated with By-products of Drinking Water Disinfection by Chlorine, Chlorine Dioxide, Ozone, and UV Irradiation. *Env. Health Persp.,* 46:197–205.

IV

Engineering Aspects

William D. Bellamy
François Damez
Bruno Langlais
Antoine Montiel
Kerwin L. Rakness
David A. Reckhow
C. Michael Robson

The three major components of an ozonation system are preparation and treatment of the feed gas to the ozone generator, the ozone generator including power supply, and dissolution of the ozonized gas into the water under treatment. Additional subcomponents and considerations include offgas ozone destruction management, control strategy system, materials of construction, and safety.

The choice of feed gas may be dictated by process, dissolution, or other requirements. The gas preparation system must be capable of consistently delivering a dry, contaminant-free, comparatively cool feed gas to the ozone generator. The components of the gas preparation system must be carefully considered to assure a system that will deliver the required quality of feed gas despite the wide range of ambient conditions that are likely to exist over an annual operating period.

The ozone generator and power supply unit are provided as a package. Although ozone generators currently on the market utilize the same generation process, the details of the systems vary in choice of type and size of the individual ozone generation modules, applied power frequency, and cooling mode. Production efficiency, O&M reliability, operational experience, and vendor support are all factors that must be evaluated when considering a specific application.

Selection of a mode or modes of ozonized gas dissolution is based on the water quality and the specific ozone application. The fine bubble diffuser, multi-compartment contactor is the most common concept for many applications, including primary disinfection. However, there are many other types that may be better suited for applications such as organics oxidation. Both initial construction and long-term operational costs must be evaluated as well as other O&M considerations. Reduction of the ozone concentration of the offgas from the dissolution units must be provided to protect personnel, plant components, and the general environment. This is achieved by several types of ozone destruction devices.

Instrumentation and control of the ozone system is addressed using familiar equipment and control concepts. However, specific provisions must be made to measure (and alarm, if necessary) specific parameters such as feed gas dewpoint, ozone-generator cooling-medium supply, and offgas ozone concentration both before and after destruction. Although the trend, particularly in large systems, is toward highly automated systems, the basic control strategies are simple and based on primary sensors, which are becoming more reliable as the market is becoming established.

Common construction materials have proved appropriate for ozone system service. Particular care, however, must be taken in selecting materials used in ozonized gas conveying and dissolution systems. Stainless steel has proven to be the material of choice. Acceptably ozone-resistant materials have been identified for valve seats, gaskets, and seals.

The ozonation system components must be carefully tested to assure operability and compliance with the specifications. Since ozonation is operational cost–intensive in terms of electrical power, many owners require the ozonation system selection to be based on both initial construction cost and long-term power cost considerations. Evaluation of the power costs will likely require ozone production power testings, which are normally performed at the owner's plant after construction is complete. Carefully planned test procedures, performed and monitored by experienced personnel, must be carried out to protect the owner's interests and to minimize subsequent disagreements.

Personnel safety must be addressed despite the intrinsic safety employed in the overall production of the chemical, since no ozone is stored and production ceases when the power supply to the ozone generator is interrupted. Also, the human nose is capable of perceiving ozone concentrations in the air at far lower levels than those which would be of safety concern. Nevertheless, the safety of personnel and equipment are addressed throughout the ozonation system by reliable ambient air ozone monitors, equipment interlocks, and safety equipment. O&M manuals must address safety issues of the ozonation system at length due to the different hazards associated with a complex system incorporating components that may be unfamiliar to the O&M staff.

IV.A TREATABILITY STUDIES

Treatability studies are used to determine the treatment characteristics of a specific water. They can range from determining the feasibility of a single treatment process to the optimization of the entire treatment train. The evaluations are normally conducted on the waters of concern. The scale of the study is dictated by the study objectives and can include bench-scale, pilot-scale, or even demonstration-scale evaluations.

During the design of ozone facilities, the design engineer must make decisions regarding the ozone dose, contact time, rate of application, and other process variables. There are several ways in which the design engineer can gain the information needed to make these decisions; treatability studies with the same or similar waters, treatment plant experience on the same or similar waters, or an educated guess based on literature review and discussions with other experienced professionals. The most desirable information would be from an operating ozone facility on the same water supply, as might be the case in an ozone facility expansion. In the absence of an existing plant, treatability studies can provide accurate estimates on which the design engineer can base the ozone facility design.

The degree of conservatism in an ozone facility design is determined in large part by the quality of the treatment process information the design engineer has to work with. If the design engineer must select design criteria with little knowledge of the amount of ozone required, the reaction time needed to meet the treatment objectives, and the effects of ozonation on downstream processes, then the design would necessarily be more conservative than if accurate estimates of the application requirements were available. This extra conservatism would result in added capital costs and may result in additional operating costs caused by inefficient operation. An example of the positive impact treatability studies can have on overall design is seen in the Los Angeles Aqueduct Filtration Plant. The results of treatability studies indicated that preozonation allowed direct filtration rates as high as 33 m/h (15 gpm/ft^2) while maintaining excellent filtered water quality (Monk et al. 1985). Without the treatability studies, the design would likely have used filtration rates of 8.8 to 13.2 m/h (4 to 6 gpm/ft^2) resulting in double the construction costs for the filtration system. The savings in capital costs for the filtration system as the result of the treatability studies can be roughly estimated at $14 million (about 96 million French Francs) for this 600-mgd (94,635 m^3/h) treatment plant.

IV.A.1 General Study Considerations

Need for treatability studies. In the development of water treatment facilities, process information and engineering must be integrated for the development of a successful design and, ultimately, a treatment facility that achieves its water quality goals. The following is a general overview of the information to be gained from treatability studies and its impact on design.

Location of ozone application(s). One of the fundamental decisions in applying ozone is to determine the best location(s) to employ the ozone in the treatment sequence. Ozone may be applied at one point or at multiple points in the process (see chapter III). The treatment objectives determine the best location(s), i.e., the best location in the process sequence for the addition of ozone is the one that allows the treatment objectives to be met at the least overall cost. For example, in a highly colored Florida groundwater, acceptable color removal required ozone doses of 4, 6, and 12 mg/L for filtered, prefiltered (postsoftened), and raw water, respectively (Beaudet et al. 1988). Based on ozone utilization, the postfiltration location would be chosen as most appropriate. However, prefilter (postsoftening) ozonation resulted in the lowest filtered water turbidities and ozonation by-products, and the hydraulics of the existing plant made addition of ozonation before filtration attractive. As a result, the prefiltration ozonation application point was selected to meet the overall treatment objectives. Treatability studies are desirable to determine the most appropriate application point(s) in the process if more than one location is available.

Ozone dose requirements. The ozone dose required to meet the treatment objectives is usually the most important input to the design of the ozone generation and contacting system. In all applications, ozone is consumed, not only in reactions with the target compounds, but in reactions with "nontarget ozone demand" substances and by self-decomposition (see sec. A, chapter II). The overall ozone demand is water quality–specific and may vary widely with season and temperature. Treatability studies allow the dose/response relationship necessary for achieving an objective to be determined. Existing application on the same water or treatability studies are the only ways of determining this dose requirement.

Contact time. Optimal contact time can vary widely depending on the intended ozone objective. Determining the minimum time necessary to accomplish the intended treatment objective can be accomplished with the aid of a treatability study.

In general, the oxidation reactions of ozone with easily oxidizable compounds occur in a much shorter time frame than does disinfection. (See sec. A, chapter II and sec. H, chapter III for detailed information.) Consequently, for applications where primary disinfection is one of the treatment objectives, contact time will most likely be controlled by the disinfection requirement. For disinfection, the dose/response relationship is determined by the necessary residual and decay, i.e., the ozone dose required to provide and maintain a biocidal residual for a sufficient period of time to achieve a stated level of disinfection. The U.S. EPA has specified the "C · t" concept to assure adequate disinfection, where C is the concentration of dissolved disinfectant (in milligrams per liter) and t is the nominal contact time (in minutes) (*Federal Register* 1989). Although the C · t approach is being used by the EPA to assess the disinfection capabilities of an ozone system, the methods associated with determining C and t have not been finalized. Draft documents for determining the C · t value have been developed, including draft Appendix O of the Surface Water Treatment Rule Guidance Manual (U.S. EPA 1989) and two papers that outline the basis for the rules of thumb for determining C and t given in Appendix O (Lev and Regli 1990a, 1990b). The concept is to give some set disinfection credit for the first dissolution stage (e.g., 0.5 log *Giardia* and 1 log virus inactivation if certain ozone residuals are maintained at the exit of the first compartment) and calculate the C and t for subsequent stages based on a reduced C and t value. Another approach is being considered where a model of the ozone system can be developed to establish a C · t credit on a case-by-case basis. Since the method for the C · t determination has not been finalized by the EPA, a model approach to estimating a C · t value for design purposes will be presented in this text that can be modified once a procedure has been established. C will be assumed to be the measured or calculated value at the point in the contactor of concern, while t will be assumed to be the t_{10} value, where t_{10} is determined by tracer study and equals the time necessary for 10 percent of the tracer mass to exit the contactor. Both the C and t can be modified as the guidelines develop, while the method for estimating and modeling the C · t credit (with appropriate modification) should remain valid.

The World Health Organization (WHO) recommendation and French practice have followed the 0.4/4 guidelines, in which adequate disinfection is assumed if a 0.4 mg/L ozone residual can be maintained over a 4-min contact time (Légeron 1984). While this may appear to be equivalent to a 1.6 mg-min/L C · t value, French practice has followed the convention of a 0.4 mg/L ozone residual during 4 min of "hydraulic" detention time (volume/flow). The EPA draft recommended procedure requires that tracer studies of the contactor be performed and that a characteristic or nominal detention time (e.g., t_{10}) be determined. The t_{10} can vary from 20 to 50 percent of the hydraulic detention time (Levenspiel 1972). This indicates that French practice corresponds to an approximate EPA equivalent C · t value of 0.3 to 0.8 mg-min/L. However, numerous French installations have been designed with a "hydraulic" detention time in the contactor of 10 to 12 min. Under these conditions, the EPA equivalent C · t value is in the range of 0.9 to 2.2 mg-min/L.

With the exception of primary disinfection (as mentioned above), the effect of contact time on the extent of most reactions between ozone and target compounds is much less important than the amount of ozone transferred during the contact time. Figure IV–1 shows a comparison between 6–, 9–, and 12-min contact times with applied ozone doses of 1 and 2 mg/L ozone for the oxidation of the taste and odor compound 2-methylisoborneol (MIB). The results indicate that for this contacting system the removal is relatively insensitive to the contact time (MWDSC 1989). This will be true for fast ozonation reactions that quite often involve the hydroxyl radical as the oxidizing species (see sec. A, chapter II). However, the contacting system can have an effect on the reaction rate; thus, it is recommended that the effect of contact time be investigated for each application as the complex ozonation chemistry precludes an evaluation of the importance of contact time *a priori*.

Contact time does not necessarily have the same meaning for different types of reactors. Mixing intensities and rate of transfer can have a dramatic effect on the relative contact time associated with a process. For example, during a pilot study of conventional contacting and an in-line contacting system (Bellamy et al. 1990), the effects of

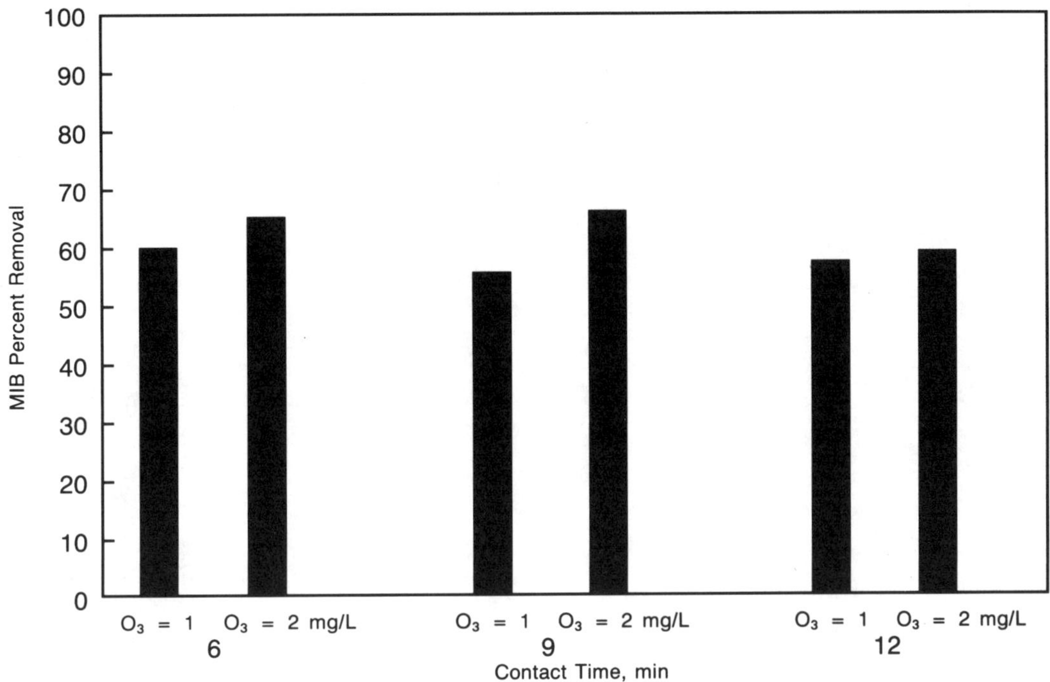

Source: MWDSC Project Status Report (1989).

Figure IV–1 MIB Percent Removal for Varying Contact Time and O_3 Doses

mixing and ozone dissolution were quite pronounced on the perceived contact time necessary to complete the color removal. For the conventional contact pilot column, detention times less than 4 min resulted in lower color removal than those over 4 min. An in-line dissolution and contacting system was able to achieve approximately the same color removal with an equivalent ozone dose in 30 sec. Figure IV–2 presents a plot of this rapid color oxidation for an in-line system. Although the conventional contactor may have been affected by short circuiting or mixing (as discussed below) at the low detention times, the results of this study reinforce the need to investigate or even demonstrate that the rate of transfer is a predominant factor.

Effect of ozone on performance of subsequent treatment processes. The application of ozone as a unit process will have an impact on the water quality and the operation of the treatment facility. The areas of most impact are coagulation/flocculation, filtration, disinfection by-products (from ozonation and from subsequent or previous addition of other disinfectants), and biodegradable organic material. (See chapters II and III for specific information.) For this reason, treatability studies must consider not only the ozonation step but also its integration into the total process sequence.

Test objectives. In the design of ozone treatability studies, the objectives of the process study and the role of ozone must be clearly defined. Ozone may be used in treatment for several purposes, including primary disinfection; disinfection by-product control; micropollutant oxidation (organic or inorganic); control of taste, odor, or color; and particulate removal. In many applications, ozone is used to achieve multiple objectives (see chapter III). Ozone treatability studies should be designed and executed to determine design parameters that encompass all treatment objectives.

Perhaps the most critical element of the successful treatability study is the experimental plan. The experimental plan includes all elements of the study's actual implementation: the apparatus to be used, the analytical techniques, the process parameters to

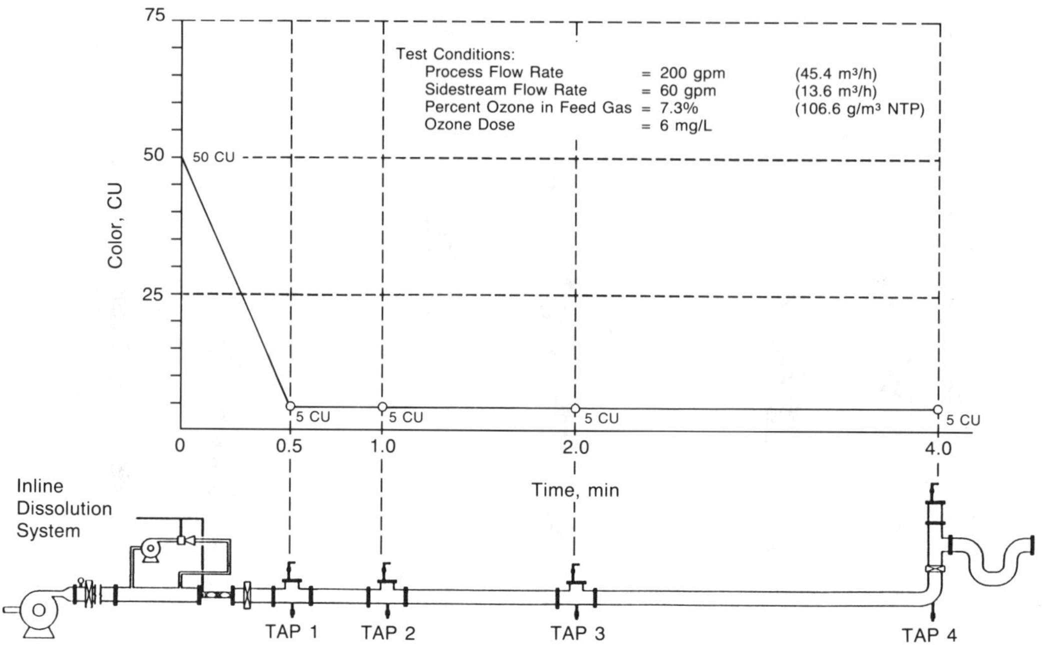

Source: Bellamy et al. (1990).

Figure IV–2 Rapid Color Oxidation: In-Line System

be varied and the levels to be investigated, the number and location of sample collection, and the expected results. One of the key provisions of the experimental plan that is frequently overlooked is data handling, analysis, and quality control. A detailed experimental plan ensures that all participants to the study understand the objectives and the techniques that will be used to meet those objectives.

Study scale (applicability). Treatability studies can be classified by scale. Bench-scale studies are generally characterized by batch reactors of 2 to 70 L (0.5 to 18.5 gal) or by semi-batch reactors with low flows in the 0.1 to 1 L/min range (0.03 to 0.3 gpm). In France and Germany, initial tests are often performed in ozone demand flasks. This methodology uses 0.5- to 1.0-L (0.13- to 0.26-gal) glass vessels to which the water and a known ozone gas concentration and volume are added. The two are then mixed and the effects noted (Légeron 1978). Pilot-scale studies are generally continuous flow systems with processes operating at flows of 5 to 100 L/min (1 to 25 gpm). Demonstration-scale projects are sized to provide project-specific information, which may include process scaleup relationships, operator training, or operating data for a nonconventional process. Their sizes can range from as small as 100 L/min (25 gpm) to any size considered necessary to appropriately represent a given application and not be subject to indeterminant scale-up factors.

The type and scale of a treatability study should be determined by the desired reliability of the results compared with the intended full-scale application. This means that the type and scale of the study should be determined by its objectives in terms of engineering information required, operational information required, the time frame in which the information is needed, the degree of involvement of the local regulatory establishment in the study and their willingness to accept the study results as a basis for design, and economics.

For example, bench-scale studies may be sufficient to determine the ozone dose required to oxidize free iron or manganese as a replacement for chlorine in pretreatment whereas a demonstration-scale project may be necessary to convince the regulatory community that ozonation followed by high rate filtration will provide sufficient turbidity and pathogenic cyst removal to protect the consuming public.

IV.A.2 Study Approach

As stated above, the study approach will be developed based on the objectives of the ozone application. Testing will normally include evaluation of location, dose, contact time, by-products, and downstream effects. Also, site-specific considerations should always be considered. For example, the source water quality may vary significantly on a seasonal basis. The treatability studies must consider the seasonal and annual variations, since the production of disinfection by-products, the residual ozone decay, and the dosage required for oxidation are all affected by variations in the source water.

The ozone facility design must balance the desire to provide optimum treatment under the worst possible conditions with the economic reality of providing for an extreme water quality event. A well-conceived and executed treatability study can provide this information and result in significant savings in the design, construction, and operation of the facility. The objectives of the treatability study should be carefully studied to select the scale and duration appropriate to meet the goals.

Although each treatability study is unique, there are some basic concepts associated with the study considerations and basic approach to testing that can be used to initiate the planning process for the study design. The following sections address these preliminary considerations and the testing approach.

Preliminary study considerations. The treatability study plan will be developed to address the specific objectives of the ozone application. The plan will have to be developed by carefully considering the study objectives and determining if the desired results can be obtained (e.g., design parameters for ozone generation, dissolution, contacting, and location of application). There is no general study plan that can be used for each anticipated application. However, in an attempt to simulate the planning process, the following list of considerations has been developed. This list includes examples of the initial considerations for the design of a treatability study (ozone considerations primarily):

1. Treatment characteristics and parameters to be determined (additional details are given for these determinations in subsequent sections)

- Oxidation characteristics (e.g., stoichiometry and pseudo-kinetics) for specific compounds (e.g., MIB, TCE) or gross characteristics (e.g., color, THMFP) versus transferred ozone dose while varying detention time, ozone concentration in gas, or pH.
- Ozone residual versus transferred dose for various detention times, application rates, and possibly pH's (ozone concentration in gas may become important for a given contactor configuration or required gas-to-liquid (G/L) ratio; consequently, it may be necessary to test various ozone concentrations in gas).
- Ozone decay rate following addition of a range of transferred doses, possibly at different pH's.
- Effect of ozone application on downstream treatment (e.g., coagulant dose requirements, disinfection by-products, filter effluent turbidity/microflocculation, filter solids holding capacity).
- Pilot contactor characteristics (e.g., t_{10}, t_{50}, and dispersion number) for various flow and gas feed rates (i.e., complete tracer studies using full range of water and gas flow rates to be determined once only for a given pilot contactor and to be applied during scale-up).
- Ozone efficiency (e.g., disinfection and oxidation) and suitability at different locations in the treatment train.
- Special contactor considerations (e.g., in-line or turbine dissolution [see sec. IV.D]).

2. Testing periods and duration
   - Determine the above treatment parameters for those seasons in which changes in raw water quality can significantly affect the ozone treatment (for example, during late summer algal blooms for taste and odor, or during high-turbidity runoff in the spring). At least two seasons are recommended (average and worst case) and four are desirable.
   - The minimum duration for testing will be approximately six weeks for the initial testing and a minimum of three to four weeks for subsequent testing during other seasons. Six weeks is required to set up the pilot facility and complete the array of testing described above. The six weeks not only includes the ozone testing but also testing of the other treatment processes associated with the treatment system, e.g., chemical selection, sedimentation, filter media selection. A more reasonable initial test period is ten weeks. Subsequent testing may require more than four weeks if the desired conditions are not present during the planned test period or if significantly different results are obtained.
3. Test scale
   - Most ozone testing can be conducted with 0.1-m (4-in.) diameter contactor columns with flows of 0.3 to 1.3 m^3/h (1 to 5 gpm) and ozone generation of 9.5 to 95 g O_3/h (0.5 to 5 lb O_3/day). Bench-scale testing can be used to assist in determining the necessary ozone production.
   - For raw water oxidation prior to sedimentation (or other specific processes), the scale of the sedimentation unit will dictate the scale of downstream processes. For example, flows as small as 0.3 m^3/h (1 gpm) can be used to study the effect of ozone on downstream processes (e.g., direct filtration) but scale-up of some of the downstream processes (e.g., sedimentation) may not be possible. For those cases where scale-up of a specific unit process (e.g., sedimentation) is desired, package units with flows of approximately 14 m^3/h (60 gpm) can be used. This will require an equivalently sized ozone test facility. French studies have been successfully performed (i.e., reliable scale-up between pilot unit and full-scale plant) with lamellar settlers treating 2 m^3/h (8.8 gpm).
   - Specific designs for nonconventional (e.g., turbine or in-line) contactors will have to be determined on a case-by-case basis.
4. Operations
   - At least one operator is required full time during the pilot plant operation. Some unattended operation may be possible for extended studies with automated pilot plants.
   - Two operators are required during startup and during unusually active test periods, e.g., during wide swings in raw water quality.
   - On the average, 1.5 operators are required for the entire testing period.
   - Laboratory assistance is required to analyze for specific parameters, e.g., MIB, trichloroethylene (TCE), or disinfection by-products.
5. Study Plan

As discussed above, a good study plan will help ensure a successful and cost-effective treatability study. The plan should cover all aspects of the treatability study. It should describe the objectives, expected (or desired) results, schedule, test sequence, equipment needs, operations, data collection, data handling, data analysis, quality control, analytical procedures, staffing, responsibilities, cost, and reporting.

Testing approach. Like the development of the study plan, the approach to testing will be dictated by the specific objectives of the project. It is important to conduct the pilot testing with a clear set of objectives and a well-developed test plan that provides for flexibility as the program progresses. One possible approach to developing an ozone testing program is:

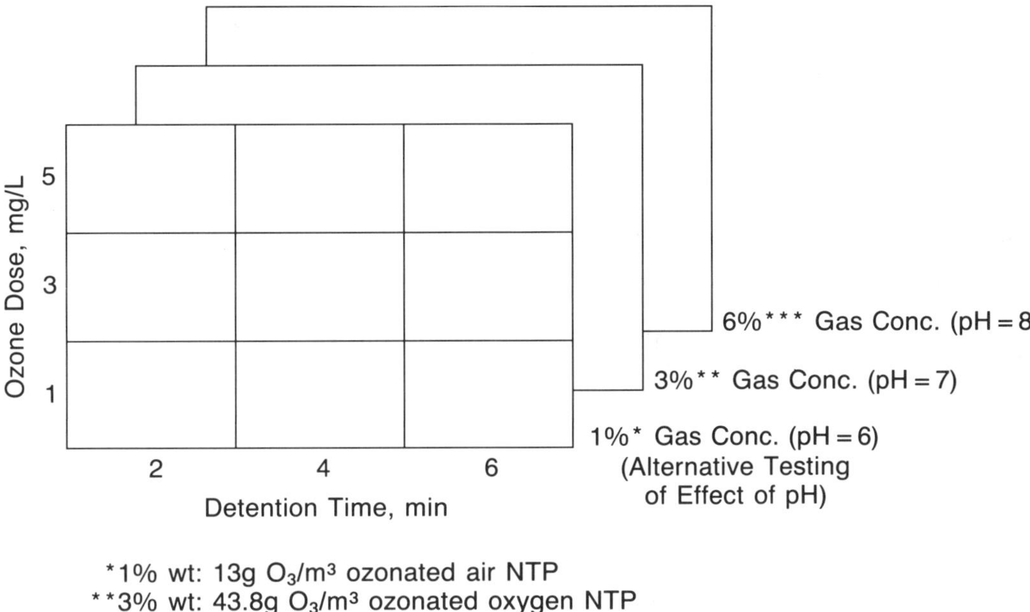

Figure IV–3 Test Matrix

- Conduct "initial testing" to determine basic ozone treatment characteristics associated with the subject waters.
- Conduct "confirmation testing" of the information developed during the initial testing.
- Conduct "seasonal testing" to verify design conditions for various water quality conditions if necessary.

Initial testing should be conducted to define the interrelationships between the raw water and ozone. The scale of testing can be bench or pilot. If bench-scale testing is used, pilot-scale verification of some of the test points will be required. This testing sequence is not intended to develop relationships between ozone and downstream processes such as ozonation and its effects on GAC or slow sand filter efficiency.

A simple test matrix can be used to accomplish this goal. A $3 \times 3 \times 3$ matrix consisting of three ozone doses, three ozone contact times, and, if appropriate, three ozone concentrations or pH values will provide the information necessary to establish the basic relationships between the raw water and the ozone. Figure IV–3 presents such a matrix. The values presented in Figure IV–3 are for discussion purposes only; specific test values will have to be determined on a case-by-case basis. This test matrix should be completed on representative waters from each of the locations where ozone may be applied. If the pilot contactor characteristics are well known to the operator, the variation on gas-phase ozone concentration could be eliminated. Appropriate ozone concentrations for optimizing the contacting process (see sec. IV.D) should be used. It is sometimes useful to test three pH values as well. This can be especially useful for high-pH waters where it is more difficult to establish an ozone residual for effective disinfection.

Parameters to be measured during this initial testing should include, at a minimum, ozone gas flow, the ozone feed gas concentration, ozone offgas concentration, ozone residual (necessary for disinfection applications), and treatment parameters. The treatment parameters to be measured will be dictated by the objectives of the planned ozone application, e.g., color removal, trihalomethane (THM) precursor removal, oxidation of taste and odor compounds, oxidation of iron and manganese, or oxidation of

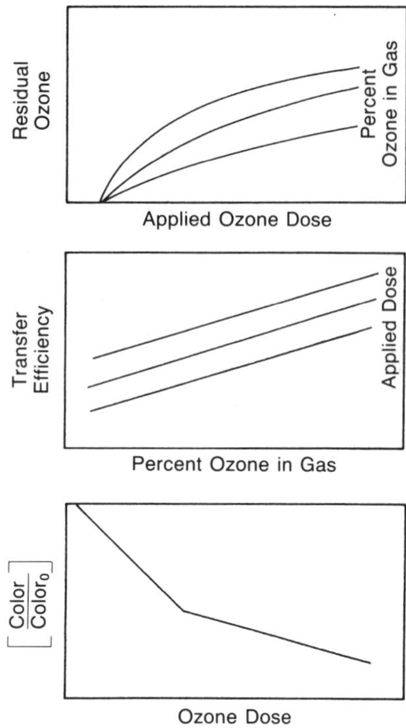

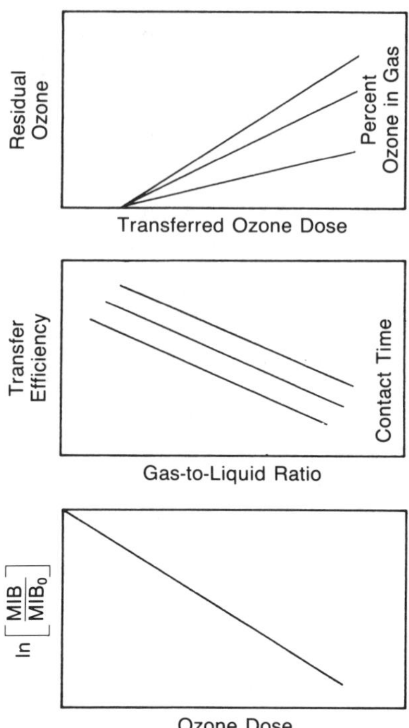

Figure IV–4 Interpretation of Results

synthetic organic chemicals (SOCs). The results from this initial matrix will establish the bounds of the remaining test program as well as providing the majority of information necessary for the design of the full-scale facility.

The reactor detention times in the matrix can be varied, but it is important to try values as low as 1 to 2 min since it may be possible to design the first stage of the full-scale reactor at this detention time.

The range in ozone doses will vary according to site-specific water quality conditions. To establish the lowest ozone dose, it will be necessary to determine the ozone dose required to accomplish the desired level of oxidation or, for the objective of disinfection, the ozone dose required to establish a residual. Values above and below the oxidation dose should be tested. For disinfection, the minimum ozone dose required to establish an ozone residual will be the lower limit for the test matrix.

One additional test that should be conducted during the initial test phase (if disinfection is an objective) is a determination of the ozone decay rate. This can be accomplished by ozonating the water, carefully collecting a large sample, and conducting ozone residual measurements over time. The sample should be collected and handled to prevent ozone loss due to causes other than its decay in the water (see chapter II, sec. C). This test should be conducted at two or three different initial residual concentrations.

Once the matrix is completed for a given water condition and location, the analyses of results can be completed and the following relationships (as presented in Figure IV–4) as well as others can be developed:

- Stoichiometry for oxidation processes (e.g., color, THM precursors, SOCs, taste and odor, metals) as a function of ozone dose, concentration of ozone in feed gas, and detention time.
- Ozone residual versus applied ozone dose, transferred ozone dose, and contact time for disinfection applications.
- Transfer efficiency versus ozone dose, percent ozone in gas, and gas-to-liquid ratio.

- Others as appropriate, such as the effect of pH or H_2O_2 addition (advanced oxidation process) on these parameters.

This short but efficient testing approach will supply the basic criteria necessary for establishing the most cost-effective confirmation testing program. As mentioned above, it will also provide basic insight into the design of the full-scale facility.

Confirmation testing is used to test the most promising of the ozone applications (e.g., optimum dosages and locations) developed during the initial test phase. Testing will be primarily concerned with determining the effects of the ozone application on downstream processes and the effect of upstream processes on the ozone application. This test phase will most probably have to be conducted at a pilot or a demonstration scale to appropriately determine the effect of or on other important processes. Completing the initial testing (i.e., matrix) presented above prior to beginning the confirmational testing will assist in optimizing the study time. The types of issues that might be considered in the confirmational testing could include:

- Determining the effect on downstream treatment for parameters such as the removal of turbidity, by-products, biodegradable organic carbon, THM precursors, color, taste and odor, iron and manganese, and SOCs.
- Testing at various pH values following an upstream treatment process, such as pH adjustment.
- Confirming the decay rate and ozone application necessary to maintain a residual in a second contactor if necessary for disinfection (two-stage contacting is discussed below). (Normally, disinfection determinations will be made based on a $C \cdot t$ model and not on actual microbial inactivation measurements.)
- Determining if cocurrent or countercurrent first- or second-stage contacting is most efficient (first stage countercurrent is the norm).
- Confirming primary objectives.
- Testing effects of advanced oxidation application.

Seasonal testing will be used to determine the effect of changing raw water conditions on the ozone application. Changes in pH, temperature, color, turbidity, inorganics, and other raw water constituents can have a significant effect on ozone performance and effectiveness. A design that does not incorporate the effect of seasonal variations and relies on the response to a single raw water condition has the distinct possibility of failure.

Seasonal testing need not be extensive. It can consist of several weeks of testing to confirm the ozone dose and the response of downstream processes. A quick preliminary matrix test sequence followed by a shortened confirmational testing protocol may be all that is necessary. If significant changes in performance are discerned, it may be necessary to extend the test to fully evaluate these differences.

IV.A.3 Bench-Scale Studies

Bench-scale studies can be utilized to develop preliminary estimates of design parameters and assist in limiting the range of parameters to be evaluated at pilot scale. These studies can also provide valuable insight into the fundamental processes involved in ozonation. A major drawback of all bench-scale process evaluations is the difficulty of determining the process effects on the downstream processes. Another important consideration is the characteristics of the water samples used for the bench-scale studies. These samples must display the different water characteristics anticipated for treatment. The following sections discuss the common reactor and ozone generator systems used for bench-scale studies.

Bench-scale reactors. Bench-scale treatability studies can be and have been performed with a variety of reactor/contactor types and scales. The term reactor will be used in this section to describe these devices. The appropriate system is determined by

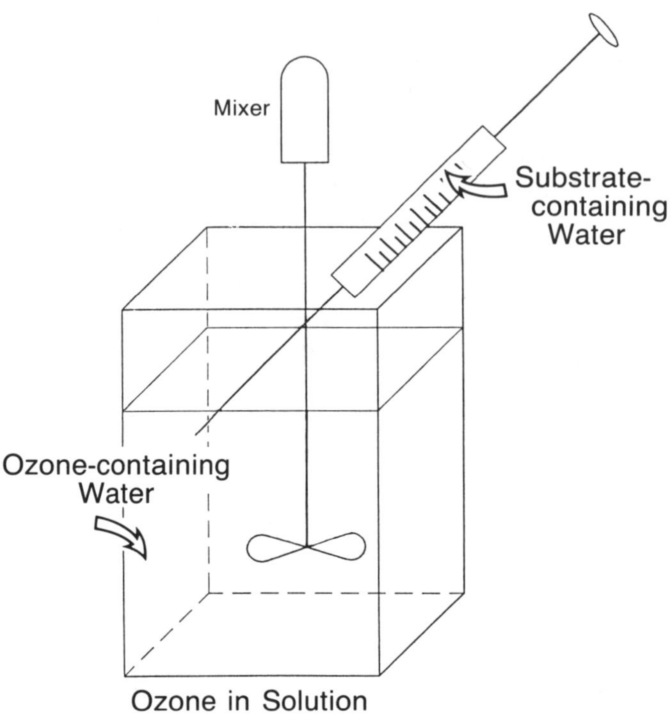

Figure IV–5 Batch Reactor

the objectives of the study. The more common reactor types are the batch reactor, the semibatch reactor, and the continuous flow-through reactor.

Batch reactor. In bench-scale evaluations, batch reactors are used in the fundamental study of ozonation reactions: the stoichiometry, kinetic constants and mechanisms. While this information is highly valuable, given our present level of understanding of ozone chemistry and its interaction with reactor hydrodynamic performance, it is difficult to translate the information gained into useful design parameters. To develop information on the required ozone doses and contact time for treatment plant design, the semibatch and continuous reactors are generally more useful.

In the batch reactor, two solutions, one containing ozone at a known concentration and one containing the substrate at a known concentration, are brought together under controlled conditions of temperature and mixing (Figure IV–5). The mixing regime is such that the time for complete mixing of the two solutions is much shorter than the characteristic time of reaction (kinetics). Since many reactions involving ozone are very rapid, the rapid mixing required to ensure homogeneous reaction conditions can be difficult to achieve.

The aqueous matrices are generally "synthetic" waters with controlled composition and reactant concentrations. The extent of reaction is monitored by sampling the reactor and measuring the reactant and product compositions over time. For ease of sampling for rapid reactions, a flow technique can be used in which the reacting solutions are rapidly mixed and then flow in small-diameter tubing to the measuring device. The time of reaction is dependent on the flow rate of the reacting solution between the point of mixing and the detector. This technique is referred to as the "stopped-flow" technique (Figure IV–6).

Semibatch reactor. The semibatch reactor may be thought of as an ideal plug-flow reactor (PFR) into which ozone is continuously introduced. It is batch in the substrate (i.e., fixed volume) but continuous in gaseous ozone feed (Figure IV–7). The ozone feed

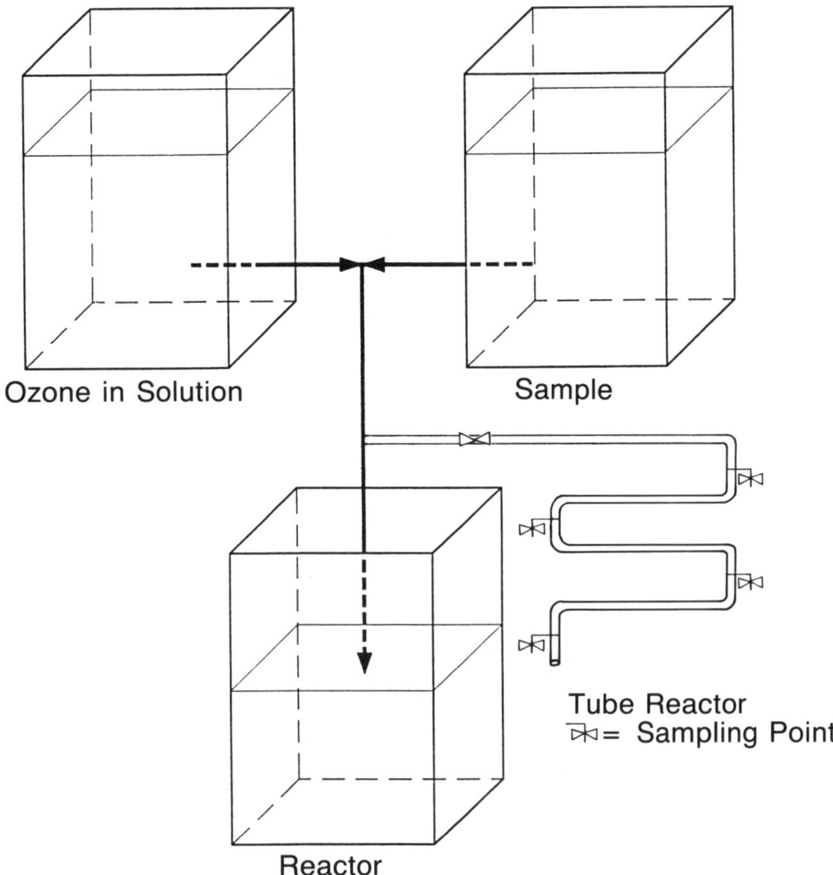

Figure IV–6 Batch Reactor: Stopped-Flow Technique

is transferred from a continuous flow of ozonated air or oxygen, so that the reactor volume remains constant over time. In general, the semibatch reactor is well mixed so that the concentration of the reactants is uniform within the reactor liquid phase. The ozone liquid-phase concentration achieves a constant value after a certain time. Changes in ozone liquid-phase concentration are accomplished by changing the gas flow rate, the gas-phase ozone concentration, or both.

There are some limitations to using a semibatch system. In real waters, ozone is consumed by a variety of substances and at a variety of rates. In a vigorously mixed semibatch system, the most readily consumed ozone-demanding substances will be rapidly exhausted, allowing the compounds more resistant to ozonation to become the only substrates available to the continuous supply of ozone, whereas the incomplete mixed continuous flow reactor will show localized differences in the degree of oxidation and have a continuous inflow of nonoxidized water. This will tend to produce less of a sequential oxidation phenomenon and more of a simultaneous competitive oxidation system. The semibatch system may also allow the buildup of reaction intermediates to levels that can influence the final reaction products found or result in ozone consumption not observed otherwise. The tendency of the semibatch reactor, or any ideal plug flow reactor, is to overestimate the extent of reaction compared to a reactor system with a significant backmixing and short circuiting (Levenspiel 1972).

The semibatch reactor's strengths are the ease of operation, small volume required, and rapid results. The semibatch system can be used to estimate the required ozone dose for an application, develop residual decay information, and estimate the extent of ozone

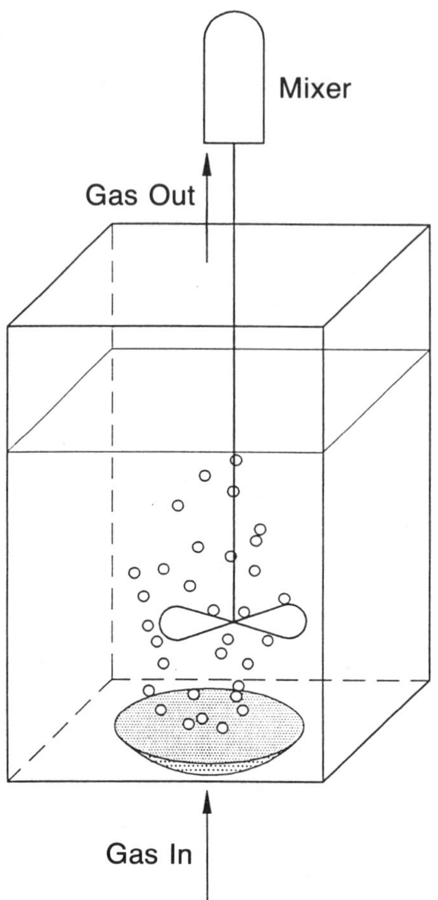

Figure IV–7 Semibatch Reactor

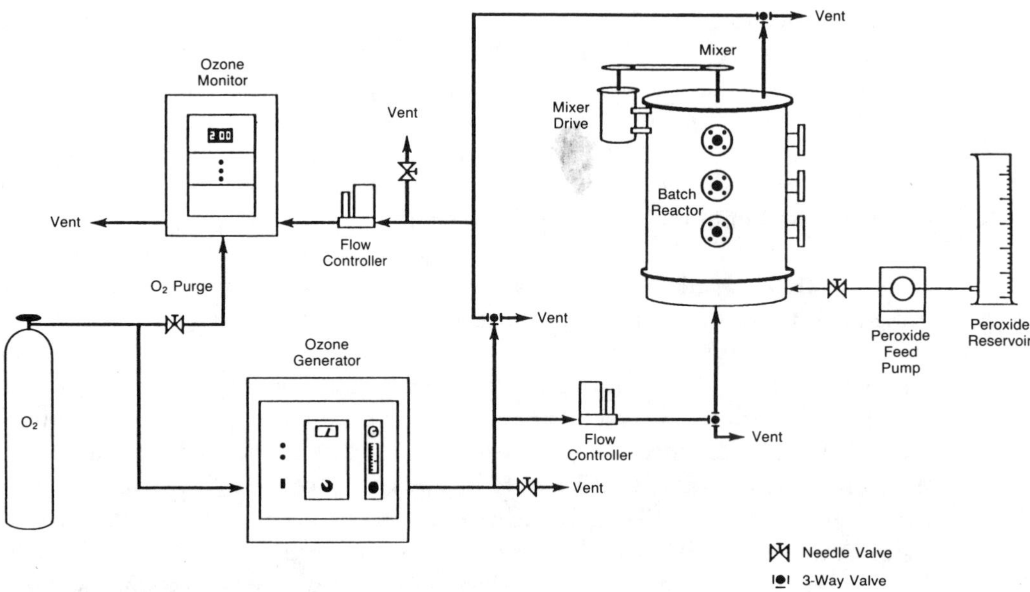

Figure IV–8 Schematic of Bench-Scale System

Figure IV–9 Reactor Used to Investigate Advanced Oxidation of Volatile Organics in the San Gabriel Basin (Southern California)

oxidation reactions. The use of this information for design purposes must be done with care.

There are any number of methods for implementing a semibatch reactor investigation. Figure IV–8 is a schematic of a semibatch reactor, bench-scale design that has been used by Glaze and Kang (1988), Bellamy et al. (1989), and others for investigations of oxidation reactions. Figure IV–9 is a photograph of the reactor used to investigate advanced oxidation of volatile organics in the San Gabriel Basin (Southern California). The reactor volume is approximately 70 L (18.5 gal). The same general configuration has been used at smaller scales, e.g., 2 L (0.53 gal) (Langlais 1979) down to 0.5 L (0.13 gal) (Reckhow and Singer 1984).

Continuous flow reactor. Bench-scale, continuous flow reactor designs are varied, but all have the common characteristic of a continuous flow ozone-containing gas stream and a continuous flow liquid stream. A laboratory-scale continuous flow reactor does not suffer from some of the limitations associated with the semibatch reactor related to backmixing. As an approximation to the nonideal flow regimes encountered in full-scale (and pilot-scale) reactors, a series of continuous flow reactors (Figure IV–10) can be utilized that will then approximate the intermediate case between plug flow reactor and completely stirred tank reactor (CSTR) behavior (Levenspiel 1972). The drawbacks of a laboratory-scale continuous flow reactor are mainly in the physical setup of the equipment: monitoring and controlling very low liquid and gas flow rates and

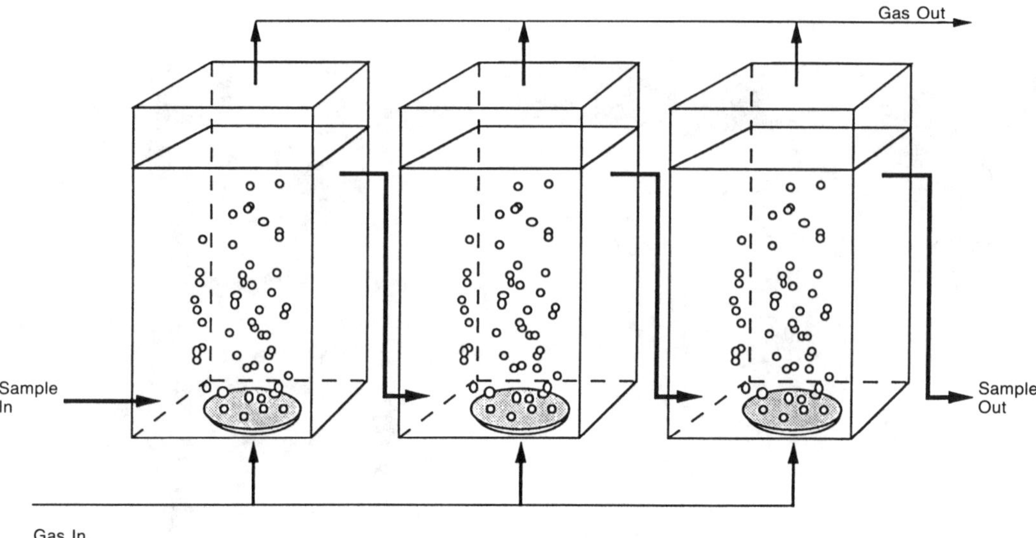

Figure IV–10 Continuous Flow Series Reactor

shipping and handling large volumes of sample. The physical configuration of the equipment can be similar to that presented for the semibatch system.

Ozone generation at bench scale. There are three basic types of ozone generating devices used in small applications: tubular corona discharge, plate-type corona discharge, and ultraviolet light ozone generation (for very small ozone production). Laboratory-scale generators mimic their full-scale counterparts in most design factors except power consumption, which is high. As with the larger generators, the most common design is the corona discharge, and it is available in many of the variations also found in large-scale equipment. The tubular form of the corona discharge unit is by far the most common, with the plate-type generators still in use in some places but not generally available. Piloting the type of generator is not important.

The concentration of ozone created by the corona discharge units ranges from 1 to 3 percent by weight (13 to 39 g O_3/m^3 NTP) at the maximum production rate of the generator when ozone is produced from air. If the flow rate is decreased in this type of generator, the ozone concentration will increase accordingly until it becomes temperature limited. A further decrease in flow rate will continue to decrease the mass ozone output of the generator but will still increase the concentration to a degree.

Concentrations of 6 percent by weight (87.6 O_3/m^3 NTP) or more of ozone have been generated using oxygen as the feed gas. At least one manufacturer claims the ability to generate ozone on the laboratory scale at concentrations up to 8 percent by weight (116.8 g O_3/m^3 NTP) from oxygen.

Ozone generation by ultraviolet light involves a photochemical reaction. The ozone concentration that can be generated by UV is generally less than 0.3 percent by weight (3.9 g O_3/m^3 NTP) in air and 0.4 percent by weight (5.8 g O_3/m^3 NTP) with oxygen as the feed gas. These concentrations are generally not applicable or appropriate for treatability studies, and this type of generator should normally not be used.

IV.A.4 Pilot- and Demonstration-Scale Studies

The information contained in this section pertains to both pilot- and demonstration-scale studies. When designing and constructing a pilot-scale facility, two basic premises should always be remembered: (1) the convenience and quality of the equipment can far outweigh the cost differential for a less easily operated system that has frequent

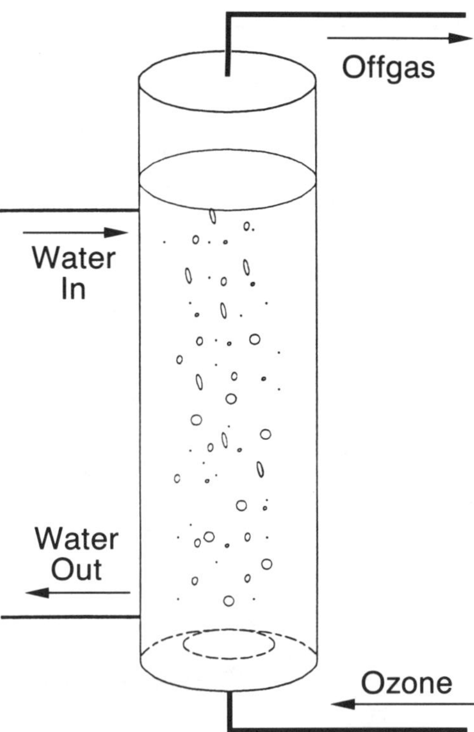

Figure IV–11 Continuous Flow Reactor

equipment malfunctions, and (2) the major study expense will often be the highly educated and skilled labor required to operate the system and collect meaningful data.

Pilot-scale reactors. The design of pilot-scale reactors should follow the general guidelines of providing good gas/liquid contact with liquid-phase hydrodynamics similar to those found in full-scale reactors. Variables to be considered are depth and volume of the reactor, liquid flow rate, gas flow rate, bubble size, and gas-to-liquid ratio. The characteristics of the reactor will have to be fully defined by a tracer evaluation. This characterization is necessary to relate the results of the pilot study to the design of the full-scale facility.

Of the pilot-scale reactors, the most commonly used system is the bubble column reactor. This reactor is typically operated with ozone gas flow countercurrent to the liquid flow (Figure IV–11). Depending on the design and operational parameters selected, the bubble column reactor can have hydrodynamic characteristics of a completely mixed stirred-tank reactor, a plug-flow reactor, or a reactor with some intermediate flow characteristics. This type of reactor finds wide application in treatability studies because it tends to simulate a full-scale reactor. Care must be taken, however, to assure that the hydrodynamics of the process study reactor meet the reality of the designed system as close as possible. Results from a PFR pilot-scale reactor may severely underestimate the required dose and/or contact time for a full-scale reactor with significant CSTR behavior.

The hydrodynamic condition of a single or multiple reactors can be characterized by conducting a tracer study of the reactor(s) for various operating conditions. A set of typical tracer curves is shown in Figure IV–12 for a single pilot reactor subjected to a pulse tracer input. This series of tracer experiments was performed on a 5.2-m (17-ft) water depth tubular reactor, 0.15 m (6 in.) in diameter (MWDSC 1989). There are a number of mathematical analyses of the tracer curves that can be used to characterize the reactor (Levenspiel 1972). For example, the mean contact time, t_{10}, t_{50}, dispersion

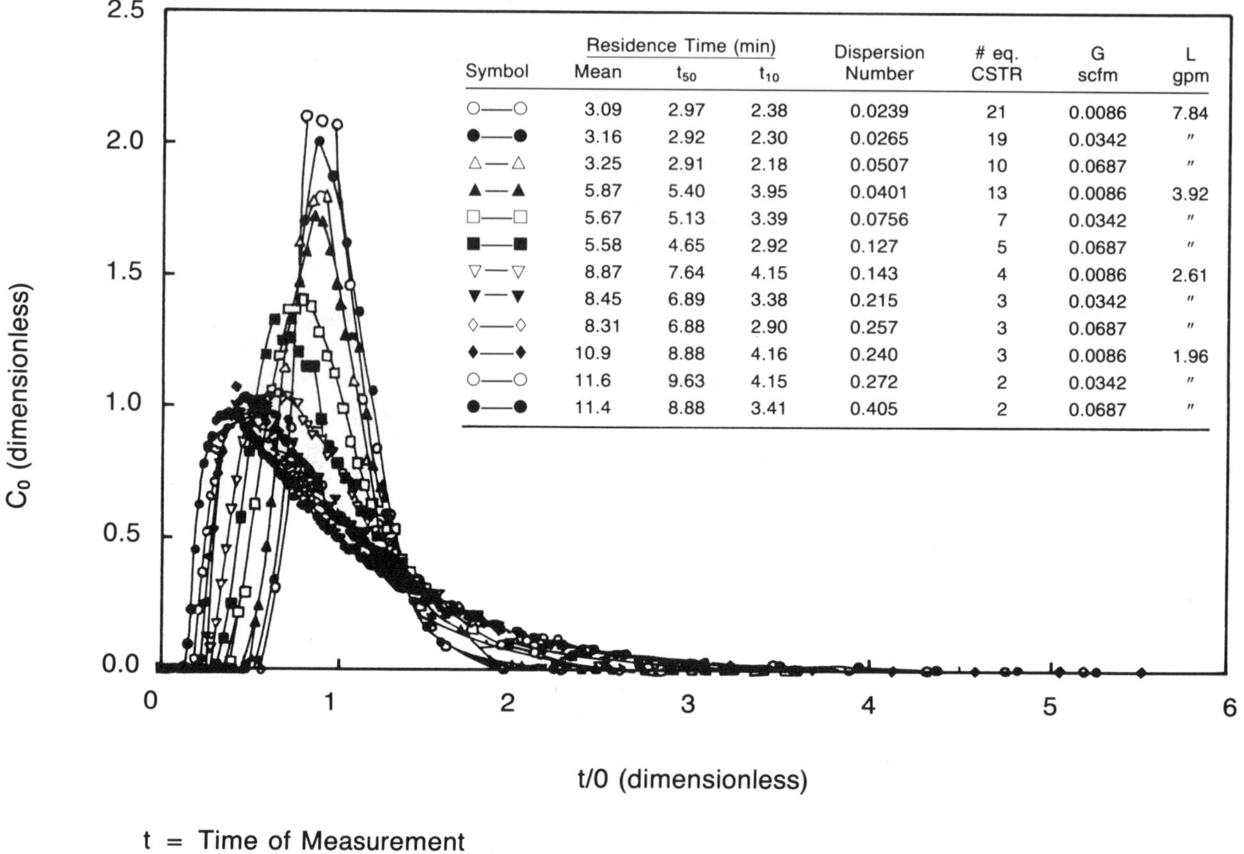

Symbol	Residence Time (min) Mean	t_{50}	t_{10}	Dispersion Number	# eq. CSTR	G scfm	L gpm
○—○	3.09	2.97	2.38	0.0239	21	0.0086	7.84
●—●	3.16	2.92	2.30	0.0265	19	0.0342	"
△—△	3.25	2.91	2.18	0.0507	10	0.0687	"
▲—▲	5.87	5.40	3.95	0.0401	13	0.0086	3.92
□—□	5.67	5.13	3.39	0.0756	7	0.0342	"
■—■	5.58	4.65	2.92	0.127	5	0.0687	"
▽—▽	8.87	7.64	4.15	0.143	4	0.0086	2.61
▼—▼	8.45	6.89	3.38	0.215	3	0.0342	"
◇—◇	8.31	6.88	2.90	0.257	3	0.0687	"
◆—◆	10.9	8.88	4.16	0.240	3	0.0086	1.96
○—○	11.6	9.63	4.15	0.272	2	0.0342	"
●—●	11.4	8.88	3.41	0.405	2	0.0687	"

t = Time of Measurement

0 = Theoretical Detention Time

$$C_0 = \frac{\text{Conc.} \times \text{Water Flow Rate} \times \text{Theoretical Detention Time}}{\text{Tracer Mass Added in Spike}}$$

Source: MWDSC (1989).

Countercurrent gas and water flows.

Figure IV–12 MWDSC Tracer Study Results

number, equivalent CSTRs in series, gas flow rate, and liquid flow rate for each tracer test are also given in the figure. These types of characterizations can be used to compare full-scale contactors to pilot- or demonstration-scale units. Tracer tests should also be conducted at various flows without gas addition.

Figure IV–13 presents an example of the use of a mathematical model to evaluate reactor hydraulics. Theoretical response curves for equivalent CSTRs in series for a continuous tracer input have been plotted against the results of an actual tracer study of the ozone contactor at Los Angeles Aqueduct Filtration Plant (LAAFP) (Stolarik and Christie 1988). These curves represent a continuous tracer input versus a slug input as presented above. The two results can be converted mathematically to accommodate either type of presentation. The LAAFP response is best represented by a 10-to-15-CSTR-in-series model. This and other types of analysis can be used to assess the characteristics of the pilot reactor and relate it to the full-scale reactor. The same kind of tracer experiments are also performed in Europe on full-scale plants (Langlais and Bablon 1990).

Reactor depth and volume. Reactor depth in a pilot study is more a function of convenience than it is of transfer efficiency. The depth in full-scale reactors is to assure

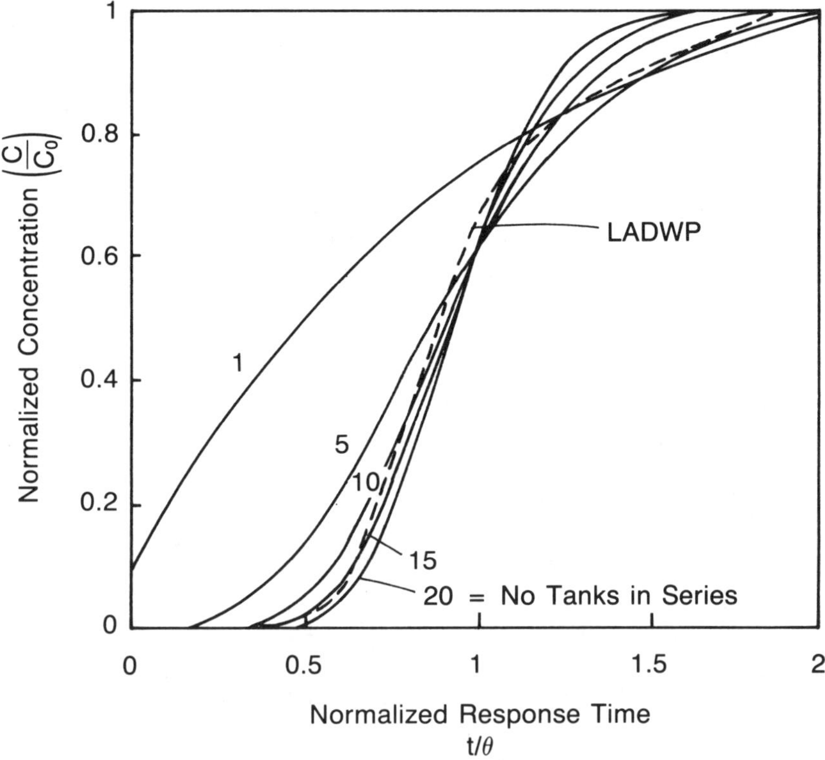

t = Time
θ = Theoretical Detention Time
C = Reactor Effluent Concentration
C_0 = Tracer Input Concentration

Source: Stolarik and Christie (1988).

Figure IV–13 Response Curve for Multiple Mixed Tanks in Series to a Continuous Tracer Input with LAAFP Tracer Data, California

high transfer efficiency to minimize ozone production costs. For process studies, high transfer efficiencies are not required to determine the design parameters of dose and contact time. While high transfer efficiencies are not necessary, measurement of the transfer efficiency to determine transferred ozone dose is required. Pilot columns of 3 to 6 m (10 to 20 ft) are common. The taller columns will tend to behave as PFRs while the shorter columns will tend to behave as CSTRs. Providing the ability to vary the water depth can be beneficial.

The choice of reactor volume is determined by the liquid flows required (see below) and the range of contact times to be investigated. If the objectives of the process study are to examine certain classical oxidation reactions that are relatively fast (e.g., some taste and odor compounds, color), contact time can be relatively short as these reactions are generally dependent on transferred ozone dose (Hoigné et al. 1987). For disinfection experiments, contact times should be chosen to provide C · t values sufficient to meet disinfection needs as estimated or established by the WHO and EPA.

Normally, columns are constructed of glass, clear acrylic, or clear polyvinyl chloride (PVC). The column diameters are standard pipe sizes normally between 0.1 and 0.3 m (4 and 12 in.). The contactors do not need to be transparent; however, it is often an advantage to see the process to determine if it is operating properly. Most flow and detention time parameters can be met with 0.1- to 0.15-m (4- to 6-in.) diameter columns. Larger columns are only necessary when larger flows are required for downstream treatment such as sedimentation. For mounting and convenience, columns of 4.6 m (15 ft) are reasonable. Support and alignment become more difficult as the columns become taller.

Liquid flow rate. The selection of liquid flow rate for a pilot-scale ozone reactor should be based on the process study needs and the hydrodynamics of the reactor itself. In terms of the process needs, the ozone reactor must be sized to provide sufficient ozonated water for the downstream processes. At given reactor dimensions and a set gas flow rate, increasing the liquid flow rate (and the fluid velocity in the reactor) increases the plug-flow characteristics in the reactor. The relationship between hydrodynamics and G/L ratio is discussed below. A general rule of piloting is to size each process for flows greater than the flow required for subsequent processes and then waste the excess flows. This is necessary to separate or decouple the processes and make them hydraulically independent in terms of operations.

Gas flow rate and bubble size. The range of gas flow rates for the pilot reactor should be matched to the diffuser system selected. Optimum bubble sizes for an ozone reactor range from about 2 to 5 mm. Diffuser manufacturers will be able to recommend a diffuser that will provide a range of gas flow rates in which this size bubble can be produced. Operationally, the ozone pilot system should be capable of varying the ozone dose rate by either (1) changing the gas flow and maintaining a constant concentration or (2) changing the ozone concentration and maintaining a constant gas flow. Once set, the system's ozone concentration and gas flow rate should remain constant.

Gas/liquid ratio. The effects of the G/L ratio on the hydrodynamics of a pilot ozone reactor system have been demonstrated in the tracer analysis of the contactor depicted in Figure IV–12. For this reactor, as the liquid rate decreases and the gas rate increases, the contactor characteristics progress from predominantly PFR to predominantly CSTR. This figure indicates the significant effect the G/L ratio can have on the pilot reactor and the need to evaluate the behavior of the reactor with a tracer evaluation. Considerable control over the G/L ratio can be obtained by planning for a wide range in ozone gas concentrations (e.g., from 1 to 6 percent). A G/L ratio that will meet the objectives of the study can be obtained if flexibility is provided in the contactor liquid volume and the ozone gas concentration.

Special considerations. It may not always be possible to operate the pilot plant when a specific water quality episode occurs. For example, if the objectives for the treatability study may include seasonal taste and odor control, it may be necessary to add the compounds of interest into the pilot plant flow. Care must be taken to assure a consistent known liquid-phase concentration of the compounds of interest. In addition, sufficient operational time must be provided to allow for process stabilization before sampling and between process changes.

Ozone generator. The ozone generator must be capable of supplying the maximum amount of ozone expected. An oversized generator may be difficult to control unless excess gas is generated and wasted. Most available small-scale systems have turndown ratios of 10:1. In general, ozone generators that include air preparation systems are more expensive and larger than those that run on bottled air or oxygen, but the operation costs are lower.

The pilot system must have a provision for measuring the ozone concentration and the gas flow rate. The system should also have provisions for measuring the reactor ozone concentration in the offgas to be able to determine transferred ozone dose. Generators should be selected that will give ozone gas-phase concentrations in the range normally encountered in practice (1 to 6 percent by weight depending on which gas is used to produce the ozone). This precludes the use of UV-based ozone generation systems unless the treatment facility anticipates the installation of a generation system

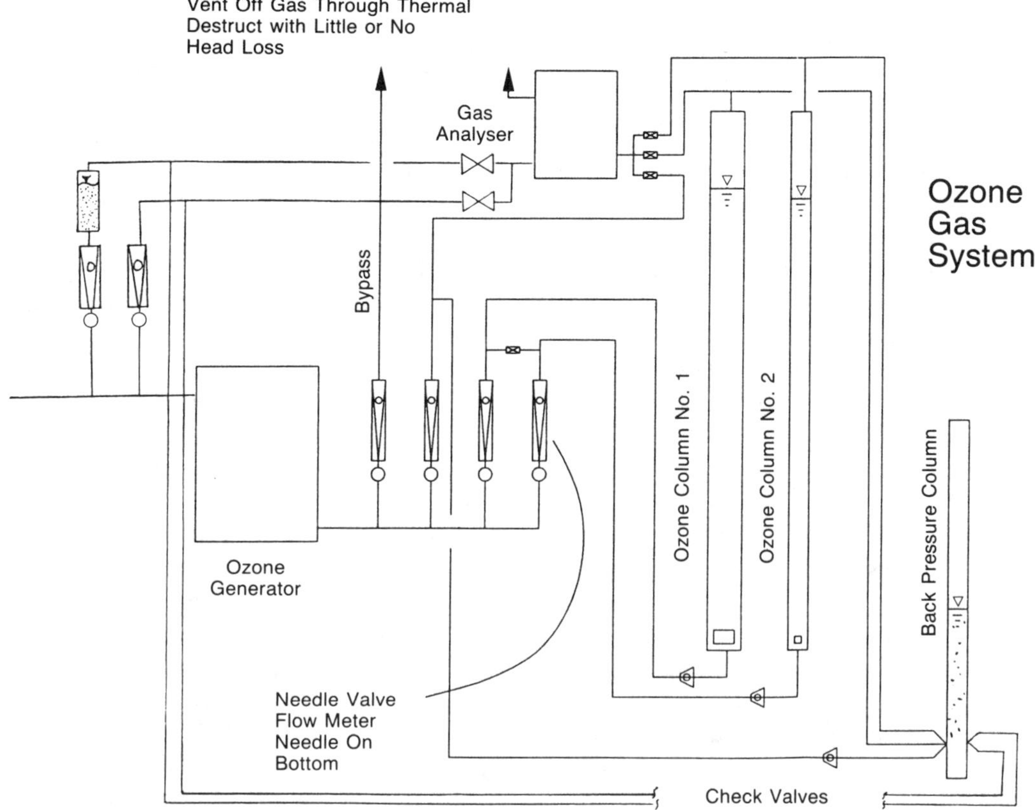

Figure IV–14 Ozone Gas Distribution System

that produces gas-phase concentrations of less than 0.5 percent by weight (6.5 g O_3/m^3 NTP).

To simplify ozone measurements and for multiple point dosing, a gas manifold system can be constructed that splits the generator output flow into multiple streams. Each stream must have its own flow meter and each reactor should have provision for offgas measurement. Balancing split stream gas flows can be difficult, depending on the mechanical and hydraulic conditions.

Figure IV–14 is a flow schematic of a manifold system that will accomplish each of these requirements. The most important characteristic of the system is that once the flows and pressures are set to each of the columns and to the gas analysis lines, they will remain constant even when switching among the analysis lines. This will prevent measurement errors caused by changes in flow or pressure. The back pressure column shown in the figure is a tube with 0.6 to 0.9 m (2 to 3 ft) of water, which will provide a consistent pressure and (consequently) flow to the gas phase analyzer from each of the offgas streams and the feed gas. The water column in the zero gas stream to the gas-phase analyzer is used to humidify the zero gas when analyzing the offgas from the reactors. The nonhumidified zero gas is used when analyzing the feed gas. Another solution, which is possibly more desirable although more expensive, is to use two ozone gas monitors, one for the feed gas and one for the offgas.

IV.A.5 Parameter Measurements

When evaluating the effects of ozonation in a treatment scheme, data from the entire process need to be collected, recorded, and analyzed. The process conditions such as coagulant dose, pH, flocculation energy input, polymer addition, and flow rates are needed along with performance measurements such as turbidity, disinfection by-products, taste and odor, and color, for each experimental condition of ozone dose, contact time, and application point. Raw, intermediate, and finished water quality parameters should also be part of the experimental database.

It is most convenient to provide detailed data sheets for all variables of interest and a computerized database for recording and analyzing the data. For extended studies, it may be desirable to provide online monitoring and data collection and storage for specific parameters, (e.g., feed gas concentration, offgas concentration, liquid and gas flow rates, pressures, and turbidity). As with all large data set manipulations, assuring the integrity of the data requires a well-planned and -executed program of quality assurance.

For the ozone system, the minimum information required is feed gas ozone concentration, offgas ozone concentration, temperature, pressure, gas flow rate, liquid flow rate, liquid volume, ozone residual, and a measurement of ozone decay. The following sections discuss the common ozone-related parameter measurements that will normally be required during the course of a treatability study. Sec. IV.A.6 presents methods of interpretation of these measurements and their relationship to design parameters.

Offgas ozone measurements. Full- and pilot-scale operations rely on the use of commercially available gas-phase ozone monitors that use ultraviolet absorption of ozone to measure the concentration in the gas stream. These monitors require a gas flow rate of at least 1 L/min. Bench-scale ozone processes can have a gas flow rate that is a fraction of this value; thus, the commercial monitors themselves cannot be utilized for measurements of ozone in the offgas. For the larger semibatch reactors (e.g., 70 L [18.5 gal]) the gas flow rates may be sufficient to be measured with a conventional monitor.

The same principle utilized by commercial ozone monitors can be used in the laboratory to measure the gas-phase concentration at low gas flow rates. The absorbance of the gas stream can be measured spectrophotometrically at 253.7 nm by passing the gas stream through a flow-through cell in a UV spectrophotometer. The molar extinction coefficient or molar absorptivity of ozone at this wavelength is known; thus, the gas-phase concentration can be calculated using the Lambert-Beer relationship (see sec. C, chapter II), considering the temperature and the pressure in the cell.

As long as the materials of construction used in the apparatus are not reactive with ozone, the gas flow rate can be reduced to less than 0.1 L/min without affecting the accuracy of the technique (Aieta 1990). If materials other than stainless steel, Teflon, and glass are used to construct the reactor and/or transfer the gas to the spectrophotometer, the measured concentration can differ from the actual concentration in the apparatus due to reaction of the ozone with the materials of construction.

Wet chemistry measurements of offgas can also be used at bench scale. The total offgas stream can be routed through a wet chemistry analysis apparatus (see sec. C, chapter II). This measurement technique will determine the total ozone mass for the selected sample period. However, this method cannot be conveniently used to differentiate changes in ozone transfer efficiency during the sample period (i.e., it is integrative rather than instantaneous).

Feed gas ozone concentration measurement. Feed gas concentrations can normally be measured with commercial monitors. This assumes that excess feed gas is generated during the experimentation. It is important to use the same gas flows, pressures, and power to the generator during the feed gas measurement and during testing. If there is not at least the 1-L/min excess gas production required by the commercial monitor, a wet chemistry method will have to be used. Consult sec. C, chapter II for a description of the analytical techniques.

Ozone residual. Sec. C, chapter II identifies a number of methods for measuring ozone residual. The most common measurement uses spectrophotometry with indigo trisulfonate or some other colorometric technique (e.g., iodometric). This approach has been established for batch-type measurements (this being the most common) and automated procedures. There are also electrochemical devices (i.e., electrodes) and gas stripping techniques for automated residual measurements.

Sampling technique is a very important factor in measuring ozone residual. Sec. C, chapter II describes the sampling procedures. It is also important that proper sample handling is exercised during the measurement of decay rate (i.e., the successive measurement of residual).

Performance parameters. In addition to ozone analyses, measurements of performance need to be monitored and related to the ozone utilized. The parameters measured depend on the objectives of the treatability study (for example, color, disinfection by-products, microorganisms, or micropollutants). These parameters may be measured with or without quenching of the ozone residual, depending on the objectives of the study and the final treatment system characteristics.

IV.A.6 Interpretation of Results

Scale. Design criteria originating from the results of bench, pilot, or demonstration testing have to be developed with an understanding of the limitations involved in the test technique. The limitations of bench-scale reactors and their deviation from real systems has been discussed above. Most bench-scale studies evaluate ozonation as a unit process only and not in an integrated treatment scheme. For example, the effects of ozone pretreatment on filtered water turbidity are difficult to predict or determine from bench-scale data alone. The production of ozone disinfection by-products in a water source can be determined at bench scale, but the biodegradation that may occur in subsequent treatment cannot be determined.

The data from batch, semibatch, or continuous flow bench-scale studies can aid in the determination of ozone's ability to meet the treatment objectives. For example, bench-scale studies (in semibatch or continuous flow reactors) can indicate whether or not ozone can effectively oxidize color in a source water. Comparisons can be made between ozone and other oxidants, and a preliminary economical analysis may indicate other treatment options should be considered in the pilot study. Initial doses and application points can be determined for pilot-scale studies, thus decreasing the time and expense of field evaluations.

Bench-scale studies do provide valuable information at a reduced cost from more extensive pilot-scale studies. The information gained represents a valuable step in the design process, but relying solely on the results of a bench-scale evaluation for the final treatment plant design parameters will require a greater degree of conservatism in the design, resulting in higher costs than would be necessary with quality pilot-scale results.

Results of pilot-scale or demonstration evaluations will more closely represent the conditions expected in full-scale evaluations. However, like the bench-scale results, they will require interpretation prior to implementation into final design criteria. For example, the test-scale reactor hydraulics will most probably not represent full-scale reactor hydraulics. As presented above, a model can be developed to assist in relating the results of the test facility to the intended full-scale application.

Ozone transfer efficiency. In all process studies, ozone concentration in the feed gas (O_{3feed}) and ozone concentration in the offgas (O_{3out}) must be measured and the transfer efficiency must be calculated. Determining transfer efficiency for the pilot system is required to relate the pilot results to full-scale applications. These results are not used to estimate full-scale transfer efficiency unless the study is of the size and appropriate design to accommodate this objective. Transferred ozone is calculated by determining the difference between the ozone applied to the reactor and the ozone in the offgas, as shown in the following equations:

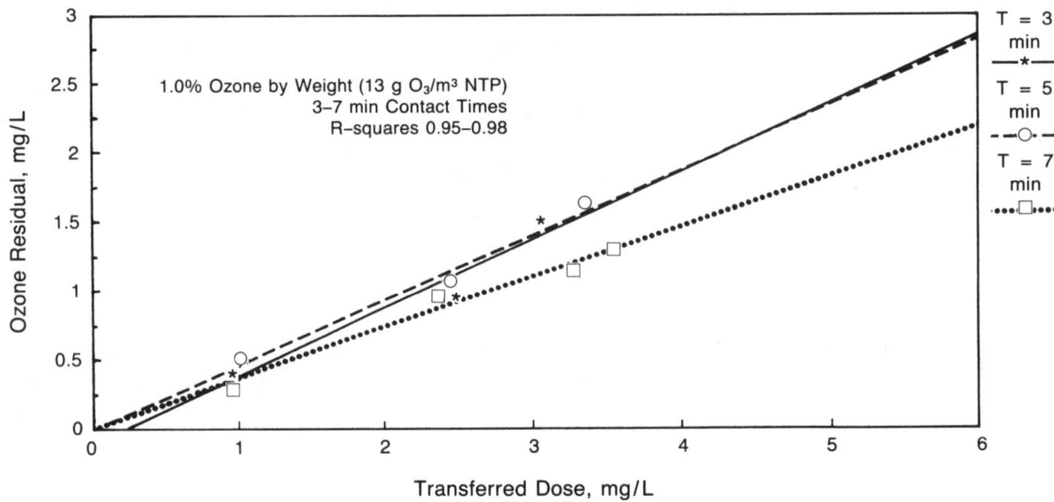

Single-Stage Contactor
pH = 7.7, Temp. = 20°C (68°F)

Source: Price et al. (1988).

Sobrante Filter Plant (northern California) raw water, tested 9/26/88, 9/27/88, and 9/30/88.

Figure IV–15 Ozone Residual vs. Transferred Dose

$$\text{Transfer efficiency (TE)} = [(O_{3feed}) - (O_{3out})]/(O_{3feed}) \times 100 = \text{percent}$$

$$\text{Transferred mass (TM)} = \text{TE} \times O_{3feed}\ (\text{mg/L}) \times \text{L/min} = \text{mg } O_3/\text{min}$$

$$\text{Transferred Dose (Batch)} = [\text{TM} \times \text{min of feed}]/\text{reactor volume (L)} = \text{mg/L}$$

$$\text{Transferred Dose (Flow)} = [\text{TM}]/\text{flow rate in L/min} = \text{mg/L}$$

Ozone demand. Ozone demand can be determined and expressed in a number of ways. The simplest method is to graph ozone residual versus transferred ozone dose (Figure IV–15). The demand is the difference between the transferred and residual concentrations. For a given water, this value will depend on the contact time and contactor design, as well as the ozone dose. This type of curve can be used to assess the requirements for disinfection. Ozone demand relative to the oxidation of a specific compound can be determined and expressed as shown in Figure IV–16. In this figure the oxidation of color versus the transferred ozone dose indicates the design requirements for a desired color reduction. The interpretation of this type of data is discussed in greater detail below.

Ozone decay. Ozone decay should be measured for each water tested (e.g., different locations in the treatment system and different seasons or testing conditions). Decay is an important parameter in disinfection determinations but is less important when considering oxidation reactions. However, it should still be measured, since decay rates indicate the type of competing reactions (i.e., decomposition rate) that occur after the immediate demand is satisfied. As presented above, decay is determined by

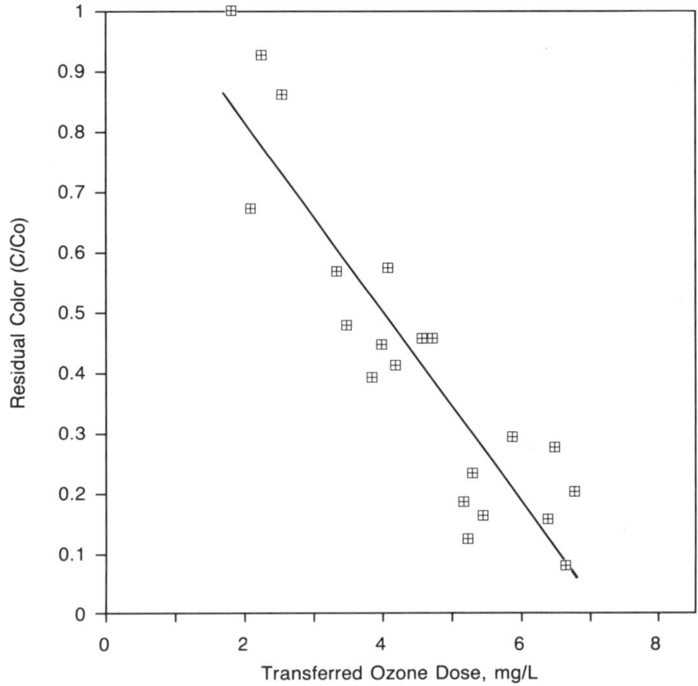

Source: Bellamy et al. (1990).

Figure IV–16 Color Reduction for Groundwater—Long Beach, California

measuring successive ozone residuals in a sample of water that has been ozonated sufficiently to satisfy the immediate demand and establish a residual.

In most cases, decay can best be interpreted and presented as a pseudo first-order kinetic reaction. Figure IV–17 shows decay data plotted as a pseudo first-order reaction. This pseudo first-order reaction is represented by the following relationship:

$$\ln \frac{[C]}{[C_0]} = -k'[t]$$
$$[C] = [C_0]\exp(-k'[t])$$

Where:

$[C]$ = ozone residual (mg/L) at the designated time (t)
$[C_0]$ = ozone residual (mg/L) at time 0
k' = decay rate constant in (min^{-1}).

The rate constant is the slope of the linear relationship between $\ln[C/C_0]$ and time. If this relationship is not relatively linear, the decay rate cannot be successfully modeled as pseudo first-order.

The utility of the rate constant can be seen in the following example. Assuming that a reaction chamber with no ozone addition has a 4-min detention time, an influent ozone concentration of 0.3 mg/L, and a k' value of 0.06 min^{-1} (Figure IV–17), the effluent residual can be approximated by:

$$\begin{aligned} [C] &= [C_0]\exp(-k'[t]) \\ &= [0.3]\exp(-0.06 \cdot 4) \\ &= 0.24 \text{ mg/L} \end{aligned}$$

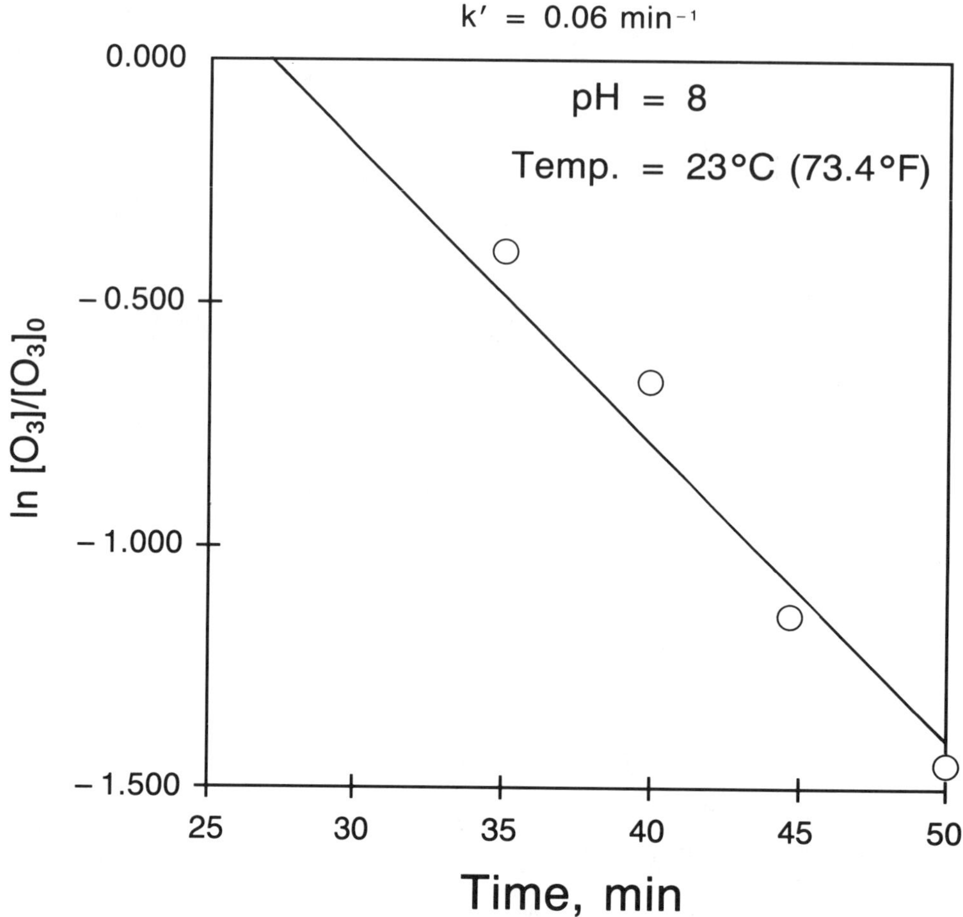

Source: Glaze et al. (1988).

Figure IV–17 Rate of Ozone Decay for Surface Water at San Leandro, California

Oxidation. For oxidation results, comparisons of doses and contact times can conveniently be represented on bar charts or graphs as appropriate (e.g., Figure IV–18). This data can also be subject to mathematical modeling or used directly. The mathematical models often used are similar in form to classical chemical kinetic models. However, they are entirely empirical and describe only the empirical stoichiometry that results from simultaneous competitive reactions. Figure IV–18 presents an empirical exponential stoichiometric model for taste and odor reduction based on the measurement of MIB.

For rapid oxidation reactions, the reaction rate can be expressed or converted to a stoichiometric relationship by plotting ozone dose versus $\ln[C/C_0]$ for exponential relationships (also presented in Figure IV–18) or $[C/C_0]$ for linear relationships (e.g., color, Figure IV–16). These relationships are represented by:

Exponential

$$\ln \frac{[C]}{[C_0]} = -k_D[O_3]$$

$$[C] = [C_0]\exp(-k_D[O_3])$$

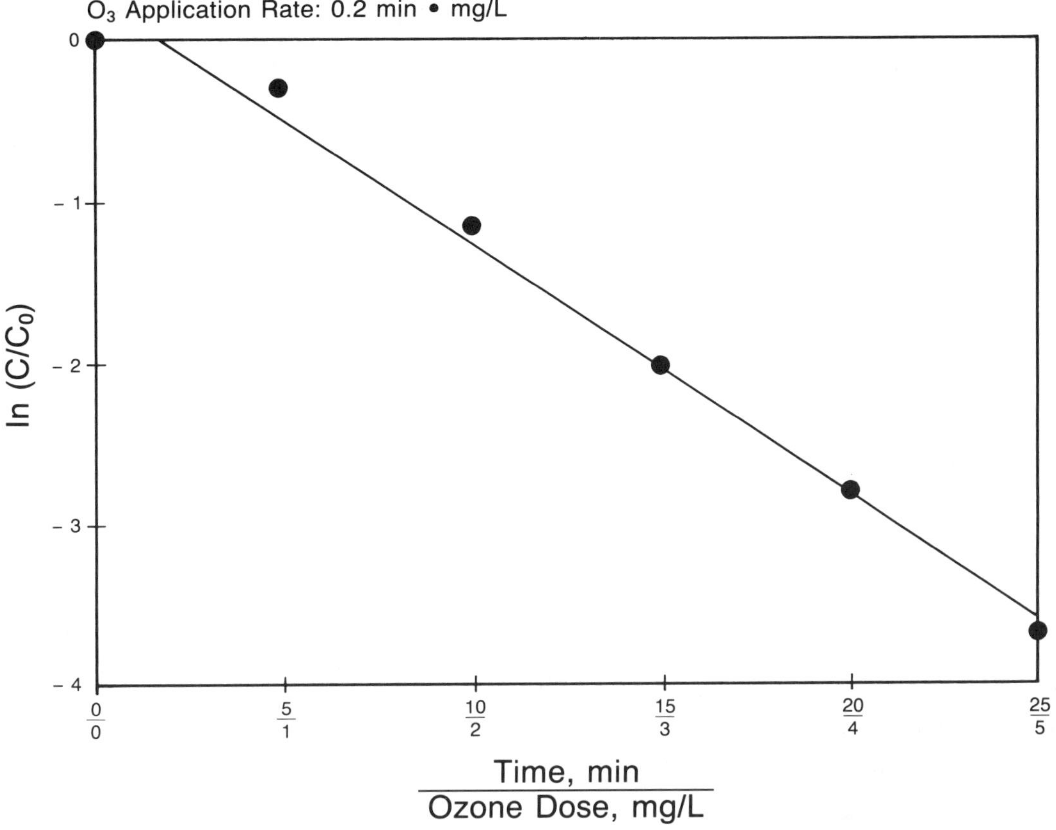

Source: Glaze et al. (1988).

Figure IV–18 Upper San Leandro Filter Plant Settled Water, MIB Removal—California

Linear

$$\frac{[C]}{[C_0]} = k_D[O_3]$$

$$[C] = [C_0]\,(k_D[O_3])$$

Where:

k_D = empirical stoichiometric coefficient (L/mg)
$[O_3]$ = ozone dose (mg/L).

These empirical stoichiometric coefficients define the relationship between transferred ozone dose and level of treatment. The coefficients can be used (with appropriate judgment and only for the waters studied) to calculate the transferred ozone dose required for a given or desired level of oxidation (as presented in the decay rate example above).

Disinfection. Disinfection can be evaluated during a treatability study in two basic ways. The first method is to assign a disinfection objective based on known inactivation data and an appropriate mathematical model, such as Chick's Law (i.e., the $C \cdot t$ concept), and determining the relationship between transferred ozone dose and the specific water in relation to the disinfection criteria. The second method is to conduct inactivation investigations. The more common approach is to use the $C \cdot t$ model. The $C \cdot t$ model should normally be used since the complications and possible errors

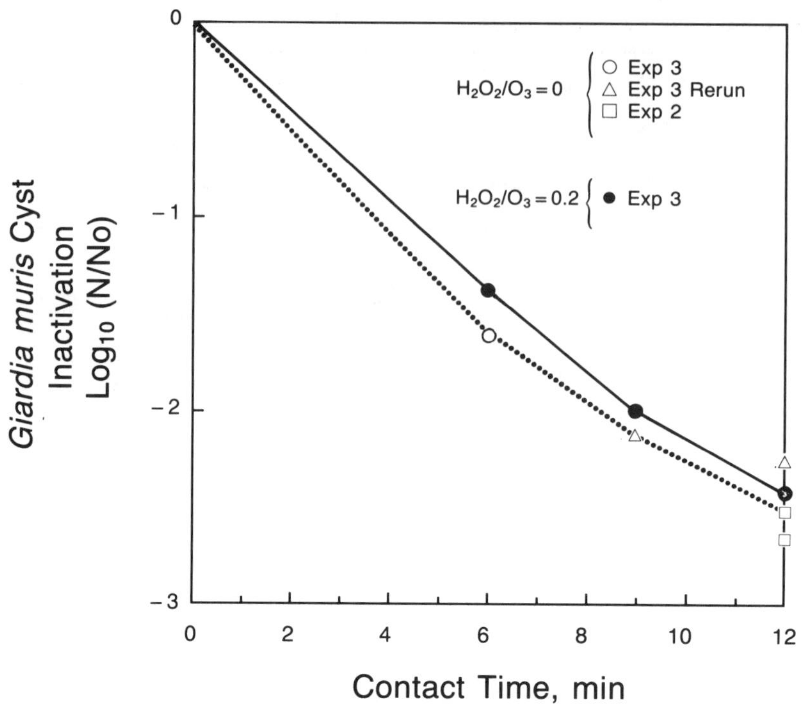

Source: MWDSC Status Report, Phase IV (1989).

Figure IV–19 ***Giardia muris*** **Cyst Inactivation by Ozone and Ozone/Hydrogen Peroxide Processes**

associated with an inactivation study can be overwhelming unless the study team has a very experienced disinfection specialist. An inactivation study is also significantly more expensive.

For disinfection determinations based on C · t values it will be necessary to determine the ozone dose required to establish a given residual and to determine the ozone decay rate. Figure IV–15 presents a plot of ozone residual versus transferred ozone dose. This plot provides the information necessary to estimate the ozone required to establish a given ozone residual (i.e., the C value). The decay curve presented above (Figure IV–17) provides the data necessary to estimate the ozone required to maintain a residual or to estimate the average residual in a reactor (no ozone addition). These two values, together with estimates of the hydraulic characteristics of the full-scale reactor, supply the information necessary to complete a contactor design for disinfection based on C · t values. Sec. D.3 of this chapter includes an example design procedure using this approach.

For disinfection results based on inactivation studies, the effects of contact time, dose, and microorganism type and inactivation should be presented. In general, disinfection results are presented as $\log_{10}$ values or survival ratio:

$$\text{Survival Ratio} = \log_{10} [N/N_0]$$

Where:

N = number of organisms at time = t
N_0 = number of organisms at time = 0

Figure IV–19 presents the results of inactivation of *Giardia* by ozone as a function of contact time (MWDSC 1989).

Performance parameters. In addition to the direct results of ozonation, the results of performance measures related to the conditions of ozone application must be presented and interpreted (e.g., effects on downstream processes). The parameters to be analyzed depend on the objectives of the treatability study and may include parameters such as disinfection by-products, production of biodegradable organic compounds (BDOC), and BDOC reduction through posttreatment, as well as posttreatment effects on those parameters already discussed (e.g., disinfection, oxidation of organic and inorganic micropollutants, and color treatment). For example, microflocculation effects are most easily seen by presenting a plot of the filtered water turbidity while cycling the ozonation process on and off, assuming the quality of the water does not change in between. Disinfection by-product results could be presented as a function of ozone dose, contact time, and final disinfectant.

In an evaluation of the effects of ozonation in a treatment scheme, data from the entire process should be analyzed and presented. Process conditions such as coagulant dose, pH, flocculation energy input, polymer addition, and flow rates are needed along with performance parameters for each experimental condition. Raw, intermediate, and finished water quality parameters, such as pH, alkalinity, hardness, total organic carbon, and dissolved gases, should also be part of the experimental database.

The selection of the type of presentation is an individual decision and relates both to personal preference and the intended audience.

IV.A.7 Process Recommendations

The results and recommendations of the treatability study will form the foundation upon which the design engineer will design the treatment facility. The process recommendations from the study should provide sufficient information so that the designer can size equipment, provide operational flexibility, and assure constructability, and do so in the most economical manner possible.

The final report from a treatability study should include a complete discussion of the objectives of the treatment process in terms of the required final water quality and the objectives of the study itself. A detailed presentation of the experimental plan, equipment utilized, analytical procedures, raw data, and chronology of the study should be included.

Separate sections on issues relating directly to design should be included to describe in detail the results observed and their impact on the overall process. These sections might include chemical optimization, flocculation, filter media evaluation, and disinfection/disinfection by-products. The ozone conclusions should include the test results for ozone demand, decay, reaction rates, and other information necessary to proceed with the design. A final section of process recommendations should include the conclusions reached concerning treatment to meet the objectives developed at the outset. This section presents the recommended process flow diagram showing the relationship between ozonation and the other unit processes.

IV.B FEED GAS PREPARATION

There are several feed gases to be evaluated for an ozonation facility. These include air, high-purity oxygen, recycled high-purity oxygen, and oxygen-enriched air. Each of these feed gases has its attributes depending upon the goals of the application of ozonation, the proposed facility, and its operating agency. Table IV–1 summarizes the advantages and disadvantages of the various feed gases.

Factors to be considered in selection of a feed gas include the following:

Table IV–1 Ozone Generation Feed Gas Selection

Feed Gas	Advantages	Disadvantages
Air	Most common system. Good track record. Applicable for both small and large systems.	High specific energy (kWh/kg O_3). Largest gas handling requirement. Maximum ozone concentration 2.5 percent by weight.
High-Purity Oxygen (General)	Approximately doubles production capacity of a given generator (Lowest specific energy). Highest ozone concentration (5 to 8 percent by weight), therefore greatest gas transfer driving force. Lowest gas handling requirement.	General sensitivity (public and operators) to application of oxygen. Requires more attention to safety and material selection. Requires special attention to contactor design due to low gas flows for mixing. Possibly patented system.
High-Purity Oxygen (Cryogenic Air Separation)	Most feasible for large ozone applications.	Capital-intensive gas preparation system. Gas preparation system is complex to operate and maintain.
High-Purity Oxygen (Pressure Swing Adsorption [PSA] Air Separation)	PSA unit is similar to "heatless dryer" air preparation system, which is a common air preparation unit.	Perceived maintenance concerns based on initial problems with wastewater applications of GOX generation.
High-Purity Liquid Oxygen (LOX)	Most simple of all feed gas systems. Least capital cost of all feed gas alternatives. More cost competitive with high-efficiency generators.	Variable operational cost because of cost of LOX purchase.
Externally Recycled High-Purity Oxygen	Savings of high-purity oxygen by recycle. Medium specific energy.	Special gas preparation considerations to manage recycled pollutants and high O_2 concentrations with associated capital and operational costs. No successful operating U.S. municipal water or wastewater applications of this concept.
Internally Recycled High-Purity Oxygen	Potentially very high ozone concentrations (>10 percent).	Unproven technology (as of 1990).
Oxygen-Enriched Air	Reduced capital cost for short-term peak ozone demand. Provides increased flexibility in satisfying a range of plant ozone demands.	No operating applications in U.S. municipal water or wastewater practice. Makes plant more complex to operate.

1. The projected operational production of the ozonation facility over the annual operational period.
2. The specific aspects of the application, which may include:
   - Raw water with a high initial ozone demand;
   - Oxidizing an easily oxidizable compound;
   - Oxidizing a comparatively refractory compound; and
   - Maintaining a dissolved ozone residual for a specific period.
3. The initial capacity and projected growth pattern of the ozonation facility.
4. The sophistication of the operational and maintenance staff of the agency.
5. The financial philosophy of the operating agency.

IV.B.1 Feed Gas Selection

Air. Dried, filtered air remains the most commonly used feed gas in ozonation systems throughout the world. Of the 40 U.S. water treatment ozone applications

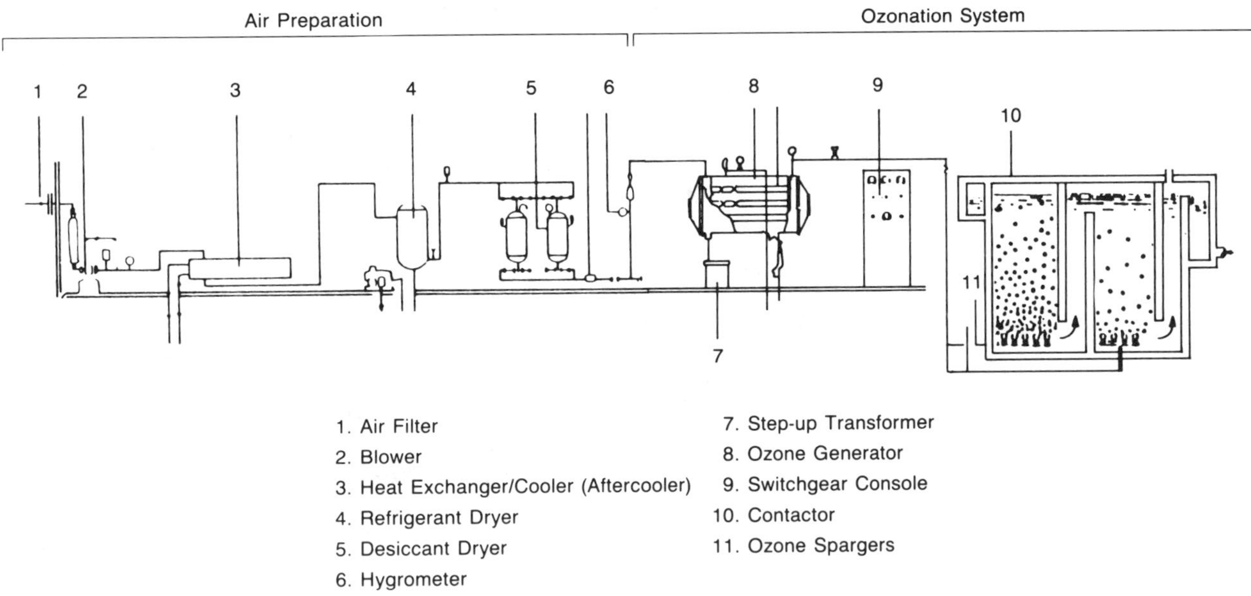

1. Air Filter
2. Blower
3. Heat Exchanger/Cooler (Aftercooler)
4. Refrigerant Dryer
5. Desiccant Dryer
6. Hygrometer
7. Step-up Transformer
8. Ozone Generator
9. Switchgear Console
10. Contactor
11. Ozone Spargers

Source: Miller et al. (1978).

Figure IV–20 Low Pressure Air Preparation System

surveyed in 1987–1988, all but two were air systems (Robson et al. 1988). In Europe, more than 99 percent of the drinking water ozonation systems use air as the feed gas (Langlais 1989). A similar situation exists throughout the remainder of the world.

The wide variation of geographic and climatic conditions throughout North America renders it necessary to generalize regarding the air feed gas character to be expected. The factors that the designer of an air feed system must take into account in the design of an air feed system include ambient temperature range, elevation, humidity range, continuous or seasonal air particulate content, and hydrocarbon concentrations.

The air feed system includes air compression equipment to deliver pressurized air to the dryer equipment. The delivery pressure is dependent on the complete design of the ozonation system (see later sections). The dryer equipment should provide air at a dew point of −60°C (−76°F) or lower (relative to standard pressure), depending upon ozone generator production and drying equipment life cycle costs. The ozonized air mixture is then utilized in the dissolution subsystem. Offgas is vented through an ozone destruction system where remaining ozone decomposes to oxygen. Obviously, there is associated monitoring instrumentation and control equipment. Figure IV–20 illustrates one of the different air preparation systems that are typical of ozonation systems throughout the world.

High-purity oxygen. Ozone yield increases with increasing oxygen concentration of the feed gas and may reduce the number of ozone generators and/or ozone generator size required for a system. However, the capital and/or operational cost for the oxygen supply equipment must also be included in the cost analysis. As the ozonation system size increases, a point may be reached where the total (capital plus operation and maintenance) cost for a high-purity oxygen feed or enriched ozone system will be less than that of an air feed ozone system. Since corona-specific energy (kWh/kg O_3) decreases with increasing oxygen concentration, the ozone generator power consumption drops. However there is an added cost for oxygen purchase or production. Those considering the use of high-purity oxygen should be aware of the existence of U.S. Patent 4,352,740, dated October 5, 1982 and entitled "Water Ozonation Method." The design conditions covered by the patent include the following:

(a) producing at least 70 percent oxygen by volume as the oxygen-enriched feed gas;

(b) generating between 4 and 8 percent by weight (58 to 116 g O_3/m^3 NTP) ozone in the silent electrical discharge such that the ozone generation figure, i.e., energy demand of merit η in 1 <2 h/lb O_3, is less than that defined by the equation (1) as follows:

$$\eta = 1/(0.34-0.033x) \qquad (1)$$

where x is the ozone concentration in the so-generated ozone containing feed, and

(c) performing said contacting of *said* water stream and said ozone-containing feed gas to dissolve between 0.5 mg/L and 8 mg/L ozone in said water stream.

The cost of application of the patent is currently not available.

The use of oxygen allows operating ozone concentrations that are two to four times greater than those normally obtained with the air feed. This affects the ozone dissolution system design. Consideration must be given to the specific ozone generator cooling characteristics, since without good heat transfer, it is difficult to achieve high ozone concentrations at reasonable corona-specific energy levels. Although the majority of state-of-the-art ozone generators have good heat exchange characteristics, the ozone generator cooling medium supply subsystem must be carefully designed.

High-purity oxygen can be provided in several manners: onsite generation by cryogenic air separation processes, pressure swing adsorption (PSA), or vacuum swing adsorption (VSA); or offsite generation of bulk liquid oxygen (LOX). In all of these methods, air is fed into the oxygen production system and nitrogen, other gases, and water vapor are removed from the gas stream. Oxygen production systems contain many of the same elements as air preparation systems discussed above, since the gas stream must be cleaned and dried in order to be utilized in the system.

In most plants utilizing onsite production of oxygen, a backup LOX storage system is included. When cryogenic oxygen production is used, the LOX can be generated onsite from the small sidestream of liquid oxygen formed during the production of the gaseous oxygen stream.

Cryogenic oxygen generation (Warakomski 1988). The basic principles behind cryogenic oxygen production are based on the physical properties of air and its components. Air liquefies at −192°C (−313.6°F) at atmospheric pressure. The key to the separation of oxygen from nitrogen by this process is based on the relationship between their respective boiling/condensation points at different pressures. Low temperature refrigeration is used to liquefy the air, then column distillation is used to separate oxygen from nitrogen. Small quantities of LOX are produced in cryogenic systems as well.

The steps involved in the process of oxygen generation by cryogenic distillation are as follows (illustrated in Figure IV–21):

- Feed air is filtered to remove particulates.
- Feed air is compressed.
- Compressed feed air is chilled, and water vapor, carbon dioxide, and hydrocarbons are removed by adsorption.
- Refrigeration is provided to liquefy air by both isothermal throttling (3 percent) and adiabatic expansion (97 percent).
- The liquid air is distilled by a two-step process (double column) into oxygen (95–99.99 percent) and nitrogen.
- Liquid oxygen can be produced in excess and stored as LOX for backup.

The systems incorporate sophisticated equipment and controls, and operational and maintenance expertise is required for their successful application. Twenty years of wastewater treatment experience shows that utility workers can be trained to successfully operate and maintain these systems. For a given capacity, cryogenic oxygen

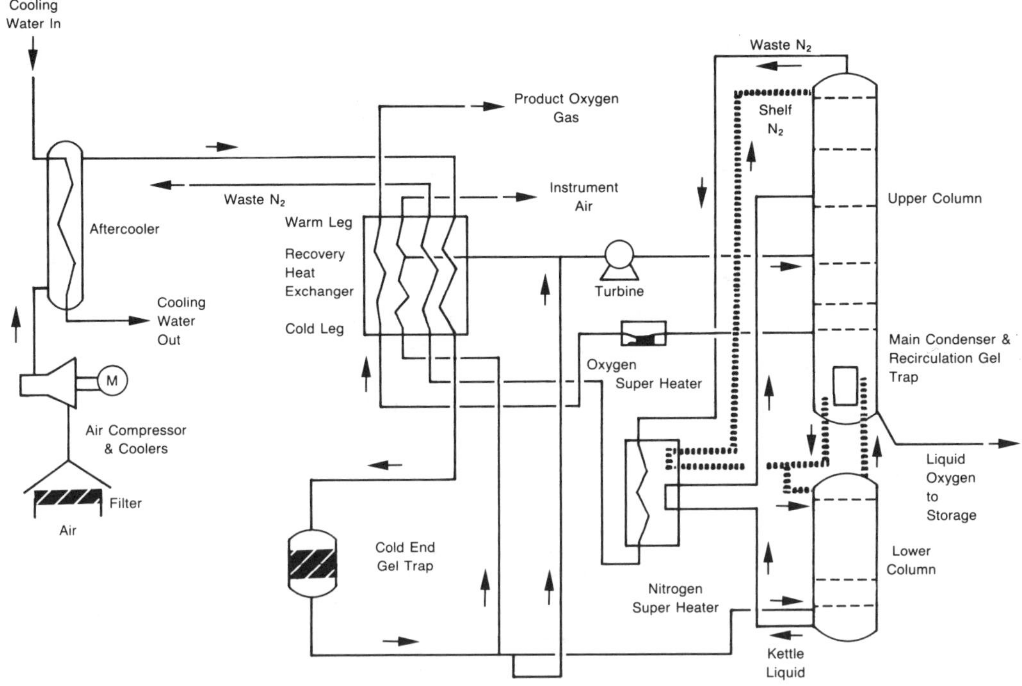

Source: Lotepro (1988).

Figure IV–21 Cryogenic Oxygen Production System

production is more capital-intensive than pressure swing adsorption systems. (See chapter VI, sec. C and D.) Cryogenic systems are feasible for large installations, generally within the range of 20 to 20,000 tons/day of oxygen production (Lotepro 1988).

Pressure swing adsorption. The separation vessels of a conventional PSA system usually contain molecular sieve media and operate at ambient temperatures within the pressure swing range of 1.0 to 3.0 bar (14.7 to 44 psi). Nitrogen adsorption by the molecular sieves is much higher than oxygen adsorption. During the high pressure phase, nitrogen, as well as water, hydrocarbons, and CO_2, is adsorbed onto the adsorption medium and oxygen is discharged from the system as product gas. When the pressure is reduced, the nitrogen and other gases and vapors desorb and are removed from the vessel by purge gas. The adsorptive capacity of the medium in each column decreases over the duration of the oxygen separation/nitrogen adsorption phase of the cycle as the medium becomes exhausted. Switching from the oxygen separation/nitrogen adsorption phase to the adsorbent regeneration phase of the cycle occurs when the oxygen concentration of the product gas from one column falls below a setpoint level, such as 90 percent. Thus, the product gas usually contains approximately 90 to 95 percent oxygen, 5 percent argon, and a small amount of nitrogen (Namba and Honda 1987). U.S. suppliers will normally guarantee 90 percent oxygen, 6 percent nitrogen, and 4 percent argon for PSA systems.

A schematic of a conventional PSA system is shown in Figure IV–22. The design of a PSA system with continuous product supply normally calls for a minimum of two adsorbers operating in parallel. One adsorber is in the adsorption phase and the other is in the regeneration phase simultaneously. In the multi-bed system shown in the schematic, the expansion gas released during the pressure reduction in one absorber can be used to pressurize and purge other adsorbers (Riquarts and Leitgeb 1985). Modular PSA systems are for installations ranging in size from 200 lb to 30 tons of oxygen production per day based on highway transport constraints (Lotepro 1988). Field erection techniques allow construction of PSA units up to 100 tons/day.

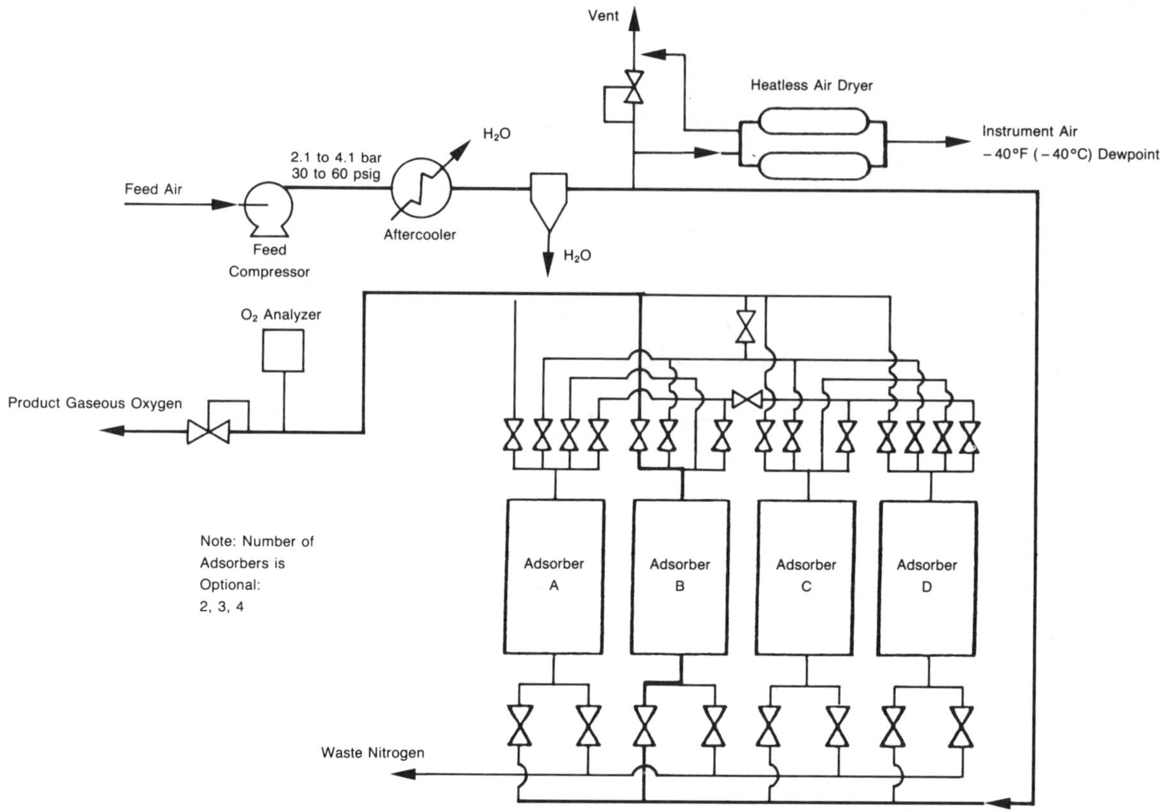

Source: Lotepro (1988).

Figure IV–22 Pressure Swing Adsorption Oxygen Production System

Warakomski (1989) provides a detailed history of the evolution of the PSA technology and its application in U.S. practice.

Vacuum swing adsorption. Vacuum swing adsorption is a comparatively recent evolution of the reversible adsorption process on zeolite media. It is similar to PSA except that regeneration is performed under a vacuum. There are currently no operating VSA systems in U.S. water or wastewater utility facilities.

Bulk liquid oxygen. The application of bulk liquid oxygen as the feed gas for ozone generation is common in laboratory-scale and pilot-scale applications and is applied at the 12-mgd (1893 m^3/h) Rocky Mount, North Carolina water treatment plant ozone installation. Similar applications are in various stages of design-construction at Springfield, Massachusetts; Laughlin, Nevada; and Valdosta, Georgia. LOX is trucked to an onsite insulated storage tank, after which it is vaporized as required for onsite application.

Use of LOX has benefits for comparatively small installations that seek the benefits of high-purity oxygen as an ozone generation feed gas with the simplicity and low capital cost of LOX. The local cost of LOX would be an important factor in considering the feasibility of this source of high-purity oxygen. (See chapter VI, sec. C and D.)

External recycle of high-purity oxygen. Another economic consideration arising from use of oxygen feed gas is the option to recycle the oxygen-rich contactor offgases. Recycling of contactor offgas requires the equivalent of a conventional air preparation system, which includes filtration and removal of moisture to a minimum −60°C (−76°F) dewpoint before returning the gas to the generator. Removal of impurities in the feed gas resulting from the stripping action of the ozone contactor may be achieved by adsorption in filters using granular activated carbon, silica gel, or molecular sieve

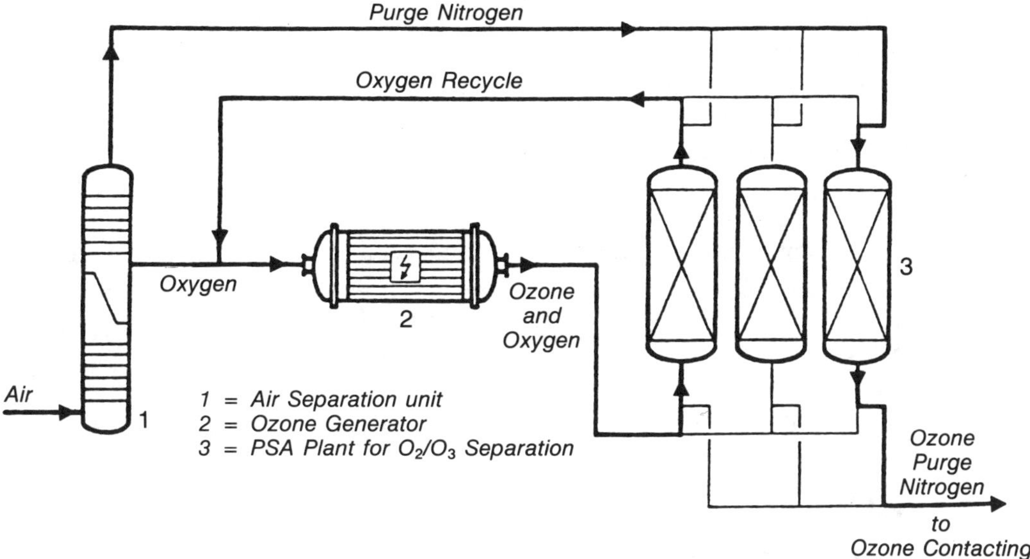

Source: Riquarts and Leitgeb (1985).

Figure IV–23 Ozone Generation Plant with O_2/O_3 Separation by Pressure Swing Adsorption

media. Recycling of oxygen-rich contactor offgas to the ozone generator must also require a mechanism for purging nitrogen, argon, and carbon dioxide from the offgas stream. The equilibrium of gas in solution as water enters the contact chamber is dictated by ambient pressure (usually atmospheric), solubility, and the composition of air, i.e., 21 percent vol. oxygen, 78 percent vol. nitrogen, and 1 percent vol. argon. If the process gas contains a higher concentration of oxygen, the nitrogen previously dissolved in the water is stripped from solution. The result is an increased concentration of nitrogen and a corresponding decrease of oxygen in the contactor offgas. For example, at the Los Angeles Aqueduct Filtration Plant, although the feed gas contains approximately 95 percent oxygen, contactor offgases typically contain as much as 20 percent nitrogen (Stolarik and Christie 1988). Recycling thus requires the bleed-off of a portion of the recycle stream and the introduction of fresh oxygen as makeup gas (Masschelein 1982b). While recycling of oxygen-rich contactor offgas has not been a success in U.S. wastewater applications (Robson, 1989) it has been satisfactorily applied at the Duisberg Wasserwerk III Wittloer III water treatment plants since 1965 (Miller et al. 1978) and at an industrial kaolin slurry bleaching facility in Georgia since 1988 (Hartwig 1989).

Internal recycle of high-purity oxygen. Increased efficiency of ozone generation using high-purity oxygen has also been achieved by passing ozonized gas generated from high-purity oxygen through an adsorbent bed. The most commonly applied adsorbent is silica gel. Feed gas may be high-purity oxygen from cryogenic air separation, pressure swing adsorption, or vaporized LOX. As illustrated by review of Figure IV–23, ozonized gas is passed through an adsorber bed in which ozone is retained and oxygen passes through. The oxygen is recycled through the ozone generator and the ozone is discharged from the bed during a desorption cycle, which may be achieved by a nitrogen purge gas, PSA, or VSA dispersion cycle, or by dry air.

These concepts may have specific applications but there are currently no operating facilities on which to make judgments about the economics or reliability of these processes. Published literature and patents on internal recycle of high-purity oxygen include the following: Cook et al. (1959), Kiffer (1959), Lowther (1976a, 1976b), Riquarts and Leitgeb (1985), and Joel (1987).

Air feed systems with LOX enrichment. The use of LOX-enriched air feed systems is presently utilized at the Tailfer Water Treatment Plant in Brussels, Belgium. This 68.7-mgd (10,833-m^3/h) plant (see chapter VI, sec. E.1 and E.2) treats Meuse River water and has been utilizing LOX enrichment since 1978. The treatment train is conventional with postozonation for primary disinfection and chloramination for final disinfection. Ozone is produced in low-frequency horizontal tube generators. LOX enrichment of the air feed gas is used on a continuous basis. An oxygen stream comprising a few percent of the total feed gas flow is added in the winter months during periods of low oxidant demand or low water consumption, and is increased up to a maximum oxygen stream of 10 percent of the total gas flow in summer months when oxidant demand and water consumption increase (Masschelein 1982a).

Several U.S. water treatment plants are currently being designed to incorporate LOX enriched air (Monk 1989). The application of this concept is justified on the basis of ability to meet infrequent short-term peak ozone demands and reduced capital investment.

IV.B.2 Feed Gas Quality

The quality of the feed gas is critical to the performance of any ozone generation system. For successful long-term operation with minimum system maintenance, the feed gas, whether air or oxygen, must be extremely dry; free of particulates, hydrocarbons, and other contaminants; and relatively cool. It is the function of the feed gas preparation system to supply a feed gas of appropriate quality and in quantities sufficient for the process.

Moisture. Of all quality parameters, the most important is moisture content. Excessive moisture will adversely impact ozone production (see chapter II, sec. D.2, Figure II–39) and may react with nitrous oxides in the generator to form nitric acid and cause subsequent damage to the generator. The moisture content of gases is usually expressed as "dewpoint," which is the temperature at which moisture begins to condense out of the gas. Since cooler gas has a lower moisture-holding capacity, the lower the dewpoint, the lower the moisture content in the gas.

Dewpoint must always be qualified as "ambient" or "line pressure," since the moisture holding capacity of air varies inversely with pressure. For most low-pressure gas preparation systems, this distinction is modest, as indicated in Figure IV–24. However, for high-pressure systems, the difference between ambient and pressure dewpoint is significant.

In ozone applications, the feed gas dewpoint must be extremely low. In fact, the design dewpoint has become ever lower with each new generation of ozone system designers and with the development of new drying materials. European systems designed in the 1950s and 1960s were based on dewpoints of −40°C (−40°F), while U.S. and European systems designed in the 1970s and early 1980s were based on −60°C (−76°F). Today, some systems are being designed and operated at −80°C (−112°F). One reason for this dewpoint level is the advent of narrower dielectric gaps, as with medium-frequency ozone generators. The narrower the gap, the lower the dewpoint must be (Langlais 1989). For current practice, −60°C (−76°F) should be considered the highest acceptable operating level, and the design dewpoint should be some conservative level below this.

Removal of moisture from the gas can be accomplished through compression, cooling, and desiccant drying. Gas compression, followed by an appropriate aftercooler, will remove some moisture, since the moisture holding capacity is reduced at increased pressures. Likewise, moisture holding capacity is reduced at lower temperatures, and direct gas cooling will remove additional moisture. In most systems, several steps are used to reduce the moisture content, but only desiccant dryers are effective in ultimately

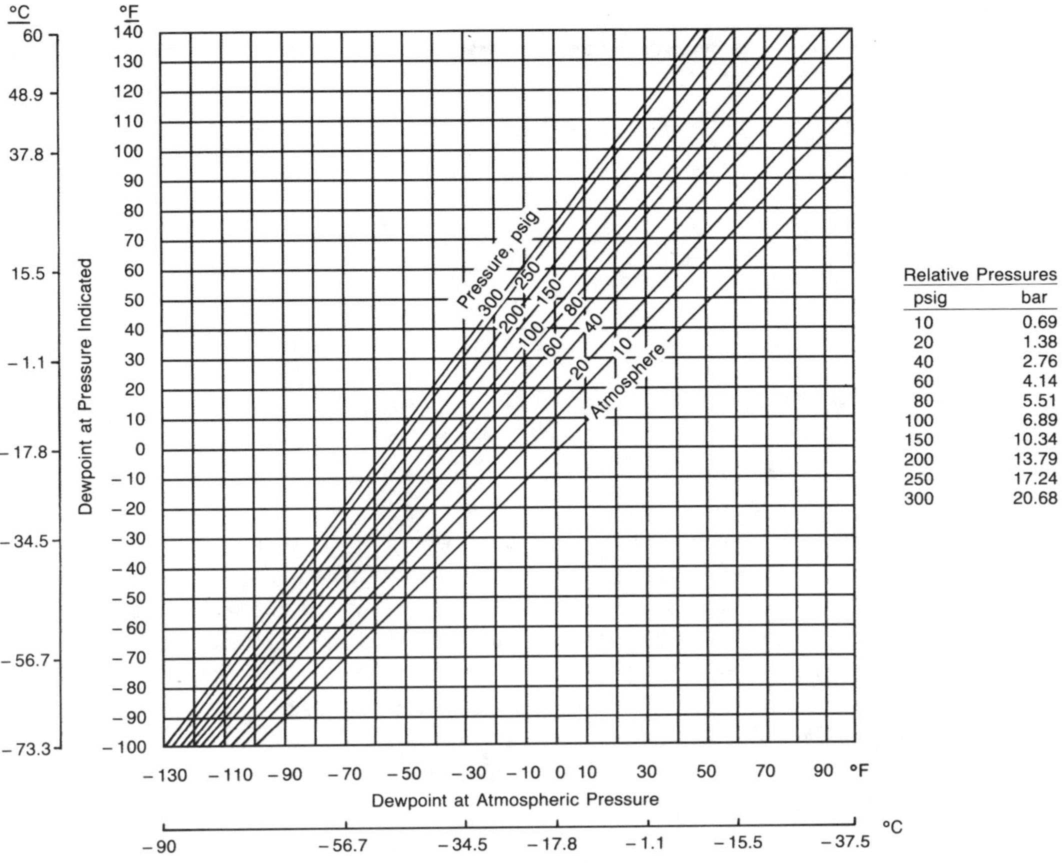

Relative Pressures

psig	bar
10	0.69
20	1.38
40	2.76
60	4.14
80	5.51
100	6.89
150	10.34
200	13.79
250	17.24
300	20.68

Source: Shumate (1989).

Figure IV–24 Relationship of Pressure and Atmospheric Dewpoint

achieving the very low dewpoints necessary for an ozone feed gas, and are always the last gas drying step prior to the generators. Oxygen produced cryogenically or with PSAs will have a dewpoint below −100°F (−73.3°C) by nature of the oxygen generation process. However, pipeline leakage and other factors may result in gases with higher dewpoints being received at the generator.

Table IV–2 presents a summary of typical moisture levels encountered in the design of air preparation systems.

Particulates and other contaminants. Particulates have the potential to cause problems at a number of locations within an ozone system. Compressors with close running clearances must be protected from particles that would score impeller, lobe, or rotor surfaces. Desiccant dryers must be protected from dust and hydrocarbons, which would block the desiccant pores, thereby reducing the desiccant moisture holding capacity. And since the ozone generator acts as an electrostatic precipitator, particulate material and hydrocarbons will attach to the dielectric surface. In some cases, hot spots will develop as a result of this coating and eventually lead to dielectric failure. Finally, fine bubble ozone diffusers must be protected from particulates that would clog the diffuser.

Particulates may be inorganic or organic in nature, and may be present in the ambient air or generated from within the ozone system itself. Typical inorganic materials include sand, dust, lime dust, coal dust, construction debris, and moisture droplets. Common organic materials include pollen, cottonwood seeds, and similar particles. From within a gas preparation system, the most common particulate is desiccant dust.

Table IV–2 Moisture Levels

Location/Condition	Dewpoint (°C)	Dewpoint (°F)	Moisture Content (ppm)
Design Inlet to Compressor	40	104	45,000
Normal Summer Inlet to Compressor	30	86	28,000
Design Discharge from Refrigerant Drier	5	41	5,000
Minimum Operating Feed Gas to Generator	−60	−76	7
Specified Design Feed Gas to Generator	−80	−112	1.5

Hydrocarbons may be present in the atmosphere in large metropolitan areas and in the oil- and gas-producing areas of the country. Although not commonly used in the United States but more common in Europe, oil lubricated–type compressors may also introduce unwanted hydrocarbons into the system.

Filters are the best method of controlling particulate material. They should be placed ahead of the compressor, ahead of the desiccant dryers, and after the desiccant driers. While filter rating is something of a black art, the goal should be a final feed gas entering the ozone generator free of all particulates larger than 0.1 μm. Hydrocarbons can be controlled with coalescing filters and granular activated carbon adsorber–type filters. For installations with oil-lubricated screw-type compressors, the coalescing filter will typically remove oil droplets larger than 0.05 μm. An adsorber incorporating granular activated carbon, silica gel, or molecular sieve can be used for capturing and removing hydrocarbon vapors.

IV.B.3 Gas Preparation Systems

Classification. Gas preparation systems have been classified according to the operating pressure at the point of compressor discharge as follows:

- ambient pressure
- low pressure (<1 bar—14.5 psi)
- medium pressure (<4 bar—<58 psi)
- high pressure (>4 bar—>58 psi)

Ambient pressure. Ambient pressure systems frequently do not use a compressor at all but rely on the vacuum created by an aspirating device to draw the air through the air preparation system and the generator and into the contactor. This approach is being marketed in the United States by a single vendor who will provide the complete system from the dryers through the contactor. There are several small ambient pressure installations of the type in the United States for drinking water treatment with a number of larger European installations. Many of these have been installed since the beginning of this century (Langlais 1989). (See sec. IV.C.5.)

Low pressure. Low-pressure systems are designed for very low pressure losses and usually operate as constant-volume systems. This allows the use of rotary lobe blowers, which are very reliable, low-maintenance units when used at pressures less than 0.7 bar gauge (10.1 psi), as well as screw-type compressors. Due to the high moisture holding capacity of the low-pressure air, drying always includes both refrigerant- and desiccant-type dryers. Desiccant dryers are limited to thermally regenerated types.

Medium pressure. Medium-pressure systems are configured much like the low-pressure systems, except the rotary lobe blower is not an acceptable compressor type, and a pressure reduction step is incorporated after the dryers. Many systems are being designed today for a rather broad range of system pressures to allow for additional pressure losses due to dirty filters, aged desiccant, and fouled diffusers or to obtain a higher pressure at the ozone generator. While "normal" system operating pressure is expected to be 22 to 27 psig (1.5 to 1.9 bar), the systems are being designed to still maintain operations at pressures up to 37 psig (2.5 bar). As with low-pressure systems, both refrigerant and desiccant drying types will normally be provided in a medium pressure system.

High pressure. High-pressure systems can usually be differentiated from the other types by the use of multi-stage positive displacement compressors and pressure swing adsorption–type desiccant dryers. Pressure swing desiccant dryers will operate at pressures as low as 65 psig (4.5 bar), but with higher pressures, the purge requirements are reduced. Typical operating pressures are 85 to 100 psig (5.9 to 6.9 bar). High-pressure systems have been used successfully in many small systems (less than 200 lb/day [3.8 kg O_3/h]). A limited number of larger installations (600 lb/day [11.3 kg O_3/h] at St. Rose, Quebec and 900 lb/day [17 kg O_3/h] at Daytona Beach, Florida) are in operation in North America today.

Gas compression. (Also see chapter VI, sec. D.1.) Gas compression in ozone gas preparation systems provides air at mass flow rates required by the process, and at pressures necessary to overcome all downstream losses. The five major types of compressors which have been used successfully in U.S. ozone systems are (1) liquid ring, (2) centrifugal, (3) rotary screw, (4) rotary lobe, and (5) reciprocating. Although comparatively unknown in the United States, a vane-type compressor has been successfully applied in approximately 200 European installations (Langlais 1989).

Liquid ring. Liquid ring compressors have been widely used in U.S. ozone systems (Robson et al. 1988). The available flow range and optimum operating discharge pressure fits well with a broad range of system sizes, and they are naturally oil-free. The liquid ring compressor is a simple, durable, low-maintenance machine that uses a rotating band of liquid sealant to accomplish the actual compression. The liquid is kept rotating by an eccentrically mounted rotor that causes the compressant to almost fill, then partially empty, the compression chambers with each revolution. This design precludes the need for fine tolerances between moving parts, or for sophisticated controls. Service life should be similar to a typical water pump. One particular advantage of the liquid ring compressor is that the heat of compression is absorbed by the compressing liquid, eliminating the need for an aftercooler downstream of the compressor. Discharge air temperature should be not more than 10°F (5.5°C) higher than the liquid compressant temperature.

The liquid ring compressor is a positive displacement type of blower and delivers a relatively constant gas flow regardless of inlet conditions. However, this should not be inferred to mean constant flow or mass flow. As the inlet conditions change from warm air to cool air, the delivered mass flow rate and horsepower requirements increase due to the increased inlet air density. Therefore, as for all compressors, the summer inlet conditions control the minimum mass flow delivery, while the winter inlet conditions control motor horsepower requirements. Liquid ring compressors are relatively power-inefficient, with horsepower draw at 16 to 24 bhp per 100 scfm at full load at 15 psig discharge pressure (70.2 to 105.3 Wh/m^3 NTP at 1.03 bar discharge pressure).

Since most of the load on the compressor is in keeping the liquid ring in motion, it is not possible to "unload the compressor" as a means to save energy during no-load conditions. While it is possible to blow off excess air, or to recycle discharge air in a bypass line to modulate the flow rate, neither results in reduced power draw. For this reason, liquid ring compressors are not typically used in variable volume systems unless multiple units are used.

Liquid ring compressors are generally suitable for systems smaller than 1500 lb/day (28.3 kg O_3/h) and with system operating pressures less than 35 psig (2.4 bar). However, one manufacturer offers a high-pressure (up to 80 psig [5.5 bar]) model in a range of capacities that may be suitable for small to medium (less than 750 lb/day [14.2 kg O_3/h]) high-pressure systems.

Centrifugal. Centrifugal compressors have not been widely used in air feed gas preparation systems for ozone since they are traditionally high-volume, low-pressure (10 psig [0.69 bar]) machines. However, multi-stage units capable of 100 psig (6.9 bar) have been used in several cryogenic oxygen generation plants for ozone systems with success. Several manufacturers now market high-pressure (30 psig [2.1 bar]) single-stage machines in a capacity range that should make them applicable in large ozone systems (3000 lb/day [56.7 kg O_3/h] and larger). The centrifugal compressor is a simple, reliable machine that is easily incorporated into either variable- or constant-volume control systems. Variable capacity with constant outlet pressure is now easily achieved through computerized control of inlet guide vanes as opposed to a throttling valve. At reduced flows, horsepower draw is also reduced. At full load, horsepower draw is 14 to 15 bhp per 100 scfm (61.4 to 65.8 Wh/m^3 NTP). Compressor turndown is limited by the compressor surge point, which is usually about 40 to 50 percent of full load or 60 to 70 percent of full power draw. For large systems ($\geq$3000 lb/day [$\geq$56.7 kg O_3/h]), the centrifugal compressor can be a very cost-effective alternative.

Rotary screw. Rotary screw compressors have been applied in air feed gas preparation systems for ozone. These compressors are positive displacement in design, consisting of two interlocking screw-shaped rotors, available in either single-stage configuration for discharge pressures up to 50 psig (3.4 bar) or two-stage for discharge pressures over 100 psig (6.9 bar). While available as either lubricated or nonlubricated, most ozone system designers may prefer nonlubricated as part of a total hydrocarbon control effort. Rotary screw compressors have a proven track record and are used extensively in a wide variety of industrial applications, including such diverse installations as food processing and microchip manufacturing.

Rotary screw compressors can be readily used in constant- or variable-volume systems. In variable-volume systems, the compressor may discharge to a receiver that is being maintained within a preset pressure range. When air in the receiver reaches maximum pressure, the compressor "unloads" by closing an inlet valve and opening a blowoff valve on the discharge. Horsepower draw at full load ranges from 12.5 to 14.0 bhp per 100 scfm (54.9 to 61.4 Wh/m^3 NTP), depending upon the specific model. When unloaded, horsepower draw is approximately 40 percent of full load power draw. Although variable speed drives have been used by some system suppliers, the selection of the specific compressor to be used with variable speed drives must be reviewed with the compressor manufacturer, since at least one manufacturer indicates its machine is not designed for reduced speeds and may experience rotor overheating if used with a variable speed drive.

The nonlubricated rotary screw compressors operate under dry conditions with relatively fine tolerances between the rotors. Although the air end casing is watercooled, the rotors of screw compressors protected by appropriate air filtration must be replaced approximately every 40,000 h at a cost of about 20 percent of the original purchase price. Recommended annual maintenance includes replacing the soft parts of the unloading valve.

Since they are not available in capacities less than 350 scfm (595 m^3/h NTP), nonlubricated rotary screw compressors are generally not suitable for systems smaller than about 1500 lb/day (28.3 kg O_3/h).

Rotary lobe. Rotary lobe blowers, with discharge pressures to about 15 psig (1.03 bar), have been successfully used in low pressure installations. Practical advantages include low capital cost and the support of this technology by numerous equipment manufacturers. At least two manufacturers have also developed nonlubricated high pressure rotary lobe compressors suitable for up to 35 psig (2.4 bar) discharge pressure in a single-stage configuration, and up to 110 psig (7.6 bar) in the two-stage configuration. These have been used only on a limited basis in ozone systems, and one of the

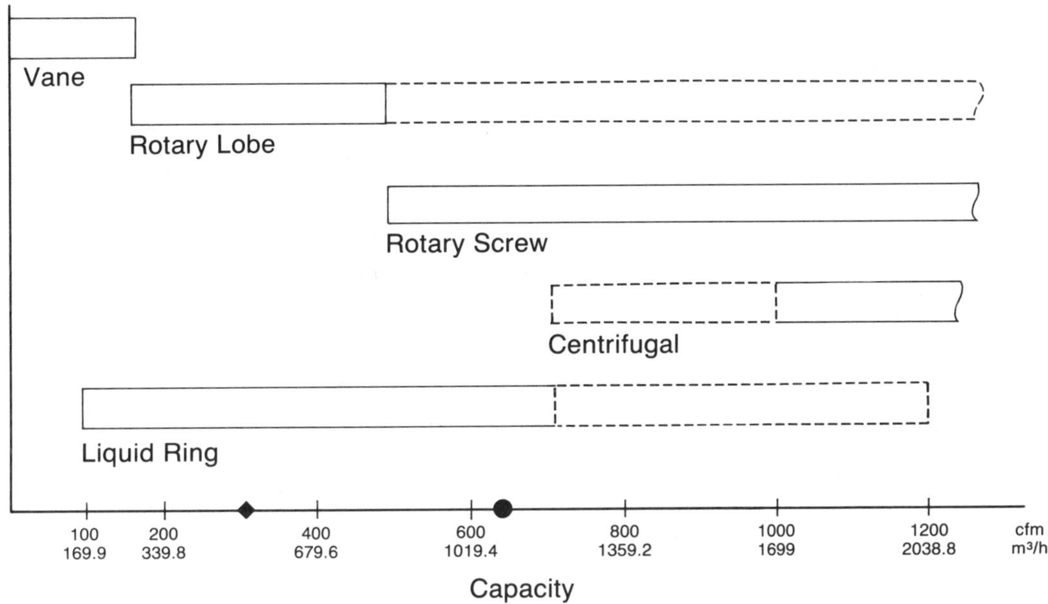

Source: Hesby (1989).

Figure IV–25 Compressor Capacity Range

manufacturers has stopped marketing the compressor due to difficulties in maintaining quality control of the lobe castings.

The rotary lobe blower is ideally suited to constant- or variable-volume low-pressure systems and will see continued use in those applications. The rotary lobe compressor is sometimes configured with an unloading valve and can be controlled similar to the rotary screw machines. Power draw for rotary lobe compressors is slightly higher than the power draw for rotary screw requirements.

Reciprocating. Nonlubricated reciprocating compressors have not generally been used in low- to medium-pressure ozone systems, but have been used with success in small (less than 200 lb/day [3.8 kg/h] ozone) high-pressure systems. Two-stage, double-acting, balanced reciprocating compressors have also been used with PSA-type oxygen-generating plants.

Of simple piston design, the reciprocating compressor is a positive displacement machine. At less than 16 bhp per 100 scfm (70.2 Wh/m^3 NTP), it is the most power-efficient high-pressure unit available. Similar to the rotary lobe and rotary screw types, it can be readily incorporated into either the variable- or constant-volume system. Like those other units, the compressor can be configured with an unloading valve to reduce the power draw during periods of no load without stopping the compressor.

Reciprocating compressors are available in the widest possible range of capacities, but should only be used where high pressures are required.

Vane. The vane compressor is widely used for high-pressure applications in Europe with constant or variable gas flows (Langlais 1989). This oil-lubricated unit is capable of operating at both fixed and variable speed with capacities ranging from 2 m^3/h NTP (1.18 scfm) to 290 m^3/h NTP (170 scfm) at 7 bar (100 psig) (Hydrovane 1989). For compressors of this type in a flow range of about 200 m^3/h NTP (118 scfm), horsepower draw at full load is up to 30 bhp per 100 scfm (131.6 Wh/m^3 NTP).

Compressor selection. The selection of a particular compressor for a given installation frequently depends upon considerations including the following: required capacity, required discharge pressure, system type (whether variable- or constant-volume), and efficiency of power utilization. Figure IV–25 indicates the generally available

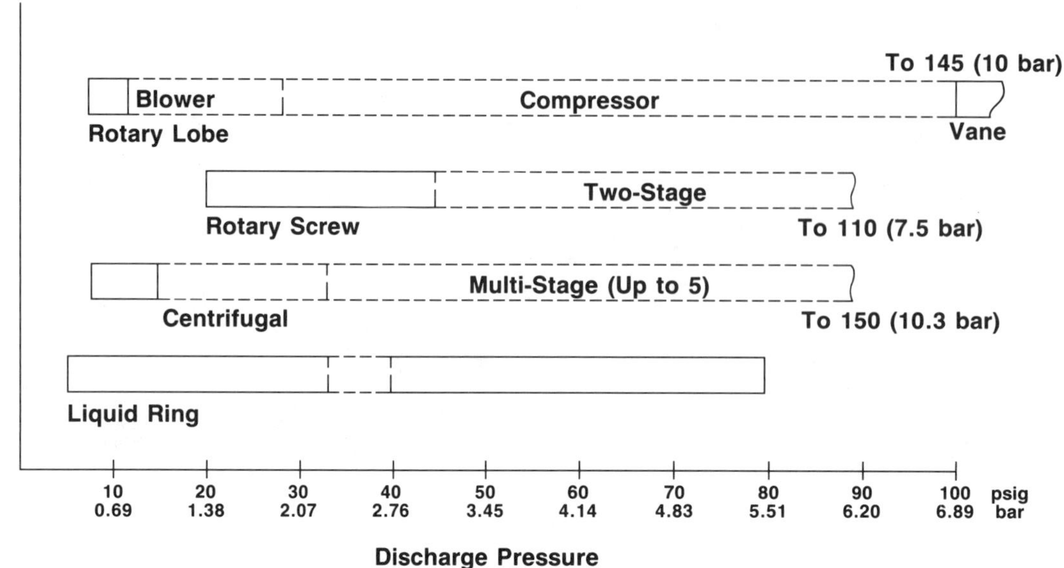

Source: Hesby (1989).

Figure IV–26 Compressor Pressure Capabilities

capacity range for each of the major compressor types. Required capacity, based on the treatment process requirements, must consider total required ozone production and design ozone concentration. Equipment longevity, maintenance cost and difficulty, and other factors associated with maintaining the equipment are more difficult to measure but should play a part in specifying this equipment.

The discharge pressure capabilities of each major compressor type are indicated on Figure IV–26. The selection of a design discharge pressure must be based on total system considerations, including desiccant dryer types. A conservative approach to estimating system pressure losses will allow the system to continue operating even during periods of unexpected high losses due, for instance, to dirty filters.

As discussed in the earlier sections, some compressors are more readily adapted to variable volume control schemes than others, and the proposed system operation should be carefully discussed with the compressor manufacturer. This will ensure appropriate use of each compressor type, resulting in optimum power utilization.

Compressor controls. Compressor controls can range from very simple to complex depending on the sophistication of the overall control package for the ozonation system. At a minimum, compressors should be equipped with controls to address the following situations:

- Internal conditions within the compressor that can cause damage to the compressor, and
- Conditions of the ozone system external to the compressor that require air flow through the system to be interrupted.

In the former case, compressors are typically protected against lubrication oil and cooling water failures. Liquid ring compressors do not have an oil lubrication system but must be protected against a loss of seal water. The best example of an external condition requiring compressor shutdown is a high dewpoint of the ozone generator feed gas. This condition can damage the ozone generators. The designer may choose to incorporate other external conditions into the compressor shutdown control scheme.

Compressor status can be monitored with local control panels and/or a central ozone system control panel. In either case, all shutdown conditions should be indicated through an alarm panel for diagnostic maintenance purposes. If desired, compressor on-off status and alarm conditions can be transmitted to a plant control system.

Compressor capacity or the discharge flow rate may or may not be controlled. An ozone system may be designed to modulate ozone production capacity in two different ways.

- The ozone concentration is varied while the feed gas flow rate is held constant (constant-flow system).
- The feed gas flow rate is varied while the ozone concentration is held constant (variable-flow system).

In general, the former is used in smaller applications while the latter is used in larger applications. If a variable-flow ozone system is desired, consideration will have to be given to efficient modulation of feed gas flow.

The primary method of compressor capacity control may involve the use of a receiver tank between the compressors and desiccant dryers. Compressors are loaded and unloaded in accordance with pressure switch setpoints on the receiver tank. The frequency of loading and unloading is dependent on the size of the receiver tank, the differential between pressure setpoints, the operating compressor's capacity, and the ozone production capacity. It is important that the compressor be designed to withstand continuous operation at the worst case cycling frequency. Compressor types that can be unloaded without stopping the motor include rotary screw, rotary lobe, and reciprocating.

Another type of capacity control utilizes a blow-off valve or recirculation line from the compressor outlet to the inlet. This is a relatively inefficient method of capacity control and should be avoided if possible. Centrifugal compressor capacity can be modulated by inlet guide vanes as mentioned previously. This method of capacity control is relatively efficient and does not result in the pressure and flow surges that loading and unloading can produce.

Some types of compressors, most notably centrifugal and rotary lobe, can be provided with variable speed drivers for capacity control. Compressors of this type are also efficient but the additional capital cost of variable speed motors and controls may be weighed against operational savings. Because of surge conditions that can result with reduction in speed of a centrifugal compressor, any type of variable speed drives should be provided by the compressor manufacturer.

Aftercoolers. Heat generated in the gas stream during compression results in elevated compressor discharge temperatures. For low- to medium-pressure systems, the discharge temperature will be between 180° and 350°F (82.2 and 176.7°C) for most compressors. As indicated previously, the liquid compressant of a liquid ring compressor will absorb most of that heat and the discharge temperature should only be 10°F (5.5°C) above the compressant temperature. For all other compressors, this temperature must be reduced using an aftercooler before the air stream enters any of the downstream dryers.

An aftercooler is typically a gas-to-water shell and tube heat exchanger. Standard design provides for a 10°F (5.5°C) approach; that is, the exit gas temperature is 10°F (5.5°C) above the entering cooling water temperature. High-efficiency units designed for a 2°F (1.1°C) approach are available. These units use thin-walled, internally finned tubes. Designs should provide for a maximum exit gas temperature of 100°F (37.8°C).

Refrigerant drying. Refrigerant drying utilizes physical cooling of the air to condense moisture out of the air stream, thereby lowering the dewpoint. Refrigerant drying in ozone systems may be of two types: (1) direct expansion or (2) chilled water–based systems. To avoid freezing of condensed moisture, refrigerant drying is limited to producing dewpoints no lower than the freezing temperature of water, 32°F (0°C). The operating gas dewpoint can range from 38° to 40°F (3.3 to 4.4°C) (Shumate 1989) or, more commonly, 40 to 50°F (4.4 to 10°C) (Langlais 1989).

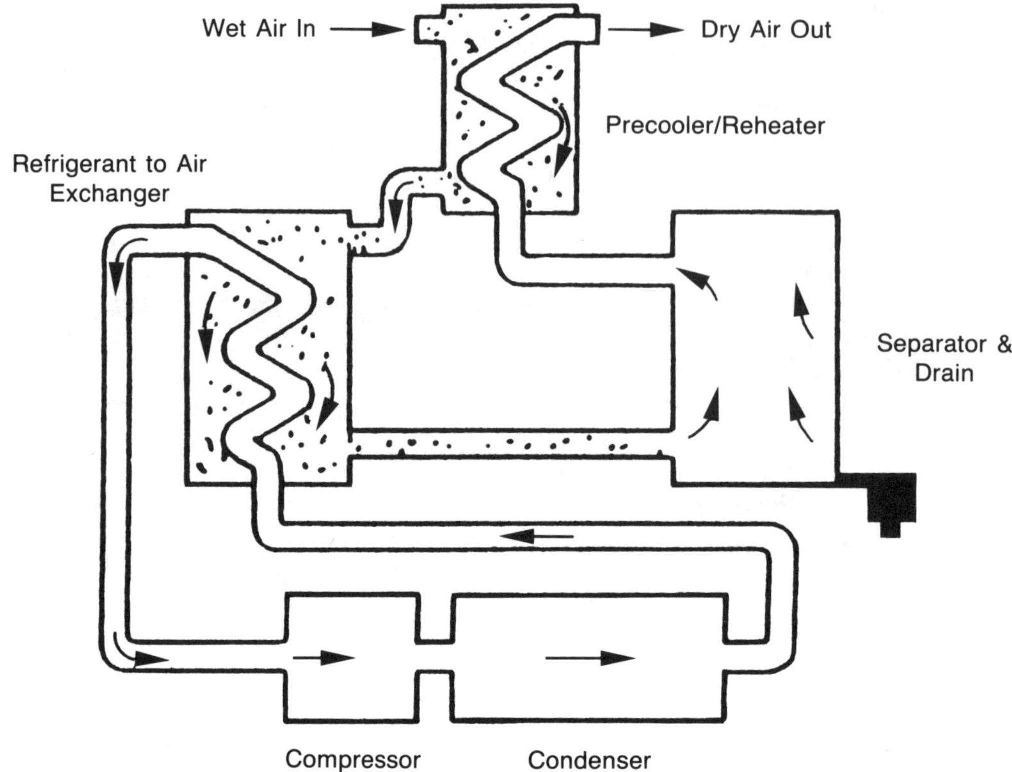

Source: Hesby (1989).

Figure IV–27 Refrigerant Dryer—Direct Expansion

Types of refrigerant dryers.

Direct expansion: Self-contained direct expansion dryers are available from several manufacturers and are provided exclusively for air drying. These dryers employ a conventional refrigeration cycle, as indicated in Figure IV–27, using a refrigerant such as Freon (R22) with a compressor, condenser, and heat exchanger for gas cooling. The heat exchanger may be configured in a refrigerant gas-to-air or liquid-to-gas (flooded evaporator) arrangement. The wetted heat exchange surfaces in the flooded evaporator are said to provide improved heat exchange efficiency. Condensers may be air- or water-cooled. A disadvantage of a direct expansion system is the potential contamination of the refrigerated feed gas by the cooling medium (Langlais 1989).

It is noted that most refrigerant dryers sold by U.S. manufacturers are for applications in high-pressure plant air systems and are rated for performance at 100 psig (6.9 bar), 100°F (37.8°C) inlet conditions, and 3.5 to 5 psig (0.24 to 0.34 bar) pressure drop through the dryer. For lower-pressure ozone gas preparation systems, dryer capacity must be derated to as little as one-third of the standard rating to maintain the design pressure dewpoint and minimize pressure drop through the dryer.

Water chiller: Chilled water–based refrigerant drying has been used successfully where chilled water is being provided for ozone generator cooling. In these systems, a separately mounted chilled water–to–gas heat exchanger is used for gas cooling, with the chilled water being provided from a closed loop chilled water system shown schematically in Figure IV–28. Design dried air dewpoint from these systems is usually between 40°F and 50°F (4.4 to 10°C), so the desiccant dryer must be sized accordingly.

Self-contained chilled water dryers provided exclusively for air drying are also available. Compared to direct expansion dryers, chilled water dryers employ an extra

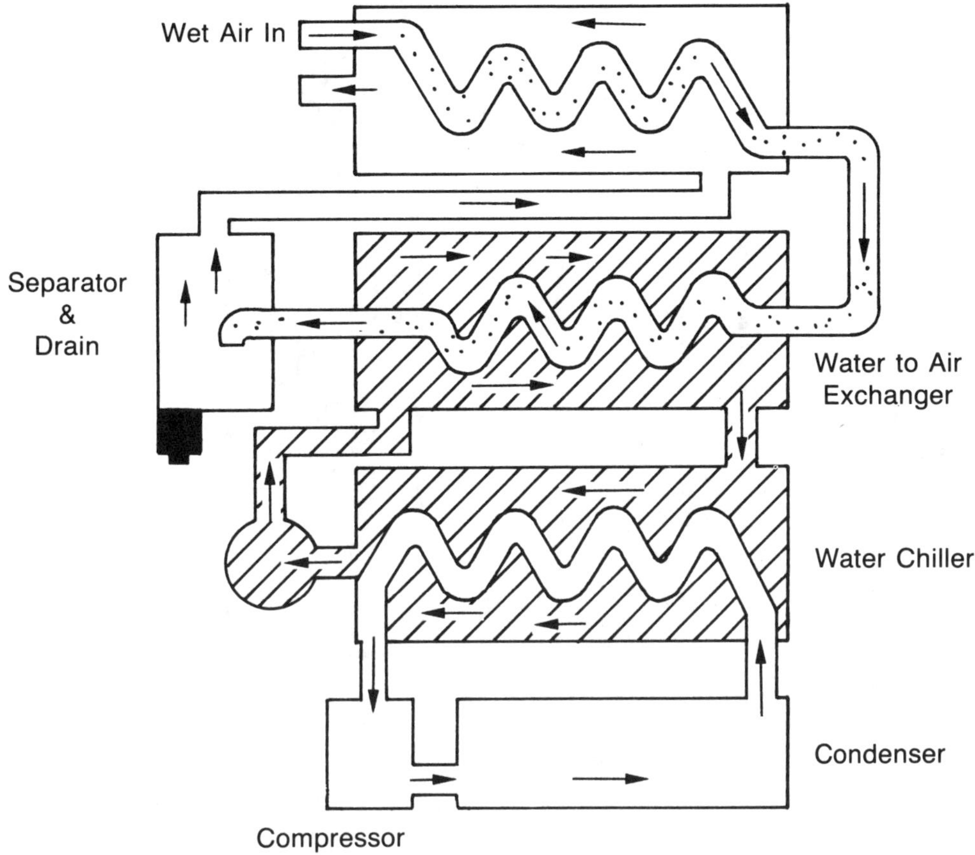

Source: Hesby (1989).

Figure IV–28 Refrigerant Dryer—Water Chiller

heat exchanger. Although normally using a water/glycol mixture, design outlet dewpoints are also limited to 50°F (10°C).

Refrigerant dryer system control. Refrigerant dryers of the type discussed here are designed for constant operation under varying flows. Internal control valves (hot gas bypass valve) modulate refrigerant flow to accommodate changing system load and to prevent freeze-up during no-load periods. Placing these dryers on line after periods of extended shutdown can be achieved by activating the crankcase heater for several hours and then operating in the normal mode (Shumate 1989). For this reason, the number of refrigerant dryers in a system should be minimized, and units should be maintained on line during changing system loads. Refrigerant dryers are available for design flow rates up to 25,000 scfm (42,475 m^3/h NTP) under standard conditions.

Discussion. While refrigerant drying is not normally able to achieve final feed gas dewpoint alone, it is a very cost-effective process for removing over 80 percent of the incoming moisture and reducing the subsequent load on the desiccant dryers. For a typical system, the total annualized capital and operating costs can be estimated at $1.50/lb (2.3 centimes/g) moisture removed for a direct expansion type of refrigerant dryer, and nearly $10.00/lb (15 centimes/g) moisture removed for a desiccant dryer (PAI 1987).

On the negative side, refrigerated dryers are relatively complicated and require specialized expertise available through a commercial refrigeration service firm for some maintenance tasks. With multiple component parts, including the condenser, compressor, and several control valves, refrigerated dryers require frequent maintenance attention for long-term reliable operation.

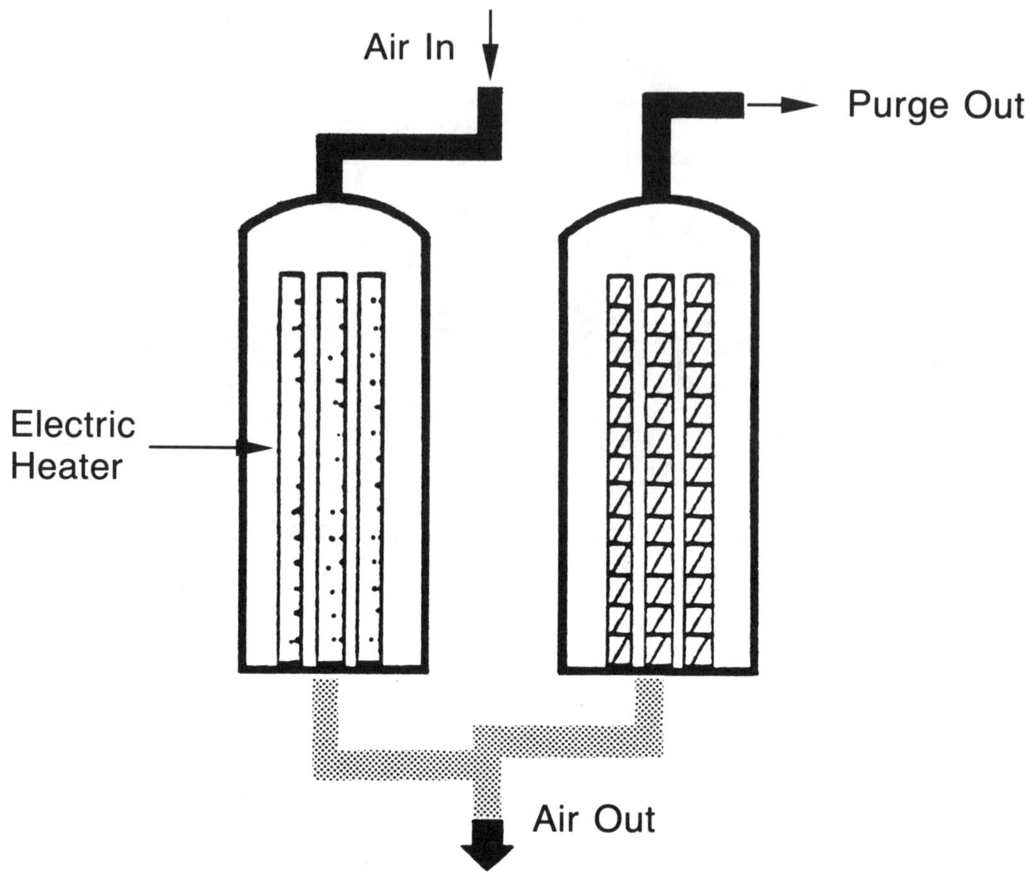

Source: Hesby (1989).

Figure IV–29 Heat-Reactivated Internally Heated Desiccant Dryer

Desiccant drying.

Principle of operation. Desiccant drying is the final significant step in the air preparation system to reduce moisture in the air and achieve the final design dewpoint. These dryers function by adsorbing moisture vapor onto a solid surface—actually submicroscopic pores in the desiccant material. During regeneration, this adsorbed moisture is driven off by high temperatures or system pressure reduction (depending upon the type) and carried out of the dryer bed by a purge gas stream.

Most desiccant dryers in use with ozone systems are of the dual tower design, shown schematically in Figure IV–29. While one tower is in drying service, the other tower is being regenerated, as shown schematically in Figure IV–30. All components are typically factory-preassembled with a four-way or other switching valve(s) to alternate flow between the chambers. Typical total cycle times are 16 h for thermally regenerated dryers (8 h drying, 4 to 5 h heating, and 3 to 4 h cooling [Manz and Shumate 1988]), and 6 or 10 min for heatless dryers (3 or 5 min each for drying and purging) plus 30 sec for repressurization.

Desiccant dryers are currently available for rated flows up to 25,000 scfm (42,475 m^3/h NTP).

Desiccant media. The three primary types of desiccant materials are silica gel, activated alumina, and molecular sieve. General physical properties for each of these types are presented in Table IV–3 and equilibrium capacity curves are presented on Figure IV–31.

Silica gel: Silica gel has been commercially available for a long time and has been

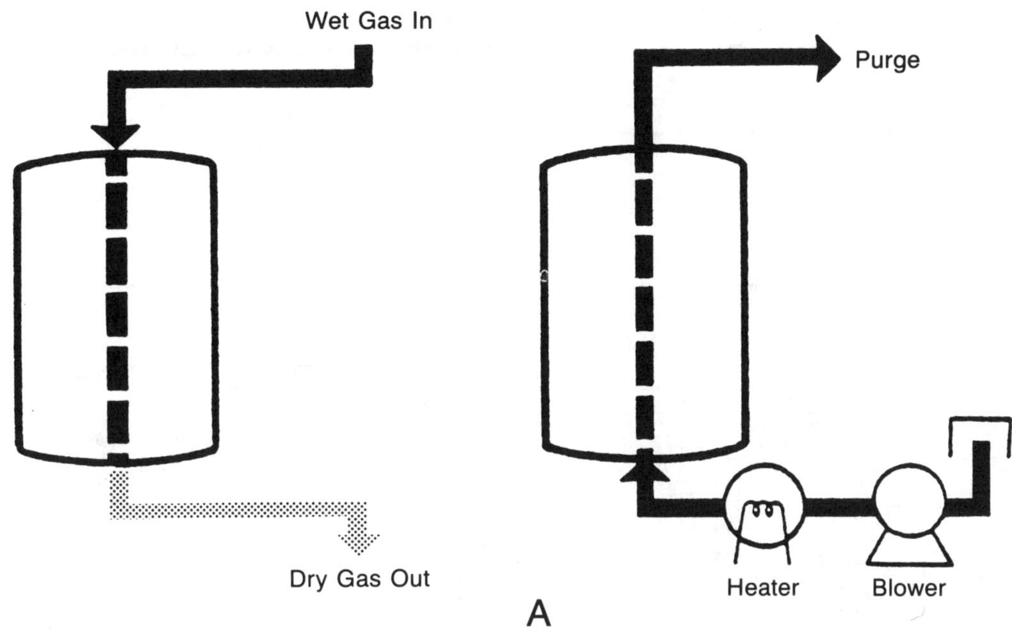

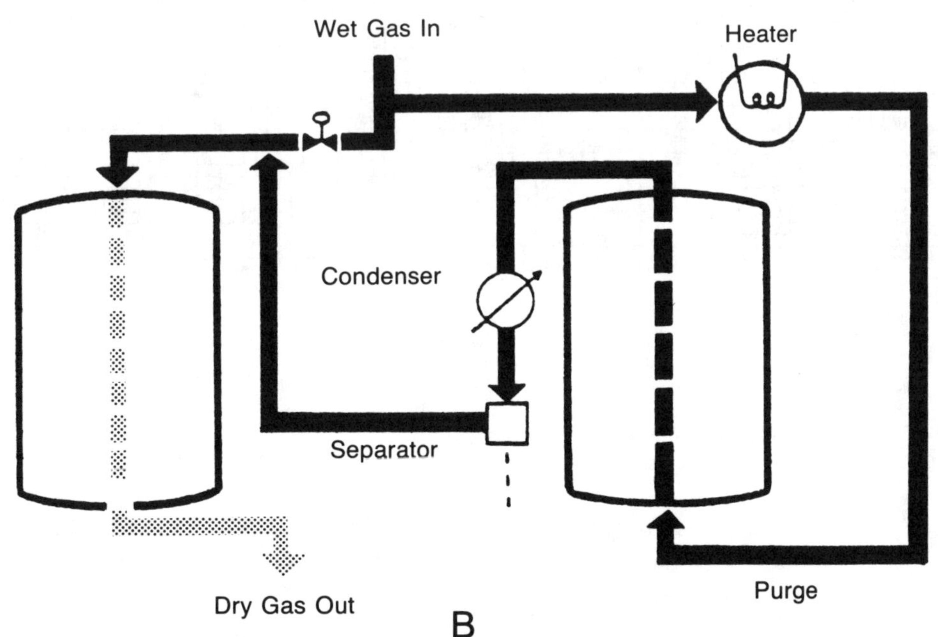

Source: Hesby (1989).

Figure IV–30 A: External Heat Reactivated Atmospheric Blowers. B: External Heat Reactivated—Line Pressure, Closed Loop for Desiccant Dryer

Table IV–3 General Characteristics of Desiccant Media

Characteristic	Silica Gel	Activated Alumina	Molecular Sieve
Bulk Density, lb/ft^3 (kg/m^3)	38–45 (609–721)	45–50 (721–801)	30–45 (480–721)
Regeneration Temp., °F (°C)	300–450 (149–232)	350–500 (177–260)	350–600 (177–315)
Surface Area, m^2/g	650	325	–
Apparent Specific Gravity	1.2	1.6	1.1
Particle Shape	Granular	Beads	Pellets

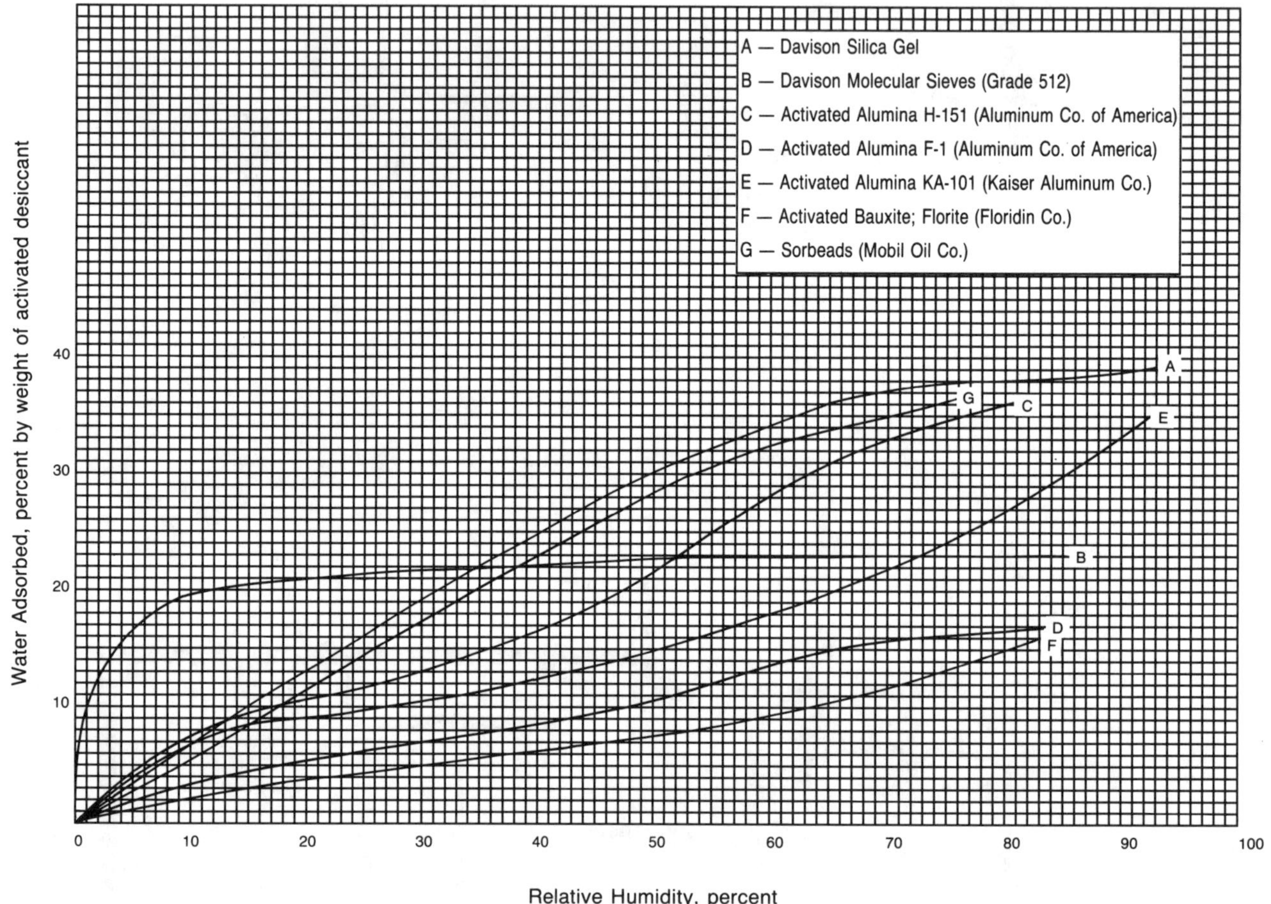

Source: Hesby (1989).

Figure IV–31 Equilibrium Capacity Curves

used in the past for air drying in ozone systems. With the development of newer desiccants, it has fallen out of favor with most ozone system designers. Silica gel may be represented by the formula $SiO_2 \cdot nH_2O$.

As noted from the equilibrium curves, silica gel has a very high adsorption capacity in feed gases with high humidity. This, combined with rather moderate regeneration temperatures, is the reason for silica gel's previous popularity. It is also capable of adsorbing hydrocarbons in the gas stream. However, regenerated beads will shatter on contact with liquid water, which may be present in incoming gas streams. Also, silica gel's effectiveness is greatly reduced at lower humidities; it is therefore less capable of producing the very low dewpoints now desired by ozone system designers.

Activated alumina: Activated alumina is a porous, amorphous form of aluminum oxide matrixed with trace quantities of other materials. It can be represented by the formula $Al_2O_3 \cdot nH_2O$. The desiccant is typically provided in a bead form and is available in several sizes, ranging in diameter from 1/10 in. to 1/2 in. (2.5 to 12.7 mm).

Activated alumina has a good moisture-holding capacity, especially at high relative humidities, and can be regenerated at rather moderate temperatures when compared to molecular sieves. Additionally, it is highly abrasion- and crush-resistant and is unaffected by liquid moisture. At $1.50/lb (22.7 FrF/kg) in bulk quantities (1989 quotation), the cost of activated alumina is less than half that of molecular sieves. For these reasons, activated alumina is now the most popular of all desiccants for general purpose air drying. However, activated alumina does not adsorb hydrocarbons well.

Molecular sieve: Molecular sieves are synthetically produced crystalline metal alumina silicates that have been activated for adsorption by removal of their water of hydration. The most common molecular sieve used in general air drying can be represented as $Na_{86}[(AlO_2)_{86}(SiO_2)_{106}] \cdot 276\ H_2O$. Unlike other desiccants, the pores of molecular sieve are precisely uniform in size and of particular molecular dimensions. Depending on the size of these pores, molecules may be readily adsorbed or completely excluded, based on molecular size differentiation or sieving. Molecular sieve can also adsorb hydrocarbons in the gas stream. However, the most useful, and unique, characteristic of molecular sieve is its high moisture-holding capacity at low inlet gas relative humidities, indicated in Figure IV–31. It is this characteristic that makes molecular sieve excellent for use as a "trim" bed at the bottom of an activated alumina desiccant bed. Some ozone system designers now require a minimum of 25 percent molecular sieve in the dryer bed to assure low dewpoints. Only rarely, except in "heatless" systems, is molecular sieve used exclusively as the desiccant medium for air drying in ozone systems. Molecular sieve is available from several manufacturers in 1/16- and 1/8-in. (1.6- to 3.2-mm) pellet forms, with approximately 20 percent clay binder. Relatively expensive, at over $3.50/lb (53 FrF/kg) (1989 quotation), molecular sieve also requires the highest regeneration temperature of all desiccants and about 500 more Btu/lb (about 278 more kcal/kg) to remove 1 lb (453.6 g) of H_2O than does activated alumina.

Regeneration. One of the major advantages of solid adsorbents is their ability to be regenerated. That is, the adsorptive process is reversible. The adsorbed moisture can be removed and the dryer placed back in service for another drying cycle. While drying techniques are similar for all designs, dryers are usually identified by the mode of regeneration: pressure swing (heatless) or temperature swing (internal and external heat).

Proper regeneration of a desiccant dryer is dependent upon several interrelated factors, including;

- regeneration temperature,
- purge gas moisture content,
- regeneration direction,
- purge flow rate and duration, and
- bed cooling.

The total heat applied to the desiccant must be sufficient to bring the desiccant material and the adsorber vessel to the regeneration temperature and to provide the heat of desorption. Since the desorption process is also a function of partial vapor pressure in the purge gas, higher temperatures are necessary for dryers using wet gas (atmospheric air) for regeneration when attempting to achieve comparable dewpoint performance.

Most desiccant manufacturers recommend that regeneration purge flow direction be countercurrent to the direction of drying. This ensures that the desiccant near the outlet of the bed is thoroughly regenerated. Since the dewpoint of the gas being dried is in equilibrium with the adsorbent at the bed outlet, this results in the best dewpoint performance. Also, since most of the water is adsorbed at the inlet of the bed, countercurrent regeneration minimizes the amount of water carried through the bed, thus minimizing hydrothermal aging of the media. Finally, bed cooldown is especially important in ozone gas preparation, since hot feed gas to the generators will adversely affect ozone production and will support a high dewpoint.

Pressure swing regeneration: For air preparation systems operating at pressures above 65 psig (4.48 bar), the desiccant can be regenerated without heat. A portion (at pressures between 70 to 100 psig [4.83 to 6.9 bar], purge ranges from 20 to 15 percent, respectively) of the dry process air is passed through the bed countercurrent to the direction of drying and expanded to the atmosphere (Shumate 1989). Since the dry low-pressure air has a lower vapor pressure than the adsorbate on the desiccant, moisture will desorb and be carried away by the purge gas.

The typical cycle time for heatless type dryers is very short—on the order of 2–3 min for drying. Heatless dryers are designed on the basis of actual bed volume gas throughput. The moisture content of the inlet air, within normal conditions, does not affect the capacity of the dryer. The actual volume of air passed through the tower should not exceed 50 percent of the gross volume of the bed during one adsorption period. With increased air throughput, a larger quantity of regenerative gas flow is required. Also, flows in excess of 50 percent remove the "heat front" (the heat of adsorption) that is used to regenerate the bed. Design on the basis of a moisture loading rate, typically 2–3 percent of the total dynamic capacity, should be approached cautiously as only the upper 4 to 6 in. (about 10 to 15 cm) of the bed is actually saturated (Shumate 1989). Heatless dryers can reliably produce −100°F (−73.3°C) dewpoints, and this performance will not suffer due to thermal aging of the desiccant material as is the case with heat-regenerated dryers.

Due to their simplicity in operation and reliable performance, pressure swing–type driers have been widely used in several industrial applications, including microchip manufacturing. However, except for small systems (less than 200 lb/day [<3.78 kg O_3/h]), they have only been used on a limited basis in ozone systems. This is easily explained by the fact that the pressure requirements of the typical ozone system are much too low for operation of a heatless dryer. The savings in power and capital costs of the heatless dryer over a heat-regenerated dryer are normally not sufficient to offset the power and capital costs of a high-pressure compressor operated strictly for the benefit of high-pressure drying.

The simplicity and potential maintenance cost reduction should be part of economic evaluations of this gas preparation mode. One instance that justifies serious investigation of the heatless dryer, regardless of system size, is where the ozone system is expected to operate at short notice without maintaining the gas preparation system on standby for limited number of hours each day. In this case, the short operating cycle facilitates system startup after an abrupt shutdown, without the concerns of a partially regenerated bed (which could be the case with a thermally regenerated system).

Internal heat regeneration: Internal heat–regenerated dryers rely on convection and radiation to transfer heat to the desiccant and drive off the moisture. A small purge flow (10 to 15 percent of the dryer capacity) is passed over the bed countercurrent to the direction of drying to carry off this moisture.

A typical internal heat–regenerated dryer has several vertical heater elements spaced throughout the desiccant bed. Each heater element should be enclosed in an individual sheath designed to avoid high skin temperatures, to allow space for thermal expansion and contraction of the heater element, and to enable removal of the heating element without removing the media from the dryer tower. Because of the difficulty in achieving even heating of the desiccant bed through this method, internal heat–regenerated dryers have normally been limited in size to approximately 1000 scfm (about 1700 m^3/h NTP) and smaller or to a 35- to 40-kW heater due to problems of expansion (Shumate 1989).

The designer, in evaluating this unit, should consider specifying the insulation of the exterior of the dryer towers to conserve energy (Joost 1989). Currently, although the internal heat units are significantly more energy-efficient than external heat regeneration units, many ozone system designers are avoiding these dryers altogether, citing uneven performance and premature aging of the desiccant. Joost (1989) considers that the maximum feed gas flow rate for this type of dryer should be 500 scfm (about 850 m^3/h NTP).

External heat regeneration: External heat–regenerated dryers rely on convection to heat the desiccant material and to drive off the adsorbed moisture. For these dryers, purge air is heated to the regeneration temperature and passed through the bed countercurrent to the direction of drying. Several variations on this basic theme have been developed, including (1) atmospheric purge with dry gas sweep, (2) closed-loop, and (3) atmospheric purge with closed-loop cooling.

In the first instance, a blower draws in ambient air, passes it over an externally mounted heater through the bed, and vents it to the atmosphere. To accomplish removal of additional residual moisture and improve dryer dewpoint performance, the desiccant bed is then purged with dry process air, also vented to the atmosphere.

In the closed-loop system, a captive volume of air is circulated by a positive displacement compressor over an externally mounted heater through the bed, through an aftercooler and separator to remove the adsorbed moisture, and then back to the compressor. The advantage of this system is that no process air is used for regeneration purge flow. This is most important for constant volume–type ozone systems where changing process gas flow rates will adversely impact system performance. With activated alumina, dewpoints of −60°F to −70°F (−51.1 to −57.6°C) can be achieved in this mode (Shumate 1989).

The atmospheric purge with closed-loop cooling alternative is a combination of the previous two systems. The atmospheric purge air is heated and passed through the desiccant bed as in the earlier alternative for the heating cycle, and a captive volume of gas is recycled in a closed-loop system for desiccant cooling. With activated alumina, dewpoints to −100°F (−73.3°C) can be achieved with this mode (Shumate 1989).

Desiccant dryer controls. Desiccant dryers are designed for constant operation under varying flows, and the dryer cycle can be operated in a fixed time mode or based on demand. In the fixed time mode (most common), the drying cycle is based on a fixed time interval—typically, 8 h for ozone systems—regardless of the load on the dryer. This system, based on the maximum volume of air (for the worst humidity conditions) that can be passed through the dryer to achieve the design dewpoint, is the proven mode of operation. A timer is used to determine when it is necessary to change towers. While very reliable, this approach may waste energy, since the desiccant may not be fully utilized during periods of low flow.

When demand-based drying is employed (less commonly), the drying cycle is terminated when the desiccant bed has reached its design loading. This termination point can be determined by monitoring, either at the outlet or at some location within the bed, with a moisture sensor. The latter approach is preferred, since the moisture wavefront is then detected before it exits the desiccant bed. The success of the system depends on the moisture sensor; therefore, demand-based drying appropriate for the larger systems where a high level of instrumentation calibration and maintenance may reasonably be expected. An absolute moisture measuring device should be used rather than a relative humidity instrument (Shumate 1989). Another way to control a demand-based drying system is the measurement of the total gas volume passing through the tower. When the volume reaches a setpoint (depending on the design of the dryer), the regeneration process is started up (Langlais 1989).

Like compressors, desiccant dryers can be provided with a full range of control systems. The control system has the function of timing the intervals for the drying,

Table IV–4 Coalescing Filter Performance Ratings

	MIL F-1079	A	B
Aerosols			
Efficiency (percent)	–	95	–
Size (μm)	–	.009	–
Efficiency (percent)	99.985	100	99.999
Size (μm)	0.3	.75	–
Solids			
Efficiency (percent)	99.985	100	100
Size (μm)	0.3	0.3	.025

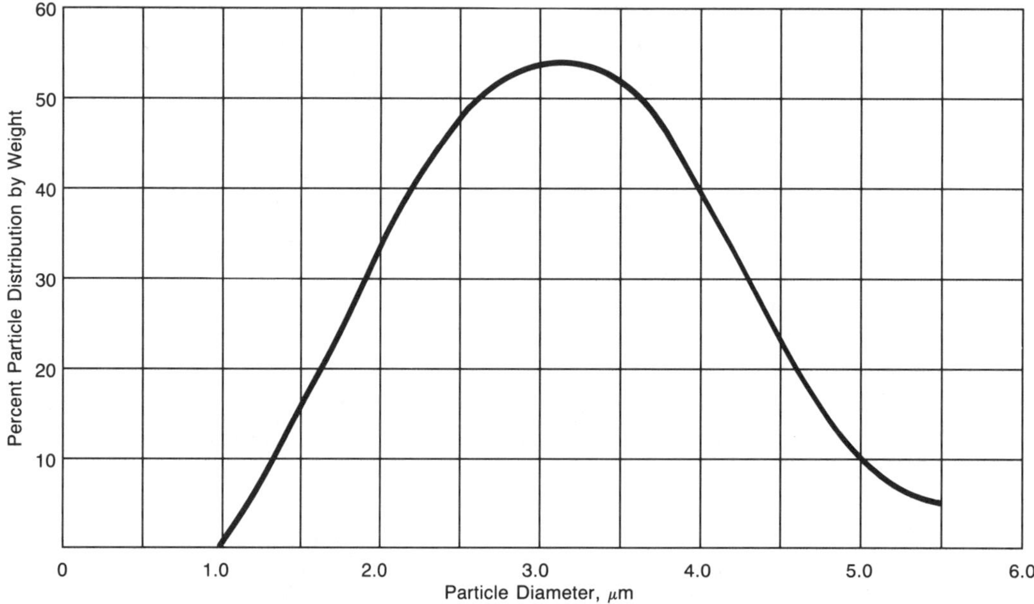

Source: Hesby (1989).

Figure IV–32 Typical Contaminant Challenge to a Particulate After-Filter

heating, and cooling portions of the cycle in addition to monitoring the status and alarms of the dryer. The controls can be microprocessor-based or realized through a cam timer and relays.

The drying interval of a desiccant bed before regeneration can be dictated by time and/or moisture content of dried gas or by the total gas volume measurement. A dryer operating with only time as a parameter is termed a fixed time mode dryer, while one that incorporates time and moisture content, or total gas volume, is termed a demand-based dryer. The demand-based dryer will have a minimum drying interval, and if the moisture content or the gas volume is below a setpoint after this interval, the dryer will not change beds unless the setpoint or maximum drying interval is exceeded. The objective of a demand-based dryer is to save energy and minimize the number of heating cycles on the desiccant material.

The most important alarm associated with a desiccant dryer is the high moisture alarm for the output gas. It is recommended that each dryer's output gas be monitored for high moisture content for diagnostic purposes. This does not necessarily imply that each dryer must have its own analyzer, as some analyzer manufacturers can provide multi-probe units. In general, a high moisture alarm should initiate an ozone system

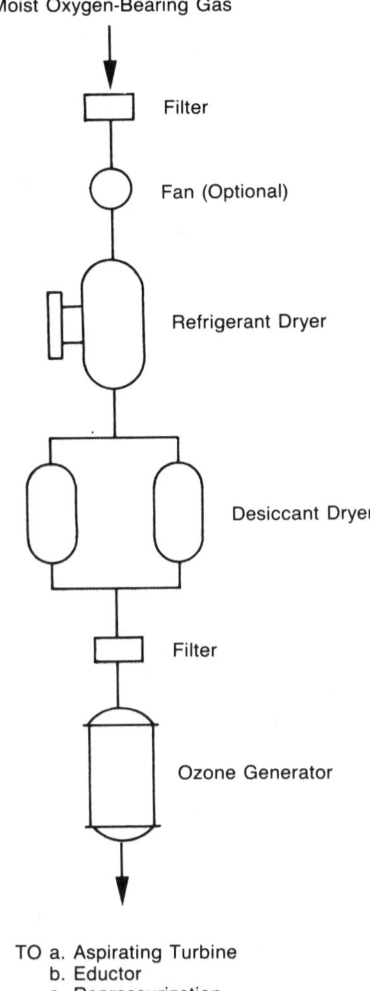

Figure IV–33 Process Diagram of Ambient Gas Preparation System

shutdown, although it may be desirable to have two relays—one as a warning and the second for shutdown.

Other alarms that may be provided with a desiccant dryer include:

- Valve switching failure (indicates if a pneumatically actuated valve fails to switch properly).
- Heater element failure (in heated dryers).
- Cooling water low flow and blower motor overload (for externally heated dryers).

Gas filtration. As indicated previously, gas filtration is an important step at several key locations to protect the components of the gas preparation system, to provide the best feed gas quality for optimum long-term performance of the ozone generators, and to prevent plugging of the diffusion system.

Raw gas. Raw gas filtration is used primarily to protect the compressor. Each type of compressor requires a different degree of protection. Many compressors, such as rotary screw and rotary lobe compressors, are typically provided with inlet filter/silencers as part of the prepackaged assembly. Since these compressors operate with very close running tolerances between lobes, filtration requirements are relatively stringent and the inlet filters are typically rated for less than 10 μm.

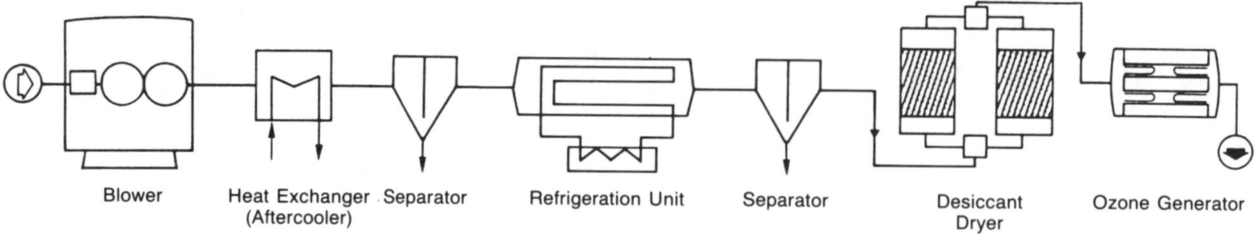

Source: Schulhof and Dyer-Smith (1988).

Figure IV–34 Low Pressure Air Preparation Ozonation System

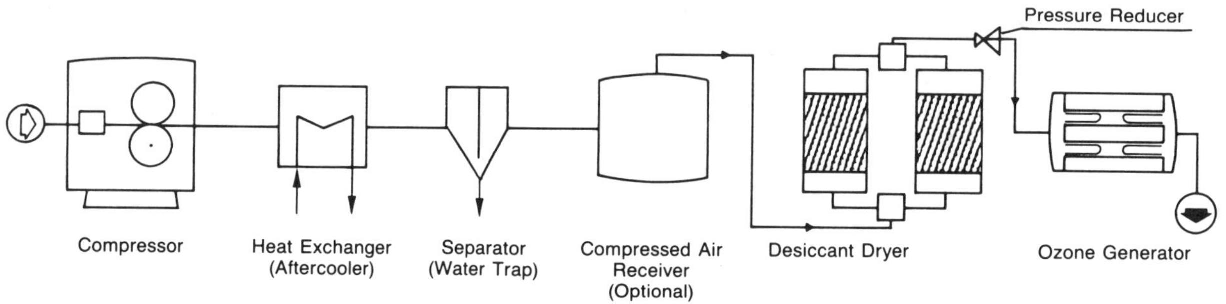

Source: Schulhof and Dyer-Smith (1988).

Figure IV–35 Medium Pressure Air Preparation Ozonation System

Most inlet filters provided with the compressors are of the cleanable/replaceable pleated cartridge type, similar to a high-quality automobile intake filter. Air can be drawn in from the room or piped to the compressor inlet from outdoors. In the latter case, it is advisable to use a 20 μm inlet dust filter at the intake louver to remove sand, leaves, and other coarse materials.

Before the dessicant dryer. Downstream of any refrigerant-type dryer, the feed gas will be saturated and will carry liquid moisture. A separator and trap should be provided to remove the liquid moisture. In order to protect the desiccant dryer from entrained water droplets, and from particulates in the air stream, the dryer must be preceded by a coalescing filter or a cyclonic separator. Coalescing filters provide an effective means of reducing the water droplets and solid aerosol burden of the base stream. Unfortunately, there is no widely accepted, meaningful rating method other than U.S. Government Specification MIL F-1079, dated April 6, 1962, for DOP penetration and pressure drop. The efficiency of the material should be 99.985 percent retention (0.015 DOP penetration) at 0.3 μm. Table IV–4 presents the rating as stated by three different manufacturers for their coalescing filter (one of them using MIL specifications).

Most coalescing filters now employ inside-to-outside flow direction for optimum coalescing. In this arrangement, the coalesced moisture is collected on a final drainage layer on the outside surface of the filter and drips to the bottom of the housing for removal through an automatic drain trap. The coalescing prefilter or cyclonic separator is usually provided skid-mounted with the refrigerant or desiccant dryer.

Final feed gas. Final feed gas filtration is provided as a particulate after-filter, also skid-mounted with the desiccant dryer. Particulate challenge to the after-filter comes primarily from dusting of the desiccant material, which typically has a particle size distribution as indicated in Figure IV–32. To accommodate this loading, particulate

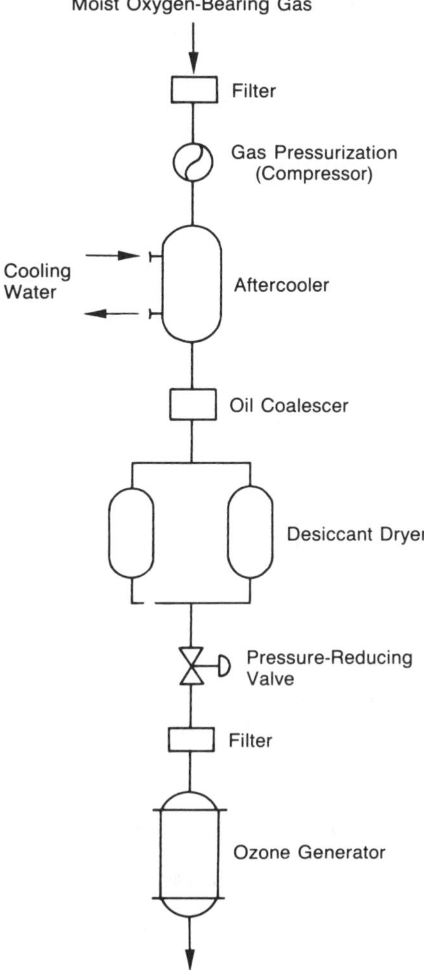

Figure IV–36 High Pressure Air Preparation Ozonation System

after-filters should be of the replaceable pleated cartridge type, with at least a 1.0-μm absolute rating. A filter rating of 0.1 μm is recommended particularly for medium- and high-frequency ozone generators, which may incorporate a smaller dielectric gap (Rakness 1989). An initial maximum pressure differential of 1 psig (0.069 bar) across the filter medium is specified to assure that sufficient filter medium is available (Rakness 1989). Operational filter cartridge lifespan is limited only by particulate loading, but six to nine months is a reasonable expectation.

Final design. The detailed design of the gas preparation system is frequently assigned to the ozone system vendor. However, the facility designer and/or operator must establish the type of overall gas preparation desired and the quality level of the individual components. Figures IV–33, IV–34, IV–35 and IV–36 illustrate the four basic air preparation systems in common use and the components normally included in them.

IV.C OZONE GENERATION

The designer is faced with selection of an ozone generator from a variety of types of units that have been applied or are currently being promoted in the United States. A number of these may be limited to highly specific applications but are identified in this subsection for completeness.

Ozone generators may be described on the basis of a number of parameters. These parameters include type of dielectric, frequency, and mode of generator cooling.

Dielectrics incorporated in commercial scale ozone generators include glass plates, metallized glass tubes, and ceramic plates, in addition to even more proprietary concepts. The type and/or specific characteristics of each vendor's ozone generator is most likely specific to that generator.

Ozone generator frequency has been generally grouped into three broad categories:

- Low frequency: 50 or 60 Hz
- Medium frequency: 60 to 1000 Hz
- High frequency: >1000 Hz

Low frequency is essentially the line or main frequency as used in the power supply distribution grid. For example, this would be 60 Hz for North America and 50 Hz throughout most of Europe. Medium and high frequencies are the frequencies selected by the specific vendor.

The mode of generator cooling is mainly water but may also be air as well as other media such as oil and Freon. Chilled water cooling loops may be provided as a matter of course by certain vendors or to meet specific project requirements, such as the high treated water temperatures experienced in the southern United States as well as in other countries.

The ozone generators covered in this subsection include the following:

- Horizontal tube, water cooled, low frequency
- Horizontal tube, water cooled, medium frequency
- Vertical tube, water cooled, low frequency
- Vertical tube, dual cooled, high frequency
- Vertical plate, water cooled, low frequency
- Vertical plate, air cooled, high frequency

Without returning in detail to the points discussed in sec. D, chapter II, it is worth pointing out that the ozone required for large drinking water treatment plants is provided by corona discharge generators operating on principles similar to or derived from those of the "Welsbach" generator. This type consists of concentric tubular electrodes, glass dielectrics, and a cooling water system. Ozone generators producing ozone from ultraviolet radiation or by electrolysis (see sec. D.1, chapter II) exist, but they are generally not used in drinking water treatment.

We shall first describe the parameters that affect the generation of ozone and the various types of generators available before going into the concepts used in designing and manufacturing generators. An ozone generator is a complex piece of equipment: to build and optimize such a unit not only requires a thorough knowledge of the electrical aspects; fabrication of the metal components and selection of the dielectric glass to optimize ozone production requires considerable technical expertise. In addition, the manufacturer needs experience in the conditions under which the equipment will operate and must be familiar with the peripheral equipment and processes such as air drying, oxygen use, refrigeration, and so on (see sec. B, this chapter).

IV.C.1 Commercial Scale Ozone Generator Operating Principles

General principles applicable to an ozone generator cell. A silent electrical discharge, frequently described as a corona discharge, is maintained between two electrodes separated by a dielectric (very often consisting of glass, sometimes of ceramic material) and an air gap by applying a high voltage across the electrodes. At the frequency used, this voltage is held at a value between the threshold at which the oxygen-bearing gas ionizes (or "fires"), producing ozone, and the value at which the glass dielectric will break down its "dielectric strength" (Gary and Moreau 1976; Filipov and

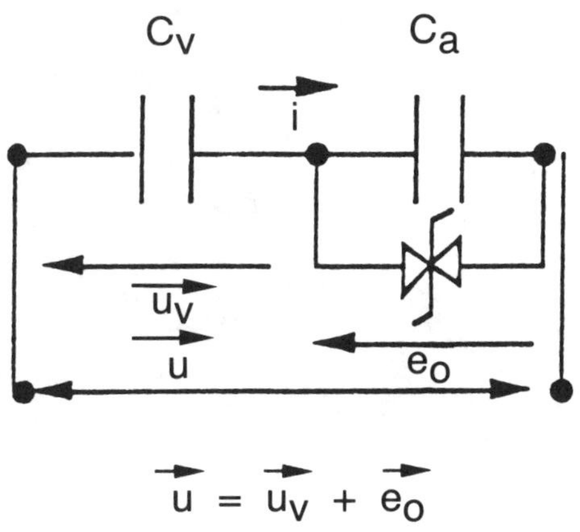

$$\vec{u} = \vec{u}_V + \vec{e}_O$$

C_V: Capacitance of the glass dielectric
C_a: Capacitance of the discharge gas
u: Voltage applied to the ozone generator
u_V: Voltage across the glass dielectric
e_O: Firing voltage across the gap
i: Current through the ozone generator

Source: Pare (1967).

Figure IV–37 Equivalent Electric Diagram of a Basic Ozone Generator Cell

Emilianov 1957). The electrical characteristics of this basic ozone generator can be represented by the equivalent electric circuit shown in Figure IV–37.

This equivalent circuit consists of two capacitances in series, representing the glass dielectric (capacitance C_v) and the oxygen-bearing gas (capacitance C_a). A voltage clipper consisting of two Zener diodes connected back-to-back is connected across this second capacitance. This clipper determines the firing voltage E_0.

At the electron discharge threshold, the absolute value of this voltage can be written (Peek 1924):

$$E_0 = 28.6 \times d \times K$$

Where:

d = the density of the oxygen-bearing gas
K = a factor that depends on the electrode geometry.

The expression for each of the capacitances can be written as follows:

- the air: $C_a = 2 \pi \ell \Sigma_0 \cdot \dfrac{1}{\ln \dfrac{r + e1}{r}}$

- the glass: $C_v = \dfrac{2 \pi \ell \Sigma_0 \Sigma_r}{\ln \dfrac{r - e2}{r}}$

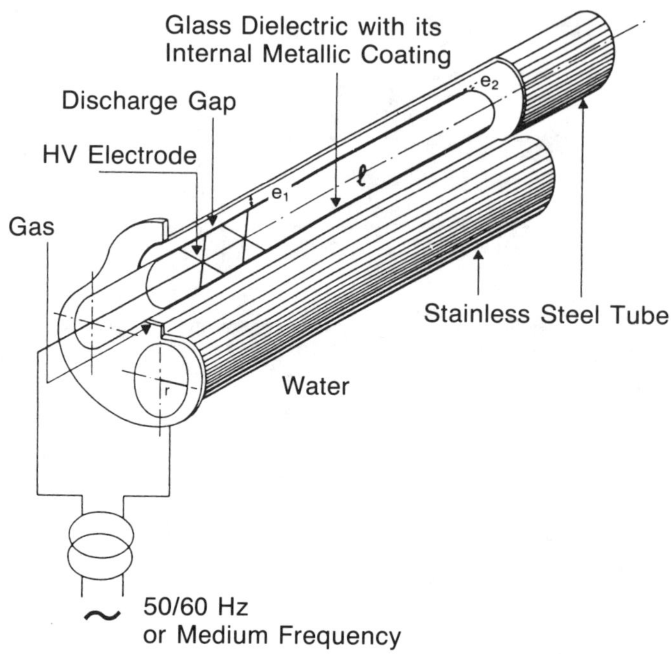

Source: Chapsal (1989).

Figure IV–38 Geometry of Tubular Ozone Generator Cell (Basic Welsbach Technology)

Where:

ℓ = the length of the cylindrical high-voltage electrode (m)
r = the outside radius of the dielectric tube (m)
e1 = the discharge gap (m)
e2 = the thickness of the glass (m)
Σ_0 = the absolute permittivity of vacuum equal to 8.85×10^{-12} F/m
Σ_r = the relative permittivity of the dielectric, i.e., the ratio of the dielectric capacitance value between two electrodes (C_x) and the value of the capacitance (C_v) of the same electrode shape under vacuum. $\Sigma_r = C_x/C_v$; this parameter is without dimension.

Figure IV–38 illustrates the geometry of a tubular ozone generator cell.

C_a, C_v, and E_0 are three electrical parameters that, as will be shown later, depend on the characteristics of the basic generator cell and the operating conditions.

The fundamental electrical variables in a basic cell are:

- the voltage across the glass dielectric (u_v)
- the current through the ozone generator: $i = C_a \cdot \frac{du_v}{dt}$
- the voltage applied to the ozone generator u and its frequency f

When the currents and voltages defined in the equivalent electric diagram (Figure IV–37) are displayed on an oscilloscope during discharge, their forms are as shown in Figure IV–39.

The power dissipated in the glass dielectric is expressed by the equation:

$$P_v = \frac{\omega}{\pi} \int_{t_0}^{\frac{\pi}{\omega} + t_0} U_v \cdot i \cdot dt$$

where t_0 = time of the start of the discharge.

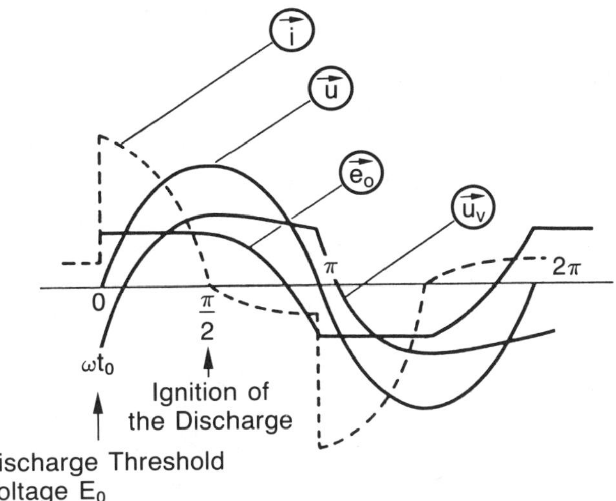

i: Current through the ozone generator

u: Voltage applied to the ozone generator

e_0: Firing voltage across the gap

u_v: Voltage across the glass dielectric

Source: Faes (1975).

Figure IV–39 Voltage and Current Curves During Discharge

Integrating this power equation over half a cycle gives the glass "loss angle" factor, defined as $\tan \delta = R \cdot C_v \cdot \omega$ where R = the transverse resistance of the capacitance. This factor is directly proportional to temperature and inversely proportional to frequency. The power dissipated in the glass is approximately 1 percent, which is totally negligible compared to the last electrical parameter to be considered: the power dissipated in the discharge (P_e).

This power is expressed by the Manley (1943) formula which, in fact, describes the total power dissipated:

$$P_v + P_e \approx P_e = 4 f E_0^{\approx} [(U \cdot C_v/E_0) - C_a - C_v]$$

All the terms in this formula have already been defined; it will be used later to show how the ozone production can be optimized. Various factors that can affect the electrical discharge and, consequently, the production of ozone and the ozone generator system efficiency (Chapsal 1976) include:

- The quality of the ozone generator feed gas, i.e., the percentage of oxygen it contains, its dryness, the mineral or organic solid or gaseous impurities it contains, its pressure, and its temperature at inlet to the generator (see II.D.1).
- The quality of the electric power supply, which depends on the balance of the electrical supply; the frequency, voltage, harmonics, and any corrections applied; and voltage surges and any corrections applied.
- The quality of the dielectric, i.e., (1) the quality of the glass (its permittivity, resistivity, and thickness); and (2) the effective dielectric length and surface area, geometry, and discharge gap used. The quality and consistency of these properties depend on the manufacturing tolerances applied.

- The effectiveness of the cooling, which depends on (1) the cooling fluid temperature and flow, and (2) the internal and external arrangement of the cooling fluid distribution system.
- The quality of the high-voltage (HV) electrode, i.e., the type of metal used, the contact obtained between the HV electrode and the dielectric, and the contact obtained between the HV power source and the HV electrode.
- The possibility of scaling up from a single-cell generator, which depends on the cell manufacturing tolerances and the possibility of adapting the electrical power supply to the increased number of cells.

These parameters interact and a variation in any of them almost inevitably means that one or several other parameters must be corrected accordingly.

Basic parameters that affect the design of a system. The criteria that must be satisfied at the outlet from any ozone generator system include the quantity of ozone to be generated (in kg O_3/h or lb/day); the ozone concentration in the product gas (percent by weight or g O_3/m^3 NTP), which itself defines the ozonized gas flow (m^3/h NTP or scfm); the minimum ozonized gas pressure at outlet from the generator (essential to the design of the contactors and their diffusion systems); and the maximum ozonated gas temperature at outlet from the generator (to avoid destruction of the ozone in the product gas).

The number and type of ozone generators will also depend on:

- The required ozone production rates (particularly the minimum, nominal and maximum values).
- The temperature and flow of the cooling water available.
- The ambient conditions: maximum ambient air temperature and humidity. These parameters are important for the gas preparation train design and for the cooling conditions of the power supply unit (transformers, frequency converters, etc.).
- The required energy consumption.
- The space available and the standby capacity required.
- The expertise, training, and availability of the prospective owner-agency's operation/maintenance department. This last aspect will affect the technology employed in the same way as the required ozone production.
- The reliability of the system. The ozone produced is used to treat drinking water; continuous operation is therefore vital. Consequently, the technology used must be well-proven. A number of examples demonstrate the risks involved in using insufficiently tried and tested technologies. It is for this reason that the manufacturer's performance claims must be carefully evaluated, since they frequently are generated from tests carried out in a laboratory or at a pilot plant. It is advisable only to choose manufacturers who have operating ozone generators of a size equivalent to the required capacity (i.e., at least the capacity of each ozone generator). Only long-term, real-life operation can demonstrate the capabilities or unsuitability of new technology. Unreliable equipment can even lead to a technology's being completely abandoned.
- The same prudent approach must also be applied when using oxygen. For example, economic and maintenance aspects should be considered (see sec. IV.B, chapter V, and chapter VI). Second, the standards for oxygen gas supply must be employed such as special cleaning of the pipes, limited velocity of the oxygen gas, and explosion-proof electrical equipment. Most oxygen-fed ozone systems operate at concentrations less than 8 percent by weight. It is possible to generate higher concentrations (Kogelschatz 1988). It is commonly agreed that it is safe to generate ozone up to 14 percent. However, there is disagreement as to the maximum safe ozone concentration.
- In fact, many of the basic parameters discussed here also apply to any part of an installation; consequently, many of these points are also discussed in the

remainder of chapter IV and in chapter V. Here we shall only consider the mechanical and electrical aspects of ozone generators and the problems of cooling them.

IV.C.2 Generator Technology and Materials Used

The types of materials used in fabrication of an ozone generator depend on the generator cooling method and the dryness of the feed gas. Theoretically, ANSI 304 or 304L stainless steel could be used, but in practice, it has proved unsatisfactory in many cases and is not recommended. Preference should be given to ANSI 316L stainless steel, which is compatible with most operating conditions. A 2 to 2.5 percent molybdenum content increases the ability of the generator components (stainless steel shell and stainless steel tubes) to resist corrosion both by nitrogen oxide compounds and by certain ions, such as chlorides, in the cooling water. Resistance to possible corrosion must be considered, particularly when using an open cooling water arrangement.

Any nonmetallic components of the generator in contact with the ozone must be made of ozone-resistant materials, such as polyfluoride compounds (e.g., Teflon or Hypalon).

Each ozone generator manufacturer uses its own manufacturing processes (e.g., welding processes). Among the available techniques, the TIG (Tungsten Inert Gas) is preferred. In accordance with the size of the generator and the operating pressure, the construction should follow the applicable code. Preference must be given to equipment in which risks of corrosion are minimized, particularly corrosion in weld zones and in the space between the tube bundle and the generator shell, which are prime areas for corrosion. The general design of the stainless steel tube bundle should provide a uniform distribution of water flow around the stainless steel tube, it is common practice to avoid internal baffles as these can encourage corrosion by creating dead zones where solids can settle, particularly metal hydroxides, which together with chlorides may lead to corrosion.

The design must also include various instruments or systems to guarantee reliable operation and indicate or correct the problems that inevitably arise. For example, the ozone generator must be protected against excessively high pressure or negative pressure to prevent permanent distortion of the cell. The feed gas pressure also affects the ozone generator operation and must therefore be monitored by instruments that trigger alarms and can shut down ozone production if the pressure varies excessively (i.e., ± 0.15 bar for a nominal pressure of 0.7 bar; ± 2 psig for a nominal pressure of 10 psig). In the same way, the cooling water flow and/or temperature must be measured by instruments that trigger alarms and shut down production if the cooling is insufficient.

When an ozone generator is taken out of service, it must be purged with dry air to eliminate any residual ozone. In the same way, when it is placed in service, it must again be purged with dry air to eliminate any humidity. Finally, the pipework between the product gas outlet and the contact chambers generally includes, from the ozone generator, a check valve to prevent any backflow of water, an isolation valve, and a goose neck. This becomes particularly important when certain dissolution systems, capable of creating a partial vacuum in the ozonator, are used.

Strict manufacturing tolerances are a vital factor in the repeatability of ozone generator performance. For example, the uniformity of the discharge gap depends directly on the metal tube and dielectric glass tube manufacturing tolerances (both diameter and straightness). The uniformity of the surface also has a significant effect on the regularity of the discharge.

Either drawn or rolled and welded steel tubes capable of satisfying the required manufacturing tolerances are available from various suppliers. However, obtaining satisfactory glass tubes is far more difficult. These tolerances can only be obtained by inspecting large numbers of tubes and rejecting those that are not satisfactory. An alternative is to refit all the tubes to comply with the tight tolerances required; however, this requires an additional manufacturing operation. The dielectric metallization process (if used) must be very tightly controlled. This metallization and the high thermal stresses thereby generated significantly affect the ability of the glass to withstand high voltages and/or high power densities, with consequent effects on the life of the tubes. Other types of internal electrodes are also used.

Finally, a uniform discharge gap can only be optimized if radial locating devices are provided to keep the glass tubes in the center of stainless steel tubes. These support devices must be capable of withstanding the ozone concentration in the gas.

IV.C.3 Electrical Parameters and Variables That Affect Ozone Production

In sec. IV.C.1, we defined the various electrical parameters and variables that affect the operation of a basic ozone generator element, along with Manley's formula, which expresses the dissipated power at different values of these parameters and variables. Here we shall consider the various possibilities of optimizing ozone production on a commercial scale.

Selection of the power supply frequency and voltage. Manley's formula (Manley 1943) shows that once the three electrical parameters (C_a, C_v, and E_0) have been fixed, the power supplied to an ozone generator depends solely on the power supply voltage and frequency. For a given power level, increasing the supply frequency reduces the supply voltage. In addition, the following parameters limit the power that can be applied to a dielectric tube:

- the dielectric breakdown voltage, and
- the failure temperature of the dielectric, which itself depends on the power applied and therefore the electron density within the discharge gap (Scholze 1980) and the type of cooling.

If, for reasons associated with the process, the ozone generator must operate with a gas other than air and with high ozone concentrations in the gas, the discharge gap and the frequency must be matched to obtain the required performances (ozone concentration, ozone production, specific energy consumption).

The frequencies commonly used at water treatment plants are low frequency (50 or 60 Hz) and medium frequency (60 to 1000 Hz). Few ozone generators run at high frequency (higher than 1000 Hz).

The voltages applied can vary for each frequency and depending on the manufacturers. For instance, at low frequency (i.e., at 60 Hz), the peak voltage is 27 kV (19 kV r.m.s.) for one manufacturer's ozone generator, and 20 kV (14.2 kV r.m.s.) for another one. These differences are due to the different types and thickness of the glass, the different gap, the different metallization techniques, etc. At medium frequency the voltage range is lower and the power density is higher (i.e., at 400 Hz: 18 kV peak, 12.6 kV r.m.s.). Anyway, the voltage cannot be an evaluation parameter by itself. The important parameter is the reliability and the lifetime of the complete ozone generation system. This parameter can be quantified for a given ozone generator by the percentage of dielectric failures per year. Low, medium, and high frequency ozone generators can be reliable if they receive properly prepared gas and if they are operated within the limits of the equipment.

Dielectric tubes and the discharge gap. First, it must be noted that (1) the dielectric tube capacitance (C_v) depends on the tube wall thickness, its diameter, and the type and temperature of the glass (Prod'homme 1960), and (2) the discharge gap capacitance and the discharge threshold voltage E_0 depend on the type of gas used to produce the ozone, its pressure, its temperature, and the thickness of the gas film within the discharge gap. Either reducing the discharge gap or increasing the power supply frequency, or both, will reduce the power supply voltage required. However, these three factors also affect many other parameters, particularly losses due to dielectric hysterisis, conduction, and dissipation. It is therefore again necessary to determine the most efficient compromise before fixing these parameters.

In selecting the glass used for the dielectric, the manufacturer's objective is to obtain a low loss angle tangent (tan δ), constant relative permittivity of the dielectric (Σ_r) and a product of these two parameters (tan $\delta \times \Sigma_r$), which does not vary with frequency.

Provided that the gas characteristics (pressure, temperature, dewpoint) remain within acceptable ranges (see sec. IV.B), a given dielectric tube may reach its breaking point for a too-high voltage peak or a too-high power density (W/m^2 or W/ft^2) or both. In general, higher power densities are applied to medium-frequency tubes, and higher voltages to the low-frequency tubes. For instance, power density is in the range 1000–1500 W/m^2 (about 93 to 139 W/ft^2) at low frequency and 2000–4000 W/m^2 (about 186 to 372 W/ft^2) at medium frequencies. The maximum r.m.s. voltage is in the range of 14 to 19 kV for low frequency and 9 to 14 kV for medium frequency.

The point of dielectric failure will vary from one dielectric tube to another, according to the glass characteristics, geometry, metallization process, or the nature of the high-voltage electrodes. For a well designed system, the life cycle of all dielectrics (working at low or medium frequencies) is similar and has extended longevity when operating within designed feed gas preparation criteria. High concentrations can be achieved with smaller discharge gaps. It should also be noted that reducing the gap implies stricter technological constraints (tighter tolerances for tube diameter, straightness, accuracy of location at assembly, etc.).

The power supply and balancing of mains phases. As was stated previously, the type of electrical power supply used depends directly on each manufacturer's experience. The power supply quality depends on the electrical components used. Low-frequency ozone generators can operate with a single phase load. In plants where it is necessary to balance the phases, the phase loads can be balanced in various ways. For example, a three-phase/single-phase system requires a special transformer and a rebalancing system consisting of inductors and capacitors. For a three-phase/two-phase system, two single-phase transformers can be coupled (Scott) or one specific transformer (Leblanc circuit) can be used. Figure IV–40 shows a basic electric diagram of a power supply for an ozone generator operating at low frequency (50 or 60 Hz).

Balanced loads can also be achieved by spreading the ozone generators within a system over the phases or by splitting down the dielectric tubes within the generator into three equal sections. In the first configuration, the balance is achieved on the low voltage; in the second, the high voltage is balanced and a step-up three-phase/three-phase transformer and three high-voltage terminals on each generator are used.

For medium-frequency ozone generators the quality of the power supply depends on the converter switching speed, the way in which the converter is connected to the transformer and the transformer to the ozone generator, the waveforms used, and the use of materials and construction techniques to minimize electrical losses. A header rectifier allows the type of current selected by the manufacturer, particularly a single-phase supply, to be easily produced from three-phase mains. Figure IV–41 shows an electric diagram of a power supply (from three-phase mains, 50 or 60 Hz) for a medium-frequency ozone generator.

Correction of the power factor and compensation for the harmonic content. To make the best use of the power supply, the current and voltage must be in phase and the waveform must be as close to a sine wave as possible. Careful selection of

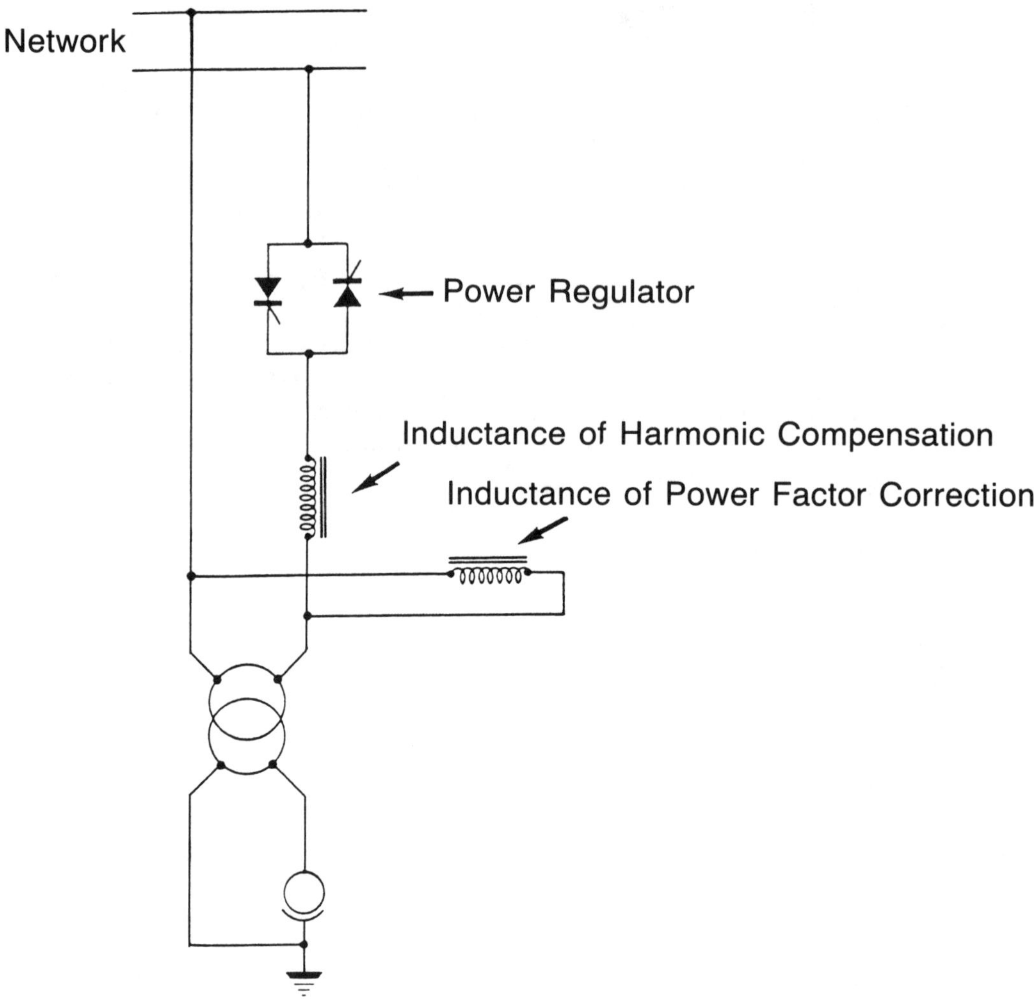

Figure IV–40 Basic Electric Diagram of a Power Supply for an Ozone Generator Working at Low Frequency (50 or 60 Hz)

the ozone generator capacitance can provide combined correction (power factor and harmonic) to optimize the complete power supply system. Figure IV–42 shows an electric diagram for combined correction.

The power factor requirements imposed by the electricity-generating authority generally required power factor correction on all ozone installations. There is as yet (1990) no international agreement on the definition of a harmonically "clean" network. However, manufacturers generally include filters in their equipment to avoid interfering with the electrical supply. The power factor is usually corrected at the power supply unit but can also be corrected at the source (ozone generator) or at both points.

Electrical protection of the dielectric tubes. There are two schools of thought on electrical protection of the dielectric tubes: one is to install fuses for each dielectric tube, or group of tubes, while the other is to provide general electronic protection. Some manufacturers provide both modes of protection. The consideration for using fuses is that the ozone generator may stay in service if one or a few dielectric tubes have been damaged; however, in this condition the remaining tubes take an extra load and this may create a chain reaction of additional dielectric problems. The consideration for general electronic protection is that the generator will always be operated at its design electrical conditions, but the generator must be immediately opened to replace a

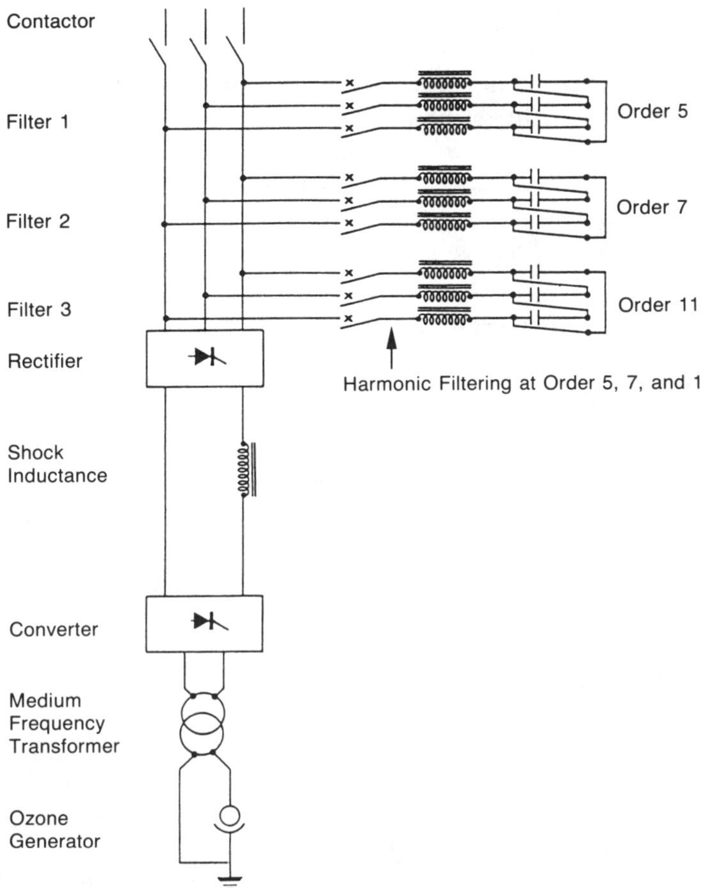

Figure IV–41 Electric Diagram of a Power Supply (From the Three-Phase Mains; 50 or 60 Hz) for a Medium-Frequency Ozone Generator

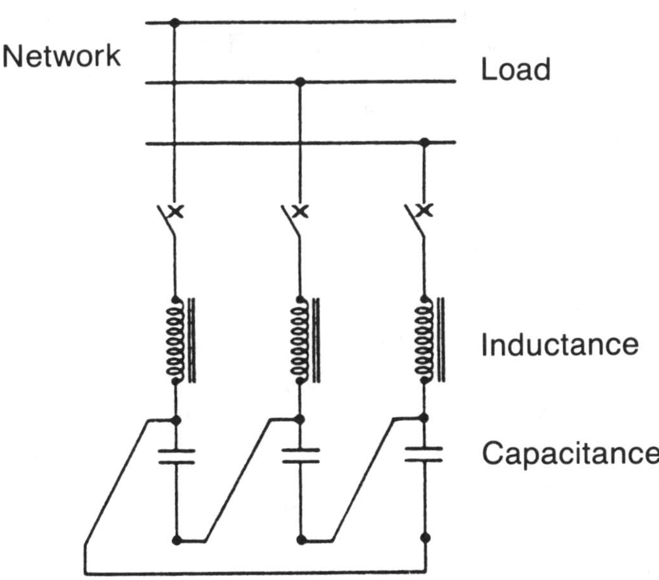

Figure IV–42 Electric Diagram for Power Factor Correction and Harmonic Content (First Order) Compensation

damaged dielectric tube. The decision process for choosing between fuses or general electrical protection should include a discussion among the owner and designer in consultation with the ozone generator manufacturer. Where fuses are installed they are designed to protect the stainless steel tube and/or the dielectric tube and keep the generator in service. It is noted that, when a fuse blows, it disconnects one or several tubes and modifies the capacitance of the generator. If the generator is a medium-frequency unit, it may cause problems with the frequency convertor such as the destruction of the thyristors if they are not well protected.

IV.C.4 Ozone Generator Cooling

The requirement for cooling. However good the performance of an ozone generator, its ozone production efficiency still remains mediocre and a large part of the energy consumed will be converted into heat which must be extracted (see sec. D.1, chapter II). The actual quantity of heat to be extracted will depend on the flow velocity of the gas through the discharge gap and, therefore, on the width of this gap and the power density. However, there will always be large quantities of heat to be extracted.

To minimize decomposition of the ozone, the common practice is to limit the temperature at the outlet from the generator to less than 50°C (122°F). Most of the cooling capacity must therefore be provided by the cooling fluid; in fact, the generator acts as an air/air or air/liquid heat exchanger, depending on the particular type of ozone generator.

Different cooling systems. As was stated earlier, water is generally used as the cooling agent in most ozone generation facilities. Depending on the circumstances, it may be readily available (filtered water, for example). The highest water temperature should be used to size the generator. A low inlet/outlet temperature differential ($\Delta\theta$) increases the generator efficiency. Generally, sufficient water flow is provided such that the difference between the inlet and outlet temperatures is 3°C (5.4°F). Systems with a $\Delta\theta$ up to 5.5°C (10°F) have been installed and successfully operated. It is mainly a question of economics.

Direct cooling with the process water (where the filtered water flows through the ozone generator and then returns to the drinking water treatment process) can be acceptable. If a direct system is used, particular care must be taken to select materials that will resist the constituents in the water (chlorides, dissolved metals, dissolved O_2, etc.) (see sec. IV.C.2).

If it is decided to use a closed cooling loop to avoid corrosion, there are two alternatives: (1) a closed circuit with a simple plate exchanger and one or several recirculation pumps, or (2) a closed circuit with a chiller. When the first system is used, the water on the closed-circuit side of the heat exchanger will be conditioned water, while that on the open side may either be raw (with an exchanger that can be dismantled and compatible material) or filtered water. One advantage of this system is that it protects the ozone generator against the corrosive constituents in the direct cooling system. It should also be noted that the additional exchanger increases the conditioned water temperature at the inlet to the ozone generator compared to the temperature of the direct application of water, and therefore reduces the generator efficiency. The overall efficiency of the system is further reduced by the power consumed by the circulation pump(s).

In instances where the cooling water temperature is quite high (80 to 95°F [26.6 to 35°C]), some manufacturers may optimize their ozone generation systems using a chiller. The cost reduction in energy consumed by the generator operated with colder water and/or the capital savings offered by using a smaller generator should be evaluated against the cost of the energy consumed by the chiller and its capital cost. Others may not use chillers at these same temperatures because for their ozone generators the chiller system is more expensive. In order to obtain the least expensive system overall, a capital cost plus life cycle cost bid evaluation must be used (see chapter VI). Generally, the chiller is designed to cool the water input to the ozone generator to between 5 and 15°C (41 and 59°F). The type and number of chillers used will depend on the number of generators and the operating conditions. Some have used a chiller for each generator, and others have provided fewer larger units. In all cases, turndown of the chiller should be matched with the turndown of the generator. The multiple chiller system offers more flexibility but may be more expensive.

Potential condensation problems must be addressed in the case of low cooling water temperature. Appropriate insulation will be provided for the cooling water circuit, including the ozone generator.

IV.C.5 Various Types of Ozone Generators

The low-frequency (50–60 Hz) horizontal-tube water-cooled ozone generator. This is currently the most common type of ozone generator. It is a derivative of the first American "Welsbach" tubular ozone generator introduced into Europe in the early 1950s (Chapsal 1989a).

The electrical discharge occurs between two concentric electrodes separated by a composite air/glass dielectric as shown in Figure IV–38. It is cooled by a direct or closed water loop.

The electrical power supply is provided by a single-phase or three-phase/two-phase transformer; the second version allows the network load to be balanced (see sec. IV.C.3). Dry transformers, cooled by natural air convection, have recently come into use, although local environmental conditions may make an oil-cooled transformer preferable. A series and a parallel inductor are sometimes installed to improve the power factor to close to 1.

The power supply to the ozone generator is generally controlled in one of two ways: either by varying the power supply voltage using an inductive controller or a voltage tap-changer with a step-up/step-down booster transformer or, again, a rotary transformer; or by varying the firing angle of the thyristors in a static power controller similar to those used for electric motor speed control.

The voltages applied to the dielectric tubes depend on the manufacturer (see sec. C.3). All tubes will withstand these values provided that:

- The glass has been selected correctly.
- The power supply system does not generate high voltage surges.
- The operating pressure in the ozone generator remains relatively constant around the setpoint.
- The dewpoint of the gas used is ≤−60°C (≤−76°F).
- The manufacturing tolerances applied ensure that each generator cell (i.e., the metal tube with its 1 or 2 [simple or double-acting generator] glass tubes) is identical to all the other basic cells in the generator (see sec. IV.C.2).

This is the simplest type of water-cooled tubular ozone generator. The ozone production efficiency is less sensitive to changes in the discharge gap than in the same type of generator operating at medium frequency. It therefore avoids the construction constraints introduced by reducing the gap (see sec. IV.C.3). In general, low-frequency generators achieve optimum performance at low operating pressure (7–14 psig [0.48 to 0.96 bar]).

The medium-frequency horizontal-tube water-cooled ozone generator. This type is a development of the Welsbach type generator described previously. The difference is that the frequency applied to the dielectric tubes ranges between 60 and 1000 Hz. This discussion addresses only the basic differences from the low-frequency unit.

The medium-frequency type first appeared toward the end of the 1960s. Its development is now facilitated to a considerable degree by the commercial availability of an increasing number of solid-state frequency converters with higher and higher power ratings (Brichant 1982). The power supply for a medium-frequency generator is generated by one or two transformers, depending on whether it is a simple or double-acting unit. Different manufacturers use various operating r.m.s. voltages between 9 and 14 kV and at frequencies unique to their specific units.

The frequency converter can be a series or parallel type centered around a single or double thyristor bridge. Some manufacturers use variable frequency through voltage or current variation.

The dielectric power density used in medium-frequency ozone generators can be more than twice that of a low-frequency unit (see sec. IV.C.3). For the same $\Delta\theta$, therefore, the flow of cooling water per dielectric through the generator must be increased (see sec. C.4). The total flow of water per gram of ozone produced may be similar to that for the low-frequency generator. Chillers are more frequently used for this type of generator to obtain optimum production efficiency. The cooling water flow must be developed relative to the $\Delta\theta$ and the maximum power applied to the dielectric tube (and therefore to the generator). This is also true for the low-frequency unit (see sec. IV.C.4).

Among the medium-frequency generators, the size of the discharge gap can be different according to the required performance. It should be noted that generators with a small discharge gap and consequent higher gas velocities in the gap are less sensitive to increases in water temperature.

The optimum operating pressure for the ozone generator depends on the discharge gap and is generally higher than that required for a low-frequency unit. In some cases it may be necessary to install a pressure control valve downstream of the generator, since the back-pressure generated by the water in the contact columns and the diffusion system might not be sufficient to maintain the required pressure in the ozone generator. In practical terms, this means that the overall least expensive system is obtained by an evaluation of capital cost plus life cycle cost for optimized medium- and low-frequency systems (see chapter VI).

The use of a medium-frequency supply allows the production capacity of each individual cell to be considerably increased. This offers the prospect of commercial ozone generators, operating at higher ozone production rates per generator (50 kg of O_3/h [about 2645 lb/day]). In addition, higher ozone concentration can be achieved. When the discharge gap is reduced, the ozone produced per energy unit consumed is slightly increased. However, it must be remembered that the minimum gap that can be achieved is limited by practical manufacturing tolerances and by costs (see sec. IV.C.3). The effects on the overall cost of the installation of using medium-frequency generators are discussed in chapter VI.

Other types of ozone generators.

Tube type ozone generators in which the glass dielectrics are in direct contact with the cooling water. These ozone generators, with horizontally or vertically assembled tubes, are based on the Van der Made ozone generator, commercially available since 1920. Each generator element consists of a closed glass dielectric tube surrounded with water, the latter serving as both cooling fluid and a low-voltage electrode, and an open concentric metal tube assembled inside the glass tube. This metal tube serves as a high-voltage electrode.

An illustration of this type of generator is shown in Figure IV–43. Air is drawn through the system by a vacuum or by means of a low-pressure blower. Ozone is produced as the carrier gas passes through the annular space enclosed between the metal tube and the glass tube. The negative pressure operation originally used for this system restricted the choice of the diffusion system. Nowadays, the ozone generators based on this principle usually function on low frequency, whereas before World War II, medium

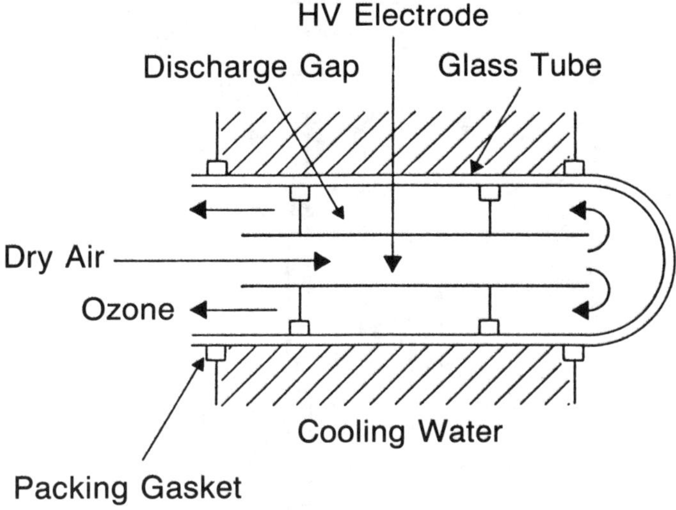

Source: "L'Ozonation des Eaux de Consommation," Trailigaz Document (1966).

Figure IV–43 Principle of the Van der Made Ozone Generator

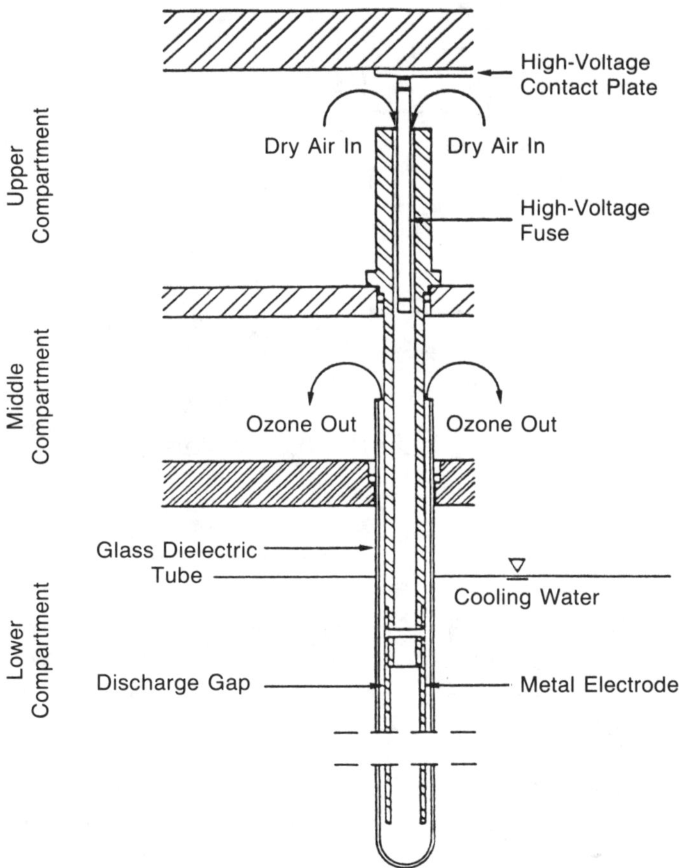

Source: Carlins and Clark (1982).

Figure IV–44 Vertical-Tube Water-Cooled Cell Derived from the Van der Made Technology

frequency (500 Hz) was used. This medium frequency was produced by electrical rotating machines. This unit was installed in large European water treatment plants. Figure IV–44 illustrates the principle of a vertical cell derived from the Van der Made technology.

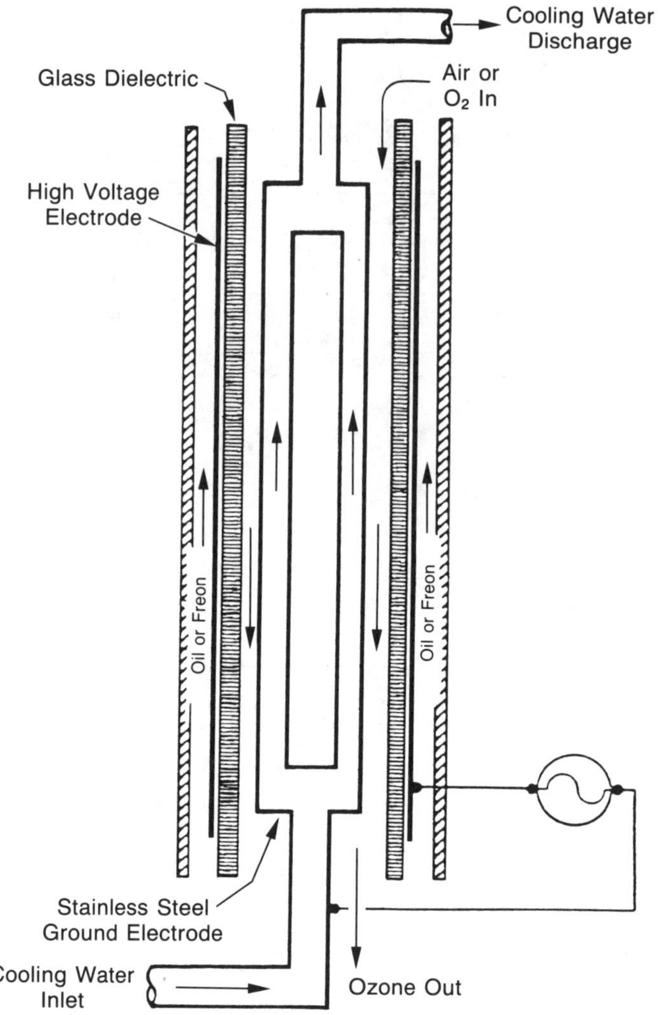

Source: Carlins and Clark (1982).

Figure IV–45 High-Frequency Vertical-Tube Double-Cooled Generator

The high-frequency vertical-tube double-cooled generator. This kind of generator usually consists of a vertically assembled generator element. The basic component of the generator is a cell made up of a low-voltage stainless steel electrode cooled by water circulating internally and a silver-coated glass dielectric tube cooled externally by oil or some other fluid (e.g., freon), in turn cooled by a shell and tube heat exchanger. The carrier air passing through the discharge space between the two electrodes is exposed to a fixed voltage of 10 kV, controlled by varying the frequency between 60 and 2000 Hz using a frequency converter (Miller et al. 1978). Operating pressure may be up to 1.4 bar of relative pressure (20 psig). Figure IV–45 illustrates a double cooled elementary ozone generator cell.

Otto-type water-cooled plate ozone generator. This was the first ozone generator to be used in a large-scale installation (in 1906, for the disinfection of water in Nice in southeast France). An elementary generator of this type is illustrated in Figure IV–46.

This apparatus consists of:

- A hollow metal plate with parallel external sides, serving as a shell electrode. Cooling water circulates in this plate.

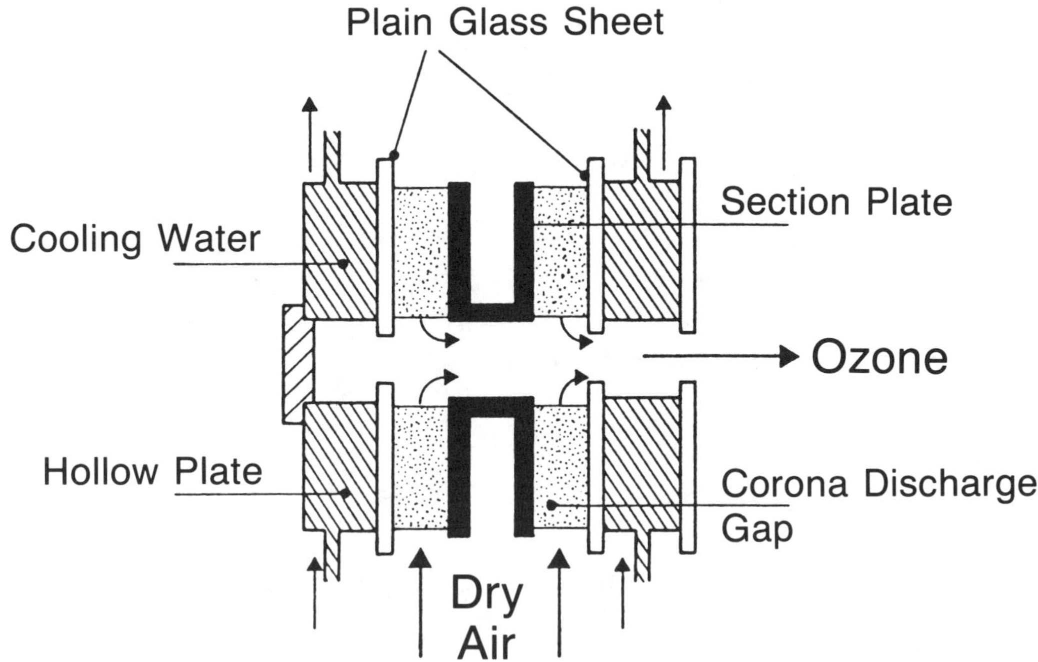

Source: "L'Ozonation des Eaux de Consommation," Trailigaz Document (1966).

Figure IV–46 Principle of OTTO'S Ozone Generator

- A thin glass plate, constituting the dielectric, closely packed beside the shell electrode.
- A high-voltage electrode in the shape of a thin section plate.
- A set of wedges creating the space between the high voltage electrode and the shell electrode.

Both the electrodes and the dielectric have a square cross-section and a central orifice.

Several of these elementary generators are hung vertically, tightly packed in a parallelepiped enclosure. They make up a solid unit through which their central orifices form a channel of communication with the discharge spaces.

The carrier gas to be ozonized is admitted to the airtight enclosure at a pressure of around 1 absolute bar (14.5 psia). It penetrates the discharge space via the periphery and reaches the center of the electrodes, where it is channeled for evacuation. The cooling circuit of the shell electrode plates consists of two collectors, for water inlet and discharge, respectively. Each one is connected to the plates via an individual piping assembly. These units were originally designed for operating at low frequency at a maximum voltage of 20 kV. It is to be noted, however, that these ozone generators have also been known to operate at medium frequency (150 to 500 Hz) obtained by using electrical rotating machines.

As in the Van der Made generator, the gas is drawn through the system by a vacuum, which limits the choice of diffusion systems. New installation of this type of equipment is no longer available in France or North America, but many installations in France, Switzerland, and Quebec, Canada, are still in operation; some have been operating for as long as 30 years.

Lowther-type air-cooled plate ozone generator. This kind of ozone generator operates on a dielectric composed of a glass or ceramic-covered steel electrode separated from a second identical electrode by silicon bracing that provides the discharge space as illustrated in Figure IV–47. Each of the electrodes is coupled with an aluminum

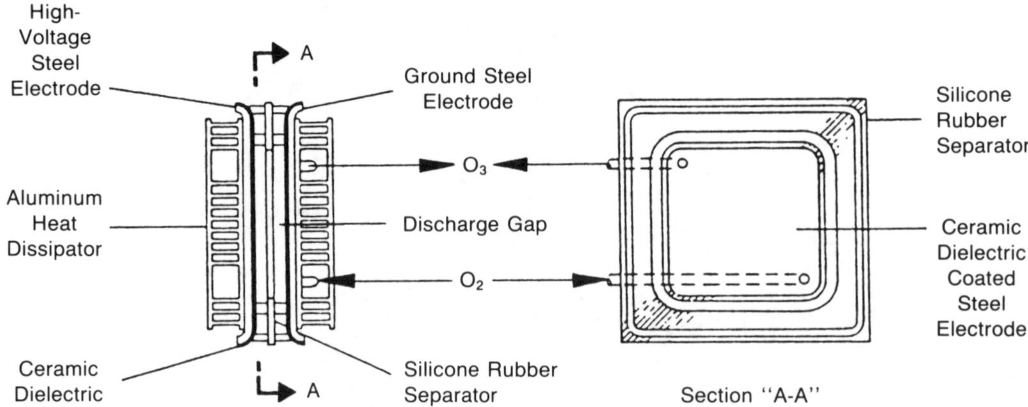

Source: Miller et al. (1978).

Figure IV–47 Lowther-Type Air-Cooled Plate Ozone Generator

ventilator designed to cool the ozonated gas with ambient air. The upper effective pressure in the cell is about 1 bar (14.5 psi). This type of ozone generator runs on 9 kV electric current at a frequency of 2000 Hz.

The air cooling systems have often been found insufficient to ensure optimum ozone generator production. These generators are no longer used in any major water treatment plants in the United States (MWDSC 1988) and have never been used in France.

Several manufacturers and licensees have promoted this technology. Units of a major U.S. manufacturer have experienced significant maintenance problems at a number of North American water and wastewater applications (Taylor et al. 1989). The Japanese licensee of this technology claims a large number of successful applications in Japan applying modifications of this unit (Yokomi et al. 1989). Also, the smaller units (1 to 25 lb O_3/day [18.9 to 472.5 g O_3/h]) of another U.S. manufacturer have operated reliably in industrial applications.

IV.D DESIGN OF CONTACT CHAMBERS AND DIFFUSION SYSTEMS

IV.D.1 Overall Considerations

Ozone, which is always at a relatively low concentration in the carrier gas, must be transferred from the gas phase to the liquid phase to perform its intended function(s) in the treatment of water. Economics normally require that the design of a dissolution unit maximize ozone transfer for the intended purpose and/or reaction.

Dissolution/contacting units to provide ozone for various ozonation applications will vary in terms of type, design, operating conditions, the specific functions of ozone at the points of application, and the design of the contact chambers. In some cases, even the type of ozone generator will dictate the contactor type. For example, some commercial ozone generators are more appropriate for use in conjunction with subatmospheric pressure contactors (aspirating turbines or injectors) than with pressurized contacting systems. Nevertheless, it is more logical that the purposes of ozonation lead to the choice of the type of contactor and, consequently, to the choice of ozone generator and feed gas.

The type of reaction involving ozone may determine the type of contactor selected. The mass transfer–limited reaction normally requires the transfer of the maximum amount of ozone in the shortest time for economical construction and operation. On the other hand, for a reaction-limited reaction aimed, for example, at maintaining an ozone residual for a specific length of time, a high ozone driving force is not as essential compared with the necessity of having a homogeneous ozone residual in the total volume of treated water.

When the reactions occurring in the liquid phase are relatively slow with respect to the ozone transfer velocity from the gas phase to the liquid, the design of the contactor could be different. If the economics of both ozonized gas concentration and contactor design are important, the choice of both ozone concentration and contactor size and geometry should be performed in concert. Diffusion of high concentrations of ozone (say 6 to 8 percent [87.6 to 116.8 g/m^3 NTP]) requires careful design of the contactor to achieve a uniform distribution of ozonated gas in the mass of water.

IV.D.2 Ozone Dissolution Techniques

A number of techniques are available for dissolution of ozone in the liquid to be treated. These include the following:

1. Conventional fine bubble diffusion
2. Turbine mixers
   - Positive gas pressure
   - Negative gas pressure (aspirating type)
3. Injectors
   - Positive gas pressure
   - Negative gas pressure
4. Packed columns
5. Spray chambers
6. Recently developed techniques
   - Deep U tube
   - Sweeping porous plate diffuser contactor
   - Submerged static radial turbine contactor

These techniques are discussed in the following subsections.

Conventional fine bubble diffusion. The fine bubble diffuser contactor is the most widely used of the available ozone transfer systems because it is operated without the further addition of energy in addition to that of initial gas compression. The technology is not new; in addition to its use in ozonation systems, fine bubble diffusers have been used for many years as a means of oxygen transfer in both water treatment and wastewater treatment aeration basins. Of the 40 potable water ozonation facilities constructed in the United States, 34 are known to have been equipped with fine bubble diffuser contact basins (Robson et al. 1988). In Europe, where ozonation has a much longer history of use than in the United States, bubble diffuser systems are also the most prevalent contactor design. A wide range of configurations have been applied for fine bubble ozone dissolution units ranging from single stage tanks to the six-stage units. Reasons for selection of the number and type of contactor stages are based on the type of application of ozone.

Several different contactor configurations, baffle arrangements, and flow patterns are used with conventional bubble diffuser systems. Figures IV–48 and IV–49 illustrate fine bubble diffuser designs of U.S. water treatment plants.

In the United States, most ozone contactors used in water treatment have previously been either two or three oxidation chambers (stages) per contactor, depending on the purpose of ozonation. The general "rules of thumb" are:

- If the objective for ozonation is a reaction that is mass transfer–limited, such as iron and manganese oxidation, a minimum of two chambers per contactor is recommended. These reactions can be quite rapid and the optimum (minimum) contact times should be determined from pilot plant testing (Renner et al. 1988).

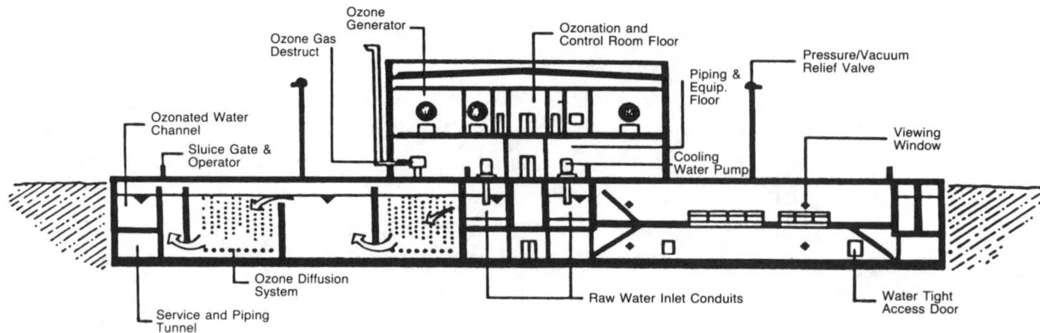

Source: MWDSC (1988).

Figure IV–48 Ozone Contactor at the Los Angeles Aqueduct Filtration Plant, California

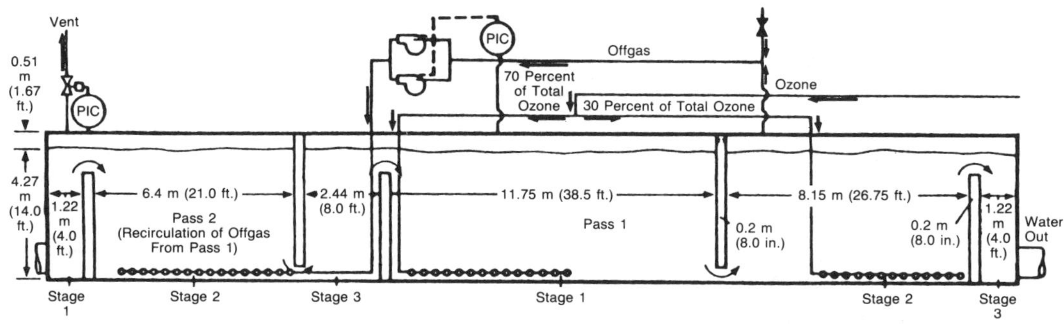

Source: LePage (1981).

Figure IV–49 Ozone Contactor at Monroe Water Works, Monroe, Michigan

- If the objective for ozonation is reaction rate–limited (e.g., primary disinfection) rather than mass transfer–limited, three or more ozone dissolution stages should be used so that the flow through the contactor will be more representative of plug flow in that short circuiting will be minimized (Renner et al. 1988). Homogeneous distribution of ozonized gas in the liquid is required.
- Also, in complying with C · t requirements (see sec. IV.A.1), the t used in the calculation increases as the reactor hydraulics approach plug flow. Thus, the closer a reactor design is to plug flow, the smaller it can be to achieve a given t.

In summary, as more chambers are used in the contactor or reaction tank, the flow regime will approach that of a plug flow reactor. Because a plug flow reactor requires a shorter contact time than a continuously stirred tank reactor (CSTR) to achieve an equivalent degree of reaction efficiency, a smaller contactor is required when more baffles are used. In light of the C · t requirements proposed in the current SWTR Guidance Manual, the more the ozone contactor resembles a plug flow reactor, the greater will be the value of t_{10} (the detention time necessary for 10 percent of the tracer mass to exit the contactor). However, there are practical limits to the number of chambers. The more extensive the baffling, the greater the concrete requirements and the capital cost of the contactor, and the greater the headloss through the contactor. At some point, the cost of adding baffles outweighs the benefits, and other contacting approaches should be evaluated.

The majority of the conventional bubble diffuser contactors in both the United States and Europe employ two or three chambers of contacting and reaction. However, some plants, such as the Neuilly-sur-Marne, Choisy-le-Roi, and Clairfont plants in France, and the plant in Bay City, Michigan, use four oxidation chambers per contactor. The Tucson, Arizona plant uses six stages.

Advantages/disadvantages of conventional bubble diffuser systems. One of the most apparent advantages of using the conventional bubble diffuser system for ozonation is that it is a widely used technology with many years of demonstrated performance. Additionally, it requires no additional energy input for effective operation. Ozonized gas is normally discharged from the ozone generator at a pressure of 10 to 15 psig (0.69 to 1.03 bar), which is sufficient to overcome the hydrostatic head plus the head loss due to gas distribution piping and diffusers.

Some further advantages and the disadvantages of this type of system are summarized below:

Advantages

- No moving parts: Because there are no moving parts associated with this type of diffuser system, maintenance requirements are minimal.
- Effective means of ozone transfer: Ozone transfer efficiencies greater than 90 percent are generally associated with bubble diffuser systems. At the Los Angeles Aqueduct Filtration Plant (LAAFP), 96 to 99 percent transfer efficiencies were measured in 17- to 20-ft (5.18- to 6.10-m)–deep contact basins. However, it should be pointed out that these transfer efficiencies are achieved with high ozone concentrations (~6 percent [87.6 g/m^3 NTP]) in the gas applied for dissolution (Stolarik and Christie 1988).
- Scale-up is not a major issue. Because the conventional fine bubble diffusers are widely used and are based on well-known technology, the scale-up of a demonstration-size contactor to full-scale (approximately 600 mgd [94,635 m^3/h]) will not be a difficult design issue. LAAFP prototype test results at a flow of 200 gpm (45.4 m^3/h) showed similar performance to its 600-mgd (94,635 m^3/h) plant. Scale-up of fine bubble diffusion systems would simply mean building a larger contactor and adding more diffusers to it.

  Scale-up may become more of an issue in the future, in light of $C \cdot t$ requirements and the need for tracer studies.
- Low hydraulic loss: Designs can be structured to accommodate a comparatively low headloss through the tanks.

Disadvantages

- Relatively deep (18- to 24-ft [5.49- to 7.31-m]) contact basins are required for effective ozone transfer.
- Clogging of diffuser pores is possible, particularly when water flows and applied ozone dosages are intermittent or when iron and manganese oxidation is required. Depending on the quality of the water, clogging of the diffuser pores by metal precipitates is a potential problem. At the ozonation facility at the Hackensack Water Company's Haworth Plant in Haworth, New Jersey, aspirating turbines rather than bubble diffusers are used. The water has a high iron and manganese content and the designers were concerned about diffuser plugging and the associated maintenance requirements.
- Vertical channeling of bubbles is possible. Depending on the gas flow through the diffusers and the diffuser spacing, vertical channeling of bubbles can occur, adversely affecting gas-liquid contact.

Rod-type diffusers. Rod or tubular ozone diffusers supported by stainless steel headers are frequently used in fine bubble ozonation contactors. Figure IV–50 shows a typical rod diffuser, while Figure IV–51 illustrates a rod diffuser ozone dissolution system. A claimed advantage of tubular diffusers is an ability to rotate them when

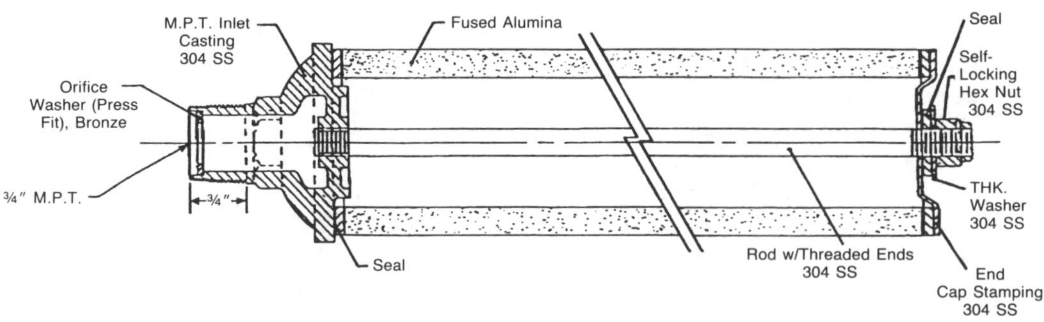

Source: Refractron Technologies Corp. (1990).

Figure IV–50 Detail of a Tube/Rod–Type Ozone Diffuser

Source: Compagnie Générale des Eaux.

Figure IV–51 Rod Diffuser Ozone Dissolution System

surface pores become plugged. European experience is that rotation of the tubular diffusers is extremely expensive due to labor costs, risk of breakage, and the necessity to change the seals of the rods. The units incorporate a number of seals as well as components that are subject to corrosion. The trend in Europe over the past 10 years has been away from the tubular units (Langlais 1989). Gas flow characteristics for the diffusers should be obtained from equipment vendors to assure good gas distribution over the floor of the contact chamber and thus minimize vertical channelling and possible short-circuiting of the liquid under treatment. Also, gas flows per diffuser should be appropriate to obtain the desired 2- to 3-mm bubble size.

Dome- or disc-type diffusers. Dome- or disc-type diffusers are the common alternative to tubular ozone diffusers. Figure IV–52 shows a typical disc-type diffuser, while Figure IV–53 illustrates a disc diffuser dissolution system. Gas flow characteristics of the typical diffusers should be obtained from equipment vendors for the same reasons discussed for rod-type diffusers.

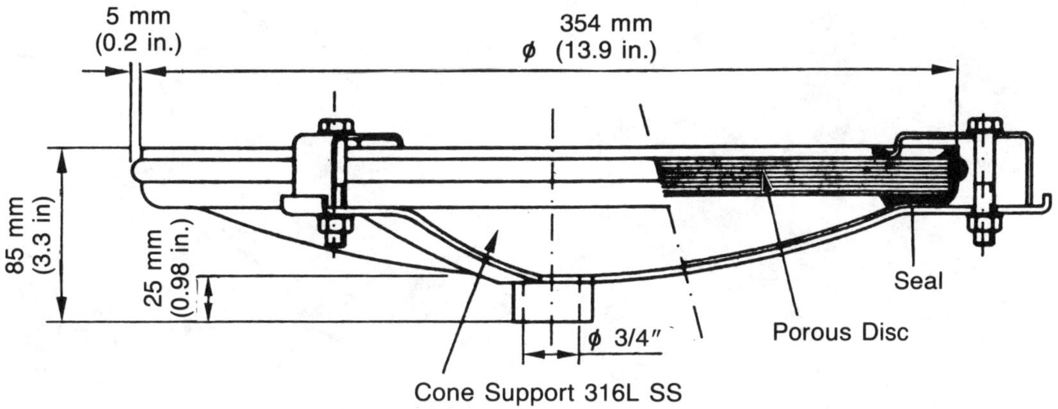

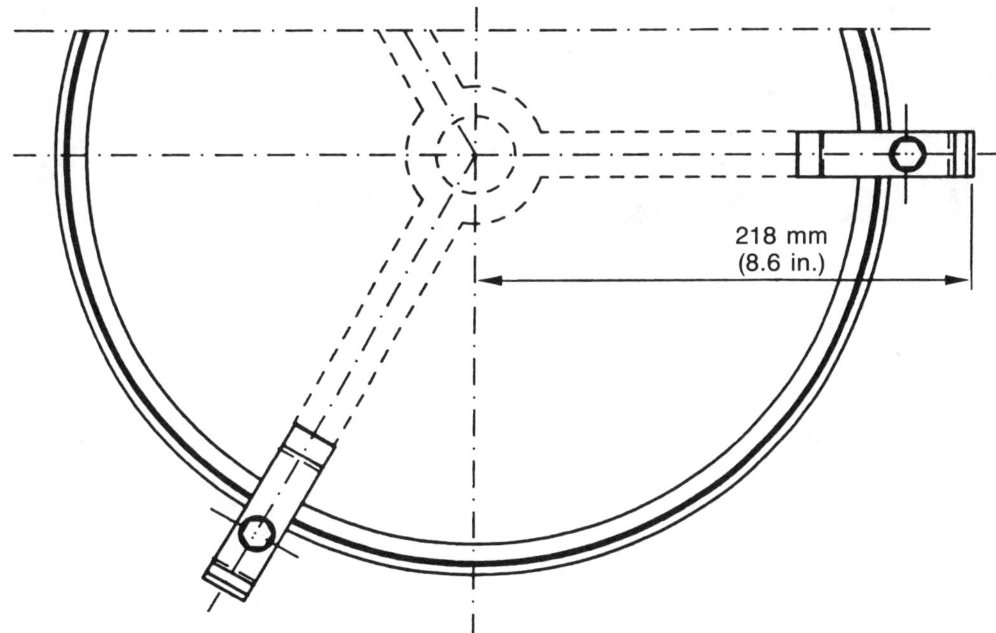

Source: Trailigaz (1990).

Figure IV–52 Detail of a Disc-Type Ozone Diffuser

Turbine contactors. Turbine contacting systems are also used quite often in water treatment plants around the world. Aspirating turbines, which draw ozonized gas into the contactor as well as mixing the ozone with the water in the contactor, are used in a number of U.S. and European plants. The most commonly used turbine types in Europe designed for use with ozone are manufactured by Frings, Asea Brown Boveri (ABB), and Trailigaz (aspirating and submerged turbines), although other manufacturers produce turbines for gas contacting as well. Figure IV–54 illustrates a typical submerged aspirating turbine.

As opposed to bubble diffusion contactors, use of turbines requires an input of energy to run the turbine. Richard (1986) estimates that rotating turbine contactors require an energy expenditure as high as ten times that of bubble diffusers. The reported energy requirement of the Frings turbine at the Tailfer, Belgium plant is 5.5 kWh/lb

Source: Compagnie Générale des Eaux.

Figure IV–53 Disc Diffuser Ozone Dissolution System

(12.1 Wh/g) applied ozone (MWDSC 1988), while other Frings turbines in Europe reportedly require energy expenditures of 2.3 to 2.7 kWh/lb (5.07 to 5.95 Wh/g) (Blankenfeld 1988).

A principal design consideration for turbine contactors is the requirement to maintain a constant gas flow rate, regardless of the water flow rate through the contactor. This requirement presents two issues: (1) the energy requirement to operate the turbine is the same regardless of the water flow rate; and (2) transfer efficiency decreases as the gas-to-liquid ratio increases. The first issue has economic implications in that the cost per unit volume of treated water increases as the flow through the plant decreases. The second issue also has economic implications since ozone is being generated and is not being utilized in the contactor. During periods of lower water flow, the same gas flow must be maintained, which means that in order to maintain the desired applied ozone dose, the concentration of gas-phase ozone must be reduced. Thus, there is a decreased driving force for transfer efficiency. Some manufacturers suggest reducing the gas flow. To reduce energy consumption, they suggest oversizing the turbine motor or providing a variable-speed motor (Langlais 1989).

Despite the drawbacks, the advantages arising from use of turbine contactors led to their use at the Hackensack Water Company's Haworth Water Treatment Plant in Haworth, New Jersey. The Haworth contactor (Figure IV–55) was designed specifically to provide rapid mixing for coagulant chemicals (alum and polymer), to promote solids removal by flotation, and to allow simultaneous oxidation of iron and manganese by ozone. The contactor is designed with special scum troughs and the downstream reaction basin is designed with skimmers for algae and foam removal. Another reason for the selection of turbine contacting over bubble diffusion was the resistance of turbines to clogging in water containing high concentrations of iron and manganese.

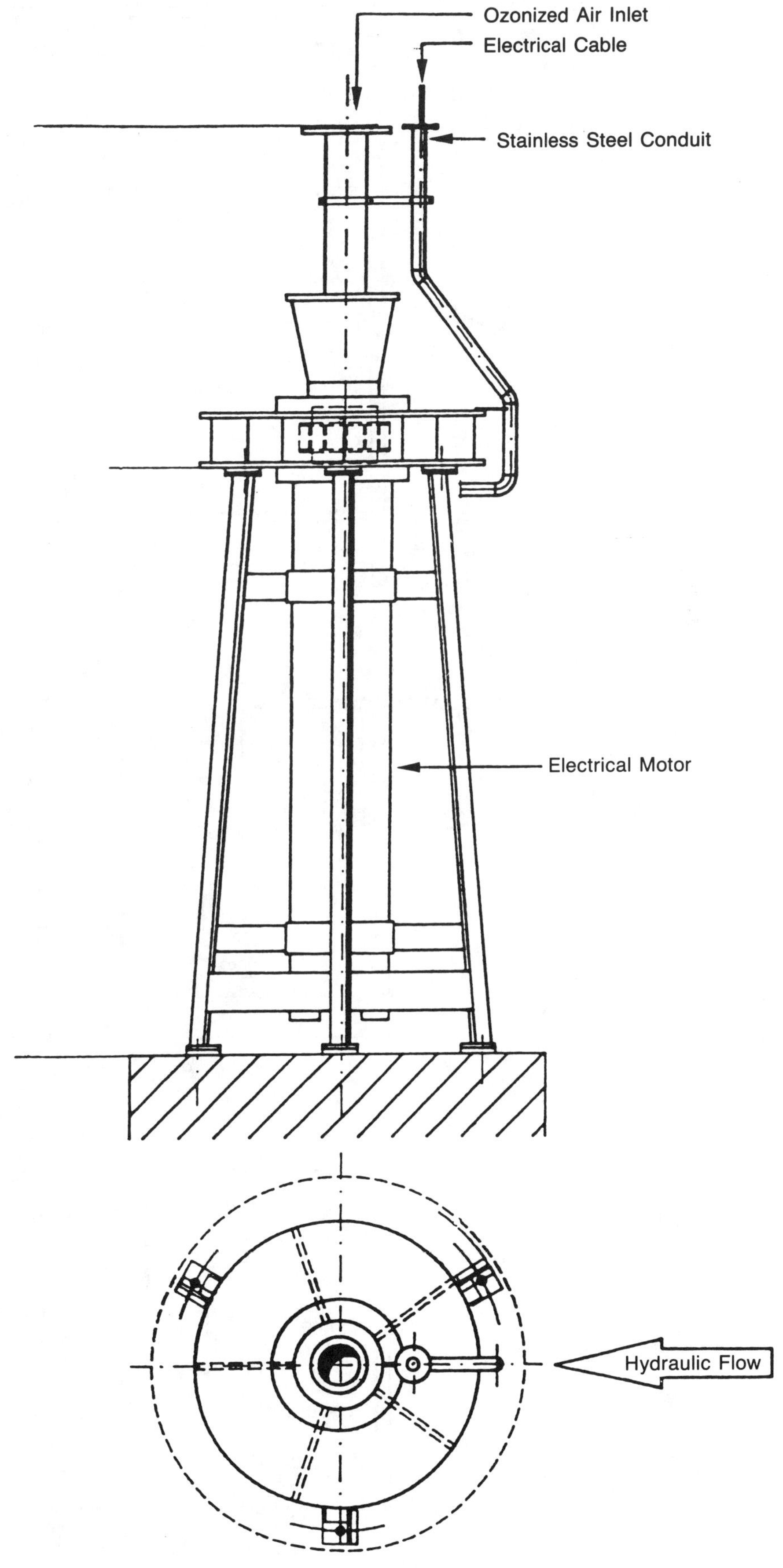

Source: Trailigaz (1990).

Figure IV–54 A Typical Submerged Aspirating Turbine

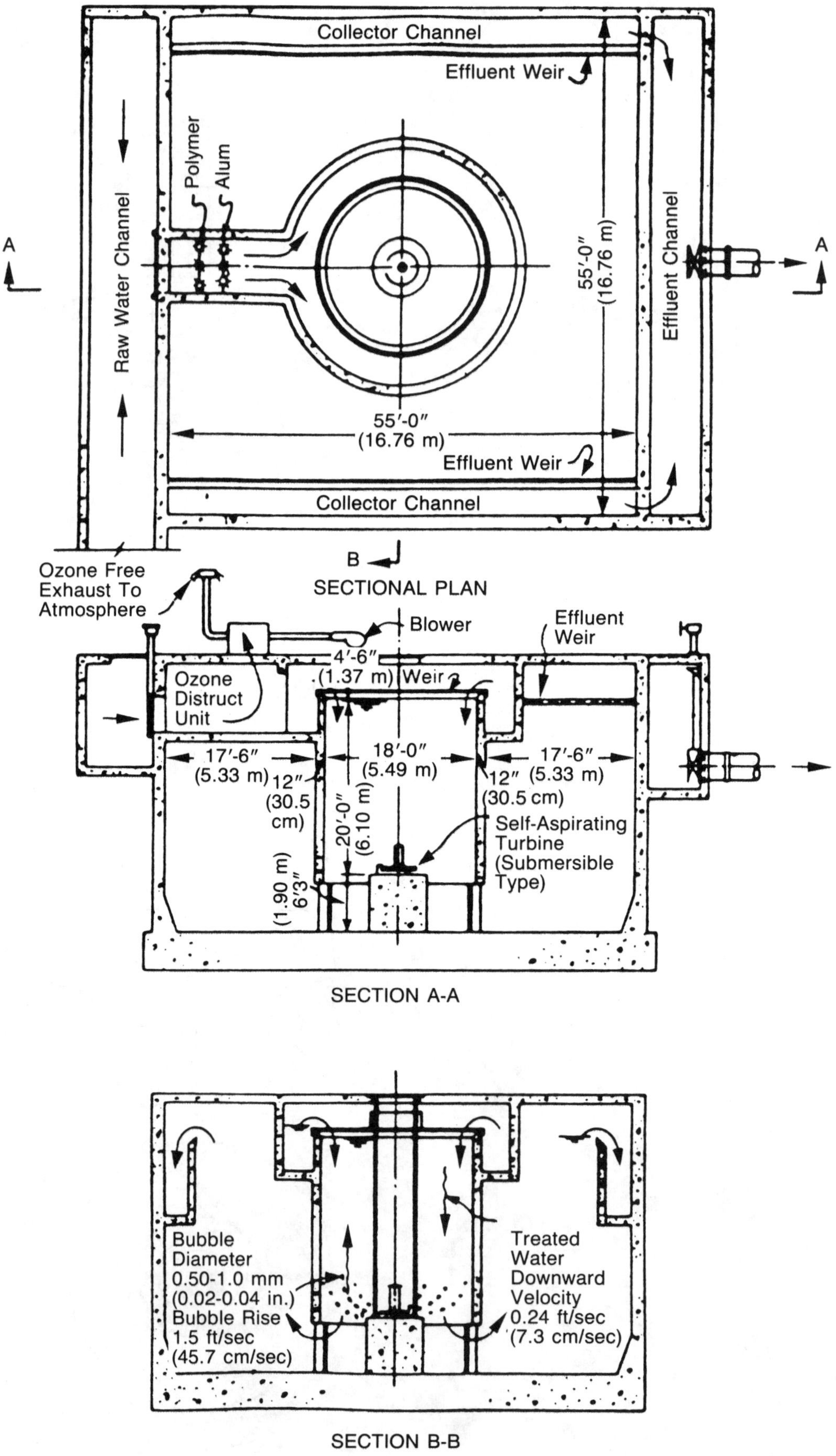

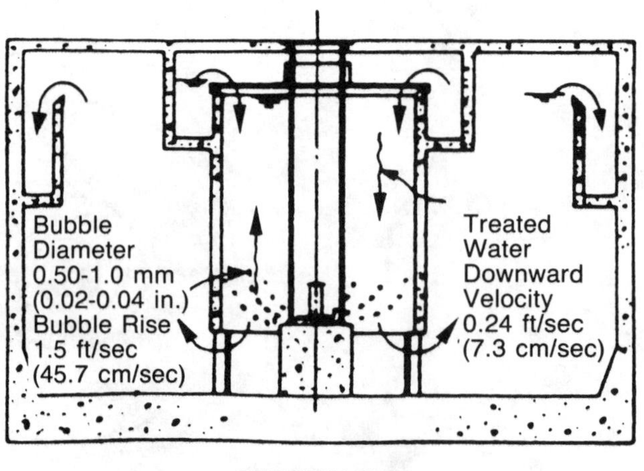

SECTION B-B

Source: Schwartz (1990).

Figure IV–55 Haworth, New Jersey Aspirating Turbine Contactor

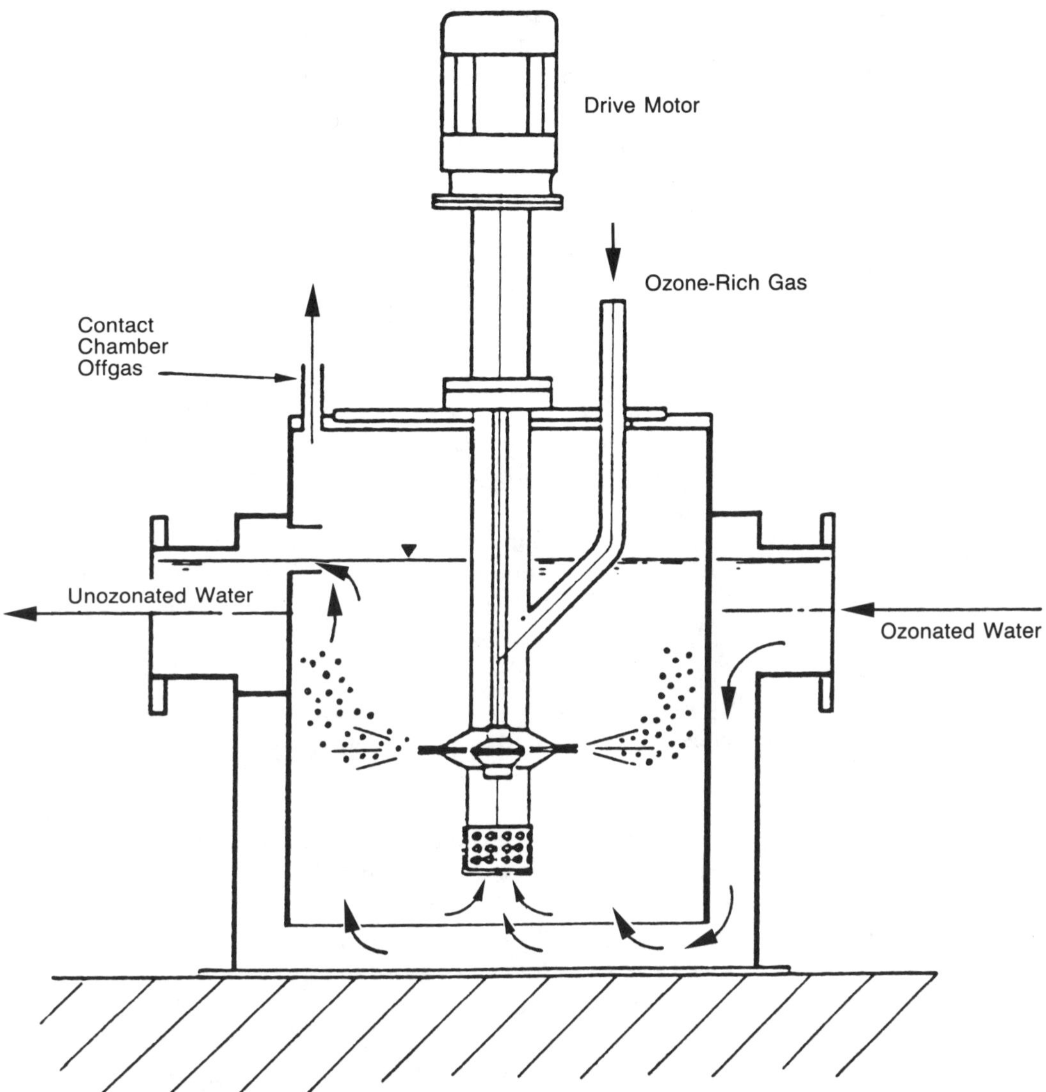

Figure IV–56 Kerag Turbine-Type Ozone Contactor

A common use of aspirating turbines is in multiple-chamber contactors to draw offgases from second, third, and fourth chambers into water having high ozone demand in the first chamber. In this way, ozone in the offgas is utilized to help satisfy ozone demand and the concentration of ozone in the offgas entering the destruct system is minimized. Figure IV–56 illustrates a type of aspirating turbine that has been installed and operated at North Andover, Massachusetts and at Sturgeon Bay, Wisconsin. Although not specifically designed for flotation, the combined use of a turbine contactor for rapid mixing of coagulants and ozone dissolution is part of the so-called "Mulheim Process" or "Essen Process." As part of this treatment technique, a turbine in the preozonation contact tank draws offgas from other ozonation stages into the basin as well as mixing polyaluminum chloride coagulant. Ozonized gas coming from the generator can also be added to the gas stream being drawn into the tank. This process is used at several plants treating Ruhr River water, including the Essen, Dohne, and Styrum West plants in the Federal Republic of Germany (Sontheimer 1978; Heilker 1979; Jekel 1983; Blankenfeld 1988).

Advantages of using turbine contactors include:

- Transfer of ozone from the gas to the liquid phase can be enhanced by the small bubble size resulting from the high turbulence and high shear produced by turbines.
- Turbine contactors can achieve high transfer efficiencies without the requirement for 18 to 24 ft (5.49 to 7.31 m) of contactor depth.
- Aspirating type turbines can be used to draw offgases from other contacting chambers into an upstream contactor to satisfy ozone demand and decrease the load on the ozone destruction system.
- Turbines can be used to achieve both ozone contacting and rapid mixing in one chamber.
- The application of turbines reduces concerns of clogging of the diffusion devices (i.e. for waters with high iron and/or manganese levels).
- Scale-up of turbine contactors from experimental scale to full scale does not present problems. Transfer efficiency results obtained in demonstration-scale facilities should reflect expected full-scale results at the same ozone gas phase concentrations and gas-to-liquid ratios (Conway 1988). The turbine contactors designed for the Haworth plant were based on data from extensive pilot studies conducted from 1980 to 1985. Based on pilot study results, anticipated transfer efficiency in the full-scale contactors is 92 percent, with turbine submergence depth in excess of 20 ft (6.1 m). Full-scale plant data were not available in early 1990 to verify expected results.

The disadvantages associated with the use of turbine contactors include the following:

- Unlike bubble diffuser contactors, turbine contactors require an input of energy. The energy requirement to run the turbine is the same regardless of fluctuations in the water flow rate through the contactor.
- A constant gas flow rate must be maintained, regardless of fluctuations in water flow rate, resulting in decreased transfer efficiency.
- Maintenance requirements for a turbine with both turbine and drive motor totally submerged may present a problem.

Injectors and static mixers. Ozonized gas can be injected into the liquid stream under positive or negative pressure. Selection of the type of injection will depend upon a number of factors, including the hydraulic head available, the ozonized gas pressure available, and downstream treatment units. Injection of gas may require downstream static mixing depending on the features of the injection method.

To date, the use of injectors and/or static mixers is rare in the United States but is relatively common in installations in Europe and Canada. The only known U.S. application is the positive pressure ozonized gas injection application at the Mahoning Valley wastewater treatment plant near Youngstown, Ohio (Jain et al. 1979).

In France, an example of a large facility using this dissolution device is the Choisy-le-Roi water treatment plant (33,330 m^3/h [about 211 mgd]). At the preozonation step, three contact columns have been built in parallel. Two of them are equipped with "emulsifiers," i.e., injectors sited at the top of vertical dissolution tubes (Langlais 1982). Each of these contactors (theoretical contact time at nominal flow: 6.7 min) is capable of treating between 5417 m^3/h (about 34.3 mgd) and 12,500 m^3/h (about 79.2 mgd). A canal distributes the raw water along the twelve emulsifiers in each contact column. Each of those emulsifiers receives between 930 m^3/h and 1200 m^3/h (5.9 to 7.6 mgd) (Langlais and Bablon 1990). Depending on the raw water flow, the number of emulsifiers put into service varies between four and twelve. Figure IV–57 shows a simplified cross-section of a typical contact column using emulsifiers as the dissolution system. At Choisy-le-Roi, for each emulsifier the gas/water ratio can vary from 3 to 10 percent with an optimum running point at about 8 percent for a water head of about 450 mm (17.7

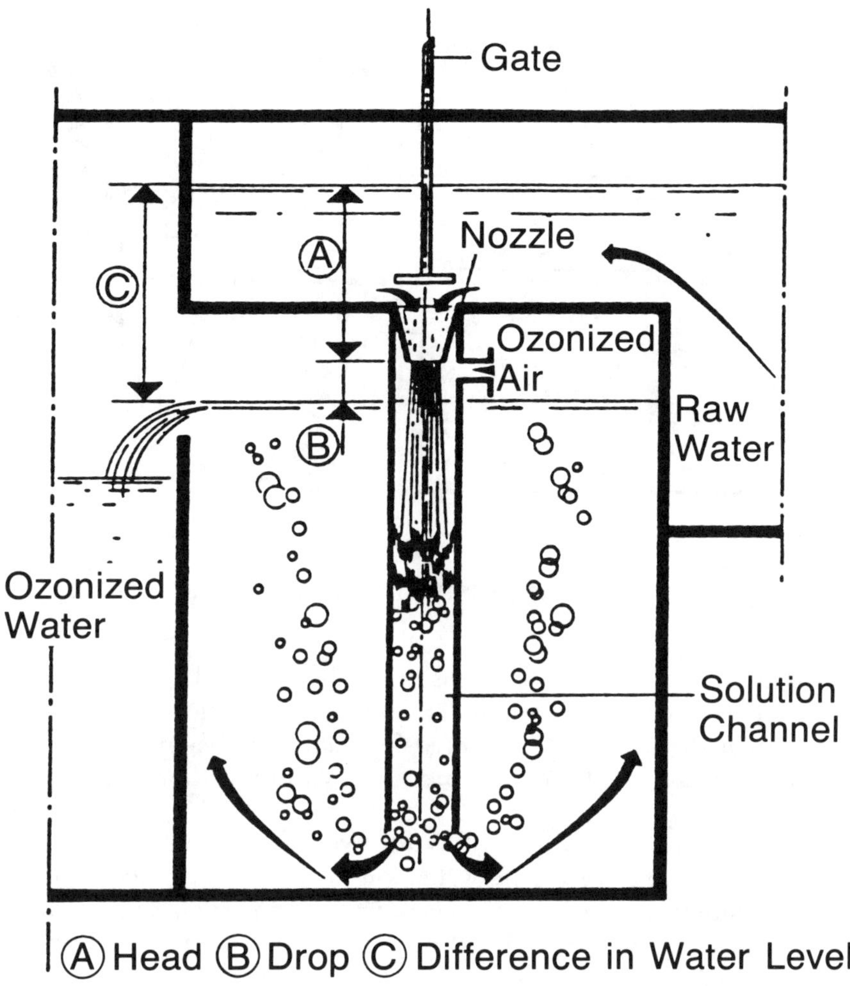

Source: Gerval and Bablon (1983).

Figure IV–57 Schematic Diagram of a Typical Preozonation Contactor Using an Emulsifier as Dissolution System

in.). The concentration of the ozonated air is about 16 g/m^3 NTP (1.2 percent by weight) and the ozone dosage is about 1 mg O_3/L. Under these conditions the transfer efficiency is very high (close to 100 percent) (Perrot 1989).

Negative pressure injection of ozone usually involves water flowing rapidly through a pipe past a small orifice, creating a Venturi effect (partial vacuum) that pulls the gas through the orifice into the water stream. Often, a sidestream of the total plant flow is pumped at elevated pressure into a separate line where the orifice is located in order to produce a better partial vacuum. The entire gas stream is injected into the sidestream, resulting in very high ozone concentrations, in conformance with Henry's Law, and then the sidestream is very quickly combined with the rest of the plant flow in a large pipe that serves as the contactor/reactor. Injection may or may not be followed by a static mixer within the pipeline. Either method results in very short contact times, on the order of a few seconds. Thus, injection with or without static mixing may need to be followed by a reaction tank to provide additional contact time.

The advantages of injection and static mixing include the following:

- Both injection and static mixing are simple systems with very low maintenance requirements due to the lack of moving parts.

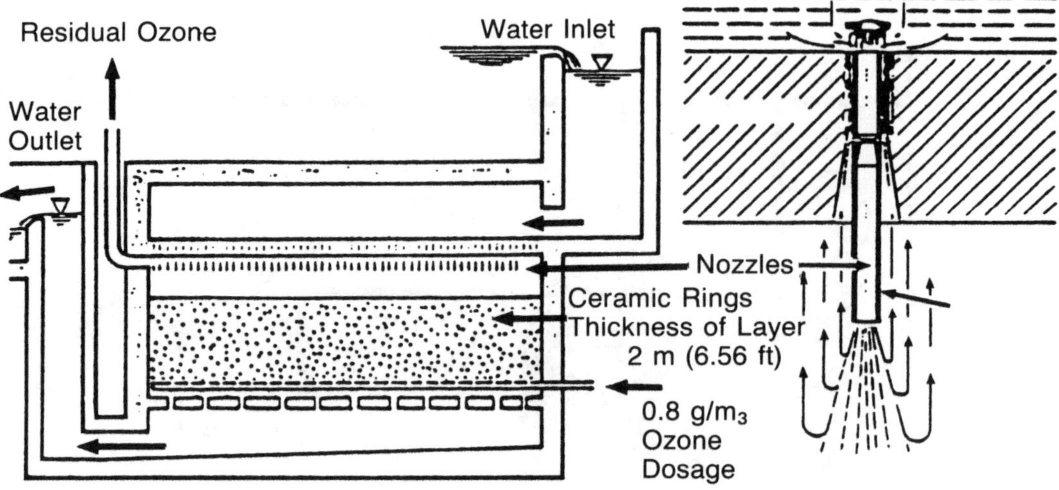

Source: Elsenhans (1989).

Figure IV–58 Schematic of the Packed Bed Ozone Contactor Used at Sipplingen (Germany)

- Good mixing and mass transfer can be achieved.
- Depending on the desired detention time, reaction tank depths and volumes can be smaller than those required for bubble diffusion.

Disadvantages of injection contacting with or without static mixing are described below:

- The major disadvantage of static mixers is head loss, which must be overcome by pumping. However, an evaluation of the energy costs associated with pumping requirements must be weighed against the capital and operating costs of other contacting methods in order to determine the most advantageous alternative. Depending on the hydraulic head available in an existing facility, pumping may be necessary in any case.
- The systems have turndown capability limited to the capacity of the specific injection device, and separate trains would have to be placed in or taken out of service depending on the flow rate of the liquid under treatment.

Packed columns. Packed columns have been used as ozone contactors in only a limited number of water treatment applications. In the United States, packed columns have not yet been used for municipal drinking water or wastewater ozone contacting, although the first installation (Los Angeles Department of Water and Power Ozone/Hydrogen Peroxide Demonstration Plant in Los Angeles, California) is currently under construction. In Germany and Switzerland, packed columns are used at several treatment facilities, most of which have treatment capacities less than 1 mgd (157.7 m^3/h). However, one European installation using packed columns, the Sipplingen Berg Plant in the Federal Republic of Germany, has a capacity of 200 mgd (about 31,540 m^3/h). Figure IV–58 shows the packed bed ozone contactor used in this plant.

Two general approaches have been taken in a packed column contactor/reactor design. The first is to add the ozone to the water ahead of the packed column and utilize the packed column as a reaction chamber. The second is to inject the ozone directly into the packed column and utilize the column as both a contactor/mixer and a reaction chamber.

Advantages of using packed columns include the following:

- Flow regime with packing is more representative of a plug flow reactor than is the flow regime without packing. However, in packed bed columns, the packing

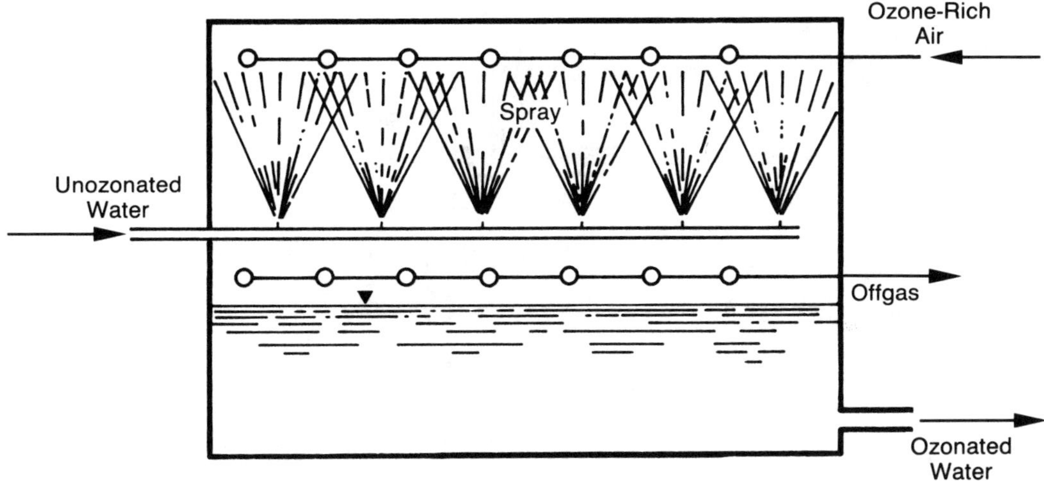

Figure IV–59 Schematic Diagram of the Spray Chamber Contactor at the Wuppertal Plant, Benrath, Germany

occupies a significant amount of volume. Although the actual retention may approach the hydraulic retention time (HRT), the HRT of the contactor is reduced by the volume of the packing, thereby reducing the benefit of the packing.

- There are no moving parts in the contactor, so mechanical maintenance should be minimal.
- Although packed column contactors have not been widely used in water treatment for ozonation, scaleup with regard to hydraulics and head loss should not be a critical issue. It is assumed that ozone transfer efficiencies and treatment efficiencies will also scale up, but this is not known due to the limited experience with packed column contactors for this treatment purpose.
- Packed column contactors have reduced inlet gas pressure requirements.

One disadvantage of using packed columns is that chemical scale may build up on the packing and cause increased head loss through the reactor. The degree of scale buildup is a function of the quality of the water being treated.

Spray contact chambers. The spray contact chamber reverses other known ozone dissolution methods. The liquid to be treated is sprayed into an ozone-rich atmosphere. The actual contact time of water with ozone-containing gas is very short; consequently, these types of contactors are used primarily for instantaneous or rapid reactions. The Wuppertal plant in Benrath, Federal Republic of Germany was using this type of ozone contact in the late 1970s (Miller et al. 1978). At the time, Rhine river sand bank–filtered water (containing iron and manganese) was being pumped to a large storage tank positioned on the roof of the building housing the rest of the water treatment plant. The amount of water head created by this large storage tank was sufficient to force the water through nozzles by gravity into the ozone/air gas mixture. Ozonized water then flowed through the filters (sand, then GAC). Waste gases from the ozone oxidation stage were collected and fed into the raw water as it was pumped into the large storage tank. In this manner, most of the excess ozone was utilized in partially satisfying the ozone demand of the raw water. This type of contactor has been installed at several other German water treatment plants by the same engineering firm which designed and installed the Benrath plant. A schematic diagram of the spray chamber contactor is shown in Figure IV–59.

Similar concepts are described by Richard and Blue (1978) for the Jonchay and Crissey water treatment plants in France.

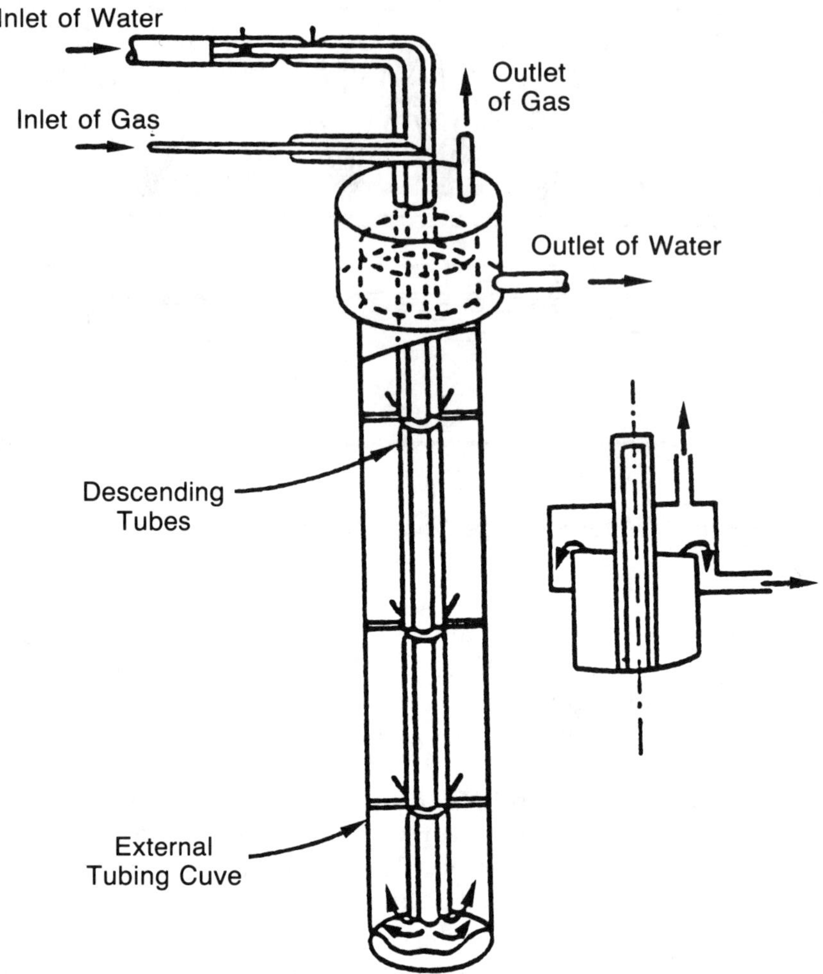

Source: Roustan et al. (1987).

Figure IV–60 Schematic of a Deep U-Tube Contactor

New concepts.

Deep U-Tube. The Deep U-Tube contactor is a unique and relatively new contactor design; as such, it should be considered as still in the developmental stages. However, it is applicable to specific contacting situations. This contactor is a patented design of Lyonnaise des Eaux, a French water treatment company. The contactor is comprised of two vertical concentric tubes of approximately 20 m (66 ft) in depth. Water flows downward through the inner tube. Ozone is injected through an orifice into the water stream near the top of the inner tube and travels downward with the flow of water. After reaching the bottom of the inner tube, the water/ozone mixture flows upward through the outer tube. A schematic diagram of this type of contactor is shown in Figure IV–60.

The design of the Deep U-Tube contactor is based on the following principles of gas transfer that enhance the transfer of ozone from the gas to the liquid phase:

- Increasing pressure: Because the depth of a full-scale Deep U-Tube is approximately 20 m (66 ft) as the water and gas flow downward through the contactor, the external pressure on the bubbles increases. In turn, this increases the mass transfer driving force (C_s-C) because it increases the partial pressure of the ozone in the bubble and the equilibrium saturation concentration (C_s) for ozone.

In contrast, most conventional bubble diffuser contactors operate at a depth of 18 to 24 ft (5.49 to 7.31 m), which provides only one-third the static pressure provided by a Deep U-Tube contactor.

- High turbulence: The Deep U-Tube operates under a flow regime of high turbulence, as do other injector contactors. This turbulence causes shearing of the bubbles, which increases the gas/liquid interfacial area and induces an increase in ozone transfer.
- Plug flow hydraulics: The Deep U-Tube contactor operates as a plug flow reactor (Brodard et al. 1984). As will be discussed, the degree to which the contactor simulates plug flow conditions depends on the diameter of the outer tube and the flow regime through that portion of the reactor. Plug flow reactors, in comparison with CSTRs (the two extremes of reactor design) require a shorter contact time to attain $C \cdot t$ values for primary disinfection. In the SWTR Guidance Manual, plug flow reactors are shown to be more advantageous than CSTRs because they increase the value of t_{10}.

The Deep U-Tube has been the subject of detailed study at the pilot scale (Speece et al. 1980; Brodard et al. 1984) and has been studied at one demonstration-scale installation (Brodard et al. 1986). Today, there are three full-scale Deep U-Tube installations around the world, two in France and one in South Africa.

The advantages associated with the Deep U-Tube contactor include the following:

- High transfer efficiencies (95 to 99 percent in full-scale applications) are obtained as a result of the high pressure and turbulence associated with this type of reactor.
- Because the U-Tube is such a deep contactor, surface area requirements are substantially less than for other contactor designs. In comparison with the conventional bubble diffuser contactor, the surface area to treatment capacity (ft^2/mgd) requirements of the Deep U-Tube are an order of magnitude less (1.3 ft^2/mgd to 4.3 ft^2/mgd [7.66×10^{-4} to 2.53×10^{-3} $m^2/m^3/h$]) for the Deep U-Tube compared with approximately 20 ft^2/mgd to 60 ft^2/mgd (1.18×10^{-2} to 3.53×10^{-2} $m^2/m^3/h$) for conventional diffuser contactors.
- Scale-up does not appear to be an issue; three full-scale facilities, scaled up from a demonstration-scale unit, have already been constructed and are operating without any noteworthy operational problems. The limitations of scale-up, such as maximum effective inner tube diameter, currently are not known due to the limited use of this type of contactor.

Disadvantages associated with the Deep U-Tube contactor are summarized below:

- The Deep U-Tube is a new technology and has been used in only a limited number of applications. Operational experience with this type of contactor is limited.
- High water velocities are required through the inner tube to pull the ozone bubbles to the bottom of the reactor. As a result, turndown capability may be limited.
- The Deep U-Tube is a proprietary system subject to payment of royalties and to sole-source specification issues.
- Costs of drilling and excavation and possible geotechnical problems may be associated with the depth of this contactor.
- Maintenance may be more difficult due to the configuration of this contactor as compared to a conventional bubble diffuser contactor.

Sweeping porous plate diffuser contactor (see chapter III, sec. G.4). The sweeping porous plate diffuser system is a proprietary system (patented design of Omnium de Traitement de Valorisation, a French water treatment company) when it is contained in a specially designed housing that combines ozonation and flotation in one unit process. It was originally developed to facilitate the removal of algae by ozonation and flotation

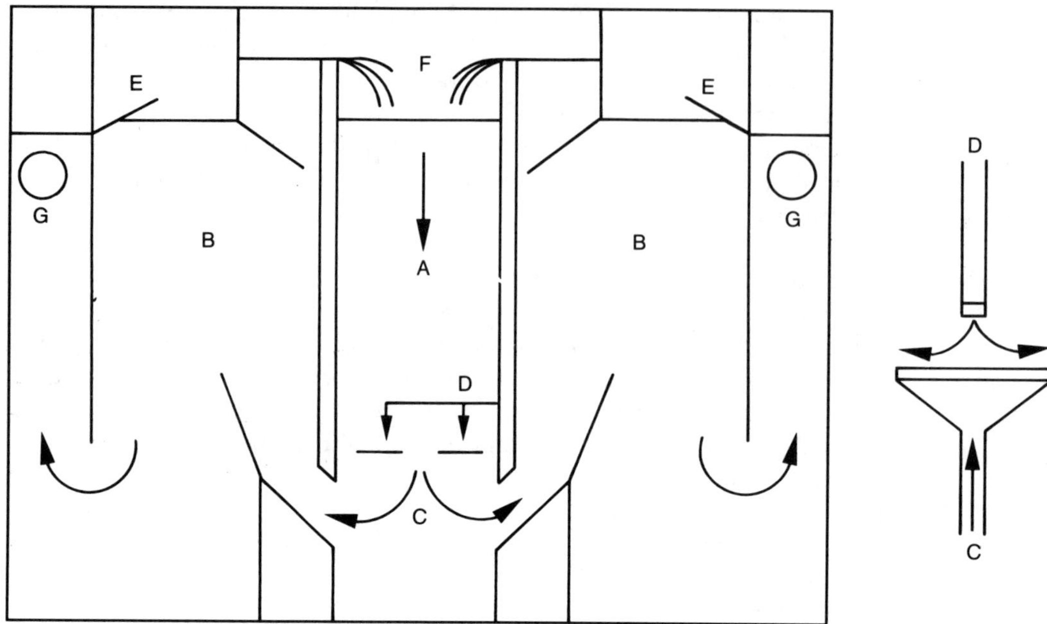

A: Downflow Ozonation Chamber

B: Flotation Chambers

C: Porous Diffusion Plates

D: Sweeping Current
(Water Injected Through Nozzle)

E: Scum Removal Channel

F: Raw Water Inlet

G: Treated Water Outlet

Source: Bourbigot et al. (1987).

Figure IV–61 Schematic of Sweeping Porous Plate Diffuser System

from eutrophic raw waters prior to filtration. This diffuser system, however, may be used for the purpose of enhancing mass transfer in addition to simply inducing flotation.

A schematic diagram of the sweeping porous plate ozonation-flotation unit is shown in Figure IV–61 (Bourbigot et al. 1987). The contact basin is divided into two chambers, an ozonation chamber and a flotation chamber. As shown in the diagram, water flows downward through the center ozonation chamber. Ozone is diffused into the water through porous plate diffusers located near the bottom of the ozonation chamber. Simultaneously, a jet of pressurized water flows downward onto the diffuser, causing shearing of the bubbles. The jet of water sweeping the plate is a small fraction (generally 5 to 10 percent) of the total raw water flow through the contactor. Some percentage of the bubbles will be too large to be carried over into the flotation chamber by the downward velocity of water. These bubbles will rise through the water column in the ozonation chamber and ozone will diffuse from the gas phase to the liquid phase just as in a conventional bubble diffuser contactor. The bubbles carried over into the flotation chamber adsorb onto the lighter floc and algae particles and rise to the surface, where they are accumulated, concentrated, and withdrawn at a desired rate. It is critical that the upper portion of the flotation chamber remain quiescent. The more dense particles remain suspended and pass through the flotation chamber and onto the filters.

In addition to enhancing ozone mass transfer, other advantages of the sweeping porous plate diffuser system include the following:

- Filter run times (in the case of direct filtration) can be substantially increased because a large percentage of the floc, algae, and other suspended matter is removed by flotation. Pilot study results showed that the ozonation-flotation system can increase filter run lengths by 20 to 40 percent (Bourbigot et al. 1987).
- Problems associated with plugging of the diffusers may be reduced because the plates are continuously cleaned by the jet of water.
- This new technology has full-scale experience with three plants in France (see chapter III, sec. G.4).

Disadvantages of the sweeping porous plate diffuser system include the following:

- Turndown may be limited because the velocity of the downward flow of water is a critical factor in assuring that the small bubbles (approximately 0.2 to 0.5 mm in diameter) are carried over into the flotation chamber.
- The "Ozoflotator" is a proprietary system subject to the payment of royalties and to sole-source specification issues.

Scale-up of the sweeping porous plate contactor from demonstration-scale to full-scale should not be difficult.

Submerged static radial turbine contactor. A schematic diagram of the ABB submerged static radial turbine is shown in Figure IV–62. This turbine does not contain moving parts, such as an impeller, below the water surface, so it may be easier to maintain than submerged turbines. A partial stream of water (approximately 4 percent of the total flow into the contactor) is pressurized and mixed with the ozonized gas in a static mixing element located in the head of the turbine (above the water level). The gas/water mixture is injected into the contactor through nozzles that form fine bubbles due to high shear. The contactor is operated in the countercurrent mode. After the bubbles rise to the water surface, some of the ozone contained in the offgas can be further utilized by the cascading water entering the mixing/contacting chamber (Dyer-Smith et al. 1988).

Because of stringent space limitations, the bubble diffusion contactor was neither large nor deep enough to achieve the desired transfer efficiencies, which led to the application of a submerged static radial turbine contactor at the Horgen, Switzerland plant (Dyer-Smith et al. 1988). Use of the submerged static radial turbine permitted design of two contactors, each with dimensions of approximately 6 by 6 ft (1.83 by 1.83 m) and 8 ft (2.44 m) in depth for a plant flow of 8 mgd (about 1260 m^3/h). Ozone is mixed/contacted with water in a small first chamber of the contactor, then allowed to react for a longer period of time in the balance of the contactor/reactor. Depending on the applied ozone dose, transfer efficiencies ranging from 90 to 97 percent are achieved. Currently (1990), operational experience with this type of contactor is limited.

IV.D.3 Contactor and Dissolution System Selection

The process evaluations described in chapter III and the treatability studies addressed in sec. A of this chapter will establish the proposed application(s) of ozone and the proposed point(s) of application in the overall water treatment system. As illustrated in sec D.2 above, there is a broad range of dissolution devices available that have been used for specific ozone applications. Review of the literature indicates that many types of contactors have been used for many applications. However, certain types of contactors appear to be more beneficially applied for certain processes.

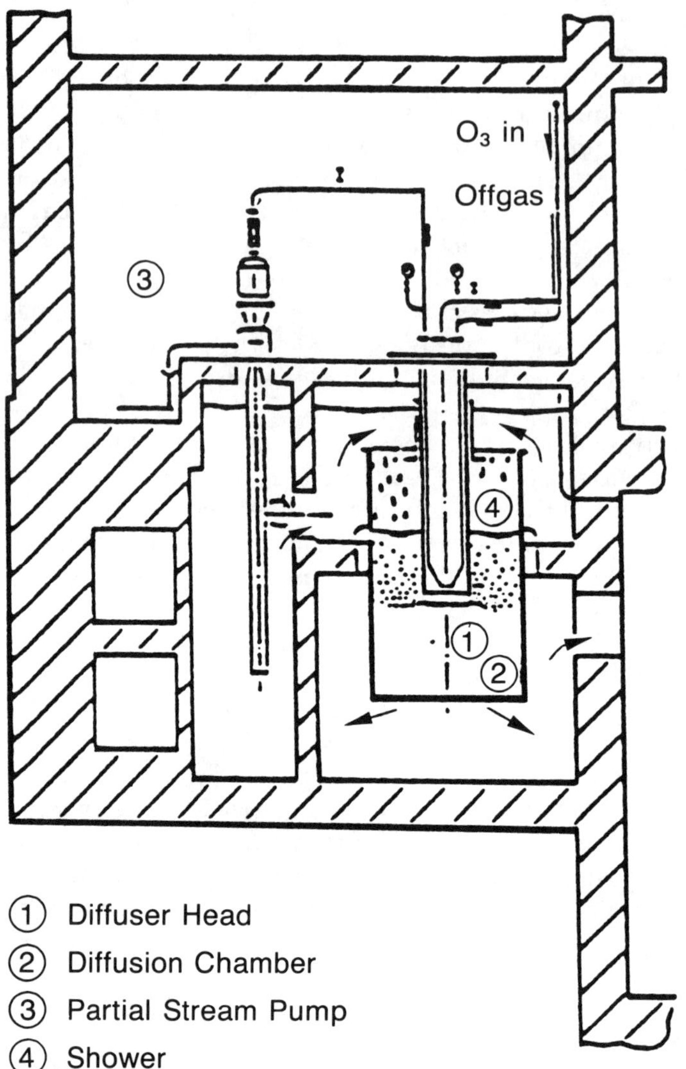

Source: Dyer-Smith (1988).

Figure IV–62 Typical Static Radial Turbine Installation

Primary disinfection. The concept of maintaining a residual concentration of ozone for a specific period of time most frequently results in the selection of a multiple-stage fine bubble diffuser–type contactor. Sufficient ozonized gas must be provided to satisfy the initial ozone demand of the liquid (which is normally a function of the organic and inorganic matrix of the water, pH, temperature, and concentration of ozonized gas) and to uniformly distribute the ozone. Once the initial ozone demand of the liquids is satisfied, then additional ozone will result in measurable residual ozone levels. It is appropriate to provide a dissolution stage designed to satisfy the immediate (initial) ozone demand, after which subsequent stages serve as retention vessels in which the required residual levels are maintained.

Iron and manganese oxidation. Oxidation of iron and manganese in a preozonation or intermediate ozonation step has been achieved in a variety of contactors. The formation of precipitates of iron and manganese and the relative speed of the reaction would appear to favor the selection of contactors other than fine bubble diffusers, although fine bubble contactors have been successfully applied.

Color removal. Ozonation for color removal can be achieved in preozonation or intermediate ozonation steps. Since contact time as well as ozone dosage may be a factor in color removal, U.S. drinking water ozonation applications have frequently applied the multiple-stage fine bubble contactor. Injectors have also been successfully tried.

Taste and odor. Multiple-stage fine bubble contactors have been successfully applied in the removal of taste and odor compounds. However, a single stage contactor of almost any type would serve the purpose.

Algae removal. The formation of high levels of entrained fine gas bubbles in the liquid under treatment provides for removal of algae microorganisms by flotation. Once the material becomes a "float," an efficient and operable means of removing the material from the contactor and from the liquid is required. The configuration of the Haworth submerged aspirating turbine contactor and the "Ozoflotator" are specifically designed to provide for flotation.

Particulate removal. Ozonation for enhanced particulate removal processes will primarily be achieved in a preozonation step. A key issue will be a capability to carefully control the transferred ozone dosage. This will likely be achieved in a single-stage contactor of any one of various types.

Oxidation. Application of ozone for the oxidation of dissolved organic compounds necessitates the selection of an effective dissolution device.

IV.D.4 Contactor Design Considerations

Conventional fine bubble diffuser contact chamber design.

Hydraulic considerations. The floor area of a stage of the contactor may be established by the number of diffusers and associated piping that can be installed as well as by considering practical provisions for required system maintenance.

The depth of water over the diffusers is an important design parameter because it has great influence on ozone transfer efficiency. Computer models are now available for determination of the optimum contactor depth for specific geographical conditions (Rakness et al. 1986) and for simulation of the ozone transfer conditions regarding the hydrodynamics of the contactors and the quality of the water being treated. Most conventional fine bubble diffuser contactors have a water depth of 16 to 20 ft (4.88 to 6.10 m) above the diffusers. Contactors at higher elevations above sea level require the greater depths to compensate for the reduction in ozone transfer efficiency related to the increase in elevation. For example, the contactor at the Avon, Colorado water treatment plant, located at an elevation of 7500 ft (2286 m) was designed with a 20-ft (6.1 m) depth of water over the diffusers (Renner et al. 1988). Higher ambient water temperatures would also require greater water depths, based on Henry's Law.

The contactor stage configuration preferred in current designs is one in which liquid flow is downward, countercurrent to the upward flow of the ozonized gas. The countercurrent reactor appears to minimize short circuiting and maximize contact of liquid and ozonized gas. Current practice would have two or more countercurrent stages in series to achieve and maintain the required residual (Figure IV–63). Flow from one countercurrent stage to another is transmitted upward through a downstream vertical channel or chimney. The "throat" dimension of the chimney may be governed by practical minimum dimensions for constructing and stripping reinforced concrete formwork. A minimum throat dimension of 3 ft (0.91 m) has been used on a number of U.S. ozone contactors that used cast-in-place reinforced concrete for the vertical baffling of the contactor. At Neuilly-sur-Marne (France), the dimension of the chimney is only 0.5 m (1.64 ft) and the baffles are made of reinforced aluminum (Gerval et al. 1989).

Chédal (1979) uses a series of photographs of hydraulic test models to illustrate the importance of uniform distribution of fine bubble diffusers across the bottom of the

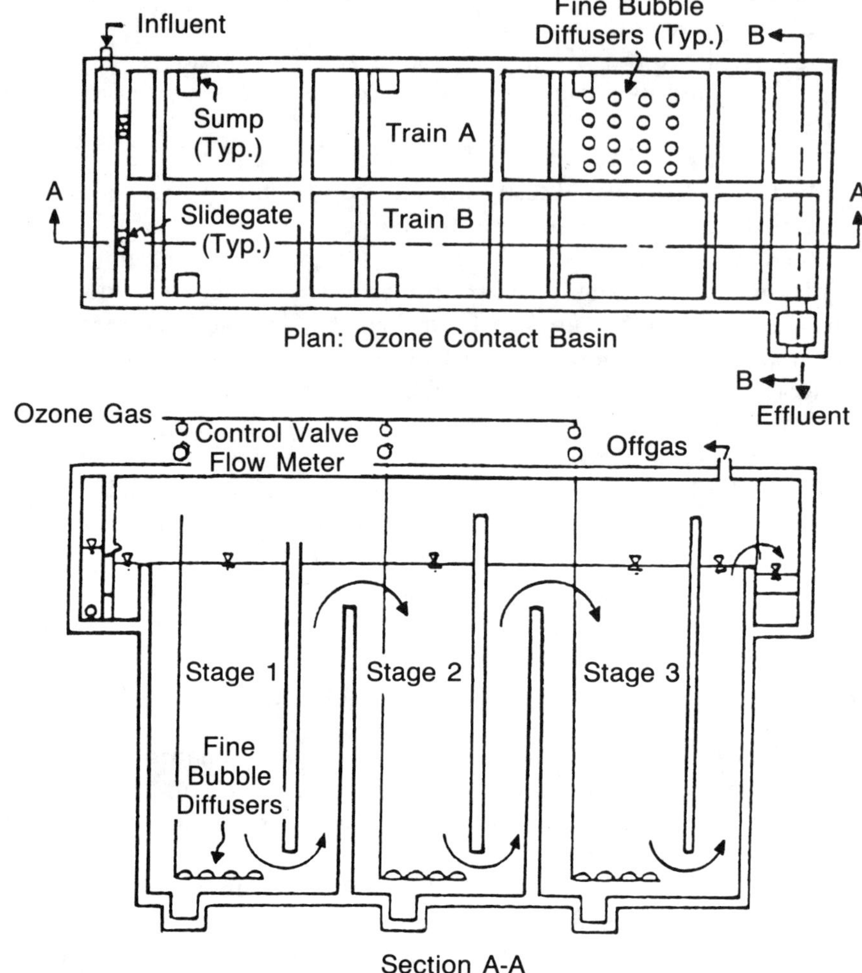

Figure IV–63 Schematic of a Three-Stage Bubble Diffuser Ozone Contactor

contactor. Liquid flowing countercurrent to the gas flow takes the path of least resistance by preferentially following channels low in or devoid of upflowing gas bubbles.

The situation becomes particularly significant if the reduced gas flow is in the downstream or exit portion of the compartment. The gas bubbles that are discharged enter a dead water area where their effectiveness is at least partially lost through decay. Chédal points out that basing transfer efficiency merely on ozone applied less ozone exhausted from the contactor may be deceiving in terms of achieving effective application of the applied ozone.

Hydraulic calculations for contactors require consideration of the situation of air being diffused into the contactor. The level drop, Δh, between the free surface of a compartment where gas is injected and that of an upflow compartment downstream may be expressed as (Gerval et al. 1989):

$$\Delta h = \frac{V'}{V + V'} h = \alpha h$$

Where:

Δh = level drop, m
V = volume of water in the contactor stage, m^3

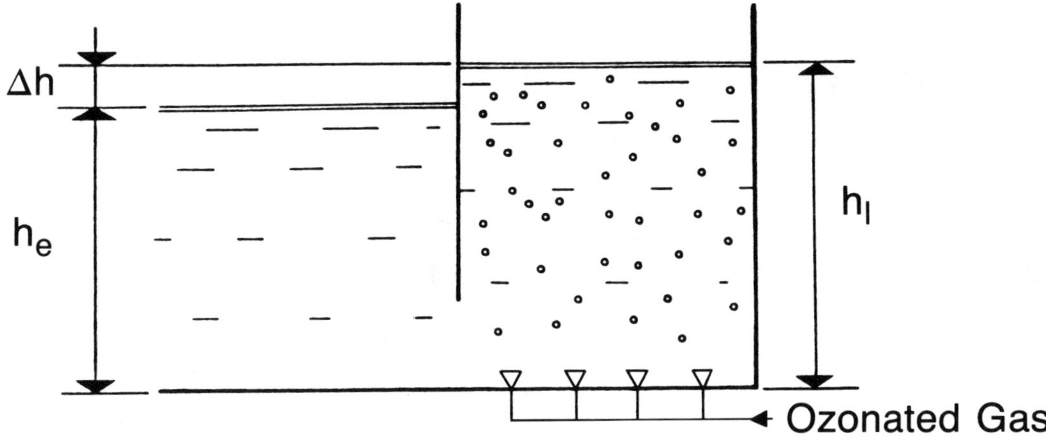

Source: Chédal (1979).

Figure IV–64 Pressure Drop Due to Gas Bubble Diffusion

V′ = volume of gas in the contactor stage, m^3
h = water depth of contactor immediately downstream, m

$$\alpha = \frac{V'}{V + V'}$$ = void factor for the compartment where the gas is injected.

Chédal (1979) expressed a similar concept as follows, keyed to Figure IV–64:

$$\Delta h = h_e \cdot \frac{\rho}{1 - \rho}$$

and

$$\rho = \frac{qa}{qe} \times \frac{Ve}{Va} = \frac{T_R}{C} \times \frac{Ve}{Va}$$

Where:

Δh = pressure drop due to gas bubbles, m
h_e = water depth of contactor immediately downstream, m
ρ = the percentage of air: volume of air in the bubbles per unit of liquid volume
qa = applied ozone gas flow rate, m^3/h
qe = applied liquid flow rate, m^3/h
Ve = liquid vertical velocity, cm/s
Va = gas bubble vertical rising velocity, cm/s
T_R = ozone treatment rate, g O_3/m^3 of water
C = concentration of utilized ozonized gas, g O_3/m^3 NTP.

Contactor hydraulic baffling. The Surface Water Treatment Rule of the Safe Drinking Water Act established C · t as a means of establishing and monitoring the degree of primary disinfection being achieved at a water treatment plant. The designer is faced with the problem of designing a contactor to provide a sufficient (but not excessive) actual contact time period in which a disinfectant residual level is maintained. After completion of construction, the actual contact time is to be determined. The t may

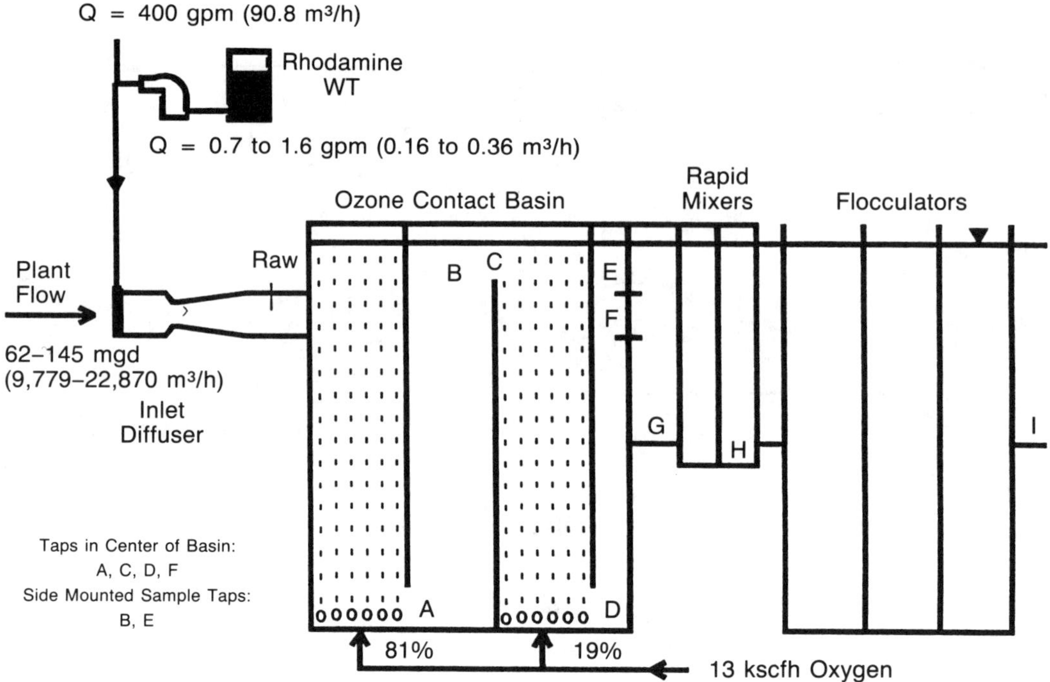

Source: Stolarik and Christie (1988).

Figure IV–65 Tracer Feed System and Sample Tap Locations Used to Test the Ozone Contactor at the Los Angeles Aqueduct Filtration Plant, California

be based on t_{10} as measured by tracer studies (see sec. IV.A.1). The relationship of t_{10} to the theoretical hydraulic retention time (volume ÷ flow) is an indication of degree of short-circuiting in the contactor.

In early 1990, there were few published data providing a comparison of actual (measured) contact times relative to the theoretical hydraulic retention times (HRT) of operating units. Two ozonation facilities in the United States that have provided data are those of the Los Angeles Aqueduct Filtration Plant of the Los Angeles Department of Water and Power and the Hagerstown Wastewater Treatment Plant of Hagerstown, Maryland. This kind of study is also performed in Europe, e.g., the studies of Compagnie Générale des Eaux at the Méry-sur-Oise and Neuilly-sur-Marne facilities (Martin and Bourbigot 1989; Martin 1989).

The LAAFP ozonation system generates ozone from high-purity oxygen for application to a multiple-stage contactor (see sec. F.2, chapter III, and sec. E.1, Chapter VI). Gas dissolution is performed in countercurrent stages 1 and 3 while no dissolution is designed for in cocurrent stage 2. Stolarik and Christie (1988) reported on testing the performances of the full-scale facility, although the initial design was not based on C · t concepts. Figure IV–65 shows the tracer feed system and sample tap locations used for the evaluation of the performance of the contactor. As illustrated in Figure IV–66, t_{10} for the entire contactor (i.e., from ozone contact basin entry to the exit of the plant's rapid mixing) ranged from 48 to 50 percent of the HRT, depending upon the design water flow rate, the oxygen flow rate, and the distribution between the 2 compartments remaining constant during the tracer studies. The t_{50} value for the entire contactor and for the same conditions approached the HRT. Evaluation of Figure IV–67 (from the same reference) provides data for a countercurrent gas diffused chamber followed by a passive chamber without gas diffusion. At sample tap B (see Figure IV–65), values of t_{10} ranged from 33 to 42 percent of the HRT while t_{50} was close to the HRT. Evaluation of

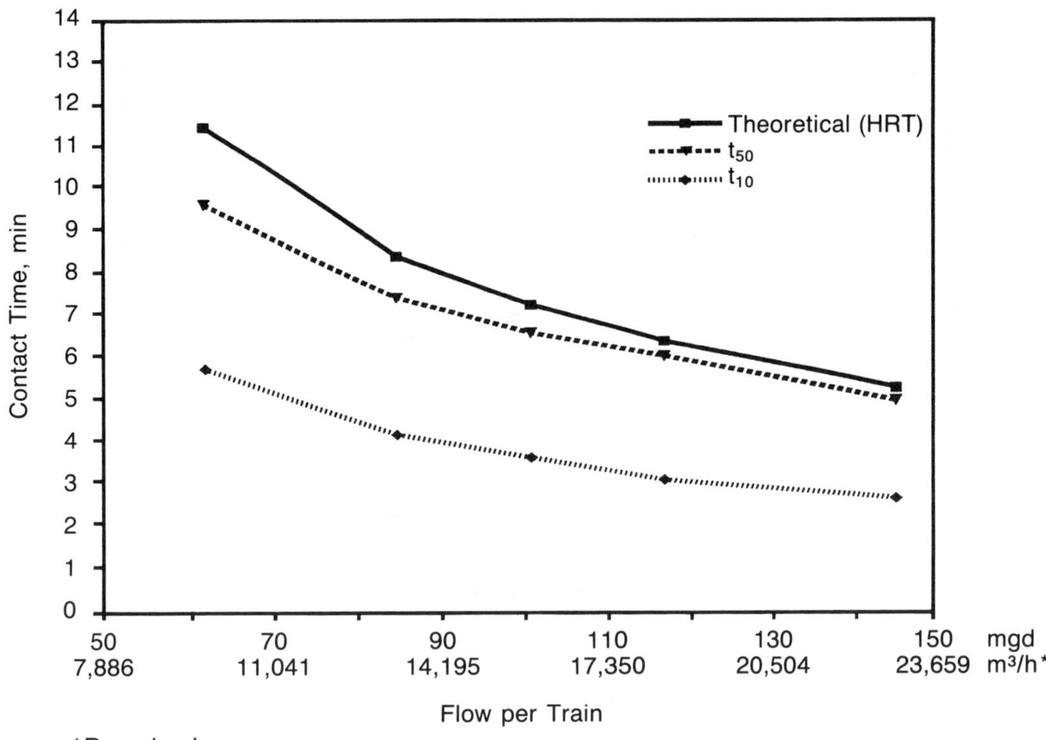

Source: Stolarik and Christie (1988).

Figure IV–66 Tracer Test Contact Times for Entire Ozonation System at the Los Angeles Aqueduct Filtration Plant, California

Figure IV–68 indicates that between the sample taps B and H (i.e., from the second compartment of the ozone contactor to the exit of the rapid mixers) values of t_{10} are always higher than 50 percent of the HRT. Another tracer study performed at Hagerstown, Maryland (Appelton et al. 1990) showed that changing the existing contactor from a double-stage (countercurrent followed by cocurrent) to a three-stage (three countercurrents) led to an increase of the t_{10} value from 29 to 52 percent of the HRT. Figure IV–69 illustrates this result.

Once the immediate or initial ozone demand has been satisfied and a residual ozone level has been satisfactorily maintained in several contactor stages, it may be practicable to incorporate a "passive" stage without ozone diffusers, which will serve to increase the required retention time without the cost of the additional diffusers and associated piping.

Contactor design must provide for periodic access for unit maintenance. The basic arrangement of the contactor will establish the provisions required for personnel access. Contactors with access galleries between the contactor trains can provide access to each individual stage by means of gasketed stainless steel plates. Access from the deck of contactor stages can be provided for tanks without galleries between the stages. Stainless steel hatches with ozone-resistant gaskets should be provided for each stage to minimize difficulty and delay of entry and exit from the contactor under regular and emergency conditions. Access between stages can be provided by openings in the walls between the stages. However, this would preclude the continuous use of safety harness by personnel in the tank, which may be unacceptable to some operating agencies. Also, apertures large enough for personnel passage may be too large for hydraulic control in the stages. Arrangements must be made to provide and exhaust ventilation air prior to the entry of maintenance personnel. This may be by means of manways or specific ports for this purpose (see sec. C.6, chapter V).

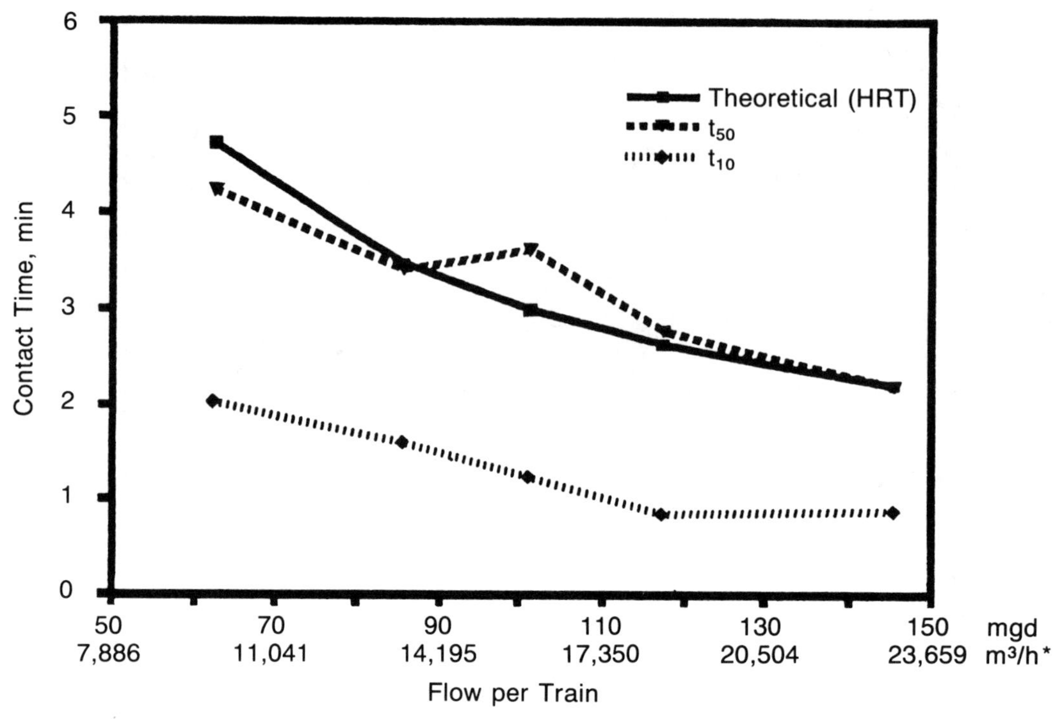

Source: Stolarik and Christie (1988).

Figure IV–67 Tracer Test Contact Times for First Stage Ozonation System at the Los Angeles Aqueduct Filtration Plant, California

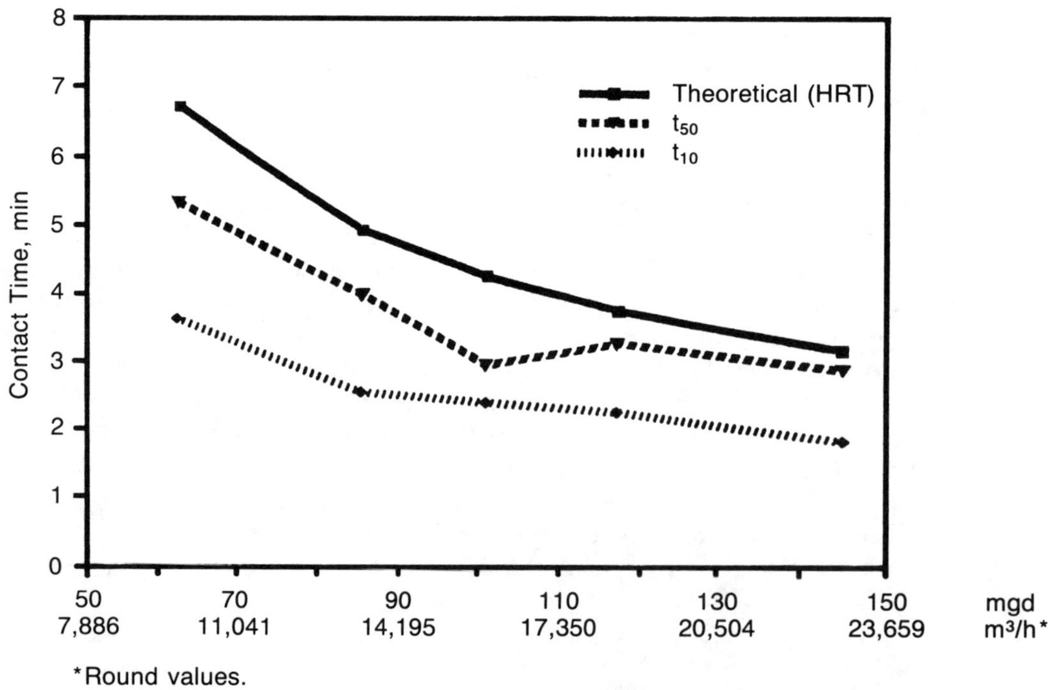

Source: Stolarik and Christie (1988).

Figure IV–68 Tracer Test Contact Times for Second Stage Ozonation System at the Los Angeles Aqueduct Filtration Plant, California

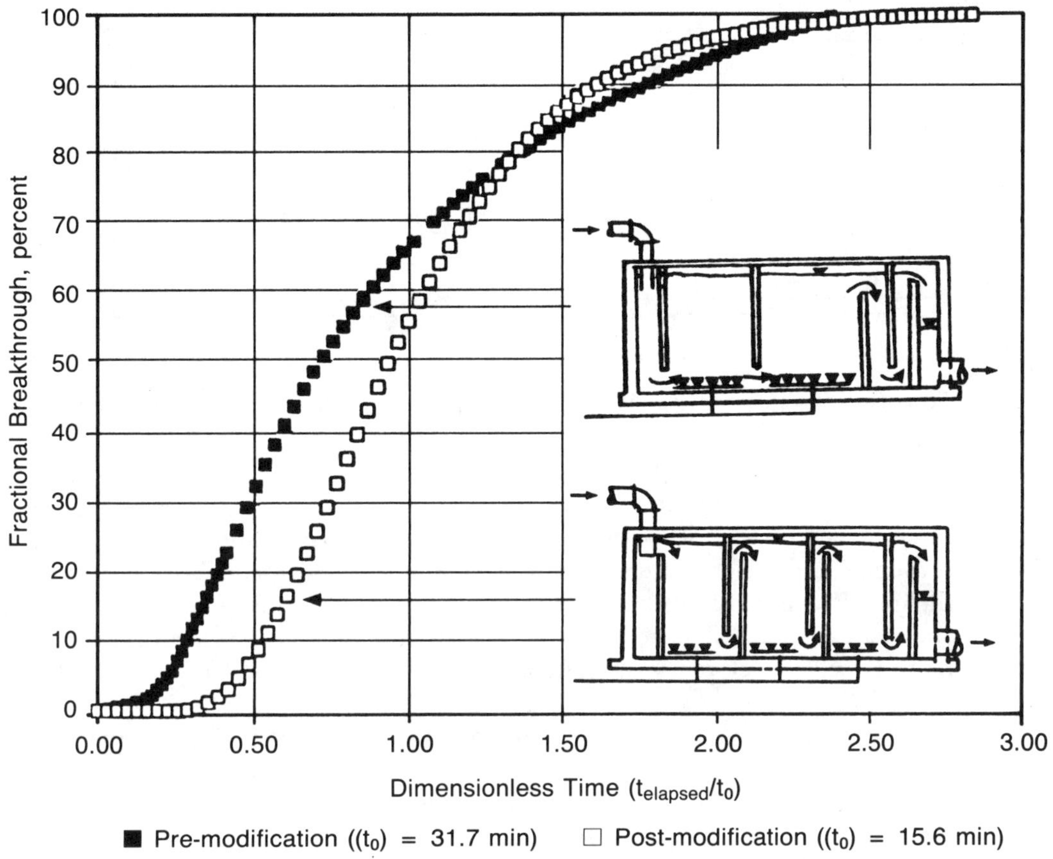

Source: Appleton et al. (1990).

Figure IV–69 Influence of Baffling Modifications on t_{10} at Hagerstown, Maryland Contactor

Headspace considerations. Sufficient headspace should be allowed to provide unimpeded gas flow between the stages. Depending on the characteristics of the water, headspace provisions may be required to reduce the potential for ozone-induced foam rising above the water surface and being carried into the offgas treatment system. A minimum of 2 ft (about 0.6 m) clearance should be provided between the bottom of the contactor cover slab and the top of an intermediate baffle wall, with 3 ft (about 0.9 m) being a preferred dimension if foaming is predictable. Spray nozzles should be provided to control froth, float, or foam in the headspace, particularly in the vicinity of the offgas vent.

Gas flow considerations.

Gas/liquid ratio: The ratio of gas flow from the gas diffusion to the flow of liquid through the chamber is important for hydrodynamic conditions. A gas/liquid ratio of 0.15 to 0.20 scfm/cfm (m^3 NTP/h/m^3/h) is used as an upper limit design guide.

Diffusers: Fine bubble diffusers (producing bubbles on the order of 2 to 3 mm in diameter), rather than coarse bubble diffusers, are used for ozonation because of bubble size considerations discussed earlier. The types most commonly employed are the tube, disc, and dome-shaped porous stone (ceramic) or stainless steel diffusers. Diffuser stones are graded by porosity and permeability which specifies the bubble sizes produced under a given condition. The most appropriate grades for ozonation systems are those with permeabilities ranging from 2 to 15 cfm/ft^2/in. of diffuser thickness at a pressure of 2 in. water column (36.6 $m^3/h/m^2$ to 274.3 $m^3/h/m^2$ for a diffuser thickness of 2.54 cm at 5 cm of water column pressure) and porosities of 35 to 45 percent (Rakness et al. 1984).

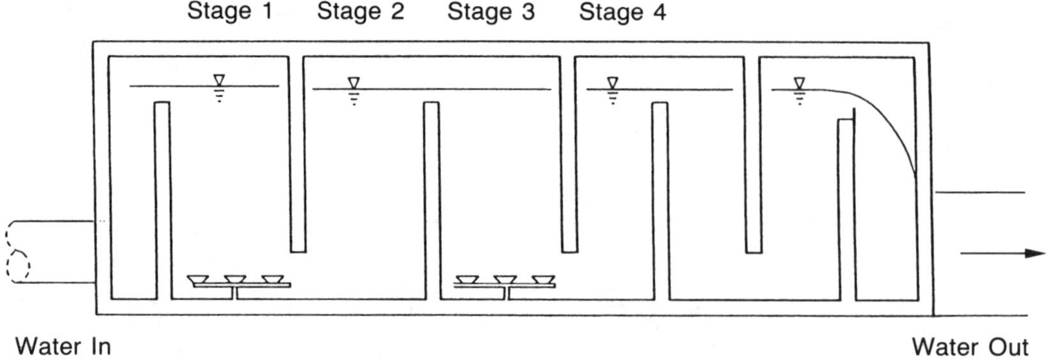

Figure IV–70 Fine Bubble Contactor Used as an Example for the Design Approach

The choice between porous tubes and discs or domes is basically a matter of personal choice for the designer or operator in terms of maintenance. The gas flow rate through the diffuser should be within the range recommended by the manufacturer. If the gas flow is too low, uniform bubble patterns will not develop across the contactor and inadequate mixing will result. If the gas flow exceeds the proper maximum, coarse bubbles will develop and the efficiency of gas transfer will be reduced (Renner et al. 1988).

Example contactor design approaches. The following examples for contactor design are generally confined to fine bubble dissolution systems in a conventional-type contactor. This contactor is the most common type and will normally be the preferred when primary disinfection is required.

Design for primary disinfection based on the SWTR (C · t) concept. Contactors based on the C · t requirement can be designed by assessing the combination of the dissolved ozone concentration and contact time in the contactor stages used for ozone dissolution and those used for additional detention time (i.e., stages with no ozone addition). Consequently, for disinfection determinations based on C · t values, it will be necessary to determine the ozone dose required to establish a given residual and the ozone decay rate. (See sec. A of this chapter.)

As stated in sec. A.1 of this chapter, the C · t evaluation procedure for ozone systems has not been established in final form by the EPA. Consequently, the procedures presented here may need to be modified once the EPA has finalized its evaluation methodology. This model approach should continue to be valid for design, although modification to reflect changes in C · t evaluation procedure may be necessary.

The contactor shown in Figure IV–70 is presented for discussion purposes. This is a fine bubble diffuser system with alternating stages for ozone dissolution and reacting. This configuration was selected because it contains each of the types of dissolution and contacting stages that will occur in a conventional contactor. The principles presented here can be used for any number of states and combinations for dissolution, reacting, and decay. The contactor has 4 stages:

- Stage 1—initial ozone dissolution and ozone residual development,
- Stage 2—reaction stage for inactivation of microorganisms,
- Stage 3—ozone dissolution stage and inactivation stage, and
- Stage 4—reaction stage for inactivation, as well as a decay stage to reduce ozone residual prior to water leaving contactor. (This stage has been baffled to improve detention time.)

The C · t value in a dissolution stage was estimated by averaging the expected ozone residual obtained across the stage based on the ozone residual curve and multiplying this by the expected t_{10} time. (See the example below.)

C · t values for stages without ozone addition can be determined based on the decay rate determination, which is normally a first-order reaction (Figure IV–17, sec. A.6 of this chapter). The first-order decay rate shown in Figure IV–17 (sec. A.6 of this chapter) is represented by the following relationship:

$$[C] = [C_0]\exp(-k'[t])$$

Where:

$[C]$ = dissolved ozone concentration in mg/L at time t, min
$[C_0]$ = initial ozone concentration, mg/L
k' = rate constant, min^{-1}.

The C · t value can be calculated based on the average residual ozone concentration. The average ozone residual $[C_{avg}]$ can be calculated by determining the area under the decay curve and dividing by the time:

$$[C_{avg}] = \frac{[C_0]}{k'[t]} \times (1 - e^{-k'[t]})$$

In addition to determining the C for the C · t, it is important to estimate the t_{10} (as defined in sec. A of this chapter). Values for t_{10} have been estimated from 30 to 50 percent of the hydraulic detention time for contactor stages that are not mixed and from 40 to 60 percent for stages that are gas-mixed (i.e., stages with ozone addition) (see Figures 4–68 and 4–69).

For this example, the following detention times were assumed for C · t calculations:

- Ozone-addition stage: 0.5 times the hydraulic retention time (HRT) (i.e., 50-percent short-circuiting)
- No-ozone-addition stage: 0.4 times the HRT (i.e., 60 percent short-circuiting)

If better data are available for a given contactor design, they should be used rather than these generic examples. Also, SWTR guidance documents should be reviewed, once finalized, to determine if some other t value should be used.

Example of C · t design procedure: C · t values in the United States are based on the requirements of the EPA and are listed in the guidance manual to the SWTR for specific temperatures and inactivation efficiencies (i.e., log removal capabilities for *Giardia* cysts and for viruses). The following design example assumes an ozone C · t requirement of 0.7 min · mg/L based on a given design water temperature and a given level of inactivation. [Note: This C · t value already takes into account a credit in the first stage of the contactor as provided for in the EPA guidance manual. Therefore, this C · t value is to be obtained in the ozone contactor subsequent to the first stage as discussed below]. The example calculations are based on the contactor presented in Figure IV–70. Alternative contactor designs can be modeled with the procedures presented here.

The C · t value for each stage (Stages 2 through 4) is determined separately. The method for determining detention time consists of iterative calculations between the C · t calculation presented here and the ozone dissolution requirement calculated during the contactor gas dissolution system design. Detention times assumed for this example for each stage are as follows:

- Stage 2—HRT = 2 min, retention time for a non-mixed stage of 0.8 min (2 min × 0.4)
- Stage 3—HRT = 2 min, retention time for a mixed stage of 1.0 min (2 min × 0.5)
- Stage 4—As required to complete C · t requirement

The design process is iterative since there are a number of assumptions made during the design that can be changed to accommodate preferences such as contactor size, contactor dimensions, and method of ozone addition. The values presented above are presented for the purposes of this example only.

Stage 1. Stage 1 is a dissolution stage where the immediate ozone demand is satisfied and a residual is established. The most important function of this stage is to establish an ozone residual in the water. The ozone residual will be determined based on the residual versus ozone dose determination presented in Figure IV–15 (sec. A.6 of this chapter). For this example, it is assumed that a residual of 0.3 mg/L will be established. A transferred dose of 0.9 mg/L will be required to establish the residual.

Studies have shown that significant disinfection can be provided in the first stage (Joret et al. 1986) even if no dissolved ozone residual is detectable at the outlet of the first stage (see sec. H.2, chapter III). However, the current EPA concept is to give a 0.5 log *Giardia* inactivation credit and a 1 log virus inactivation credit if specified residual levels of 0.3 mg/L and 0.1 mg/L, respectively, are established in the first stage. As stated previously, the credit for the first stage was considered in determining the C · t requirement in this design example of 0.7 min · mg/L. This C · t is developed in stages 2 through 4.

Stage 2. The C · t value in Stage 2 is a simple determination based on the averaged C determination at the decay rate (k') of 0.06 min^{-1} (see Figure IV–17 [sec. A.6 of this chapter]) and the t_{10} time. The t_{10} time is 0.4 of the HRT, since this is a non-mixed stage. The C · t value is determined as follows:

$$C \cdot t_{\text{stage 2}} = (\text{HRT} \times \text{SCF}) \times \frac{[C_0] \times (1 - e^{-k'[t]})}{k'[t]}$$

$$= (2 \times 0.4) \times \frac{[0.3] \times (1 - e^{-0.06 \times 2}) \text{ min} \cdot \text{mg/L}}{0.06 \times 2}$$

$$= 0.226 \text{ min} \cdot \text{mg/L}$$

The remaining C · t required is 0.474 min · mg/L (i.e., 0.7 − 0.226).

The ozone concentration leaving Stage 2 is calculated as follows:

$$[C] = [C_0] \exp(-k'[t])$$
$$[C] = [0.3] \exp(-0.06 \times [2])$$
$$[C] = 0.266 \text{ mg/L}$$

Stage 3. Stage 3 is used to reestablish the ozone residual. For this example it is assumed that the residual will be replenished to 0.3 mg/L. It can be assumed that the average ozone residual in Stage 3 will be half the effluent ozone concentration unless special studies are performed that indicate a different relationship. The resultant C · t value is:

$$C \cdot t_{\text{stage 3}} = (\text{detention time} \times 0.5) \frac{C_{\text{out}}}{2}$$
$$= 2 \times 0.5 \times 0.3/2$$
$$= 0.150 \text{ min} \cdot \text{mg/L}$$

The remaining C · t to be obtained is 0.324 min · mg/L (i.e., 0.474 − 0.150).

The ozone required to reestablish the ozone residual is equal to the ozone necessary to go from a residual of 0.266 to 0.3 mg/L. The ozone required to increase the residual is approximately 0.15 mg/L transferred (see Figure IV–15 [sec. A.6 of this chapter]).

Stage 4. The remaining C · t value of 0.324 mg/L·min has to be achieved in the last stage of the contactor. The influent residual will be 0.3 mg/L. The required detention time for the stage can be calculated with the following equation:

$$
\begin{aligned}
t_{stage\ 4} &= -\frac{1}{k'} \cdot \ln\left(1 - \frac{k' \cdot C \cdot t}{SCF \cdot C_0}\right) \\
&= (-1/.06) \times \ln(1 - [0.06 \times 0.324/0.4]/0.3) \\
&= 2.95 \text{ min}
\end{aligned}
$$

The C · t then should equal 0.324 min · mg/L

$$
\begin{aligned}
C \cdot t_{stage\ 4} &= (HRT \times SCF) \times \frac{[C_0] \times (1 - e^{-k'[t]})}{k'[t]} \\
&= (2.95 \times 0.4) \times \frac{[0.3] \times (1 - e^{-0.06 \times 2.95})}{0.06 \times 2.95} \\
&= 0.324 \text{ min} \cdot \text{mg/L}
\end{aligned}
$$

This confirms the C · t calculation for the fourth stage.

The actual retention time in the fourth stage should also provide for ozone decay prior to leaving the contactor. Assuming an outlet residual below 0.1 mg/L is desired (inlet residual is 0.3 mg/L), the retention time will be:

$$
\begin{aligned}
t &= \frac{-1}{k'} \ln \frac{C}{C_0} \\
t &= (-1/.06) \times \ln(.1/.3) \\
t &= 18 \text{ min}
\end{aligned}
$$

A retention time of 18 min is probably excessive for the fourth stage. It would be reasonable at this point to reassess the design criteria and determine if a different residual or ozone addition pattern should be used. The design needs based on the dissolution requirements should also be evaluated at this point.

Contactor design for oxidation reactions. Calculated ozone dose and reactor design for oxidation reactions can be based on the empirical stoichiometric relationship determined for the desired reaction. Figure IV–18 (see sec. A.6 of this chapter) shows an exponential rate relationship for taste and odor reduction in terms of 2-methylisoborneol (MIB) compounds. The empirical stoichiometric coefficient can define the relationship between ozone dose and level of treatment. This value can be used (with appropriate judgment) to calculate the ozone dose required for a given application.

For example, the exponential relationship presented in Figure IV–18 (see sec. A.6, this chapter) can be reordered to calculate the required ozone dose for a given level of oxidation:

$$
[O_3] = -\frac{1}{k_D} \ln \frac{[C]}{[C_0]}
$$

Where:

k_D = empirical stoichiometric coefficient, L/mg (0.783 mg/L from Figure IV–18 [sec. A.6 of this chapter])

$[O_3]$ = ozone dose, mg/L.

If a 95 percent reduction (e.g., from 100 μg/L to 5 μg/L) of MIB is desired, a transferred ozone dose can be determined as follows:

$$[O_3] = -\frac{1}{0.783} \cdot \ln(5/100)$$
$$= 3.83 \text{ mg/L}$$

The advantage of this interpretation and design approach is the ability to interpolate data easily during the inevitable iterative design process.

The contactor for oxidation can be any one of the types commonly used for ozone dissolution and contacting. Since most oxidation reactions are comparatively rapid, it is reasonable to use contactors other than the fine bubble systems normally used for primary disinfection. For example, either an in-line or turbine-type system would be applicable. The objective of the contactor for this example would be to transfer 3.83 mg/L ozone into the water. The contact time could be as short as 1 min. Demonstration-scale studies with in-line contacting at Long Beach, California for color removal indicated that the color oxidation reaction was completed in less than 30 sec (Bellamy et al. 1990). Experimentation is required to verify the lower limit of the reaction time.

Contactor sizing based on ozone dissolution requirements. The design of an ozone contactor requires the design parameters of ozone dose, dissolution characteristics, and intended application. Some of the methods for determining the ozone dose were presented above.

The intended application will normally be either primary disinfection or oxidation. Primary disinfection requires establishing and maintaining a residual for a given retention time. Fine bubble diffuser type contactors are suitable for this type of application. Oxidation requires transferring a specific ozone dose, and given sufficient mixing, the retention time is not as critical as for disinfection.

Dissolution criteria that affect sizing of a contactor include a variety of parameters; for example, the effect of side water depth on transfer efficiency, the desirability of equally spaced diffusers with complete floor coverage, minimum gas flow rates to prevent diffuser plugging, and others. Many of these subjects are discussed in texts and papers on design of ozone systems. This section deals with the criteria of minimum gas feed rates for mixing and gas-to-liquid ratio for fine bubble diffuser contactors.

Minimum gas feed rate for mixing: There is a lower gas feed rate where the mixing energy within the contactor drops and the water and gas tend to channelize and pass through the dissolution stage without proper dissolution and/or contacting with the ozone. This condition can be caused by too few diffusers or too low a gas feed rate. Low gas flow will most probably occur in contactors designed for disinfection under conditions of reduced flow rate and high ozone concentrations (e.g., a pure oxygen feed gas system).

A minimum gas flow rate has not been identified in the literature. There are two sources of information that suggest what this rate might be. One manufacturer suggests that a gas flow rate of 0.13 cfm/ft^2 (2.38 $m^3/h/m^2$) (of contactor area) was the lower limit practiced in France at their water treatment plants (Delcominette 1988). In addition, fine bubble systems for activated sludge have been studied and a minimum rate of 0.12 cfm/ft^2 (2.19 $m^3/h/m^2$) has been suggested as being the lower limit for proper mixing (Design Manual for Fine Pore Aeration Systems, EPA 1989).

Based on these values, a minimum gas flow rate of 0.12 cfm/ft^2 (2.19 $m^3/h/m^2$) will be assumed to be the lowest desirable gas flow rate sufficient to create proper mixing and homogeneous ozone dissolution. An example for applying this value is presented below.

Gas-to-liquid ratio: There is an upper limit for the gas-to-liquid ratio (cfm_{gas}/

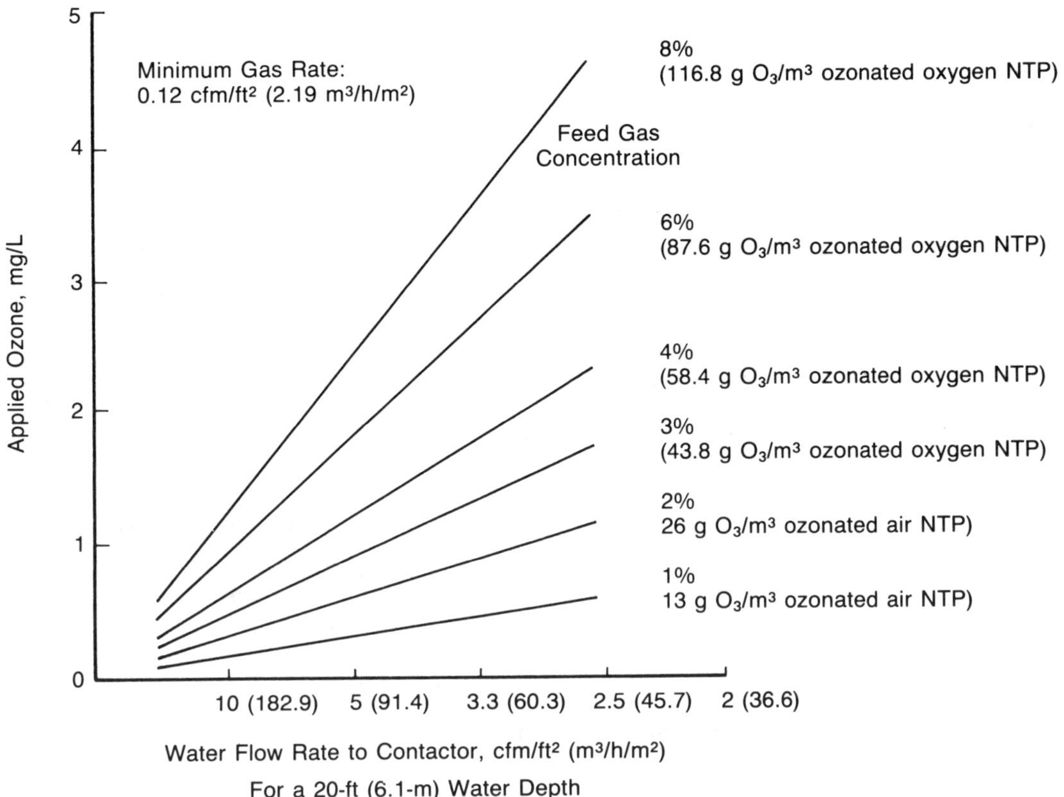

Figure IV–71 Possible Applied Ozone Dose in Relation to Unit Water Flow Rate Versus Gas Concentration Considering a Minimum Gas Flow Rate

cfm_{water} [m^3_{gas}/m^3_{water}]) where ozone transfer efficiency is sufficiently reduced to limit the quantity of ozone that can be reasonably transferred. An exact determination of the upper limit has not been and probably cannot be made. Values have been suggested between 0.15 to 0.20 (Rice and Netzer 1984). The 0.15 ratio was indicated as a preferred limit by an ozone equipment manufacturer (Dimitriou 1989).

The reduction in transfer efficiency for high gas-to-liquid ratios is thought to be caused by interference of bubble patterns, which causes the fine bubbles to coalesce. This reduces the total surface area of the bubbles, hence the transfer efficiency. This coalescing can occur at the diffuser or in the water column.

Example of contactor sizing based on dissolution characteristics: As presented above, the dissolution characteristics of minimum gas rate and maximum gas-to-liquid ratio will affect the design of an ozone contactor. The design process will be iterative and compromises will be necessary before an acceptable balance of all the design criteria can be made.

Example for minimum gas flow. Figure IV–71 presents the relationship among the minimum feed gas rate (assumed to be 0.12 cfm/ft^2 [2.19 m^3/h/m^2]), ozone feed gas concentration, and the applied ozone dose versus the liquid flow rate to the contactor. This figure can be used in a number of ways.

Assume that a utility is going to design a system around purchased liquid oxygen (LOX) and generate a high ozone concentration. The intended application is disinfection and the system criteria are:

- Applied ozone dose—minimum 1 mg/L and maximum 3 mg/L
- Ozonized gas concentration—6 percent (87.6 g O_3/m^3 NTP)
- Liquid flow range—3 to 7 mgd (473 to 1104 m^3/h)

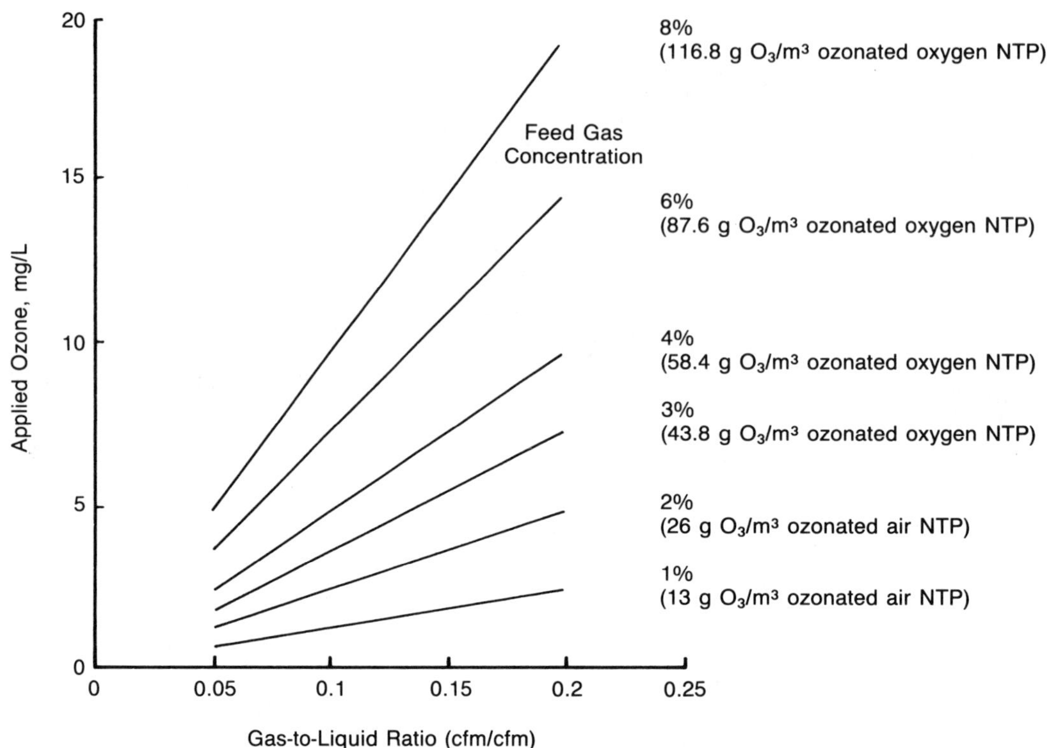

Figure IV–72 Possible Applied Ozone Dose in Relation to Gas-to-Liquid Ratio versus Feed Gas Concentration

The minimum gas feed rate will occur when the minimum water flow rate and minimum ozone application rate correspond. From Figure IV–71, a 1-mg/L ozone dose and 6 percent ozone (87.6 g O_3/m^3 NTP) results in a liquid floor loading rate of approximately 8 cfm/ft^2 (146.3 m^3/h/m^2). For a minimum flow of 3 mgd (280 cfm) the contactor area would be:

$$(280 \text{ cfm})/(8 \text{ cfm/ft}^2) = 35 \text{ ft}^2 \ (3.25 \text{ m}^2)$$

Alternatively, Figure IV–71 can be used to estimate the gas concentration necessary to maintain gas flows for a given contactor area and desired ozone dose. For example, a contactor with a floor area of 100 ft^2 (9.29 m^2) and a flow rate of 3.6 mgd (330 cfm [567.8 m^3/h]) will have a floor loading of 3.3 cfm/ft^2 (60.35 m^3/h/m^2). If a dose of 1 mg/L is required, the percent ozone should be approximately 2.3 percent (29.9 g O_3/m^3 ozonated air NTP).

Example for maximum gas-to-liquid ratio. Figure IV–72 presents the relationship among the ozone feed gas concentration and the applied ozone dose versus the gas-to-liquid ratio.

For the same system described above, the upper limit for the gas-to-liquid ratio will be assumed to be 0.15 cfm/cfm (m^3/m^3). For a maximum applied dose of 3 mg/L, the 0.15 ratio can be easily achieved. With a 6 percent ozone (87.6 g O_3/m^3 NTP) source, a dose of approximately 11 mg/L can be applied before the gas-to-liquid ratio would become a problem.

It is apparent from Figure IV–72 that problems with the gas-to-liquid ratio will not occur unless a large ozone dose is required with a low ozone concentration in the feed gas. For low feed gas ozone concentrations (e.g., 1.5 to 2 percent [19.5 to 26 g O_3/m^3

NTP]), the upper limit for applied ozone dose will be from 3 to 5 mg/L for a maximum gas-to-liquid ratio of 0.2. For high–ozone demand waters, this may cause problems in injecting the gas into a single chamber.

Suggestions for the types of compromises that can be made to accommodate the gas-to-liquid ratio and minimum gas flow rate are:

- If the lower gas flow rate cannot be satisfied, decrease the minimum allowable ozone concentration, i.e., keep a constant gas flow and decrease the ozone concentration.
- If the upper gas-to-liquid ratio is exceeded, subdivide the dissolution stage and apply the ozone in successive stages (in series).
- If a balance between these two parameters cannot be met because there is too wide an operating range in ozone feed rate and water feed rate, divide up the contactor into parallel units and take a unit off line under low flow conditions.
- If no compromise is acceptable, a different type of contactor may be indicated.

In-line static mixer. Oldshue (1983) provides design guidance on static in-line mixers. Although the most efficient design is that which provides the necessary result with the least power consumption, the most energy-efficient design is not always the most practical nor most economical (e.g., if the lowest pressure drop for a desired mixing result requires an extraordinary mixer diameter). In using power from existing process flow (derived from a pump, gravity, or pressure source, for example), the motionless mixer consumes some of that power, which is expressed in terms of the pressure drop (ΔP) across the mixer. The energy consumed, then, is expressed as the horsepower required to overcome that pressure drop:

$$\text{Theoretical horsepower} = 5.83 \times 10^{-4}\, Q\Delta P$$

Where:

Q = flow rate, gal/min
ΔP = pressure drop, psig

Thus, pressure drop is a limiting factor in motionless mixer design.

Injector and static mixer design. The use of injector design requires the evaluation of the gas saturation solubility in the water to be treated (Joel 1990). The saturation limits of gases are temperature and pressure related.

Figure IV–73 shows the saturation limits of ozone in water relative to the ozone concentration in oxygen. The graphs indicate that as the ozone/oxygen concentration increases, so does the saturation limit.

The designer of an injector dissolution system should develop, based on their specific requirements, an ozone/oxygen mixture solubility curve. All ozone/oxygen points below the solubility curve indicate dosages and concentrations of ozone that can be fed via injection and remain in solution if the pressure does not change along the system. Once the mass transfer has been accomplished, the designer must provide sufficient contact time. This can be accomplished in-line (in a pipe) or in a contactor. The following example reports on a study where the ozone was transferred into solution with an injector and a pipe was used as a contactor.

A 1989 study by the Long Beach Water Department (Bellamy et al. 1990) provides initial design guidance for prospective applications of static in-line mixers. Color removal and ozone transfer efficiency were used as measures of treatment performance (i.e., the dependent variables). (See Figure IV–2, sec. A.1 of this chapter.) These parameters were evaluated to assess the effect on treatment of a change in an independent variable.

Sidestream injection of ozone was determined to be more effective than direct

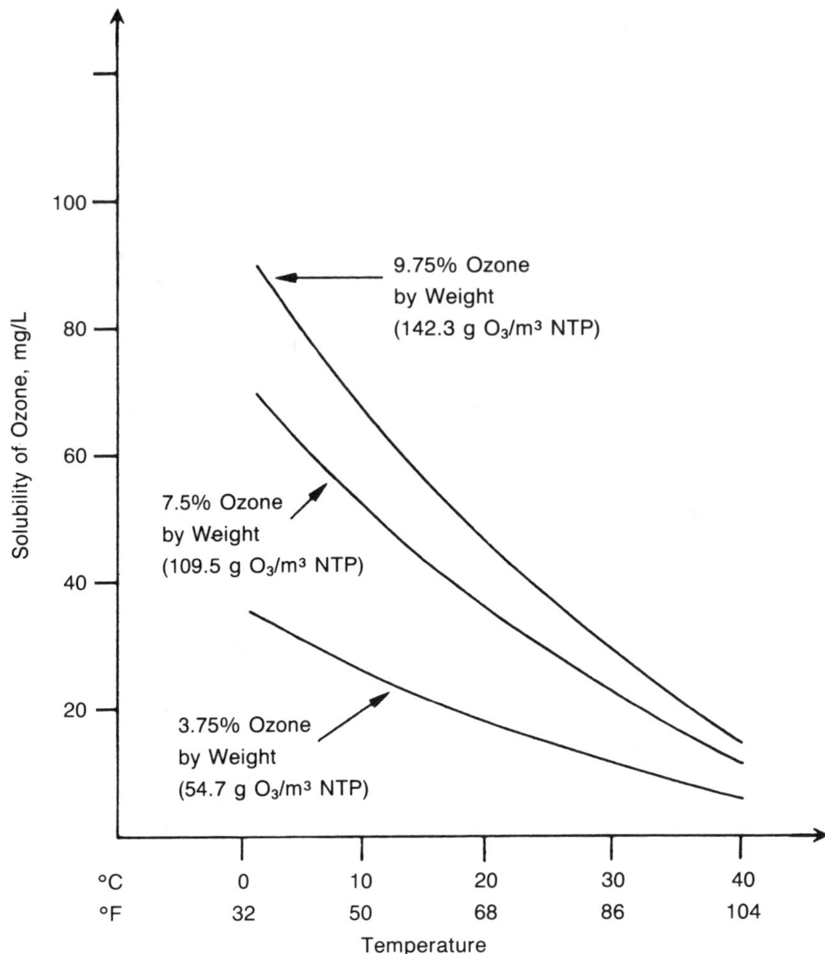

Figure IV–73 Solubility of Ozone in Water at One Atmosphere for Different Ozone Concentrations in Oxygen

injection into the process flow. Sidestream injection resulted in higher ozone transfer efficiency as a result of increased mixing at the eductor, a higher ozone utilization rate (color units removed per mg/L of ozone applied), and a reduced contact time requirement to obtain a specific level of treatment.

The pressure in the dissolution and mixing system (i.e., the injector and static mixer) and the pressure in the contact pipe did not have a significant effect on ozone residual or color removal. Pressures were evaluated at between 2 and 7 psi (0.14 to 0.48 bar). Ozone doses up to 8.1 mg/L were evaluated. Increasing the applied ozone dose increased color removal. Increasing the ozone dosage also increased the sidestream gas-to-liquid ratio, which resulted in a lower ozone transfer efficiency. The ozone utilization rate for in-line ozone treatment was higher than for conventional ozone treatment (i.e., fine bubble diffusion in a countercurrent contactor).

The injector nozzle and throat sizes are not independent variables. Changing the sizes of nozzle and throat will cause a change in the pressure drop across the injector, which is directly related to the energy input provided at the injector. Increased energy input increases the ozone transfer efficiency and color removal. Each injector has a specific operating range, and operating outside of that range results in ineffective ozone injection. For a specific injector size, sidestream flow rate determines the level of dissolution energy implemented across the injector. At the test conditions examined during this study (i.e., injector size and 10 percent sidestream flow to total process flow [300 gpm—about 68 m^3/h]), a 0.5-hp (0.37-kW) dissolution energy was determined adequate to achieve optimum ozone utilization rates.

Changes in flow rate (velocity) through the static mixer did not significantly affect color removal. However, flow rate is directly related to contact time, and the two variables have to be evaluated together during design. The results of testing of combined hydrogen peroxide and ozone oxidation testing indicated that as the H_2O_2/O_3 ratio increased, the color removal decreased, the ozone transfer efficiency increased, and the ozone residual decreased.

The measure of the effect of the sidestream–to–main flow ratio on treatment performance was complicated by the effects of the gas-to-liquid ratio at the injector. As the size of the injector decreases or the pressure drop across the injector decreases, there is no significant effect on treatment efficiency as long as the flow rate is sufficient and the injector is sized to maintain the minimum dissolution energy required (i.e., 0.5 hp [0.37 kW]). These results suggest that the sidestream flow rate should be designed so that:

- Energy input is minimized
- The sidestream gas-to-liquid ratio will not be limiting (i.e., the ratio is kept below 0.067 cfm/gpm [0.5 cfm/cfm or $m^3/h/m^3/h$])
- The design flow rate is within the specified operating range of the eductor to maintain the minimum dissolution energy required (e.g., 0.5 hp [0.37 kW] for the eductors tested)

The effect of sidestream gas-to-liquid ratio on color removal is evident only when the ratio becomes a limiting factor (i.e., when the ratio is above 0.067 cfm/gpm or 0.5 cfm/cfm [$m^3/h/m^3/h$]). Above the critical value, increasing the sidestream gas-to-liquid ratio will reduce the color removal and the ozone transfer efficiency.

The effect of percent ozone in feed gas on the in-line system performance is complicated by the effects of the sidestream gas-to-liquid ratio. Increasing percent ozone in feed gas will decrease the gas-to-liquid ratio. Below the critical gas-to-liquid ratio, percent ozone in feed gas did not affect the in-line system performance.

Packed column design. A paper by van Leeuwen (1980) provides guidance as to the design considerations for packed column ozone contactors based on the design of a wastewater reclamation facility. Design considerations included:

- Packing material: Available packing material included stainless steel pall rings, porcelain Raschig rings, and porcelain Intalox saddles. Polypropylene cascade rings were not considered because of hardening of the rings when exposed to ozone during a material compatability test.
- Flooding velocity in columns: The correlation developed by Sherwood and Holloway (1938) can be used for dumped packing:

$$(U_0^2 S\, g^{-1} F^{-3})\,(\rho_G \rho_L^{-1})\mu^{0.2} = f[(L/G)\,\rho_G \rho_L^{-1}]^{0.5}$$

Where:

U_0 = superficial gas velocity (based on empty column), ft/s
S = surface area of packing, ft^2/ft^3
g = acceleration of gravity, ft/s^2
F = fraction of free volume in packing, ft^3/ft^3
ρ_G, ρ_L = densities of gas and liquid, respectively, lb/ft^3
μ = viscosity of liquid, cP
L/G = ratio of liquid to gas mass flow rates.

- Mass transfer coefficients for the packed tower.
- Depth of packing material.

The design described by van Leeuwen (1980) required modification in the form of providing "redistributors" of 2/3 depth, 1/3 depth, and the top of the bed to improve gas and liquid distribution. Ozone transfer efficiencies of 88 to 95 percent were measured.

Spray contact chamber design. Design considerations for spray contact chambers are difficult to define because of the broad scope of this category. Information on the application of mechanical surface "aerators" as a dissolution device may be derived from the design guidance documents of oxygen/ozone system licensors (Lotepro 1988).

Deep U-Tube design.

Important design parameters for sizing the Deep U-Tube: When sizing the Deep U-Tube to optimize ozone transfer efficiency, the important design parameters are gas-to-liquid flow ratio, depth of the tube, and gas-phase ozone concentration (Brodard et al. 1986). The reasons are:

- The water flow rate must be great enough to propel the gas bubbles downward through the inner tube, yet low enough to minimize head loss through the contactor. Thus, the tube diameter, water velocity, and gas-to-liquid ratio must be optimized to maximize transfer efficiency, minimize head loss through the tube, retain plug flow characteristics, and meet the treatment plant's design capacity.
- The depth of the tube affects both the head loss through the contactor and the transfer efficiency resulting from increased pressure. According to Brodard et al. (1986), head loss is less sensitive to tube depth than to water velocity.
- The gas-phase ozone concentration affects transfer efficiency because of its influence on the gas transfer driving force.

Gas-to-liquid ratio. Based on an optimization study conducted on a demonstration-scale Deep U-Tube installation in Aubergenville, France, the design parameters that influence ozone transfer efficiency in this type of contactor are, in order of increasing importance, gas-to-liquid flow ratio, tube depth, and gas-phase ozone concentration (Brodard et al. 1986). The optimal gas-to-liquid ratio is less than 15 percent (Brodard et al. 1986), and 10 percent is generally best (MWDSC 1988). For example, at the Western Transvaal Plant (South Africa), the largest of the full-scale installations, a water velocity of approximately 6 ft/sec (1.83 m/sec) and a gas-to-liquid ratio of 8 percent are optimum.

Depth. Results of this optimization study (Brodard et al. 1986) also demonstrated that a 20-m (66-ft) contactor depth is sufficient to provide optimal transfer efficiency.

Head loss. Head loss through a Deep U-Tube is a function of tube diameter, water velocity, and gas flow rate. Head loss increases as the tube diameter decreases and as the water velocity increases. The gas-to-liquid ratio also influences head loss; substantial increases in head loss will be observed if the water velocity is too low and the gas flow rate is too high. The relationship between liquid velocity, gas-to-liquid ratio, and head loss through a demonstration-scale Deep U-Tube contactor at Aubergenville, France is shown in Figure IV–74. As indicated, head loss increases sharply as a function of both water velocity and gas-to-liquid ratio (Brodard et al. 1985).

Head loss through the Deep U-Tube as a function of both liquid and gas flow rate has been measured through the 31.7-mgd (5000-m^3/h) installation in South Africa. The results indicate that for gas-to-liquid ratios ranging from 0 to 16 percent, the head loss through the tube varies from 1.4 ft (of water column) to 3.9 ft (4.18×10^{-2} to 11.66×10^{-2} bar), respectively. In comparison, head losses through large conventional bubble diffuser contactors (without degassing considerations) range from 1.5 ft to 2.5 ft (4.48×10^{-2} to 7.47×10^{-2} bar) (based on data from the LAAFP; the North Bay Regional Water Treatment Plant in Fairfield, California; and the Kennewick, Washington Water Treatment Plant (MWDSC 1988).

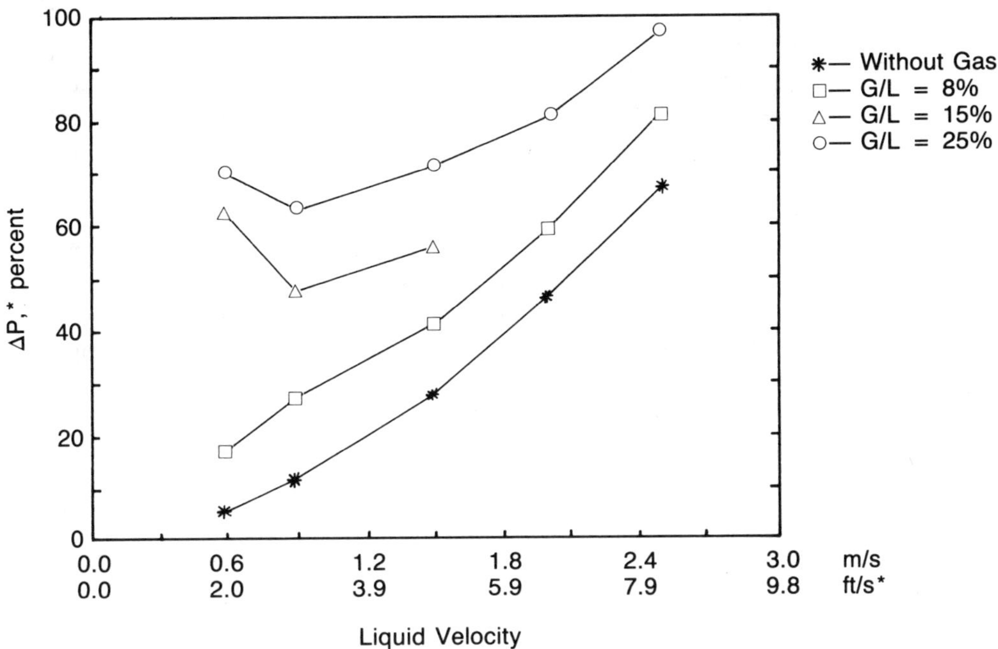

Source: Brodard et al. (1985).

Figure IV–74 Influence of Liquid Velocity and Gas Flow/Liquid Flow Ratio on Head Loss Development in a Deep U-Tube

Reactor hydrodynamics. Dye tracer tests have been conducted through two of the existing Deep U-Tube installations, Aubergenville (52 gpm [11.8 m^3/h]) and Le Pecq (5.1 to 9.5 mgd [804 to 1498 m^3/h]).

The shape of the tracer curve (concentration vs. time, or "C" curve) for the test conducted at the Aubergenville installation indicates that the contactor operates as a plug flow reactor. The dispersion number calculated for the operating conditions used during the test was 0.0068, which, according to Levenspiel (1972), also indicates that the contactor operates as a plug flow reactor. The theoretical (hydraulic) retention time and calculated mean retention time were 4.20 min and 3.67 min respectively.

Materials of construction. The materials of construction recommended for Deep U-Tube ozone contactors are stainless steel and concrete (MWDSC 1988), since both are resistant to oxidation by ozone. Although concrete is less expensive than stainless steel, Deep U-Tubes have been constructed of both materials. At the South African installation, both the inner and outer tubes are constructed of concrete; at the French installations, stainless steel is used for both the inner and outer tubes.

Sweeping porous plate diffuser contactor design. In designing and sizing a sweeping porous plate contactor, several parameters require optimization:

- The percentage of total raw water flow making up the jet of pressurized water which sweeps the plates,
- The distance between the nozzle of the jet and the porous plate,
- The water pressure of the jet,
- The depth of the water column, and
- The downward flow rate of the water through the ozonation chamber.

Based on pilot-scale optimization studies (Baetz et al. 1986; Faivre et al. 1987), in which (1) the flow of the jet was between 5 and 10 percent of the total influent flow rate, (2) the pressure of the jet ranged from 0.5 to 1.0 atm (7.35 to 14.70 psig) and (3) the depth of the water column was 2 m (6.6 ft), the liquid velocity at the surface of the plate was approximately 0.33 ft/s (10 cm/s) and the bubbles generated ranged from 0.6 to 0.8 mm in diameter. The bubbles produced by the sweeping porous plate diffuser system enhance ozone mass transfer. Results of this study showed that sweeping, compared with not sweeping, enhanced the mass transfer coefficient (for oxygen), K_La, by 50 percent. The study also showed that, for the same volume of water, oxygen transfer in a 2-m (6.6-ft) column of water with sweeping exceeded the amount of oxygen transferred in a 4-m (13.2-ft) column of water with non-sweeping porous plate diffusers. Results of similar tests using ozone were not available; it is not known if similar differences in mass transfer coefficients (for sweeping versus nonsweeping systems) would be measured for ozone.

Results of this same study also showed that (1) entrainment of the fine bubbles in the downward flow of water is proportional to the water velocity but essentially independent of the gas injection rate and (2) the optimum distance between the nozzle and the diffuser plate is a function of the plate diameter (Faivre et al. 1987).

Turbine contactor design. The design for these contactors depends on the type of turbines used. Design parameters for this dissolution system should be obtained from the system vendors.

IV.E INSTRUMENTATION AND CONTROL SYSTEMS

IV.E.1 General Considerations

Ozone has a number of applications, as has already been mentioned in chapter III. Control methods based on ozone measurement techniques and devices described in, must therefore be selected to suit the purpose. For particular applications, some control systems will be more sophisticated than others. In the field of continuous control systems to ensure fully automated operation, users have a number of alternatives.

Control of ozone production by treated water flow rate. The most basic control system is based exclusively on the flow rate of the treated water, with the ozone dosage previously having been determined during pilot plant tests, in the laboratory, or onsite (see Figure IV–75). The system is applied in cases where ozonation is being used for:

- Removal of manganese;
- Improvement of clariflocculation stages;
- Direct filtration;
- Oxidation of micropollutants; and
- Ozonation prior to activated carbon filtration.

This is due to the fact that, for such applications, it is often hard to identify sensors on the market suitable for the continuous measurement of one of the parameters for which the ozonation is being applied. This system does not provide automated control that takes into account variations in water quality, which may occur suddenly and on a large scale.

Two possibilities exist for varying ozone dosages proportional to flow:

- The power of the ozone generator can be varied to pace on flow without altering the gas throughput. The ozone concentration in the gas will, however, vary according to the operating conditions.
- The flow of ozonized gas can be changed at the same time as the ozone generator power in order to maintain a constant operating ozone concentration. In addition to a flow rate indicator, this method requires continuous metering of the ozone concentration in the gas at the outlet of the ozone generator. It is a solution that enables constant ozonized gas transfer conditions to be maintained in the contact chamber (MacKay 1985).

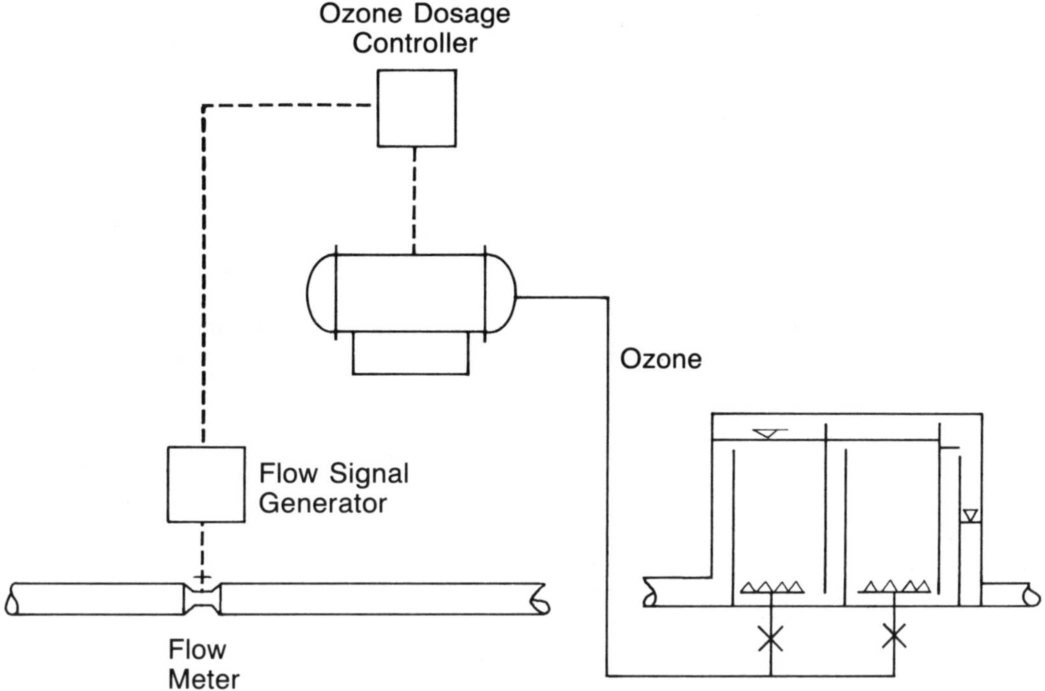

Figure IV–75 Ozone Dosage Control Based on Water Flow Rate

Control of ozone production by ozone residual in water. This system is used mainly to assure that water is suitably disinfected with respect to bacteria and viruses. The key control sensor will be a dissolved ozone analyzer. The residual recorded by the sensor is compared to a set value and the ozone dosage is adjusted (Louboutin 1985). Here again, the plant can operate at a fixed or variable ozonized gas concentration. This method of control, illustrated by Figure IV–76, allows for fluctuations in the quality of the water to be disinfected (the residual ozone is maintained independent to the ozone demand of the water). It can only be used if the facility is operating at a relatively constant flow rate. If this is not the case, cascade control, as described below, must be provided.

Control of ozone by water flow rate and residual ozone. In this case, illustrated by Figure IV–77, mainly applied to primary disinfection, the two control modes described above are applied in a cascade sequence. The secondary control, based on the residual dissolved ozone level, accurately adjusts the treatment rate after an initial regulation of the ozone production in accordance with the flow rate (i.e., pace on flow and trim on ozone residual).

Closed loop control based on offgas concentration. A similar kind of ozone dosage adjustment as presented for residual control can also be obtained by measuring the residual ozone present in the offgases issuing from the contact columns as illustrated by Figure IV–78. This system is applicable if the operators are confronted with problems due to excessive fouling of the residual ozone sensors. It would not be usable when residual ozone control is required to satisfy regulatory requirements except as a secondary measurement/control. This adjustment involves working in constant hydrodynamic conditions (constant ozone concentration in the ozonized gas). The variations of residual ozone in the gas to be taken into account should be solely those due to fluctuations in the quality of the water and not to variations of the conditions in which the ozonized gas is injected into the water.

The concept was originally developed for wastewater disinfection wherein

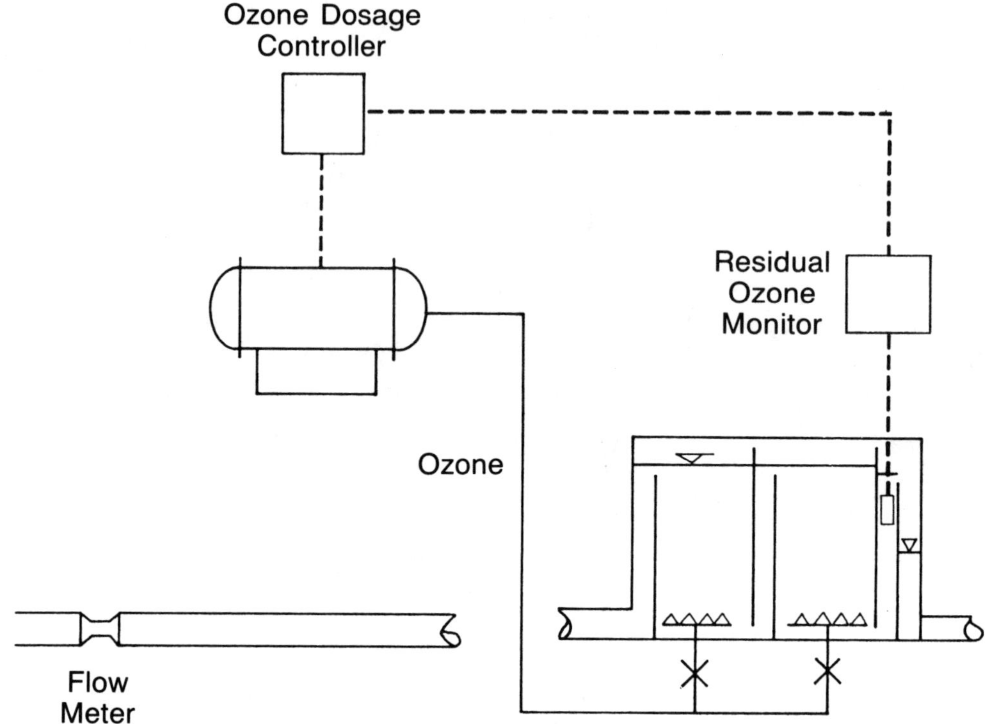

Figure IV–76 Ozone Dosage Control Based on Dissolved Residual Ozone Measurement

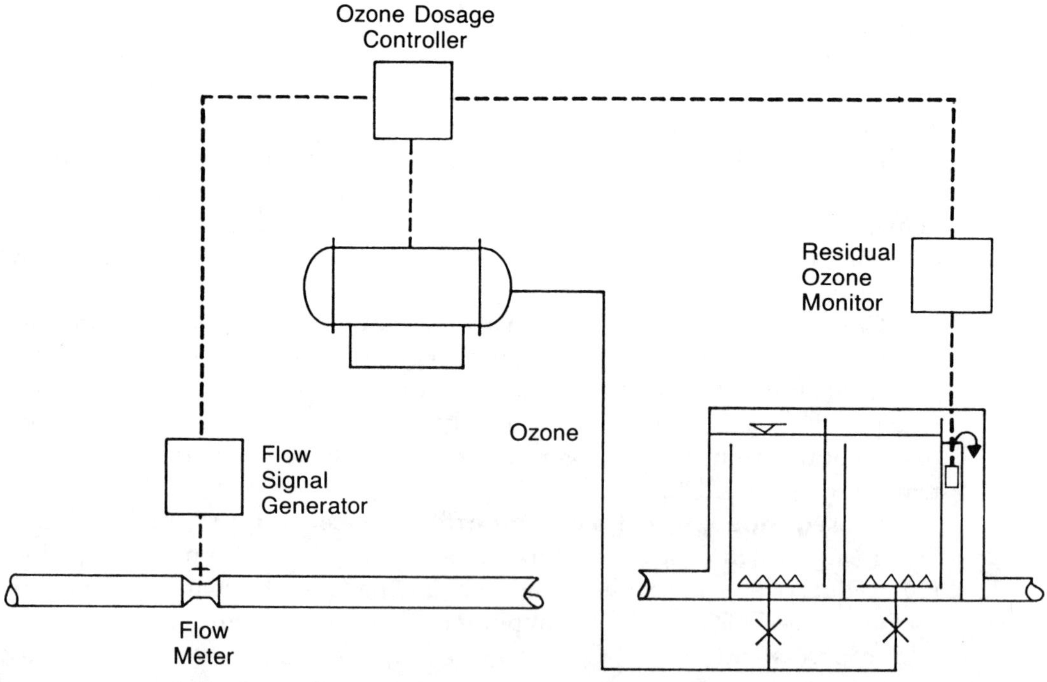

Figure IV–77 Ozone Dosage Cascade Control Based on Water Flow and Dissolved Ozone Residual

maintaining an ozone residual is difficult and fouling of the residual ozone sensors frequently occurs. On the other hand, measurement of offgas ozone concentrations was comparatively simple and reliable with chemiluminescent or UV-absorbent monitors. It

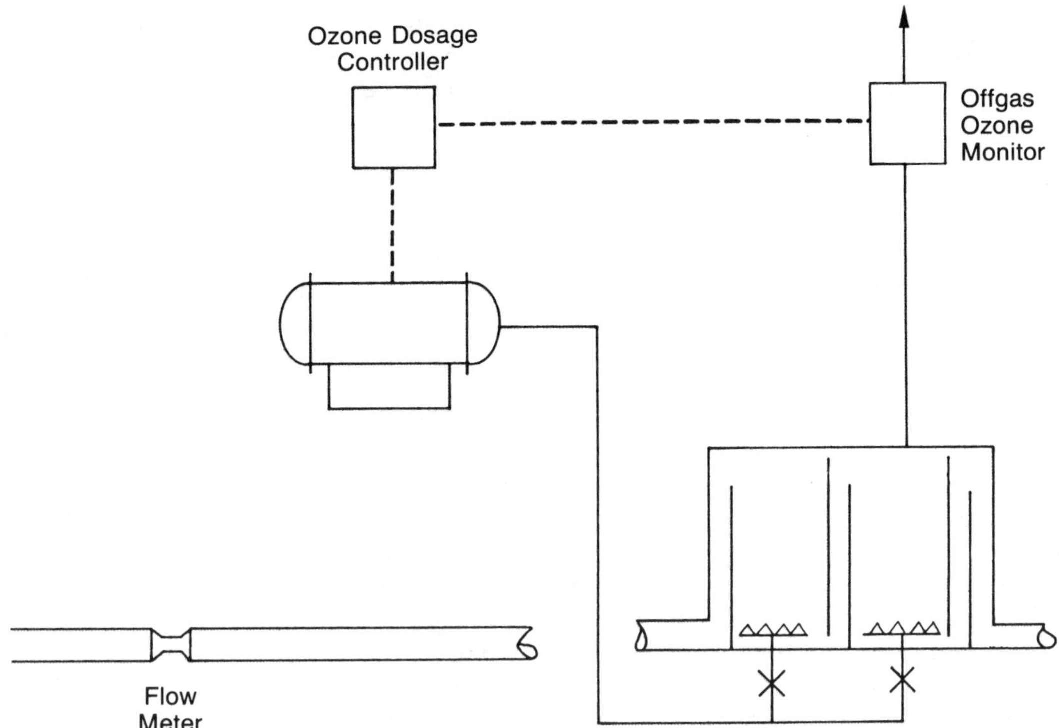

Figure IV–78 Ozone Dosage Control Based on Offgas Ozone Levels

was originally believed necessary to operate on a cascade closed loop mode with liquid flow being the primary signal and offgas ozone concentration as the trimming signal. It was subsequently found that offgas concentration was sufficient as a primary signal (Venosa and Meckes 1986). In wastewater treatment strict control over ozonized feed gas flow to liquid flow rate ratio seems not to be necessary if it is maintained in the range of 0.1 to 0.5 (Venosa et al. 1985). However, a conservative design approach would be to provide for cascade control in the original construction bid documents, thereby allowing the operational staff to evaluate several modes of control for their specific application.

This concept is being applied for automated control of several wastewater ozonation facilities: Bahrein; Indianapolis, Indiana; Springfield, Missouri; and El Paso, Texas. The latter facility is a wastewater reclamation facility that utilizes ozonation on a liquid stream that is nominally finished water. This concept is also being incorporated in recently designed drinking water plants such as at Tucson, Arizona.

Control based on specific or overall representative water quality parameters. In this case, and according to the objective to be achieved through ozonation (other than disinfection), the basic sensor in the control sequence is an analyzer capable of measuring, in a continuous operation, one or several specific compounds whose removal is one of the purposes of ozonation (e.g., iron and manganese removal, color control).

In cases where ozone is used to improve the removal of organics by biological filtration as discussed in chapter III sec. K, a sensor measuring a parameter representative of the water's organic content can be used (for example, the absorption of ultraviolet radiation at 254 nm [Dietrich et al. 1981] or fluorometric analyses).

This type of control, based on a sensor that either is specific to a given pollutant or reflects an overall parameter, is not widely used in connection with control of systems designed for the removal of a specific organic parameter. This situation is primarily due to the absence on the market of a sensor designed for the continuous measurement of several specific parameters (e.g., iron, manganese) or overall parameters (e.g., oxidizability test, biodegradable organic carbon measurement). Conversely, some continuous sensors that are available, such as the TOC meter, can be used more or less easily depending on TOC levels in the water, but they are costly.

Among feasible short-term solutions, the only two that can be considered are UV absorption at 254 nm (after care is taken to eliminate any possible ozone from the samples, ozone being a reagent that is intensely absorbent at that particular wavelength, as discussed in sec. B, Chapter II) or fluorometry. The control system chosen can be identical to the one described under "Control of ozone by water flow rate and residual ozone," the residual ozone analyzer being replaced by a spectrophotometer.

Use of mathematical models for managing ozone facilities. In view of the development of computers, it is quite foreseeable to work with systems that include curves or nomographs obtained through mathematical modeling and giving the correct rate of ozonation in order to achieve the required level of abatement for any particular parameter, irrespective of seasonable variations in influent quality.

This type of system has been described in connection with the use of ozone before filtration on activated carbon (Duguet et al. 1986) and with preozonation as a means of improving clarification efficiency in the Choisy-le-Roi waterworks (Bablon and Gugenheim 1985). At the latter facility, designing the system involved an onsite test program lasting one year, during which 22 parameters were monitored in terms of four variable plant operating factors. A statistical analysis of the results enabled evaluation of the most representative analytical parameters for this particular application of ozone (Pascal et al. 1984). These parameters served as a basis for the construction of nomographs now in operational use. This example shows that such systems take a long time to design and can only be used on very large facilities.

IV.E.2 Dewpoint Measurement and Drying System Control

Dewpoint is a measure of the absolute moisture content of the ozone generator feed gas and is commonly used to denote the temperature at which moisture will begin to condense from a flow of gas. Dewpoint is monitored continuously by electronic moisture-measurement devices that provide direct readings and are continuous in their operation (Weiner 1974). Some devices contain a sensing element consisting of a hygroscopic cell, which is usually lithium chloride. Since there is a dynamic equilibrium between the moisture in the pores of the cell and the sample fluid, the electronic circuit measures the capacitance. The higher the moisture content, the higher the capacitance. Manufacturers of these moisture analyzers claim dewpoint ranges down to −80°C (−112°F). Various sensing elements are used for different true ranges.

Several electronic analyzers measure surface conductivity of an inert nonconductor. Others use a thermoelectric technique that combines optical detection and temperature control. When the system is cooled, water vapor condenses on the mirror; an optical sensor then detects a change in the light reflected from the surface. This change in reflected light generates a signal, as does the temperature sensor. The temperature at which condensation forms is the dewpoint. The dewpoint monitor can be calibrated by means of the dewpoint cup or the Alnor analyzer (Weiner 1974).

Dewpoint cups function by determining the temperature at which the first sign of vapor condensation is visible. This apparatus consists of a small, polished stainless-steel cup placed in a container into which the gas sample is passed. The container has a transparent window, where the gas impinges, that makes the outer surface of the cup visible. A thermometer in the acetone-containing cup, into which dry ice is slowly introduced, measures the temperature of the mixture. As the gas to be analyzed is passed into the container, the polished surface is observed. As soon as condensation is visible, the temperature of the bath is recorded as the dewpoint of the gas sample. Accuracies of ±3°C (± 1.7°F) are obtainable.

The Alnor analyzer is a device to measure water vapor content in gas, which relies on the relationship between pressure and temperature during adiabatic expansion. A sample of gas at atmospheric pressure is drawn into a small chamber, where it is compressed to some pressure above atmospheric (gas temperature prior to compression must be noted). The sample is then rapidly vented to atmospheric pressure, resulting in sudden expansion and cooling.

If the cooling is sufficient to drop the temperature to or below the dewpoint, a fog will be visible. To be certain that the correct measurement is made, the pressure of the sample is raised in increasing steps, so that the minimum pressure ratio at which the fog appears is observed. Mathematically, this is defined as:

$$T_f = T_i \left(\frac{P_f}{P_i}\right)^{(K-1)/K}$$

Where:

T_f and T_i = final and initial temperatures
P_f and P_i = final and initial pressures
K = (specific heat at constant pressure) / (specific heat at constant volume).

The final temperature is regarded as the dewpoint. The device measures dewpoints of gases down to −80°F (−62.2°C), and is reported to be accurate within ±2°F (±3.6°C).

The dewpoint monitor may be used in several control strategies in the gas preparation system, including measurement of the moisture content of the prepared gas immediately upstream of the ozone generator. (See sec. B.3 of this chapter and sec. B and C, chapter V.) The dewpoint measurement is used to assure that the prepared gas dewpoint does not exceed the specified level. Electronic equipment used to sound alarms, initiate system shutdown, and provide a signal for recordkeeping purposes should be provided for all ozone systems.

An additional use of dewpoint measurement is identified in the preceding gas preparation section (See sec. B.3 of this chapter). Desiccant dryer regeneration may be controlled by the dewpoint of the gas being discharged from a specific tower. Regeneration of the tower would be initiated by the gas dewpoint reaching a predetermined level. This level of complexity is normally applied only on larger installations or on systems which are highly motivated in energy conservation.

IV.E.3 Gas Flow Meters

Measurement of gas flows throughout the system is important for control and monitoring purposes. Selection of types of meters and materials of construction depends on the application and the gas being measured. The material of that portion of the meter in contact with the gas being measured must be minimum Schedule 10, AISI stainless steel. These meters are discussed as follows:

1. Ambient or prepared air flow meters: Diaphragm differential pressure type.
2. High-purity oxygen: "Pitot" tube type with annular flow sensor with internal interpolating tube producing a differential signal, the square root of which is directly proportional to flow.
3. "Dry" ozonized air: Flow tube with differential pressure measurement with

temperature and pressure compensation. Alternatively, a rotameter can be used, of the appropriate construction materials.

4. "Dry" ozonized high-purity oxygen: Flow tube with differential pressure measurement with temperature and pressure compensation. Alternatively, a rotameter can be used, of the appropriate construction materials.

5. "Wet" ozonized air: "Pitot" tube type with annular flow sensor with internal interpolating tube producing a differential signal, the square root of which is directly proportional to flow.

6. "Wet" ozonized high-purity oxygen: "Pitot" tube type with annular flow sensor with internal interpolating tube producing a differential signal, the square root of which is directly proportional to flow.

IV.E.4 Alarms

A number of conditions can exist that could be detrimental to plant personnel, the operating equipment, the ozonation process, or the environment of the plant. It is good practice to establish acceptable limits of specific parameters above or below which sensor measurements will generate alarm signals requiring action by operating personnel and/or initiating independent automatic action of the equipment, which in certain situations might be shutdown of the entire system.

Alarms that could be required (depending on the size and sophistication of the ozonation system) include the following:

1. Gas preparation:
   - Water flow to units requiring cooling water.
   - High discharge water temperature from units requiring cooling water.
   - High differential pressure across gas filters.
   - High discharge regeneration gas temperature from thermally regenerated gas desiccant dryers.
   - High gas pressure downstream of gas pressure reducing valves.
   - High feed gas dewpoint upstream of ozone generators.
2. Ozone generators:
   - Cooling water flow and gas flow to ozone generators.
   - High discharge water temperature from ozone generator.
   - Low or high ozone concentration in ozonized gas discharged from generator.
   - High ozonized gas temperature from ozone generator.
   - All the necessary electrical safety devices.
3. Ozone dissolution system:
   - High pressure in contactor head space.
   - Low pressure in contactor head space.
4. Ozone destruction system:
   - Gas flow to destruction units.
   - High temperature in ozone destructor.
   - High ozone levels in exhaust gas from destructor.
5. General environment:
   - High ozone levels in ozone generation system enclosed space.

IV.F OZONE DESTRUCTION SYSTEMS

IV.F.1 Reasons for Ozone Destruction

Personnel, equipment, structural components, and the general environment must be protected from exposure to high levels of ozone. Even when contactor offgas is reinjected into an upstream contact chamber, the offgas from this upstream chamber may contain as much as 0.2 to 0.5 g/m^3 NTP [0.015 to 0.038 percent] (Masschelein

1982c). These concentrations far exceed the maximum allowable ozone concentration for an 8-h work day of 0.0002 g/m^3 or 0.1 ppm by volume. This standard is applicable in the United States (set by the U.S. Occupational Safety and Health Administration) as well as in European countries.

IV.F.2 Current Design Practice

Thermal ozone destruction. Thermal ozone destruction systems rely on the decomposition of ozone at increased temperatures. In a clean vessel at room temperature, the half-life of ozone may range from 20 to 100 h in dry air (Manley and Niegowski 1967). At 120°C (248°F), the half-life is reduced to the range of 11 to 112 min, and further reduced at 250°C (482°F) to only 0.04 to 0.4 s. In a thermal ozone destruction unit, the offgas is heated to a prescribed temperature, typically between 300°C and 350°C (572 to 662°F) for a short period of time, usually less than 5 s. Since discharge of gases at these elevated temperatures also has environmental concerns, a heat recovery heat exchanger is usually provided.

Heating in a single-pass system (without heat exchange) is a simple process. Head loss is quite low, typically less than 2 in. of water column (4.98×10^{-3} bar) (Masschelein 1982c). Gas is discharged from the system at temperatures of 250° to 300°C (482 to 572°F), requiring use of refractory materials for exhaust system construction. This system is energy-intensive, however, because the energy imparted to the gas stream by heating is lost to the atmosphere on discharge. For this reason, heat recovery units are usually included in thermal destruction systems. Different types of heat exchangers are used, with cross-flow, gas/gas shell and tube, and gas/gas plate type units the most common (Coste 1982). Heating the offgas permits recovery of a large portion (up to 80 percent) of the input energy by preheating the incoming gas.

The heat exchanger allows the heat from the gas stream passing by the heating coils to preheat the gas stream about to enter the heating coils. Head loss for thermal systems with heat exchangers can be as high as 3 ft of water column (8.97×10^{-2} bar). Gas discharged from the system normally has a temperature of 70 to 100°C (158 to 212°F) (Masschelein 1982d; Orgler 1980). Figure IV–79 illustrates a thermal ozone destruction system.

Water sprays are normally provided to prevent foam from entering the heat exchanger and heating elements. A demister consisting of a stainless steel mesh screen may be provided to maximize water droplet removal. A fan is provided to move gas through the system when the contactor is operated under slightly negative pressure, which is the recommended case.

Thermal ozone destruction is the most commonly used method in Europe (Langlais 1989).

Catalytic and thermal/catalytic destruction systems. The use of catalysts for ozone destruction is a fairly recent development, but this method is the most commonly used in the United States today (1990). Specific information regarding the composition of catalysts is frequently treated as proprietary. However, metals, metal oxides, hydroxides, and peroxides are known for their ability to decompose ozone (Horst 1982). Many catalysts are based on palladium, manganese, or nickel oxides. Sometimes the catalyst is coated on a support medium, such as palladium-based coating on aluminum granules (Masschelein 1982c). According to European catalyst manufacturers, 0.25 to 0.4 kg (0.55 to 0.88 lb) of catalyst is required to treat 1 standard m^3/h of offgas (Chapsal 1989).

A schematic of a thermal/catalytic ozone destruct system is shown in Figure IV–80. A demister minimizes moisture from coming onto contact with the catalyst. Moisture condensation on the catalyst can greatly reduce its effectiveness and life expectancy. A heating chamber is usually an element of catalytic systems to maintain the temperature

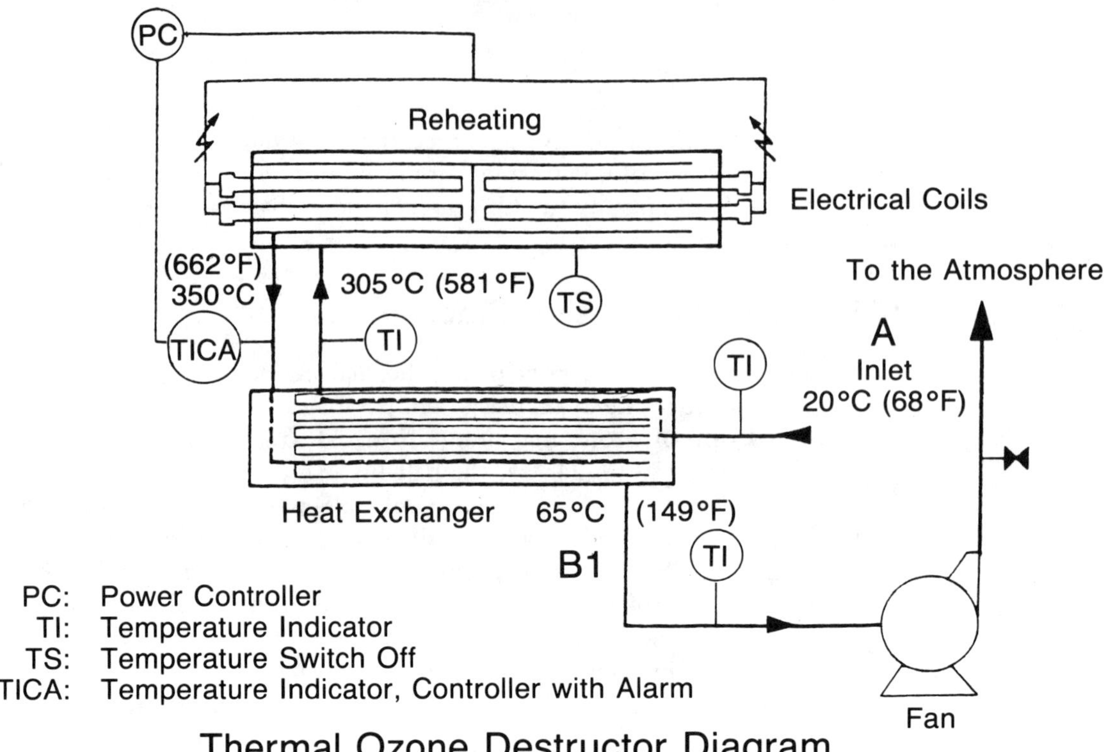

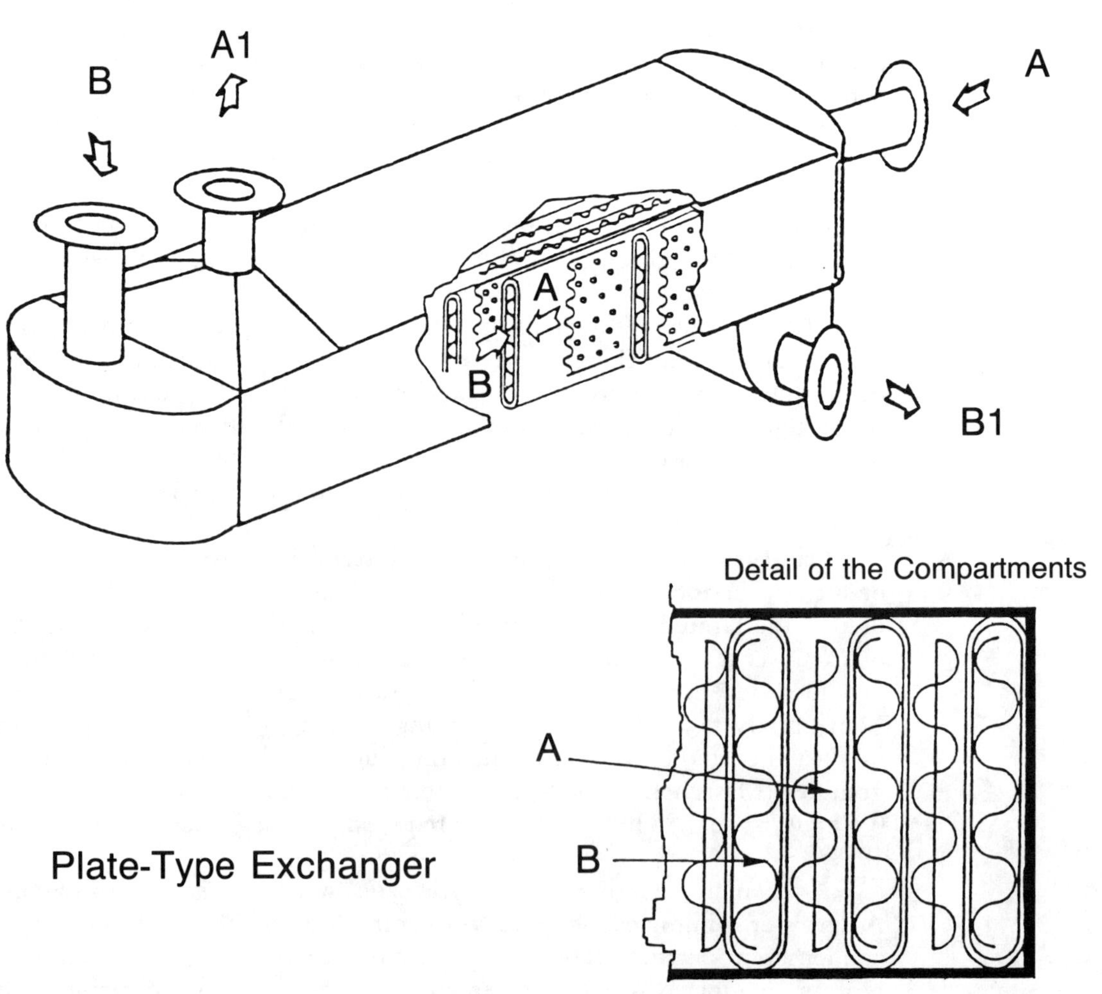

Source: Chapsal (1989).

Figure IV–79 Thermal Ozone Destruct System with Heat Recovery. A: Diagram; B: Plate-Type Exchanger

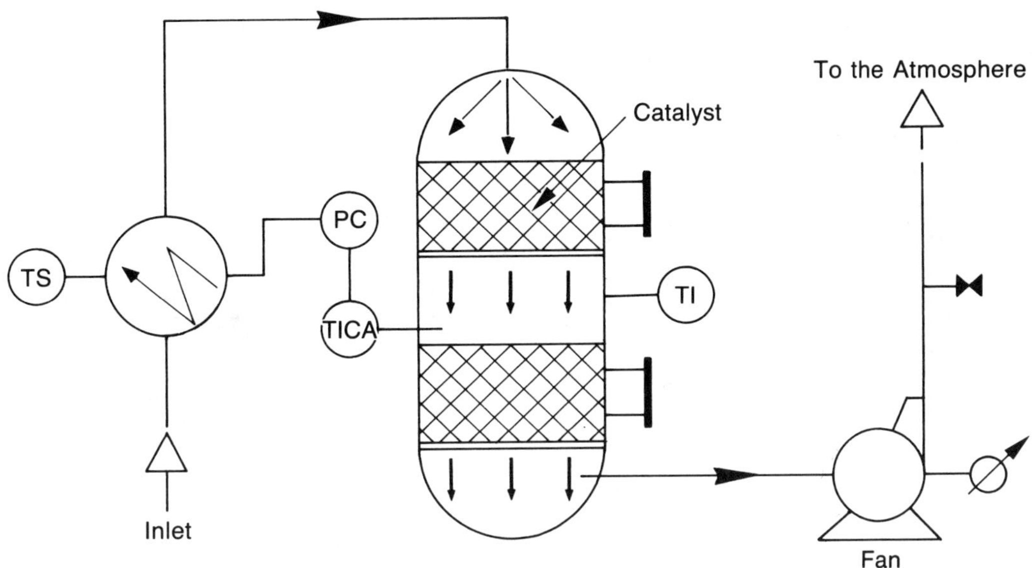

TI: Temperature Indicator
TS: Temperature Switch Off
TICA: Temperature Indicator, Controller with Alarm
PC: Power Controller

Source: Chapsal (1989b).

Figure IV–80 Thermal-Catalytic Destruct System

of the water-saturated contactor offgas above its dewpoint (Orgler 1982). Operation without preheating of the gas prior to catalytic destruction is not recommended due to the continuing danger of possible moisture accumulation on the catalytic medium. The required temperature increase depends on the nature of the catalyst. Metal catalysts may be operated at temperatures as low as 29°C (84.2°F), while metal oxide catalysts may operate at 50° to 70°C (122 to 158°F) (Horst 1982; Coste 1982).

Catalysts may be reversibly or irreversibly deactivated by nitrogen oxides, chlorine compounds, or sulfides, depending on the nature of the catalyst (Orgler 1982). To illustrate this point, a thermal ozone destruction system was chosen by the designers of the 8-mgd (about 1260 m^3/h) Horgen plant in Switzerland because it was feared that use of prechlorination in the plant could damage the catalyst (Dyer-Smith et al. 1988). Improvement in catalyst composition has been rapid since catalysts for ozone destruction were introduced to the market. The newer catalysts can operate at higher gas flow rates and lower heating temperatures than their predecessors, have longer service lifetimes, and are more tolerant to sulfur, nitrogen oxides, and small quantities of chlorine (Coste 1982). Use of thermal catalytic destruction systems results in a slightly lower energy expenditure than thermal systems because the required temperature increase is lower when catalysts are used. Catalyst life expectancy can be on the order of 5 years when the system is properly operated (Horst 1982). Catalysts can be regenerated by heating to the temperature recommended by the catalyst supplier. This is frequently in the range of 500°C (932°F) for a period of 6 to 7 h (Coste 1982).

IV.F.3 Other Ozone Destruction Methods

Adsorption and reaction on GAC. Activated carbon adsorption has been used in the past, primarily in small ozonation systems, but its use is not presently

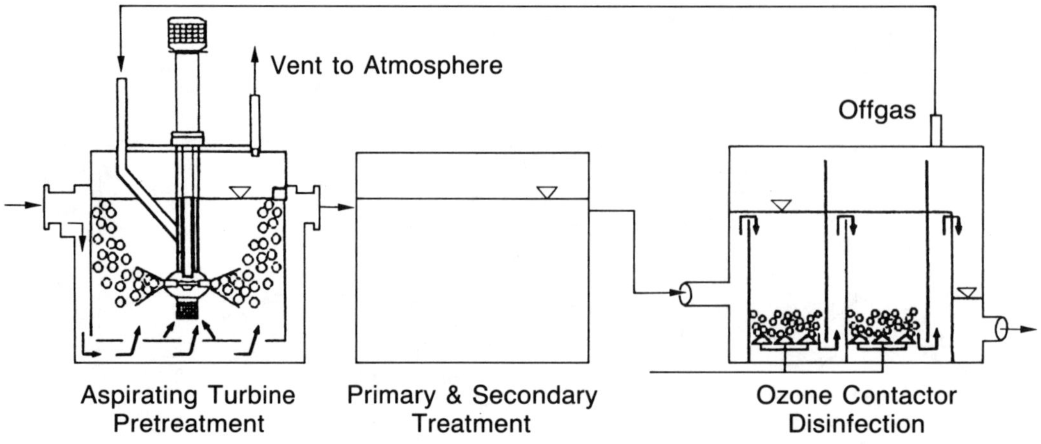

Figure IV–81 Offgas Treatment by Recycle

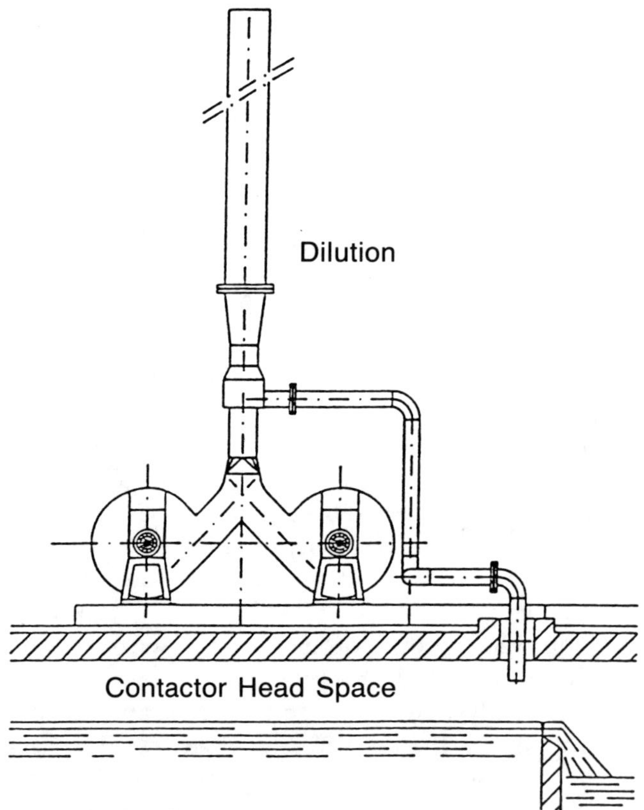

Source: Schulhof and Dyer-Smith (1988).

Figure IV–82 Air Dilution System

recommended. Reaction of ozone with dry activated carbon generates considerable heat. The carbon is consumed by slow combustion, resulting in the formation of carbon fines which may be explosive under the conditions existing in the destruction unit (U.S. EPA 1986). Some of the risks associated with explosiveness are avoided by moistening of the carbon bed (Horst 1982). This method of ozone destruction is difficult to monitor and is dangerous.

Recycling of offgas to ozone generator. Recycling of offgas to the ozone generator may be practical when oxygen is used for ozone generation. However, use of this technique prior to conditioning and drying the gas is limited due to the need to destroy the remaining ozone in the offgas and condition and redry the oxygen returning in the gas stream to the generator (see sec. IV.B.1). Also, all impurities that may have been stripped from the gas must be removed from the gas stream. The only water treatment plant reporting use of this technique is the Wittlaer Water Works Duisberg plant in Germany (Miller et al. 1978).

Reuse of offgas in an upstream contactor. A common practice in Europe is the reuse of contactor offgas in an upstream contact chamber, so that the excess ozone is utilized by the demand of untreated water. Offgas is typically drawn into the upstream chamber by use of an aspirating turbine or "emulsifiers" (see sec. IV.D.2). The Monroe, Michigan plant reuses the offgas by recompressing the offgas from the second chamber of a two-chamber contactor and feeding it through bubble diffusers to the first chamber of the contactor as illustrated in Figure IV–49 (sec. IV.D.1) (LePage 1981). There is no ozone destruction system associated with the first chamber of the Monroe contactor, which is very unusual for a plant in the United States. Plants designed for offgas reuse in this manner generally dispose of offgas from the upstream chamber by dilution and venting or by a thermal or thermal/catalytic system. These systems are recommended. Figure IV–81 shows an offgas treatment recycling from a postozonation to a preozonation step through an aspirating turbine.

Dilution and venting. Dilution and venting as a method of emissions control was common in the past, but its use is generally limited to older or rural water treatment plants in Europe. An exception in the United States is the plant at Monroe, Michigan, as discussed above. Dilution is only practicable after extensive reuse of the offgas (for example, by recontacting as described above) and by ensuring an appropriate rate of atmospheric dilution. A dilution ration near 100:1 by forced venting is usually sufficient to allow atmospheric discharge (Masschelein 1982c). Langlais (1989) recommends a dilution ratio near 10 with fresh air. Forced venting allowing a discharge velocity of the diluted gas up to 15 m/sec is usually sufficient (the high velocity of the offgas at the outlet of the stack leads to a secondary dilution nearly as important as the primary dilution). Nevertheless, the dilution method is not likely to be sufficient under severe inversion conditions. Figure IV–82 illustrates an ozone offgas dilution system.

Chemical destruction. Dyer-Smith and Schulhoff (1988) describe the use of a chemical solution for ozone destruction. They reported laboratory work using a solution of NaOH, Na_2SO_3, and $Na_2S_2O_3$ for wet scrubbing of offgas. To date (1990), no full-scale installations applying this concept have been constructed.

IV.G CORROSION CONSIDERATIONS AND OZONE-RESISTANT MATERIALS

The power of ozone as an oxidant requires careful selection of construction material in design to provide a facility that is resistant to the attacks of gas-phase ozone (both "wet" and "dry") and ozone dissolved in the liquid being treated. Much of the information is basically rule-of-thumb, but there are several published papers on the subject (Fonlupt 1979; Mrazek 1981; Zawierucha and Charleson 1982; Damez 1988; Walton 1983). Operational experience in U.S. installations has provided information that provides design guidance for the various components of an ozonation facility.

IV.G.1 Gas Preparation System

Air feed gas. The material selection process should include materials that would be used for conventional compression, drying, and conveyance for ambient air. Carbon steel, cast iron, aluminum, conventional gasketing, and piping coupling techniques are all acceptable for the air preparation system to a point prior to the discharge pipe from the desiccant dryer. Experience indicates that ozonized gas from the ozone generators may backflow or migrate upstream of the generators with deleterious effect on that portion of the gas preparation system. Therefore, it is recommended that the gas preparation piping system downstream of the desiccant dryer be treated as that for "dry" ozonized gas as recommended below. As a minimum, the gas piping system should be designed for "dry" ozone service from the isolation valve immediately upstream of the ozone generator.

Operational experience at southern U.S. ozonation facilities such as Belle Glade, Florida and Shreveport, Louisiana indicates that concern for corrosion (not ozone-induced) should extend to the gas preparation system. Use of stainless steel piping should be considered for the gas preparation systems in warm, humid climates.

High-purity oxygen feed gas. Experience with operational high-purity oxygen piping in the feed gas system in water and wastewater service indicates that the piping system should be addressed for "Wet Oxygen-Rich Gas Service." Piping should be ASTM A-312 TP 304L for welded services and ASTM A-312 TP 304 when not welded. Fittings and flanges should be compatible with basic material. Gate valves should be 150-lb rating with bronze trim and virgin Teflon packing. Butterfly valves should be 125-lb rating, wafer type, with cast iron body, stainless steel or bronze seat, stainless steel stem, bronze bushings, and Buna "N" seat.

European standards require the use of carbon steel for dry oxygen service and an upper limit of 8 m/s (26.25 ft/s) for gas flow in the pipeline. All pipe welds are to be polished.

IV.G.2 Material Selection For Ozonation Systems

Ozone generators (see sec. IV.C.2). Materials for surfaces in contact with dried and ozonized gas should be selected for "dry" ozonized gas service. Stainless steel in the AISI series 316 and 321 are acceptable for this application, with 316L and 321 being used for welded applications. Tungsten Inert Gas (TIG) welding procedures are to be used. Although not recommended for general service, aluminum 6061 piping has demonstrated satisfactory service for dried air and high-purity oxygen influent piping and ozonized gas discharge piping of the Lowther ozone generator.

"Dry" ozonized gas piping systems. Piping systems for "dry" ozone service should be a minimum stainless AISI 304 and 304L (316L being most common) with TIG welding and Teflon-filled gaskets. Valves should be cast iron body, Viton A body liner, 316 stainless steel disc, and Viton A seat and seals or 316 stainless steel body, disc, and shaft, with TFE-filled seat and seal. It should be clearly stated that unplasticized polyvinyl chloride (UPVC) or "hard" plastic piping systems have demonstrated unsatisfactory service in a number of U.S. wastewater ozonation applications and should not be used (RQAW 1981).

"Wet" ozonized gas piping systems. The most rigorous service in the system is "wet" ozonized gas service, which is that downstream of the contactors. Therefore, piping systems and associated valves and fittings should be constructed of 316 and 316L stainless steel. System components such as demisters and ozone destructor media cabinets should also be constructed of 316 or 316L stainless steel.

Contact tanks. There is comparatively little published literature offering formal recommendations or recording the results of inspection of existing facilities. Mrazek (1981) provides details of the original structural design of the ozone contactor of the Southwest Wastewater Treatment Plant of Springfield, Missouri. The ozone contactor, which used mechanical surface aerators as the ozone dissolution devices, has been in continuous operation since 1978.

Submerged concrete and water stops. Type 2 or Type 5 Portland cement concrete is recommended for the water-bearing structures. Conventional water stops can be used. Hypalon or stainless steel gas seals should be used where the contactor walls connect with the contactor deck.

Gas space concrete, gas stops, gas sealing. To date (1990), there have been no identified problems with existing reinforced concrete structures exposed to "wet" ozone in the contactor head space. The gas space concrete of the structure should be of Type 2 or Type 5 Portland cement using a low water/cement ratio mix design. Consideration may be given to the use of galvanized steel reinforcing steel or at least to providing an additional 10 percent reinforcing steel if uncoated carbon steel reinforcing bar is used. A minimum of 3 in. (7.6 cm) of concrete cover for the reinforcing steel should be provided. According to Fonlupt (1979), 4 cm (1.13 in.) of concrete cover is sufficient. Stainless steel or Teflon gas space seals should be used. All fittings and concrete penetrations such as hatchways, vents, and valves should be 316L stainless.

Ozone destruction systems. The ozone destruction system from the demister through the ozone destructor unit should be treated as "wet" ozone service.

Electrical power cable protection. A rigid or flexible polyvinyl chloride or stainless steel shroud or covering should be provided to isolate the electrical cable from the ozonized liquid of the contactor. The void space between the shroud and the coating of the electrical cable should be filled with water.

IV.H RETROFIT OF OZONE SYSTEMS

IV.H.1 Introduction

The design of an ozone facility that is to be constructed at an existing plant site (i.e., a retrofit) can present complexities not normally encountered in a new plant design. There are different and additional study, design, construction, and operational details to be considered. The final facility arrangement has to be a compromise between the desired objectives of the ozone application and the physical and process constraints of the existing plant. The following are some of the considerations that will have to be addressed during the project development and design process:

1. Process requirements or goals:
   - Treatment and design objective development that is compatible with the existing system.
   - Downstream effects on existing treatment units (e.g., sedimentation or filtration and the possible need for recoagulation or modifications to existing processes).
   - Ozone emissions from downstream processes (e.g., release of ozone in enclosed filter area).
   - Bacterial regrowth potential in downstream facilities, especially reservoirs and distribution systems.
   - Foaming in downstream facilities.
   - Control system (e.g., level of sophistication and technique for integration into the existing system or for upgrading the entire system).
2. Site considerations:
   - Hydraulic considerations and constraints, between processes and into the plant.
   - Size of available sites for ozone facilities.
   - Arrangement of existing facilities and possible interferences (e.g., conduits, buildings, pipelines, roads, drainage).
   - Power to and from the grid and distribution within the site, including emergency power.

- Proximity of proposed ozone facilities to existing operations, buildings, and control interfaces (i.e., plant operations).

3. Design considerations:
   - Ozone system design—equipment, controls, and contactor.
   - Compatibility of existing material of construction in facilities immediately downstream of ozone contactor (i.e., corrosion considerations).
   - Compatible architecture of ozone building with existing facility and visual impacts on the surrounding area.
   - Process interconnections to existing channels, conduits, and control system.
   - Ability to maintain operations and to continue producing water during construction.

The process of designing a retrofit ozone facility includes all the considerations of a new plant facility as well as the interface concerns of a retrofit design. The following discussions expand on some of the factors that should be considered during a retrofit and presents an approach to conducting the project development.

IV.H.2 Approach to Retrofit Project Development

Figure IV–83 is an example project development flow schematic that might be used during the development of an ozone retrofit project. The most important part of this procedure is the iterative process of comparing and revising objectives, constraints, and process capabilities until there is a satisfactory balance reached between these three factors. Each of the procedures is discussed in greater detail in the following subsections. The general approach is presented here.

The first two tasks to be performed are (1) the establishment of the treatment objectives and (2) identifying the existing facility requirements and constraints. Both these tasks should be considered preliminary, since each may have to be reconsidered and/or revised if the constraints and the process criteria necessary to meet a treatment objective are not compatible. Both the primary and secondary treatment objectives (defined below) should be established at this point.

Once the treatment objectives are set, it will be necessary to determine the process criteria (e.g., the ozone dose and detention time) necessary to accomplish these objectives. This will require bench-scale or pilot-scale testing (see sec. IV.A) and a review of existing applications that are accomplishing similar objectives. Some testing is required, since ozone interactions are water chemistry–specific. These chemical interactions are not sufficiently understood to reliably predict the interactions based on an analysis of the water chemistry. The testing should be limited at this stage to a level of detail sufficient for a comparison of existing facility requirements to the process criteria and treatment objectives. The criteria and requirements may not be compatible and may have to be revised.

Once the preliminary process criteria are set, a comparison of facility requirements and constraints to the process criteria should be made. This stage will determine if the project should progress or if the treatment objectives and facility constraints should be redefined.

The interactive evaluation continues until a balance or compromise has been reached. For example, it may be determined that the objective of taste and odor control can best be accomplished by applying ozone to settled water. During bench-scale testing, it was determined that the ozone dose would be 30 percent less if applied to settled water when compared to raw water. However, facility constraints indicate that there is no space available in the area of the sedimentation basins and filters to allow the retrofit of a settled water contactor within the hydraulic gradient. This will require a decision to be made: whether to pump the settled water to a contactor and route the water back to the filters or to accept the additional operating cost associated with ozonating raw water for taste and odor control.

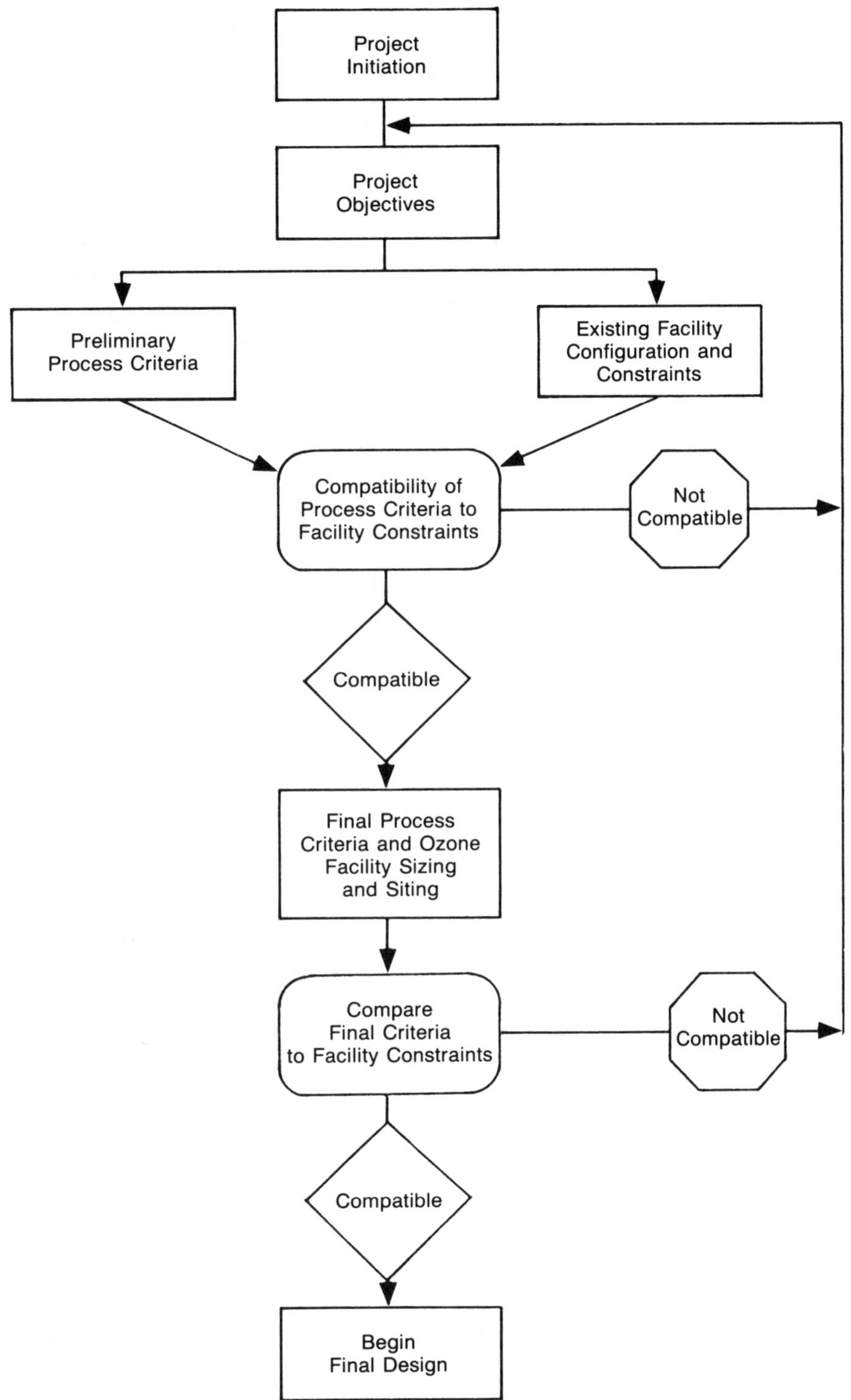

Figure IV–83 Example Project Flow Schematic

Once a balance between the facility requirements and the preliminary processes criteria have been reached, additional process development will be required. The preliminary process criteria were best estimates of the treatment requirements based on preliminary testing and experience. Additional testing is required to verify the feasibility of the treatment process and possibly determine effects of ozonation on downstream facilities. For example, if the decision was made above to pump from the sedimentation basin to the ozone contactor, it will be necessary to evaluate the proposed treatment train to determine if the pumping and ozonating of the settled water will interfere with filtration efficiency. The same test sequence could also be used to determine if recoagulation is possible and if filter efficiency could be maintained.

After extensive process criteria development, it is necessary to reevaluate the process criteria and the facility constraints. If the process criteria, facility requirements, and the treatment objects are at odds, it will be necessary to go back to the beginning of the process and establish each of the requirements and criteria based on new information. Once all of these factors are balanced (usually through compromise), it will be possible to progress to detail design activities.

IV.H.3 Process Considerations

There are a number of process decisions to be made during the development of an ozone system retrofit to an existing facility. As stated above, the development of the process will have to be balanced with the project objectives and fit into the site constraints. The following process issues will be discussed in this section: design objectives and criteria, process criteria development, downstream effects of ozone application, and process control.

Design objectives and criteria. As presented in previous sections, the objectives of the ozone project must be clearly defined prior to beginning the study and design activities. In the case of a retrofit to an existing facility, the primary objective should be evident, since the project would not have been initiated without a clear reason to consider an ozone retrofit. However, once the primary objective has been established, there are usually a number of additional benefits that are desired.

The primary objective for a retrofit installation will probably fall into one of four categories:

- Maintaining or increasing disinfection capabilities while lowering disinfection by-products,
- Controlling taste and odor or biological stabilization,
- Treating for color, or
- Oxidizing synthetic organic chemicals (SOCs) or volatile organic chemicals (VOCs).

The secondary benefits to be derived are usually one of the primary objectives relegated to secondary importance and/or one of the following:

- Microflocculation, either of raw or settled water.
- Oxidation of iron and manganese.

For example, if taste and odor control is the primary objective of an ozone retrofit, the secondary benefits could very well be disinfection, microflocculation, and reduction of disinfection by-products (especially [THMs]).

Establish preliminary process criteria. Once the primary and secondary treatment objectives have been established and the facility requirements and constraints have been defined, it is necessary to determine the feasibility of a system to fit within these objectives, requirements, and constraints. Process criteria (e.g., required ozone dose and detention time) provide a link among these factors and are necessary to determine the facility feasibility. Process criteria can be established to a limited degree based on experience with existing systems; however, some level of treatability study (see sec. IV.A) will also be required.

At this stage of the project, it is probably best to limit the treatability study to bench-scale testing or preliminary pilot evaluations. The testing should be limited until the preliminary process criteria have been determined and it has been demonstrated that these criteria fit within the framework of the treatment objectives and facility requirements. Once this has been accomplished, additional treatability testing can be used to verify the preliminary criteria, establish treatment feasibility, and provide the basis for the final process design criteria.

Develop final process criteria. The final process criteria provide a description of the process requirements necessary to accomplish the treatment objectives. These criteria have to be developed within the constraints of the existing facilities. As stated above, a compromise may be necessary between the facility requirements and the process criteria, for example, achieving a given disinfection $C \cdot t$ by applying a larger ozone dose to accommodate a shorter detention time caused by limited space availability for the contactor, or allowing for reduced ozone transfer efficiency to accommodate a contactor retrofit in an existing shallow basin.

The final process criteria will include items such as:

- Application points, single or multiple points.
- Required ozone dose and design transfer efficiency range.
- Ozone feed gas concentration and generation technique.
- Range in ozone concentration in feed gas.
- Contactor design parameters: depth, width, length, number of cells, cells for ozone addition.
- Gas-to-liquid ratio and gas loading rate to each cell.
- Contact time.
- Dissolution system.
- Basic control philosophy.

Each factor is discussed in greater detail in the appropriate sections of this book. Once the final process criteria have been established, the final design can begin. There will continue to be reevaluations of the process criteria as the design details are developed. This results in the need for a process review to continue through detailed design.

Downstream effects. In a retrofit situation, the importance of the downstream effects resulting from the ozone application becomes important, since the existing facilities were not designed with an ozone process in mind. There are process, operating, and physical considerations to be addressed. The following are some of the items to be considered.

Filtration. If an ozone system is placed between a sedimentation process and a filtration process, there is a possibility that this could cause problems with floc filterability and filter efficiency. Due to the ozone and the agitation in the contactor, the floc carried over from the sedimentation basin can shear or age to the point that it is not as readily removed in the filters. This possibility should be investigated during a treatability study. If this phenomena should occur, it may be possible to correct the problem with a change in coagulants (or any improvement in the sedimentation step) or with the addition of coagulants or a filter aid after the ozone contactor. This may include a new mixing process after the contactor.

Ozone addition prior to the filters could also cause gas binding of the filter media. This can be more of a problem for a retrofit, since there may be very limited head available to degas the water with a free-fall at the effluent of the contactor. Also, in the United States, filtration rates can vary anywhere from 2 to 10 gpm/ft^2 (4.84 to 24.19 m/h) or greater. The filter configuration should be evaluated to determine if there is a possibility for developing negative head across the filters.

Sedimentation. Some of the same changes in floc characteristics mentioned in the filtration section can occur in the sedimentation process. Changes in floc characteristics prior to sedimentation are not as great a concern as those for the filter process. Changes in coagulants and the mixing system can be made to compensate for these changes. Treatability studies should be conducted to determine what changes may be necessary.

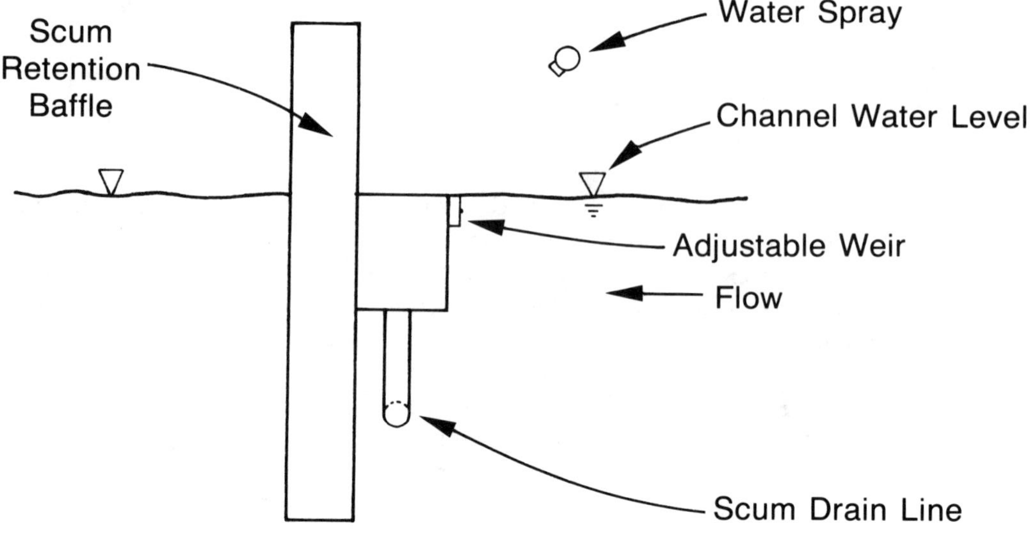

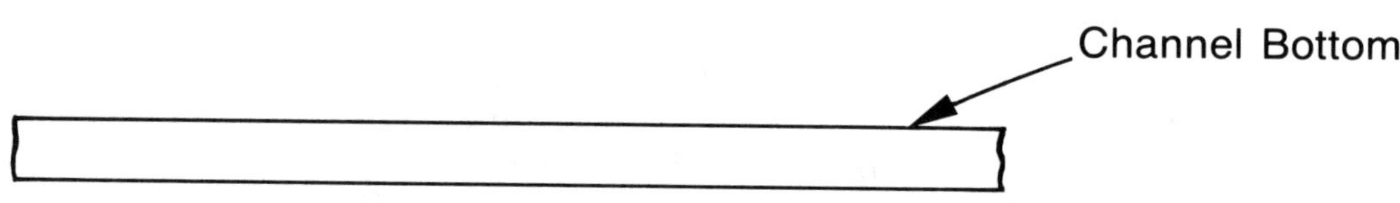

Figure IV–84 Section of Skimmer for Scum and Foam Removal

Ozone emissions. Both limited hydraulic head and site constraints can contribute to ozone being carried out of the contactor and into the next treatment process. For an existing plant, the emissions of ozone in a downstream process could be very undesirable. If ozone is released in a covered filter building, the ambient concentration could well exceed the exposure limit.

In a new plant design, ozone reduction is usually addressed in one or all of the following ways: (1) the outlet design of the contactor has sufficient free-fall to assist in degassing and removing ozone; (2) the contactor has detention time allocated to the dissipation of residual; and (3) the conduits and facilities downstream can be covered and emissions contained (e.g., a covered rapid mix).

These approaches may not be available to the designer of a retrofit facility due to hydraulic head limitations. Possible remedies for a retrofit are addition of hydrogen peroxide at the contactor effluent to accelerate the ozone destruction in the water, addition of a mechanical or vacuum degassing device, or covering downstream facilities and collecting gas.

Scum and foam removal. Scum and foam removal facilities should be placed in or after most ozone contactors where there is reason to believe this will become an operational or aesthetics problem. The retrofit of a scum removal system in an existing process may not be practicable without some form of modification to the structure.

Scum and foam removal is not a common practice; consequently, there is not much design information available. The concept used for the retrofit at East Bay Municipal Utility District was to provide skimmers in the effluent channels of the contactors and skimmers in the channels after the contactors and prior to the filters (see Figure IV–84).

A skimmer is set just below the water level in the channel. A thin film of water flows over the skimmer weir and into the collection trough with the help of sprays that direct the scum into the skimmers. The skimmers have to be mounted upstream of a scum retention baffle. The collected scum is drained or pumped into the plant waste collection system for treatment or preferably disposal.

Where foaming is minimal, sprays could be used to continually knock the scum back into solution, and it could be allowed to pass through the system. It may be best to plan for the skimmers and sprays but not install them until a problem develops. Pipeways, drainage, and mounting considerations should be taken into account.

Distribution system concerns. The issues of ozone by-products, biodegradable organic carbon (BDOC), and regrowth are addressed in chapter II (sec. A.3 to A.6), chapter III (sec. K), and sec. A of this chapter. For a retrofit facility, special consideration should be given to these concerns. It may not be possible in a retrofit situation to implement the techniques employed in new facilities to control these parameters. Alternative approaches may have to be considered. For example, the retrofit of ozone into a direct filtration facility with high-TOC raw water may require a change in the filter operations or design to accommodate BDOC removal. The filtration rate can be reduced or the media can be changed to granular activated carbon (GAC) to accentuate biological treatment.

Controls. The level of control for the ozone equipment and the ozone process can be based on compatibility with the existing system or on renovating the entire control system. A decision as to the control philosophy and complexity will have to be made taking into consideration the level of operations at the plant and the desires of the owner. A decision should be made as to whether the control system will remain at the same level or be upgraded during the ozone retrofit. This will not only include the controls for the ozone equipment and process but could include the entire plant control system. Methods of integration into the existing system should be considered carefully.

Site considerations. Existing facilities and the available site play a major role in the development of a retrofit design. Since these matters are so site-specific, only a general discussion covering the areas of concern can be presented here.

Hydraulics. The location of the ozone contactor could be restricted by the plant hydraulic profile. Most plants have been designed with minimum head loss between processes (e.g., between the sedimentation basin and the filters). If satisfying the ozone process objectives requires the contactor to be located in a hydraulically restricted area, an analysis of the following types of issues will have to be made: Can the upstream or downstream facilities be modified to accept the contactor? Is pumping to the contactor an acceptable alternative? Can the objectives or process criteria be modified to allow the installation of the contactor at a different location? Even a raw water contactor may be restricted by site limitations or available head.

The contactor hydraulics are discussed in sec. D of this chapter.

Site availability and constraints. The most obvious of the site considerations is determining where the contactor, ozone building, and power facilities can be located. Issues associated with these selections will include the following:

- Size and geometry of the site.
- Proximity to process interfaces.
- Interferences with existing structures and utilities (e.g., power, water, and chemicals).

In addition to the basic siting criteria, early consideration should be given to factors such as:

- Proximity to control building and operator accessibility.
- Access and drainage (site development).
- Interference with existing structures (e.g., foundations).
- Aesthetics, compatible architecture, and limited visual impact to neighbors.

Considering all these issues as the site is selected will assist in preventing an unexpected design change.

Power. Power distribution both onsite and offsite must be considered. Existing transformers and switch gear will have to be evaluated together with the emergency power requirements. The need to have the ozone system operating during emergency conditions must be addressed. It may be acceptable to operate on an emergency basis without ozone, for example, by using chlorine for disinfection.

Design considerations. Most of the design considerations for an ozone retrofit are included in the preceding sections. The basics of the ozone equipment and contactor design will be the same for a retrofit as a new facility. However, there are some additional items that should be considered for a retrofit that may not have to be considered for a new facility design.

Ozone system. Most of the ozone equipment and contactor design considerations will be basically the same for a retrofit as a new design once the site and process criteria are set. The most significant differences in design will be caused by the site and hydraulic constraints. There will be little or no difference in the ozone equipment. However, the application of high-purity oxygen as well as medium- or high-frequency ozone generators might result from limited site area.

The major impacts will occur during the contactor design. Hydraulic constraints can significantly limit the head loss across the contactor. This will require a design of influent distribution, baffling, and effluent weir that will minimize head loss while maintaining positive flow control for an equal flow split among contactors and providing for proper hydraulic distribution in the contactor.

Site constraints can create a situation where the contactors are limited in depth or in size. Special consideration must be given to differences in ozone transfer efficiency or in developing acceptable detention times and preventing short-circuiting.

Material compatibility. Construction materials downstream of the ozone contactor should be surveyed to determine the potential for corrosion being caused by residual ozone. The potential for this occurring can be greater for a retrofit design if the design conditions limit the possibility of reducing the ozone residual at the effluent of the contactor.

Architecture. Architectural consistency with existing facilities will have to be considered. Beyond matching the architectural style, it may be important to screen or enclose facilities such as the ozone destruction units and additional electrical equipment.

Process interconnections. Process connections to the existing channels, conduits, and control system will require a special design effort. Special consideration has to be given to these connections to assure that the process objectives will be met (e.g., assuring that there is sufficient capacity in the control system or data highways). These interconnections should be designed and planned to occur with minimal impact on the existing plant and its operations, as discussed below. Special consideration has to be given to existing facility interferences and the possible disruption of operations due to the rerouting of plant utilities or process streams.

Scheduling of construction. It may be necessary for the treatment plant to continue operations during the construction of the ozone facilities. Careful scheduling and planning of connections to process streams, power connections, and other interfaces will have to be made during the design and bid document preparation.

IV.I PERFORMANCE EVALUATION

IV.I.1 Introduction

Electrical power is used when producing ozone in commercial ozone generators. It is the major expense component of an operating budget, comprising about 75 percent of the total ozone system's operation and maintenance costs (Rakness et al. 1984, 1988). In some cases, energy consumption has been included in the bid evaluation in order to achieve the lowest-cost system on a present worth basis (Rakness and Stolarik 1988). However, energy cost alone should not be the sole factor in system selection. Equipment maintenance and reliability considerations should also be assessed (Rakness et al. 1989). (See chapter VI.)

Energy consumption of an ozone system is dependent upon several items, including generator type, operating pressure, and feed gas character, plus ambient temperature and pressure. Due to the individuality of the installation and difficulty in safely destroying the ozone produced during a factory test, power production evaluations are generally conducted after the ozone equipment has been placed in service. Since it is normally impractical to remove installed equipment, provisions for noncompliance, other than replacement, should be included in the specifications. An example of a noncompliance clause is a specified monetary value for each unit of additional power required to produce the specified quantity of ozone. In unique instances a factory test may be completed, particularly for small installations using skid-mounted units. In those cases, only the equipment that meets the energy requirements of the specification is installed.

An energy assessment of the ozone system is most useful when it includes an evaluation at different operating points, such as 25, 50, 75, and 100 percent of the design production rate. This multiple-point data is useful because the unit energy consumption (Wh/g or kWh/lb) may be considerably higher at the lower production rates, unless precautions are taken in equipment selection. It is also helpful in showing the operators how to obtain efficient operation at less than design production requirements.

IV.I.2 Specification Requirements

A successful power evaluation begins with a comprehensive specification. Some of the elements that should be addressed are described below.

Design criteria. Power required to produce ozone can be separated into generator power, system auxiliary power, and nonsystem power. Generator power is that used by the ozone generator alone. It is primarily influenced by the production rate, ozone concentration, and cooling water temperature. Auxiliary, production-related power is used by the feed gas preparation and offgas treatment equipment. Their power is primarily affected by ozone production, ozone concentration (i.e., gas flows), operating pressure, and ambient temperature. Nonsystem power includes building lighting and heating and is usually excluded from the assessment.

The particulars contained in the specifications depend on the purpose of the evaluation. In all cases, it is necessary to properly measure ozone concentration and gas flow, these two parameters being used to calculate the ozone production rate for given operating conditions. Moreover, the purpose of the specification is often to obtain a measured power value for the total ozone system (ozone generation plus gas preparation equipment) at a specified production rate. When total system power is assessed, the power measurements can be made at slightly under or over the specified ozone concentration because power saved by the generator from operating at a lower ozone concentration is part of the auxiliary equipment measurements and includes the additional feed gas required. By allowing a range for ozone concentration, the ozone generator manufacturer can adjust feed gas flows and achieve optimum total system performance for their equipment. Occasionally, the power of the ozone generator only is assessed if the generator manufacturer is not responsible for the feed gas treatment system. In this case, a precise ozone concentration should be specified.

Cooling water temperature will vary from season to season. Generator power requirements are greatest at the warmest temperature. Specified criteria should include the sustained (e.g., weekly or biweekly) warmest temperature of the cooling water.

Practically speaking, a temperature range of ±2°C (± 3.6°F) should be stipulated in order to allow flexibility in scheduling the power test. Other parameters also may be defined as a range of values, such as operating pressure, feed gas dewpoint, and ambient temperature. The approach to be used in interpolating the data for the variable operating conditions that may be encountered in the test should be addressed in the specifications. Optional techniques are discussed below.

The specified production rate should be a singular value. It should not be adjustable based on operating variables such as ozone transfer efficiency. Ozone transfer is sensitive to water quality conditions that affect ozone demand. Water characteristics are difficult to control during a power test. By specifying an exact production rate, the power assessment is more straightfoward.

Interpretation of test results. The actual test environment likely will not match the specified conditions. Adjustments to measured data may be necessary. The methodology for reconciling test results should be addressed in the specifications. Some of the issues to be addressed are further described.

Ozone production rate. Adjustments to measured power for slight under- or over-production should be completed. Two sensible techniques are: (1) determine the specific energy at power settings as close as possible to the subject test production rate (e.g., ±3 percent of desired rate) and use that number to calculate the power for the specified condition, or (2) obtain power and production data above and below the specified criteria and interpolate from the data. The favorite approach is the first technique because it minimizes testing time yet attains a valid outcome. It is based on the fact that specific energy is not highly sensitive to the ozone production rate. Specific energy is mostly affected by ozone concentration and cooling water temperature.

Ozone concentration. Adjustments to measured power for test ozone concentrations that are slightly under or over the specified ozone concentration (which will be measured in all cases) are not necessary when the total system power (ozone generator and gas preparation system) is measured because total system power is being assessed. However, if generator power only is evaluated, then ozone concentration should be incorporated into the appraisal. Adjustments for ozone concentration can be developed by obtaining specific energy data at concentrations above and below the specified number. The resultant specific energy for the specified concentration can then be interpolated from the data. Final power values are obtained in the product of measured, adjusted specific energy times design production rate.

Cooling water temperature. Specific energy for the ozone system is affected by cooling water temperature. An empirical relationship between temperature and production can be developed for each ozone installation, but the process is very time-con suming because it involves formalized testing activities at multiple temperatures. A less complicated approach is to perform only one power assessment, but conduct the evaluation when the cooling water temperature is within the specified range. This approach must be so stated in the specifications. It addresses the concerns of temperature effects on performance, but allows testing to be done only during the warmer months of the year. A concern of this technique is that equipment may be operating for several weeks or months before the test can be initiated.

Equipment operation before testing. Operating problems can affect energy efficiency, such as a high dewpoint of the feed gas. However, the system can be expected to maintain its specified power value for more than one full year if operating parameters have been maintained within warranty provisions. Addressing potential operating responsibility issues that may develop before the power test is started, and their effect on the power evaluation, must be undertaken in the specifications. The objective should be to restrict the need for any adjustments to measured results due to operating time. Both the owner's and manufacturer's concerns should be recognized.

Testing design production rates. Design ozone production rates may not be required until a future date when the demand for water is greater or the raw water quality is poor (i.e., ozone demand is great). At the same time, it is advisable to test the ozone system at its design production rate within the first year of operation. Excess ozone production may overload offgas destruction equipment, and operation at the full output of *all* generators may not be possible. A sensible solution is to test individual ozone generators at their design capacities, and add the results. The auxiliary equipment would be tested separately. The possibility of using this procedure should be included in the specifications.

The number of generators to be tested simultaneously could be determined based on plant operating conditions. Individual generator feed gas flows would be controlled to attain a certain ozone concentration. Auxiliary equipment would be tested separately at its full output relative to the tested ozone concentration. Splitting the power assessment into the two parts of generator and auxiliary equipment tests also accommodates energy measurements for equipment that operates on a periodic basis.

Intermittent equipment operation. Ozone generators operate continuously. They reach operating stability in about one to two hours. Valid, reproducible specific energy results for the generators can be obtained in a 3- to 4-h time period by collecting data at half-hour intervals after the first 1 or 2 h of operation. This approach allows multiple test points to be evaluated in a relatively short time period. However, the auxiliary equipment cannot be assessed simultaneously because equipment like heat-reactivated desiccant dryers draws power periodically and may not be operating during a shortened test.

Intermittent operation of equipment can be addressed by one of two techniques: (1) determine the instantaneous power demand of the equipment and proportion the power number with respect to measured on-time; or (2) operate the equipment for a long period of time (e.g., 24 h), cumulate the energy consumption, and calculate average power from the quotient of energy consumption divided by measurement time. The approach to be used should be stated in the specifications. A conservative approach is to use the second technique of long-time collection of energy consumption data as the measurement of record, and the first technique of instantaneous power and measured on-time as a quality control check.

Measurement of record. Readings for ozone concentration, feed gas flow, and power demand are minimum data that are required to assess the specific energy requirement of the ozone generation system. When high-purity oxygen feed gas is used, the concentration of individual gas constituents should be identified in order to ascertain feed gas density. Obtaining other data is also important in order to judge compliance with specified operational ranges, and may include cooling water inlet and outlet temperature, flow, and pressure; feed gas temperature and pressure; ozone product gas temperature and pressure; and ambient temperature, pressure, and relative humidity.

Most measurements can be obtained using permanently installed equipment, such as UV monitors for ozone concentration, venturi meters for flow measurement, and temperature and pressure gauges. Power measurements are usually obtained by renting equipment for the duration of the test. All measuring equipment utilized should be "precision"-type machinery. Permanently installed equipment that is to be used for the power tests should be noted in the specifications. Where duplicate monitoring devices are available, only one should be designated as the meter of record.

Instrument calibration. The instruments used for measurements are customarily correct to within a certain percentage of the "true" value. This accuracy is generally expressed by the instrument's manufacturer as the inherent error of the device. For the determination of specific energy, the readings of several instruments are combined into a single equation to obtain a finite number. The calculation of specific energy is fairly simple, but the approach to combining the inherent errors of related equipment is not straightforward. The subject of instrument calibration and error analysis could lead to compliance arguments unless they are clearly defined in the specifications.

Instrument calibration does not lead to error elimination; it does allow the equipment to provide representative numbers for the subject measurement to the best ability of the machinery. Permanently installed equipment used for measurements of record should be jointly calibrated by at least two instrument technicians, one representing the owner and the other representing the manufacturer. The manufacturer's technician may take the lead role. Supplemental equipment, such as the power meters, may be calibrated by an independent party. The equipment should be calibrated before the test and checked after the test. If drift occurs beyond a specified percentage (e.g., ±3 percent), then the test should be repeated. This approach, or a variation thereof, should be documented in the specifications. It should be conscientiously implemented during the test so that all ozone equipment/system suppliers are satisfied that the test is a fair one, even if they did not win the bid.

The reading from each calibrated instrument should be accepted as correct. Temperature, pressure, and gas density adjustments may be completed as necessary, but adjustments due to instrument calibration inaccuracies should be avoided. The question of measurement inaccuracies should be addressed separately in the context of "error analysis."

Error analysis. Error analysis encompasses the approach to interpreting the measured results of the test data when assessing compliance with the specifications. It addresses the concern of inherent instrument inaccuracies. Presumably, the instruments are calibrated as discussed above. The matter of error analysis must be addressed in the specifications. Almost any approach would be defensible if it is clearly addressed in the specifications, but a reasonable approach should be stipulated.

Techniques for addressing error analysis include the following: (1) a calculation based on accepted statistical procedures, (2) a specific predefined number, such as ±3 percent of the final criteria being assessed (i.e., ±3 percent of the final criteria and not ±3 percent of intermediate measurements), or (3) no error analysis (i.e., the measured numbers are the final assessment numbers). The first option of specifying a statistical procedure is cumbersome and time-consuming, which reduces its benefit in relation to the effort involved. The second option of selecting a specific, predefined number establishes a value for instrument inaccuracies and is easy to implement; however, it potentially allows the installed equipment to be slightly undersized. The third option of disallowing error analysis ensures that the installed equipment will perform at or above its nameplate ratings. It requires the manufacturer to slightly oversize their equipment as insurance that they can meet the power evaluations tests, or to risk the consequences. In the United States, excluding error analysis is recommended by the Hydraulics Institute Standards for water and wastewater pumps (Hydraulics Institute 1983). It should also be seriously considered for ozone generation equipment. Whichever approach is used, it must be clearly addressed in the specifications to minimize testing arguments.

IV.I.3 Example Power Evaluation Test

The Los Angeles Department of Water and Power (LADWP) operates a 600-mgd (94,635 m^3/h) direct filtration water treatment plant which includes a 7900-lb/day (149-kg/h) ozonation system. Ozone is applied as a preoxidant for the purposes of disinfection and microflocculation. High-purity oxygen is generated onsite and is used as the feed gas in a once-through system. The process was selected through competitive bids and evaluated for total present worth from 20-year life cycle costs, which included capital plus energy expenses. Power consumption (and penalty) was valued at $6500 (about 44,600 FrF) per kW. System power demand was evaluated at nine ozone production rates. Each rate was weighted based on a specified percent of time expected at the subject operating condition.

The specifications stated that the power evaluation tests would be conducted when the cooling water temperature was between 18°C and 22°C (64.4 to 71.6°F). The cooling water temperature was 20°C (68°F) during October 1987, when the evaluation tests were completed. The tests were conducted jointly by plant staff personnel, manufacturer's representatives, and a technical consultant hired by the LADWP. The active participation by the plant staff increased their awareness of techniques available to optimize performance without forfeiting maintenance interests. Through this awareness, a greater opportunity was provided to perpetuate the optimum operating conditions that were experienced during the performance testing.

A summary of selected, important readings and calculations from the power tests at 70 percent of the maximum production rate are listed in Table IV–5. For this test, five readings were taken at approximately 30-min intervals; these included time of the reading, energy consumption, ozone concentration, feed gas flow, feed gas purity, and cooling water temperature. Results for each 30-min period were immediately analyzed to assess system stability. Power for each half-hour time interval was determined by the quotient of Δ energy consumption divided by Δ time. Production was assessed for each of the five readings. Final results for the test incorporated the results of all readings. Power was determined by the quotient of Δ energy consumption for the total time of the test divided by Δ time of the test. Production was assessed from the average of the results of the five individual readings.

Quality control assessment. During the power tests there were three measured parameters that were prominent in interpreting the results; namely, the power demand measurements, the oxygen feed gas flow measurements, and the ozone concentration measurements. A quality control assessment of each parameter was completed.

Power measurement quality control. The power meters were rented from a power meter manufacturer, who had performed a calibration check and had certified the meters prior to shipment. The meters were installed by an LADWP electrical engineer. The installation was checked by a project engineer representing the manufacturer. Both parties were convinced that the power meters were appropriately recording the energy consumption of the equipment.

Feed gas ozone concentration measurement quality control. Two UV monitor (PCI Ozone Corp.) ozone concentration readings from each ozone generator, each reading at a different purge cycle of the UV meter, were used in determining the resultant ozone concentration. The monitor span was determined based on an assessment of sample cell temperature, pressure, and gas density. The unbuffered potassium iodide (KI) and "European" neutral buffered potassium iodide ($NBKI_e$) methods of ozone concentration testing were used to support the credibility of the UV monitor results (Birdsall et al. 1952; Masschelein 1987). The KI method was used because it was so stated in the specifications. The $NBKI_e$ method was used because it is a standard proposed by the European Committee of the International Ozone Association, and it was felt that interested parties from that part of the world would relate better to that quality control comparison (see chapter II, sec. C.6).

The UV meter reading was to be accepted as the representative ozone concentration number if its results were within ±3 percent of the unbuffered KI wet chemistry results. It was expected that the UV meter would achieve this precision if it was properly installed and if accurate sample cell temperature, pressure, and feed gas density readings were taken and subsequently used to adjust the span. A comparison of the UV versus wet chemistry readings is shown in Figure IV–85. Both the KI and $NBKI_e$ results were within ±3 percent of the UV meter's numbers.

Oxygen feed gas flow measurement quality control. A schematic of the plant's oxygen/ozone system is shown in Figure IV–86. Feed gas flow rate to the ozone generators is measured by two instruments, flow element 197 (FE-197) and flow element 53 (FE-53). FE-197 was pressure and temperature compensated by a Micon programmable controller. FE-53 was temperature and pressure compensated by Flow

Table IV–5 Summary of Results for One Power Evaluation Test

Standard Temperature	70°F	Molecular Weight of Ozone	48 g/mol
% of "Other" gas = Argon	80.0%	Ozone wt/vol at 100% volume	1,988.1 mg/L
% of "Other" gas = Nitrogen	20.0%	Molecular Weight of Argon	39.948 g/mol
VOLUME (L/mol) at STD TEMP	24.144 L/mol	Molecular Weight of Oxygen	32 g/mol
Molar Volume	0.08205 L/mol/K	Molecular Weight of Nitrogen	28 g/mol
Conversion Factor	28.317 L/ft³	Standard Temperature	21.11°C
Conversion Factor	453.592 g/lb	Absolute Temperature	273.15°K
Conversion Factor	0.746 kW/hp	Estimated MAC Motor Efficiency	95%

Reading Number	Unit	1	2	3	4	5	AVERAGE
Target Production	lb/day	5,530	5,530	5,530	5,530	5,530	5,530
Ambient Pressure	mm Hg	736.1	736.1	735.6	735.3	735.6	735.7
Cooling Water Temp.	°C	20.0	20.0	19.8	20.0	20.0	20.0
GENERATOR "A"							
Time	hr:min:s	09:25:46	09:56:53	10:29:16	10:56:39	11:25:58	
Energy	kWh	377.4	512.2	652.1	770.7	902.7	
Ozone Concentration 1	%wt	5.217	5.231	5.436	5.398	5.378	
Ozone Concentration 2	%wt	5.216	5.242	5.432	5.390	5.368	
Δ Time	min	31.12	32.38	27.38	29.32		120.20
Δ Energy	kWh	134.8	139.9	118.6	132.0		525.30
Avg. Power Demand	kW	259.9	259.1	260.0	270.2		262.2
GENERATOR "B"							
Time	hr:min:s	09:26:50	09:57:42	10:31:12	10:57:48	11:26:40	
Energy	kWh	458.1	616.7	786.1	925.2	1,074.0	
Ozone Concentration 1	%wt	5.162	5.221	5.341	5.354	5.320	
Ozone Concentration 2	%wt	5.151	5.220	5.341	5.338	5.318	
Δ Time	min	30.87	33.50	26.60	28.87		119.83
Δ Energy	kWh	158.6	169.4	139.1	148.8		615.90
Avg. Power Demand	kW	308.3	303.4	313.8	309.3		308.4
GENERATOR "E"							
Time	hr:min:s	09:30:08	09:59:09	10:32:47	10:58:43	11:28:52	
Energy	kWh	477.0	627.2	803.7	938.9	1,095.0	
Ozone Concentration 1	%wt	5.162	5.221	5.341	5.354	5.320	
Ozone Concentration 2	%wt	5.151	5.220	5.341	5.338	5.317	
Δ Time	min	29.02	33.63	25.93	30.15		118.73
Δ Energy	kWh	150.2	176.5	135.2	156.1		618.00
Avg. Power Demand	kW	310.6	314.9	312.8	310.6		312.3
TOTAL OF GENERATORS							
Total Power Demand	kW	878.8	877.4	886.5	890.1		882.9
"Micon" Flow Reading	kscfh	52.78	52.11	51.67	52.50	52.70	52.35
Oxygen Purity	% Vol.	96.8	96.5	96.3	96.2	95.0	96.2
Feed Gas Density	lb/ft³	0.08320	0.08324	0.08327	0.08329	0.08346	0.08329
Feed Gas Mass Flow	lb/day	105,391.5	104,107.6	103,264.2	104,941.1	105,559.0	104,652.4
Average Ozone Conc.	%wt	5.177	5.226	5.372	5.362	5.337	5.295
Measured Production	lb/day	5,455.6	5,440.5	5,547.4	5,626.9	5,633.5	5,541.0
Measured Spec. Energy	kWh/lb	3.866	3.871	3.835	3.796		3.824
Adjusted Gen. Power	kW	890.8	891.8	883.7	874.8		881.1
MAIN AIR COMPRESSOR							
Time	hr:min:s	09:34:23	10:02:01	10:39:04	11:05:13	11:31:17	
Energy	kWh	1,050.0	1,354.0	1,755.0	2,038.0	2,321.0	
Δ Time	min	27.63	37.05	26.15	26.07		116.90
Δ Energy	kWh	304.0	401.0	283.0	283.0		1271.00
Avg. Power Demand	kW	660.1	649.4	649.3	651.4		652.4
MAC Air Flow	kscfh	258	258	258	258	258	258.0
MAC Operating Pres.	psi	70	70	71	71	69	70.2
MAC Unit Power	BHP/Cscfm	19.55	19.23	19.23	19.29		19.32
Adjusted MAC Flow	kscfh	261.5	262.2	257.2	253.6	253.3	257.5
Adjusted MAC Power	kW	669.1	660.1	647.3	640.2		651.1

(continued)

Table IV–5 (Continued)

Standard Temperature	70°F	Molecular Weight of Ozone	48 g/mol
% of "Other" gas = Argon	80.0%	Ozone wt/vol at 100% volume	1,988.1 mg/L
% of "Other" gas = Nitrogen	20.0%	Molecular Weight of Argon	39.948 g/mol
VOLUME (L/mol) at STD TEMP	24.144 L/mol	Molecular Weight of Oxygen	32 g/mol
Molar Volume	0.08205 L/mol/K	Molecular Weight of Nitrogen	28 g/mol
Conversion Factor	28.317 L/ft^3	Standard Temperature	21.11°C
Conversion Factor	453.592 g/lb	Absolute Temperature	273.15°K
Conversion Factor	0.746 kW/hp	Estimated MAC Motor Efficiency	95%

Reading Number	Unit	1	2	3	4	5	AVERAGE
MCC-12-2							
Time	hr:min:s	09:33:23	10:00:37	10:35:57	11:00:51	11:30:20	
Energy	kWh	11.91	14.89	19.44	22.63	26.59	
Δ Time	min	27.23	35.33	24.90	29.48		116.95
Δ Energy	kWh	3.0	4.6	3.2	4.0		14.68
Avg. Power Demand	kw	6.6	7.7	7.7	8.1		7.5
MCC-11-1							
Time	hr:min:s	09:32:36	10:00:20	10:37:17	11:02:30	11:31:18	
Energy	kWh	82.7	109.1	141.7	165.3	190.3	
Δ Time	min	27.73	36.95	25.22	28.80		118.70
Δ Energy	kWh	26.4	32.6	23.6	25.0		107.63
Avg. Power Demand	kW	57.2	52.9	56.2	52.1		54.4
TOTAL SYSTEM POWER SUMMARY							
Adjusted Generator Power	kW	890.8	891.8	883.7	874.8		881.1
Cooling Water Power	kW	34.0	34.0	34.0	34.0	34.0	34.0
MCC 12-2 Power	kW	6.6	7.7	7.7	8.1		7.5
MCC 11-1 Power	kW	57.2	52.9	56.2	52.1		54.4
Adjusted MAC Power	kW	669.1	660.1	647.3	640.2		651.1
TOTAL ADJ. POWER	kW	1,657.6	1,646.6	1,628.9	1,609.1		1,628.1

Source: Rakness and Stolarik (1988).

Indicating Controller 53-1 (FIC 53-1). For all test points (except the 90 percent and 100 percent test points, which introduced supplemental liquid oxygen [LOX]), the same flow stream was being measured by both meters. However, FIC 53-1 consistently indicated a slightly different flow rate than the Micon, ranging from 0.8 percent to 4 percent lower. It could not be determined which reading was superior. However, prior to the tests the Micon was chosen as the flow monitoring device of record. More time was spent on calibration of its pressure, temperature, and flow element signals. The Micon results were used as the number of record.

The second approach to confirming the accuracy of the feed gas flow reading was to perform an oxygen mass balance around the cryogenic oxygen production system. Toward this end, an examination was completed of the mass of oxygen applied to the cryogenic oxygen production facility using the main air compressor flow reading. The mass applied was then compared to the mass of oxygen captured, based on the Micon flow reading plus LOX measurements. The comparison indicated that 93 to 96 percent of the oxygen mass supplied to the cryogenic process was accounted for in the Micon flow plus oxygen concentration reading. The manufacturer indicated that this capture percentage in the cryo plant is reasonable, since some loss of the main air compressor oxygen mass supply occurs due to instrument air usage, valve switch-over, and "nitrogen" venting.

Power evaluation results. After each test was completed, the raw data were transcribed into a computer spreadsheet. This immediate analysis in the field allowed process stability to be assessed, questionable readings to be recognized and resolved, additional measurements to be made as desired, and compliance projections to be

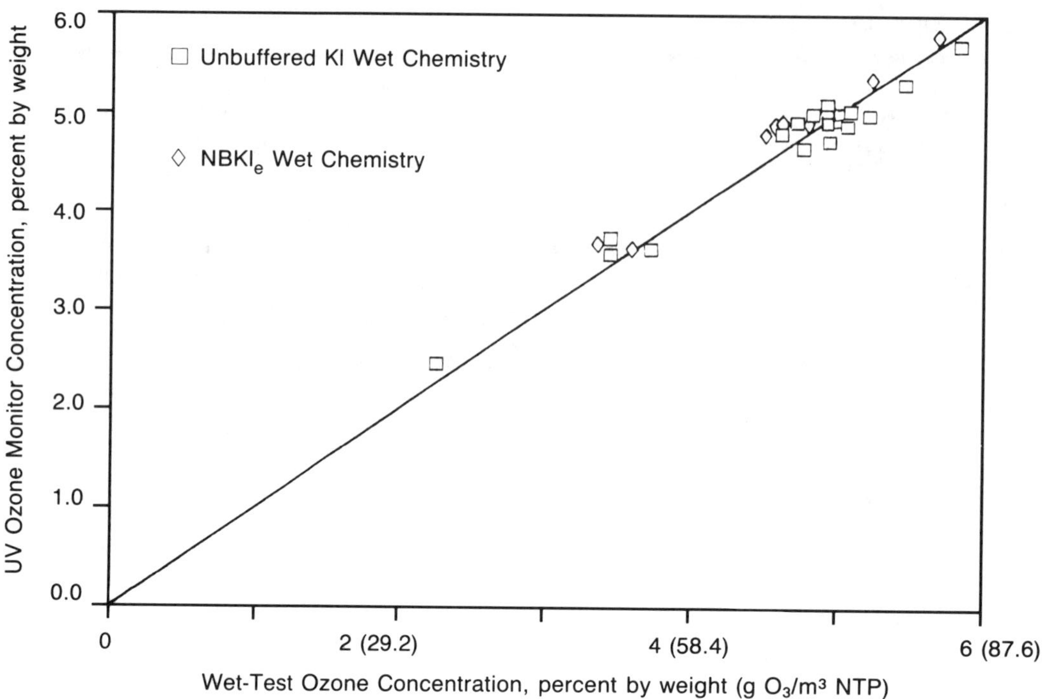

Source: Rakness and Stolarik (1988).

Figure IV–85 UV Monitor Results Compared Favorably with Unbuffered KI and NBKI$_e$ Wet Chemistry Results

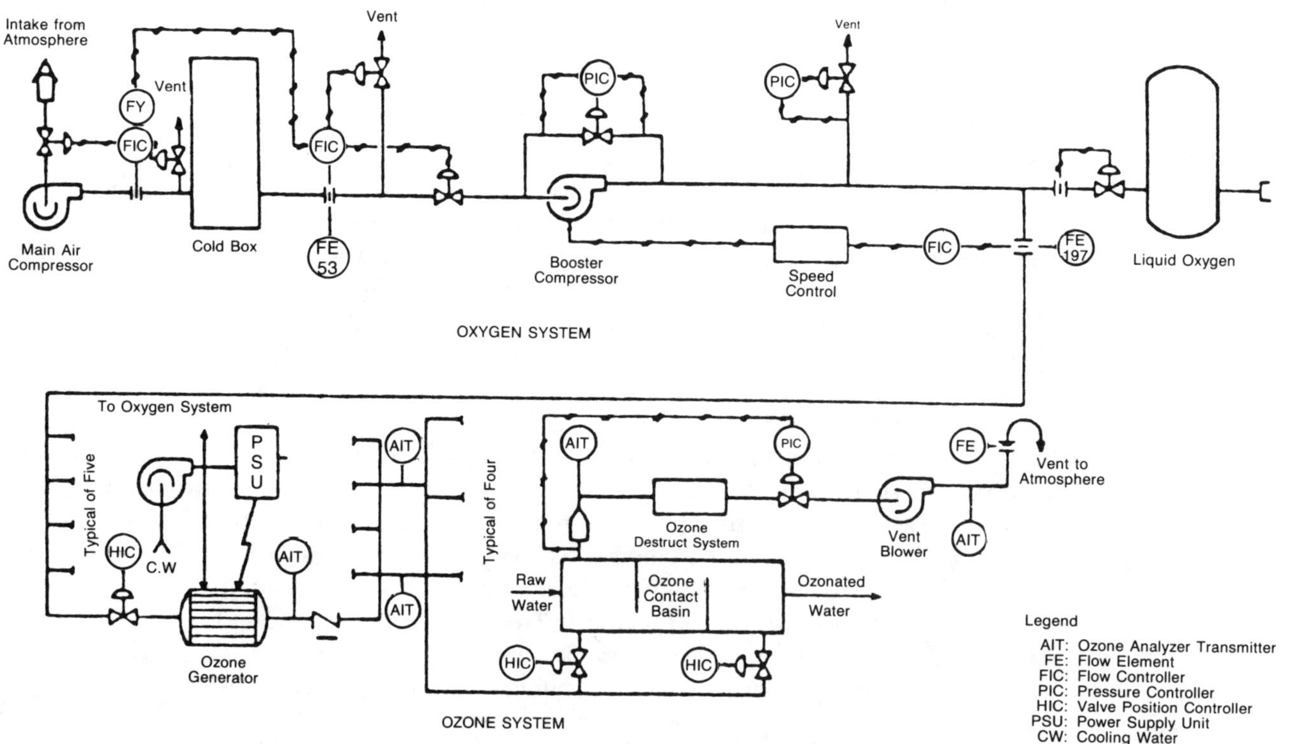

Source: Rakness and Stolarik (1988).

Figure IV–86 Oxygen/Ozone System Simplified Process Schematic—Los Angeles Aqueduct Filtration Plant, California

Table IV–6 Adjusted Power for Ozone Generators

Test No.	Measured Power	Measured Specific Energy	Measured Ozone Conc.	Measured Oxygen Flow	Measured Ozone Production	Design Prod.	Percent of Design	Adjusted Power	Bid Power
	kW	kWh/lb	%wt	lb/h	lb/day	lb/day	%	kW	kW
20-1	168.5	2.590	2.445	2,659.6	1,560.8	1,580	98.79	170.5	197.0
30-1	324.8	3.388	3.610	2,656.4	2,301.2	2,320	99.19	327.5	329.0
30-2	276.2	2.838	3.662	2,657.5	2,335.8	2,320	100.68	274.4	329.0
Avg. 30	300.5	3.113	3.636	2,657.0	2,318.5	2,320	99.94	300.9	329.0
40-1	414.8	3.243	4.810	2,658.9	3,069.6	3,160	97.14	427.0	494.0
40-2	438.9	3.370	4.894	2,661.2	3,125.5	3,160	98.91	443.8	494.0
40-3	429.0	3.317	4.858	2,662.4	3,104.0	3,160	98.23	436.8	494.0
Avg. 40	427.6	3.310	4.854	2,660.8	3,099.7	3,160	98.09	435.9	494.0
50-1	551.4	3.359	4.920	3,336.1	3,939.5	3,950	99.74	552.8	614.0
60-1	729.4	3.685	4.976	3,978.0	4,750.7	4,740	100.23	727.8	740.0
70-1	882.9	3.824	5.295	4,360.5	5,541.0	5,530	100.20	881.1	889.0
80-1	1,076.2	4.060	6.008	4,412.5	6,361.9	6,320	100.66	1,069.1	1,125.0
90-1	1,291.2	4.323	5.947	5,021.9	7,167.7	7,110	100.81	1,280.8	1,286.0
100-1	1,323.8	4.141	5.947	5,375.2	7,672.0	7,900	97.11	1,363.1	1,472.0
					Sum of Individual Test Points:			6,782.1	7,146.0
									Δ = (363.9)

Source: Rakness and Stolarik (1988).

examined. Compliance projections are believed to be very beneficial because they allow equipment adjustments to be fine-tuned, if possible, and potential conflicts in data interpretation to be immediately addressed. Additional equipment fine-tuning adjustments were not necessary during the LAAFP tests because initial power measurements indicated that compliance most likely would be achieved. However, the spreadsheets were not considered as the final power values until an exhaustive review of the data entries was completed.

Ozone generator power. The ozone generator power measurements and production results for each test are shown in Table IV–6. For each test, the measured ozone production rate was close to, but not exactly equal to, the required test point production rate. An adjustment to measured power was made to reflect the power that would be consumed at the exact, specified test-point production rate.

Adjustments were made to the dominant power-consuming equipment, namely, to the ozone generators and the main air compressors. The approach to adjusting the ozone generator power was to use the measured specific energy for the subject test point, coupled with the specified ozone production rate for that test point. The resulting adjusted ozone generator power is summarized in Table IV–6, along with the manufacturer's projected power numbers for the ozone generators. The measured/adjusted power demand for the ozone generators was consistently less than the generator power number presented in the bid. In summing the data for all nine test points, the ozone generators used 363.9 kW less power than was presented in the bid documents.

Cryogenic oxygen production power. An important element in completing the ozone generator power adjustments was the condition that the ozone concentration remain unchanged. Therefore, adjustments to the feed gas flow rate were needed. The power of the main air compressors (MAC) was adjusted to reflect the required changes in feed gas flow rate. The MAC power adjustments were determined by calculating the change in MAC air flow that corresponded to the change in oxygen feed gas flow that kept the ozone concentration equal to the measured concentration at the specified ozone

Table IV–7 Adjusted Power for Main Air Compressors*

Test No.	Name MAC Used	MAC Air Flow	Meas. Required Prod.	% of Adjust Flow	MAC Meas. Power	MAC Unit Power	Meas. Adjust. Power	MAC Vapor. Meas.	Cryogenic MAC + Vapor. Total	MAC Bid Power
		kscfh	%	BHP/kscfh	kW	Cscfm	kW	kW	kW	kW
20-1	B	160.0	98.79	162.0	377.3	18.02	381.9	0	381.9	379.0
30-1	B	160.0	99.19	161.3	376.1	17.96	379.1	0	379.1	379.0
30-2	B	160.0	100.68	158.9	376.5	17.98	374.0	0	374.0	379.0
Avg. 30	B	160.0	99.94	160.1	376.3	17.97	376.5	0	376.5	379.0
40-1	B	160.0	97.14	164.7	378.2	18.06	389.3	0	389.3	390.0
40-2	B	160.0	98.91	161.8	380.8	18.18	385.0	0	385.0	390.0
40-3	B	160.0	98.23	162.9	377.7	18.04	384.5	0	384.5	390.0
Avg. 40	B	160.0	98.09	163.1	378.9	18.09	386.3	0	386.3	390.0
50-1	B	195.0	99.74	195.5	456.4	17.88	457.6	0	457.6	489.0
60-1	A	248.0	100.23	247.4	627.3	19.33	625.8	0	625.8	641.0
70-1	A	258.0	100.20	257.5	652.4	19.32	651.1	0	651.1	642.0
80-1	A	261.4	100.66	259.7	665.3	19.45	660.9	0	660.9	643.0
90-1	A	260.8	100.81	258.7	664.4	19.47	659.1	35.9	694.9	669.0
100-1	A	261.0	97.11	268.8	662.2	19.39	681.9	53.8	735.7	696.0
					Sum of Individual Test Points:				4,970.8	4,928.0
										Δ = (42.8)

Source: Rakness and Stolarik (1988).
*Assume MAC motor efficiency is 95 percent and standard conditions are 70°F (21.11°C) and one atmosphere.

production test point. The adjusted MAC air flow rate was coupled with the subject test-point measured unit power requirement for the MAC to calculate the adjusted power. The results for the adjustments to MAC power to achieve a constant ozone concentration at each test point are summarized in Table IV–7.

The overall power for the cryogenic oxygen system included the power for the MAC, power for the heating elements of the vaporizer, which converts LOX to gaseous oxygen (GOX) when supplemental LOX is required, and power for reactivation of the hydrocarbon adsorbers. In Table IV–7, the total power for cryogenic process is shown as the sum of the MAC plus vaporizer power. The power for reactivating the hydrocarbon adsorbers is included with the "other" power numbers, which are discussed later in this section. Power for the vaporizer was required at the 90 and 100 percent test points when supplemental LOX was used to meet gas flow requirements. The power values anticipated by the manufacturer for the cryogenic equipment are also shown in Table IV–7. In all cases, the measured/adjusted cryogenic power was quite close to the manufacturer's anticipated power number. In total, the cryogenic facility used 42.8 kW more power than was proposed in the bid documents.

Other ozone equipment power. The measured power requirement for the remaining equipment of the ozonation system, such as booster compressor, offgas destruct unit heater, and lobe oil pumps and heaters, is reported in Table IV–8 under column headings for Motor Control Center (MCC) 11 and 12. The power values for the cryogenic plant hydrocarbon adsorber reactivation unit and the cooling water pumps are listed separately. It was recognized that the power of some of this "other" equipment is also affected by ozone production adjustments, such as gas flow effect on the power demand of the ozone destruct unit heater; however, the adjustments would be very small, would have taken considerable time and exhaustive measurements to document, and would result in only minor changes to the total system power requirement. These adjustments

Table IV–8 Summary of Measured Power For All "Other" Equipment

Test No.	MCC-11 Meas.	MCC-12 Meas.	React Hydro. Ads. Meas.	Cool Water Bid	Total "Other" Meas.	Total "Other" Bid
	kW	kW	kW	kW	kW	kW
20-1	23.4	21.7	0.7	17.0	62.8	31.0
30-1	18.5	27.8	0.7	17.0	64.0	31.0
30-2	22.2	21.6	0.7	17.0	61.5	31.0
Avg. 30	20.3	24.7	0.7	17.0	62.7	31.0
40-1	18.9	26.7	0.7	17.0	63.3	34.0
40-2	17.4	26.2	0.7	17.0	61.2	34.0
40-3	25.2	20.9	0.7	17.0	63.8	34.0
Avg. 40	20.5	24.6	0.7	17.0	62.8	34.0
50-1	25.9	25.6	0.7	17.0	69.2	38.0
60-1	48.6	7.0	0.7	17.0	73.2	41.0
70-1	54.4	7.5	0.7	34.0	96.6	60.0
80-1	53.9	7.6	0.7	34.0	96.3	61.0
90-1	55.7	8.1	0.7	34.0	98.5	64.0
100-1	57.8	7.0	0.7	34.0	99.5	68.0
			Sum of Individual Test Points:		721.6	428.0
						Δ = (293.6)

Source: Rakness and Stolarik (1988).

Table IV–9 Summary of Measured/Adjusted Power vs Bid Power

Test No.	Test Point Weight	Direct Power		Weighted Power			
		Adjust/Meas. Total	Bid Total	Adjust/ Meas.	Bid	Δ W/ Bid Amount	Δ W/ Bid Percent
	%	kW	kW	kW	kW	kW	% of Bid
20-1	5.0	615.3	602.0	30.77	30.10	0.67	2.21
30-1	18.0	770.6	734.0	138.70	132.12	6.58	4.98
30-2	18.0	709.8	734.0	127.77	132.12	(4.35)	−3.29
Avg. 30	18.0	740.2	734.0	133.24	132.12	1.12	0.85
40-1	29.0	879.6	918.0	255.09	266.22	(11.13)	−4.18
40-2	29.0	890.0	918.0	258.10	266.22	(8.12)	−3.05
40-3	29.0	885.0	918.0	256.65	266.22	(9.57)	−3.59
Avg. 40	29.0	884.9	918.0	256.62	266.22	(9.60)	−3.61
50-1	24.0	1,079.6	1,141.0	259.11	273.84	(14.73)	−5.38
60-1	12.0	1,426.9	1,422.0	171.22	170.64	0.58	0.34
70-1	6.0	1,628.8	1,591.0	97.73	95.46	2.27	2.38
80-1	3.0	1,826.2	1,829.0	54.79	54.87	(0.08)	−0.15
90-1	2.0	2,074.2	2,019.0	41.48	40.38	1.10	2.74
100-1	1.0	2,198.3	2,236.0	21.98	22.36	(0.38)	−1.68
Overall Totals		12,474.5	12,492.0	1,066.9	1,086.0	(19.1)	−1.75

Source: Rakness and Stolarik (1988).

were not made because they were judged to be insignificant in the determination of compliance with the bid power values.

The "other" equipment represented a relatively small percentage (5.78 percent) of the power consumed by the total ozonation system (i.e., sum of power for individual test points for the "other" equipment was 721.6 kW, and for all ozonation equipment was 12,474.5 kW). However, "other" equipment power consumption could have been a critical constituent in the bid power assessment. In Table IV–8 the measured power for the "other" equipment is greater for every test point than the power number anticipated in the bid documents. In summing the results for all nine test points, the "other" equipment used 293.6 kW more power than was reflected in the bid submittals. This higher-than-bid power consumption of the "other" equipment could have caused a problem in meeting the total system bid number. It did not cause a problem because the power of the ozone generators was lower than anticipated.

Total ozone system power. In Table IV–9 the aggregate measured/adjusted power numbers, using the applicable weighting factors, are compared to the manufacturer's final bid power number. For most test points, the measured/adjusted weighted power is less than the bid power number. The sum of weighted bid power numbers for each test point was 1086.0 kW. The sum of the weighted, measured power for each test point was slightly lower, at 1066.9 kW. Based on the power evaluation tests, the LAAFP ozonation system was judged to be in compliance with the bidding documents.

IV.J OTHER CONSIDERATIONS

The preceding sections provide design information specific to the ozonation system. Other factors must be considered for a system that is readily operable by plant personnel.

IV.J.1 Noise Control

In the United States, Occupational Safety and Health Administration requirements establish acceptable noise levels for municipal as well as industrial applications. Acceptable noise levels can be achieved by limiting the period of employee exposure and/or providing ear protection. However, careful consideration of potential noise control issues can result in less noise-intensive areas for all or the majority of the ozonation system.

Gas preparation components should be evaluated for potential noise generation levels. Typically, these include the gas pressurization device. However, refrigerant dryer components (compressor), "heatless" desiccant dryer, and high-frequency ozone generator electronic control systems also exhibit elevated noise levels.

Specification of maximum noise levels for each system component is desirable to establish a noise-oriented standard for the component. A maximum allowable noise level of 85 dBA is common. This may not be achievable for certain operating equipment (such as a screw-type compressor, which may have rotational speeds as high as 3600 rpm).

Isolation of equipment with high noise levels (most commonly the gas compression equipment) is good practice. This may be achieved by provision of a unit soundproofing enclosure or isolation of noisy equipment in a separate sound-treated room of the facility. The success of equipment enclosures is subject to the willingness of plant maintenance staff to reinstall the enclosures after maintenance activities. Within the limitations of plant size and associated construction budget, it is suggested that consideration be given to providing separate equipment rooms for the following:

- Gas compression devices.
- Gas dryers (refrigerant and desiccant).
- Ozone generators.

Architectural treatment of the sound-reflecting surfaces of the rooms of the ozonation system components should be appropriate for the equipment housed in the rooms in question. A small investment in sound-absorbing materials during the facility construction will pay dividends in operation and maintenance personnel comfort throughout the operational life of the facility.

IV.J.2 Structural Arrangement

Various arrangements of the gas preparation equipment and ozone generation, ozone dissolution, and offgas treatment units are possible. Many of these arrangements are a result of critical climatic conditions (e.g., maintenance of equipment in cold weather). However, other arrangements are based on designer or operator preference.

The relationship between ozone generation equipment and ozone dissolution equipment is an example of what has been, to date, a result of designer preference. Two possible arrangements are (1) ozone generation equipment in room(s) over ozone contactor(s), or (2) ozone generation equipment in a structure separate from the ozone dissolution structure(s).

Common ozone generation/dissolution structure. Provision of a common structure for ozone generation and ozone dissolution is typical of European ozonation facilities and is illustrated by the Los Angeles Aqueduct Filtration Plant of the Los Angeles Department of Water and Power (see Figure IV–48, sec. IV.D.1).

The benefits of combining ozone generation and dissolution in a single structure include the following:

1. Site area conservation.
2. Employee protection from inclement weather.
3. If galleries are provided:
   - Easier access to contactor stages through watertight access panels in gallery side walls.
   - Easier location and access to contactor monitoring instrumentation.
   - Ability to view (depending on water characteristics) the interior of the contactor to evaluate bubble patterns and diffuser conditions.
   - Possible installation of gas handling piping in the gallery to reduce piping in the ozone generation room(s).

Disadvantages of this type of arrangement include the following:

1. Increased concern for the leakage of ozonized gas from the contactor.
2. Increased complexity and cost of the contactor to accommodate galleries and associated equipment.

Separate ozone generation/dissolution structures. Provision of separate ozone generation and dissolution structures is the common arrangement in U.S. wastewater ozonation practice and is illustrated by the Myrtle Beach, South Carolina water treatment plant installation. Benefits of this arrangement includes the following:

1. Lower construction cost, since the ozone generation building may be a single story slab-on-grade installation, with the contactor consisting of a number of trains with common walls.
2. Reduced concern for possible leakage of ozonized gas from the contactor into an occupied work space.

Disadvantages include the following:

1. A higher level of personnel exposure to inclement weather.
2. More difficult access to contactor stages for maintenance.
3. No easy means of visual inspection of bubble patterns or diffuser condition.
4. Need to weatherproof or otherwise protect contactor instrumentation and control equipment and conduit.

IV.K CHECKLISTS FOR OZONE SYSTEM DESIGN

The following checklists are adapted from a document of the European Committee of the International Ozone Association (Masschelein 1989). They are intended to serve as a review aid to identify issues that should be addressed during certain phases of a project.

1. Preliminary planning phase:

a. Primary and secondary purpose(s) or objective(s) of the ozonation process. (See chapter III.)

b. Ozone application point(s) (pre, intermediate, and/or post).

c. Seasonal ozone usage design criteria (determined from bench- and pilot-scale treatability studies [see sec. IV.A]).
- Initial or immediate ozone demand.
- Rate of ozone decomposition.
- Minimum contact time.
- Minimum transferred ozone dosage.

d. Preliminary level cost estimates.
- Capital costs.
- Operating and maintenance costs.

2. Pre-design phase:

a. Ozone production requirement (includes water flow rate and ozone dosage variations).
- Startup, minimum conditions.
- Average or normal conditions.
- Design point with standby considerations.
- Peak, maximum conditions.
- Projected future conditions.

b. Feasibility of available system options.
- Site constraints.
  - Available site area.
  - Available hydraulic head.
  - Location of main power supply.
  - Unique topographic features.
- Contactor design criteria.
  - Type.
  - Number and size.
  - Internal configuration (baffling).
  - Constraints with respect to gas or liquid flow range.
- Ozone generation design criteria.
  - Allowable ozone concentration (i.e., gas flow).
    - * Impact on ozone contactor design criteria.
    - * Impact on available equipment options.
    - * Impact on feed gas selection (i.e., air or oxygen).
  - Bidding approach.
  - Number of units (include standby considerations).
  - Plant energy unit costs ($/kWh or FrF/kWh).

c. System cost.
- Capital cost estimate.
- Energy cost estimate.
- System maintenance considerations and cost.
- Cost of consumables.
- Present worth evaluation (include operating personnel skill levels in final selection process).

3. Design phase

a. Air feed gas design considerations.
- Compressor system.
  - Type (oil-lubricated or oil-free).
  - Model (piston, rotating piston, screw, liquid ring).
  - Energy consumption and system maintenance requirements.
  - Size (flow, pressure and energy turndown considerations).
  - Monitoring and control.
  - Filter system for ambient particulate or other pollutant dust removal.
  - Filter system for oil removal, if applicable.
  - Acceptable noise level and control.
  - Temperature control, after-cooler systems (type, materials of construction and performance criteria).
  - Shutdown alarms (temperature and pressure).
  - Standby (backup) considerations.
- Gas dryer system design considerations.
  - Size and degree of backup.
  - Energy consumption and system maintenance considerations.
  - Acceptable operating conditions (inlet and outlet temperature, pressure, flow, and dewpoint).
  - Shutdown alarms (dewpoint temperature, pressure).
  - Utilization of refrigerant dryer.
  - Desiccant dryer.
    - ⋆ Type and adsorption capacity.
    - ⋆ Danger of poisoning from flooding.
    - ⋆ Danger of dust formation (filtration required).
    - ⋆ Regeneration method.
    - ⋆ Gas flow used in regeneration.
    - ⋆ Cycle length and control.
    - ⋆ Possibility of visual inspection of dessicant media.

b. Oxygen feed gas design considerations.
- Storage or supply facilities.
- Built-in safety provisions.
- Recycling of offgas.
- Materials of construction.
- Gas flow velocity in pipes.
- Lower explosion limit criteria and monitoring.
- System alarms and safety provisions.

c. Ozone generator design considerations.
- Type of generator.
- Cooling media requirements.
- Maximum operating voltage (rms and peak voltage).
- Minimum and maximum operating frequency.
- Sustained maximum cooling water temperature (several days).
- Shutdown alarms (gas and liquid temperature, gas and liquid pressure, gas dewpoint, and gas and liquid flow).
- Minimum turndown capability.
- Provisions for dielectric removal for cleaning.
- Construction materials (shell, gaskets, and piping).
- Specific energy versus ozone concentration characteristics.
- Operating pressure.
- Possibility of visual inspection (dielectrics side and cooling water side).
- Gas flow measurement (pressure, temperature, and density correction).

- Electrical power supply considerations.
  - Voltage and amperage requirements (low and high voltage considerations).
  - Ambient temperature.
  - Cooling fluid in voltage transformers.
  - Electrical load balance for single-phase units (if required).
  - Electronics cabinet cooling and dust contamination considerations.
  - Quality of electronic components.
  - Shutdown alarms.
- Water backflow prevention (i.e., backup of humidity from the ozone contacting system).
- Dielectric protection.
- Provisions for purging (i.e., drying) with feed gas.
- Safety provisions when using oxygen feed gas.
- Recommended stock and availability of spare parts.

d. Ozone contactor design considerations.
- Bubble diffuser contactor.
  - Type of diffusers (rod or disc).
  - Flow range of diffusers.
  - Gas to volume flow for mixing (consider low gas flow conditions).
  - Gas flow per unit area of contact column (consider low gas flow conditions).
  - Materials of construction.
  - Velocity in pipes.
  - Liquid detention time.
  - Degree of short-circuiting (t_{10}) and residual concentration ($C \cdot t$).
  - O&M considerations.
  - Transfer efficiency at different conditions and in relation to water quality.
- Other contact systems.
  - Type (turbine, injector, eductor, other).
  - Principal of operation.
  - Flow and energy turndown.
  - Liquid detention time.
  - Degree of short-circuiting (t_{10}) and residual concentration ($C \cdot t$).
  - O&M considerations.
  - Transfer efficiency at different conditions and in relation to water quality.
  - Power consumptions.

e. Control systems and automation.
- Gas pressure control systems.
- Gas flow measurements (equipment, units, and accuracy).
- Monitoring of ozone concentration (in the inlet and outlet gas, residual in the water).
- Safety control systems in generator rooms.
- Manual sampling points for control.
- Possibility of manual operation in all instances.
- Built-in safety provision of the contacting and monitoring systems.

f. Offgas control and removal system.
- Monitoring system for controlling ozone in the offgas.
- Offgas treatment integration in the system.
- Degree of automation of the removal system.
- Explosivity of the system under control.
- Presence of gases other than ozone, oxygen, and nitrogen (i.e., poisoning of catalyst).

- Potential for flooding with water.
- Temperature and pressure control of the offgas system.
- Reliability of the materials of construction.

g. Process alarms, control and automation.
- Ambient ozone alarms.
- Equipment protection alarms (temperature, pressure, and dewpoint; electrical safety devices).
- Allowance for manual operation (monitoring as well as control).
- Sampling points.
- Type and degree of automation.

h. Materials of construction.
- Nature.
- Corrosion.
- Gas-tight systems.
- Coatings.
- Pressure or vacuum conditions.

i. Safety measures.
- Room ventilation.
- Fire protection.
- Electrical hazards.
- Explosion hazards (oxygen feed gas).
- Excessive pressure/vacuum relief.
- Offgas ozone destructor discharge point.

References

ANNARELLI, D. 1988. FMC Corporation. Personal communication.

APPELTON ET AL. 1990. Final Effluent Disinfection Ozonation System Modifications at the Hagerstown, MD WPCP. Presented at the International Ozone Association PAC Spring Conference, Shreveport, LA, March 27–29.

Asea Brown Boveri (ABB). 1988. Product information.

BABLON, G. & GUGENHEIM, A. 1985. Use of a Mathematical Model to Control the Influence of Preozonation During Coagulation Process. Presented at the Workshop on Instrumentation and Control of Water and Wastewater Treatment and Transport Systems, Denver, CO.

BADER, H. & HOIGNE, J. 1979. Analysis of Ozone in Water and Wastewater by an Indigo Method. 4th Ozone World Congress, International Ozone Association, Houston, Texas.

_____. 1981. Determination of Ozone in Water by the Indigo Method. *Wtr. Res.,* 15:449–454.

BAETZ, J. ET AL. 1986. An Ozone Diffusion Technique Using Fine Bubbles: Examples of its Application in Treatment Processes. Presented at the I.O.A. Conference of Perrysburg. April 28–29.

BEAUDET, B.A. ET AL. 1988. "Ozone/Lime Softening Treatment of a Highly Colored Florida Groundwater: From Bench Studies to Design." Proc. Intl. Ozone Assn. Symp., Pan American Congress, Myrtle Beach, SC, December, 1988.

BELLAMY, W.D. ET AL. 1989. Evaluation of Volatile Organic Compound Treatment by Advanced Oxidation for Drinking Water. National Conf. Water Pollut. Contr. Fed., San Francisco, CA.

_____. 1990a. Ozone Demonstration and Pilot Evaluation For Color Removal at Long Beach. Proc. AWWA Ann. Conf., Cincinnati, OH.

_____. 1990b. In-line Ozone Dissolution Demonstration-Scale Evaluation. Presented at the International Ozone Association Pan American Committee Spring Conference: New Developments: Ozone in Water and Wastewater Treatment. Shreveport, LA, March 27–29.

BIRDSALL, C.M. ET AL. 1952. Iodometric Determination of Ozone, *Anal. Chem.,* 24:62.

BLANKENFELD, D. 1988. Schmidding-Werke. Personal communication.

BOURBIGOT, M.M. ET AL. 1987. Combined Ozonation and Flotation—Application to the Treatment of Highly Algae-Loaded Water. Proceedings 8th Ozone World Congress, International Ozone Association, Zurich, Switzerland B-91.

BRICHANT, F. 1982. *Les Onduleurs Autonomes. Conception et Applications Industrielles.* Edited by Dunod, Paris, France.

BRODARD, E. ET AL. 1984. Use of a Deep U-Tube Ozone Reactor for the Disinfection of Potable Waters. Proc. AWWA Ann. Conf., p. 1543.

_____. 1985. Development of Full Scale Deep U-Tube: South African and French Experiences. Presented at the 7th Ozone World Congress. International Ozone Association, Tokyo, Japan, Sept. 9–12.

_____. 1986. Tube Diameter and Height Influence on the Ozonation Operating Conditions with a U-Tube. *Ozone Sci. Engrg.*, 74(4):235.

Capital Controls Company, Inc. 1988. Product information.

Carlins, J.J. & Clark, R.G. 1982. "Ozone Generation by Corona Discharge." In *Handbook of Ozone Technology and Applications,* edited by R.G. Rice and A. Netzer. Ann Arbor Science Publishers, Inc., Ann Arbor, MI, pp. 41–75.

Chapsal, P. 1976. Traitement par l'Ozone des Eaux d'Alimentation. Conference presented at Interbytmach exhibition, Moscow, USSR.

_____. 1989a. Technologie des Générateurs d'Ozone TRAILIGAZ. In *L'Ozonation des Eaux. Manuel Pratique,* edited by W.J. Masschelein (in press). Technique et Documentation, Paris, France.

_____. 1989b. Destruction de l'Ozone dans l'Air à l'Évent. In *L'Ozonation des Eaux. Manuel Pratique,* edited by W.J. Masschelein (in press). Technique et Documentation, Paris, France.

Chédal, J. 1979. Examination of Some of the Problems in Design of an Ozonation Tank. Presented at the International Ozone Association European Conference on the Application of Ozone in Water Treatment, January 25–26.

_____. 1984. Raw Water Preozonation. In *Handbook of Ozone Technology and Applications,* Vol. 2, edited by R.G. Rice and A. Netzer. Ann Arbor Science Publishers, Inc., Ann Arbor, MI.

Clark, J.M. 1988. Ozone Technology, Inc. Personal communication.

Conway, W.T. 1988. Framco Aeration & Mixing Company. Personal communication.

Cook, G.A. et al. 1959. Separation of Ozone from Oxygen by a Sorption Process. Advances in Chemistry Series 21, 44–52.

Coste, C. 1982. Excess Ozone Disposal. In *Ozonation Manual for Water and Wastewater Treatment,* edited by W.J. Masschelein. John Wiley and Sons, Inc., New York, pp. 204–208.

Coste, C. & Fiessinger, F. 1986. Recent Advances in Ozone Generation. Presented at AWWA Sunday Seminar, Denver, CO, June 22.

Damez, F. 1988. "Materials Resistant to Corrosion and Degradation In Contact With Ozone." In *Ozonation Manual for Water and Wastewater Treatment,* edited by W.J. Masschelein. John Wiley and Sons, Inc., New York.

Delcominette, A. Trailigaz. 1988. Personal communication.

Dietrich, A.M. 1981. Physical Chemical Mechanisms of Aqueous Ozonation. Proc. Natl. Conf. Env. Eng. ASCE, Atlanta, GA.

Duguet, J.P. et al. 1986. An Automated Procedure for Monitoring the Effectiveness of Ozonation Processes. In *Analytical Aspects of Ozone Treatment of Water and Wastewater,* edited by R.G. Rice et al. Lewis Publishers, Inc., Chelsea, MI, p. 357.

Dyer-Smith P., et al. 1988. Uprating Existing Facilities with Medium Frequency Ozone Generators. Asea Brown Boveri product information.

Elsenhans, K.H. 1989. Experience With the Maintenance of Ozone Systems. Proc. Ninth Ozone World Congress International Ozone Association, edited by L.J. Bollyky. New York. Vol 2:697.

Faes, Y. 1975. Ozoneurs: Leur Théorie et Application des Techniques Nouvelles à Semi-Conducteurs à Leur Alimentation. *Revue Générale de l'Electricité,* 84(1):13–24.

Federal Register. 1989. National Primary Drinking Water Regulations; Filtration; Disinfection; Turbidity, *Giardia Lamblia,* Viruses, *Legionella,* and Heterotrophic Bacteria; Final Rule. 40 CFR Parts 141 and 142, Vol. 54, Number 124, June 29.

Filipov, V. & Emilianov, M. 1957. Théorie Électrique des Ozoneurs. Théorie des Caractéristiques Dynamiques des Ozoneurs. *Zh. Fiz. Kim.* (USSR), 31(7):1629–1635.

Fischer, M. et al. 1987. Technical and Economical Advantages of Producing and Applying Ozone at High Concentrations. Proc. Second Intern. Conf. The Role of Ozone in Water and Wastewater Treatment, edited by D.W. Smith and G.R. Finch. Tektran International. Kitchener, Ontario.

Fontalinant, P. et al. 1977. An Example of Automatic Regulation of Ozone Production: The Plant at Nantes LaRoche. Presented at 3rd International Ozone Congress, Paris, France.

Fontlupt, J. 1979. Comportement de Différent Matériaux en Contact avec l'Ozone. Presented at the International Ozone Association European Committee Workshop: Les Applications de l'Ozone au Traitement des Eaux, January 25–26.

Gary, C. & Moreau, M. 1976. L'Effet de Couronne en Tension Alternative: Pertes et Perturbations Radioélectriques Engendrées par les Lignes de Transport d'Énergie Électrique. Edition Eyrolles. Paris, France.

Gerval. R. & Bablon, G. 1983. Optimization of the Ozonation-Floculation Stage. Proc. of the 6th Ozone World Congress, International Ozone Association, Washington, D.C., May 23–26, 115.

Geuther, D.A. & Pierson, S.S. 1988. Advantages of Ozone Production at High Concentrations. Publication of Capital Controls Company, Inc. Colmar, PA.

GLAZE, W.H. & KANG, J.W. 1988. Advanced Oxidation Processes for Treating Groundwater Contaminated With TCE and PCE: Laboratory Studies. *Jour. AWWA,* 80:5.

GLAZE, W.H. ET AL. 1980. Oxidation of Water Supply Refractory Species by Ozone and Ultraviolet Radiation. EPA-6002-30-110, U.S. EPA, Cincinnati, OH.

_____. 1982. Destruction of Pollutants in Water with Ozone in Combination with Ultraviolet Radiation. 2. Natural Trihalomethane Precursors. *Envir. Sci. and Technol.,* 16:454.

_____. 1988. Evaluation of Ozone and Ozone-Peroxide Processes for Oxidation of Taste and Odor Compounds in East Bay Municipal Utility District Source Waters. Test Results Submitted to EBMUD, August.

GRADER, R.J. & JOHANSON, R.P. 1982. U.S. Patent 4352740: Water Ozonation Method. Assignee Linde AG/Lotepro.

Griffin Technics, Inc. 1988. Product information.

HEILKER, E. 1979. The Mulheim Process for Treating Ruhr River Water. *Jour. AWWA,* 71(11):623.

HESBY, J.C. 1989. Black & Veatch. Personal Communication.

HESS, T.F. 1980. Removal of Entrained Air From a Source Water. Master of Science Thesis, University of Colorado, Boulder, CO.

HOIGNE, J. & BADER, H. Determination of Ozone and Chlorine Dioxide in Water by Indigo Method. *Vom Wasser,* 55:261.

HOIGNÉ, J. ET AL. 1987. Rate Constants for OH Radical Scavenging by Aquatic Humic Substances: Role in Ozonation and in a Few Photochemical Processes for the Elimination of Micropollutants. 193 National ACS Division Environmental Chemistry, April, 27:1.

HORST, M. 1982. Removal of the Residual Ozone in the Air After the Application of Ozone. In *Ozonation Manual for Water and Wastewater Treatment,* edited by W.J. Masschelein. John Wiley and Sons, Inc. New York.

Hydraulics Institute. 1983. *Standards for Centrifugal, Rotary and Reciprocating Pumps.* Cleveland, Ohio (14th ed.), p. 174.

Hydrovane Compressor Company Limited. 1989. Product information. Redditch, Worcestershire, England.

JAIN, S.J. ET AL. 1978. Field-Scale Evaluation of Wastewater Disinfection by Ozone Generated from Oxygen, Prog. Wastewater Disinfection Technology, Proc. Natl. Symp. Cincinnati, Ohio, Sept. 18–20, edited by A.D. Venosa. EPA-600/9-79-018, U.S. EPA, Cincinnati, OH.

JOEL, A.R. 1990. Capitals Controls Co., Inc. Personal Communication.

JOHNSON, J.D. & DUNN, J.F. 1976. Ozone Amperometric Membrane Technology. Proc. Second Int. Symp. Ozone Technol., edited by R.G. Rice et al.

JOOST, R. 1989. John Corrollo Engineers, personal communication.

JORET, J.C. ET AL. 1986. Inactivation des Virus dans l'Eau sur une Filière de Production à Ozonation Étagée. *Wtr. Res.* 20:871.

KIFFER, A.D. 1959. Method and Apparatus for Producing Ozone-Carrier Gas Mixture. Patented in U.S. on the Feb. 3, 1959, no. 2872397.

KOGELSCHATZ, U. 1988. In *Advanced Ozone Generation: Process Technologies for Water Treatment,* edited by S. Stucki. Plenum Press, New York, pp. 87–120.

KOGELSCHATZ, U. ET AL. 1987. Ozone Generation from Oxygen and Air: Discharge Physics and Reaction Mechanisms. 8th Ozone World Congress, Zurich, Switzerland, Vol 1:A-1.

KUO, C.H. & YOCUM, F.H. 1982. Mass Transfer of Ozone into Aqueous Systems. In *Handbook of Ozone Technology and Applications,* Vol. 1, edited by R.G. Rice and A. Netzer. Ann Arbor Science Publishers, Inc., Ann Arbor, MI.

LANGLAIS, B. 1979. Mécanisme de l'Ozonation de Quelques Micropolluants Aromatiques en Milieu Aqueux Dilué. Incidence sur la Toxicité et la Réactivité des Sous Produits vis à vis du Chlore. Thèse No. 749. Université de Poitiers, France.

_____. 1982. Schéma Technologique de l'Ozonation de l'Eau en Vue de sa Potabilisation. Presented at the International Seminar about Ozone, Torino, Italy, June 28–29.

_____. 1989. Anjou Recherche, Centre de Recherche de la Compagnie Generale des Eaux. Personal Communication.

LANGLAIS, B. & BABLON, G. 1990. Les Différentes Applications de l'Ozone et Techniques Associées en Traitement d'Eau: Un Moyen pour Répondre aux Normes da Potabilité. *Sci. Tech. de l'Eau,* 23(1):33.

LÉGERON, J.P. 1978. Chemical Ozone Demand of a Water Sample by Laboratory Evaluation. *Ozonews,* International Ozone Institute, August, 5:8, Part 2.

_____. 1984. "Ozone Disinfection of Drinking Water." In *Handbook of Ozone Technology and Applications,* vol. II, edited by R.G. Rice & A. Netzer. Ann Arbor Science Publishers, Inc., Ann Arbor, MI.

LEPAGE, W.L. 1981. The Anatomy of an Ozone Plant. *Jour. AWWA,* 73(2):105.

LEV, O. & REGLI, S. 1990a. Compliance of Ozone Disinfection Systems with the Surface

Water Treatment Rule: Selection of Characteristic Concentration "C". Submitted to *J. Env. Eng. ASCE,* New York, Nov.

_____. 1990b. Compliance of Ozone Disinfection Systems with the Surface Water Treatment Rule: Selection of Characteristic Time "T". Submitted to *J. Env. Eng. ASCE,* New York, Nov.

LEVENSPIEL, O. 1972. *Chemical Reaction Engineering,* 2nd ed., John Wiley and Sons, Inc., New York.

LEWIS, W.K. & WHITMAN, W.E. 1981. Principles of Gas Absorption. *Ind. Eng. Chem.* 16(12): 1215.

LOTEPRO CORPORATION. 1988. Product information.

LOWTHER, F.E. 1976a. Procédé et Appareil de Création et de Récupération Continues d'Ozone. Brevet Déposé en France le 11 Mars 1976, numéro 7607028. Patented in U.S. on March 12, 1975, no. 557 594 and 557 595.

_____. 1976b. Procédé et Appareil de Création et de Séparation d'Ozone. Brevet Déposé en France le 16 Avril 1976, numéro 7611422. Patented in U.S. on April 17, 1975, no. 588934.

MACKAY, D.R. 1985. Control of a Full Automated Ozone Application System, *Ozone Sci. Engrg.,* 7(1):77.

_____. 1987. Ozone Dosage Control. In *Analytical Aspects of Ozone Treatment of Water and Wastewater,* edited by R.G. Rice et al. Lewis Publishers, Inc., Chelsea, MI.

MAIER, D. ET AL. 1988. Ozone Gas Scrubbing for Air-Operated Ozone Generation Systems. *Ozone Sci. Engrg.,* 10:241.

MALLEVIALLE, J. ET AL. 1979. Fluorescence des Eaux, Etude de la Signification, Recherche des Corrélations avec d'Autres Paramètres Globaux (COT, Absorption à 254 nm, Oxydabilité au $KMnO_4$). *TSM l'Eau Special Hydrologie,* Mars: 74–182.

MANLEY, T.C. 1943. The Electric Characteristics of the Ozonator Discharge. *Trans. Electrochemical Society,* 84:83–96.

MANLEY, T.C. & NIEGOWSKI, S.J. 1967. *Kirk-Othmer Encyclopedia of Chemical Technology,* Vol. 14, John Wiley and Sons, Inc., New York (2nd ed.), pp. 410–432.

MANZ, C.W. & SHUMATE, A.A. 1988. Dryer Packages Available for Drying Air in the Production of Ozone for Sewage/Wastewater Treatment Plants. Proc. of 1988 Autumn Conf. on Ozonation Systems and Drinking Water Treatment, International Ozone Association, Pan Am Com., Myrtle Beach, SC.

MARTIN, N. 1989. Mise en Oeuvre de l'Ozone dans le Traitement des Eaux Potables. Diplôme d'Etude Approfondie: Génie des Procédés Industriels. Université de Compiègne, France. Juin.

MARTIN, N. & BOURBIGOT, M.M. 1989. Critères de Conception d'une Cuve d'Ozonation. In *L'Ozonation des Eaux. Manuel Pratique,* edited by W.J. Masschelein. Technique et Documentation, Paris, France (in press).

MASSCHELEIN, W.J. 1982a. "The Use of an Oxygen-Enriched Process Gas." In *Ozonation Manual for Water and Wastewater Treatment,* edited by W.J. Masschelein. John Wiley and Sons, Inc., New York.

_____. 1982b. "Practical Aspects of the Recycling of Effluent Gas into Generation Systems." In *Ozonation Manual for Water and Wastewater Treatment,* edited by W.J. Masschelein. John Wiley and Sons, Inc., New York.

_____. 1982c. "Contacting of Ozone with Water and Contactor Offgas Treatment." In *Handbook of Ozone Technology and Applications,* Vol. 1, edited by R.G. Rice and A. Netzer. Ann Arbor Science Publishers, Inc., Ann Arbor, MI.

_____. 1982d. "The Ozonation Scheme at the Tailfer Plant." In *Ozonation Manual for Water and Wastewater Treatment,* edited by W.J. Masschelein. John Wiley and Sons, Inc., New York.

_____. 1982e. "The Tailfer Plant and New Developments in Ozone Destruction Techniques." In *Ozonation Manual for Water and Wastewater Treatment,* edited by W.J. Masschelein. John Wiley and Sons, Inc., New York.

_____. 1987. Iodometric Method for the Determination of Ozone in a Process Gas. *IOA Ozone News,* 15:4 (July/August).

_____. 1989. Present State of Standardization of Ozone Measurements in Europe. Proc. Ninth Ozone World Congress, edited by L.J. Bollyky. International Ozone Association, Norwalk, CT, vol. 2:562.

MASSCHELEIN, W.J. & FRANSOLET, G. 1977. Spectrophotometric Determination of Residual Ozone in Water with ACVK, *Jour. AWWA,* 61: 461.

_____. 1979. Technique of Continuous Electrochemical Measurement of Residual Active Oxidants (RAO) in Water. *Tribune du Cebedeau,* 422–423, pp. 31–38.

MCKEON, T. & GROSS, L. 1987. A Major Ozonation/Direct Filtration Water Treatment Facility. *Public Works,* Sept.

METCALF & EDDY, INC. 1979. *Wastewater Engineering: Treatment, Disposal, Reuse.* McGraw-Hill, New York.

Metropolitan Water District of Southern California (MWDSC). 1988. Oxidation Demonstration Project Scoping Study. Ozone/PEROXONE Technology Status Report. Document prepared by James M. Montgomery Consulting Engineers.

_____. 1989. Evaluation of Ozone and PEROXONE for Disinfection and Control of Disinfection By-Products and Taste and Odor Compounds. Phase IV. Experiments 1 and 2. Project Status Report. June.

MILLER, G.W. ET AL. 1978. An Assessment of Ozone and Chlorine Dioxide Technologies for Treatment of Municipal Water Supplies. U.S. EPA, EPA-600/2-78-147, NTIS PB-285972, Cincinnati, OH.

MONK, D.G. ET AL. 1985. Prepurchasing Ozone Equipment. *Jour. AWWA*, 77:8.

MONK, R. 1989. Camp Dresser & McKee, personal communication.

MRAZEK, L.G. 1981. Resistance of Concrete Structures to Ozone Penetration. *Concrete International*, April, pp. 69–74.

NAKAYAMA, S. ET AL. 1979. Improved Ozonation in Aqueous Systems. *Ozone Sci. Engrg.*, 119.

NAMBA, K. & HONDA, T. 1987. A High Efficiency Ozone Generation System. Eighth Ozone World Congress, Zurich, Switzerland, Vol. 1:A11.

NONNENMACHER, K. Analyseur d'Ozone Pour la Mesure Continue et Selective de l'Ozone dans la Phase Aqueuse. *Documentation Technique Société Anseros*, Görtringen, Germany.

OLDSHUE, J.Y. 1983. *Fluid Mixing Technology, Chemical Engineering*. McGraw-Hill Publishing Co., New York.

ORGLER, K. 1980. Procédés et Frais d'Exploitation de Destruction de l'Ozone dans l'Air d'Event. In *L'Ozonation des Eaux: Manuel Pratique*, edited by W.J. Masschelein. Technique et Documentation, Paris, France.

_____. 1982. "Methods and Operating Costs of Ozone Destruction in Offgas." In *Ozonation Manual for Water and Wastewater Treatment*, edited by W.J. Masschelein. John Wiley and Sons, Inc., New York.

PALIN, A.T. Simplified DPD Procedure for the Determination of Residual Ozone and Mixtures of Ozone with Free Combined Chlorine. Intl. Ozone Assoc., Symp. on Adv. Ozone Tech.

Pan American Committee of the International Ozone Association. 1990. *Design Guidance Manual for Ozone Systems*, edited by M.A. Dimitriou. Norwalk, CT.

PARÉ, M. 1967. Contribution à l'Étude du Comportement Électrique d'un Appareil Industriel de Production d'Ozone. Chaire d'Électricité Industrielle. Thesis, Conservatoire National des Arts et Métiers, Paris, France.

PASCAL, O. ET AL. 1984. La Chimiométrie pour Piloter un Traitement Combiné de Préozonation-Coagulation. Exemple de l'Usine d'Eau potable de Choisy-le-Roi (800,000 m^3/jour). Presented at the International Ozone Association Montreal Workshop, Montreal, Canada.

PEEK, F.W. 1924. Phénomènes Diélectriques dans la Technique des Hautes Tensions. *Edition Delagrave*, Paris, France.

PERROT, J.Y. 1989. Trailigaz. Personal communication.

PRICE, M. ET AL. 1988. Pilot Plant Test Results. East Bay Municipal Utility District. Preliminary Design of Ozonation Facilities.

Process Application Inc. 1987. Utilizing Ozone: Design and Generating Requirements. Seminar presented to James M. Montgomery Consulting Engineers Inc. Pasadena, California.

PROD'HOMME, L. 1960. Action de la Température sur les Propriétés Diélectriques des Verres. *Verres et Réfractaires*, 2 and 3:1–12.

RAKNESS, K.L. 1989. Process Applications, Inc., personal communication.

RAKNESS, K.L. & STOLARIK, G.F. 1988. Power Evaluation of the Los Angeles Oxygen-Fed Ozone System. Presented at the International Ozone Association Pan American Committee Spring Conference, Monroe, Michigan (April).

RAKNESS, K.L. ET AL. 1984. Design, Start-up, and Operation of an Ozone Disinfection Unit. *Jour. WPCF*, 56:1152.

_____. 1987. Ozone System Design for Water and Wastewater. *Proc. Second Intl. Conf. The Role of Ozone in Water and Wastewater Treatment*, edited by D.W. Smith and G.R. Finch. Tektran International. Kitchener, Ontario, pp. 135–152.

_____. 1988. Practical Design Model for Calculating Bubble Diffuser Contactor Ozone Transfer Efficiency. *Ozone Sci. Engrg.*, 10:2:173.

_____. 1988. Start-Up and Operation of the Indianapolis Ozone Disinfection Wastewater Systems. *Ozone Sci. Engrg.*, 10:3:214.

_____. 1989. Design Considerations in Alternative Ozone Feed-Gas Systems, Proc. Ninth Ozone World Congress, edited by L.J. Bollyky. Int. Ozone Assoc., Norwalk, CT. Vol 2:634.

RECKHOW, D.A. & SINGER, P.C. 1984. The Removal of Organic Halide Precursors by Preozonation and Alum Coagulation. *Jour. AWWA*, 76:4.

Refractron Technologies Corp. 1990. Product Information.

Reid Quebe, Allison and Associates. 1981. Use of UPVC Piping in Ozone Diffuser Header Piping.

RENNER, R.C. ET AL. 1988. Ozone in Water Treatment—The Designer's Role. *Ozone Sci. Engrg.*, 10:55.

RICE, R.G. ET AL. 1981. Uses of Ozone in Drinking Water Treatment. *Jour. AWWA*, 73:1.

RICHARD, Y.R. 1986. Improvement of Ozone Oxidation and Disinfection Design. *Ozone Sci. Engrg.,* 8:261.

RICHARD, Y. & BLUE, P. 1978. Ozone Pretreatment of Drinking Water. Ozone Technology Symp., International Ozone Institute. Los Angeles.

RIQUARTS, H.P. & LEITGEB, P. 1985. Gas Separation Using Pressing Swing Adsorption Plants. Linde Reports No. 40. Linde AG, West Germany.

ROBSON, C.M. ET AL. 1988. History and Current Status of U.S.A. Applications of Ozonation for Drinking Water Treatment. Proc. 1988 Autumn Conf.: Ozonation Systems and Drinking Water Treatment. International Ozone Association, Myrtle Beach, SC.

ROSEN, H.M. 1973. Use of Ozone and Oxygen in Advanced Wastewater Treatment. *Jour. WPCF* 45:(12)2521–2536.

ROUSTAN ET AL. 1987. Mass Transfer Considerations for the Design of Ozone Contactor. Proc. Second Intl. Conf. The Role of Ozone in Water and Wastewater Treatment, edited by D.W. Smith and G.R. Finch. Tektran International, Kitchener, Ontario, pp. 125–133.

SCHECHTER, A. 1973. Spectrophotometric Method for Determination of Ozone in Aqueous Solutions. *Wtr. Res.* 7:729.

SCHOLZE, H. 1980. *Le Verre: Nature, Structure et Propriétés.* Institut du Verre, Paris, France (2nd édition), pp. 250–270.

SCHULHOF, P. & DYER-SMITH, P. 1988. Generation of Ozone. Proc. of the International Ozone Symposium at Rio de Janeiro, Brazil. Sept. 16, pp. 27–45.

SCHWARTZ, B.J. 1990. Hackensack Water Company. Personal Communication.

SHUMATE, A.A. 1989. Lectrodryer of Ajax Magnethermic, Richmond, KY. Personal Communication.

SPEECE, R.E. ET AL. 1980. Pilot Performance of Deep U-Tubes. *Prog. Water Tech.,* 12:395.

Standard Methods for the Examination of Water and Wastewater. 1985. APHA, AWWA, and WPCF. Washington, D.C. (16th ed.).

STOLARIK, G.F. & CHRISTIE, J.D. 1988. Projection of Ozone C · t Values—Los Angeles Aqueduct Filtration Plant. Conf. Intl. Ozone Assn., Pan American Congress, Monroe, MI.

TALBOT, E.M. 1985. *Compressed Air Systems.* Fairmont Press, Inc., Lilburn, GA.

Trailigaz. 1990. Personal Communication.

Trailigaz Documentation. 1966. L'Ozonation des Eaux de Consommation. Edited by E.R.I.C., Paris, France.

UHLIG, G. 1984. Closed-Loop System for Generating Ozone from Oxygen and its Application to the Treatment of Drinking Water at Duisburg, Federal Republic of Germany. *Handbook of Ozone Technology and Applications,* Vol. 2, edited by R.G. Rice and A. Netzer. Ann Arbor Science Publishers, Inc., Ann Arbor, MI, pp. 141–150.

U.S. Environmental Protection Agency (U.S. EPA). 1986. *Design Manual: Municipal Wastewater Disinfection.* EPA-625/1-86/021.

_____. 1989. Appendix O, Guidelines to Predict Performance of Ozone Disinfection Systems. In *Guidance Manual for Compliance with the Filtration and Disinfection Requirements for Public Water Systems Using Surface Water Sources.* Science and Technology Branch, Criteria and Standards Division, Office of Drinking Water, October.

_____. 1989. Design Manual for Fine Pore Aeration Systems.

Vade Mecum du Chef d'Usine de Traitement d'Eau Destinée à la Consommation. 1987. *Technique et Documentation Lavoisier,* p. 54–56.

VAN LEEUWEN, J. 1980. The Design and Application of a Packed Column Ozone Absorber in Water Reclamation. *Ozone Sci. Engrg.,* 2:283.

VENOSA, A. & MECKES, M.C. 1986. Control of Ozone Disinfection (of Wastewaters) by Exhaust Gas Monitoring. In *Analytical Aspects of Ozone Treatment of Water and Wastewater,* edited by R.G. Rice and W.J. Lacy. Lewis Publishers, Inc., Chelsea, MI, pp. 303–314.

VENOSA, A.D. ET AL. 1985. Reliable Ozone Disinfection Using Offgas Control. *Jour. WPCF,* 57:929.

WALTON, J.R. 1983. The Effect of Ozone on the Corrosion Rate of Metals. Presented at the 6th Ozone World Congress. International Ozone Association, Washington, D.C., May 23–26.

WARAKOMSKI, A. 1989. Ozone Generation Utilizing High Purity Oxygen Feed Gas. Proc. 9th Ozone World Congress. International Ozone Association, edited by L.J. Bollyky. New York, vol. 2:596.

WEINER, A.L. 1974. Drying Gases and Liquids. *Chem. Eng.,* Sept. 16.

ZAWIERUCHA, R. & CHARLESON, H. 1982. Recent Experimental Studies Dealing with Corrosion and Degradation of Materials in Ozone-Containing Environments. In *Handbook of Ozone Technology and Applications,* Vol. 1, edited by R.G. Rice and A. Netzer. Ann Arbor Science Publishers, Inc., Ann Arbor, MI, pp. 227–252.

V

Operating an Ozonation Facility

François Damez
Bruno Langlais
Kerwin L. Rakness
C. Michael Robson

The most common type of ozone system uses air for the feed gas. In this case, the operation and maintenance (O&M) of the air treatment equipment, as well as the ozone generation and contacting equipment, rest solely with the plant operator.

If liquid oxygen (LOX) is used, the oxygen feed gas is produced offsite and the onsite O&M activities are associated with storage in insulated tanks and the LOX vaporization facilities. When gaseous oxygen (GOX) is used, it may be produced onsite and operated by the water treatment plant staff, or produced offsite and delivered via pipeline.

The use of LOX is gaining popularity, especially in areas where its price is low. The onsite production of GOX is feasible for large plants (i.e., 5,000 lb O_3/day—about 100 kg O_3/h); GOX delivery has not been used to date.

Selection of specific equipment for an ozone system usually requires that a balance be achieved between the opposing issues of capital cost, energy usage, and O&M considerations. Chapters IV and VI contain additional discussion on these items. In this chapter, O&M considerations are discussed more specifically for air-feed ozone systems in order to give the reader who may be new to ozone an overview of the operation of an ozone system. The reader should consult with ozone manufacturers and operators of other ozone systems for additional information.

V.A OPERATING METHODS

V.A.1 Air as Feed Gas

In facilities where air is the feed gas, the air treatment systems can be categorized into high-pressure or low/medium-pressure air treatment systems (see IV.B). This

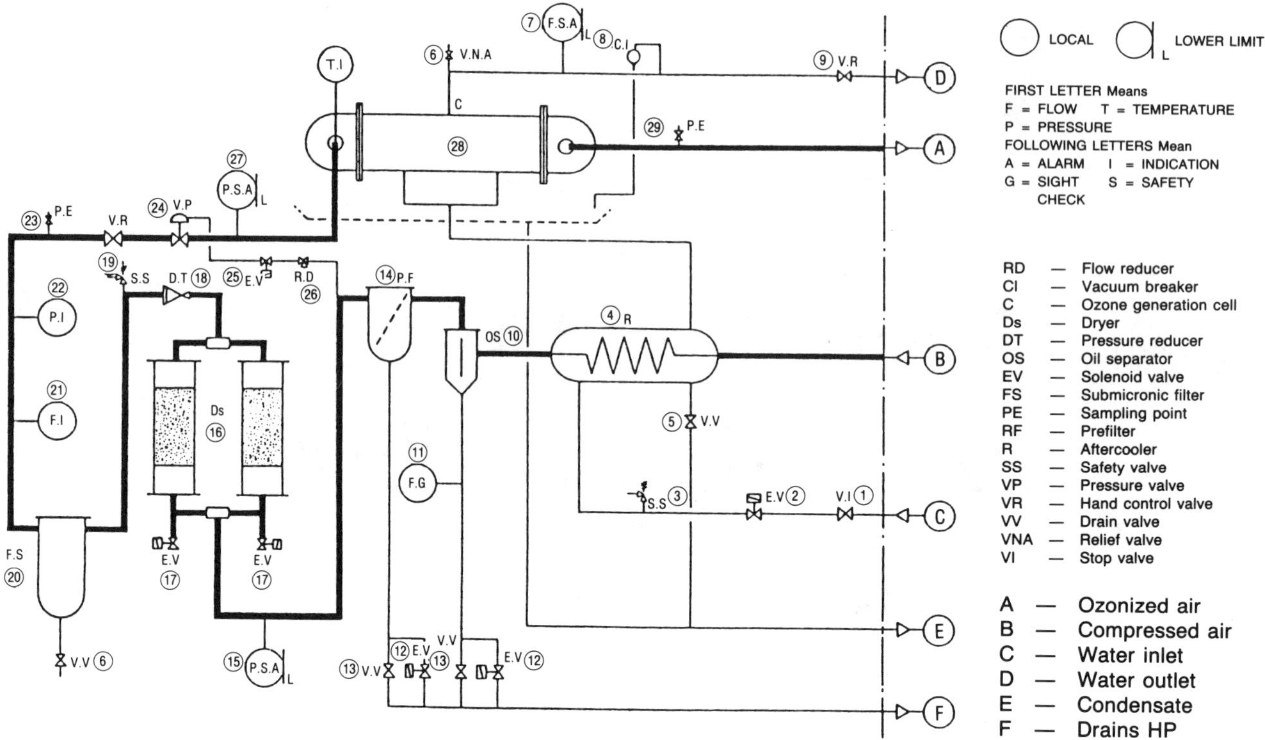

Figure V–1 Ozone Production Flow Sheet. High-Pressure Air Treatment System

section will focus on two types of operating systems—high- and low-pressure air-feed systems. It should be noted that the specific operating details apply only to the types of equipment presented here. Details may change if other types of equipment are installed. However, the overall "concept" of the equipment operation is similar for all facilities.

High-pressure air treatment system. Figure V–1 is a schematic of a high-pressure air treatment system, which is relatively common in small-capacity facilities and also used in medium-capacity facilities. It is noted that other equipment choices are available (see sec. IV.B and sec. VI.C, D and E).

In the case described here, the air compressor is the only apparatus that is separated from the rest of the equipment. It is housed in a soundproof room that is well-ventilated to enable the heat generated by the compressor to escape. The remainder of the ozone production unit is mounted on a welded structural steel frame that supports the machinery required for air treatment and ozone production and the completely automated controls.

The air discharged from the oil-lubricated compressor is at a relative pressure of 7 bar (100 psig) and a temperature of 80°C (176°F). This air contains water and oil vapors that must be removed. The compressor is connected to the remainder of the air preparation system at point B.

The air passes through an aftercooler (4) in which most of the oil and water is condensed. The air then reaches the cyclone separator (10) where it moves in a spiral motion, causing more oil and water droplets to separate out, seep along the walls of the separator, and settle to the bottom. The oil and water droplets are periodically drained from the separator via a solenoid valve.

The air passes through a prefilter (14) that removes solid and liquid particles measuring up to 1 μm in size. The maximum residual oil content at the outlet of the prefilter is 0.5 ppm by weight. The condensates, like those in the cyclone separator, are periodically drawn off via a solenoid valve with a manual backup valve. A pressure switch controls the pressure at the set value of 7 bar (100 psig).

The air is then conveyed to the desiccant dryers (containing activated alumina in this case). The dryer is comprised of two tanks, one tank is in service while the other is undergoing regeneration by a countercurrent flow of dry air. The inlet and outlet of each tank are fitted with a sintered bronze filter, designed to trap the desiccant dust. At the outlet of the dryer, the dewpoint is at a maximum of −60°C (−76°F) at standard pressure of 1 atm (see chapter IV for the value of the dewpoint at different pressures).

The pressure of the dry air is then reduced to an operating value of approximately 1 bar (15 psig) by means of a pressure-reducing/control valve (18). A safety valve (19), located downstream, protects the low-pressure circuit. A submicronic filter (20) traps any final traces of oil that may have passed through the desiccant dryer in the form of aerosols. The maximum residual oil content should be less than 0.01 ppm by weight.

Upstream of the ozone generator, the gas preparation subsystem is equipped with a flowmeter (21), a pressure gauge (22), a sampling intake (23), a control valve (24) that regulates the flow of air to be ozonated, a solenoid valve (25), and a pressure switch (27) that stops ozone generation if the pressure drops below a set value established for the ozone generator.

Finally, the prepared air passes through the ozone generator (28), which is fitted at the outlet with a sampling device (29) to measure the ozone concentration of the ozonized air. The air then leaves the ozone generator via point A to the dissolution system.

The cooling water system is connected at point C. The water, at a pressure of 1 to 2 bar (15 to 30 psig), flows through an isolation valve (1), a water inlet solenoid valve (2) that opens on a startup order, and a safety valve (3) that protects the piping.

The water passes through the aftercooler (4). It then passes through the ozone generator (28), cooling the generator as it does so.

At the outlet of the ozone generator the following controls are installed: an air-relief valve (6) for purging the air from the water circuit when it is first filled; a water flow rate controller (7); a vacuum breaker (8); and a manual control valve (9) located before the outlet [point (D)]. The ozone generator and aftercooler are drained via a manually controlled valve (5) connected to the drain system, discharging at point E.

The automatic startup procedure, initiated by closing the startup switch, takes place as follows:

- the water solenoid valve opens and admits water to the compressor aftercooler,
- the compressor starts operating,
- the drain system is switched on and controls the opening of the solenoid drain valves for 0.5 sec every 3 min, and
- after a time lag the air solenoid valve opens.

The air preparation system should be operated through several cycles to ensure that the required dewpoint is being provided to the ozone generator.

With the air and water solenoid valves open and when the contacts on the various water and air circuit safety devices are in order, as described above, the ozone generator is energized and the power is gradually stepped up until it reaches a preset level.

When ozone dissolution is performed by diffusers, a preliminary test to check that the diffusers are in working order must be performed. This is done by filling the contactor with water to a level of about 0.30 m (12 in.) above the diffusers and then feeding them with unozonized air. It is essential that all the diffusers are working.

Ozone production can be adjusted based on a given treatment rate or, in the case of disinfection, to achieve a constant dissolved residual ozone concentration over a certain time ($C \cdot t$ concept). In both cases, adequate production is achieved by adjusting the power applied to the ozone generator and by adjusting the input of air to be ozonated. This is done by using the two graphs given in Figure V–2, which depict

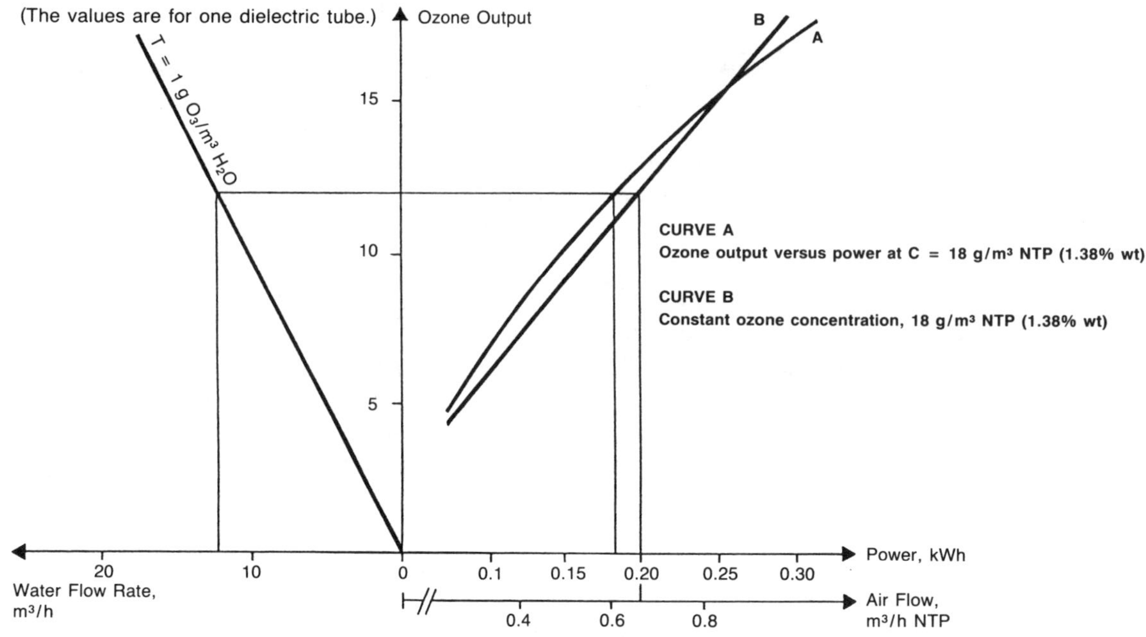

Figure V–2 Curves for Adjusting the Air Flow Rate and the Electric Power

- production of ozone versus power at optimum concentration, and
- the constant concentration curve at its optimum value versus water flow rate.

These two curves enable the operator to determine for each ozone production run the power to be applied to the ozone generator and the amount of compressed air required to maintain the concentration at its optimum level.

This relatively simple operating method can be automated. For treatment at a constant dose rate, the quantity of ozone to be generated is calculated based on the flow rate of water to be ozonated. When there is a set residual concentration, it is the variations in this residual that indicate if production must be increased or decreased. For treatment at a constant dose rate, proportional regulation is sufficient; if a set residual is used, control must be proportional, integral, and derivative.

Low/medium-pressure air treatment system. Low/medium-pressure systems are common for large-capacity plants. Large installations are often designed in trains, that is, each contactor is fed by a separate ozone generator and an air treatment system. Obviously, it is important that these trains can be interconnected at different points, primarily to meet standby supply requirements.

The example described here is for a low-pressure air treatment system (see chapter IV.B for other available equipment). The air blowers are of the rotary-piston, variable-speed type, driven by asynchronous electric motors controlled by a microprocessor speed converter. The air drawn into the compressor is treated using a dust filter and is drawn through a silencer. It is then discharged at a relative pressure of 0.8 bar (12 psig) and a temperature between 80°C (176°F) and 130°C (266°F) via a second silencer. The blower is protected by a pressure switch and safety valve.

The air then passes through an aftercooler that is fed by treated water. The air outlet temperature depends on that of the water; the maximum temperature is approximately 25°C (77°F). The treated water flows through the aftercooler at a constant rate (maximum pressure: 4 bar [57 psig]) and is recirculated back through the treatment plant.

Air then flows through a refrigerant dryer, where it is cooled with glycolated water that is kept at a temperature of 0°C (32°F) by a Freon-based cooling device. A thermostat at the outlet of this refrigerant dryer maintains the air temperature between 5 and 7°C (41 and 45°F). The condensates formed during the cooling process are trapped in a purifying separator and discharged via a solenoid-operated drain valve that opens for 2 sec every 3 min.

Downstream of the separator, the treated air flow rate is measured by a vacuum flowmeter and a 4-20 mA differential-pressure transmitter. This flow measurement activates an appropriate number of refrigerant dryers.

The refrigerant dryer consists of two heat exchangers. One is the air/glycolated water heat exchanger and the second is the glycolated water/freon unit. According to the air flow, and considering a preset value, one or two Freon compressors are started. These two compressors are installed on two independent circuits connected to one evaporator and one condenser. The quantity of water circulating in the condenser is controlled by means of two pressostatic valves operated by the Freon pressure in the circuits. Each Freon circuit includes a pressure switch with a high-level alarm and a second pressure switch with a low-level alarm, which stops the compressor in case of misoperation. The starting and stopping of the circuits also depends on the main start order of the ozonation system, and on indications from air temperature controllers with a high- and low-level alarm at the outlet of the refrigerant dryer. Such a system maintains the required temperature of the glycolated water at around 0°C (32°F) for producing air at a temperature between 5 and 7°C (41 and 45°F). On the other hand, the glycolated water circuit also includes a temperature controller with low-level alarm, which stops the Freon unit to avoid circuit freezing.

The air to be treated then enters the desiccant dryer, which is composed of two chambers filled with desiccant (activated alumina in this case) and a regeneration circuit. The two vessels are connected by two lubricated four-way valves and, in turn, coupled via a common lever-operated shaft to a pneumatic cylinder. The regeneration circuit includes a heat exchanger that is composed of enclosed immersion heaters, steel tubes, and safety thermostat regulators. The system is designed for automated operation, with each chamber acting alternately in order to obtain a dewpoint of less than −60°C (−76°F) at standard pressure while the other is undergoing regeneration.

Regeneration consists of two operating stages that function differently. First, there is a period of heating, during which time the regeneration air (dry because it is tapped at the outlet of the desiccant dryer) is heated to the temperature required for satisfactory desorption of the activated alumina [180 to 200°C (356 to 392°F)]. The dry regeneration air flows from top to bottom. The heating phase ends when the regeneration outlet temperature reaches 140°C (284°F). The time required for the heating phase (typically 4 h) is controlled by a time switch.

A cooling sequence (typically 4 h) is necessary after the reheating stage to restore the alumina to ambient temperature. This is done by continuing to tap the dry air at the outlet of the operating vessel after the electric heating system has been turned off. The necessary time lag is controlled by another time switch. The dry air is then filtered in order to capture the desiccant dust that may have passed through the desiccant dryer.

Upstream of the ozone generator, the dry air circuit is fitted with a control valve plus flow meter, a sampling point for dewpoint measurement, a pressure gauge, a solenoid valve to prevent backflow of ozone, and a pressure switch that stops the ozone generator if there is an abnormal loss of pressure. At the outlet of the ozone generator, a sampling device is installed, as well as a check valve to prevent a return of wet air to the ozone generator.

The ozonized air is sent to the dissolution units where it is distributed within the contactor. The offgas from the contactor is destroyed by a thermal ozone destructor, consisting of an extraction blower, a primary heat exchanger that raises the temperature of the air to 300°C (572°F), and a heater that is heated to 350°C (662°F), the temperature at which ozone is destroyed rapidly (see sec. F.2 chapter IV).

The ozone generator is cooled by a water circuit that is pressurized by a booster pump. For safety reasons, a flow switch shuts down the generator if there is a lack of water.

Such a facility, particularly if it consists of several identical trains, must include, at a minimum, remote control from a central control room. The facility should also be equipped with a maximum number of local controls, for example, controls for emptying the drainage system, changing over desiccator chambers, and managing all alarms on air and water systems. This involves the use of several continuously operating sensors that transmit essential control and monitoring data, including air flow rate at different points along the treatment train, dewpoint, concentration of ozone in the product gas, and ozone generator power.

For facilities with a certain degree of sophistication, and whenever it is possible to determine a set rate of treatment (such as the ozone dose rate or the residual ozone content in the water), it is advantageous to install a computerized control system to control the entire ozonation sequence.

V.A.2 Oxygen as Feed Gas

The methods for supplying oxygen as a feed gas include gaseous oxygen (GOX) production by cryogenic or molecular sieve adsorption (pressure swing adsorption [PSA] or vacuum swing adsorption [VSA] systems) or purchasing liquid oxygen (LOX), as described in chapter IV.B.1.

Gaseous oxygen production facilities. The operating considerations for the GOX facilities (usually for large or very large plants) are specific to the particular type of equipment used to generate the oxygen. For example, a cryogenic plant requires sophisticated training that is usually provided by the oxygen plant suppliers. In some instances, an ongoing maintenance contract is developed; in other cases, special technical support is on an on-call basis. Several oxygen plants (cryogenic and molecular sieve adsorption) have been successfully used at municipal wastewater treatment plants and operated by the plant operators.

Liquid oxygen supply. Liquid oxygen (LOX) is the simplest ozone feed gas system. LOX storage tanks, liquid to gaseous oxygen evaporators (either ambient or electrically heated) and flow and pressure control valves are the equipment that is necessary.

The dewpoint of the oxygen supplied through GOX or LOX system is normally below −60°C (−76°F), which is compatible with the needs of the ozone generators. The primary operational condition is to control the oxygen flow rate as the ozone production rate changes in order to minimize operating cost. This is generally accomplished by maintaining a certain range in ozone concentration. The ozone concentration monitor must be reliable and very well maintained. Ozone generators are equipped upstream with a flow meter and a control valve that is activated by the ozone concentration monitor. A back-up oxygen flow control device, in case of an ozone concentration meter malfunction, is also used. In all cases, it is wise to install a second fast-closing safety switch, which will enable the oxygen to be cut off immediately in the unlikely event of fire.

V.B OZONE SYSTEM RELIABILITY

Equipment reliability and its influence on water quality is fundamental to the field of water treatment. The slightest problem with any treatment unit can jeopardize the quality of the product water.

Each element of a water treatment facility is dependent on the other treatment elements to meet the overall treatment goal. Therefore, each element must have standby capacity to enable it to complete its assigned role. This is particularly true for ozonation, which has a very important role in water treatment, particularly when primary disinfection is concerned.

V.B.1 Design and Standby Capacities

When determining the standby equipment to be used, two primary factors must be taken into account:

- the range of ozone production, especially as it relates to the design capacity; and
- the degree of modularity as it relates to ozone production equipment and contact systems.

The primary factor in designing standby equipment is the range of ozone production. During the initial design phases, variations in water flow and ozone dose must be studied. From these data, a histogram of the quantities of ozone to be generated can be drawn. Using the histogram, it is then possible to calculate the probability of the facility being used at low, average, and peak ozone production rates.

The consideration of standby capacity is most important at peak ozone production rates. In most instances, standby units need to be available when operating at peak ozone production, particularly in order to meet disinfection requirements. The degree of standby equipment needed should be determined on a case-by-case basis, should consider all equipment components (e.g., air compressor, refrigerant dryer, desiccant dryer, ozone generator, and ozone destructor) and should address all local regulatory standards.

In many applications, facilities operate at their peak ozone production requirement only rarely. In these cases, it may be possible to eliminate the provision of a standby unit by performing all necessary service work during the low-demand periods. The difference between the average ozone requirement and the installed capacity becomes the standby capacity. This approach may be a result of a limited construction budget, agency policy, or designer's philosophy. As discussed in the preceding paragraph, regulatory agency standards may preclude this approach.

The second factor in determining the size of the standby equipment is the number of units or trains that are installed. In order to provide standby capacity, it is advisable to provide the required production capacity in the form of several trains or units, which serve as backups for each other. Providing multiple units or trains is also useful when optimizing plant efficiency for operation at its most common rate of production.

For contactors, the provision of several trains or units allows for operation of only the number of units corresponding to the flow rate; each unit is operated at its design output. As a result, the contact time remains constant at its design value (see sec. IV.D). At a low flow rate, the contact time would always be longer. This extra time may lead to excessive self-decomposition of the ozone and, when treating according to the $C \cdot t$ criterion (see sec. III.H), to surplus ozone consumption. The provision of a number of contactors is therefore necessary in order to obtain optimum compliance with ozone treatment requirements.

Provision of multiple units also makes economic operation of the ozone production process possible. For each ozone generation technique (low or medium frequency; low-, medium-, or high-pressure air preparation) there is an optimum operating concentration that is a compromise between the cost of the energy required for feed gas production and ozone generation system production efficiency (assuming that minimum conditions for an adequate mixing and transfer are respected [see sec. IV.D]). It is possible to determine the capacity of the gas preparation equipment that is most suitable to the ozone generator in question.

Moreover, it is preferable to design these components so that they function under optimum conditions most of the time. To do so, the histogram of ozone production rates can be used.

The installed production capacity can be divided into a number of trains or elements that have a capacity such that the most frequent production rates are ensured by a combination of certain elements.

Most importantly, the number of air compressors that are used (or speed of the compressor) should be adjusted to operate the ozone system in a range of ozone concentration that maintains the best overall system efficiency.

V.B.2 Reliability of Equipment

Even if standby capacity is appropriate, the proper operating condition of each unit must be ensured. This is particularly true for ozonation equipment, in which even minor disruptions can result in shutdown or even serious damage that will affect a large part of the plant. For instance, poor air drying can lead to the breakage of multiple dielectric tubes in the ozone generator. The reliability of specific pieces of ozone generator equipment will be discussed in the following sections.

Gas treatment line. Compressors must be installed in a properly ventilated room at an ambient temperature of less than 35°C (95°F). Because compressors are often installed in closed, soundproof rooms, the ventilation must be carefully designed. The safety device on a compressor is either a check valve or a safety valve. If the maintenance operations described in sec. V.C.1 are properly performed and the manufacturer's instructions are followed carefully, compressors will not present any special difficulties.

Aftercoolers that function on a water circuit (which presents some scaling problems) may be easily addressed by following the maintenance steps and manufacturer's instructions.

On the other hand, refrigerant dryers with dual cooling fluid circuits (glycolated water plus Freon, for instance) are much more sophisticated machines. Fast and efficient troubleshooting can be achieved if the unit has a monitoring system or panel that displays major faults.

In order for the generators to operate under their optimum conditions, the dewpoint at the generator inlet must be kept below −60°C (−76°F) at standard pressure, requiring that the desiccant dryers continually perform their required function. When a refrigerant dryer is used, regardless of inlet temperature and flow rate, the temperature of the air reaching the desiccant dryer must be kept at 5°C (41°F).

Inside the desiccant dryer, a sufficient amount of adsorbant must be maintained. Regeneration cycles should be started at fixed periods rather than at varying intervals based on the discharge gas dewpoint. After the dryer has been in service for some time, the adsorbant will become saturated and a front will advance until it suddenly reaches the end of the chamber. Consequently, fixed timing with a sufficient safety margin is recommended to initiate regeneration before the dewpoint at the outlet of the dryer exceeds the design dewpoint.

For regeneration by dry air heating and cooling, it is preferable to adopt fixed timing rather than initiating regeneration based on temperature. A sufficient safety margin must be provided; if it is not provided, the medium gradually loses a part of its adsorbing power. It is noted than even under optimum operating conditions the desiccant will gradually lose its efficiency and have to be replaced (see sec. IV.B). The dryer regeneration heater must supply heated air at a temperature between 180 and 200°C (356 and 392°F) when activated alumina is used. Generally speaking, desiccant dryers are complex units and must be fitted with adequate fault indicators, enabling the trouble source to be located quickly.

In conclusion, all filters and drain systems of the air circuit must be in good working condition. This is necessary because they influence air quality (in particular, moisture content and dust).

Ozone generator. Ozone generators must be protected from:

- high dewpoint (greater than −60°C [−76°F]),
- lack of air, and
- lack of cooling water.

The risk of moist air in the ozone generator can stem from a fault in the air drying system or a backflow of air from the dissolution chambers, which usually arises when one generator is out of service while the others are in operation.

A failure in the desiccation stage of treatment must be identified as quickly as possible using capacitance hygrometers; each desiccant dryer must be equipped with a hygrometer. Moreover, a hygrometer must be installed directly upstream of the ozone generator, after the different air inlets converge to one line. Not only must the hygrometer sound the alarm, it must also induce automatic shutdown of the ozone generator.

A check valve downstream of the ozone generator is not sufficient to protect the generator from the backflow of moist air from the contactor. Isolation valves should be provided in addition to the check valve provided at the ozone contactor.

The automatic ozone generator shutdown sequence must include initial purging with a high flow of dried air to clear the ozone generator of residual ozone traces, followed by a continuous purge with a low flow of dry air (e.g., 10 percent of the design flow) for as long as the ozone generator stands idle, in order to avoid condensation inside the generator.

In order to prevent internal condensation, it is wise to stop the cooling water flow every time the generator is switched off. When the generator is likely to be out of service for several weeks, it is advisable to drain all the water.

When the ozone generator has been stopped or perhaps opened for cleaning, it must be purged for several hours with a high flow of dry air that will remove all traces of moisture. This operation should be continued until the dewpoint at the outlet of the ozone generator remains below −60°C (−76°F) for several hours (say, 12).

These precautions will minimize the conversion of the nitrogen oxides, which are formed during ozone generation (with prepared air as a feed gas), to nitric acid (see sec. II.D). Formation of nitric acid affects both operation and maintenance. Nitric acid forms on the glass dielectric surfaces and reduces ozone production. Cleaning of the tubes requires removing the tubes from the generators, cleaning the tubes, and reinstalling them. It is noted that the multiple handling of the glass tubes results in breakage of tubes or, even more serious, creates planes of weakness in the glass that will result in subsequent failures of the tubes during service, requiring additional opening of the generators. There is comparatively little annual loss of ozone capacity with good quality air feed gas. Nevertheless, yearly cleaning is common.

Lastly, a loss of feed air or cooling water in the generators must be avoided. If the flow of feed air or cooling water is interrupted, the apparatus will overheat, which may lead to breakage of all the generator tubes. The air and water feed lines must therefore be fitted with flow detectors that automatically switch off the generator when air or water flows are interrupted. The most reliable device is an orifice-plate flow meter equipped with a safety pressure switch.

Contactor and dissolution systems. Any trouble with a contactor negatively impacts treatment quality. The trouble must be rectified as quickly as possible, since treatment capacity is typically reduced. Consequently, in addition to having good ozone transfer efficiency and mixing, dissolution systems must have a simple and reliable design, must be easy to maintain so as to minimize downtime, and must include the devices necessary for fast troubleshooting.

The choice of a dissolution system must be made on a case-by-case basis. Such a choice will depend first and foremost on treatment data, such as contact time, variations in influent flow rate and ozonation rate, and the hydraulic profile of the water treatment plant. Numerous design techniques are available to the designer for making this choice (see sec. IV.D). Reliability criteria will then be considered when a dissolution system has been chosen. Only the three most common systems will be discussed here: porous diffusers, turbines, and injectors.

Porous diffusers. The use of porous diffusers, installed on the bottom of a contact vessel that consists of several baffled compartments, is the most widespread method of dissolution. Its reliability depends primarily on the facility design. The optimum system, from the point of view of efficient transfer of the ozonized air to the influent water under treatment, is injection of ozonized gas in a countercurrent flow pattern in the compartments.

The ozonized air manifolds and the diffusers must be able to withstand the corrosive conditions. Corrosion can be caused by the damp ozonized air and dissolved ozone in the water, and also by the formation of low-pH by-products inside the diffusers and at the surface of the water inside the vessel. The common materials used to withstand this environment are stainless steel (316L), ceramics and hypalon, or ethylene-propylene type elastomers. Moreover, the diffusers must be able to resist difficult hydraulic conditions, particularly while the vessels are being filled. Additionally, their service life is limited by the fact that the diffusers gradually become clogged due to the precipitation of iron oxides and manganese if these substances are present in the influent water. As a result, the discs or rods should be inspected annually for proper operation. They may have to be replaced every three or four years. The take-down and reassembly operations must be done as fast as possible to avoid keeping the ozone vessel idle for an excessive length of time.

For all these reasons, one-piece jointless porous diffusers with the simplest assembly system possible, for example, screwed on the ozonized air inlet pipe, are preferred.

Turbines. Dissolution by turbine is another frequently applied means of transferring and mixing ozone. The turbines are immersed aspirating or pressure units. A small portion of the water flow is driven out by the impeller. The energy supplied by the impeller generates an emulsion of ozonized air and water in the vessel, at the bottom of which the turbine is installed. For the turbine to continue operating economically, the flow of ozonized air must be kept as constant as possible. If construction materials are suitable for exposure to ozone and if the immersed equipment meets the criteria for watertightness, the reliability of the turbine will depend on its motor, which will generally give satisfactory performance.

Injectors. Contact by injection is another highly reliable method of dissolution. This is due to the fact that it does not involve the use of moving parts. Materials used for construction (concrete, stainless steel) must be designed to withstand the wet ozone environment (see sec. IV.D.2).

Ozone destruction units. The offgas from the contactors contains about 1 g/m^3 NTP (0.08 percent by weight) of ozone. Two methods of destruction are feasible at the present time (1990). One method involves thermal destruction with heat recovery. The second is thermal-catalytic destruction. For both methods, the offgas entering the destructors must be separated from the foam floating on the water inside the contactor. This highly acidic foam substance would otherwise attack even the stainless steel machinery parts. A water-spraying device in the contactor section of the air vent plus a stainless steel demister generally achieve the necessary results.

The thermal destructor is composed of an air/air pre-heating exchanger, followed by a furnace that raises the temperature of the offgas to 350°C (662°F), the temperature for rapidly destroying ozone. When it leaves the furnace, the offgas enters the preheating exchanger where it is exhausted by a fan. The system is very simple, and therefore reliable, providing that the parts in contact with the very wet ozonized gas are constructed of stainless steel (316L).

The thermal-catalytic type destructor consists first of a pre-heating oven that dries the offgas and heats it to avoid condensation and to optimize the catalyst reaction—50 to 70°C (122°F to 158°F) for a metal oxide and 30°C (86°F) for a metal catalyst. The offgas that is heated above its dewpoint temperature is brought into contact with the catalyst and then discharged via an exhaust fan. Like the thermal destructor, the system is simple and reliable if the proper construction materials are used and the catalyst remains active (see sec. IV.F). For safety reasons, it is advisable to equip the discharge end of the destructor with a continuous ozone analyzer that can detect very low ozone concentrations. The analyzer will indicate any gradual loss of catalyst effectiveness, as well as trouble with the catalyst, such as poisoning.

Even if the offgas is recovered and reinjected at the head of the treatment train, the concentration in the exhaust from the reinjector contactors will be higher than the maximum safety standard. Consequently, the residual ozone in the offgas must be destroyed (thermal or thermal catalyst destruct unit) prior to discharge to the environment.

Protection against corrosion. The choice of the materials suitable for each air treatment stage, ozone generation, contact dissolution, and destruction of offgas is explained in sec. IV.G. All recommendations, even the slightest detail, must be observed in order to avoid serious corrosion which could dramatically affect the reliability of the entire treatment facility.

As a general rule, stainless steel (316L) should be used wherever ozone is present. Despite its resistance, even this material can corrode if wet ozone is present under low-pH conditions (pH of 1 to 2).

In small quantities, nitric acid reduces the efficiency of the generator; in larger amounts it can corrode the stainless steel tubes inside the generator. For this reason, it is wise to install a hygrometric alarm at the inlet to the ozone generator, which is designed to switch off the generator when the dew point rise above −60°C at standard pressure (−76°F).

The combined presence of wet ozone and acid compounds may also internally damage diffusers. Extra-thick stainless steel (316L) (thickness: 1.5 mm [0.06 in.]) must therefore be used so that the metal portion of the diffusers remains intact for at least as long as it takes for the porous ceramic to become clogged.

The last point in the treatment process where wet ozone may exist in an acidic environment is in the contactor end of the offgas vents, which may carry both wet residual gas and fragments of very acidic foam found at the water surface in the dissolution chambers. The offgas pipe section situated between the contactor exit and the foam removal unit should be constructed of stainless steel (316L), since low ozone concentration and low pressure are the prevailing conditions at this point. In addition, at the lowest point of the offgas line between the contactor and the destruct unit, a drain should be provided to remove condensation.

V.C REQUIRED PLANT MAINTENANCE

Generally speaking, ozonation equipment should be designed with maintenance in mind. The equipment should require the minimum amount of maintenance and, when maintenance is necessary, it should be possible to perform the maintenance quickly and easily. If ozonation equipment is designed in this way, downtime will be reduced, as will the need for standby capacity and an excessive number of maintenance personnel. All of these reductions represent a cost savings for the utility.

V.C.1 Air Compressors

Maintenance tasks for the compressors are generally more complicated for high-pressure systems that include oil-lubricated compressors than for low- and medium-pressure systems that operate using "oil-free" compressors.

Table V–1 Maintenance Schedule for High-Pressure Oil Compressor

	Maintenance Schedule			
Maintenance Function	Daily	Weekly or Every 50 h	Every 6 Months or 1750 h	Yearly or Every 35,000 h
Check oil level	X			
Clean air filter		X		
Replace air filter				X
Clean air-cooler fins		X		
Change oil			1st time after plant startup	X
Change oil return valves			X	
Change separator oil			X	
Clean oil filter	1st time after plant startup		X	

A high-pressure system, which is commonly used by small-scale facilities, may include an eccentric rotor compressor with sliding vanes. For example purposes, the maintenance for this type of compressor will be discussed here.

The maintenance schedule is given in Table V–1. Maintenance of the electric motor consists of normal servicing and occasional lubrication of the bearings. In most cases, bearings are permanently lubricated when the motor is assembled.

For larger installations using low-pressure systems, a Roots type compressor with an asynchronous variable-speed motor can be used as an example. This type of compressor does not require a great deal of checking on a daily basis since its rotary lobes revolve without coming into contact with each other and are not lubricated. Maintenance involves changing the oil (which lubricates the sharts)—the first oil change should be after 500 h of operation, then every 2000 h or once yearly. Additional maintenance includes checking the safety valve, cleaning the filter once a week and replacing it every year, and general overhauling (by the manufacturer) every three years or after 20,000 h of operation.

Other types of compressors are also available for the air treatment system (see sec. IV.B).

V.C.2 Aftercooler and Refrigerant Dryer

Minimum maintenance should be required for the aftercoolers under normal operating conditions. In the event of extended shutdown, it is wise to drain the water from the unit. If there is a wide gap between the temperature of the water and that of the air outlet, scaling may have occurred. If that is the case, flush the exchanger tubes for a few hours with a 20 percent hydrochloric acid solution.

Refrigerant dryers are a more sophisticated type of machinery. The most important things to watch for are:

- clogging of the drying filter,
- obstruction of the pressure-reduction valve opening with ice,
- fouling of the evaporator
- concentration of glycolated water,
- compressor check-valve seals, and
- electric safeties (fuses and relays).

A maintenance schedule for a specific type of refrigerant dryer is given in Table V–2. Other types also exist (see sec. IV.B).

V.C.3 Moisture Separators and Air Filters

The performance of moisture separators as well as oil separators and air filters in high-pressure systems is of primary importance. Although a cyclonic-type air separator

Table V–2 Refrigerant Dryer Maintenance Schedule

Maintenance Function	Maintenance Schedule					
	Daily	Weekly	Once per Month	Every 3–4 Months	Yearly	Every 3–4 Years
Check air temperature (inlet/outlet)	X					
Check glycolated water level		X				
Check glycolated water makeup			X			
Check glycolated water circulating pump				X		
Check pump packing leakage				X		
Check Freon circuit					X	
Replace pump packing						X
Inspect cooling tubes and clean if necessary						X

requires, in itself, no special maintenance, the condensate controller must be checked daily. The important things to check for include making sure that the controller is working properly and efficiently; regulating the draining sequence (check interval and timing of each draining operation); and making sure that no solenoid valves are faulty. In addition, the condensate drain valve must be periodically cleaned with kerosene and soapy water, and the cartridges of the prefilters and submicron filters must be routinely replaced (for example, every six months if the plant is operating around the clock or annually in the case of an 8- to 12-h working day).

V.C.4 Desiccant Dryer

Two main types of dryers are used depending on system pressure. Pressure swing heatless desiccant dryers are used in high-pressure air preparation systems, and heat-reactivated desiccant dryers are used in low- and medium-pressure systems.

Pressure swing heatless desiccant dryers. Pressure swing heatless desiccant dryers have a relatively simple design and require little daily checking. The switch valves must be serviced periodically (e.g., once a year). It is also necessary to ensure that the sintered bronze filters are not clogged; declogging them requires dipping the filters in a 25 percent acid bath for a few hours. This will also provide the opportunity to check that the desiccant remains in good quality. The desiccant should be changed periodically (e.g., every 3 to 5 years), unless it has been damaged by an excessive oil carry-over due to filtration problems upstream. It is also necessary to check the workings of the pressure-reducing valve downstream of the desiccant dryer, which is designed to reduce the relative pressure from 7 bar (100 psig) to the operating level. The regulator must be lubricated whenever one of the following symptoms appears: excessive discharge to the atmosphere, impossibility of reaching a high secondary pressure, irregular secondary pressure, or too much lag time.

Heat-regenerated desiccant dryers. Heat-regenerated desiccant dryers require more maintenance than the pressure swing unit just discussed. From an O&M perspective it is preferable to use heat-regenerated systems that use externally heated dry air rather than internal heating elements, which may damage the desiccants, resulting in more frequent replacement. However, the external heating desiccant units are generally more expensive. This is an example of the many trade-offs between cost and maintenance of ozone equipment (see chapter VI).

Maintenance operations on external heat-regenerated dryers should be performed when the apparatus is not operating, both for the safety of operating personnel and to avoid putting the automatic operations out of sequence.

Table V–3 Example Heat-Reactivated Desiccant Dryer Maintenance Schedule

	Maintenance Schedule					
Maintenance Function	Daily	Weekly	Monthly	Every 3–4 Months	Yearly	Every 3–4 Years
Check dewpoint	X					
Check temperature, heating startup and finish		X				
Check valve greasing			X			
Check head loss at alumina filters		X				
Check oil level in lubricator			X			
Check regeneration flow rate				X		
Clean alumina filter packing				X		
Test cylinder seal tightness					X	
Replace alumina filter packing					X	
Replace cylinder seals						X

When the dryer is first put into operation, the gaskets between flanges may become more compact and cause leakage, especially those exposed to the effects of heat. Therefore, flange bolts should be tightened after a few weeks of operation. To stop any possible leaky cocks, the grease separators must be adjusted and faucets regularly lubricated.

After several years of service (e.g., 3 years), the condition of the desiccant must be checked. This must be done when the machinery is shut down and after venting the system to the atmosphere. The cover is first removed from the top of the chamber. A 5-L sample is then taken from the outlet layer. A similar sample from the inlet layer is obtained by opening the chamber drainage system. The grain size distribution is then checked. If desiccant graduation is different than the initial one, the desiccant must be replaced.

The occurrence of automatic changeover and regeneration operations must be checked periodically. The chamber changeover sequence, which is performed by the pneumatic cylinders, must take only a few seconds. The electric heater must be energized shortly after the changeover, and the temperature of the air must be between 180 and 200°C (356 and 392°F) when activated alumina is used as desiccant. At the end of the heating sequence, the temperature at the outlet of the regeneration chamber must reach approximately 140°C (284°F). It must then fall to near ambient temperature by the end of the cooling period. A schedule for these operations is given in Table V–3. The dewpoint monitor must be calibrated weekly.

V.C.5 Ozone Generator

If the air at the ozone generator inlet is of high quality, that is, the dewpoint is at −60°C (−76°F) and the air is perfectly filtered, ozone generator maintenance operations are reduced to a minimum. Maintenance does require checking the following items at the time intervals specified:

- daily temperature, flow rate, and pressure control of both air and cooling water. An analyzer will be required for checking the ozone concentration in the product gas at the generator outlet, in order to control the quantity of ozone being produced;
- operating control of flow rate and pressure-relief valves on air and cooling-water circuits every month, with recalibration of the sensors;
- measurement of high-voltage transformer temperature every three months; and

- evaluation of ozone generator performance every year—ozone production must be measured for different applied power levels. This makes it possible to check the efficiency of the ozone generator and, in turn, to determine the cleaning frequency for dielectric tubes if a drop of 5 percent in production efficiency is observed.

Dielectric tubes. Cleaning the dielectric tubes is a complicated operation requiring specially trained personnel. In all cases, the ozone generator must be shut down by operators experienced with high-voltage electricity. The specifics of this procedure are for a particular type of ozone generator, but the "concepts" are similar for other types as well.

Before opening the ozone generator, it must be purged with dry air for at least one half hour. After the ozone generator is de-energized, its main isolating switch must be moved to the open position. This operation releases a key that opens the doors of the high-voltage isolating switch cubicles. At this point in the operation, great care must be taken. A notice must be hung on the cubicle door indicating that work is in process on the ozone generator. The operator, wearing insulated gloves, should fasten one end of the ground wire to the end of the grounding rod. The operator will fix the end of the ground wire, which is attached to the pole, to the grounding terminal. The operator then releases the ground wire, pulling it towards the body and fixing the opposite end to the pole. The ground wire can then be fixed to the high-voltage terminal and the pole withdrawn.

The ozone generator can be opened by unscrewing the bolts located around the edge of the cover, after having secured the latter with taut slings. The cover must be loosened and removed carefully to avoid damaging the seal. If the seal is damaged despite the precautions taken, it must be replaced by a new seal to ensure gas-tight closure of the ozone generator.

The knurled nuts located at the end of the contact brush extensions are then removed. Next, the aluminium wire conductors that interconnect the dielectric tubes are removed. The tubes can then be easily withdrawn by first pulling them towards the body, then removing the contact brushes and the center straps. The tubes must be stored where there is no risk of impact, preferably on a rack or specially fitted trolley.

Tube maintenance includes the following steps:

- clean the inside walls of the stainless steel ozone generator tubes with a rag dipped in methanol attached to a handle. Then wipe the tubes with a dry, lintless cloth to remove the deposits more easily.
- clean the dielectric tubes immediately after they have been removed. Clean the outside surface of the dielectric tubes with a rag dipped in methanol, then wipe with a dry lintless cloth.
- clean the dielectric tubes in this way until the metal-coated portion of the tube is reached. It is very important to avoid brushing of the metal-coated portion of the tube.

While this work is being done, great care must be taken not to strike the dielectric tubes. A tube with the slightest imperfection, particularly in the active portion, must never be reused, because a fractured tube may induce cracking or perforation by dielectric arcing and short circuit the transformer.

The dielectric tubes must be constructed of glass that is specially designed to withstand the high voltages used and to give the best ozone production rate (see sec. IV.C). The glass may be subjected to "stress corrosion" if the tubes are to be used under adverse conditions, especially in a wet or humid environment that would cause cracks or break the tubes. To avoid such accidents, the following precautions must be taken:

- tubes must be stored in a very dry, air conditioned atmosphere.

- air issuing from the ozone generator must always be at a dewpoint below −60°C (−76°F).
- no trace of moisture must be left in the generator after shutdown of the ozone generator (condensation while the cover is open, washing, etc.).
- before starting, the opened ozone generator must be flushed with dry air for 12 h.
- the inside of the tubes must never be washed or dampened. When cleaning the parts without metal coating, use a dry rag or a cloth slightly dampened with methanol if the parts are extremely dirty. If a damp cloth is used, the tubes must be wiped several times after cleaning with dry cloths.
- while the ozone generators are being cleaned, the cooling water must be turned off to avoid all condensation inside the generator.
- in the event of extended shutdown (several days), no cooling water must be allowed to stagnate in the ozone generator.

Depending on the downtime involved, the ozone generator can be either exhausted or left with a small flow of dry air (e.g., 10 percent of the normal flow rate).

Startup. After being opened, the ozone generator must be gradually restarted. After the 12-h dry-air flushing operation and reinitiation of the cooling circuit, the apparatus must be energized at its minimum level and stepped up in three or four phases lasting 1 h each.

When removing and reinstalling the dielectric tubes, it is important to closely observe all manufacturer's instructions. Tubes must be handled with great care to avoid fractures. Any tube with the slightest flaw must be discarded; a flawed tube would soon break after the ozone generator is put back into service, making it necessary to stop and open it for repair.

If these operations are not performed by the ozone generator supplier, personnel must be specially trained to maintain the ozone facility. These employees must be qualified in electrical maintenance.

V.C.6 Contactor and Dissolution System

Generally speaking, regardless of the contact process chosen, it must be possible to check bubble formation by sight, measure the dissolution efficiency periodically, have access to the vessel interior, and use the proper materials required for maintenance inside the vessel. All maintenance work must be done as promptly as possible.

A periodic check on dissolution efficiency is essential every six months in order to monitor dissolution and detect possible anomalies, such as plugged or broken diffusers or plugged or broken ozone dissolution system piping. For evaluation purposes, a number of continuously operating sensors are required. The amount of ozone entering the contactor is measured by a flow meter that measures the quantity of product gas entering the chamber and a sensor that monitors the concentration of ozone in that gas. The range of this sensor should be between 0 and 50 g/m^3 NTP (0 and 3.8 percent by weight) if the carrier is air and 0 to 100 g/m^3 NTP (0 and 8 percent by weight) in the case of oxygen. To be able to compare at any time the dissolution efficiency, it is necessary to perform the tests in the same conditions (flows, water temperature, and dissolved ozone concentration). Measurement of the quantity of residual ozone in the water issuing from the contact vessel involves measuring the ozonated water flow rate and the concentration of residual ozone in the water at the contactor outlet. Measuring the amount of residual ozone in the contactor offgas outlet requires a gaseous ozone analyzer that is operable in a range of 0.1 to 13.0 g/m^3 NTP (0.008 to 1 percent by weight) plus a flow meter that can measure the quantity of air at the outlet, which is also the amount taken in by the destructor. This quantity may be greater than the volume of product gas injected, due to the presence of vent holes in the contact chamber.

All analyses must be performed continuously by automatic sensors. Some analyzers, such as the one used to measure the ozone concentration in the offgas and residual ozone in the water at the outlet of the contact chamber, must be fitted with an alarm system. The ozone residual analyzers must be recalibrated each week, the flow meters every month. When this is done, a balance of all the dissolution parameters will be established and the overall transfer efficiency can be calculated. This calculation can then be compared to the reference values determined at the time of startup. These values reflect the forecast range of influent flow rates to be treated, the different carrier gas flow rates, and the various concentrations of ozone. If, under identical treatment conditions, a drop in output of over 5 percent is noticed, the dissolution system must be visually inspected.

These controls are particularly necessary when contacting with porous diffusers, because they gradually become clogged with time. Experience shows that clogging is extremely variable for porous diffusers that function basically at the same flow rate and pressure. A diffusion balance must be determined, either as soon as a drop in transfer efficiency of over 5 percent is observed or systematically once a year.

The visual inspection consists of examining the porous diffusers, through which unozonated air is diffused under 0.3 m (12 in.) of water above the diffusers at various flow rates: minimum, average, and maximum. This quickly identifies the diffusers that diffuse insufficiently or, on the contrary, too profusely, generating very large bubbles of air. All the faulty diffusers should be replaced and the evenness of diffusion should be checked throughout the assembly using the same process. On average, the operating life of the diffusers is generally five years. Their replacement can be one of the most costly items in the maintenance budget.

In order to avoid undue downtime due to maintenance activities, the water in the contactor must be quickly drained. For safety reasons, it is not advisable to enter a chamber containing an ozone concentration above the standard of 0.2 mg/m^3 NTP (0.1 ppm by volume) even when wearing an individual self-contained breathing apparatus. It is far better to equip the chamber with sufficient ventilation to quickly obtain and sustain a concentration of ozone below the safety level. In this way, operators can enter the chamber and work without a mask, achieving enhanced safety and comfort and enabling the maintenance work to be done faster.

The chamber must be designed for an easy diffusion sight check. There must also be walkways above the water level to allow for inspection of the entire surface. Enough windows and adequate electric lighting facilities should be provided to ensure good visibility. All access ways, doors, body-rails, and steps must be able to withstand wet ozone and must be constructed of stainless steel (316L).

These criteria also apply to other dissolution systems, such as turbines. Because turbines can offset relatively high power costs and a lack of flexibility when adapting to changes in water and ozonized air flow rates, they are very economical with regards to maintenance, providing they are suitably designed. First, turbines must be made of materials capable of resisting wet ozone and, secondly, in the case of immersed motors, they must comply with the prescribed watertightness conditions. Maintenance for turbines is similar to that for an electric motor.

Maintenance is even easier in the case of diffusion by injection. The construction materials for injectors must be able to withstand wet ozone.

V.C.7 Destructor

Destructor maintenance primarily involves monitoring of the output. A continuous ozone analyzer that functions within a range corresponding to the standard of 0.1 ppm by volume (0.2 mg/m^3 NTP) must be fitted to the air line leaving the destructor. Any overshooting of this value will trip a general alarm.

Achieving this standard is of fundamental importance; failure to do so will cause contamination of the ambient air surrounding the ozone plant. If the ozone generators and contact vessels are completely airtight, the destructor outlet will be the only possible source of offgas in the atmosphere and, as such, must be closely watched.

The proper operation of the destructors will depend on the following two precautions:

- there must be no entrainment of the acidic foams that form on the surface of the contactor. The foams must be arrested by a water-spraying device or equivalent.
- standards on construction materials for contact with wet ozone must be strictly observed.

Providing these conditions are met, thermal destructor maintenance is simple. For thermal destructors, maintenance involves checking the exhaust fan and cleaning the heating oven and the preceding air-recovery heat exchanger once every five years.

Catalytic ozone destruction involves the problem of the effective life span of the catalyst. Installing an ozone control analyzer at the outlet of the destructor will reveal any loss of catalyst efficiency. Projected life spans of the catalyst are 2 to 4 years. However, carryover of froth, flood, or foam from the contactor can blind or poison the catalyst very quickly, thereby requiring its replacement, which will be comparatively expensive.

V.C.8 Control Boards and Electronic Equipment

Current maintenance of control boards and electronic equipment poses no particular problem, providing the equipment is located in an ozone-free atmosphere. This constitutes yet another argument in favor of designing a facility so that no leakage can possibly occur or fitting the housing of the ozonation plant with a sufficient number of ambient air ozone sensors that will operate within the range of 0.1 ppm by volume (0.2 mg/m^3 NTP) and trip an instant alarm if the concentration rises above that threshold.

The only precaution to be taken concerns the cables that are found in an ozone-containing environment, particularly cables inside the contact vessels for lighting or feeding an immersed turbine. The insulated cable sheathing must be made of a material that withstands ozone.

If these conditions are met and the conventional procedures are followed with respect to the ventilation of electrical and electronic equipment, there is no reason to install this equipment in an insulated building. It should be noted that ozone plant control and monitoring boxes, in particular, can be installed close to the generators, which is more practical for periodic control.

V.D EMPLOYEE TRAINING

An ozonation facility is, in itself, a complete production unit that is contained in a drinking water treatment plant. Because ozonation involves specialized techniques, personnel training is required at several levels.

In general, operation of an ozone facility requires skilled personnel who have been trained to operate the equipment and the actual process. Currently, this training is generally not a part of the curriculum of most plant operator schools in the countries where ozone is a relatively new process, such as the United States. Therefore, the facility owner should provide, as part of the facilities construction budget, operations-oriented outside consultation or in-house activities that deal with the operation of the process. Such activities should result in (1) an operation plan (operation budget, staffing plans, organizational charts, and activities schedule) that is prepared at the same time the plans and specifications for the ozone facility are prepared; (2) an operator training program that is set up to assist the plant staff in their preparation of the operating policies and procedures (the program should be conducted during the construction phase of the project, and completed when construction is finished); and (3) a system startup training program that helps the plant staff prioritize operational responses, especially during the first 6 to 12 months of process operation.

V.D.1 Operator Participation in Pilot Tests

Pilot testing prior to installation of an ozonation system has several objectives (see sec. IV.A):

- to determine the role(s) and position(s) of the ozonation process in the water treatment plant,
- to define the conditions under which the treatment should be applied, including dosages and contact time characteristics; and
- to establish the technical specifications of the ozone production and dissolution equipment.

Familiarizing operators with this pilot work is advisable. First, they can be made to realize the contribution of ozone to the general treatment process. In exchange, the operators bring their practical knowledge regarding treatment problems and the improvements that are needed, which is extremely valuable for pinpointing the correct test conditions. Factors such as seasonal variations in raw water quality and the likely pollution hazards can be more easily accomodated in the pilot test plan when operators are involved.

Moreover, operators are best able to supply information on ozone application characteristics as they relate to the flow rate of the influent to be treated and its variations in reponse to demand and the management of treatment plant reservoirs. This knowledge is fundamental for the optimum design of ozone production and contact equipment, which is greatly influenced by operating rates and the expected dosages. This has already been discussed in this chapter (see sec. V.B.1).

Finally, it is essential that operators take part in preparation of the bid documents for the ozone system, because the choice of equipment depends on, among other things, operating and maintenance criteria. The degree of automation will be determined according to the available personnel and their levels of qualification. The technical standards and performance requirements of the specifications also have a significant effect on the choice of equipment. If these conditions are not considered when the specifications are being drafted, the equipment suppliers will propose solutions with the lowest investment cost that may have a negative impact on the treatment plant as a whole (for example, increased maintenance costs and reduced reliability for the machinery concerned).

V.D.2 Training Personnel During Facility Startup

In order to obtain the maximum efficiency, initial operator training should be provided by the supplier of the ozone system. This training should, therefore, be included in the list of services to be supplied and should be written into the system specifications.

An effective training technique is to have the operators take part in the startup of the plant. This should not, however, release the supplier from the responsibilities of delivering the facility in perfect working order and within the contractual time limits specified in the bid documents.

Subject to this proviso, operator participation at the time the facility is commissioned enables the operator to acquire detailed and practical knowledge about operating the facility. Operators can also learn the system check-out procedures, performance testing, and various operating methods for each piece of equipment, including manual, automatic, and remote-controlled units. In addition, the ability to solve the normal problems that occur during initial start-up will give the operator useful maintenance information.

Formal classroom training periods should be specified to ensure that all operational and maintenance staff are provided with a degree of knowledge of the theory of the system and the system subunits.

V.D.3 Plant O&M Manuals

The supplier is obligated to supply the plant with O&M literature when the plant is commissioned.

These documents must be written for use by operating personnel and adapted to the particular features of the equipment installed. Simple descriptive literature and, at the other extreme, in-house brochures that are only understandable by specialists, are both equally unsuitable. The manuals must include:

- a precise description of the equipment;
- practical operating procedures in the different operating circumstances;
- a complete inventory of control and maintenance methods, especially when the maintenance involves obtaining specific products such as oil, grease, and spare parts (the manual should include the suppliers' requirements); and
- a list of faults and failures for each piece of equipment, giving short- and long-term troubleshooting methods.

Final acceptance of the plant should not be given until these documents have been supplied, reviewed, and approved.

V.D.4 Troubleshooting Schedules

In order to plan all the check-up and maintenance operations required for plant operation, a special troubleshooting schedule is necessary. The schedule should discuss:

- systematic check-up operations for each piece of equipment;
- recalibration of sensors;
- periodic reports regarding such topics as energy consumption by ozone generators and other units vital to production, ozone dissolution efficiency, and ozone destruction efficiency; and
- all systematic plant maintenance jobs.

The schedule should also indicate:

- the personnel required for each job;
- the time required for each job; and
- the equipment to be shut down during maintenance and repair.

The entire schedule and the results of each job should be recorded in a computer database, at least at medium and large plants. This provides for easy analysis when it is time to draw up the operating costs, O&M statistics, and figures to be included in operating cost determinations.

V.E SAFETY

V.E.1 Gaseous Ozone Exposure—The Labor Code

Ozone becomes a toxic gas above a certain concentration. Standards on maximum exposure have been defined by health and occupational safety agencies. In the United States, the organizations responsible for these standards include

- Occupational Safety and Health Administration (OSHA);
- American National Standards Institute/American Society for Testing and Materials (ANSI/ASTM);
- American Conference of Government Industrial Hygienists (ACGIH).

In accordance with United States regulations, an individual must not be exposed to a concentration of ozone higher than:

- 0.1 ppm by volume (0.2 mg/m^3 NTP), determined as a time-weighted average (TWA) over a period corresponding to a full working day (maximum 8 h) (it should be noted that this value is also the limit in most countries); and
- 0.2 ppm by volume (0.4 mg/m^3 NTP), as ceiling limit for an exposure time of 10 min.

In addition, personnel working in an ozonation plant must undergo regular medical check-ups.

A notice must be posted at the entrance of the ozonation facility, which states (in effect) the following:

OZONE WARNING!
IRRITANT GAS
ADEQUATE VENTILATION REQUIRED
AVOID PROLONGED OR REPEATED
BREATHING OF OZONE.

The notice must also include instructions on what to do in the case of accidental ozone leakage.

Protective canister-type respirators must be kept available. Rubber masks can be used for concentrations up to 5 ppm (10 mg/m^3 NTP). Beyond that level, respirators with compressed-air cylinders must be used. Personnel must practice handling and wearing masks and respirators.

V.E.2 Protection of Personnel

Two kinds of measures can be taken to protect personnel who may be exposed to ozone: preventive and remedial.

Preventive measures. These measures involve the personnel directly and the ozonation equipment indirectly. As with any other chemical, personnel must first be informed of the dangers of ozone. Training should cover:

- the nature and dangers of ozone;
- precautions to be taken on premises where an exposure hazard exists; and
- first aid in case ozone gas is accidentally inhaled.

To be effective and practical, this training must be established by safety personnel, industrial hygienists, members of the plant operations team, and their representatives. Training courses should be conducted at the work site to make demonstrations on actual machinery possible. To maintain training effectiveness, refresher courses should be held regularly.

Outside contractors who may be called in to work on ozonation facilities must also be informed of the dangers of ozone. A notice that includes safety instructions must be enclosed with contractor orders. Before any job can begin, the safety manager should issue a work permit that includes all the precautions to be taken when working on ozonation equipment. The person in charge should ensure that these safety measures are observed.

Onsite notices must be posted, reminding workers of the dangers of ozone, all appropriate safety measures, and the location of respirators and how to use them.

Preventive measures also include warnings to personnel of the presence of ozone. In order to do this, the premises must be fitted with continuous analyzers that detect ozone in the ambient atmosphere. At the present time (1990), one of the most accurate and reliable instruments (but also the most expensive) is an ultraviolet analyzer that is equipped with a large measuring cell adapted to a range of 0.01 to 100 ppm by vol (0.02 to 200 mg/m^3 NTP). The analyzers must be installed in ozonation rooms at regular intervals along the ozonized gas distribution pipes, in contactor access galleries, and at ozone destructor discharge points. The analyzer must trigger both a displayed and an acoustic warning signal as soon as the ozone content in the ambient air exceeds 0.1 ppm (0.2 mg/m^3 NTP). The alarm must be generated at the measuring site and transmitted to the entrance of the facility and to the central control unit.

To guarantee that there is no ozone in the ambient atmosphere, there are two things to do:

- strictly observe instructions concerning the types of materials to be used in contact with ozone in order to minimize the risk of leakage; and
- ensure proper operation of offgas destructors in order to avoid continuous discharge to the atmosphere.

It can be said that all efforts to select high-quality materials when designing ozone facilities have a twofold effect:

- they are instrumental in achieving better O&M with regard to ozonation structures; and
- they ensure improved occupational safety.

Preventive measures indirectly involve the ozonation machinery, since it is designed to give maximum reliability with a minimum ambient atmospheric ozone hazard. Occupational safety, as it relates to ozone hazards, can be considered part of equipment safety, and this subject has been dealt with in detail in sec. V.B.

Remedial action. Remedial actions are taken if an accident occurs. When an accident occurs, an alarm is triggered by the continuous analyzers, which detect ozone in the surrounding air.

The production of ozone must be stopped immediately by cutting off the supply of electricity. At the same time, all available ventilation must be put into operation in order to adjust the concentration of ozone in the ambient atmosphere to below the standard level. The layout of ventilators must allow for the fact that ozone is denser than air. Consequently, ozone tends to accumulate near the floor. Natural ventilation is therefore of limited effectiveness, and a forced draft must be created by blowers that are installed as mobile or permanent fixtures.

Respirators with cartridges and air cylinders should be used when personnel are locating the leak, although the ozone system may be shut down and the room sufficiently ventilated before the operators enter the room. It may be necessary to locate the leak by purging the ozone system with nonozonated gas and "soap testing" the piping, valving, and pipe joints to find the leak.

Response procedures for accidental ozone inhalation. Accidental inhalation of ozone will cause breathing trouble, the degree of which will depend on the ozone content and exposure time. The patient must rest immediately after the accident in order to reduce the suffocation effect, which is caused by the irritation of the respiratory tract, while waiting to be taken to the hospital as quickly as possible.

The discussion concerning ozone safety presented herein gives some initial safety tips. The reader should consult other documents that contain a more extensive discussion, such as O&M manuals supplied by the ozone system manufacturers, when developing specific design or operation criteria for a particular ozone facility.

VI

Economics of Ozone Systems: New Installations and Retrofits

William D. Bellamy
Bruno Langlais
Benjamin Lykins, Jr.
Kerwin L. Rakness
C. Michael Robson
Pierre Schulhof

Before examining the cost of ozonation, one should consider two points that have already been discussed in chapters IV and V because they are of importance relative to their influence on costs: (1) optimization of ozone use at a facility and (2) facility maintenance. For each gram or pound of ozone produced, the best possible water quality should be obtained for minimum cost. This implies that contact chamber design should enable the minimum amount of ozone to be used while meeting its intended use. This is the only method that facilitates a reliable comparison between the costs of ozonation and the different production and application techniques. Cost can be discussed in terms of French francs or dollars per cubic meter or per thousand gallons of treated water, or in grams per hour or pounds per day of ozone produced. In addition, maintenance of the facilities must be adequate to assure that ozone utilization efficiency is at its maximum and that the ozone generating process is maintained to the optimum ozone yield (i.e., minimizing energy consumption).

The ozonation facilities should also be designed for a capacity not too far from the actual ozone production planned. If this is not done, the amortized cost of the capital relative to the amount of ozone used will be excessive.

Even if the above objectives are obtained, the fact still remains that costs are difficult to assess. Nevertheless, an "order of magnitude" cost estimate of an ozone system can still be useful. The intent of the first section of this chapter is to assist the reader in obtaining such an estimation for initial planning and budgeting purposes. Subsequent sections address the problems that must be faced when trying to develop more precise estimations of capital and operation and maintenance (O&M) costs of an ozonation facility and help the reader to ask the right questions during the design period of a specific plant. Finally, case studies based on the different types of existing facilities and on projects where the ozone plant has been fully optimized will enable the reader to better understand the cost of ozonation systems in potable water treatment.

VI.A "ORDER-OF-MAGNITUDE" ESTIMATES OF OZONE SYSTEM COSTS

This section cannot be used as a means of differentiation between choices available for ozone systems or equipment, such as between air-fed or oxygen-fed processes or between medium- or low-frequency ozone generation equipment. Information for such a detailed assessment may be found in chapters IV and V and in the other sections of this chapter during preparation of a preliminary layout of the proposed facilities, and by contacting equipment vendors with specific questions. Since process differentiation based on cost was not intended, only prices for the air-fed ozone systems are included in this section because they provide the greatest source of reference information. Actual costs at an individual plant may vary according to choices made during design (such as for more capital-intensive equipment that provides lower energy or maintenance cost) or for other reasons (see sec. B and sec. C of this chapter).

Both capital and O&M cost estimates may be attained by filling out the forms in Figures VI–1 and VI–2, respectively. The procedures listed below can be used when determining the cost estimate, along with the specific price information shown in Figures VI–3 and VI–4. (Representative calculations for an example plant are shown in Figures VI–7 and VI–8 and are discussed in more detail in each step of the proposed procedures.)

VI.A.1 Procedure for Obtaining Estimated Capital Cost

Overview of ozone pricing information.

Equipment prices. Estimated costs for ozone generation equipment are shown in Figure VI–3. Two sources of information have been used to plot these curves. The black squares are bid numbers at U.S. plants where the information was available, as shown in Table VI–1. No attempt was made to adjust prices to a cost index because of the current nature of the industry. (Historically, only a few ozone systems were being installed annually in the United States, and bid prices may have been nonrepresentative and unrealistically low. Currently, several ozone systems are being designed, bid, and constructed every year; while competition is still keen, market price stability is undetermined.) These bid numbers include the prices of the air treatment equipment (low-pressure system—see sec. B.3, Chapter IV), the ozone generators, and the electrical gear. Also included are prices for the diffusers and the destruction equipment.

The information represented by the black circles was obtained from a French manufacturer. The prices quoted pertain to equipment producing ozone from air at low or medium frequency for an ozone capacity between 100 and 1000 lb/day (about 1.9 to 19 kg/h). Between 100 and 190 lb/day (about 1.9 to 3.6 kg/h) the air preparation is under

Plant
Name: ____________________

1. Design Water Flow Rate	________	mgd	
Ozone Dosage	________	mg/L	
Ozone Production	________	lb/day	
Standby Capacity	________	lb/day	
Installed Generation Capacity[1]	________	lb/day	
2. Ozone Equipment Unit Cost[2]	________	\$/lb/day	
3. Projected Equipment Cost			\$________
4. Anticipated Hydraulic Detention Time:	________	min	
5. Projected Contactor Cost[3]			\$________
6. Estimated Unit Building	________	ft^2/lb/day	
7. Calculated Building Size	________	ft^2	
8. Estimated Unit Building Cost	________	\$/ft^2	
9. Projected Housing Cost[4]			\$________
10. TOTAL ESTIMATED OZONE SYSTEM COSTS			\$________

NOTE: The prices above include installation, but do not include other costs such as "site work" (highly variable depending on site conditions and whether or not the ozone facilities are constructed as a stand-alone project or are part of other plant works); design and construction management services; or owner's administration, legal and management expenses.

[1]Includes operating plus standby units.

[2]Includes ozone generators, air treatment equipment, offgas treatment units, electrical gear, and miscellaneous.

[3]Includes two bubble diffuser ozone contactors in parallel. Each contactor has six compartments in series (see Figure VI-5). BE SURE TO USE THE FLOW RATE TO EACH CONTACTOR WHEN USING FIGURE VI-4.

[4]Includes space for the ozone generators, air treatment equipment, offgas treatment units, electrical gear, and miscellaneous.

Figure VI–1 Procedure for Estimation of Capital Cost of Ozonation System

high pressure (i.e., 7 relative bar or about 101 psig—very common in France); from 200 lb/day to 1000 lb/day (3.8 to 19 kg/h), a low-pressure air preparation system is used (0.8 to 1.5 relative bar or 11.6 to 21.7 psig).

Costs represented by the black circles include air treatment, ozone generation equipment, ozone destruction equipment (thermal unit with heat recovery—see sec. F, Chapter IV), piping, valves, instrumentation, disc-type diffusers, control system, and installation. The prices of the diffusion system and of the control system have been estimated at 5 and 20 percent of the total cost, respectively. The ozone production rates have been estimated based on an ozone concentration of 1.5 percent by weight (19.5 g O_3/m^3 NTP) at a cooling water temperature of 15°C (59°F).

These two curves form the band of ozone equipment prices seen in Figure VI–3. Only a case-by-case study can lead to a more precise price.

Plant Name: ______________________________

1. Operating Ozone Production Rate

   Annual Average Flow ________ mgd

   Annual Average Dose ________ mg/L

   Annual Average Production ________ lb/day

2. Estimated System Specific Energy[1] ____________kWh/lb

3. Daily Average Energy Consumption[2] ____________kWh/day

4. Estimated Unit Energy Cost[3] ____________$/kWh

5. Daily Average Energy Cost ____________$/day

6. Estimated Annual Energy Cost .... ________ $/yr

7. Est. Cost Factor for Energy Alone[4] ____________%

8. Estimated "Other" O&M Cost .... ________ $/yr

9. TOTAL ESTIMATED O&M COST .... ________ $/yr

[1]Typically ranges from 10 to 12 kWh/lb for low-pressure air treatment ozone generation systems.

[2]Formula: Cost = O_3 production (lb/day) × specific energy consumption (kWh/lb).

[3]For purposes of this estimate, all electrical company charges, such as power demand, fuel cost adjustment, and energy consumption charges, should be incorporated into a single number relative to the kWh of energy consumed.

[4]Energy costs typically range from 65% to 80% of the total system O&M cost and generally are 75%. The "other" costs include activities or items such as operator time for daily equipment readings, spare parts replacement cost, preventive and emergency maintenance activities, etc.

Figure VI–2 Procedure for Estimation of Operating Cost of an Air-Feed Ozone System

Contactor prices. The estimated costs of the contactor are shown in Figure VI–4. These costs are based on the basin layout presented in Figure VI–5 (See sec. D, chapter IV for information on contactor design.) As shown, two contactors are constructed in parallel, with each contactor treating one-half of the plant design flow. Each contactor has six compartments, three with ozone addition and three without. The estimated costs in Figure VI–4 are for both contactors and include concrete structure, stainless steel piping and pipe installation, sample ports and lines and their installation, drains, baffles, and miscellaneous expenses. The ozone diffusion and destruction systems were included in the ozone generation system cost. Theoretical hydraulic detention times of 5, 10, and 15 min have been considered.

Housing costs. The estimated area requirement for ozone facilities is shown in Figure VI–6. Data for selected plants are plotted on the curve and are shown in

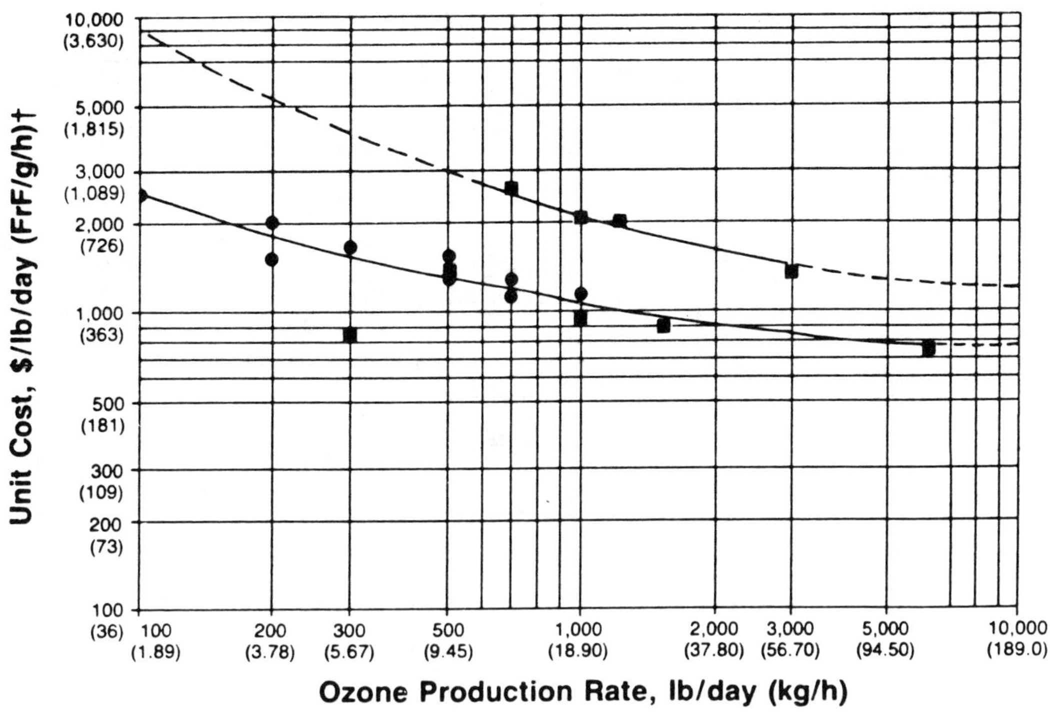

Figure VI–3 Estimated Cost for Ozone Generation System Equipment, Excluding Contactor and Housing

Table VI–1 Summary of Information from Selected Ozone Installations (U.S. Plants)

	Building Information			Bid Cost Information			
Name of Plant	Ozone Capacity (lb/day)	Floor Area (ft²)	Unit Floor Area (ft²/lb/day)	Year Built	Bid Ozone Capacity (lb/day)	Bid Price ($)	Unit Bid Price ($/lb/day)
Lincoln, NB	1,050	9,770	9.30		700	1,940,000	2,771
Lawton ,OK	470	4,221	8.98				
Fort Worth, TX	556	4,525	8.14		1,200	2,390,000	1,952
Martinez, CA	750	5,610	7.48		1,000	2,030,000	2,030
East Bay MUD, Oakland, CA				1989	3,000	4,000,000	1,333
—Sobrante	2,250	12,543	5.57				
—Upper San Leandro	2,250	12,343	5.57				
Stillwater, OK	500	2,304	4.61				
Belle Glade, FL	750	3,080	4.11		500	670,000	1,340
Hackensack, NJ	2,700	11,000	4.07				
Monroe, MI	450	1,684	3.74				
Fairfield, CA—North Bay	2,000	6,263	3.13				
Myrtle Beach, SC	3,750	11,610	3.10				
North Andover, MA	300	882	2.94				
Tucson, AZ	9,000	19,200	2.13	1989	6,000	4,510,000	752
Shreveport, LA—Ammiss	3,750	7,488	2.00				
—McNeil	825	2,610	3.16				
Central Lake Co., IL					1,500	1,310,000	873
Palm Beach Co. No. 8				1989	1,000	950,000	950
Manasquan, NJ					300	250,000	833

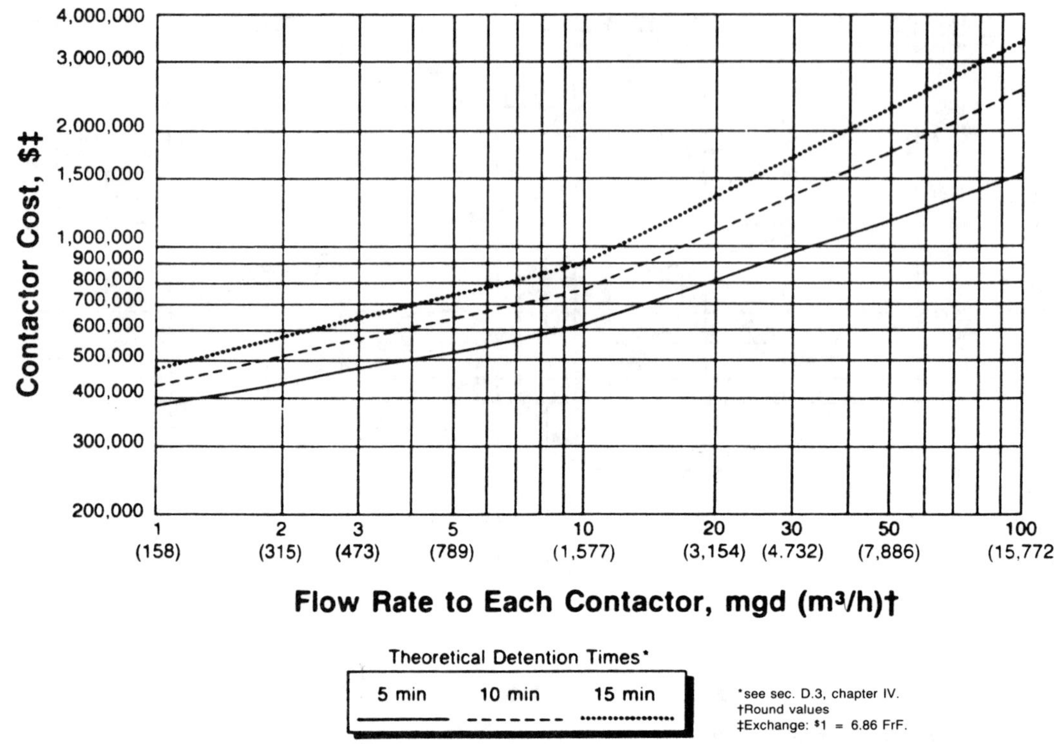

Figure VI–4 Estimated Cost for Ozone Contactor Assuming Two Parallel Contactors, Each Treating One-Half of the Design Flow

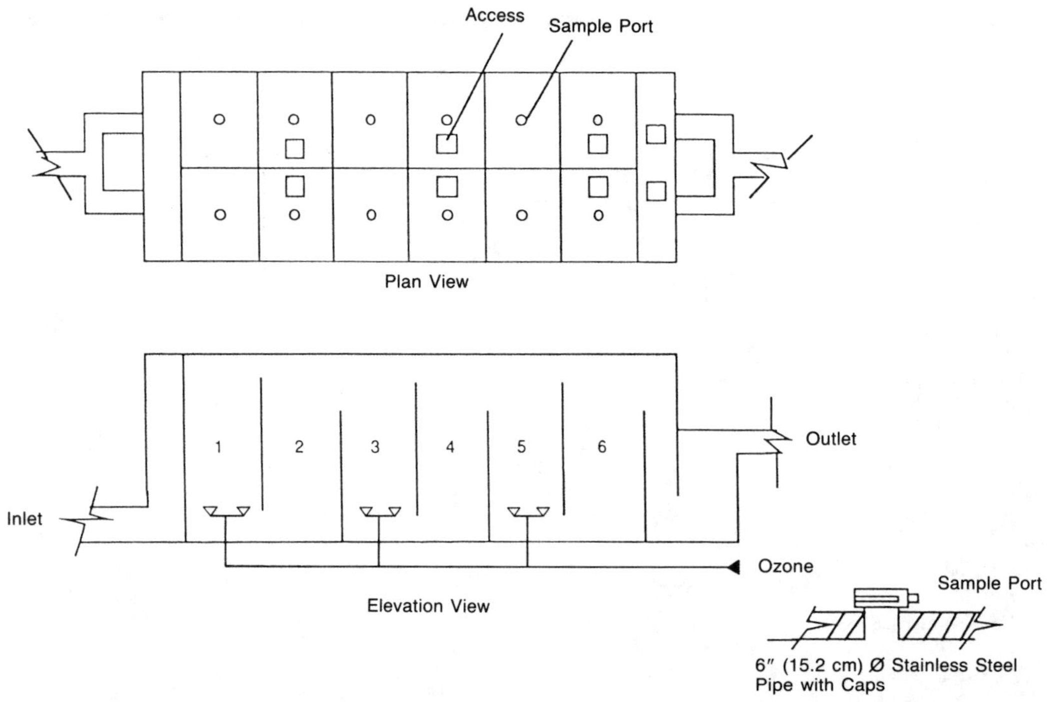

Figure VI–5 Schematic of the Example Contactor Used for Cost Estimation

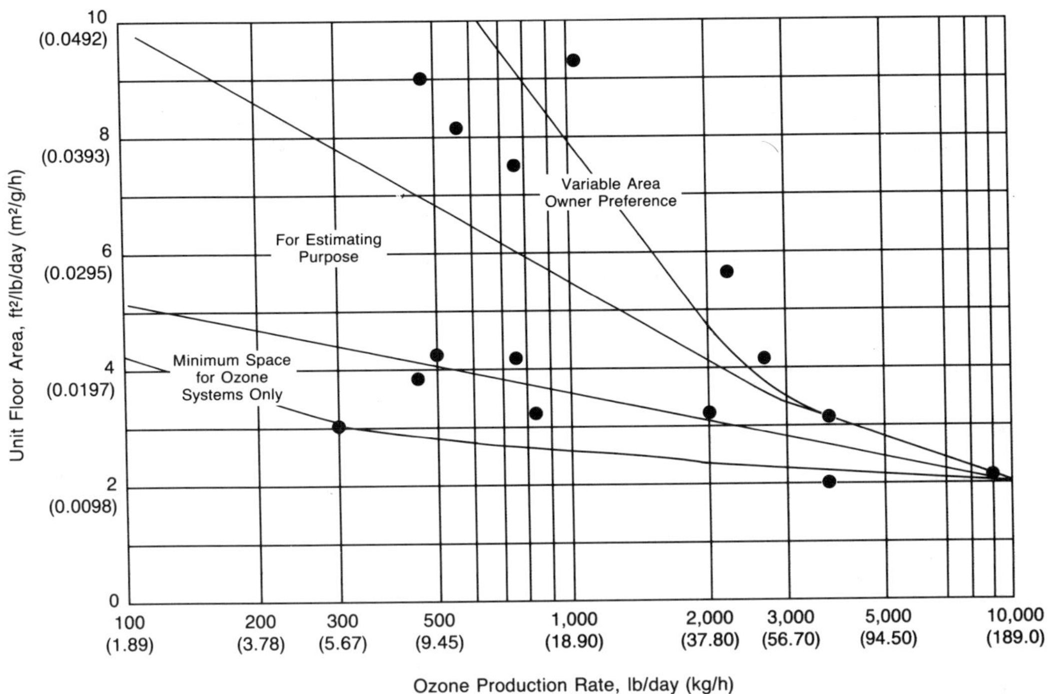

Figure VI–6 Estimated Building Area Requirement for Ozone Equipment

Table VI–1. As indicated, there is a broad range in unit area for a given ozone production rate, probably due more to design engineer and owner preference than to the needs of the equipment. A greater floor space would provide extra room between units. The smaller unit area would have equipment closer together, but may be manageable with respect to access for maintenance.

An "envelope" for building space is shown in Figure VI–6. An exact area provision may be obtained in the design phase of the project, when the choices of equipment are finalized.

Proposed procedure for obtaining estimate of capital cost. (see Figure VI–7).

Step 1: Estimate the installed capacity of the ozone system. The installed capacity includes production plus standby equipment. Production needs may be obtained by first selecting the design water flow rate, say 30 mgd (about 4730 m^3/h) and applied ozone dosage, say 3 mg/L. The resulting production rate is 750 lb/day (about 14.2 kg O_3/h). The degree of standby capacity required is highly site-specific; special considerations are discussed in more detail in sec. B.1 of chapter V. For this example, it is assumed that the ozone generators are operated at 75 percent of their rated capacity to meet the design production rate. This means that installed capacity is 1000 lb/day (about 18.9 kg/h) and "standby" capacity is 250 lb/day (about 4.7 kg/h).

Step 2: Using the installed capacity number and Figure VI–3, pick the unit cost of the ozone generation system. For this example, the installed capacity is 1000 lb/day (18.9 kg/h). Using the higher part of the price zone, the projected unit cost is $2000/lb/day (726 FrF/g/h).

Step 3: Using the estimated unit cost from step 2 and the installed capacity number from step 1, calculate the cost of the ozone generation equipment. For this example, the proposed installed capacity is 1000 lb/day (18.9 kg/h) and the anticipated unit cost is $2000/lb/day (about 726 FrF/g/h). The resulting ozone generation equipment cost is $2,000,000 (13,720,000 FrF).

Plant Name: Example Plant

Item	Value	Units	Cost
1. Design Water Flow Rate	30	mgd (4,730 m³/h)	
Ozone Dosage	3	mg/L	
Ozone Production	750	lb/day (14.2 kg/h)	
Standby Capacity	250	lb/day (4.7 kg/h)	
Installed Generation Capacity[1]	1,000	lb/day (18.9 kg/h)	
2. Ozone Equipment Unit Cost[2]	2,000	$/lb/day (726 FrF/g/h)	
3. Projected Equipment Cost			$ 2,000,000 (13,720,000 FrF)
4. Anticipated Hydraulic Detention Time:	10	min	
5. Projected Contactor Cost[3]			$ 1,000,000 (6,860,000 FrF)
6. Estimated Unit Building	4.5	ft²/lb/day (0.022 m²/g/h)	
7. Calculated Building Size	4,500	ft² (418 m²)	
8. Estimated Unit Building Cost	100	$/ft² (7400 FrF/m²)	
9. Projected Housing Cost[4]			$ 450,000 (3,090,000 FrF)
10. TOTAL ESTIMATED OZONE SYSTEM COSTS			$ 3,450,000 (23,667,000 FrF)

NOTE: The prices above include installation, but do not include other costs such as "site work" (highly variable depending on site conditions and whether or not the ozone facilities are constructed as a stand-alone project or are part of other plant works); design and construction management services; or owner's administration, legal and management expenses.

[1]Includes operating plus standby units.

[2]Includes ozone generators, air treatment equipment, offgas treatment units, electrical gear, and miscellaneous.

[3]Includes two bubble diffuser ozone contactors in parallel. Each contactor has six compartments in series (see Figure VI-5). BE SURE TO USE THE FLOW RATE TO EACH CONTACTOR WHEN USING FIGURE VI-4.

[4]Includes space for the ozone generators, air treatment equipment, offgas treatment units, electrical gear, and miscellaneous.

Figure VI–7 Estimation of Capital Cost of Ozonation System for Example Plant

Step 4: Record the anticipated theoretical hydraulic detention time in the ozone contactor. The theoretical hydraulic detention time is a function of the purpose of the ozonation and is discussed in more detail (chapter III and sec. D.3, chapter IV). For this example, it is assumed that the theoretical hydraulic detention time is 10 min.

Step 5: Using the design water flow rate from step 1 and anticipated theoretical hydraulic detention time from step 4, interpolate, if necessary, between the price curves shown in Figure VI–4 and identify the cost of the ozone contacting system. For this example, the design water flow rate is 30 mgd (4730 m^3/h) and the anticipated theoretical hydraulic detention time is 10 min. The cost curve is based on two basins that call for flow to each basin at 15 mgd (2365 m^3/h). From Figure VI–4, the estimated contactor cost is $1,000,000 (about 6,860,000 FrF).

Step 6: Using the installed ozone capacity number from step 1 and Figure VI–6, select the unit area for housing the ozone equipment. For this example, the installed capacity is 1000 lb/day (18.9 kg/h). Using Figure VI–6, the proposed building area is 4.5 ft^2/lb/day (0.022 m^2/g/h). Some installations have used slightly smaller buildings, but for budgeting purposes a higher number is more appropriate.

Step 7: Using the unit building size from step 6 and the installed ozone capacity from step 1, calculate the building size. For this example, the installed capacity is 1000 lb/day (18.9 kg/h) and proposed building area is 4.5 ft^2/lb/day (0.022 m^2/g/h). The resulting building area is 4500 ft^2 (about 418 m^2).

Step 8: Based on local building costs for industrial-type facilities, select the unit price for construction. For this example, it is assumed that the price for constructing a building is $100/$ft^2$ (about 7400 FrF/m^2).

Step 9: Using the building size from step 7 and unit building price from step 8, calculate the ozone housing cost. For this example, the building area is 4500 ft^2 (about 418 m^2) and unit building price is $100/$ft^2$ (about 7400 FrF/m^2). The resulting building cost is $450,000 (about 3,090,000 FrF).

Step 10: Using the generation system cost from step 3, the contactor cost from step 5, and the housing cost from step 9, calculate the total estimated ozone system cost. For this example, the generation system cost was $2,000,000 (13,720,000 FrF), the contactor cost was $1,000,000 (about 6,860,000 FrF), and the housing cost was $450,000 (3,090,000 FrF). The resulting total estimated ozone system cost is $3,450,000 (about 23,667,000 FrF).

Figure VI–7 summarizes the results obtained for this example.

VI.A.2 Proposed Procedure for Obtaining Estimated O&M Costs (See Figure VI–8)

Step 1: Estimate the average annual ozone production rate for the installed ozone system. This may be obtained by first selecting the average annual water flow rate, say 20 mgd (3150 m^3/h), and applied ozone dosage, say 2.5 mg/L. It is noted that these numbers are less than the design rate and account for fluctuating needs during different seasons. The resulting annual average production rate is 417 lb/day (about 7.9 kg O_3/h).

Step 2: Pick the ozone generation system specific energy number. The system specific energy for a low-pressure air treatment system is generally 10 to 12 kWh/lb (22.0 to 26.4 Wh/g). This range is appropriate for budgeting purposes. A low-pressure system typically has air compressors, aftercoolers, refrigerant dryers, and heat-reactivated desiccant dryers (see sec. B, chapter IV). For this example, the selected specific energy is 11.5 kWh/lb (25.3 Wh/g). Note that the costs for ozone generation equipment are based on the U.S. bid numbers from Figure VI–3 (the upper cost curve), which assume a low-pressure air treatment system.

Step 3: Using the annual average production rate from step 1 and system specific energy from step 2, calculate the daily average energy consumption. For this example, the annual average ozone production rate is 417 lb/day (about 7.9 kg O_3/h) and system specific energy is 11.5 kWh/lb (25.3 Wh/g). The resulting daily average energy consumption is 4795 kWh/day.

Step 4: Based on local electrical pricing information, estimate the unit cost of electrical energy, expressed in terms of $/kWh or FrF/kWh. All electrical company charges, such as power demand costs, fuel cost adjustments, and energy consumption charges should be incorporated into this estimate. For this example, it is assumed that the overall price of energy is $0.06/kWh (about 41 centimes/kWh).

Step 5: Using the energy consumption number from step 3 and unit energy cost from step 4, calculate the annual energy cost. For this example, the estimated daily energy consumption is 4795 kWh/day and unit price of energy is $0.06/kWh (41 centimes/kWh). The resulting daily energy cost is $287.70/day (about 1970 FrF/day).

Plant Name: Example Plant

1. Operating Ozone Production Rate

Annual Average Flow	20	mgd (3,150 m^3/h)
Annual Average Dose	2.5	mg/L
Annual Average Production	417	lb/day (7.9 kg/h)

2. Estimated System Specific Energy[1]	11.5	kWh/lb (25.3 Wh/g)
3. Daily Average Energy Consumption[2]	4,795	kWh/day
4. Estimated Unit Energy Cost[3]	0.06	$/kWh (41 centimes/kWh)
5. Daily Average Energy Cost	287.70	$/day (1,970 FrF/day)
6. Estimated Annual Energy Cost		105,010 $/yr (720,400 FrF/yr)
7. Est. Cost Factor for Energy Alone[4]	75	%
8. Estimated "Other" O&M Cost		35,000 $/yr (240,000 FrF/yr)
9. TOTAL ESTIMATED O&M COST		140,010 $/yr (960,400 FrF/yr)

[1]Typically ranges from 10 to 12 kWh/lb for low-pressure air treatment ozone generation systems.

[2]Formula: Cost = O_3 production (lb/day) × specific energy consumption (kWh/lb).

[3]For purposes of this estimate, all electrical company charges, such as power demand, fuel cost adjustment, and energy consumption charges, should be incorporated into a single number relative to the kWh of energy consumed.

[4]Energy costs typically range from 65% to 80% of the total system O&M cost and generally are 75%. The "other" costs include activities or items such as operator time for daily equipment readings, spare parts replacement cost, preventive and emergency maintenance activities, etc.

Figure VI–8 Estimation of Operating Cost of an Air-Feed Ozone System for Example Plant

Step 6: Using the daily energy cost, calculate the annual energy cost. For this example, the daily energy cost is $287.70/day. Assuming 365 working days, the resulting annual energy cost is $105,010 per year (about 720,400 FrF/year).

Step 7: Estimate the percentage of the total annual O&M cost that is attributed to energy cost. At operating installations, the cost of energy has been 65 to 80 percent of the total system O&M cost (see sec. D.5 of this chapter). Other costs, such as operator time for daily equipment readings, spare parts replacement costs, preventive and emergency maintenance activities, etc. run from 20 to 35 percent of the total cost. The percentage for energy cost may be lower when the price of energy is relatively low and cost of labor is relatively high. Most often, the energy cost represents 75 percent of the total system O&M cost. This is the percentage used for this example.

Step 8: Using the energy cost factor from step 7 and cost of energy from step 6, determine the cost for "other" ozone O&M activities. The energy cost is $105,010 per year (about 720,400 FrF/year) and energy cost factor is 75 percent of the total system O&M cost. (*Note:* The 75 percent factor is a rough rule of thumb; for more information on determining costs, see Razcko et al. 1990.) The resulting O&M cost, excluding energy, is $35,000 per year (about 240,000 FrF/year).

Step 9: Using the estimated annual energy cost from step 6 and estimated "other" O&M cost from step 8, calculate the total system O&M cost. The energy cost is $105,010 per year (about 720,400 FrF/year) and the "other-than-energy" cost is $35,000 per year (about 240,000 FrF). The resulting total annual system O&M cost is $140,010 (about 960,400 FrF).

Figure VI–8 summarizes the O&M costs for this example.

Another example regarding the energy consumption of an ozone facility, including more detailed data on average ozone production throughout the year, is presented in sec. E.2 of this chapter.

VI.B CONSIDERATIONS FOR ESTIMATING THE COST OF AN OZONATION FACILITY

Detailed cost estimates can only be developed for specific applications after specifications have been established. Therefore, one should use cost estimates prepared from cost estimating curves and other information with caution. Preliminary cost estimates should only be used to develop a general idea of what an ozone system will cost. Prices from different contractors will vary even when they have the same information base. Decision-makers need to examine closely the items covered by the cost estimates, as well as the methodology on which they are based, prior to implementing a financial plan for a project.

Problems connected with estimating O&M-related costs will be even more extensive than with capital costs. The costs quoted for electricity, a cost-intensive factor in ozonation plant operating expenses, often correspond to the optimum working conditions according to plant design. However, real conditions are rarely close to the optimum; the more sophisticated the original design, the more this is true. Hence one has to consider that actual operation and maintenance costs can only be obtained from users of these systems, i.e., water treatment operators or managers.

Another problem that faces decision-makers is comparing the costs of different treatment trains. An example that briefly illustrates this point is the preoxidation step that can be effected with various oxidizing reagents. For the same use, such as oxidation of iron or manganese or improving flocculation, different reagents applied will simultaneously induce effects other than the one originally intended (e.g., creation of reaction by-products, influence on taste and odor control, etc.). These side effects, whether positive or negative, will have to be taken into account when the final decision-making occurs. Yet it is impossible, when studying the relative cost of oxidants, to give an accounting coefficient that reflects an impartial view of the secondary advantages and disadvantages that the oxidants used will produce.

VI.B.1 Considerations for Assessing Capital Costs

Capital costs considered here include the cost of ozonation equipment (i.e, ozone generators, feed gas preparation equipment, and the associated instrumentation for monitoring and automatically controlling the plant), the necessary ancillaries including ozone destruction or recycling systems, the power supplies, and, when available, construction costs (contactor and housing), installation costs, and contingencies.

Capital costs depend on a number of factors. For a new plant these factors include:

- Ozone requirement, including standby capacity and redundancy equipment (see sec. B.1, chapter V);
- System size (depending on gas preparation and ozone production technologies) and housing (see secs. B.1 and J.2, chapter IV);
- Availability of power to the site;
- Contactors and associated diffusion systems (see sec. D, chapter IV); and
- Degree of automation (see sec. E.1, chapter IV).

To retrofit an existing plant, one needs to consider additional cost factors relative to the implementation of contactors in the hydraulic profile and to the modification or construction of the building for equipment housing (see sec. H, chapter IV). Other considerations also need to be taken into account, such as:

- The compatibility of existing materials of construction in facilities immediately downstream of the ozonation step (possible corrosion due to ozone);
- Process interconnections to existing channels, conduits, and control system; and
- Ability to maintain operations and to continue producing water during construction.

Moreover, to have a real idea of the capital cost it is also necessary to know the rate and period of depreciation for buildings, contactors, and equipment as well as financial costs (interest). Other costs, like "site work," design and construction management services, owners' administration, and legal and other expertises, need to be added.

All these different costs were not taken into account in sec. A of this chapter because they are highly variable depending on site conditions, type of plant (new or retrofitted), countries, and laws of the different countries.

A glance at this list of the different technological and site factors affecting capital cost is enough to see why costs and their breakdown could be very different for two facilities with the same design capacity.

VI.B.2 Considerations for Assessing Operating and Maintenance Costs

Fixed operation and maintenance costs consist of costs that do not vary with the hourly effective production of ozone. These fixed O&M costs include:

- Labor for supervising the plant during operation, if this has not been entered as overhead charges; and
- Amortization if not included as capital cost.

Others costs are proportionate or variable costs, including:

- Utilities (electricity, water, liquid oxygen [if used]); and
- Maintenance (cleaning), spare parts, maintenance personnel.

Note: Water used for cooling is often recycled via the water treatment lines. In this case, the water-associated cost is the cost of pumping.

At least for the large ozonation systems, the most important O&M cost is the cost of electricity (Raczko et al. 1990). Therefore, a major consideration is to develop a realistic energy cost estimate. Ozone generator specific energy consumption changes at different production levels. Ozone production requirements change throughout the year (see sec. E.2, this chapter). The operating cost estimate must consider both of these factors.

Other operating and maintenance costs will depend primarily on:

- The technologies selected for feed gas preparation and ozone generation (air, enriched air, O_2), which may affect the maintenance and the consumable material;
- Ozone dosage rate;
- Ozone dissolution and destruction systems; and
- Degree of automation.

VI.B.3 Considerations for Assessing Secondary Benefits and Costs

Chapter III points out the attractive aspects of ozone for several applications (for example, improvements in later stages of treatment [sec. I and K], or a savings in the amount of reagents consumed [sec. F]). The following examples show some of the benefits of using ozone:

- The possibility of decreasing the amount of flocculant needed in certain cases, or according to the season;
- Longer filtration cycles;
- Transformation of a conventional filtration process into a biological one for removal of biodegradable organics which, in the case of activated carbon beds, results in a much longer useful life for the carbon; and
- Decrease in trihalomethane formation potential (THMFP).

Logically, it should be possible to account for all these advantages, but they are not always easy to interpret in terms of cost.

Extra costs may also arise by design assumptions that can lead to inefficiencies of the treatment train, for example:

- Increase in chlorine demand;
- Consumption of chlorine dioxide by ozone when the chlorine dioxide used to protect the bacteriological quality of the utility network is added without taking care to check that the water does not contain dissolved ozone;
- The need to install extra filtration with granular activated carbon downstream of ozonation at the end of a treatment train handling water heavily loaded in organics in order to remove biodegradable organic carbon. (This should not be considered as a secondary cost but as an important improvement to the treatment system, which will have its own beneficial effects on the final chlorine demand.)

VI.C INFLUENCE OF THE DESIGNED OZONATION SYSTEM ON THE CAPITAL COST

VI.C.1 Ozone Requirement Including Standby

The amount of ozone required (see chapter III) can be accurately defined by means of treatability studies (see sec. A, chapter IV). There is no general rule for a standby ozone supply and its advisable capacity. A drinking water treatment facility must normally be in a position to supply, without interruption, water of an acceptable quality as determined by the health department.

One approach is to provide extra generators sufficient to supply 100 percent of the design production rate. Another approach is to assume that the installed ozone generators would operate at a given percentage (x) of their rated maximum capacity (i.e., 75 percent) to meet the design production rate. The standby capacity would then be $100 - x$. The approach chosen depends on different factors such as the specific regulations of each country, the ozone application, and the economics (see sec. B.1, chapter V). The final decision on standby capacity should always consider locality-specific matters. For example, if there is a problem with one utility, is there another one that can instantly provide water of a high quality (as is the case for the Paris suburban water works, which are all interconnected [Tardieu 1988])? Should an alternative oxidant be kept on hand to replace ozone temporarily, if necessary?

The same type of problem exists for the gas preparation system. In cases where air is used for producing ozone, the preparation system is often slightly oversized compared with the flow rate of desiccated air needed to generate the requisite amount of ozone. This is caused by the fact that the manufacturers, concerned with production costs, have standardized each of the component parts contained in the preparation line. When each of the components is chosen, manufacturers generally choose, from a range of available material, the units capable of treating slightly more feed air than the design flow rate. The criteria for standby capacity may be the same for the gas preparation equipment as for the ozone generator.

VI.C.2 System Size and Housing

Three important factors will affect the capital cost of the equipment and the cost of the structures in which it will be housed:

- The kind of feed gas used to produce ozone,
- The preparation technique adopted, and
- The ozone production technology.

Influence of ozone feed gas. The various feed gas systems are discussed in sec. B, chapter IV. The use of air involves consideration of area for the air preparation system and ozone generators.

The use of air with a possibility of oxygen enrichment can substantially reduce the size of the equipment for both air preparation and ozone generation. The main advantage of this combination is that it results in facilities of a reasonable size capable of producing sufficient ozone for the needs of the water works during the greater part of the year. The oxygen is needed only for ozone demand peaks, which usually occur over short periods (Masschelein 1980a).

The use of oxygen alone generally reduces the housing area required. When liquid oxygen is used, space onsite (usually outside) is needed for the liquid oxygen storage tanks. When gaseous oxygen is produced onsite, some housing is needed for the oxygen preparation equipment. In all cases, building area is needed for the ozone generators.

Influence of gas preparation technology. When using air to generate ozone, the higher the pressure during preparation, the smaller the size of the preparation plant. Hence, there will be a savings on capital cost, but an increase in operating costs.

In the case of enriched air or liquid oxygen, capital costs are also significantly reduced, but operating costs will be influenced by the cost of oxygen and the storage vessels and evaporators. The cost of oxygen is extremely variable, depending on freight charges and the quantities used. Merz and Gaia quote costs for oxygen ranging from 4.1 cents/lb to 16.8 cents/lb (0.61 FrF/kg to 2.53 FrF/kg) (Merz and Gaia 1989).

When oxygen is produced onsite (pressure swing adsorption or cryogenic generation), the complete system capital costs are generally higher, but operating costs are generally lower than for an ozone system using an air preparation train. In order to obtain the lowest life cycle cost, an evaluation of the capital costs plus the operating costs must be completed. It should be noted that for operating reliability, the oxygen production facility will generally have to be equipped with liquid oxygen storage capacity (Rakness et al. 1988).

Table VI–2 compares the advantages and disadvantages of the cryogenic or non-cryogenic systems producing oxygen onsite (Labrousse 1989). Figures VI–9 and VI–10, from two different oxygen producers, give an indication of the oxygen prices by the way it is supplied.

Lastly, recycling oxygen (offgases or oxygen upstream of the contactors, —see sec. B.1, chapter IV) is capital intensive, so capital costs are high. This is probably why the recirculation of oxygen downstream of the contactors has not been widely used. As for upstream recycling, no industrial system has yet been developed (1990), though chances are that such a system will be designed in the future for facilities with a capacity of several hundred kilograms of ozone per hour (5300 lb/day or more). See sec. B.1 in chapter IV and sec. E.1 in this chapter for more discussion about recycling systems.

Table VI–2 Qualitative Comparison Between Cryogenic and Noncryogenic Plant

Process	Noncryogenic Cycle VSA or PSA*	Cryogenic Cycle
Operating temperature	Ambient 30°C (86°F)	Cryogenics −190°C (−310°F)
Oxygen purity	Up to 95 percent	Up to 99.9 percent
Normal industrial purity	93 percent	99.5 percent
Startup time	5–10 min	For cold startup 30–45 min; for warm startup 1–2 days
Number of complex machinery	2	3
Liquid oxygen production	No	Possible

Source: Labrousse (1989).
*VSA: regeneration of molecular sieves through vacuum.
PSA: regeneration of molecular sieves at atmospheric pressure.

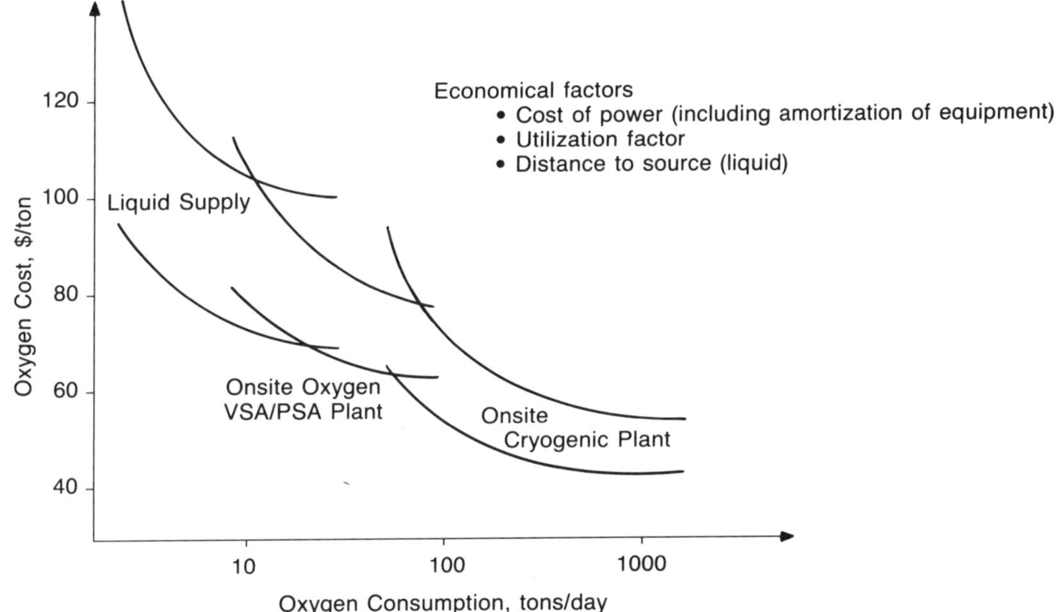

Source: Labrousse (1989).

Figure VI–9 Oxygen Costs from a French Manufacturer

Influence of ozone generation technology. The two main industrial options currently on the market for the production of ozone use low electrical frequencies (50 or 60 Hz) or higher frequencies through the use of converters (see secs. C.3 and C.5 in chapter IV). The use of the higher electrical frequencies and of small gap generators (see sec. C.5, chapter IV) makes it possible to obtain greater concentrations of ozone in the product gas and so to use a smaller ozone generator for the same rate of production. In addition, the capacity of an individual ozone generator may be increased, reducing the number of ozone generators required. In this case, the building size can be reduced.

An example of cost variations tied to the above technical options. It is very difficult to find literature on the effects of various technical options on investment costs (see Tables VI–3 and VI–4). For example, for a system using air, the unit cost is sometimes higher for medium frequency (MWDSC, 1988a), sometimes lower (Schulhof 1989).

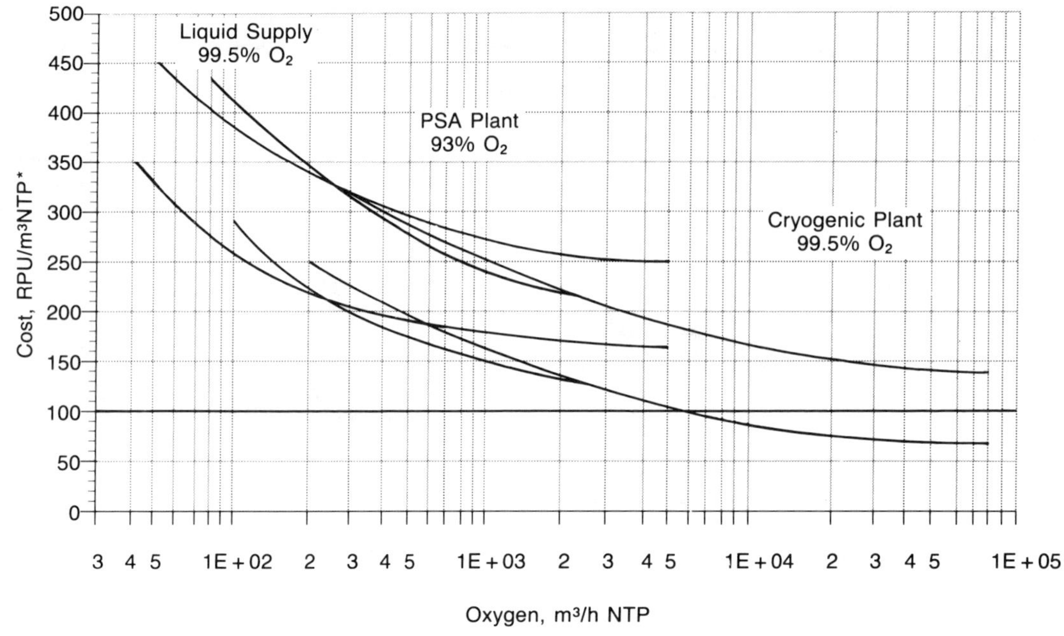

Source: Beysel (1989).

Figure VI–10 Oxygen Costs from a German Manufacturer

Table VI–3 Comparative Costs of Equipment for Various Ozone Production Methods*

Feed gas	Air	Air	GOX[†]	LOX[‡]
Frequency	Low	Medium	Medium	Low
Electrode spacing	Normal	Normal	Narrow	Normal
Investment cost				
$/lb/day	1475	1324	1683	1135
FrF[§]/g/h	535	480	610	412

Source: Schulhof (1989).
*Ozone capacity: 50–100 kg O_3/h (about 2465–5291 lb O_3/day).
[†]GOX: gaseous oxygen produced onsite.
[‡]LOX: liquid oxygen.
[§]Exchange: $1 = 6.86 FrF.

Table VI–4 Preliminary Capital Cost Estimate of Equipment for Various Ozone Production Methods*

Feed gas	Air	Air	LOX[†]	GOX[‡]	Air/LOX
Frequency	Low	Medium	Medium	Medium	Medium
Investment cost					
$/lb/day	870	895	525	990	765
FrF[§]/g/h	316	325	191	359	278

Source: Metropolitan Water District of Southern California (1988a).
*Ozone capacity: 20,000 lb/day (378 kg O_3/h).
[†]LOX: liquid oxygen.
[‡]GOX: gaseous oxygen produced onsite.
[§]Exchange: $1 = 6.86 FrF.

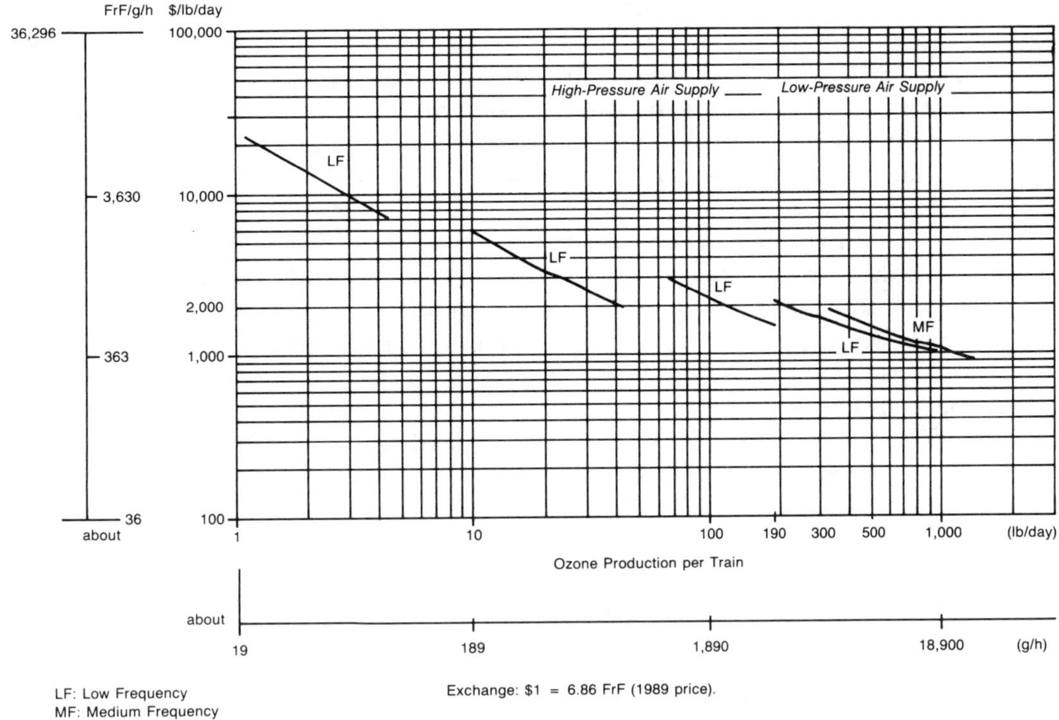

Figure VI–11 Ozone Equipment Cost (Air Feed Gas) from a French Manufacturer

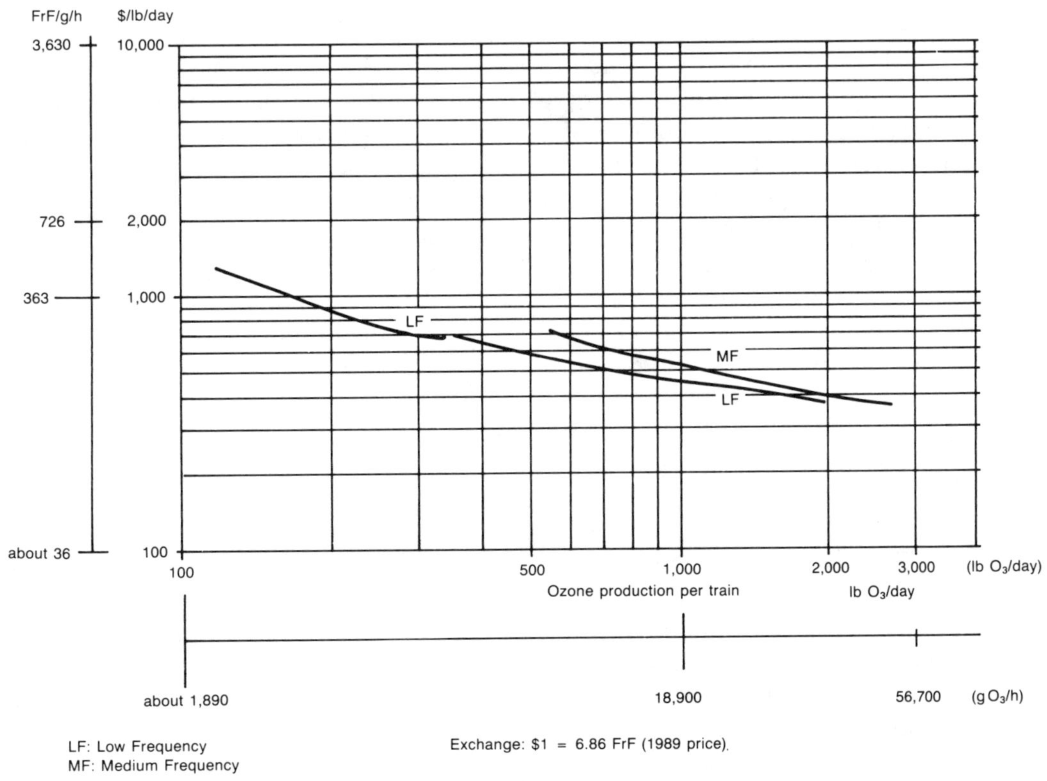

Figure VI–12 Ozone Equipment Cost (Oxygen Feed Gas) from a French Manufacturer

For its estimate, the MWDSC study takes into account the cost of the feed gas system, the generators, the electrical control package, and miscellaneous equipment. The Schulhof study includes in its estimate all the necessary equipment for an ozonation system (including ozone destruction equipment) except the dissolution device.

Rice (1989) compared the unit costs of five U.S. plants capable of treating between 13 and 600 mgd (about 2,000 to 94,600 m^3/h). He came to the conclusion that it is difficult to derive a "rule of thumb" from such studies.

It is interesting to compare the unit costs for different types of equipment coming from the same manufacturer. Figures VI–11 and VI–12 present the philosophy of one French manufacturer who generally recommends (at least in Europe) low-frequency ozone generators for low ozone production capacity, when ozone is produced from air, or for medium ozone production capacity, when ozone is produced from oxygen. For large ozone capacity (from air or oxygen), low or medium frequencies are available. The equipment cost for medium frequency is higher than the cost for low frequency. The cost difference is more pronounced for smaller ozone capacity.

The curves in Figures VI–11 and VI–12 take into consideration the following costs:

- Air preparation line (Figure VI–11),
- Ozone generation equipment,
- Ozone destruction equipment,
- Local control panels for all equipment,
- Piping, valves, and instrumentation (except for ozone distribution and dissolution),
- Control system, and
- Installation and startup.

The following additional costs are excluded:

- Equipment for oxygen supply (Figure VI–12),
- Dissolution equipment, and
- Main electrical distribution panel.

For small or medium ozone production capacity (1 to 190 lb/day [about 19 to 3,590 g O_3/h]) the example with air feed gas (Figure VI–11) uses a preparation line at high pressure (relative pressure: 7 bar [101 psig]). For larger ozone capacity (200 to about 1800 lb/day [i.e., 3,780 to about 34,000 g O_3/h]), low-pressure air preparation is used (relative pressure: 0.8–1.5 bar [11.6–21.7 psig]). These choices are obviously not the only available solutions. For instance, low-pressure air preparation for an ozone capacity of about 100 lb/day (about 1900 g/h) is feasible.

The control system cost depends mainly on the degree of automation required. It has been estimated to be 20 percent of the total equipment cost.

For this example, the ozone generators' production rates have been estimated based on the following operating conditions:

- Ozone concentration in air: 1.5 percent by weight (19.5 g O_3/m^3 NTP).
- Ozone concentration in oxygen: 5.0 percent by weight (73.0 g O_3/m^3 NTP).
- Cooling water temperature: 15 °C (59°F).

These costs are available for one ozone production train (e.g., of 10, 100, 1000 or 2000 lb/day [about 190, 1,890, 18,900 or 37,800 g O_3/h]).

Scale effect on costs. The capital costs of ozonation equipment show significant economies of scale (unit costs decreasing as ozone output increases). An example of the cost of ozonation equipment using a high-pressure air preparation line over a range of ozonation capacities is given in Table VI–5.

One can note that this scale effect is also shown in Figures VI–3 (see sec. A.1 of this chapter), VI–11, and VI–12.

In comparing two U.S. water treatment plants (treating water from the same source by the same process with ozonation equipment) Rice (1989) found a scale effect, i.e., a unit cost for the smaller plant (13 mgd [2050 m^3/h]; 825 lb O_3/day [15.3 kg O_3/h]) of

Table VI–5 Costs★ for Equipment Only, Excluding Housing and Installation (Using Air as Feed Gas)

System Size	Approximate Cost	
	$/lb/day	FrF†/g/h
Laboratory unit (0.5–2 lb/day [9–38 g/h])	7,585–12,250	2,750–4,445
Very small system (2–8 lb/day [38–151 g/h])	4,580–7,585	1,660–2,750
Small systems (8–37 lb/day [151–700 g/h])	1,480–4,580	540–1,660

★1989 prices.
†Exchange: $1 = 6.86 FrF.

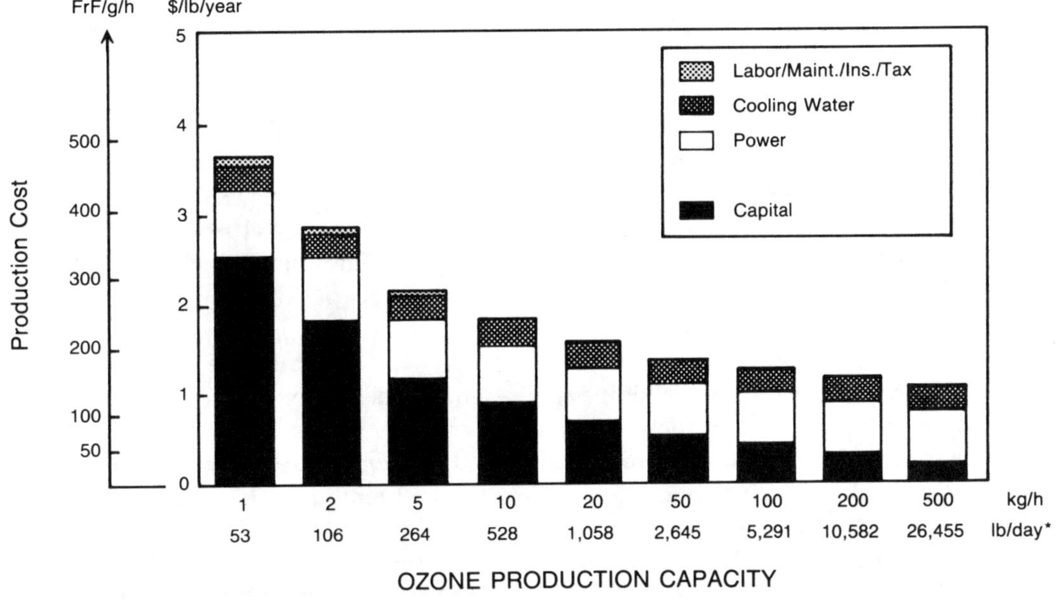

Source: Merz and Gaia (1989).

1.5 percent by weight (19.5 g O_3/m^3 ozonated air NTP, 8760 operating h/year)

Figure VI–13 Ozone Production Cost ($/lb/year or FrF/g/h) from Air

$2,935/lb/day (about 1,065 FrF/g/h) and a unit cost of $1,660/lb/day (about 602 FrF/g/h) for the larger one (90 mgd [14,195 m^3/h]; 3,750 lb O_3/day [about 70.9 kg O_3/h]).

Lastly, Figure VI–13 (Merz and Gaia 1989) shows that the scale effect achieves a limit for high ozone production rates. The cost indicated in this study seems to be underestimated when compared with the previous one (Rice 1989). Moreover, the same principle of ozonation production will not be applied for such a large scale of ozone production.

Nevertheless, some studies, such as the Metropolitan Water District of Southern California study conducted by James M. Montgomery Consulting Engineers, Inc. (MWDSC 1988b), indicate that the scale effect no longer exists for facilities from 14 kg O_3/h (approximately 740 lb/day) to 142 kg O_3/h (about 7,500 lb/day). In fact, the upper limit of ozone production capacity at which the scale effect is reduced or disappears is difficult to precisely identify. This is due to the fact that at some ozone production rates (about a few hundred pounds per day or a few kilograms per hour) many different ways of producing ozone and different degrees of automation become available.

VI.C.3 Availability of Power to the Site

Significant alterations to the production capacity of the electrical system supplying the water treatment plant are generally necessary because of the electrical power needed to operate an ozone generator (KVA) and the power consumed in connection with ozone dissolution and destruction. Hence, contracts with the suppliers of electricity will often have to be renegotiated. These factors must be taken into account in estimating both capital costs and operating costs.

VI.C.4 Contactors and Associated Dissolution System

Cost of construction. The costs of construction are difficult to firmly establish; estimates can widely vary.

In a comparative study on the cost of disinfecting potable water (by UV radiation or ozone, for flow rates less than 1,000 m^3/h [about 6.3 mgd]), Paillard (1986) considers that the cost of the ozonation chamber was about 1,080 FrF/m^2 of surface area (1986 prices). The 1989 price (excluding taxes) is 1,300 FrF/m^2 (about $18/$ft^2$).

Rice (1989) shows a figure from Gumerman et al. (1986) for the price of small contactors (concrete contactors that use baffled flow in an upflow/downflow configuration), which includes the dissolution system (porous diffusers) and manways that can be used for access to the interior of the contactor. This information is shown in Figure VI–14.

A study was done by Compagnie Générale des Eaux (Bablon and Van Elsland 1989) relative to the cost of the necessary construction (excluding tax, baseline 1989) for the latest postozonation column installed at the Neuilly-sur-Marne water works (France), designed to disinfect about 8330 m^3/h (about 52.8 mgd) and described by Gerval (Gerval et al. 1989). This analysis estimated a cost of about 11,188 FrF [$1630]/$m^3$ of concrete, or about 8,809 FrF/m^2 for the developed area (about $119.30/$ft^2$). The same study showed a cost of about 7,761 FrF/m^3 of concrete (about $1,131/$m^3$), or 3,203 FrF/$m^2$ for the developed area (about $43.30/$ft^2$) for the preozonation contactor at the Choisy-le-Roi water works (France), designed for the treatment of 33,330 m^3/h (211 mgd) of raw water (see sec. D.2 in chapter IV).

The U.S. EPA considered in its 1979 report (U.S. EPA 1979) that the cost of the structures and contact columns represents from 10 to 30 percent of the total capital cost. Another study (MWDSC 1988b) estimates that this same cost represents from 40 to 50 percent of the total capital cost, in which the contactors account for 30 to 50 percent more than the housing. Process Applications Inc. (PAI 1988) describes a unit in which all the civil works represent 43 percent of the total cost. The contact tanks cost 13 percent less than the structures themselves. Both these companies quote a unit price between $90 and $100/$ft^2$ (about 6,650 to 7,400 FrF/m^2).

One can observe a large discrepancy in prices from one plant to another in the same country, or from one country to another, regarding the volume and the type of contactors.

Cost of dissolution system. It is obvious that the dissolution method (see sec. D, chapter IV) strongly influences the cost of the contact system, at two levels:

- the cost of the dissolution system, which can only be determined via case studies;
- the size of the contact chamber, which has to be adapted to the ozonation goal and to the method by which the ozonated gas is diffused in the water to be treated (width and length of compartments, depth of the dissolution system).

The layout of the buildings housing the ozone production equipment, determined

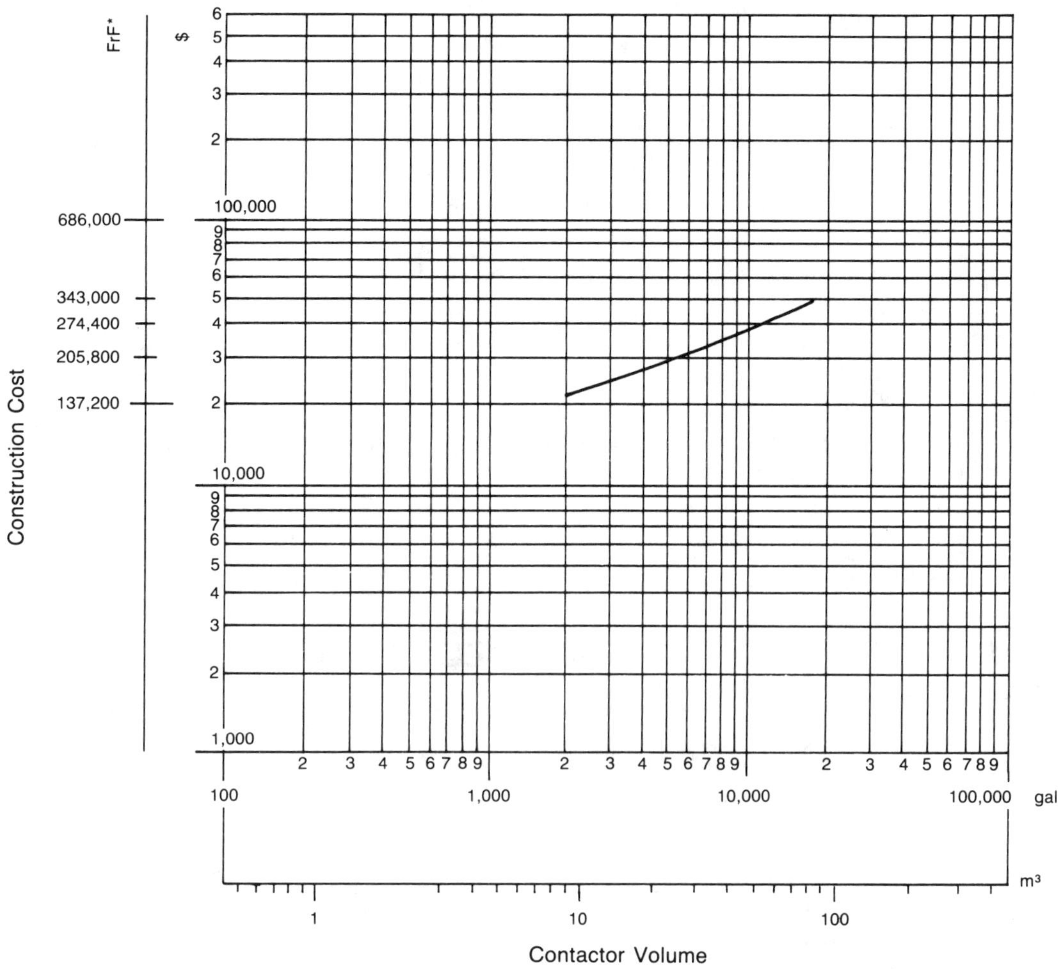

*Exchange: $1 = 6.86 FrF

Source: Gumerman et al. (1986).

Figure VI–14 Construction Cost for Ozone Contactors

by the contact columns (see sec. J, chapter IV), can cause very substantial variations in construction costs.

VI.C.5 Degree of Automation

The different types of process monitoring and control and their degree of sophistication were discussed in sec. E, chapter IV. The system requiring the least investment is a fully manual one, requiring manned intervention each time the influent water quality or flow rate varies. Although this kind of facility is conceivable and even necessary in some countries, where process monitoring and control are not yet technical realities either regarding equipment or trained labor, it does tend to be abandoned, at least in the case of larger systems. Any ozonation plant needs to be equipped with a minimum of automation in order to perform adequate startup and shutdown and provide all the requisite alarms connected with the handling of a high-voltage plant and oxygen (if any). These safety alarm measures are of the utmost importance for both personnel and equipment.

The overall management of drinking water treatment plants is becoming more and more automated, and ozonation will not be able to escape this evolution. Manufacturers are already capable of designing and installing fully automated ozonation facilities. Even now, expert systems providing the possibility of interpreting and acting on monitoring results obtained at the plant are being developed. Their purpose is to optimize the mathematical models initially incorporated in the system, improving in turn the operating conditions, through assisted decision-making at the plant control level (Wable et al. 1989; Pascal 1984; Bablon et al. 1986).

Many intermediate cases exist between the two extreme situations (manual or fully automated plants) and the capital cost of ozonation facilities with a given hourly capacity can vary by a factor of 1 to 4, depending on the degree of automation desired.

It is important to point out that if a more sophisticated process monitoring and control system is installed to theoretically result in savings on operating costs (fewer manned positions, optimal electric power consumption, etc.), this will only happen in practice if very highly qualified personnel are present onsite and if maintenance is scrupulously performed. Therefore, if the decision has been made to highly automate a plant, provisions must be made for investing in the hiring and training of qualified personnel (see sec. D, chapter V).

VI.D INFLUENCE OF THE DESIGNED OZONATION SYSTEM ON O&M COSTS

VI.D.1 O&M Costs of Feed Gas Train

We have already seen in sec. B.2 of this chapter that (1) the O&M costs depend on the ozone dosage actually applied, and (2) for a given ozone production they depend on the technologies selected for feed gas preparation, ozone generation, and ozone dissolution and destruction systems, and on the degree of automation. Moreover, operating costs also depend on the local costs of electricity, oxygen, and water used for cooling (if it is not recycled via the water treatment plant).

Different cost items must be considered depending on whether air, oxygen-enriched air, liquid oxygen, or oxygen onsite produced are used as the feed gas.

O&M cost of feed air preparation. In this case, the operating cost includes the energy consumed by the compressors or blowers supplying the air, as well as that required for air drying. The different techniques suitable for producing air with the necessary dewpoint are studied in sec. B.3, chapter IV.

Energy consumption for air pressurization. Two extreme cases can be considered: the use of blowers or of compressors. The reader may wish to refer to Figures IV–25 and IV–26 in sec. B.3, chapter IV for an example of these two cases. The various types of blowers and compressors available have widely different levels of efficiency. Table VI–6 gives a qualitative idea of the energy consumption and maintenance requirements for different types of compressors. Rakness et al. (1989) give an energy consumption for a liquid ring compressor of 80.8 W/m^3 NTP to 133.6 W/m^3 NTP (18.4 to 30.4 bhp per 100 scfm), corresponding, respectively, to operating flow rates and pressure ratings of 3 absolute bar (43.5 psia), 1,081 m^3 of air/h NTP, and 5.5 absolute bar (79.8 psia), 1,217 m^3 of air/h NTP.

Different manufacturers quoted the following specific energy consumption, according to the type of compressor:

- Screw compressor operating at 2.25 absolute bar (32.6 psia) for an air flow of about 1,000 m^3/h (NTP): specific energy consumption = 48 W/m^3 (10.9 bhp per 100 scfm).
- Roots blower operating at 1.9 absolute bar (19.9 psia) for an air flow of about 1,000 m^3/h (NTP): specific energy consumption = 35 W/m^3 (8 bhp per 100 scfm).

Table VI–6 Summary of Qualitative Comparisons for Alternative Air Pressurization

Equipment Option	Energy Consumption			Maintenance Requirements		
	Low Pressure	Medium Pressure	High Pressure	Low Pressure	Medium Pressure	High Pressure
Air compressors						
Oil-lubricated						
Reciprocating			Low			Moderate/High
Oil-flooded			Low			High
Oil-less						
Liquid ring	High	High	High	Low	Low	Low
Reciprocating	Low			High		
Rotary lobe	Low			High		
Twin screw	Low	Low	Low	Moderate/High	Moderate/High	Moderate/High

Source: Rakness et al. (1989).

- Vane compressor operating between 7 and 8 absolute bar (73.5–84.0 psia), for an air flow of about 150 m^3/h (NTP): specific energy consumption = 150 W/m^3 (34.2 bhp per 100 scfm).

Moreover, for a given piece of equipment, consumption of energy can vary drastically with regard to the operational point on the characteristic curve of the machine. For this reason, a well-designed study is necessary to be able to choose the type of equipment and operational conditions in relation to the overall operating requirements of the plant. For a given plant, a compromise between the capital cost and the O&M cost will lead to choosing more blowers or compressors at a fixed speed or fewer variable-speed units in order to meet the variation in the production of ozone.

Energy consumption for air drying. It is necessary to distinguish low-pressure, high-pressure, or in-between air preparation systems. The preparation line components that consume electric energy are the refrigerant dryer and the desiccant dryer. For the aftercooler, generally located downstream of the compressor or blowers (see sec. B.3, chapter IV), the operating cost depends on the energy required to pump the water through the heat exchanger considering the flow, the pressure loss, and the efficiency of the pump. The water is usually recycled back to the water treatment plant, so there is no associated cost with loss of water.

For the refrigerant dryer, Rakness et al. (1989) give (in the above-mentioned example) an energy consumption of about 5.5 W/m^3 NTP (1.26 bhp per 100 scfm). French manufacturers, however, give a specific energy consumption of between 8.3 and 10.4 W/m^3 NTP (1.90 to 2.38 bhp per 100 scfm), for an air flow of 1800 and 300 m^3/h NTP. This difference is perhaps due to the type of refrigerant dryer (see sec. B.3, chapter IV).

Robson estimates that the total capital and operating cost of such equipment, on an annual basis, is around $1.50/lb of moisture, i.e., 2.3 centimes/g of water (PAI 1987).

For the desiccant dryers, which can be regenerated by different means (heatless, heat-reactivation processes, or an in-between regeneration system), the energy consumed can vary widely. Table VI–7 gives a qualitative comparison for alternative air drying systems.

This energy consumption also depends on the nature of the desiccant media that needs to be heated in order to be regenerated (see sec. B.3, chapter IV). For the heatless dryers, there is no extra energy consumed except for the compressors. For dryers using a sudden depressurization effect in the tower and simultaneous heating, the energy consumption of the latter can be quoted at 37.5 W/m^3 NTP (8.6 bhp per 100 scfm).

For dryers regenerated by heating, Rakness et al. (1989) give an energy consumption of 8.6 W/m^3 NTP (1.96 bhp per 100 scfm) for a desiccant dryer working at low pressure and regenerated with an internal heating system, and of 17.1 W/m^3 NTP (3.91

Table VI–7 Summary of Qualitative Comparisons for Alternative Air Drying Systems

	Energy Consumption			Maintenance Requirements		
Equipment Option	Low Pressure	Medium Pressure	High Pressure	Low Pressure	Medium Pressure	High Pressure
Refrigerated dryer	Low			High*		
Desiccant dryers						
Heat-reactivated						
Internal heated	Low	Low		High*	High*	
External heated	High/Moderate	High/Moderate		Moderate	Moderate	
Heatless	N/A			Low		Low/Moderate

Source: Rakness et al. (1989); Rakness (1990).
*The maintenance requirements for refrigerated and internal heated desiccant dryers are considered moderate by some authors.

bhp per 100 scfm) if the regeneration is performed using external heating. A French manufacturer indicates, however, a consumption of electricity equal to 40.9 and 50.6 W/m^3 NTP (9.35 and 11.56 bhp per 100 scfm, respectively) for a range of air flow between 1800 and 300 m^3/h NTP (low-pressure system). These discrepancies can be explained by the difference in the desiccant dryers (nature of desiccant, external or internal regeneration heat) and also (perhaps mainly) by the fact that a large number of manufacturers producing desiccant dryers for purposes other than ozonation very often undersize this type of equipment. The same observation can also be made regarding the refrigerant dryers.

Because of the necessity of having a very efficient and reliable air preparation system to avoid serious trouble during ozone production, the specifications written for the manufacturing of such equipment must be closely detailed and restrictive.

Robson estimates that the total capital and operating cost on a yearly basis of the desiccation step is about $10/lb of moisture i.e., 15 centimes/g of water (PAI 1987).

Maintenance costs of feed air preparation line. These costs include the price of the usual spare parts that have to be regularly replaced, such as the different filters for oil (if any) or water droplets and dust, the oil change when oil compressors are used, and the price of larger components like the desiccant media. These prices can only be obtained from the manufacturers. Costs also include labor to replace parts and to make necessary routine checks. In sec. C of chapter V (Tables V–1, V–2, and V–3), the frequency for the routine maintenance and checking of the different components of one specific air preparation line are estimated. From this information, one can estimate the working time at about 2 h/week. For more detailed checkups on the performance of valves, drains, any type of indicators, or standby equipment; replacement of filters and/or oil (if any); cleaning and lubrication of the different mechanical parts; calibration of dewpoint meters; and so on, a working time of about 2 h/month also seems reasonable.

The necessary checking and replacement of certain parts of the equipment such as seals, gaskets, and the like can be estimated at 3 more h/quarter and 1 more day/year. Lastly, it is recommended that every three or four years a complete assessment be made of the state of the blower or compressor, the refrigerant dryer, and the desiccant dryer; that the cooling tubes be cleaned; and that parts like pump packing and desiccant tower seals be replaced. This represents between 4 and 5 working days.

O&M cost of oxygen feed line. Two different types of systems need to be considered: use of liquid oxygen or production of oxygen onsite with a PSA/VSA or cryogenic system (see sec. B.1, chapter IV and sec. A.2, chapter V).

O&M cost of an offsite feed oxygen supply. This is the least maintenance-intensive solution because the feed gas line includes only a storage capacity and an evaporator with a pressure-reducing system upstream of the ozone generator. Nevertheless, the

Table VI–8 Characteristics and Power Consumption of Different PSA Systems for Oxygen Production

	2-Bed	3-Bed	4-Bed
Size range, tons/day (kg/h)	1–10 (37.8–378.0)	1–35 (37.8–1323.0)	22–35 (831.6–1323.0)
O_2 purity, percent	90–93	90–93	90–95
Product relative pressure, psig (bar)	2.5 (about 0.17)	15 (about 1.03)	40 (about 2.76)
Power consumption, kW/ton/day	22	15	25
At max. capacity, kW/kg/h	0.582	0.397	0.661
Turndown capability, percent	0	70	20

Source: Warakomski (1989).

operating costs, which generally include the rent of the storage tank and evaporator plus the cost of the oxygen, are very dependent on the local conditions and on the annual consumption of oxygen. In sec. C.2 of this chapter, the cost of oxygen was shown to vary from 4.1 to 16.8 cents/lb (0.61 to 2.53 FrF/kg), depending on freight charges and the quantities employed (Merz and Gaia 1989). Moreover, it is necessary to take into account the waste of oxygen when the plant is stopped. Since the storage tank cannot be completely insulated, constant evaporation of the oxygen occurs. When the pressure inside the vessel becomes too high, an escape valve discharges oxygen to the atmosphere. This waste of LOX is difficult to estimate and largely depends on the operating conditions of each plant. This means that the only way to get an accurate idea of the cost of oxygen is to perform a case-by-case study.

O&M cost of an onsite oxygen supply. The two existing conventional systems for producing oxygen onsite (pressure or vacuum swing adsorption and cryogenic oxygen generation) are described in sec. B.1, chapter IV (Figures IV–21 and IV–22).

For a PSA or VSA system, the maintenance operation consists of the classical method regarding the feed air compressors and the oxygen compressors (if any) as well as the vacuum pump unit if the regeneration of the molecular sieve is performed under slight vacuum (Leitgeb 1989). Moreover, it is necessary to clean sequentially or replace the filters downstream of the air feed compressors that remove the dust, oil, and water droplets contained in the compressed air (Xorbox 1989). Nevertheless, emphasis should be placed on the maintenance of the valves through which the pressure is successively increased or decreased in the molecular sieve bed. Indeed, the efficiency of the process mainly depends on the reliability of these valves.

Regarding the oxygen production cost, a manufacturer claims a price, based on a cost of 5 cents/kWh, of between 39.7 and 21.4 cents/100 ft^3 of gas (oxygen: 90 percent), according to the size of the PSA (15 scfh at a maximum output pressure of 10 psig to 1000 scfh at a maximum output pressure of 45 psig), i.e., 0.96 to 0.52 FrF/m^3 NTP for an oxygen production between 0.4 m^3/h NTP (maximum relative output pressure: 0.69 bar) and 26.4 m^3/h NTP (maximum relative output pressure 3.1 bar). For larger oxygen production rates, Warakomski (1989) gives the power consumption of different PSA systems using 2 to 4 towers containing molecular sieves. These values are presented in Table VI–8.

For optimum flexibility, it seems better to use a PSA system design with 3 molecular sieve towers. Warakomski (1989) cites 10 U.S. patents connected with the development of PSA and estimates that this kind of process will cost in the future about $50/ton of oxygen (fully amortized onsite equipment), i.e., 378 FrF/metric ton. Leitgeb (1989) gives a specific energy consumption for an oxygen (93 percent purity) production at almost ambient pressure of 0.3 kWh/kg O_2 (0.14 kWh/lb O_2). The reader can refer to Figures VI–9 and VI–10 (see sec. C.2 of this chapter) for information about oxygen prices according to manner supplied.

With a cryogenic system, maintenance is certainly more expensive. Indeed, such a system includes 5 different steps (Warakomski 1989): air compression, contaminant removal, heat exchange, air expansion, and cryogenic distillation (see Figure IV–21 chapter IV). However, cryogenic separation plants can differ in design.

The maintenance requirements to be considered are on the filters upstream of the air compressors, the air compressors, the surge tank valves, the heat exchanger reversing valves, all the different valves and controllers along the different gas streams, liquid super-heaters, lower and upper distillation columns, and the air expansion turbine. This means that for a system to perform successfully, operating and maintenance expertise is required.

VI.D.2 O&M Cost of Ozone Generation

Section D.1 of this chapter shows that the O&M costs depend on the gas selection. The same applies to ozone generation. Moreover, for a given gas, the ozone generation technology will also have an influence on cost.

Operation cost of ozone generation. Whatever the ozone generation system used (low or medium electrical frequency), the specific energy consumption for producing ozone varies according to the nature of the feed gas. Figure VI–15 gives an idea of these rates of consumption relative to the use of air or oxygen and for different ozone concentrations in the gas. Moreover, for a given ozone generator, the consumption of energy depends on the cooling water temperature. The loss in efficiency (specific energy consumption) varies with each type of generator (see sec. C.4, chapter IV).

It is interesting to note the considerable evolution of ozone generator performance levels over the past 35 years, both for specific consumption of electricity and ozone production capacity (Chapsal 1989a), as shown in Table VI–9.

Maintenance cost of ozone generation. The major maintenance costs are as-

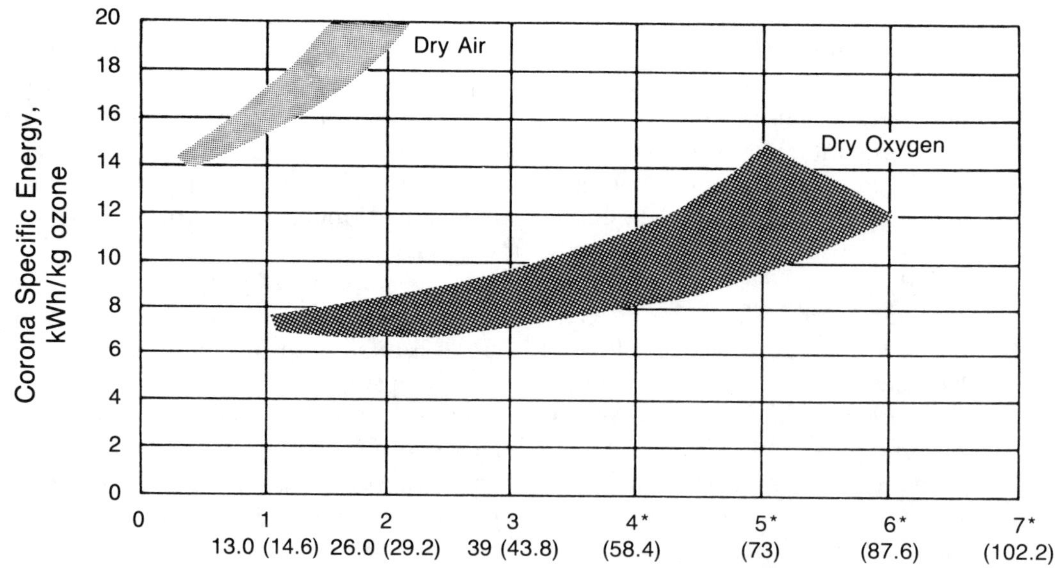

Source: Warakomski (1989).

Figure VI–15 Representative Corona Specific Energy for Dry Air and Oxygen

Table VI–9 Evolution of Ozone Generation Performances (Production from Air)

	Ozone Generation Capacity per unit		Electrodes Area		Total Dielectric Weight		Dielectrics		Applied Power,	Electrical Specific Consumption and Applied Power*				
Year	kg O_3/h	lb O_3/day	m^2	ft^2	kg	lb	kg/m^2	lb/ft^2	kW	Wh/g	kWh/lb	W/m^2	W/ft^2	W/kg
1955	1	52.9	48	516.7	296	652.6	6.1	1.25	24	24	10.89	500	46.45	81
1960	5.5	291.0	137	1,474.7	725	1,598.3	5.2	1.06	120	21.8	9.89	875	81.29	165
1970	8.3	439.1	137	1,474.7	698	1,538.8	5.0	1.02	160	19.3	8.75	1,168	108.51	230
	25†	1,322.7	140	1,507.0	698	1,538.8	5.0	1.02	500	20.0	9.07	3,570	331.66	716
1980	16	846.5	208	2,238.9	990	2,182.6	4.7	0.97	290	18.0	8.16	1,300	120.77	288
	36†	1,904.7	208	2,238.9	990	2,182.6	4.7	0.97	650	18.0	8.16	3,100	288.00	650
1990	50*	2,645.5	312	3,358.4	1,375	3,031.4	4.4	0.90	800	16.0	7.26	2,600	241.55	580

Source: Chapsal (1989a).
*Applied power per dielectric area or per dielectric weight.
†Use of medium frequency.

sociated with checking the power supply unit and control panel and, above all, with cleaning of the dielectric tubes (see secs. C.5 and C.8, chapter V). The safety of air or gas circuits, as well as cooling water and electric circuits, should be checked on a quarterly basis. Elsenhans (1989) checks the performance of the ozone generators at 33, 66, and 100 percent of the electric power applied, verifying the reactive effect on the power supply by oscillography, monitoring the harmonic content, and measuring the power on the low-voltage side. Simultaneously, the ozone concentrations at the outlet of the ozone generator are measured.

It is recommended that the ozone generator be cleaned annually when ozone is produced from air and where small plants are used. For larger facilities, where the air preparation system works at low pressure, it is reasonable to clean the ozone generator once every two years. Cleaning methods and the time required to clean an ozone generator are different for each type of equipment on the market. For the conventional dielectric tubes and discharge gaps (see sec. C.5, chapter V), one can consider the necessary working time to be between 10 and 15 min per tube. This time includes opening the generator; dismantling, cleaning, and reassembling the dielectric tubes; and closing the generator. It should be noted that labor time is not necessarily proportional to the number of dielectrics. When the discharge gap between the dielectric tube and the stainless steel tube is smaller, additional precautions and time are needed.

Because of the differences between the various ozone generators on the market, it is difficult to form a general and reliable idea as to what the cost of labor will be for cleaning an ozone generator. The reader may wish to refer to the operation and maintenance manuals of the different manufacturers.

VI.D.3 O&M Cost of the Dissolution Systems

The various dissolution systems were shown in sec. D.2, chapter IV. Among these systems, only porous diffusers do not require specific energy for dissolution because dissolution is ensured by the pressure of the ozonated gas alone, the rate of which is determined upstream of the dissolution system. Masschelein (1977) estimates that the energy required for the transfer of ozonated gas to the water corresponds to 2 to 3 Wh/g (907 to 1,361 Wh/lb) of O_3 injected. In the case of swept porous plates (see sec. D.2, chapter IV), the energy required to pump the stream of sweeping water must also be taken into account. The pressure applied is approximately 1.4 bar (20.3 psi) for a sweeping water flow rate of 700 to 1,200 L/h/porous plate (3.1 to 5.3 gpm/porous plate).

For maintenance, it appears reasonable to check the condition of the diffuser once a year. In addition, the overall effectiveness of dissolution in the contactors must be checked by measuring the transfer efficiency every six months (see sec. C.6, chapter V).

Experience shows that the rate of gradual clogging of the diffuser varies from one facility to another and depends principally on the efficiency of the water treatment upstream of the ozonation step (except in the case of preozonation).

The static mixer and the emulsifiers (see sec. D.2, chapter IV) do not need any special maintenance since they are, by definition, static components. Masschelein (1977) states that energy consumption will be from 4 to 5 Wh/g O_3 (1.81 to 2.27 kWh/lb) for the static mixers and 4 to 45 Wh/g O_3 (1.81 to 20.4 kWh/lb) according to the type of emulsifier under study (high or low load emulsifiers).

For turbines, Masschelein states (without defining the type of turbine) that the average specific energy consumption is between 5 and 7 Wh/g O_3 (2.27 to 3.17 kWh/lb). In the case of those manufactured by a French manufacturer, self-suction turbines designed for reinjecting offgas represent a specific energy consumption of 65 to 70 Wh/m^3 of air NTP (14.8 to 15.9 bhp per 100 scfm). The turbines under pressure, designed for dissolution of the ozonized gas coming from the ozone generators, consume 35 to 40 Wh/m^3 of gas NTP (8.0 to 9.1 bhp per 100 scfm), i.e., a specific consumption of 1.94 to 2.22 Wh/g O_3 (880 to 1,007 kWh/lb) for a concentration of 18 g O_3/m^3 NTP (1.38 percent by weight) ozone in the air. Miller et al. (1978) found during a case study at the Tailfer plant in Brussels, Belgium (see sec. VI.E.2) that the dissolution per turbine results in a consumption of energy corresponding to 75 percent of the amount required to produce the ozone. In the case of the Lengg plant (Zürich, Switzerland), the consumption due to the turbines is estimated at 13 Wh/g (5.9 kWh/lb) (Miller et al. 1978). All these widely divergent data depend essentially on the type of turbine used and the operating conditions (turbine running at nominal output or at a significantly lower rate, which can considerably increase the specific consumption of energy). As far as maintenance is concerned, conditions are similar to those of any equipment comprising moving parts. It must, however, be remembered that turbines allow good mixing and a high bubble contacting rate.

Lastly, in cases where packed or plate columns are used as contactors (Elsenhans 1989), which occurs rarely, Masschelein (1977) states that the related energy consumption is around 15 to 40 Wh/g O_3 (6.8 to 18.1 kWh/lb).

After having indicated certain orders of magnitude regarding energy consumption rates in relation to the dissolution systems, it can be assumed that the costs of ozone-to-water contacting can be strongly dependent on local conditions and the overall hydraulic profile of the plant.

VI.D.4 O&M Cost of Ozone-Containing Offgas Destruction Systems

The different offgas destruction systems that exist are described in sec. F, Chapter IV. If one considers a simple dilution of the offgases (which is not likely to be sufficient under severe conditions of temperature inversion and is not accepted in North America, the energy to be applied will correspond to the amount required for the extraction and mixing of the offgases with 10 volumes of fresh air, plus the discharge of this mixture to the atmosphere at a velocity around 15 m/sec (about 2,953 ft/min) at the level of the stack. The specific energy consumption will be that of the blower designed to handle these conditions, i.e., about 6 to 10 Wh/m^3 NTP (1.37 to 2.28 bhp per 100 scfm). It is not advisable to use activated carbon (see sec. F.3, chapter IV). As an example, the water works on Lake Constance (Elsenhans 1989) abandoned the activated carbon process after 10 years of use in favor of thermocatalytic destruction. The former process had needed a change of activated carbon every three months (but had been geared to a two-year schedule).

For the two most reliable processes at the present time (1990) (thermal destruction with heat recovery and thermocatalytic destruction), the following information is found in the literature:

- The average consumption of electricity is between 27 and 35 Wh/m^3 NTP (6.15 to 7.97 bhp per 100 scfm) of treated gas (Chapsal 1989b). This consumption applies almost without regard to the ozone concentrations in the offgas to be treated. Because of the efficiency of the heat recovery loop in the thermal process, the energy balance is practically the same as that obtained in the thermocatalytic method. The advantage of the thermal process is that it eliminates the risk of catalyst poisoning.
- The thermocatalytic destructor installed at the Lake Constance plant (Elsenhans 1989), where the air issuing from the contact columns is reheated to 60°C (140°F) and over 50 percent of the heat is recovered, costs about 0.07 cents (0.48 centimes) to operate per gram of ozone present in the vents. The system has been operating free of trouble for eight years.
- It is very hard to make a fair estimate for the life span of the catalysts. Some authors consider that a reasonable time is about two years (Miller et al. 1978); others forsee a useful life of five years (Horst 1982). (See also sec. C.7, chapter V.)

Lastly, the following maintenance is required: checking the extractor fans at regular intervals (lubrication of ball and roller bearing of fan motor); measuring the loss of head through the system; checking the condition of the electric resistors in the air reheating reactor; and checking the analyzer by which the efficiency of the process can be monitored continuously and the alarm triggered if the residual ozone content is overshot in the stack gases discharged to the atmosphere (in the event of trouble on the thermal or thermocatalytic destructor).

VI.D.5 Overview of Total O&M Costs

After examining the operation and maintenance costs relative to the various components of the ozonation facility, it is interesting to attempt to determine a few general ideas on overall operating and maintenance costs.

In a study conducted for the Metropolitan Water District of Southern California (MWDSC 1988b), James M. Montgomery Consulting Engineers, Inc. gives some data on the subject. By using the EPA Cost Estimating Manual (U.S. EPA 1979), these authors developed energy and labor costs of \$0.08/kWh (0.55 FrF/kWh) and \$20/h (206 FrF/h), respectively, and give an operating and maintenance cost unit of \$1.07/lb of O_3 (0.016 FrF/g of O_3), (1988 price), assuming that the plant under consideration is capable of producing 2,450 lb O_3/day (46.3 kg O_3/h).

The breakdown of this cost concerns four items, as follows:

- Energy for building, corresponding to 1 percent of the total O&M cost;
- Energy for ozone generation and dissolution, corresponding to 80 percent of the total cost;
- Maintenance (material), accounting for 10 percent of the total cost; and
- Labor, amounting to 9 percent of the total.

In the same document, the cost, based on data obtained from Process Applications Inc. (PAI 1987) and for a facility capable of producing 5,000 lb O_3 /day (94.5 kg O_3/h) using on average only 1,790 lb/day (33.1 kg/h), is \$1.29/lb O_3 (0.019 FrF/g O_3). The breakdown of this cost gives 60 to 80 percent for energy and 20 to 40 percent for maintenance.

Miller et al. (1978) consider that as a rule of thumb, maintenance cost should amount to about 15 to 20 percent of the production cost.

The experience of Compagnie Générale des Eaux at the large-scale Paris suburban water works (see sec. E.2 of this chapter) results in an estimated O&M cost of 1.56

Table VI–10 Estimated O&M Cost for Ozonation Plant

Design Ozone Capacity		Capital*		O&M†		Total‡	
lb/day	kg/h	$/lb/day	FrF/g/h	$/lb	centimes/g	$/lb	centimes/g
100	1.9	7,000	2541	2	3.0	6	9.1
1,000	18.9	3,500	1270	1	1.5	3	4.5
10,000	189	2,000	726	0.6	0.9	1.5	2.3

Source: Raczko et al. (1990).
*$/lb/day or FrF/g/h of installed design capacity. Costs escalated to May, 1990 using ENR CCI (20-city average) = 4,696.8; $1 = 6.86 FrF. Costs include ozone generation (air preparation, generation, destruction), contactor, and building.
†O&M costs divided by average ozone production, which is approximately 50 percent of design capacity. Costs include power ($0.086/kWh—0.59 FrF/kWh), labor ($14.30/h—98 FrF/h), maintenance, and coolant.
‡The sum of capital cost amortized at 10 percent, 20 years plus O&M costs, divided by the average ozone production.

cents/1000 gal (2.22 centimes/m^3), or taking into account the average ozone dosage of 3.04 mg/L: 61 cents/lb O_3 (0.93 centimes/g) (price as of 1989). Bearing in mind amortization and depreciation, this cost is 8.17 centimes/m^3, or 4.52 cents/1000 gal (i.e., 2.69 centimes/g O_3 produced or $1.78/lb). These rough estimates must be used with some reserve, however, since costs can vary widely from one facility to another, depending on the treatment train. For example, the total production cost, including amortization and depreciation, is 5.21 centimes/m^3 (2.87 cents/1000 gal) at Neuilly-sur-Marne, as compared to 11 centimes/m^3 (6.07 cents/1000 gal) at Méry-sur-Oise (both facilities described in sec. H.2, chapter III).

Merz and Gaia (1989) estimate that the annual cost for labor and maintenance accounts for 3 percent of the investment cost of the plant.

Table VI–10 summarizes the results of the study done by Raczko et al. (1990) about the cost of existing U.S. ozone installations. The total estimated cost decreases from $6 to $1.50/lb for design ozonation capacities varying from 100 to 10,000 lb/day, respectively (9.1 to 2.3 centimes/g for design ozonation capacities from 1.9 to 189 kg O_3/h).

The reader can refer to sec. A.2 of this chapter for a quick estimate of O & M costs for a future ozone installation. It should be noted that other approaches for estimating total annual cost of an ozonation facility have been published, such as the methodology described by Wunsch and Darpin (1989).

VI.E CASE STUDIES

The case studies described in this section come from Trailigaz projects or contracts, from Compagnie Générale des Eaux experiences, and from relevant literature.

VI.E.1 Case Studies of Capital Costs

Cost of new plants.

The Tailfer water treatment plant (Brussels, Belgium). This plant is capable of treating 10,833 m^3/h (about 68.7 mgd) from the River Meuse and is located about 75 km northeast of Brussels at Tailfer. The treated water is pumped to Brussels. Designed in 1966–1968 (Masschelein 1980b), the treatment train is as follows: adjustment of pH, prechlorination, oxidation by chlorine dioxide, coagulation with an aluminium salt, addition of powdered activated carbon if necessary, flocculation-sedimentation, addition of caustic, intermediate chlorination, sedimentation (second stage), high-rate sand filtration, ozonation for disinfection, and postchlorination/ammonification to protect the bacteriological quality of the water in the long distribution system.

The ozone production capacity is 24 kg O_3/h (about 1,267 lb/day). Ozone is produced through six ozone generators (each unit producing 4 kg O_3/h [about 211 lb/day]),

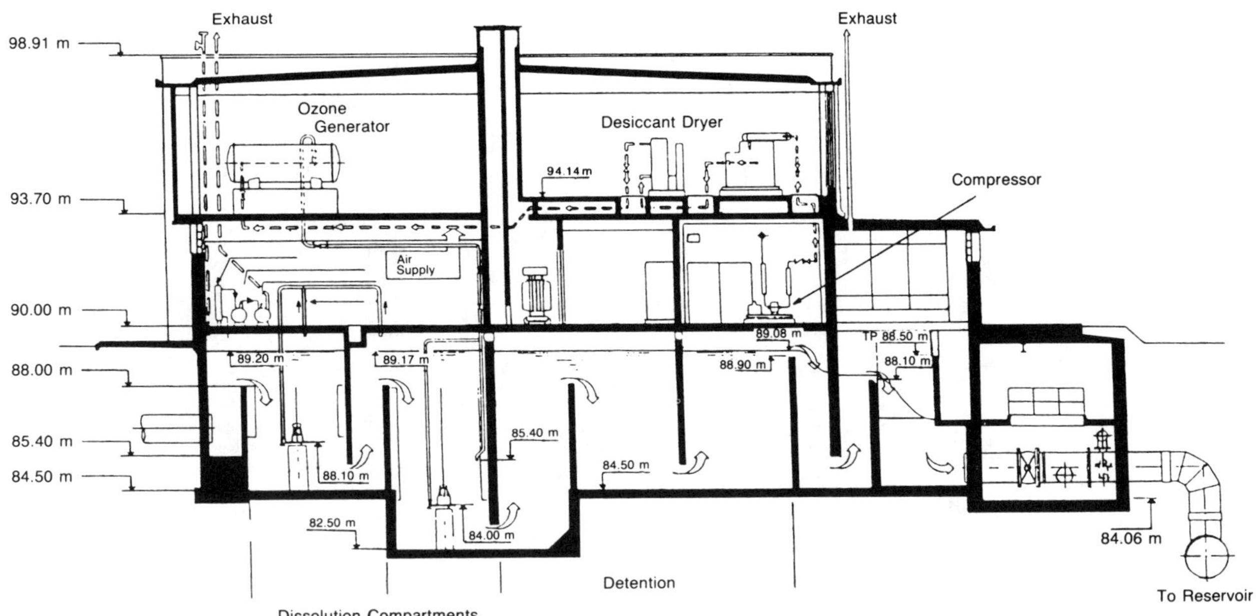

Source: Masschelein (1980b).

Figure VI–16 Ozonation Building at Tailfer, Brussels, Belgium

working on a low frequency of 50 Hz. The plant has been designed to be able to produce ozone from oxygen-enriched air, necessary during periods of peak ozone demand in summer. The oxygen is stored in liquid form. Oxygen is added to the air downstream of the desiccant dryers (See sec. IV.B.1)

The building that houses the ozonation plant and contact chambers (bubble columns with two compartments for ozone dissolution through turbines [theoretical hydraulic contact time about 5 min, 30 sec] and one retention compartment [about 6 min theoretical contact time]) has been built larger than necessary for future expansion. Figure VI–16 shows this building.

Miller (1978) reported that the complete cost (1976 price) of the current system, including the building and equipment, was 148,000,000 Belgian francs (based on an exchange rate of 1 BF = 0.16 FrF = \$0.023: \$3,404,000 or 23,680,000 FrF). The cost of housing, including the contact chambers was approximately 50,000,000 Belgian francs (\$1,150,000 or 8,000,000 FrF). Equipment costs, which include the transformer, the air preparation line, the ozone generators, the thermal destruction of ozone in the vents, venting of building, safety analyzers and process analyzers with transmitters, were 98,000,000 Belgian francs (\$2,254,000 or 15,680,000 FrF). Building costs represented 33.8 percent of the total cost; equipment costs represented 66.2 percent.

Masschelein (1989) states that today, the total investment for this ozonation facility (whose maximum ozone production when oxygen is used is 42 kg O_3/h [about 2222 lb/day]) would be 212,000,000 Belgian francs (\$4,876,000 or 33,920,000 FrF). The difference in price between this plant and a plant that could produce the same amount of ozone from air is estimated at about 124,000,000 Belgian francs; the amount saved by outfitting the plant for ozone production from oxygen is \$2,850,000 or 19,000,000 FrF.

Proposed facility for a water treatment plant in the Middle East. The description of this 35–kg O_3/h (about 1,852 lb/day) system proposed as a disinfection plant in the Middle East will enable the reader to compare a medium-frequency situation (400 Hz) with a plant running on a low frequency of 50 Hz. The air preparation systems proposed

Proposed Systems 1 and 2

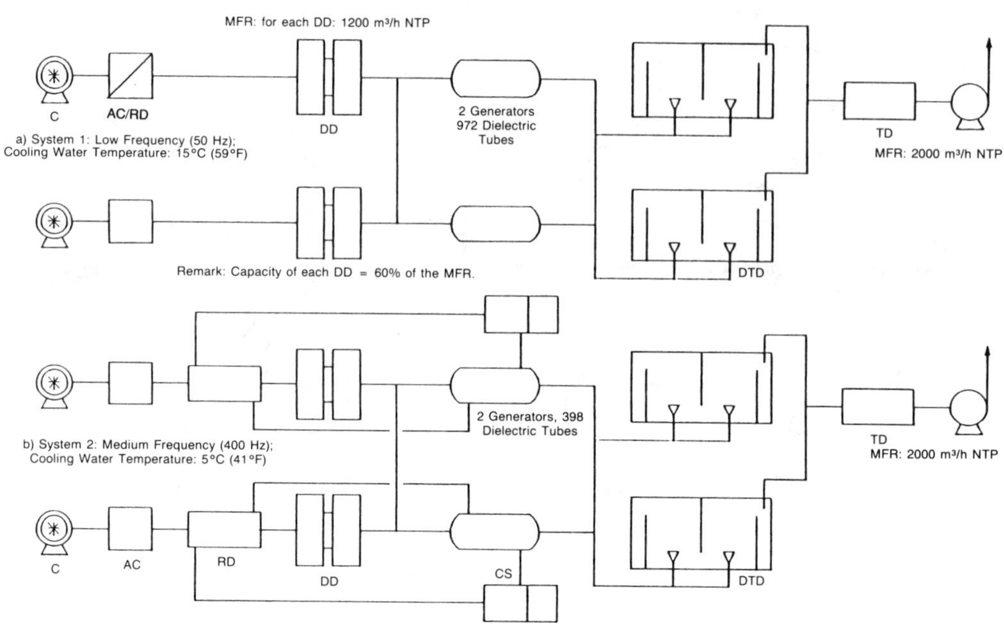

C: Compressor; AC: Aftercooler; RD: Refrigerant Dryer; DD: Desiccant Dryer; CS: Cooled Source; DTD: Disc-Type Diffusor; TD: Thermal Destructor; MFR: Maximum Flow Rate.

Ozone production at low (50 Hz) or medium frequency (400 Hz). Plant proposed in the Middle East.

Figure VI–17 Flow Diagram of Two Trains in a 35–kg O_3/h (about 1,852-lb/day) Ozonation Facility Using Air as a Feed Gas

are both operated at low pressure. The medium-frequency ozone generators are cooled with water at 5°C (41°F) (closed circuit), whereas the low frequency (50 Hz) ozone generators are cooled with water at 15°C (59°F) (open circuit). Figure VI–17 shows the schematic of the two systems that were proposed.

The 35 kg O_3/h (1,852 lb/day) is obtained with two ozone generators and associated feed air preparation units (one per generator). The complete facility is interchangeable. These two systems differ by the type of ozone generator, the power supply unit, and the air and generator cooling equipment.

Dissolution is accomplished by disc-type diffusers. The residual ozone in the off-gases is destroyed by thermal process with heat recovery. The servo control of ozone dosing is regulated on the influent flow rate and a reference dose rate with modulated feedback control based on maintaining a dissolved residual ozone concentration of 0.4 mg/L. Power applied to the generators is step-regulated to produce an average 17.5 g O_3/m^3 NTP concentration of ozone in the air (1.35 percent by weight), with variations between 13 and 26 g O_3/m^3 NTP (1 and 2 percent by weight).

Table VI–11 gives a breakdown of costs per component for the two trains described.

The low-frequency (50-Hz) system, using cooling water at 15°C (59°F), results in a lower capital cost for the air preparation line than in medium-frequency (400-Hz) systems. But savings in air-feed systems are outweighed by the higher cost of the ozone generators and power supply equipment. Indeed, in order to obtain the same overall ozone production capacity, the low-frequency system requires larger ozone generators that are more costly, despite the fact that the medium-frequency system requires more electrical equipment. Taken as a whole, the cost difference between these two systems is only about 3 percent but may be different for other configurations.

Table VI–11 Capital Cost* of an Ozonation Facility† Using Air as Feed Gas; Ozone Production at Low (50 Hz) or Medium (400 Hz) Frequency (Plant Proposed in Middle East)

	Capital Cost			
	System 1 (Low Frequency) Cooling Water at 15°C (59°F)		System 2 (Medium Frequency) Cooling Water at 5°C (41°F)	
Equipment	FrF‡	$	FrF	$
Air preparation train	2,473,000	360,496	2,962,000	431,778
Power supply unit and generators	3,700,000	539,359	2,920,000	425,656
Valving, measuring and control	2,669,000	389,067	2,669,000	389,067
Dissolution	325,000	47,376	325,000	47,376
Thermal destruction	575,000	83,819	575,000	83,819
Spare parts for two years	570,000	83,090	570,000	83,090
Total cost of equipment	10,312,000	1,503,207	10,021,000	1,460,000
Freight, assembly, commissioning	1,370,000	199,708	1,370,000	199,708
Turnkey cost (excluding construction)	11,682,000	1,702,915	11,391,000	1,660,496

*1989 prices.
†35 kg O_3/h (about 1,852 lb/day).
‡Exchange: $1 = 6.86 FrF.

Proposed facility for a water treatment plant in southern Europe. This plant was proposed in Europe for disinfection in a drinking water treatment plant. The flow rate to be treated can vary between 9,600 m^3/h (60.9 mgd) and 28,800 m^3/h (about 182.6 mgd), with an average flow rate of 20,160 m^3/h (about 127.8 mgd). The ozonation rate to be implemented is between 2 and 4 mg O_3/L. The capacity of the ozonation unit was calculated at the maximum treatment rate and flow rate (115.2 kg O_3/h [about 6,095 lb O_3/day]).

Three systems have been proposed:

- System 1: eight low-frequency ozone generators (50 Hz), each with a capacity of 14.4 kg O_3/h (about 762 lb O_3/day).
- System 2: eight medium-frequency (400 Hz) ozone generators with a unit capacity of 14.4 kg O_3/h (about 762 lb/day).
- System 3: four medium-frequency (400 Hz) generators with a capacity of 28.8 kg O_3/h (about 1,524 lb O_3/day).

In all of these cases, the ozone is produced with feed air containing a nominal concentration of 18 g O_3/m^3 NTP (1.38 percent by weight) for a maximum cooling water temperature of 29.5°C (85.1°F). Each air line (four in all, irrespective of the system selected) is capable of treating 33.3 percent of the total air flow required to produce the requisite 115.2 kg O_3/h, i.e., 1600 m^3/h NTP, per air preparation train. The maximum production capacity is 128.0 kg O_3/h (about 6,773 lb/day) for a cooling water temperature of 15°C (59°F).

The facility is manually operated and the ozone concentration in the air varies, according to the necessary ozone production, between 9 and 18 g O_3/m^3 NTP (0.69 and 1.38 percent by weight). The ozonated air is diffused through disc type diffusers. The offgases are treated by a thermal process with heat recovery.

Figure VI–18 is a schematic diagram of proposed systems 1 and 2. Figure VI–19 is a schematic of proposed system 3.

Table VI–12 gives the investment costs of equipment related to the three proposed systems. The difference in price between the least expensive system (four medium-frequency generators) and the most expensive (eight low-frequency generators) is about 20 percent.

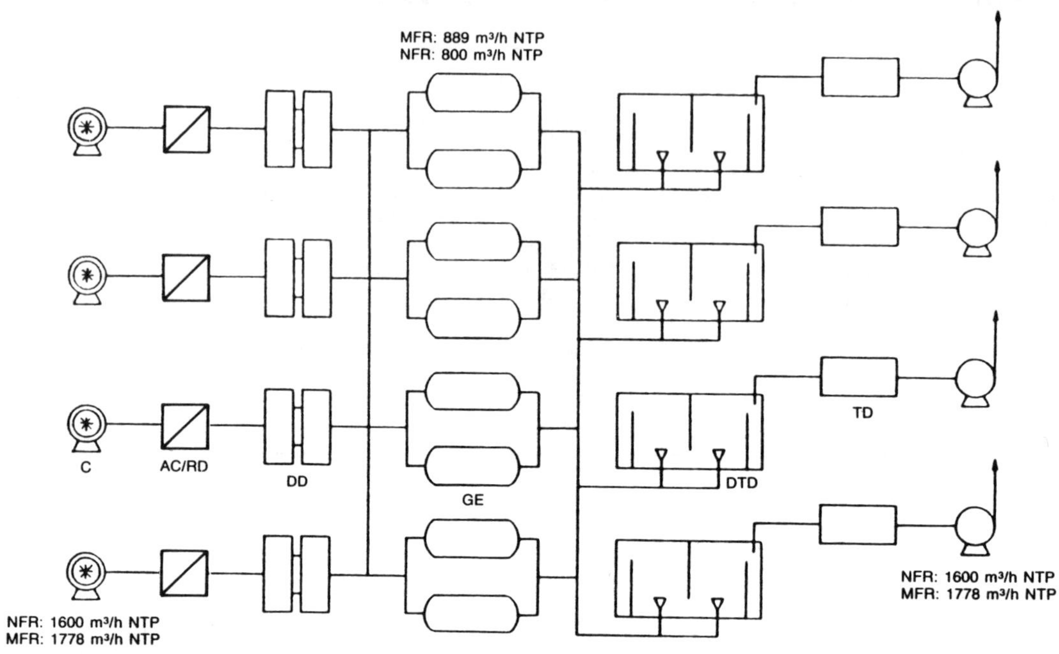

Maximum capacity: 128.0 kg O_3/h (6,773 lb/O_3/day). Ozone production at low (50 Hz) or medium (400 Hz) frequency. Plant proposed in southern Europe.

Figure VI–18 Diagram of Proposed Installation of 115.2 kg O_3/h (about 6,095 lb O_3/day): Proposed Systems 1 and 2 (Ozonation Facility Using Air as Feed Gas)

Table VI–13 shows the breakdown of the costs of the units previously defined for the three proposed systems. The largest cost difference is attributable to the ozone generator and associated equipment. Between the least expensive and the most expensive system, this difference can be as much as 50 percent. However, the system with only four ozonators is far less flexible to operate than the others.

Los Angeles Aqueduct Filtration Plant (LAAFP): 7,900 lb/day (about 149 kg O_3/h). This plant, which has completed its second year of operation, is described in sec. F.2 of Chapter III and in various publications (Georgeson and Karimi 1987; Hodges 1988). The LAAFP treats surface water brought by the Los Angeles Aqueduct from the Owens Valley, located in east central California. The average annual flow is 420 mgd (66,245 m^3/h) with a maximum capacity of 600 mgd (94,635 m^3/h). The treatment process includes an ozonation step and coagulation with ferric chloride and cationic polymer, plus direct filtration on anthracite at a very high filtration velocity (13.5 gpm/ft^2 [33 m/h]). Chlorination takes place at the end of treatment. Ozone is produced from oxygen (95 percent purity) through five ozone generators running at 600 Hz (one on standby), each unit being capable of producing up to 1,668 lb/day (about 31.5 kg O_3/h) at 6 percent concentration by weight (87.6 g O_3/m^3 NTP), i.e., a maximum capacity of 8340 lb/day (157.4 kg O_3/h) (Georgeson and Karimi 1987). Oxygen is produced from air through a cryogenic system with a capacity of 45 metric tons/day (1875 kg/h) and is sent to the generators or to a storage tank (in liquid form) for later use.

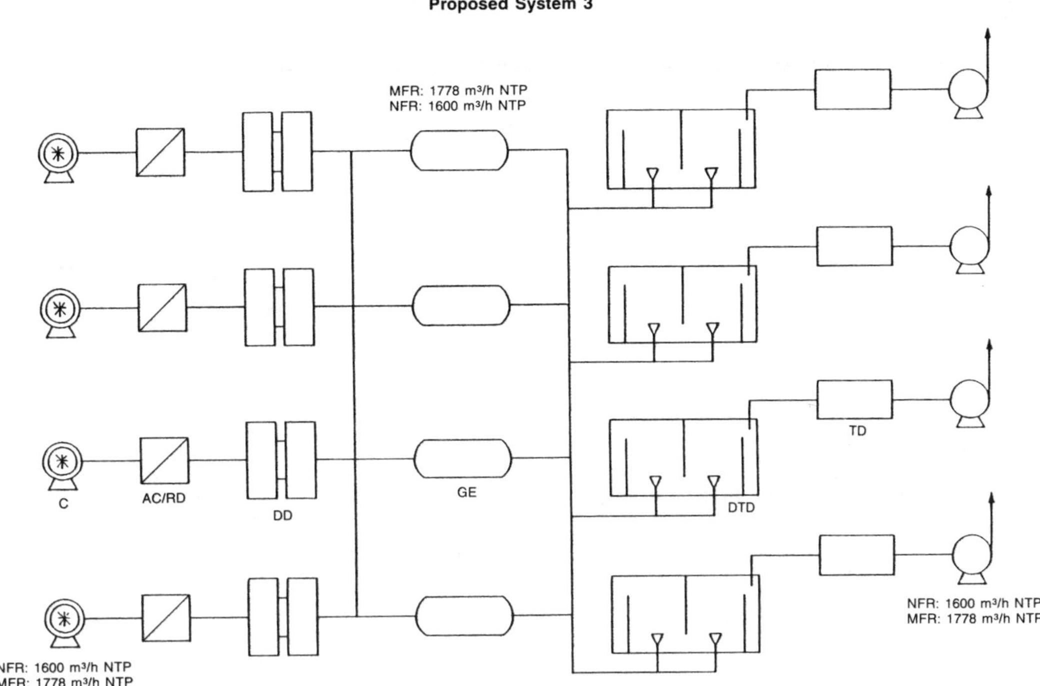

Maximum capacity: 128.0 kg O_3/h (6,773 lb O_3/day). Ozone production at low (50 Hz) or medium (400 Hz) frequency. Plant proposed in southern Europe.

Figure VI–19 Diagram of Proposed Installation of 115.2 kg O_3/h (about 6095 lb O_3/day): Proposed System 3 (Ozonation Facility Using Air as Feed Gas)

There are four treatment trains (one ozone generator per train plus one on standby) and four ozonation contact columns with a minimum theoretical hydraulic detention time of 4.9 min, equipped with porous ceramic diffusers. The ozone from the vents is destroyed in catalytic destruction units. This system is fully computerized for process monitoring and control. Updated costs for this system were given in the Metropolitan Water District of Southern California document previously cited (MWDSC 1988b) and are reported in Table VI–14.

The equipment cost breakdown is 48 percent for the ozone generation system and 52 percent for the contactor building.

During the bid phase of the LAAFP project, several air-fed systems using low-frequency (60-Hz) generators with the possibility of oxygen enrichment (from liquid oxygen) were also considered in comparison with the oxygen-fed system on medium-frequency (400- or 600-Hz) generators. The initial equipment cost of the lowest bid air-fed system was 37.5 percent less than the oxygen-fed system. However, the life-cycle costs, assuming a 20-year life, were 4.3 percent higher for the air-fed system. The oxygen system was therefore selected. Only a long period of running time will make it possible to verify whether the actual life-cycle costs are in agreement with forecasts.

Proposed facility for a water treatment plant in eastern Europe. This facility was proposed in Europe for the treatment of reservoir water containing manganese. The treatment train consists of preozonation, flocculation with aluminium sulfate, lamellar sedimentation, ozonation, double bed filtration on sand and activated carbon, pH adjustment, and an addition of chlorine before discharge to the distribution network. The treatment capacity of this water works is 54,000 m^3/h (about 342.3 mgd) with a minimum output of half that amount.

Table VI–12 Respective Investment Costs* of Three Systems Proposed for a Plant in Europe†

Equipment	System 1: 8 Low-Frequency (50-Hz) Ozone Generators FrF‡	System 1 $	System 2: 8 Medium-Frequency (400 Hz) Ozone Generators FrF	System 2 $	System 3: 4 Medium-Frequency (400 Hz) Ozone Generators FrF	System 3 $
Air preparation train	4,973,300	724,971	4,973,300	724,971	4,973,300	724,971
Generators with closed cooling circuit§	16,156,000	2,355,102	13,086,700	1,907,682	10,620,000	1,548,105
Valves, fittings and measurement	3,758,700	547,915	3,758,700	547,915	3,758,700	547,915
Dissolution	1,773,300	258,499	1,773,300	258,499	1,773,300	258,499
Destruction	2,773,300	404,271	2,773,300	404,271	2,773,300	404,271
Spare parts package for two years	1,133,300	165,204	1,133,300	165,204	1,133,300	165,204
Total cost of equipment	30,567,900	4,455,962	27,498,600	4,008,854	25,031,900	3,648,965
Transport, erection, and commissioning	3,973,827	579,275	3,574,818	521,110	3,254,147	474,365
Turnkey cost of facility (construction excluded)	34,541,727	5,035,273	31,073,418	4,529,964	28,286,047	4,123,330

*1989 prices.
†Nominal capacity = 115.2 kg O_3/h (about 6,095 lb/day).
‡Exchange: $1 = 6.86 FrF.
§The generators are cooled with a closed circuit to avoid chloride-induced corrosion, these ions being found in relatively high concentrations in the water to be treated.

Table VI–13 Investment Costs* for an Ozonation Facility Proposed in Europe† Compared with the Total Cost of Equipment or with the Turnkey Cost of the Water Works

	System 1: 8 Low-Frequency (50 Hz) Ozone Generators FrF‡	System 1 $	System 2: 8 Medium-Frequency (400 Hz) Ozone Generators FrF	System 2 $	System 3: 4 Medium-Frequency (400 Hz) Ozone Generators FrF	System 3 $
Investment costs of equipment:	30,567,900	4,455,962	27,498,600	4,008,854	25,031,900	3,648,965
Turnkey cost of facility (construction excluded):	34,541,727	5,035,237	31,073,418	4,529,964	28,286,047	4,123,330
Cost of unit considered as a percentage:	Percent with Regard to the Equipment Cost	Percent with Regard to the Turnkey Cost	Percent with Regard to the Equipment Cost	Percent with Regard to the Turnkey Cost	Percent with Regard to the Equipment Cost	Percent with Regard to the Turnkey Cost
Air preparation train	16.3	14.4	18.1	16.0	19.9	17.6
Ozone generators	52.8	46.8	47.6	42.1	42.4	37.5
Valves, fittings, and measurement	12.3	10.9	13.7	12.0	15.0	13.3
Dissolution	5.8	5.1	6.4	5.7	7.1	6.3
Destruction	9.1	8.0	10.1	8.9	11.1	9.8
Spare parts package for two years	3.7	3.3	4.1	3.6	4.5	4.0
Transport, erection, and commissioning	–	11.5	–	11.5	–	11.5

*1989 prices.
†Nominal capacity: 115.2 kg O_3/h (about 6,095 lb/day).
‡Exchange: $1 = 6.86 FrF.

The ozonation rates were 1 mg O_3/L for preozonation and 3 mg/L for ozonation prior to filtration. This ozonation plant, designed for maximum ozonation dosage and maximum flow rate, will provide a capacity of 216 kg O_3/h (about 11,429 lb O_3/day).

Table VI–14 Capital Cost of the Ozonation Plant at Los Angeles, California (LAAFP)*

Equipment	Capital Cost FrF†	Capital Cost $	Percent of Total Equipment Cost
Generators	9,329,600	1,360,000	10
Other equipment	38,484,600	5,610,000	38
Contactors/building	40,405,400	5,890,000	40
Equipment installation	11,936,400	1,740,000	12
Total cost	100,156,000	14,600,000	100

Source: Metropolitan Water District of Southern California (1988).
*7,900 lb O_3/day (about 149 kg/h).
†Exchange: $1 = 6.86 FrF.

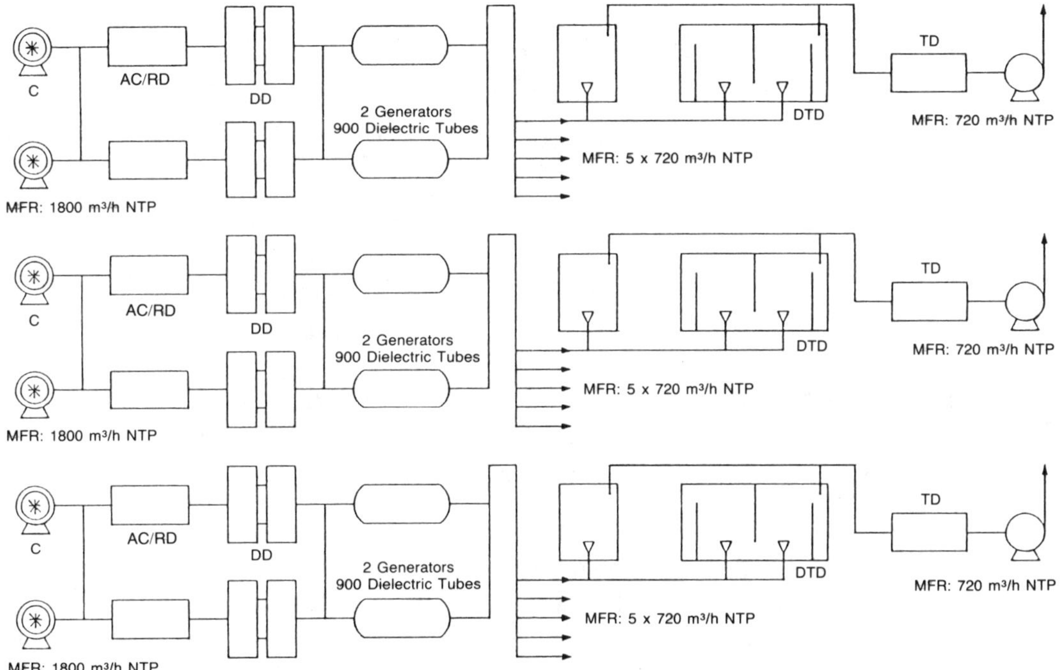

Plant proposed in eastern Europe.

Figure VI–20 Diagram of Proposed Installation of 216 kg O_3/h (about 11,429 lb O_3/day): Ozone Production from Air

The preozonation step is regulated on the flow rate to ensure a constant ozone dosage of 1 mg O_3/L. Prefiltration ozonation, where the feedback response is based on the residual dissolved ozone content at the outlet of the contact column, is regulated the same way.

The ozonized air is diffused via disc-type diffusers during both stages of ozonation. The description of the two systems proposed will enable the reader to compare two facilities operating on air or on oxygen with recycling according to a system patented by Linde in Germany (Benkmann 1983) (See sec. IV.B.1).

- System 1: Ozone is generated from air at a nominal concentration of 18 g O_3/m^3 NTP (1.38 percent by weight) and may vary between 10 and 25 g O_3/m^3 NTP (0.77 to 1.92 percent by weight) depending on the quantity of ozone to be

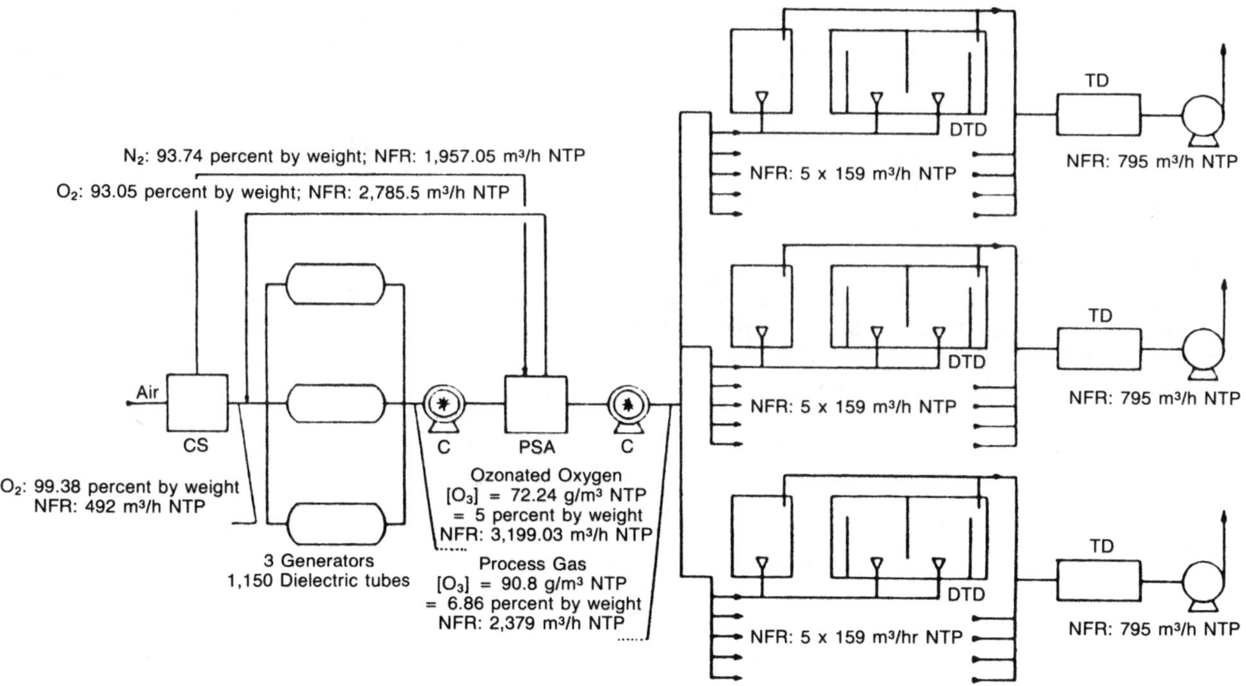

CS: Cryogenic System; C: Compressor; PSA: System for the Adsorption, then Desorption, of O_3; DTD: Disc-Type Diffusor; TD: Thermal Destructor; NFR: Nominal Flow Rate

Plant proposed in eastern Europe.

Figure VI–21 Diagram of Proposed Installation of 216 kg O_3/h (about 11,429 lb O_3/day): Ozone Production from Oxygen with Recycling Through an Adsorption/Desorption System Patented by Linde Germany

produced. The ozone generators will operate at medium frequency (400 Hz). This system is schematically diagrammed in Figure VI–20.

- System 2: The ozone is produced from oxygen obtained by cryogenic processing. The ozonated oxygen is then processed in a separation step where ozone is specifically adsorbed, the carrier oxygen being recycled to the ozone generators; at the same time, the ozone is desorbed by nitrogen (a by-product of the cryogenic stage). This process, patented by Linde Germany, is called the PSA-O_3 system. The ozone is then directed to the contact chambers. This system is schematically illustrated in Figure VI–21, where ozone concentrations are specified.

Table VI–15 gives the capital cost breakdown of the two systems in terms of total equipment cost, or turnkey cost with other construction excluded. Table VI–16 shows the percentage of cost attributable to each of the above-mentioned systems.

For system 2, where the ozone is produced from oxygen generated onsite and recycled according to the method patented by Linde Germany, the investment is less than the traditional system using air as a carrier gas (12 percent less expensive than system 1 if the cost of equipment is taken into account). In fact, this percentage is even underestimated, since the cost quoted for the cryogenic system and PSA-O_3 includes a part of the erection costs. The difference in expenditure in the case of a turnkey delivery (other construction excluded) is as much as 15 percent.

In system 1, the air treatment line represents 15.3 percent of the total equipment investment; the generators and their power supply units account for 42.9 percent. On the other hand, the cost of the equipment required in system 2 to produce onsite and

Table VI–15 Capital Cost* of an Ozonation Facility† Using Air as a Feed Gas or Oxygen with Recycling; Ozone Production at Medium Frequency (400 Hz); Plant Proposed in Eastern Europe

	Capital Cost			
	System 1: Ozone Production from Air, Cooling Water at 15°C(59°F)		System 2: Ozone Production from Oxygen, Cooling Water at 15°C (59°F)	
Equipment	FrF‡	$	FrF	$
Air preparation train or cryogenic system + PSA/O_3	8,425,488	1,228,205	25,171,200	3,669,271
Power supply unit and ozone generators	23,598,153	3,439,964	13,566,269	1,977,590
Valving, measuring, and regulation	14,070,807	2,051,138	5,681,756	828,244
Dissolution	3,746,262	546,102	1,878,910	273,894
Thermal destruction	4,811,508	701,386	1,783,589	259,998
Spare parts package for two years	367,074	53,509	329,279	48,000
Total equipment cost	55,019,292	8,020,305	48,411,003	7,056,997
Freight, assembly, commissioning	7,152,508	1,042,640	4,657,302	678,907
Turnkey cost (excluding construction)	62,171,800	9,062,945	53,068,305	7,735,904

*1989 prices.
†216 kg O_3/h (about 11,429 lb/day).
‡Exchange: $1 = 6.86 FrF.

Table VI–16 Costs* for Two Different Systems for an Ozone Plant† Proposed in Eastern Europe

	System 1: Ozone Production from Air		System 2: Ozone Production from Oxygen	
	FrF†	$	FrF	$
Investment costs of equipment:	55,019,292	8,020,305	48,411,003	7,056,997
Turnkey cost of facility (construction excluded):	62,171,800	9,062,945	53,068,305	7,735,904
Cost of unit considered as a percentage:	Percent with Regard to the Equipment Cost	Percent with Regard to the Turnkey Cost	Percent with Regard to the Equipment Cost	Percent with Regard to the Turnkey Cost
Air prep train or cryogenic system plus PSA O_3	15.3	13.6	52.0	47.4
Power unit and ozone generators	42.9	38.0	28.0	25.6
Valves, fittings, measurement and regulation	25.6	22.6	11.7	10.7
Dissolution	6.8	6.0	3.9	3.5
Ozone destruction	8.7	7.7	3.7	3.4
Spare parts package for two years	0.7	0.6	0.7	0.6
Transport, erection, and commissioning	–	11.5	–	8.8

*1989 prices.
†Capacity: 216 kg O_3/h (about 11,429 lb/day).
‡Exchange: $1 = 6.86 FrF.

recycle oxygen represents by far the larger investment (52 percent of the total amount of equipment expenditure, or 47.4 percent of the total cost of a turnkey facility). The cost of the ozone generators and related power supply unit accounts for only 28 percent of the equipment cost. Moreover, the generators running on oxygen are far less expensive for the same overall production than those required in a facility using air as a carrier gas. (The difference in the cost of the ozone generators for systems 1 and 2 is 42.5 percent.)

Lastly, less gas is used in system 2. Hence, the costs related to the dissolution of the gas followed by thermal treatment (destruction of the residual ozone present in offgases) are also far less (about 50 percent for dissolution and 63 percent for thermal destruction).

This study confirms the advantages of system 2 for large-capacity ozonation facilities. A similar study concerning an ozone output of 52.1 kg O_3/h (2,756 lb/day) shows that the cost of this process is still higher than for a system using air to produce ozone at a concentration of 27 g O_3/m^3 NTP (2.08 percent by weight) (10 percent extra investment costs) despite the choice of a system with recycling, which makes it possible to produce only half the maximum ozone capacity required (i.e., 26.05 kg O_3/h—about 1,378 lb/day) with the extra production being ensured by make up with liquid oxygen. Linde Germany (1982) considers that this recycling process becomes economically attractive for ozonation capacities above 100 kg O_3/h (about 5,291 lb/day).

The estimated power consumption is shown below for a 216–kg O_3/h (about 11,429-lb/day) facility:

- System 1 (using air as a carrier gas): The total electrical consumption including treatment of offgases is estimated at 20.5 kW/kg O_3 (9.3 kW/lb O_3).
- System 2 (ozone produced from oxygen with recycling): The total power consumption including treatment of offgases is 16.35 kW/kg O_3 (7.42 kW/lb O_3). From this amount 10.78 kW/kg O_3 (4.9 kW/lb O_3) corresponds to the generator consumption and 4.5 kW/kg O_3 (2.0 kW/lb O_3) is for oxygen production and recycling.

Because this is a new process, information is lacking for maintenance costs, but it is probable that they will be significantly higher than for a facility using air due to the relative complexity of the technology.

In all projects where this type of plant with recycling is to be installed, one has to be sure that the concentration of ozone in the gas fed to the contact columns is appropriate for obtaining a high enough gas flow rate to ensure homogeneous treatment of the water for operating conditions corresponding to the minimum water flow rate and ozone dosage (see sec. D.4, chapter IV).

Cost of retrofitted plants.

Proposed facility for a water treatment plant in the Far East. This facility was proposed as part of a plan to retrofit a water treatment plant in Asia. It includes a preozonation step for the purpose of oxidizing iron, manganese, and detergents contained in the raw water, plus an intermediate ozonation stage, located between the settling tanks and the sand filters. This second ozonation stage is designed to complete the oxidation of the detergents and increase the biodegradability of the organics.

The influent flow rate varies between 2,270 m^3/h (14.4 mgd) and 3,315 m^3/h (21 mgd), for an average production of 2,840 m^3/h (18 mgd). The ozone dosage is between 1 and 1.5 mg O_3/L for preozonation and from 1.5 to 2 mg O_3/L at the intermediate ozonation stage, i.e., a maximum overall rate of 3.5 mg O_3/L. The total capacity of the ozonation facility was designed for a maximum dosage at maximum flow rate, i.e., 11.6 kg O_3/h (about 614 lb/day).

The plant was designed for local conditions, namely an ambient atmospheric temperature of 36°C (96.8°F), a moisture content of 95 percent and a maximum cooling water temperature of 31°C (83.8°F). Four of the various systems studied that were capable of achieving the above objective are described below.

- System 1: Ozone production from air, including a low pressure feed air preparation system, and two 372–dielectric tube ozone generators running on low frequency (50 Hz); preozonation without servo control; intermediate ozonation controlled by the influent flow rate with reference to a set dose rate; and feedback modulated by the concentration of dissolved ozone at the outlet of the contact column. The power applied to each ozone generator can be varied in successive steps. The resulting concentration of ozone in the air is therefore between 7 and

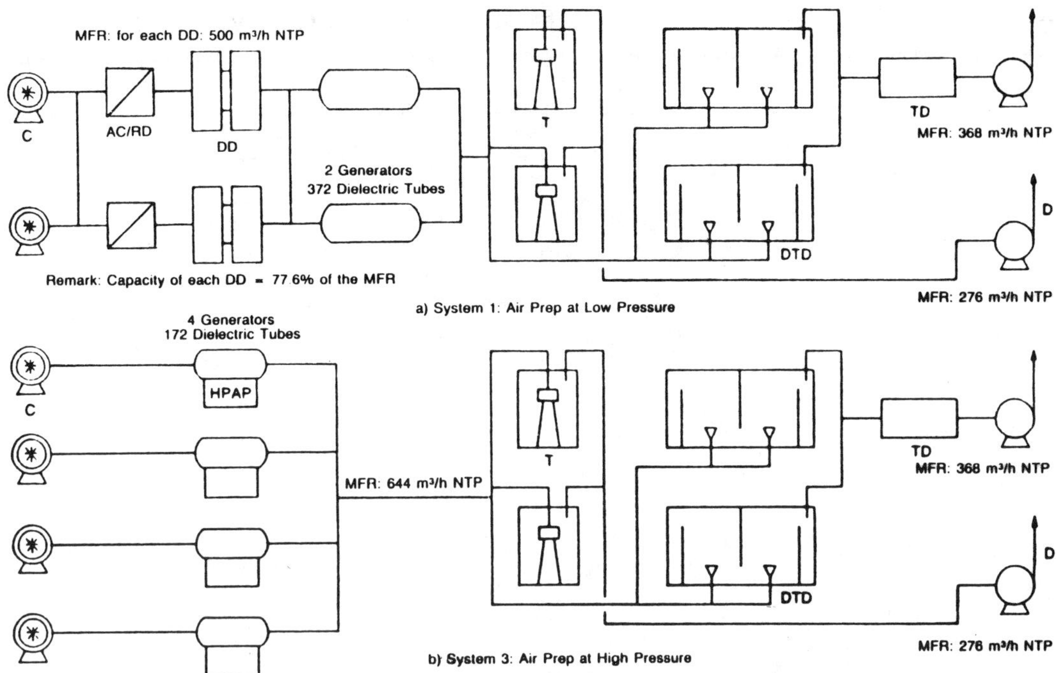

Project proposed in Asia.

Figure VI–22 Schematic of Proposed Systems for an Ozonation Facility Using Air as Feed Gas (Capacity: 11.6 kg O_3/h [about 614 lb O_3/day])

23 g of O_3/m^3 NTP (0.54 to 1.77 percent by weight), with an optimum concentration figure of 18 g O_3/m^3 NTP (1.38 percent by weight).

- System 2: Identical to system 1 above except that it is entirely manually operated.
- System 3: Ozone production from air with a high-pressure gas preparation system and four ozone generators working on low frequency (50 Hz) with 172–dielectric tube units. Ozone generation regulated identically to system 1.
- System 4: Identical to system 3, except that it is entirely manually operated.

At the outlet of the preozonation column, the residual ozone in the offgases is diluted before being discharged to the atmosphere. Offgas from intermediate ozonation undergoes thermal destruction with heat recovery.

In the preozonation step, ozone is diffused by pressurized turbines. In the second ozonation stage, this is achieved by disc-type diffusers.

Figure VI–22 is a flow diagram of systems 1 and 3 (or 2 and 4).

Table VI–17 is a breakdown of capital costs for the four systems described. Table VI–18 gives a breakdown of the capital cost for the five headings detailed in Table VI–17, plus spare parts, as a percentage of total equipment cost and turnkey delivery cost (excluding construction).

Some important observations can be drawn from these last two tables:

- The capital cost of the low-pressure system is greater than that of a high-pressure one (an extra 13.7 percent in the case of servo control and 16.8 percent for a manually controlled plant). Moreover, the cost of transport, erection, and commissioning, which in both cases accounts for 10 to 13 percent of the total turnkey cost, is significantly higher for the low-pressure option (from 48.6 percent to 64.3 percent, depending on whether manual or automated control is used).
- Although relatively simple for this kind of plant, automation increases the cost by 10 to 12 percent whether a high- or a low-pressure system is chosen.

Table VI–17 Capital Costs* of Four Systems for an Ozonation Facility† Proposed in Asia

Equipment	Capital Cost: System 1 FrF	System 1 $	System 2 FrF	System 2 $	System 3‡ FrF	System 3‡ $	System 4‡ FrF	System 4‡ $
Air preparation train	1,430,000	208,455	1,430,000	208,455				
Ozone generators	1,780,000	259,475	1,780,000	259,475	2,750,000	400,875	2,750,000	400,875
Dissolution	197,000	28,717	197,000	28,717	197,000	28,717	197,000	28,717
Destruction	340,000	49,563	340,000	49,563	340,000	49,563	340,000	49,563
Valves, fittings, measurement, and control (if any)	1,156,000	168,513	600,000	87,464	1,004,000	146,356	410,000	59,767
Spare parts package for two years	118,000	17,201	118,000	17,201	125,000	18,222	125,000	18,222
Total cost of equipment	5,021,000	731,924	4,465,000	650,875	4,416,000	643,732	3,822,000	557,143
Transport, erection, and commissioning	780,000	113,703	690,000	100,583	525,000	76,530	420,000	61,224
Turnkey cost of facility (construction excluded)	5,801,000	845,627	5,155,000	751,458	4,941,000	720,622	4,242,000	618,367

*1989 prices. Exchange rate: $1 = 6.86 FrF.
†11.6 kg O_3/h (about 61 lb/day).
‡Systems 3 and 4: the high pressure feed air preparation system is built on the same underframe as the ozone generator (compact unit). The cost corresponds to the complete unit.

Table VI–18 Capital Costs* of Four Systems for an Ozonation Facility† Compared with the Total Cost of the Equipment or with the Turnkey Cost of the Installation (Construction Excluded); Plant Proposed in Asia

	System 1 FrF‡	System 1 $	System 2 FrF	System 2 $	System 3§ FrF	System 3§ $	System 4§ FrF	System 4§ $
Total cost of equipment:	5,021,000	731,924	4,465,000	650,875	4,416,000	643,732	3,822,000	557,143
Turnkey cost of facility (construction excluded):	5,801,000	845,627	5,155,000	751,458	4,941,000	720,262	4,242,000	618,367
Cost of unit considered as a percentage:	Percent with Regard to the Equipment Cost	Percent with Regard to the Turnkey Cost	Percent with Regard to the Equipment Cost	Percent with Regard to the Turnkey Cost	Percent with Regard to the Equipment Cost	Percent with Regard to the Turnkey Cost	Percent with Regard to the Equipment Cost	Percent with Regard to the Turnkey Cost
Air preparation train	28.5	24.6	32.0	27.7				
Ozone generators	35.4	30.7	40.0	34.6	62.3	55.7	72.0	64.9
Dissolution	4.0	3.4	4.4	3.8	4.5	4.0	5.1	4.6
Destruction	6.8	5.9	7.6	6.6	7.7	6.9	8.9	8.0
Valves, fittings, measurement and control	23.0	19.9	13.4	11.6	22.7	20.3	10.7	9.7
Spare parts package for two years	2.3	2.0	2.6	2.3	2.8	2.5	3.3	2.9
Transport, erection, and commissioning	–	13.5	–	13.4	–	10.6	–	9.9

*1989 prices.
†Nominal capacity: 11.6 kg O_3/h (about 614 lb/day).
‡Exchange: $1 = 6.86 FrF.
§Systems 3 and 4: the high-pressure feed air preparation system is built on the same underframe as the ozone generator (compact unit). The percentage of the cost corresponds to the complete unit.

Table VI–19 Capital Costs* for the Haworth Plant, Hackensack, New Jersey†

Equipment	Capital Cost FrF‡	Capital Cost $	Percent of Total Equipment Cost
Air preparation line (4)	5,145,000	750,000	18
Generators (4)	3,944,500	575,000	13
Turbines (4)	3,773,000	550,000	13
Offgas destructors (4)	754,600	110,000	3
Instrumentation/controls	1,955,100	285,000	7
Equipment installation	13,720,000	2,000,000	46
Total cost	29,292,200	4,270,000	100

Source: Metropolitan Water District of Southern California (1988b).
*1988 prices, construction excluded.
†51 kg O_3/h (2700 lb/day).
‡Exchange: $1 = 6.86 FrF.

The cost of dissolution equipment and ozonated air destructors represents 10.8 to 14 percent of the equipment costs, depending on the system under consideration.

Haworth plant, Hackensack, New Jersey. The Haworth plant is owned by the Hackensack Water Company. This plant has been expanded from 50 to 220 mgd (about 7,890 to 34,700 m^3/h) with an ozonation step added as a retrofit during expansion. The new treatment process was described in sec. B.2, chapter III. The ozone capacity is 2,700 lb/day (51 kg O_3/h), produced from air at low frequency (60 Hz). Four generators supply four treatment trains of 55 mgd each (about 8,675 m^3/h). The four contactors are square, with a total depth of 26 ft (7.92 m). After treatment with alum and polymer, the raw water is forced through the center of the contactor and the turbine. The water flows down through the middle as the bubbles rise, then flows under a baffle and upward to exit from the contactor (See Figure IV–55, sec D.2, chapter IV). The ozone from the vents is catalytically destroyed. The costs are given in Table VI–19 (MWDSC 1988b). The costs of the buildings and the contactors are not included since they came under a separate contract.

At 46 percent, the installation of equipment is a much greater percentage of the total cost of the ozonation plant (other construction excluded). This large value is due to the dissolution system (Fring's turbine) and also to the fact that fitting the ozonation step into the existing plant was not easy. It is interesting to note that in the same report (MWDSC 1988b), the installation percentage for another retrofitted plant in the United States (Ammiss Plant, Shreveport, Louisiana; 90 mgd; 3,750 lb O_3/day [about 14,195 m^3/h; 70.9 kg O_3/h]) is very similar (49 percent).

Proposed retrofit of a plant in France. This facility was proposed as part of a plan to modernize a water works. The treatment process includes a preozonation stage and a postozonation stage for disinfection. The description of two of the options submitted will allow the reader to compare two medium-frequency (400-Hz) ozone facilities, using air feed gas and operating at concentrations of 18 g O_3 /m^3 NTP (1.39 percent by weight) and 27 g O_3/m^3 NTP (2.08 percent by weight).

- System 1: Ozone is produced from air at a concentration of 18 g O_3/m^3 NTP (1.39 percent by weight). This concentration is maintained at a constant level at all rates of production, i.e., 8.4 to 78.2 kg O_3/h (about 444 to 4,135 lb/day).
- System 2: The same basic system is applied as for system 1 but the facility runs at a constant ozone concentration of 27 g O_3/m^3 NTP (2.08 percent by weight).

The preozonation step is manually operated. Air is diffused via immersed pressurized turbines at variable speed. Dissolution in the postozonation contactors is performed by disc-type porous diffusers. For system 1, where higher gas flow rates are handled, two independent diffuser feed racks are provided to make sure that dissolution is

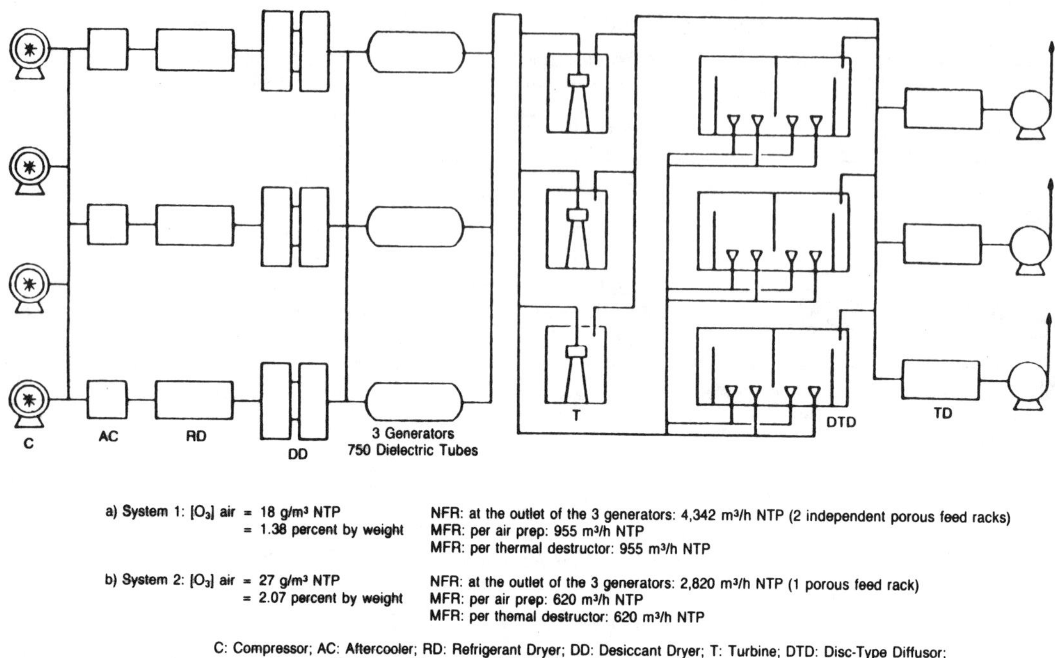

Plant proposed in France.

Figure VI–23 Diagram of Proposed Installation of 78.2 kg O_3/h (about 4,135 lb O_3/day): Ozone Production from Air at Two Different Concentrations

satisfactory whatever the rate of injection. The second rack is supplied only when the flow of ozonized gas is high. For the postozonation stage, the ozone dosage is adjusted to the water flow and to a set treatment rate, with feedback regulation by the dissolved residual ozone measured at the outlet of the contactor. The pre- and postozonation offgases are collected and transferred to the thermal destructors with heat recovery.

The feed gas preparation lines and ozone generators were designed to produce the maximum ozone demand (78.2 kg O_3/h [4,135 lb/day]) with cooling water at 22°C (71.6°F). Figure VI–23 shows the flow charts of the two proposed systems.

The air preparation trains are interchangeable. At the outlet of the ozone generators, the ozonized air is directed to the same collector, where it is distributed to the three preozonation and postozonation contactors.

Table VI–20 gives the capital cost breakdown for the two systems, in terms of total equipment cost or turnkey cost with other construction excluded. Table VI–21 shows the percentage of cost attributable to each of the two units. The ozone concentration in air influences the cost of the feed gas preparation train. By choosing a concentration of 27 g O_3/m^3 NTP (2.08 percent by weight), a 25 percent reduction in the cost of the preparation equipment can be achieved. Nevertheless, one needs to note that the contact chamber and the dissolution system will have to be designed for obtaining good mixing between the water and the ozonated air at an ozone concentration of 27 g O_3/m^3 NTP (see sec. D.4, chapter IV). Likewise, the choice of concentration affects the cost of ozone generators, destructors, and valving (respectively 3.3 percent, 18 percent, and 7.9 percent less expensive than in a solution with an ozone concentration of 18 g O_3/m^3 NTP [1.39 percent by weight]). This trend is even more prevalent with dissolution equipment (difference in cost is 34.4 percent). In this particular case, because of the wide range of ozone production rates, system 1 involves the installation of a double dissolution rack in order to ensure good transfer conditions. The total cost difference between the two systems is 12 percent.

Table VI–20 Capital Cost* of Equipment and Turnkey Plant for Two Systems for an Ozone Facility† (Plant Proposed in France)

	Capital Cost			
	System 1: O_3 Production from Air = 18 g/m^3 NTP (1.39 % by wt)		System 2: O_3 Production from Air = 27 g/m^3 NTP (2.08 % by wt)	
Equipment	FrF‡	$	FrF	$
Air preparation train	4,703,630	685,660	3,531,730	514,829
Power supply unit§ and ozone generators	8,902,720	1,297,772	8,604,590	1,254,313
Valving, measuring, and regulation	4,749,490	692,345	4,373,390	637,520
Dissolution	1,013,650	147,762	665,070	96,949
Thermal destruction	2,155,730	314,246	1,758,990	256,412
Spare parts package**	236,210	34,433	236,210	34,433
Total equipment cost	21,761,430	3,172,220	19,169,980	2,794,458
Transport, erection, and commissioning	2,983,170	434,864	2,751,540	401,099
Turnkey cost of equipment (construction excluded)	24,744,600	3,607,084	21,921,520	3,195,557

*1989 Prices.
†78.15 kg O_3/h (4,135 lb/day).
‡Exchange: $1 = 6.86 FrF.
§The cost of the power supply unit, the control panel, and the generators also includes the cost of the motor control center plus the variable speed control for the compressors.
**The contract specifications stipulate the delivery of spare parts only for the programmable logic controller for the facility's motorized devices.

Table VI–21 Costs* for Two Systems for an Ozone Facility† (Plant Proposed in France)

	System 1: O_3—18 g/m^3 NTP (1.39 % by wt)		System 2: O_3—27 g/m^3 NTP (2.08 % by wt)	
	FrF‡	$	FrF	$
Investment costs of equipment	21,761,430	3,172,220	19,169,980	2,794,460
Turnkey cost of facility (construction excluded)	24,744,600	3,607,090	21,921,520	3,195,560
Cost of unit considered as a percentage	Percent with Regard to the Equipment Cost	Percent with Regard to the Turnkey Cost	Percent with Regard to the Equipment Cost	Percent with Regard to the Turnkey Cost
Air preparation train	21.6	19.0	18.4	16.1
Power supply unit§ and ozone generators	40.9	36.0	44.9	39.3
Valves, fittings, measurement, and regulation	21.8	19.2	22.8	19.9
Dissolution	4.7	4.1	3.5	3.0
Ozone destruction	9.9	8.7	9.2	8.0
Spare parts**	1.1	0.9	1.2	1.1
Transport, erection, and commissioning	–	12.1	–	12.6

*1989 Prices.
†78.15 kg O_3/h (4135 lb/day).
‡Exchange: $1 = 6.86 FrF.
§The cost of the power supply unit, the control panel, and the generators also includes the cost of the motor control center plus the variable speed control for the compressor.
**The contract specifications stipulate the delivery of spare parts only for the programmable logic controller for the facility's motorized devices.

Table VI–22 Ozonation Cost* Compared to the Total Cost of Water Production at Tailfer, Brussels, Belgium

	Centimes/m^3	Cents/1000 gal	Percent of the Total Cost
Total cost of water, production (94,900,000 m^3/year)	46.95	25.91	100
Cost of ozone production	0.56	0.31	1.2
Cost of ozone dissolution	0.42	0.23	0.9
Cost of maintenance	0.17	0.09	0.4
Cost of spare parts	0.07	0.04	0.1
Cost of buildings and equipment amortization	2.53	1.38	5.4
Ozonation cost of water	3.75	2.05	8.0

Source: Miller et al. (1978).
*1976 price. Exchange: 1 Belgian Franc = 0.16 FrF = $0.023.

VI.E.2 Case Studies of Operating Cost

The Tailfer water treatment plant (Brussels, Belgium) (ozone capacity; 24 kg O_3/h—about 1,267 lb/day). This plant has been described in sec. E.1 of this chapter, and operating costs were also presented by Miller et al. (1978).

The energy required for the ozone production varies between 17 and 22 kWh/kg O_3 produced (7.71 to 9.98 kWh/lb O_3). In 1976, the total energy cost of ozonation was 3,340,000 Belgian francs (534,400 FrF or $78,820) per year. To this amount it is necessary to add the energy consumed by the turbines diffusing the ozonated gas, which corresponds to 2,487,000 Belgian francs (397,920 FrF or $57,201) per year. These costs are based on an average treatment rate of 1.7 mg O_3/L and a total amount of treated water of 94,900,000 m^3/year (25,070 mg/year). The energy bill for production and dissolution of ozonated gas was 932,320 FrF or $136,021/year (i.e., 0.0098 FrF/m^3 of treated water or $0.0054/1000 gal).

The labor cost for maintenance was $22,992/year (157,725 FrF). This cost was based on the twice-yearly cleaning of the ozone generators. The price of spare parts, mostly the tubes, which are systematically replaced every two years (something very unusual) amounts to about $9,391/year (64,422 FrF/year). The maintenance bill for the ozonation plant was $32,383/year or 222,147 FrF/year (i.e., 0.0023 FrF/m^3 of treated water or $0.0013/1000 gal). The total operating cost of the ozonation plant was $168,404/year or 1,154,467 FrF/year (i.e., 1.22 centimes/m^3 or 0.67 cents/1000 gal).

In amortization of buildings and equipment over a 20-year period at an annual fixed cost of 15,000,000 Belgian francs (i.e., 2,400,000 FrF or $345,000), it is necessary to add 2.53 centimes/m^3 or 1.38 cents/1000 gal to the price of ozonating the water.

The total cost of ozonation in this plant was 2.05 cents/1,000 gallons or 3.75 centimes/m^3 (i.e., 8 percent of the water production cost) in 1976. Table VI–22 shows the breakdown of the ozonation cost with regard to the total cost of water production.

In 1989, for an electricity price of 1.4 Belgian francs/kWh (i.e., 22.4 centimes or 3.2 cents/kWh) which is a special price valid for the water treatment plant capable of producing its own electricity (if necessary) and an oxygen cost of 7.25 Belgian francs/kg O_2 (i.e., 1.16 FrF or 16.7 cents/kg O_2), Masschelein (1989) verified that for an ozone production equal to 75 percent of the nominal capacity, the production cost of one kilogram of ozone is 60 Belgian francs (i.e., 9.6 FrF or $1.38) if the plant is operated with air, and 130 Belgian francs (i.e., 20.8 FrF or $2.99) if the plant is operated with oxygen.

This is due to the fact that the ozone generators have been designed basically to operate on air. This difference is partially balanced by a savings in amortization, considering the extra investment that could be necessary for a plant of the same maximum ozone capacity operated from air. These calculated amortization costs, based on the

Table VI–23 Operating Conditions for Estimating Power Consumption at an Ozonation Facility* (Plant Proposed in France)

Month	Water Temperature °C	Water Temperature °F	Number of Working Hours of the Facility	Flow Rates Treated per Time Segment m^3/h	Estimated Ozone Demand of the Water, mg/L	Ozone Production Required per Time Segment, kg	Monthly Ozone Production kg/month	Monthly Ozone Production lb/month†
January	5	41	79	15,000	1.8	27.0	2,133	4,702
			660	7,000		12.6	8,316	18,333
February	3	37.4	79	15,000	1.6	24.0	1,896	4,180
			580	9,000		14.4	8,352	18,413
March	8	46.4	78	15,000	1.7	25.5	1,989	4,385
			660	8,000		13.6	8,976	19,788
April	10	50	78	15,000	1.9	28.5	2,223	4,901
			630	9,000		17.1	10,773	23,750
May	19	66.2	79	15,000	2.4	36.0	2,844	6,270
			660	9,000		21.6	14,256	31,429
June	22	71.6	78	15,000	2.3	34.5	2,691	5,933
			630	9,000		20.7	13,041	28,750
July	23	73.4	79	15,000	2.4	36.0	2,844	6,270
			660	7,000		16.8	11,088	24,444
August	23	73.4	79	15,000	2.4	36.0	2,844	6,270
			660	9,000		21.6	14,256	31,428
September	18	64.4	79	15,000	2.2	33.0	2,607	5,747
			630	11,000		24.2	15,246	33,611
October	14	57.2	79	15,000	2.4	36.0	2,844	6,270
			660	11,000		26.4	17,424	38,412
November	8	46.4	78	15,000	1.9	28.5	2,223	4,901
			630	9,000		17.1	10,773	23,750
December	5	41	78	15,000	1.7	25.5	1,989	4,385
			660	5,000		8.5	5,610	12,368

*78.2 kg O_3/h (4,135 lb/day).
†Round capacity.

actual ozone production (75 percent of the capacity), are 126 Belgian francs (i.e., 20.2 FrF or $2.90) per kilogram of ozone for a plant operated from air, and 84 Belgian francs (i.e., 13.4 FrF or $1.93) per kilogram of ozone for the existing plant.

Operating cost for the proposed retrofit of a plant in France (capacity: 78.2 kg O_3/h—about 4,135 lb/day). This facility was described in sec. E.1 of this chapter. Although actual data are not available for this facility, the estimated energy consumption rates from the contract specifications have been calculated. Table VI–23 shows the operating conditions on which the estimate was based.

These conditions correspond to an annual production of ozone equal to 167,238 kg (about 368,697 lb) of ozone per year for an installed capacity of 684,594 kg (about 1,509,271 lb) per year, i.e., an effective operating rate of 24.4 percent, similar to that for the Ile de France Water Board plants (see below). Taking these operating data into account, the estimate of the energy cost should be calculated on the basis of electricity at the price of 0.30 FrF/kWh (4.37 cents/kWh) for a period of 10 years.

Figure VI–24 shows the guaranteed monthly rates of energy consumption based on the operating conditions adopted in the contract specifications. These rates are for the quantities of energy required to produce ozone, including air preparation and the destruction of offgas (the latter is a function of the electric furnace and the extractor). Based on this information, energy consumption rates have been calculated over 10 years (Table VI–24).

Operating cost for the three facilities of the Paris suburbs: Choisy-le-Roi, Neuilly-sur-Marne, and Méry-sur-Oise (France). These three plants, which

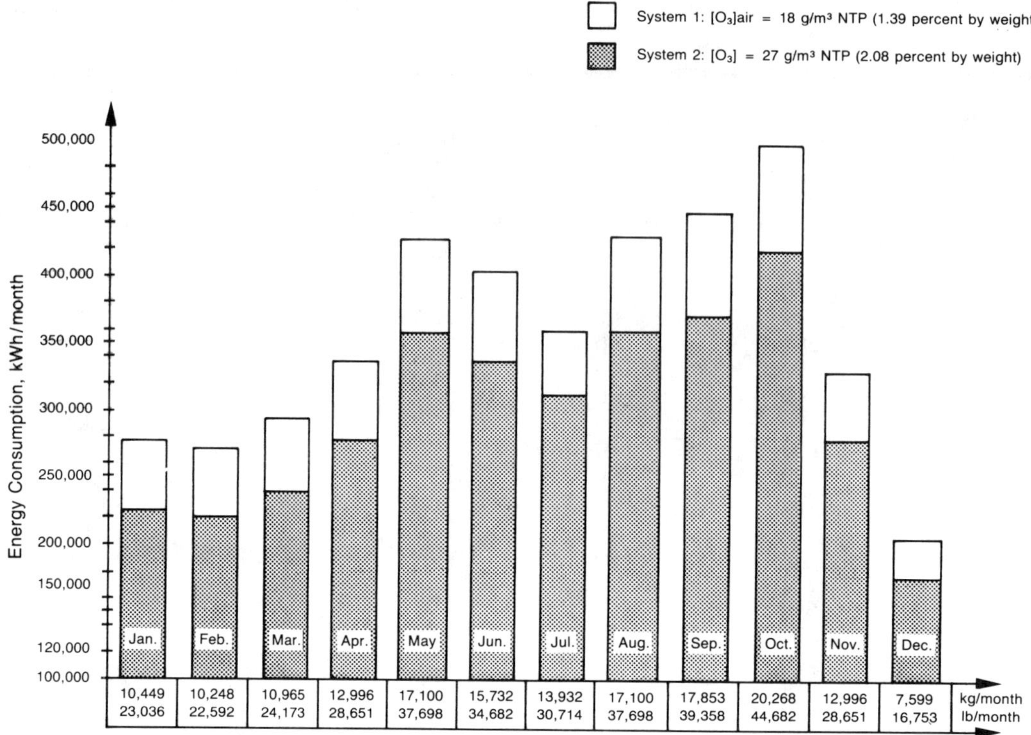

Plant proposed in France.

Figure VI–24 Monthly Energy Consumption Estimated for the Proposed Systems (Ozonation Capacity: 78.2 kg O_3/h—about 4,135 lb/day)

Table VI–24 Energy Consumption Rates Estimated over 10 Years and Corresponding Costs for Two Proposed Systems★ (Plant Proposed in France)

	System 1: Ozone Concentration in Air = 18 g/m^3 NTP (1.39 % by wt)		System 2: Ozone Concentration in Air = 27 g/m^3 NTP (2.08 % by wt)	
Aggregate power consumption over 10 years, kW	42,426,972		35,544,374	
Aggregate power cost† over 10 years, FrF ($)	12,728,092	(1,854,059)	10,663,312	(1,553,289)
Cost per unit in terms of ozone production capacity, FrF/g O_3/yr ($/lb O_3/yr)	0.0018	(0.12)	0.0015	(0.10)
Cost per unit in terms of used quantity of ozone, FrF/g O_3/yr ($/lb O_3/yr)	0.0076	(0.50)	0.0064	(0.42)
Cost per unit in terms of treated water flow rate‡ FrF/m^3/yr (¢/1000 gal/yr)	0.016	(0.87)	0.013	(0.73)

★Facility of 78.2 kg O_3/h (4135 lb/day).
†Price of kWh: 0.30 FrF = 4.37¢; exchange: $1 = 6.86 FrF.
‡Treated water flow rate over 10 years: 802,650,000 m^3.

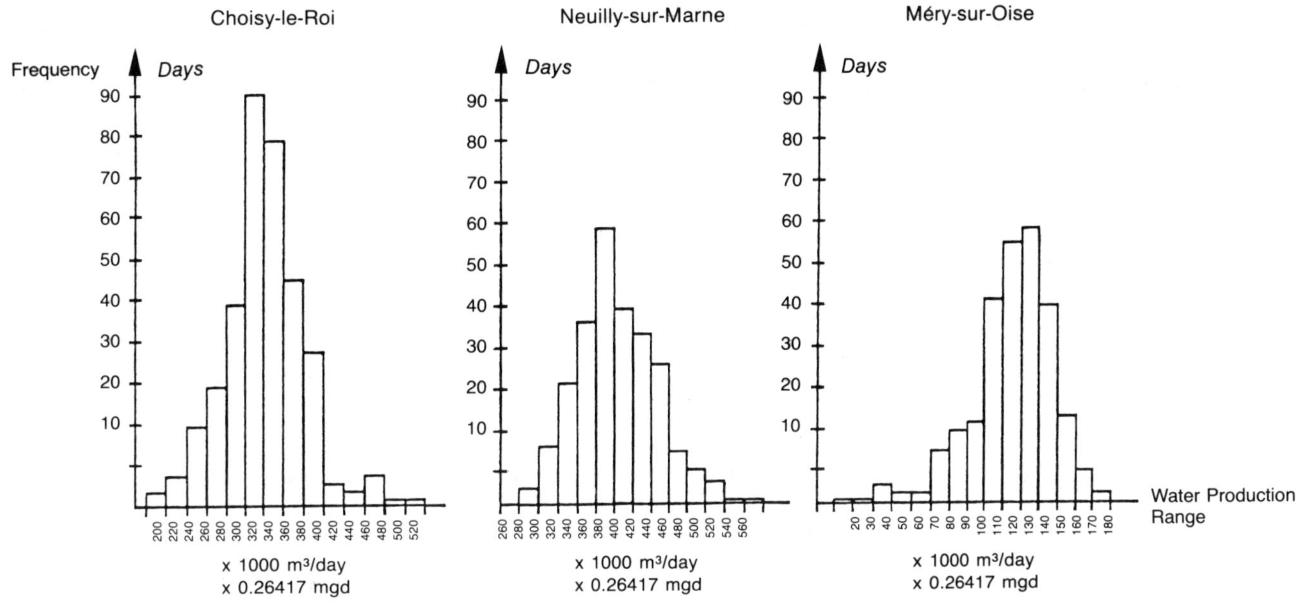

Source: Schulhof (1989).

Figure VI–25 Treated Water Daily Output Histograms of Three Plants Belonging to the Ile de France Water Board, Operated by Compagnie Générale des Eaux

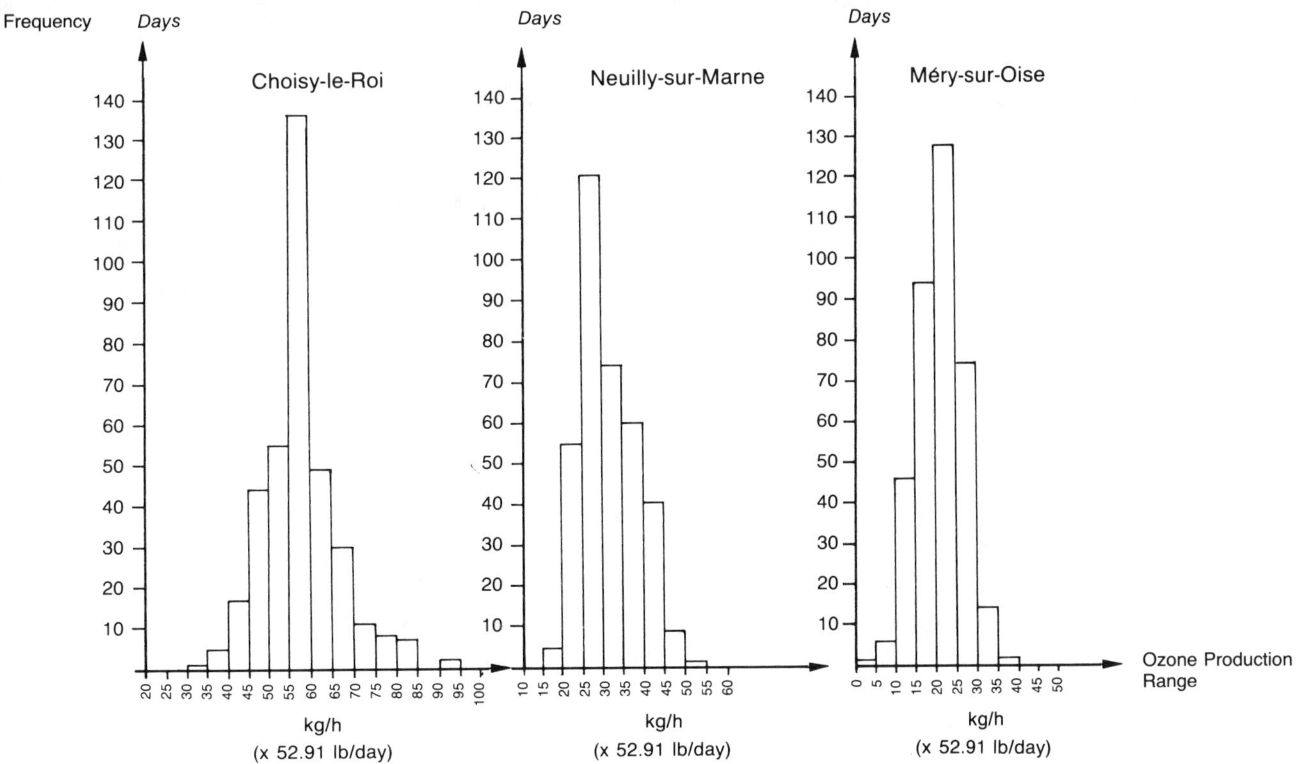

Source: Schulhof (1989).

Figure VI–26 Ozone Production Daily Output Histograms of Three Plants Belonging to the Ile de France Water Board, Operated by Compagnie Générale des Eaux

Table VI–25 Annual Operating Information for Ozonation at Three Plants Belonging to the Ile de France Water Board

	Drinking Water Treatment Plant							
	Choisy-le-Roi		Neuilly-sur-Marne		Mery-sur-Oise		Total	
Average annual flow rate, m^3/day (mgd)	337,152	(89.06)	402,375	(106.29)	122,714	(32.42)	862,241	(227.78)
Average annual ozone production, kg/year (lb/year)	497,370	(1,096,513)	274,615	(605,422)	186,433	(411,014)	958,418	(2,112,950)
Average annual power consumed, kWh	11,936,879		7,140,000		4,474,390		23,551,269	
Average annual cost of energy, FrF ($)*	4,094,349	(596,844)	2,449,020	(357,000)	1,534,716	(223,720)	8,078,085	(1,177,564)
Average annual cost of maintenance, FrF ($)	313,022	(45,630)	373,595	(54,460)	113,945	(16,610)	800,562	(116,700)
Total average annual cost of ozonation, FrF ($)	4,407,371	(642,474)	2,822,615	(411,460)	1,648,661	(240,330)	8,878,647	(1,294,264)
Ozonation cost, centimes/m^3 (cents/1000 gal)	3.58	(1.98)	1.92	(1.06)	3.68	(2.03)	2.82	(1.56)
Ozonation cost, centimes/g O_3 ($/lb O_3)	0.89	(0.586)	1.03	(0.680)	0.88	(0.585)	0.93	(0.612)

Source: Schulhof (1989).
*Exchange: $1 = 6.86 FrF.

Table VI–26 Total Cost of Ozonation (Operating Cost plus Amortization) for the Three Plants Belonging to the Ile de France Water Board

	Drinking Water Treatment Plant			
	Choisy-le-Roi	Neuilly-sur-Marne	Mery-sur-Oise	Total
Average ozone dosage, mg O_3/L	4.04	1.87	4.16	3.04
Operating ozonation cost, centimes/m^3	3.58	1.92	3.68	2.82
(cents/1000 gal)*	(1.98)	(1.06)	(2.03)	(1.56)
Amortization ozonation cost, centimes/m^3	7.11	3.29	7.32	5.35
(cents/1000 gal)	(3.92)	(1.81)	(4.04)	(2.96)
Total ozonation cost, centimes/m^3	10.69	5.21	11.00	8.17
(cents/1000 gal)	(5.9)	(2.87)	(6.07)	(4.52)

Source: Schulhof (1989).
*Exchange: $1 = 6.86 FrF.

belong to the Ile de France Water Board, have been equipped with ozone generators for the last 20 years and are operated by the Compagnie Générale des Eaux. They are describedinchaptersIII.F.2(Choisy-le-Roi),andIII.H.2(Neuilly-sur-MarneandMéry-sur-Oise). Figure VI–25 presents the histograms of treated water daily outputs and shows that the water works often operate very far from their maximum capacity (Schulhof 1989).

In the same way, Figure VI–26 shows the associated ozone output, also far below the maximum capacity.

Considering that the maximum ozone production capacity is 350 kg O_3/h (about 18, 518 lb O_3/day), the average production actually used in the course of one year is only 958,418 kg O_3 (2,112,950 lb O_3) i.e., 31.2 percent of the maximum capacity (3,066,000 kg [6,759,373 lb] O_3). This is due to the fact that the ozonation plants have been designed to cope with the accidental pollution episodes that occur on the Seine, Marne, and Oise Rivers, with ozone being considered a "crisis reagent." Table VI–25 gives the ozonation costs for these three plants, taking into account the actual amounts of ozone produced and the treated water flows.

Actual amortization costs are difficult to present. The ozone generators were commissioned at different times between 1960 and 1982. Nevertheless, it is possible to calculate the repaired value of the 32 ozone generators and their appurtenances, and then to amortize them over a period of 20 years at 7 percent per year. This average rate of amortization is 0.549 centimes/installed gram of ozone (36.30 cents/installed lb of ozone), excluding other construction. Assuming that the utilization factor is 31.2 percent, the average amortization cost is 1.76 centimes/used gram of ozone or $1.16/used lb of ozone. Table VI–26 gives the total cost of ozonation, taking these different considerations into account.

Excluding the amortization of construction, the average total ozonation cost at the Ile de France water works is 8.17 centimes/m^3 or 4.52 cents/1000 gal. If the amortization cost per gram or pound of installed ozone is considered, the average total ozonation cost decreases to 4.49 centimes/m^3 or 2.48 cents/1000 gal.

References

BABLON, G. ET AL. 1986. Mise au Point d'un Modèle Mathématique pour Contrôler le Traitement de Préozonation/Coagulation. *Water Supply*, 4:117–127.

BABLON, G. & VAN ELSLAND, E.P. 1989. Compagnie Générale des Eaux. Personal communication.

BENKMANN, C. 1983. Druckwechsel-Adsorptionsverfahren und seine Anwendung bei der Herstellung organischer Saüren. European Patent 80102162.7.

BEYSEL, G. 1989. Air Separation Plant using Cryogenic, PSA or Membrane Technology: Development and Future Application. Presented at the Symposium PSA 3, International Institute of Refrigeration, Paris, France, October 24–25.

CHAPSAL, P. 1989a. "Technologie des Générateurs d'Ozone Trailigaz." In *L'Ozonation des Eaux. Manuel Pratique,* edited by W.J. Masschelein (in press). Technique et Documentation, Paris, France.

_____. 1989b. "Destruction de l'Ozone dans l'Air à l'Évent." In *L'Ozonation des Eaux. Manuel Pratique,* edited by W.J. Masschelein (in press). Technique et Documentation, Paris, France.

ELSENHANS, K.H. 1989. Experience with the Maintenance of Ozone Systems. Proc. 9th Ozone World Congress, New York, edited by L.J. Bollyky, pp. 697–712 (vol. 2).

GEORGESON, D.L. & KARIMI, A.A. 1987. Preozonation Effects on THMs and Turbidity at Los Angeles Water Treatment Plant. Proc. 8th Ozone World Congress, Zürich, Switzerland, pp. B28–B58 (vol. 1).

GERVAL, R. ET AL. 1989. State of the Art in the Design of Ozone Contact Columns. Proc. 9th Ozone World Congress, New York, edited by L.J. Bollyky, pp. 486–495 (vol. 2).

GUMERMAN, R.C. ET AL. 1986. *Small Water System Treatment Costs.* Noyes Data Corp., Park Ridge, N.J.

HODGES, W.E. 1988. Los Angeles Aqueduct Filtration Plant: First-Year Performance. Proc. AWWA Ann. Conf., Orlando, Fla., pp. 843–860 (part 1).

HORST, M. 1982. "Removal of the Residual Ozone in Air after the Application of Ozone." In *Ozonation Manual for Water and Wastewater Treatment,* edited by W.J. Masschelein. John Wiley and Sons, Inc., New York.

HYDROVANE. 1986. Technical Leaflet: Dossier Technique Hydraulique-Pneumatique.

LABROUSSE, M. 1989. Non-cryogenic Oxygen Production—Largest Plant in Europe. Future of This Type of Production. Presented at the Symposium PSA 3, International Institute of Refrigeration, Paris, France, October 24–25.

LEITGEB, P. 1989. Key Process Parameters for Optimizing Oxygen-VSA Plants. Presented at the Symposium PSA 3, International Institute of Refrigeration, Paris, France, October 24–25.

LINDE GERMANY. 1982. Technical Leaflet: Unités Standard de Séparation d'Air.

MASSCHELEIN, W.J. 1977. Techniques de Dispersion et de Diffusion d'Ozone dans l'Eau. Presented at Wasser Berlin 77, Colloquium Verlag O.H. Hess, Berlin, Germany.

_____. 1980a. L'utilisation d'un Gaz de Processus Enrichi à l'Oxygène. In *L'Ozonation des Eaux. Manuel Pratique.* edited by W.J. Masschelein. Technique et Documentation, Paris, France, pp. 13–16.

_____. 1980b. Filière de l'Ozonation à l'Usine de Tailfer. In *L'Ozonation des Eaux. Manuel*

Pratique, edited by W.J. Masschelein. Technique et Documentation, Paris, France, pp. 214–215.

_____. 1989. Personal communication.

MERZ, E. & GAIA, F. 1989. Comparison of Economics of Various Ozone Generation Systems. Proc. 9th Ozone World Congress, New York, edited by L.J. Bollyky, pp. 69–93 (vol. 2).

Metropolitan Water District of Southern California (MWDSC). 1988a. Oxidation Demonstration Project Scoping Study. Equipment Selection Report. Appendix B. Document prepared by James M. Montgomery Consulting Engineers, Inc.

_____. 1988b. Oxidation Demonstration Project Scoping Study. Ozone/Peroxone Technology Status Report. Section 3. Document prepared by James M. Montgomery Consulting Engineers, Inc.

MILLER, G.W. ET AL. 1978. An Assessment of Ozone and Chlorine Dioxide Technologies for Treatment of Municipal Water Supplies. U.S. EPA 600/2-78-147, section 9, pp. 145–166.

PAILLARD, H. 1986. Désinfection par les Rayonnements Ultraviolet. Anjou Recherche. Internal document.

PASCAL, O. 1984. Chemometric Control for a Coupled Preozonation/Coagulation Process. The Example of Choisy-le-Roi Water Works: 800,000 m^3/day. Presented at the International Ozone Association Workshop, Montreal, Canada.

PROCESS APPLICATIONS INC. (PAI) 1987. Utilizing Ozone: Design and Generating Requirements. Seminar presented to James M. Montgomery Consulting Engineers, Inc. Pasadena, Cal.

RACZKO, R.F. ET AL. 1990. Cost of Existing Ozone Installations. Presented at the Sunday Seminar: Practical Experiences with Ozone for Organics Control and Disinfection. AWWA Ann. Conf., Cincinnati, Ohio, June 17–21.

RAKNESS, K.L. 1990. Personal communication.

RAKNESS, K.L. ET AL. 1989. Design Considerations in Alternative Ozone Feed Gas Systems. Proc. of the 9th Ozone World Congress, New York, edited by L.J. Bollyky, pp. 634–649 (vol. 2).

RICE, R.G. 1989. Ozone for Treatment of Drinking Water. EPA Contract No. 68.01.7289 with ICF Incorporated, Fairfax, VA (Work Assignment No. 88), Chapter VII.

SCHULHOF, P. 1989. The Price of Ozonation. Proc. 9th Ozone World Congress, New York, edited by L.J. Bollyky, pp. 37–47 (vol. 2).

TARDIEU, J.P. 1988. L'eau. L'impératif Sécurité. *P.C.M. Le Pont,* 86:11:39–42.

U.S. EPA. 1977. Economic Impact Analysis of a Trihalomethane Regulation for Drinking Water. NTIS PB-271 813/8SL.

_____. 1979. Estimating Water Treatment Costs, vol. 2: Cost Curves Applicable to 1-to-200 mgd Treatment Plants. 600/2-79:162b.

WABLE, O. ET AL. 1989. Computer-Aided Decision in Ozone Contactor Design and Operation: An Expert System. Proc. 9th Ozone World Congress, New York, edited by L.J. Bollyky, pp. 496–503 (vol. 2).

WARAKOMSKI, A. 1989. Ozone Generation Utilizing High-Purity Oxygen Feed Gas. Proc. 9th Ozone World Congress, New York, edited by L.J. Bollyky, pp. 596–633 (vol. 2).

WUNSCH, A.K. & DARPIN, C. 1989. The Cost-Effectiveness of Ozone Systems. Proc. 9th Ozone World Congress, New York, edited by L.J. Bollyky, pp. 48–68 (vol. 2).

XORBOX. 1989. The Xorbox Oxygen Generator. Technical Leaflet.

Appendix A

Ozone System Terminology, Measurements, and Conversions

Deborah R. Brink
Bruno Langlais
Kerwin L. Rakness

INTRODUCTION

Ozone science is moderately complex because it includes both gas and liquid flows. Also, gas is a compressible fluid, and its density, temperature, and pressure must be considered in associated calculations. If these parameters are not addressed, greater than 20 percent error could occur. Methods to account for variable gas characteristics are discussed in this appendix.

A significant amount of ozone literature is available from European and Asian countries that use metric units of expression. In the United States, a mixture of English and metric units is often employed. This book uses primarily metric units, followed by their English equivalents in parentheses. Metric-to-English conversion factors for the most common "ozone" terms are presented in Table A–1.

A block diagram of a typical ozone system is shown in Figure A–1. The liquid flow stream is displayed as a dotted line. The discharge residual ozone concentration (C) is determined in order to monitor system performance. Sometimes the residual concentration is used in the calculation of $C \cdot t$ (i.e., ozone residual concentration, in milligrams per liter, times liquid detention time, in minutes); however, the detention time representative of this residual is difficult to ascertain. Measurements of C at multiple locations within the contactor may be required to correctly identify the $C \cdot t$ number.

The ozone gas and liquid water mixes within the contact basin. The generator feed gas (G_1) must be particle-free and exceedingly dry; it may be air, oxygen, or oxygen-enriched air. The product gas (G_2) ozone concentration (Y_1) is a function of the oxygen content in the feed gas and class of ozone generator, and generally ranges from 1 to 2.5 percent by weight ozone in air and 3 to 10 percent by weight ozone in high-purity oxygen. In Figure A–1, system parameters that are commonly measured are shown in bold type.

The water flow (L) into the contact basin is measured in order to calculate ozone dosage, hydraulic detention time, transferred dosage, etc. The terminology, symbols, units of expression, and formulas for selected calculated ozone parameters are shown in Table A–2. Different conversion factors would be used if alternate units of expression were employed.

Table A–1 Useful Conversion Factors

Parameter	Metric Units	English Conversions
Flow	m^3/min	= cfm × 0.02832
		= mgd × 2.629
		= gpm × 0.003785
Temperature	°C	= (°F − 32) × 5 ÷ 9
		= K − 273.15
Absolute temperature	K	= °C + 273.15
Pressure	kPa	= psi × 6.895
		= atm × 101.325
		= ft of water × 2.989
		= in. of water × 0.249
		= in. of Hg × 3.386
		= mm of Hg × 0.133
Energy	kWh	= Wh × 1000
Power	kW	= W × 1000
		= HP × 0.7457
Specific energy	Wh/g	= kWh/lb × 2.205
		= kWh/kg × 1.0
Mass flow	g/min	= lb/day × 0.3145
		= lb/h × 7.56
Ozone concentration: approximate conversion factors		
Standard temperature = 20°C		
Air feed gas	g/m^3 or mg/L	= percent by weight × 12.1
Oxygen feed gas	g/m^3 or mg/L	= percent by weight × 13.6
Standard temperature = 0°C		
Air feed gas	g/m^3 or mg/L	= percent by weight × 13.0
Oxygen feed gas	g/m^3 or mg/L	= percent by weight × 14.6
Ozone concentration: precise conversion factor (see text and table A–3).		

STANDARD CONDITIONS

The feed gas flow (G_1) is customarily denoted as scfm (standard cubic feet per minute) or m^3/h NTP (normal temperature and pressure). The terms standard and normal refer to specific temperature and pressure conditions for the volumetric flow readings. The reference pressure is usually one atmosphere (i.e., 14.696 psi, 760 mm Hg or 101.325 kPa). The reference temperature is variable and is explained below.

"Normal" temperature is a metric expression and is 0°C. "Standard" temperature is an English expression and is usually 68°F, although in certain instances it may be 32°F, 60°F, 70°F, or other values. When evaluating data from vendor's literature or technical reports, the "standard" temperature should be noted. Also, all design documents must indicate the reference or "standard" temperature.

OZONE CONCENTRATION

Ozone concentration information is necessary for research, design, operation of ozone systems, and control of ozone generator performance. In this section, methods of measuring ozone concentration are discussed. Often, results from alternate measurement techniques are expressed differently, and conversions are necessary. However, conversion factors are variable because gas densities and "standard" conditions can vary. Precise and approximate conversion factors may be used. Each is presented herein.

Measuring Ozone Concentration

Ozone concentration is generally measured by a UV meter or by wet chemistry procedures (see sec. II.C). The meter is based on the principle of ultraviolet light

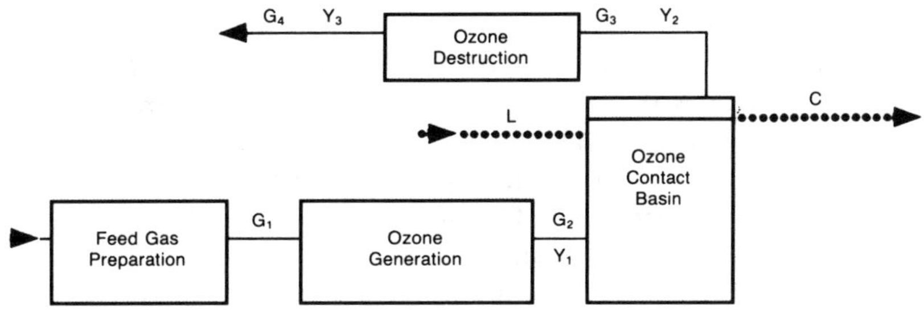

G_1 = Feed Gas Flow Rate .... **(m^3/h NTP or scfm)**
G_2 = Ozone Product Gas Flow Rate .... (m^3/h NTP or scfm)
G_3 = Offgas Flow Rate .... (m^3/h NTP or scfm)
G_4 = Exhaust Gas Flow Rate .... (m^3/h NTP or scfm)
Y_1 = Ozone Product Concentration .... **(g/m^3 NTP or percent by weight)**
Y_2 = Offgas Ozone Concentration .... **(g/m^3 NTP or percent by weight)**
Y_3 = Exhaust Gas Ozone Concentration .... **(g/m^3 NTP or ppm/vol)**
L = Water Flow Rate .... **(m^3/h or mgd)**
C = Contactor Effluent Residual Ozone Concentration .... **(mg/L or g/m^3) NTP**

NOTE: Parameters in bold are those that are typically measured.
NTP: Normal temperature and pressure — 273 K, 1013 mb.
scfm: American standard — temperature 20°C (68°F), pressure 1013 mb.
(Also see section entitled Standard Conditions in this appendix.)

Figure A–1 Block Diagram of a Typical Ozonation System

Table A–2 Terminology for Calculated Ozone Parameters

Parameter	Symbol	Units	Calculation Formula
Ozone production	P	kg/h	$G_2 \times Y_1 \div 1000$ G_2, m^3/h NTP Y_1, g/m^3 NTP
		lb/day	$G_1 \times Y_1 \times 52{,}911$ G_1, m^3/h NTP Y_1, g/m^3 NTP
Applied ozone dose	D	g/m^3	$P \div L \times 1000$ P, kg/h L, m^3/h
		mg/L	$P \div L \div 8.34$ P, lb/day L, mgd 8.34 = conversion of gal to lb (fluid is water)
Transfer efficiency Approximate	TE	percent	$(Y_1 - Y_2) \div Y_1 \times 100$
Transferred ozone dose	T	mg/L	$D \times TE \div 100$ TE in percent, not a ratio (i.e., 95.0, not 0.95)

absorbance by ozone. The product gas (Y_1) and offgas (Y_2) ozone concentration monitor's results are generally expressed in percent by weight. The exhaust gas (Y_3) and ambient UV monitors register the concentration as parts per million by volume.

Commercially available monitors utilize a reference gas as part of their measurement technique, which eliminates "drift" in the instrument's readings. The referencing technique enables the instrument to account for the gradual deterioration in lamp intensity and dirt accumulation on the observation windows. The monitors are generally accurate to within ±3 percent of the unbuffered KI (Birdsall et al. 1952) and European Committee of International Ozone Association NBKI (Masschelein 1988) wet chemistry techniques (Gordon et al. 1989; Rakness and Stolarik 1988). A wet chemistry procedure that uses a heavily buffered KI solution was shown to give significantly higher ozone concentration numbers than either a "calibrated" UV meter or the other wet chemistry procedures mentioned above (Gordon et al. 1989; Rakness et al. 1986). Refer to chapter II, sec. II.C, for more discussion of ozone measurements in the gas phase.

Commercially available UV monitors are constructed to exhibit an ozone concentration at a preset span number when the sample cell temperature and pressure are certain values, such as 25°C (77°F) and 760 mm Hg, respectively. The instruments that express the reading in percent by weight also are calibrated for a certain gas molecular weight, such as 29 g/mol for an air feed gas. The monitors allow the user to change the span number to account for different temperatures and pressures in the sample cell and for varying gas molecular weights. In order to display accurate ozone concentration readings from the monitor, the span number must be determined based on field conditions.

An inaccurate ozone concentration reading will be displayed when the span is incorrect. The error in the reading is proportional to the calibration temperature, pressure, and molecular weight. For example, a 1 percent error occurs for each 7.6 mm Hg pressure differential. The sample cell pressure for the product gas (Y_1) monitor is usually slightly greater than the ambient pressure (i.e., about 1 to 2 mm Hg), unless pressure losses are extremely high in the monitor exhaust piping. For precise ozone concentration numbers, the actual barometric pressure plus discharge pressure losses must be incorporated into the span calculation.

A negative pressure often exists in the offgas monitor's sample cell because the contact basin is usually operated under a partial vacuum. When the offgas meter is correctly installed and operating properly, only a slight negative pressure occurs (i.e., −10 to −20 mm Hg) because the offgas is drawn through the monitor with a "vacuum" blower. However, a significant negative pressure can exist (i.e., −100 to −300 mm Hg) if the meter is not working properly; this may not be correctable by a span adjustment. In order to insure that accurate offgas ozone concentration readings are being registered, *the gas pressure from the discharge of the offgas monitor should be measured and incorporated into the span calculation and adjustment.* The user should consult the UV monitor's operation and maintenance manual for the appropriate procedure to calculate the span setting and to make the span adjustments.

Changes in sample cell temperature can also dictate span adjustments. Sample cell temperature is difficult to measure directly, but it may be closely approximated by determining the temperature inside the monitor's cabinet. Temperature changes and associated span adjustments are dependent on the location of the instrument and ambient temperature conditions. Daily, weekly, monthly, or seasonal adjustments may be required. In some monitors, a thermocouple is installed to measure temperature and automatically correct the displayed ozone reading, provided the thermocouple is operating correctly. In that case, the monitor span would be adjusted only for pressure and gas molecular weight.

A span correction due to molecular weight is required if the monitor is originally set up to measure ozone in an air feed gas and oxygen is used, or vice versa. An adjustment can also be made to incorporate actual measurements for gas composition, such as for an oxygen-enriched air feed gas. It is noted that for monitors that express the result as parts per million by volume, such as the exhaust gas (Y_3) and ambient ozone concentration monitors, a molecular weight adjustment is unnecessary. For the volume-related display, the span would be adjusted only for the field temperature and pressure conditions.

Ozone Concentration Conversion Factors

The offgas (Y_2) and product gas ozone concentrations (Y_1) from the UV meters are generally expressed as percent by weight. Wet chemistry test results are denoted in the weight/volume term of milligrams per liter (mg/L NTP) or grams per cubic meter (g/m^3 NTP). (*Note:* 1 mg/L NTP = 1 g/m^3 NTP.) The precise conversion from mg/L to percent by weight is obtained using a fairly complex equation (explained below). An approximate conversion factor is shown in Table A–1.

Weight/volume to percent by weight. In the wet chemistry test result weight-to-volume concentration, the "volume" is associated with the product gas volume (G_2). It should be noted that this volume is less than the feed gas volume (G_1) because three oxygen molecules form two ozone molecules. The volume (L) of feed gas required per unit volume (m^3) of ozone product gas can be determined using Equation 1. It is noted that the standard temperature (i.e., molar volume) and actual ozone concentration are variables in the equation.

$$U_{1-2} = 1000 + (1/2 \times Y_1 \times V_m \div 48) \qquad (1)$$

Where:

U_{1-2} = Liters of feed gas per cubic meter of product gas
1000 = Nominal volume of feed gas required, L/m^3 NTP
$1/2 \times Y_1 \times V_m \div 48$ = Additional volume of feed gas needed, L/m^3 NTP
Y_1 = Ozone concentration, g/m^3 NTP
V_m = Molar volume at standard temperature, L/mol
48 = Gram molecular weight of ozone, g/mol.

The formula for calculating percent ozone by weight using a weight/volume ozone concentration (g/m^3 NTP) number is shown in Equation 2. It is noted that the density of the product gas is required, because volume (m^3 NTP) in the Y_1 expression is related to the product gas volume.

$$Y_1' = \frac{Y_1 \times 100}{W_{pg}} \qquad (2)$$

Where:

Y_1' = Ozone concentration, percent by weight
Y_1 = Ozone concentration, g/m^3 NTP
100 = Conversion of mass ratio to percent expression
W_{pg} = Density of product gas, g/m^3 NTP (see Equation 3).

The product gas density for Equation 2 can be calculated using Equation 3. Equation 3 enables feed gas data to be utilized to determine product gas information.

$$W_{pg} = U_{1-2} \times W_{fg} \qquad (3)$$

Table A–3 Conversion Factors for Ozone Concentration at Selected Conditions

Gram molecular weight gas	
Ozone	48 g/mol
Argon	39.948 g/mol
Oxygen	32 g/mol
Nitrogen	28 g/mol
Molar volume	0.08205 L/mol/K
Absolute temperature	273.15 K

Conditions

Feed gas makeup				
Oxygen	20.94% vol.	20.94% vol.	95.00% vol.	95.00% vol.
Nitrogen	78.12% vol.	78.12% vol.	1.00% vol.	1.00% vol.
Argon	0.94% vol.	0.94% vol.	4.00% vol.	4.00% vol.
Feed gas MW	28.95 g/mol	28.95 g/mol	32.28 g/mol	32.28 g/mol
Standard temperature	0°C	20°C	0°C	20°C
Molar volume	22.41 L/mol	24.05 L/mol	22.41 L/mol	24.05 L/mol
Feed gas density	1.292 g/L	1.204 g/L	1.440 g/L	1.342 g/L

Conversion Factors

O_3 Conc., (% wt)*	O_3 Conc., wt/vol (mg/L)	Unit Conv. (mg/L/%wt)	O_3 Conc., wt/vol. (mg/L)	Unit Conv. mg/L/%wt)	O_3 Conc., wt/vol. (mg/L)	Unit Conv. (mg/L/%wt)	O_3 Conc., wt/vol. (mg/L)	Unit Conv. (mg/L/%wt)
0.5%	6.47	12.94	6.03	12.06	7.21	14.42	6.72	13.44
1.0%	12.96	12.96	12.07	12.07	14.45	14.45	13.47	13.47
1.5%	19.47	12.98	18.14	12.09	21.72	14.48	20.24	13.49
2.0%	25.99	13.00	24.22	12.11	29.00	14.50	27.03	13.52
2.5%	32.54	13.02	30.32	12.13	36.32	14.53	33.84	13.54
3.0%	39.11	13.04	36.44	12.15	43.65	14.55	40.68	13.56
3.5%	45.70	13.06	42.58	12.17	51.02	14.58	47.54	13.58
4.0%	52.30	13.08	48.74	12.19	58.40	14.60	54.42	13.61
4.5%	58.93	13.10	54.91	12.20	65.82	14.63	61.33	13.63
5.0%	65.58	13.12	61.11	12.22	73.25	14.65	68.26	13.65
5.5%	72.25	13.14	67.32	12.24	80.72	14.68	75.21	13.67
6.0%	78.94	13.16	73.56	12.26	88.21	14.70	82.19	13.70
6.5%	85.65	13.18	79.81	12.28	95.72	14.73	89.19	13.72
7.0%	92.38	13.20	86.08	12.30	103.26	14.75	96.22	13.75
7.5%	99.13	13.22	92.37	12.32	110.83	14.78	103.27	13.77
8.0%	105.90	13.24	98.68	12.34	118.42	14.80	110.34	13.79
8.5%	112.69	13.26	105.01	12.35	126.04	14.83	117.44	13.82
9.0%	119.51	13.28	111.36	12.37	133.68	14.85	124.57	13.84
9.5%	126.34	13.30	117.73	12.39	141.36	14.88	131.72	13.87
10.0%	133.20	13.32	124.12	12.41	149.05	14.91	138.89	13.89

*%wt = percent by weight.

Where:

U_{1-2} = Liters of feed gas per cubic meter of product gas NTP
W_{fg} = Density of feed gas, g/L NTP (see Table A–3).

The formula for calculating percent by weight ozone concentration when the weight/volume (mg/L or g/m^3 under NTP conditions) concentration is known is obtained by combining Equations 1, 2, and 3, and is shown in Equation 4. Example results for selected conditions are presented in Table A–3. The unit conversion factor of mg/L NTP per weight percent is also shown. It should be noted that the unit conversions change for the different conditions, since they are dependent on the standard temperature and actual ozone concentration. By rearranging Equation 4, the reverse expression for converting weight percent to mg/L NTP is obtained and is shown in Equation 5.

$$Y_1' = \frac{Y_1 \times 100}{\{1000 + (1/2 \times Y_1 \times V_m \div 48)\} \times W_{fg}} \qquad (4)$$

$$Y_1 = \frac{1000 \times Y_1' \times W_{fg}}{100 - \{MW_{fg} \div (2 \times 48) \times Y_1'\}} \qquad (5)$$

Where:

Y_1' = Ozone concentration, percent by weight (expressed as percent and not as a ratio, i.e., 1.5 and not 0.015)
MW_{fg} = Molecular weight of feed gas, g/mol (see Table A–3).

Weight/volume to percent by volume. The percent by volume conversion factor is simpler than the percent weight conversion because it is not influenced by the composition of the feed gas. It is affected only by the standard temperature. The conversion factor at different standard temperatures is shown in Table A–4, and can be calculated using Equation 6.

$$Y_u = \frac{1000 \times 48}{V_m \times 100} \qquad (6)$$

Where:

Y_u = Unit ozone conversion, mg/L per percent volume
48 = Gram molecular weight of ozone, g/mol
1000 = Conversion from g to mg
V_m = Molar volume at standard temperature, L/mol
100 = Conversion for 100 percent volume concentration to 1 percent volume.

Note: Percent by volume is expressed as a percent and not as a ratio (e.g., 1.612 % vol., not 0.01612).

Table A–4 Volume Conversion for Ozone Concentration at Selected Temperatures

MW ozone: 48 g/mol Molar volume: 0.08205 L/mol/K Absolute temperature: 273.15 K	Standard Temperature 0°C	Standard Temperature 20°C
Molar volume at STP	22.41 L/mol	24.05 L/mol
Ozone wt/vol at 100% volume	2,141.90 mg/L	1,995.84 mg/L
Ozone wt/vol at 1% volume	21.42 mg/L/% vol.	19.96 mg/L/% vol.

PRODUCT GAS OZONE PRODUCTION

Ozone concentration information and gas flow rates are combined to determine the ozone production rate. One of the approaches is to multiply the product gas ozone concentration, in percent by weight, by the feed gas mass flow. Feed gas mass flow can be used in this instance because mass flow is constant through the ozone generator. Feed gas volumetric flow (corrected for temperature and pressure) is typically obtained instead of product gas flow because ozone-resistant materials are unnecessary at this location. The feed gas density must also be determined.

An incorrect production rate will result if the feed gas volumetric flow rate is multiplied by the weight/volume ozone concentration because the feed gas flow (G_1) will be slightly greater than the product gas flow. The volumetric flow changes as ozone is formed within the ozone generator because three oxygen molecules are converted to two ozone molecules.

Ozone production can be calculated using Equation 7. Solutions to Equation 7 are available whether metric or English units are employed.

$$P = G_1 \times W_{fg} \times Y_1 \div 100 \quad (7)$$

Where:

P = Mass flow rate of ozone, kg/h or lb/day
G_1 = Gas flow rate of feed gas, Nm^3/h or scfm
W_{fg} = Density of feed gas, g/L (same as kg/m^3) (see Table A–3)

Note: Density must be converted to lb/ft^3 for the applicable "standard" temperature. Also, the product must be multiplied by 1440 min/day to get the answer in lb/day.

Y_1 = Product gas ozone concentration, expressed as percent by weight and not a ratio (e.g., 1.5 and not 0.015)
100 = Conversion from percent to ratio.

OZONE TRANSFER EFFICIENCY

The offgas ozone concentration (Y_2) should be expressed in the same units as the feed gas concentration (Y_1), such as percent by weight, since both are used in the calculation of ozone transfer efficiency (TE). An approximate TE number is obtained using ozone concentration data only. A precise TE may be calculated when the feed gas and offgas ozone mass flows are determined (i.e., when the respective flow rates are also measured). Usually, offgas flows are not measured, and only the approximate TE is obtained.

The precise and approximate TEs are equal when the offgas flow rate equals the feed gas flow rate. However, offgas flows may be slightly higher due to air leakage into the contact basin from operation under a slight negative pressure. This difference is usually minor, especially when the pressure in the contactor is in the typical range of -0.25 to -0.50 kPa (-1 to -2 in. water column). The offgas flows may be lower when high-purity oxygen feed gas is used and oxygen plus ozone gas is transferred to the liquid. Offgas flow losses are typically less than 10 percent of feed gas flows when the product gas ozone concentrations are low to moderate (i.e., 2 to 6 percent by weight). More than 10 percent loss can occur at higher ozone concentrations, especially when applied dosages are also low (i.e., gas-to-liquid ratio is low).

The discrepancy in the precise and approximate TE numbers is dependent on the actual TE and the loss or gain in offgas flow. The precise TE is greater than the approximate TE when the offgas flow is less than the feed gas flow. Normally, TEs exceed 90 percent and gas flow losses are less than 10 percent. Under these conditions, the precise TE is about 1 percent greater than the approximate TE. In most applications, the approximate calculation is adequate and eliminates the need for installing a separate flow meter for offgas flow measurement. However, when exact data are required, the precise method of calculating TE should be used. Also, the precise method should be used when the ozone concentrations are high (e.g., greater than 6 percent by weight).

The ozone concentration (Y_3) from the destruct unit is measured to determine compliance with mandated ambient levels, such as 0.1 ppm by volume. Exhaust gas flow rates are seldom measured.

References

GORDON, G. ET AL. 1989. Complications in the Iodometric Determination of Ozone. *Jour. AWWA,* June: 72–76.

BIRDSALL, C.M. ET AL. 1952. Iodometric Determination of Ozone. *Anal. Chem.,* 24:662–664.

MASSCHELEIN, W.J. 1988. Personal communication regarding International Ozone Association Standardization Committee (March).

RAKNESS, K.L. & STOLARIK, G.F. 1988. Power Evaluation of the Los Angeles Oxygen-Fed Ozone System. Presented at the International Ozone Association Pan American Committee Spring Conference, Monroe, Mich. (April).

RAKNESS, K.L. ET AL. 1986. Start-Up and Operation of the Indianapolis Ozone Disinfection Wastewater Systems. Presented at the 59th Ann. Water Poll. Control Fed. Conf., Los Angeles, Cal. (October). Also published in *IOA Jour.,* 10:3 (1988).

Index